PLANT BIOTECHNOLOGY

Emerging Issues and Applications

PLANT BIOTECHNOLOGY

Emerging Issues and Applications

Pratibha Pradhan

AUTHORSPRESS

Publishers of Scholarly Books

Worldwide Circulation through Authorspress Global Network
First Published 2010
by
Authorspress
E-35/103, Jawahar Park
Laxmi Nagar, Delhi-110 092
e-mail: authorspress@yahoo.com

Plant Biotechnology: Emerging Issues and Applications
ISBN 978-81-7273-543-2

Printed in India at Tarun Offset, Delhi.

PREFACE

'Biotechnology' is neither a new concept nor a new application. Ancient fermented food processes, such as making bread, wine, cheese, curds, etc., some of which are over 6,000 year old and developed long before man had any knowledge of the existence of the micro-organisms, also genuinely constitute biotechnology. Conventional agriculture is a well-developed biotechnological industry in its own right. However, for the sake of convenience, the traditional processes are excluded from the realm of 'modern biotechnology'. In simpler words, biotechnology is the industry-scale use of organisms and/or their products.

Plant biotechnology employs a range of biological, chemical, and engineering disciplines to develop the use of plant organisms in agricultural and industrial processes. The most obvious use of plant products in traditional biotechnology was for food. Over the centuries, farmers have learned to select edible crops for yield and quality. New and even better varieties were produced when early humans learned to cross-pollinate plants and became plant breeders.

Although biotechnology has been with us for a long time, it is only now receiving the interest of the general public. This new interest in biotechnology is due to the revolution in molecular biology that has occurred over the last 30 to 40 years. Advances in DNA technology have made it possible to move genes between widely divergent organisms, creating new combinations of trits never before possible in nature. These advances were originally developed for use in the laboratory where they revolutionised our ability to ask questions and obtain meaningful answers about the fundamental biological problems. However, this new technology is now being exploited by industry, where the economic potential is immense.

The hope, the hype and the vehement opposition modern biotechnology has generated in a short-time are immense and unprecedented. Both the technical and non-technical literature on biotechnology is vast and diverse, often frustrating even the biologists. The students and teachers of biological sciences are also at a serious disadvantage in accessing and understanding the full measure of the implications and complications of the new technology. The diverse and often conflicting views that appear in the media leave the general public in a state of confusion. There is a clear need to disseminate factual science-based information that facilitates informed decisions on the acceptance or rejection of the products and services biotechnology endlessly offers.

Against this background, this textbook is conceived as a prefect resource to serve the needs of a wide range of audience. It provides an overview of the production of GM plants, highlighting the key scientific and technical advances that underpin their development. Each chapter is well-structured, containing tables and drawings that clearly illustrate concepts in the text. The book is designed for advanced undergraduates with a serious interest in this attractive topic and graduate students wishing to pursue a career in the field.

Pratibha Pradhan

Contents

1

INTRODUCTION TO PLANT BIOTECHNOLOGY

1.1. What is Plant Biotechnology ?

Biotechnology is a branch of technology that is concerned with the forms of industrial production which use micro-organisms and their biological processes. The impact of this technology are felt in agriculture, industry, medical treatment and environment. New species of plants, manure, insecticides, cloning, genetics, etc. are also influenced by this interdisciplinary field of study and research.

One definition of biotechnology is "the deliberate manipulation of DNA molecules to produce commercial products from living organisms." All life is composed of cells that contain genes, and genes are made of DNA molecules. DNA contains information used by cells as a "blueprint" or plan for the animal, plant, or other organism. All the characteristics, or traits, of any living thing are determined by the information in the DNA plan. Scientists are learning how to transfer genes from one animal, plant, or other organism into another. Theoretically, almost any trait found in nature can be transferred into any chosen organism, even if it has never had this trait before. This process, called genetic engineering or recombinant DNA technology, is an important aspect of biotechnology. It can be used to produce modified crop plants, livestock, proteins, vaccines, and drugs.

The field of plant biotechnology is concerned with developing ways to improve the production of plants in order to supply the world's needs for food, fibre and fuel. In addition, plants provide us with many pharmaceuticals and industrial compounds. As our population grows, our needs also grow. To increase the quantity of crop production as well as to produce specific characteristics in plants, biotechnologists are using selective gene techniques. The two major methods of propagation are:

i) Plant tissue culture

ii) Genetic engineering

Plant tissue culture is the sterile, *in vitro* cultivation of plant parts. Plants have the ability for differentiated cells revert to an undifferentiated state called callus. These cells will then divide and then differentiate back to somatic embryo cells that will regenerate the entire plant. Plants cultured *in vitro* yield thousands of genetically identical plants (clones) from a single plant. This process is called micropropagation and is used to commercially propagate plants asexually. The rapid multiplication allows breeders and growers to introduce new cultivars much earlier than they could by using conventional propagation techniques, such as cuttings.

Through the use of biotechnology, desirable genetic traits can be transferred from one organism to another by transfer of DNA. Many more plants with the desirable DNA can be regenerated from small pieces of the transformed plant tissue. Examples of plants produced using tissue culture include the large variety of ornamental plants; agricultural crops such as strawberry, banana, potato, and tomato; and a variety of medicinal plants. Commercial tissue culture involves exposing plant tissue to a specific regimen of nutrients, hormones, and light under sterile conditions to produce many new plants over a very short period of time.

Our ancestors have been improving crops and livestock for thousands of years through selective breeding or crossbreeding to produce desired traits. Biotechnology is just an extension of this process. Genes are added, deleted or temporarily silenced to produce desired results. Genetic engineering involves cutting and moving snippets of DNA from one plant to another. Permanently integrating new DNA into a plant's original DNA forms what's known as a transgenic plant or genetically modified organism (GMO).

Using the techniques of modern biotechnology, one or two genes may be transferred to a highly developed crop variety to impart a new character that would increase its yield. However, while increases in crop yield are the most obvious applications of modern biotechnology in agriculture, it is also the most difficult one. Current genetic engineering techniques work best for effects that are controlled by a single gene. Many of the genetic characteristics associated with yield are controlled by a large number of genes, each of which has a minimal effect on the overall yield.

While the genomics revolution has considerably expanded our understanding of plant biology, and will continue at an ever increasing pace, we are only now starting to fully appreciate the complexity of the challenges ahead, namely translating this knowledge into added value in agriculture. The apparent paradox of plant biotechnology is that the resources currently available for research – both public and private - amount to only a fraction of that of medical/ pharmaceutical research, while its scientific challenges are comparable if not greater. Not only is the field dealing with a broad diversity of plant species, the more rapid evolution of their genomes complicates the translation of molecular knowledge across species.

Despite the apparent diversity of applications of plant biotechnology around the world, the "grand" objectives of global agriculture—improved yield and quality, sustainability and diversification of products—are strikingly universal. Resources required for genomics and large scale biology programmes often exceed the budgets of national funding agencies, including those of the wealthier developed countries. Consequently, there is a rational and economic basis to start reflecting on coordinating plant biotechnology research on a global scale, and for setting the stage for a global long term strategy.

1.2. Plant Tissue Culture

Plant tissue culture refers to the technique of growing plant cells, organs in a sterile environment in a nutrient medium. It has applications in research and commerce, primarily for plant propagation and is often referred to as micropropagation. Micropropagation involves the production of whole plants from cell cultures derived from explants that usually consist of tissues containing or developing into meristem cells.

1.2.1. Ideal Conditions for Plant Cells to Multiply *In Vitro*

1.2.1.1. Sterile Conditions

In the absence of maintenance of sterile conditions, many early tissue culture experiments partly failed. Bacteria, fungi, and other organisms which can be compacted to a certain extent by a whole plant can easily outcompete an isolated piece of tissue from the plant in the relatively nutrient-rich environment of a culture flask. Therefore it is becomes imperative to remove competitor organisms from the culture and isolate them in sterile conditions. Chemical surface sterilisation of the explant with an agent such as bleach at a concentration and for a duration will destroy pathogens without injuring the plant cells beyond recovery. The medium and culture flasks used must also be sterile.

1.2.1.2. Provision of Nutrients and Waste Products Removal

A small part of a plant, when isolated, is unable to receive nutrients or hormones from the plant, which are vital for growth in vitro. The culture must be able to excrete the waste products of cell metabolism by culturing on or in a defined culture medium, that is periodically replenished.

1.2.1.3. Suitable Climate

Tissue cultures require a stable and suitable climate to sustain by the nutritive medium and confine them in a protective vessel, require. Thus there should be regularisation of light and temperature.

1.2.2. Benefits of Tissue Culture

Tissue culture offers numerous significant benefits over traditional propagation methods. Compared to traditional means tissue culture propagation can be much more rapid. Plants that are very difficult to propagate by cuttings or other traditional methods can be multiplied in vitro. Large numbers of genetically identical clones may be produced and seeds can be germinated with no risk of damping off/predation. Under certain conditions, plant material can be stored in vitro for considerable periods of time with low maintenance. Tissue culture techniques are used for virus eradication, genetic manipulation, somatic hybridisation and other procedures that benefit propagation, plant improvement, and basic research. It is an essential part of many genetic transformation protocols.

Plant tissue culture techniques include seed culture, embryo culture, ovary or ovule culture, anther and microspore culture, In vitro pollination, organ culture, shoot apical meristem culture, somatic embryogenesis, organogenesis, enhanced auxiliary budding callus cultures, *in vitro* mutagenesis, protoplast isolation, culture and fusion, *in vitro* flowering, micrografting and genetic transformation.

Table 1.1. Plant tissue culture techniques and applications in plant improvement

Technique	*Applications*
Seed culture	Increasing efficiency of germination of seeds that are difficult to germinate in *vivo*
	Precocious germination by application of plant growth regulators
	Production of clean seedlings for explants or meristem culture
Embryo culture	Overcoming embryo abortion due to incompatibility barriers
	Overcoming seed dormancy and self-sterility of seeds
	Embryo rescue in distant hybridisation where endosperm development is poor
	Shortening of breeding cycle
Ovary or ovule culture	Production of haploid plants
	A common explant for the initiation of somatic embryogenic cultures
	Overcoming abortion of embryos of wide hybrids at very early stages of development due to incompatibility barriers
	In vitro fertilisation for the production of distant hybrids avoiding style and stigmatic incompatibility that inhibits pollen germination and pollen tube growth

Anther and microspore culture	Production of haploid plants Production of homozygous diploid lines through chromosome doubling, thus reducing the time required to produce inbred lines Uncovering mutations or recessive phenotypes
In vitro pollination	Production of hybrids difficult to produce by embryo rescue
Organ culture	Any plant organ can serve as an explant to initiate cultures
Shoot apical meristem culture	Production of virus free germplasm Mass production of desirable genotypes Facilitation of exchange between locations (production of clean material) Cryopreservation (cold storage) or *in vitro* conservation of germplasm
Somatic embryogenesis	One major path of regeneration Mass multiplication Production of artificial seeds As source material for embryogenic protoplasts Amenable to mechanisation and for bioreactors
Organogenesis	One major path of regeneration Mass multiplication Conservation of germplasm at either normal or sub-zero temperatures
Enhanced axillary budding	Micropropagation
Callus Cultures	In some instances it is necessary to go through a callus phase prior to regeneration via somatic embryogenesis or organogenesis For generation of useful somaclonal variants (genetic or epigenetic) As a source of protoplasts and suspension cultures For production of metabolites Used in *in vitro* selection

In vitro mutagenesis	Induction of polyploidy
	Introduction of genetic variability
Protoplast isolation, culture and fusion	Combining genomes to produce somatic hybrids, asymmetric hybrids or cybrids
	Production of organelle recombinants
	Transfer of cytoplasmic male sterility
In vitro flowering	This can be done in some instances but with little practical applications
Micrografting	Overcoming graft incompatibility
	Rapid mass propagation of elite scions by grafting onto rootstocks that have desirable traits like resistance to soilborne pathogens and diseases
	To allow survival of difficult to root shoots
	Development of virus free plants
Genetic transformation	Many different explants can be used, depending on the plant species and its favoured method of regeneration as well as the method of transformation
	Introduction of foreign DNA to generate novel (and typically desirable) genetic combinations
	Used to study the function of genes.

1.3. Plant Genetic Engineering

Plant genetic engineering has emerged as an important tool in crop improvement. It has improved agricultural productivity, has the potential to reduce the use of herbicides and pesticides, to increase yields, and is needed to breed the next generation of crops food security. Some scientists fear-that genetically engineered crop plants and products may harm the environment and human health and may cause unintended genetic changes in other plants.

1.3.1. History of Crop Plants

The major cereal crops that we use today, providing the bulk of our calories, are descended from plants selected in prehistory for their useful traits.

Maize was introduced in Central America from a wild relation called teosinte approximately 8,000 years ago. The Middle East witness the cultivation of wild grasses with large seeds as annual crops, and bread wheat was propagated from hybrids of these wild grasses perhaps as recently as 10,000 years ago. The incidental hybridisation of the wheat species resulted in a larger grain with a soft encasing making it easier to harvest and mill. Generations of selection have led to modern wheat varieties. Over centuries a wide range of crop plants have been domesticated for food, fibre and oil production, and a host of other uses. Through continuous adaptations and refinements of local varieties and the movement of species around the globe. This led to the development of new industries for the production of sugar, cotton and rubber.

The principles of genetics were established by studying pea seed and flower characteristics, and incorporated into the new science of plant breeding during the early 20th century. Today, almost all our crops that have been carefully selected for desirable genetic changes, have genetically changed over-generations and breeders continue to make new genetic combinations to improve performance and to meet new requirements. With the discovery of DNA as the genetic material breeders are free to select desired progeny from crosses of parents that are increasingly diverse. Thus, a wider range of characteristics from wild varied crops such as disease resistance and fruit size, can be incorporated into crops.

1.3.2. Crop Plant Breeding

Plant breeding involves changing the genetics of plants for the benefit of mankind. To understand its relationship to genetic engineering, the function of DNA and genes needs to be considered. The genetic material of plants and most other organisms is a very long molecule called DNA that carries units of instructions in packages called genes. The gene usually makes one specific protein that performs a specific function in cells.

Cells and complex organisms composed of many cells requires many proteins, and hence genes, to function. For example, bacteria have between 2,000-8,000 genes, while plants, mice, humans and flies have between 15,000 and 50,000 genes. Genes occur in a specific order on long stretches of DNA called chromosomes that are visible in cells. Though DNA is extraordinarily stable, over very many generations changes in its code occur and accumulate. These changes, called mutations can affect the functions of proteins, such as, they might not perform their function properly, they may change the way the protein works, or they may interfere with another protein's functions. Though they may have no outward effects, some small changes in the DNA code can lead to large changes in the appearance of organisms.

Genetic variation is the building material of plant breeding, and understanding how this variation is inherited, and what effects it has on plant performance, are complex and interesting problems for plant scientists. In plant breeding, progeny with new combinations of genes from each parent are screened for desired characteristics over long periods up to 10 years by sifting out combinations of genes giving unwanted characters while preserving sets that give the desired character, for a new variety to make it to the field. Wild species related to modern crop plants are a very important source of new genetic material for breeding as they have often evolved useful adaptive traits, such as disease resistance or tolerance of extreme environmental conditions. For example, wild grasses from the Middle East are useful sources of drought and salt tolerance. Is the loss of habitat for wild species. A great concern to plant breeders, and much effort is spent on collecting and cataloguing them.

Plant breeders can also hybridise plants that have the entire set of genetic information from each parent. For example, oilseed rape (*Brassica napus*) is a combination of Chinese cabbage (*Brassica rapa*) and broccoli (*Brassica oleracea*). Such plants called polyploids (meaning several sets of genomes) hybridise at a low rate to generate new species and increase diversity. These polyploids can then be bred as described above to select desired progeny. Breeders have also experimented in creating mixtures of genes from unrelated plants from which useful new crop varieties can be selected in breeding programmes.

1.3.3. New Methods of Breeding Plants

Armed with the knowledge of the DNA code breeders are speeding up the breeding process and including new sources of genetic variation into breeding programmes. For example, crossing a cultivated tomato with a wild tomato has produced larger fruit, improved processing quality and pest resistance. The difficulty with crosses between more distantly related plants is the larger degree of variation among the offspring of the cross.

A technology called marker assisted breeding while sorting out the few desired characteristics from the many unwanted ones, uses landmarks on DNA to track the inheritance of different regions of DNA in the progeny. By knowing which landmarks are associated with desired and undesirable characteristics, breeders are now able to sort through the complex genetic make-up of progeny of distantly related plants directly to select improved lines.

Hybridisation is another type of breeding that involves increased vigour and increased yield. Several important crops such as maize and tomato are usually grown from F1 hybrid seed showing improved performance over that of each of the parents, but this advantage is lost in subsequent generations.

1.3.4. Genetic Engineering in Plants

In genetic engineering genes with sets of information in their DNA provide the code to make a specific protein (one gene usually makes one type of protein), and have genetic switches that dictate the amount, location and timing of protein production by the genes.

DNA molecules are far from inert and can be moved about by a variety of natural processes; for example, bacteria can exchange certain types of DNA very readily by a mating process called conjugation. About two decades ago it was found that some bacteria naturally use this mechanism to transfer DNA into plant cells. The "transferred DNA" becomes stably incorporated into the plant DNA and is then stably inherited as part of the plant DNA.

The technology of DNA cloning is useful in "genetic engineering" of plants. Whereby the combination of a plant genetic switch with the coding information of a gene in a piece of DNA can be transformed into a plant. This new technology makes it possible for the coding region of any gene to be used successfully from bacteria to human, because the genetic switch functions in plants to make the new protein. Essentially what was originally a bacterial or human gene has now become a functioning plant gene, and many new opportunities for changing the genetic makeup of crop plants has become viable.

Many types of plants can be now genetically engineered. Novel methods of introducing DNA into plants have been invented, such as the biolistic gun that shoots tiny DNA- coated gold spheres directly into plants.

1.3.5. Genetic Engineering vs. Plant Breeding

In plant breeding genetic differences between related plants are identified and combinations of genes that give desired traits, such as increased disease resistance or drought tolerance, are selected. The range of genetic variation available limits breeding to characteristics already present or latent in crops and their near relations, but it permits very complex characteristics to be selected for without any prior knowledge of the genes involved apart from their approximate location on chromosomes.

Genetic engineering differs from plant breeding in that it attempts to achieve the same objectives as plant breeding but more directly by identifying a gene of interest that confers a desired trait and transfers it to the crop plant to be improved. Genetic engineering can use any gene in any organism, and this can be transferred and switched on when required, making it suitable for genetic engineering methods. The main limitations to the genetic engineering approach to plant breeding though involves identifying appropriate genes to transfer.

1.3.6. Benefits of Genetically Engineered Crop Plants

When genetic engineering was first developed for plants there were two main challenges. The first was to find ways of making gene transfer work effectively in the major crop plants. This has been achieved for a wide range of plants, from pine to wheat. The second was to identify genes that might confer a useful characteristic in a crop. So far only a very limited number of potentially useful genes have made it into commercially successful crop varieties. Among these are genes isolated from bacteria that confer tolerance to herbicides and resistance to harmful pests.

Genetically engineered crops have different attributes depending on the functions of the introduced genes. So far three main crops have been engineered for two characteristics-herbicide tolerance and insect pest resistance.

Insect resistant cotton and maize plants, when grown with appropriate refuges to permit pests to propagate normally and avoid excessive selection pressure, do not require so many applications of insecticide and have on average higher yields. Herbicide tolerant crops require fewer treatments with herbicide and tilling is reduced, which helps to maintain soil quality. Biotech papaya has been grown in Hawaii since 1998 in order to combat a virus that devastated the crop.

Currently about half of the fresh fruit crop is biotech, while the other half is maintained as non-biotech to supply export markets. Interestingly the biotech papaya plans appear to act as a buffer that stops virus spread to the susceptible crop. About 30 per cent of the 81 million hectares of global transgenic crops were grown in developing countries in 2004. Table 1.2 shows the 2004 GM Crop Areas for the major growers.

Table 1.2. Areas of Biotech Crop Growth (2004)

Country	*Area (m HA)*	*Crop*
USA	47.6	Cotton, Maize, Soybean
Argentina	16.2	Soybean, Maize, Cotton
Canada	5.4	Canola, Maize, Soybean
Brazil	5.0	Soybean
China	3.7	Cotton
South Africa	0.5	Maize, Soybean, Cotton
India	0.5	Cotton
Paraguay	1.2	Cotton
Uruguay	0.3	Soybean, Maize
Australia	0.2	Soybean
Romania	0.1	Soybean
Spain	0.1	Maize
Mexico	0.1	Cotton, Soybean

1.3.7. Ethical Concerns

The growth of genetically engineered plants is viewed as an In comparison to the obvious importance attached to ethical issues surrounding human genetic research, the ethical issues surrounding plant genetics have not yet been systematically explored. Current concerns focus on the growth of genetically engineered plants as an intrusion on nature that may have unknown and irreversible consequences, and on the amassing of unprecedented power by corporations through ownership and control of the supply of food from seed to plate. These and related issues have generated substantial debate in some countries, including those where the crops are not grown.

Crop plant genetic engineering has the potential to introduce genes from animals and even humans into food, therefore important ethical concerns arise. For example, would it be appropriate to eat food containing a human gene? Would vegetarians consider crops containing animal genes to be non-vegetable? How do these concerns relate to the fact that plants and animals have many genes in common? Genetically engineered crops so far only contain genes from bacteria, plant viruses and other plants so these particular issues remain hypothetical for the moment. Increased experience of genetic modification including new products and more comprehensive regulatory procedures may contribute to focusing arguments and resolving ethical and social concerns over our food supply.

1.3.8. Environmental Concerns

Environmentalists are sceptical about the possibility of exchanging genetically engineered characteristics between crop plants and wild plants. Pollen movement can transfer the genetic material of flowering plants over substantial distances to sexually compatible species. Hence plant breeders need to ensure the proper genetic composition by producing seeds from plants that are an appropriate distance apart that results in minimally acceptable cross- pollination, usually about 1 per cent.

In the field pollen exchange between sexually compatible crop and weed species also has the potential to intermingle traits from crops and weeds. Pollen exchange is generally not an important issue in crops harvested from single plantings of genetically pure seedlings. In farm- kept seeds any traits derived from infrequent crosses with wild relatives generally only have a minor effect in the population of crop seeds, but over time favourable characteristics of the crop may start to be lost. Similarly weedy relatives of crop plants may inherit characteristics from crop plants, and in principle this has the potential to enhance weediness, although this has not been observed. Because genetically engineered genes are stably inherited they can also be spread by pollen from crops to weedy species, and this has the potential to confer traits such as herbicide and virus tolerance on weeds. It is conceivable that these traits may confer a selective advantage on populations of weeds, but this requires the close cultivation of genetically engineered crops with sexually compatible weedy species over time, and also requires a relatively undisturbed cultivation of weeds. Current agricultural practices generally mitigate these conditions.

Largescale experiments to measure the movement of genetically engineered traits between crop plants have helped to determine minimum separation distances so as to reduce cross-pollination to levels that are below the limits needed to maintain genetic purity. Relatively simple agronomic measures to limit any potential effects of crosses between weedy species and crops can be implemented, such as through weed control. For example, if herbicide tolerant weeds grow profusely in fields, they can be destroyed by using another type of herbicide.

The so-called "terminator technology" was developed, but not applied, for stopping the flow of pollen from GM plants. The technology makes crop plants sterile and was originally invented as a "technology protection system" to prevent farmers from keeping seeds and growing these in the next season. This technology has the potential to ensure GM plants producing non-food products are genetically isolated in the field.

Cross-pollination between crop species and their wild relatives raises the possibility that this process may erase essential differences between wild relations and their cultivated relatives. The loss of these differences has the potential to reduce genetic variation that may be useful for future breeding programmes.

The erosion of genetic diversity is among the most serious threats to agriculture and the environment because of the increasing scale of irreversible loss of genetic variation. For example, maize has been cultivated in regions of Mexico where wild varieties have been flourishing for many thousands of years. In spite of widespread cross-pollination in overlapping populations of cultivated maize and wild relatives, the wild populations have maintained their distinctive identities over many generations, denoting that cross-pollination is not really responsibie ioss of genetic diversity. Instead, habitat destruction is the principal and obvious cause of the irretrievable loss of wild species.

1.3.9. Safety of Genetically Engineered Crops

Consumption of soyabean or maize genetically engineered for herbicide tolerance or insect resistance have so far not affected humans or animals. This is because the novel genes present in these crops. We have all been eating DNA from for generations and there is no evidence of transfer of DNA from food to our DNA.

To ensure greater standardisation and conformity to best scientific practice, new biosafety laws may in olve appropriate tests comparing the composition, nutritional value, toxicity and allergenicity of the genetically engineered crop to its non-engineered counterpart.

Thailand, Brazil, Paraguay, New Zealand, Australia and England have all recently decided on limited case-by-case introduction of genetically engineered crops after the formulation of appropriate biosafety regulations.

In India, important local crops such as millet, chickpea and pigeon-pea are being genetically engineered. In Uganda biosafety laws are being established so that genetic engineering for resistance of banana to wilt disease can be performed. In China, Mexico, Australia and New Zealand and elsewhere researches are be in carried out developing local crops by genetic engineering.

There has been vast progress in experimentation with genetically engineered crops. These include engineering crop plants with improved nutrition such as vitamin- and iron-enhanced rice, alteration of the fatty acid composition of oilseed crops such as sunflower and canola to improve human nutrition and provide special fats for fish meal, etc. Farmed fish require specific polyunsaturated fatty acids currently obtained by harvesting. other fish, and finding a plant-based source may alleviate loss of marine wildlife. Extensive research conducted in industry and universities aim to provide resistance to pests in grape, cereals, potato and trees, identify genes conferring salt tolerance in tomato and cold tolerance in canola; produce antibodies and vaccines in plants which can yield high quantities of specifically modified therapeutic proteins; and use genetically engineered plants to make commodities such as specialist starches and oils for industrial applications.

1.3.10. Challenges of Plant Biotechnology

The availability of fresh water and fertile soils has always dictated agricultural production and these factors still determine global food security today. For example, unsustainable use of key aquifers is reducing the water table in key agricultural areas and lakes are drying up. Previously fertile soils, especially in irrigated areas of high production, are becoming more polluted with salt by irrigation. Moreover, the postulated increased unpredictability in weather systems due to global warming and an ever- increasing world population seriously exacerbate these threats to food security. The most serious current concern of plant scientists is the irretrievable loss of biodiversity, which reduces the scope for breeding crops to tolerate more extreme climates, to resist pests and diseases and produce more nutritious food.

Until recently the prevailing strategy in crop plant breeding and research was to increase yield. The environmental costs of this strategy, such as elevated nitrate and phosphate levels in ground water, have led to the embracing of a new strategy of sustainable agricultural production in the UK. This strategy is being framed within larger concerns about the balance between farming practices, production subsidies and the environment. Current systems supporting production are thought by many to be inconsistent with both international trade agreements and with the long- term stewardship of the farmed environment for food production and promoting wildlife.

How can plant genetics contribute to the related objectives of achieving sustainable food production with a reduced environmental impact and at the same time provide a livelihood for rural communities in the face of reduced production subsidies? There are several examples. Reduced insecticide use on insect-resistant cotton crops has already substantially reduced the market for agricultural insecticides in the US. The use of herbicide tolerant crops has improved soil condition by increasing low-till agriculture and reducing tractor usage. Further reductions in the environmental impact of agriculture can be expected from novel management systems and increased knowledge of plant nutrition and plant interactions with the soil ecosystem. Clearly many disciplines will need to be mobilised and integrated with different production systems to help address objectives set for achieving greater sustainability. Genetic engineering is only one of many methods that may contribute to maintaining productivity and food quality while reducing inputs and improving agricultural ecosystems.

1.4. Applications of Transgenic Plants

By using a variety of techniques, it has become possible to transform plants with foreign genes, thus producing a very wide range of new plant varieties. Transgenic plants have been developed to be resistant to a range of environmental stresses, including insects, viruses, herbicides, pathogens, and salt stress; to have flowers with modified colours; to a modified nutritional content, including modifications in amino acids, lipids, discolouration, and sweetness.

1.4.1. Virus-resistant Plants

Latest developments in genetic engineering by scientists have seen results in virus-resistant transgenic plants, using "immunisation" methods with viral coat protein genes, other viral genes, or viral gene antisense sequences to confer resistance.

Potatoes, cereals and pulses are the main cross that require the immunisation method for virus-resistances. Potatoes are vegetatively propagated, therefore, seed material (tubers) for planting must be virus-free.

It is very difficult to improve potato through breeding techniques as it is a tetraploid. Most of the cultivars are susceptible to various diseases caused by fungi, nematodes, and viruses. Potatoes are vegetatively propagated, therefore, seed material (tubers) for planting must be virus-free. Potatoes suffer from three important virus diseases called Photo virus X, Photo virus Y, (PVX, PVY) and potato leafroll virus.

The phenomenon of 'cross-protection' or immunisation of plant is not clearly understood. It is similar to immunisation of human-being for bacterial diseases. When a plant is inoculated with a mild strain of a virus that causes almost no disease symptoms. it will be protected against a virulent form of the virus. This cross-protection is in some way related to the synthesis of the coat protein by the plant cell. When virus infects plant cell, a plant cell starts synthesising coat proteins instead of its own proteins.

In 1986, Roger Beachy and colleagues at Washington University introduced the gene that encodes the coat protein of TMV into tobacco plants, resultantly each and every cell of the transgenic plant start producing coat protein. These plants showed considerable resistance to infection by TMV. The virus was unable to multiply rapidly in the cells already containing some coat protein. Therefore, the number of virus particles per cell remained low in transgenic plants as compared to normal control plants.

Scientists at Mogen International in the Netherlands used the same approach to make potatoes resistant to PVX. The gene encoding the coat protein of PVX was introduced into two cultivars. The transgenics showed 100 times less virus particles as compared to control plants after two weeks of inoculation. The yield performance of most cultivars was same but potatoes produced were elongated. The viral coat protein gene approach has been used to transfer tolerance to a number of different plant viruses. Using this approach, researchers have developed virus-resistant transgenic plants for a number of different crops. Although complete protection is not usually achieved, high levels of virus-resistance have been reported. Moreover, a coat protein gene from one virus sometimes provides tolerance to a number of unrelated viruses.

In both eukaryotes and prokaryotes, an RNA molecule that is complementary to a normal gene transcript (mRNA) is called antisense RNA. The mRNA, being translatable, is considered to be a sense RNA. The presence of antisenses RNA can decrease the synthesis of the gene product by forming a duplex molecule with the normal sense mRNA, thereby, preventing it from being translated. The antisense RNA-mRNA duplex is also rapidly degraded, a response that diminishes the amount of that particular mRNA in the cell. Therefore, in principle it should be possible to prevent plant viruses from replicating and subsequently damaging plant tissues by creating transgenic plants that synthesize antisense RNA that is complementary to virus coat protein mRNA.

The Ti binary vector system was used to transfer both protein producing sense and anti-sense RNA-producing cDNA sequences to separate tobacco cells, from which transgenic plants were regenerated. The transgenic tobacco plants that expressed the cucumber mosaic virus (cuMV) coat protein were protected from viral particle accumulation and did not show symptoms of viral infection, irrespective of whether the inoculum of the challenge virus was high or low. However, the transgenic tobacco plants expressing the CuMV coat protein antisense RNA were protected only when the concentration of the challenge virus in the inoculum was low. Therefore, this approach is not successful when virus infection is high.

1.4.2. Herbicide-resistant Plants

Certain herbicides can be used as pre-emergence herbicides to kill weeds before the crops are planted. If crop plants are resistant to these chemicals then they can be used with crop plants (post emergence). By understanding the mechanism of action of these herbicides and development of resistance by certain bacteria to such chemicals can provide clone for developing herbicide tolerant plants. Some plants or bacteria are resistant because they have an enzyme that detoxifies the herbicide. In other words, they possess a gene for this action. Transfer of this gene to a crop plant should protect the crop plant by same action or mechanism. Some plants or bacteria become resistant to herbicide because of mutation in the target enzyme (or gene), and because of this change they are no more sensitive to herbicide or are not damaged by herbicides. The enzyme can work in presence of herbicide. Therefore, the detoxifying mechanism or change in affected enzyme can make the organism herbicide tolerant.

Glyphosate (a herbicide) acts by inhibiting one of the enzymes that is necessary for the synthesis of amino acids in the chloroplast. Glyphosate, initially produced and marketed by Monsanto under the trade name 'Roundup', is a widely used non-selective herbicide, it effectively kills 76 of the world's 78 worst weed species.

Scientists at Monsanto isolated a gene for an enzyme involved in amino acid biosynthesis enzyme EPSP-synthase (5-enol pyruvinyl shikimate 3-phosphate synthase) from resistant E.coli bacteria. They modified the gene in such a way that it could be expressed in plants, and then transferred it to plants e.g., tobacco, tomato and soybean. Expression of bacterial gene in plant required a control region that would direct the expression at the gene in the plant (because bacterial control regions do not work in plants). In addition to this, the gene had to be modified in such a way that the enzyme, which is synthesized in cytoplasm, would be transported to chloroplast. This is important that when gene of prokaryotic origin is used, the product should be transported to right cellular compartment in the plant. This should not affect the quality or quantity of yield. The gene has been successfully transferred in soybean, where the plants showed resistance without change in yield.

Phosphinothricin is a herbicide that acts by inhibiting another enzyme necessary for amino acid biosynthesis (glutamine synthetase) and nitrogen metabolism. This enzyme converts ammonia to glutamate. Inhibiting the activity of this enzyme leads to rapid accumulation of ammonia within the plant cells. Higher concentrations of ammonia are toxic to the cell. Phosphinothricin produced and marketed by Hoechst AG under the trade name 'Basta', is also a very effective non-selective herbicide. This product is related to an antibiotic that is also a herbicide, produced by the fungus Streptomyces hygroscopicus. Scientists at Plant Genetic systems, Belgium obtained a gene from this fungus that encodes an enzyme that converts phosphinothricin to a non-herbicidal derivative by combining it with a cell metabolite. This gene, known as bar-gene, has been transferred in tobacco and potato, where it is expressed showing herbicide tolerance in these plants. The yield performance of the plants remained unchanged.

Herbicides are simply chemical compounds that kill or inhibit the growth of plants without deleterious effects on animals.

Herbicides usually inhibit processes that are unique to plants, e.g., photosynthesis. Mostly herbicides act as inhibitors of essential enzyme reactions. Any change which can reduce the inhibitory effect of herbicide, will provide increased herbicide tolerance.

Glyphosate acts by inhibiting the enzyme 5-enol pyruvinylshikimate 3-phosphate synthase (EPSP synthase), an essential enzyme in the biosynthesis of the aromatic amino acids, tyrosine, phenylalanine and tryptophan. These are essential components in the diets of higher animals since the enzymes that catalyse the biosynthesis of these amino acids are not present in higher animals. Therefore, higher animals do not contain EPSP synthase, and are not affected by glyphosate.

Glyphosate does inhibit the EPSP synthase of microorganisms as well as those of plants. Selection of organisms is made on inhibitory concentrations of herbicide by growing them in presence of herbicide. This way researchers isolated glyphosate tolerant mutant of *Salmonella typhimurium, Aerobacter aerogenes* and *Escherichia coli*. In bacteria, EPSP synthase is encoded by the aeroA gene. When aeroA genes (with plant promoter and adenylation signals) were transferred in plants, the transgenic plants showed increased tolerance to glyphosate. In plants, aromatic amino acids are synthesized in chloroplasts, but gene for EPSP is localised in nucleus. Therefore, a protein is attached to EPSP synthase, which translocates the EPSP synthase into chloroplast, where the protein is removed by clearage. It has been shown that the petunia transit peptide will target the E.coli aeroA gene-product into tobacco chloroplasts and will impart glyphosate tolerance.

In another method, glyphosate-tolerant plants have also been produced by using an EPSP synthase cDNA isolated from a glyphosate tolerant petunia cell culture line. Such lines can be selected by growing cells on medium containing increasing concentration of selection factor, e.g., glyphosate. In this cell line, tolerance resulted from amplification of the EPSP synthase gene, resulting in over production of EPSP synthase in these cells.

The EPSP synthase cDNA isolated from this cell line was joined to the CaMV 35s promoter and to the Ti nos-polyadenylation signal. The strong CaMV 35s promoter (from cauliflower mosaic virus) results in over production of EPSP synthase. When this chimeric (35s + EPSP synthase + nos) gene was introduced into petunia plants on a Ti vector, the transgenic developed were tolerant to the four times higher concentration, which kills control plants.

1.5. Biotechnology and Plant Productivity

Biotechnology began with the domestication of plants and animals. Modern biotechnology encompasses a wide diversity of activities such as gene splicing, cell and tissue culture, immobilised cell and enzyme technology, and fermentation technology. The use of recombinant DNA methodology has raised the possibility of the isolation, manipulation, and transfer of genes from any living system into bacteria, fungi, animals and plants. The applications of this technology have developed at different rates in the various systems, with the practical advances being most limited in higher plants although there are great potential benefits in this area.

Grain yields of the major U.S. crops have increased continually since about 1930. This increase has been due to genetic improvements in yield and to other factors not directly related to the breeding gains. Non-breeding inputs such as nitrogen fertiliser, timeliness of planting, better pest control, and improved harvesting machinery have been important causes of increased yield. However, variety improvement, interacting with changes in agronomic practices, has probably been the most important cause of higher yields.

Genetic improvements in yield have increased linearly for the past 25 years and have shown no signs of levelling off. Plant breeders intuitively agree with the conclusion that the gains in genetic yield can continue with present-day breeding methods. Much time, work, and ingenuity will be needed, however, to bring about these changes which will be at the expense of ever increasing research inputs. It has been estimated that maize yield potential has increased at an average rate of 1.4 per cent per year during the period of 1930-1980. For the same period the number of maize breeders has increased at an average rate of 4 per cent per year.

With this as background, what roles can the new technologies play in plant breeding? It is important to identify the genetic needs for crop improvement as judged by plant breeders. A list of the most important factors has been drawn up. These include the basic physiology and genetics of durable pest resistance, the large number of years and locations needed to evaluate and identify stress tolerance, and the long time needed to break up undesirable genetic linkages or, conversely, to assemble desirable traits. The last factor is particularly important in the utilisation of new germ plasm. Where hybrid crops are concerned, the inability to predict accurately the amounts of heterosis among crosses of inbred lines, and the complete failure of genetics and physiology to explain, except in very general, essentially descriptive terms, the basis for heterosis constrains the selection of base material in breeding programs.

The basis for genetic engineering has been established in model systems. This includes the ability to isolate genes, tailor them, transfer them between species, and subsequently recover them in mature plants that can then be used directly or incorporated into an established breeding program. However, the present methods are designed to transfer single genes which can be identified and isolated. Many traits that are connected with the final productivity are controlled by multiple genes that are only recognised in the context of the ability to manipulate them in a breeding program, but with little or no knowledge of the biochemical basis of their action. Under these circumstances it is not possible to devise a strategy to isolate and transfer the complexes of genes responsible for these quantitative traits in a single step. Thus, it must be emphasized that biotechnology will have to be used in conjunction with 'conventional' plant breeding to manipulate these gene complexes. While the new technology is very powerful, it does depend on the amount and quality of the biological knowledge that has already been accumulated.

The potential applications of biotechnology to plant productivity cover a wide range of subjects. Two areas in which advances have been made recently are in the introduction of novel genes to control virus susceptibility and pest resistance. Here we will consider two aspects of recombinant DNA technology related to identifying new genetic markers for aiding in the selection of material to be evaluated and the detection of desirable combinations. These are the potential use of restriction fragment length polymorphisms to increase the number of genetic markers and an attempt to understand the basis of heterosis.

1.5.1. Restriction Fragment Length Polymorphisms

When the genomic DNAs from two genetically distinct individuals are digested with a restriction enzyme, electrophoresed, blotted onto a nitrocellulose membrane, and probed with a radioactively-labelled DNA clone, polymorphisms in the hybridisation patterns sometimes result due to differences between the individuals. The term restriction fragment length polymorphism (RFLP) has been coined to describe this variation. The appearance of these polymorphisms is not surprising in view of the number of supposedly neutral mutations encountered at the protein level in recent years. Randomly dispersed RFLPs were first proposed as a new source of genetic markers in humans, where the low amount of genomic variability, as well as other factors, complicate the application of genetic analyses. The actual feasibility of using RFLPs as genetic markers for the diagnosis of human diseases/disorders has since been demonstrated.

The potential usefulness of RFLP markers in basic plant genetic studies, as well as in plant improvement programs, has been reviewed by Tanksley and Beckmann and Soller. In many cases the RFLP markers would essentially replace the existing isozyme variants as the molecular markers. Although isozyme loci have proved useful in several instances, their application is limited by insufficient numbers and general lack of informativeness. In contrast, a potentially unlimited number of RFLPs exist, which should allow a much wider use of the molecular marker approach.

The degree of genetic variability detectable by RFLPs in plants has been examined. Rivin et al., Burr et al. and Helentjaris et al. have reported significant variability in maize using either repetitive or single-copy DNA sequences. Restriction fragment length polymorphisms have also been detected in the tomato, barley, and peas.

The DNAs from two flax *(Linum usitatissimum)* lines were digested with a restriction endonuclease, separated on an agarose gel, transferred to nitrocellulose, and hybridised with a labelled ribosomal DNA probe. The variant fragments are identified with an arrow and are due to a variation in the length of the spacer region in the ribosomal DNA repeat unit. In a cross between the two lines, the Fl contained equimolar amounts of the two variants, and they segregated out in selfed F2. Thus, they can be used as genetic markers since they behave as normal Mendelian markers.

The molecular basis for the generation of RFLPs is best demonstrated in genes where the complete sequence is known for what is considered the wild-type allele, as well as for a number of variants. One such case is the alcohol dehydrogenase gene isolated from maize. In this case, many of the changes appear to be due to the insertion and excision of transposable elements. However, single base substitutions that alter the restriction site can also be the cause of RFLPs. Thus, any change that alters the length of a restriction fragment effectively creates a new allele. The only requirement to define this "locus" is an appropriate DNA probe. The pervasiveness of RFLPs in the genomes of higher organisms means that the number of loci that can be ultimately obtained is limited only by the number of informative probes which can be generated.

The use of RFLPs as genetic markers will greatly increase the precision of genetic maps. They also have a number of characteristics which increase their potential usefulness. They vastly increase the number of potential markers, are unaffected by the developmental state of the organism, as are many proteins that are expressed in some form of developmental time scale; and can be characterised using a small amount of material obtained from seedlings.

In general, RFLPs are associated with random sequences, so that their mapping requires considerable effort. However, they would be most useful in material where a preliminary genetic linkage map has already been established and can be used to provide a more comprehensive map. Many hundreds of new markers have been added to the already extensive maps of maize and tomato. Forest tree species are examples where RFLPs will be used to generate a much more extensive map. In some of these species, linkage maps have been established on the basis of isozymes and secondary products, and the original material is still available. From this material it would be possible to construct a more detailed map from the RFLPs. There is already evidence for considerable genetic heterogeneity among conifers, in particular, so that RFLPs should abound.

There are many potential applications of RFLPs in plant breeding. Examples of these are strain identification, mapping and monitoring of quantitatively inherited traits, measurement of genetic diversity, and controlling the level of heterozygosity/ homozygosity.

Strain identification is likely to be most useful in non-true breeding plants that are propagated asexually. These are the plants that are most frequently covered by plant patents. In many cases, the sequences used to generate RFLPs are those which are of low copy number in the genome. In one case, however, it has been shown that a dispersed, repeated sequence shows many RFLPs. Hence, the overall pattern shown is equivalent to a fingerprint, and a small number of such sequences may be sufficient to identify an individual unambiguously. These repetitive sequences may be very useful for the identification of individuals, but their complex patterns make interpretation in crosses difficult. For genetic markers, low copy number sequences are likely to be the ones of choice.

The evaluation of selections in a plant breeding program requires considerable effort. Field testing limits the size of breeding programs, and in the case of fruit or forest trees, which do not come to maturity in a single growing season, the field evaluation becomes even more important. If linkages between marker loci and factors for quantitative traits could be tight enough, then meaningful correlations could be established between the two. Since it appears possible to saturate the genome with RFLP markers, these relationships could be established for a chosen population. If the linkage was sufficiently close, then it would also be possible to use the RFLPs in the selection of material to include in the construction of an elite population as a source for future breeding programs. A start on the correlation of quantitative traits and RFLPs has been made in maize. Using the large number of RFLPs mapped and the wealth of genetic information in maize, a number of loci have been identified, which have major effects on quantitative traits such as plant height.

Widespread epidemics, such as the southern corn leaf blight in the U.S. in 1970 and the potato leaf blight in Ireland in the 1800s, can be traced to the use of a limited genetic base. Genetic diversity does not necessarily overcome the vulnerability to disease, but the maintenance of a wide variety of germplasms is increasingly recognised as important for future breeding. However, natural habitats are being lost, and the resources for maintaining extensive collections are finite. Thus, the use of RFLPs may give an indication of the extent of diversity among collections. In addition, since they are based on random sequences, they should not be biased for any particular trait. This is important as the genetic trait next in demand cannot be known.

1.5.2. Hybrid Vigour

Hybrid vigour, is also known as heterosis, is seen in the improved performance of Fl hybrids as compared with the inbred parents. However, there is no adequate theoretical, genetic, or physiological explanation for this phenomenon and it is hoped that the new technologies could help to unravel this phenomenon.

The extent of heterozygosity in an Fl is one of the suggestions for the heterotic effect observed in hybrids. The use of RFLPs, since they can measure directly the extent of heterozygosity, will be ideal for testing this hypothesis. A study of quantitative traits in maize has started to approach this question. In this study, the regression of single cross yield on the percentage heterozygosity at 34 RFLP loci was highly significant. The relationship appeared to break down at levels of heterozygosity of more than fifty percent. Thus, the combining ability of maize inbreds may be predicted by levels of heterozygosity at RFLP loci to a point, after which other factors may become an important consideration. The use of this RFLP data base grouped the maize inbreds into heterotic groups, based on whether or not they were derived from a common ancestor. The predominant allele at a given RFLP locus differed among the heterotic groups for most of the loci studied, thus providing a basis for further testing of the association between particular RFLP loci and heterotic effects.

Another phenomenon shown to occur in F1 hybrids is that of genomic instability. In certain instances it has been shown that the nuclear DNA in the F1 of a cross between two inbred parents is not necessarily equal to the sum of half of each of the two parents. The observations were that the amount of DNA in the Fl for either a specific sequence, or for the total DNA amount, was not equal to the expected value. It could be either more or less than the expected value. This effect has been demonstrated in three different systems, namely maize, flax, and *Microseris* species.

In maize, it was shown that wide crosses could activate previously quiescent, transposable elements. These elements can cause the restructuring of the genome and appear to be normal components of the genome. The net effect of the activation of these transposable elements is an increased mutation rate, particularly with respect to the generation of unstable alleles. It has been shown recently that wide crosses may not be necessary to get genomic alterations, but may occur in the combinations of more closely related inbreds without the concomitant induction of transposon activity. Thus, genomic restructuring is not the sole preserve of transposable elements, although they may play a prominent role in many species.

This non-Mendelian inheritance of DNA amount does not occur in all crosses, but is dependent on the particular parents involved. The genetic basis for the control of the change in DNA is not known. However, it appears that the system may have at least two separate parts, one controlling whether or not there is any alteration in the Fl and the other controlling the extent of that alteration. It is not clear what effect this form of genomic rearrangement has on the phenotype. One possible way it could alter the phenotype would be by some form of variegated position effect.

Part of the heterosis observed in some crosses could be the result of an interaction at the genomic level with the restructuring of specific parts of the genome being responsible for the altered growth pattern. If this is the case, then the appearance of non-Mendelian effects at the DNA level in Fls may be diagnostic of heterotic crosses. This would be a direct determination of which inbred combinations would show heterosis. Such a factor would be extremely valuable in selecting the initial base population for a breeding program, especially in species where there is insufficient genetic data to identify the loci directly responsible for heterosis.

Another alternative way of approaching the molecular basis of heterosis would be to characterise the genes expressed in the heterotic hybrids and compare them to those expressed in their inbred parents. The technology to isolate the genes that are expressed in particular organs and tissues exists. The characterisation of those genes that differ in their expression in the hybrids compared to their parents will be the genes involved in the metabolic pathways important in the expression of hybrid vigour. Eventually it should be possible to isolate the genes responsible for the control of hybrid vigour. With the gene transfer technology already in place, a further advance would be the transfer of these genes into species where natural hybrid vigour is limited. New insights into the processes contributing to the phenomenon of heterosis and an ability to predict which individuals would be good combiners and show extensive heterosis would be extremely important. This information would alter the basis for selecting the parents for hybrid crops, and perhaps alter the range of crops in which hybrids are considerably superior to inbred varieties.

Two agronomically important crops in which heterosis is highly important are maize and *Pinus* species. The life cycle of the former is short and the plants are relatively small, so that effective screening can be carried out to determine which lines will be useful parents in the production of hybrids. However, in the case of conifers, the time scales are very different; thus any method of predicting heterosis between individuals would dramatically affect the selection of elite material.

1.6. Agricultural Impacts of Plant Biotechnology

The genetically modified (GM) crops were first widely grown in 1996 and since then many milestones have been crossed in this field. This section examines specific global socioeconomic impacts on farm income and environmental impacts with respect to the use of biotechnology.

1.6.1. Farm Use of Biotech Crops

The adoption of biotechnology has had a very positive impact on farm income derived mainly from a combination of enhanced productivity and efficiency gains (Table 1.3). In 2005, the direct global farm income benefit from GM crops was $5 billion. If the additional income stemming from second crop soybeans in

Argentina is considered, this income gain rises to $5.6 billion. This is equivalent to having added between 3.6 per cent and 4.0 per cent to the value of global production of the four main crops of soybean, maize, canola, and cotton, a substantial impact. Since 1996, worldwide farm incomes have increased by $24.2 billion or $27 billion inclusive of second-crop soybean gains in Argentina directly because of the adoption of GM crop technology.

Table 1.3. Global Farm Income Benefits from Growing GM Crops 1996–2005 (million US $)

Trait and Crop	*Increase Farm Income, 2005*	*Increase in Farm Income, 1996–2005*	*Farm Income Benefit in 2005 as % of Total Value of Production of These Crops in GM-Adopting Countries*	*Farm Income Benefit in in 2005 as % of Total Value of Global Production of These Crops*
GM herbicide-tolerant soybean	2281 (2842)	11,686 (14,417)	5.72 (7.1)	4.86 (6.05)
GM herbicide-tolerant maize	212	795	0.82	0.39
GM herbicide-tolerant cotton	166	927	1.16	0.64
GM herbicide-tolerant canola	195	893	9.45	1.86
GM insect-resistant maize	416	2,367	1.57	0.77
GM insect-resistant cotton	1,732	7,510	12.1	6.68
Others	25	66	N/A	N/A
Totals	5027 (5588)	24,244 (26,975)	6.0 (6.7)	3.6 (4.0)

The largest gains in farm income have arisen in the soybean sector, largely from cost savings, where the $2.84 billion additional income generated by GM HT soybean in 2005 has been equivalent to adding 7.1 per cent to the value of the crop in the GM-growing countries, or adding the equivalent of 6.05 per cent to the $47 billion value of the global soybean crop in 2005. These economic benefits should, however, be placed within the context of a significant increase in the level of soybean production in the main GM-adopting countries. Since 1996, the soybean area and production in the leading soybean producing countries of the United States, Brazil, and Argentina increased by 58 per cent and 65 per cent, respectively.

Substantial gains have also arisen in the cotton sector through a combination of higher yields and lower costs. In 2005, cotton farm income levels in the GM-adopting countries increased by $1.9 billion and since 1996, the sector has benefited from an additional $8.44 billion. The 2005 income gains are equivalent to adding 13.3 per cent to the value of the cotton crop in these countries, or 7.3 per cent to the $26 billion value of total global cotton production. This is a substantial increase in value-added terms for two new cotton seed technologies.

Significant increases to farm incomes have also resulted in the maize and canola sectors. The combination of GM IR and GM HT technology in maize has boosted farm incomes by over $3.1 billion since 1996. An additional $893 million has been generated in the North American canola sector. Overall, the economic gains derived from planting GM crops have been of two main types: (1) increased yields and (2) reduced costs of production derived from less expenditure on crop protection (insecticides and herbicides) products and fuel.

In terms of the division of the economic benefits, it is interesting to note that farmers in developing countries derived the majority of the farm income benefits in 2005 (55 per cent) relative to farmers in developed countries (Table 1. 4). The vast majority of these income gains for developing country farmers have been from GM IR cotton and GM HT soybean. Examination of the cost farmers pay for accessing GM technology relative to the total gains derived shows (Table 1. 5) that across the four main GM crops, the total cost was equal to about 26 per cent of the total farm income gains. For farmers in developing countries the total cost is equal to about 13 per cent of total farm income gains, while for farmers in developed countries the cost is about 38 per cent of the total farm income gain. Although circumstances vary among countries, the higher share of total gains derived by farmers in developing countries relative to farmers in developed countries reflects factors such as weaker provision and enforcement of intellectual property rights.

Table 1.4. GM Crop Farm Income Benefits, 2005: Developing Versus Developed Countries (million US $)

Crop	*Developed*	*Developing*	*% Developed*	*% Developing*
GM HT soybean	1183	1658	41.6	58.4
GM IR maize	364	53	86.5	13.5
GM HT maize	212	0.3	99.9	0.1
GM IR cotton	354	1378	20.4	79.6
GM HT cotton	163	3	98.4	1.6
GM HT canola	195	0	100	0
GM VR papaya and squash	25	0	100	0
Totals	2496	3092	45	55

Table 1. 5. Cost of Accessing GM Technology (in % Terms) Relative to Total Farm Income Benefits, 2005

Crop	*All Farmers*	*Developed Countries*	*Developing Countries*
GM HT soybean	21	32	10
GM IR maize	44	43	48
GM HT maize	38	38	81
GM IR cotton	21	41	13
GM HT cotton	44	43	65
GM HT canola	47	47	N/A
Totals	26	38	13

In addition to the tangible and quantifiable impacts on farm profitability presented above, there are other important, more intangible (difficult to quantify) impacts of an economic nature. Many studies on the impact of GM crops have identified the factors listed below as being important influences for adoption of the technology.

1.6.1.1. *Herbicide-tolerant Crops*

- The GM technology provides increased management flexibility with the ease of use broad-spectrum, post-emergent herbicides like glyphosate and the increased/longer time window for spraying.
- In a conventional crop, post-emergent weed control relies on herbicide applications before the weeds and crop are well established. As a result, the crop suffering "knockback" to its growth from the effects of the herbicide. Using the GM technology the problem of growth retardation due to the herbicide effects can be avoided because the crop is tolerant to the herbicide and spraying can occur at a later stage when the crop is better able to withstand any possible knockback effects.
- This method facilitates the adoption of conservation or no-tillage systems and for additional cost savings, such as reduced labour and fuel costs associated with plowing.
- Improved weed control has contributed to reduced harvesting costs, and time leading and also improved harvest quality leading to higher levels of quality price bonuses in some regions.
- There is no potential damage than can be caused by soil-incorporated residual herbicides in follow-on crops has been eliminated.

1.6.1.2. *Insect-resistant Crops*

- The GM technology method eliminates the risk of significant pest damage.
- Time is saved as there is less need to walk through the crop fields to survey insects and insect damage and/or apply insecticides.
- Less frequent aerial spraying saves energy use.
- There are savings in machinery use.
- The farmers and farmworkers have better health as the handling and use of pesticides are reduced.
- The crop yield is faster (e.g., for some cotton growers in India), which allows some farmers to plant a second crop in the same season.

– Beekeepers too benefit as fewer bees are now lost to insecticide spraying.

1.6.2. Environmental Impacts of Plant Biotechnology

The two key aspects of environmental impact of biotech crops examined below are decreased insecticide and herbicide use, and the impact on carbon emissions and soil conservation.

1.6.2.1. Impacts from Changes in Insecticide and Herbicide Use

Usually, changes in pesticide use with GM crops have traditionally been presented in terms of the volume (quantity) of pesticide applied. While comparisons of total pesticide volume used in GM and non-GM crop production systems can be a useful indicator of environmental impacts, it is an imperfect measure because it does not account for differences in the specific pest control programs used in GM and non-GM cropping systems. For example, different specific chemical products used in GM versus conventional crop systems, differences in the rate of pesticides used for efficacy, and differences in the environmental characteristics (mobility, persistence, etc.) are masked in general comparisons of total pesticide volumes used.

To provide a more robust measurement of the environmental impact of GM crops, the analysis presented below includes an assessment of both pesticide active-ingredient use and the specific pesticides used via an indicator known as the environmental impact quotient (EIQ). This universal indicator, developed by Kovach et al. 1992 and updated annually, effectively integrates the various environmental impacts of individual pesticides into a single field value per hectare. This index provides a more balanced assessment of the impact of GM crops on the environment as it draws on all of the key toxicity and environmental exposure data related to individual products, as applicable to impacts on farmworkers, consumers, and ecology, and provides a consistent and comprehensive measure of environmental impact. Readers should, however, note that the EIQ is an indicator only and therefore does not account for all environmental issues and impacts.

The EIQ value is multiplied by the amount of pesticide active ingredient (AI) used per hectare to produce a field EIQ value. For example, the EIQ rating for glyphosate is 15.3. By using this rating multiplied by the amount of glyphosate used per hectare (e.g., a hypothetical example of 1.1 kg applied per hectare), the field EIQ value for glyphosate would be equivalent to 16.83/ha. In comparison, the field EIQ/ha value for a commonly used herbicide on corn crops (atrazine) is 22.9/ha.

The EIQ indicator is therefore used for comparison of the field EIQ/ha values for conventional versus GM crop production systems, with the total environmental impact or load of each system, a direct function of respective field EIQ/ha values, and the area planted to each type of production.

The EIQ methodology is used below to calculate and compare typical EIQ values for conventional and GM crops and then aggregate these values to a national level. The level of pesticide use in the respective areas planted for conventional and GM crops in each year was compared with the level of pesticide use that probably would otherwise have occurred if the whole crop, in each year, had been produced using conventional technology. This approach addresses gaps in the availability of herbicide or insecticide usage data in most countries and differentiates between GM and conventional crops. Additionally, it allows for comparisons between GM and non-GM cropping systems when GM accounts for a large proportion of the total crop planted area. For example, in the case of soybean in several countries, GM represents over 60 per cent of the total soybean crop planted area.

GM crops have contributed to a significant reduction in the global environmental impact of production agriculture. Since 1996, the use of pesticides was reduced by 224 million kg of active ingredient, constituting a 6.9 per cent reduction, and the overall environmental impact associated with pesticide use on these crops was reduced by 15.3 per cent. In absolute terms, the largest environmental gain has been associated with the adoption of GM HT soybean and reflects the large share of global soybean plantings accounted for by GM soybean. The volume of herbicide use in GM soybean decreased by 51 million kg since 1996, a 4.1 per cent reduction, and the overall environmental impact decreased by 20 per cent. It should be noted that in some countries, such as in Argentina and Brazil, the adoption of GM HT soybean has coincided with increases in the volume of herbicides used relative to historic levels. This net increase largely reflects the facilitating role of the GM HT technology in accelerating and maintaining the switch away from conventional tillage to no/ low-tillage production systems, along with their inherent environmental benefits. This net increase in the volume of herbicides used should, therefore, be placed in the context of the reduced GHG emissions arising from this production system change and the general dynamics of agricultural production system changes.

Major environmental gains have also been derived from the adoption of GM insectresistant (IR) cotton. These gains were the largest of any crop on a per hectare basis. Since 1996, farmers have used 95.5 million kg less insecticide in GM IR cotton crops (a 19.4 per cent reduction), and reduced the environmental impact by 24.3 per cent. Important environmental gains have also arisen in the maize and canola sectors. In the maize sector, pesticide use decreased by 43 million kg and the environmental impact decreased because of reduced insecticide use (4.6 per cent) and a switch to more environmentally benign herbicides (4 per cent). In the canola sector, farmers reduced herbicide use by 6.3 million kg (an 11 per cent reduction) and the environmental impact has fallen by 23 per cent because of a switch to more environmentally benign herbicides.

Table 1.6. GM Crop Environmental Benefits from Lower Insecticide and Herbicide Use in 2005: Developing versus Developed Countries

	Percent of Total Reduction in EI	
Crop	*Developed Countries*	*Developing Countries*
GM HT soybean	53	47
GM IR maize	92	8
GM HT maize	99	1
GM IR cotton	15	85
GM HT cotton	99	1
GM HT canola	100	0
Totals	46	54

In terms of the division of the environmental benefits associated with less insecticide and herbicide use for farmers in developing countries relative to farmers in developed countries, Table 1.6 shows that in 2005, the majority of the environmental benefits associated with lower insecticide and herbicide use have been for developing-country farmers. The vast majority of these environmental gains have been from the use of GM IR cotton and GM HT soybeans.

1.6.2.2. Impact on Greenhouse Gas (GHG) Emissions

Reductions in the level of GHG emissions from GM crops are from two principal sources:

i) GM crops contribute to a reduction in fuel use from less frequent herbicide or insecticide applications and a reduction in the energy use in soil cultivation. For example, Lazarus and Selley estimated that one pesticide spray application uses 1.045 liters (L) of fuel, which is equivalent to 2.87 kg/ha of carbon dioxide emissions. In this analysis we used the conservative assumption that only GM IR crops reduced spray applications and ultimately GHG emissions. In addition to the reduction in the number of herbicide applications there has been a shift from conventional tillage to no/reduced tillage. This has had a marked effect on tractor fuel consumption because energy-intensive cultivation methods have been replaced with no/reduced tillage and herbicide-based weed control systems. The GM HT crop where this is most evident is GM HT soybean. Here, adoption of the technology has made an important contribution to facilitating the adoption of reduced/no-tillage (NT) farming. Before the introduction of GM HT soybean cultivars, NT systems were practised by some farmers using a number of herbicides and with varying degrees of success. The opportunity for growers to control weeds with a nonresidual foliar herbicide as a "burndown" preseeding treatment, followed by a post-emergent treatment when the soybean crop became established, has made the NT system more reliable, technically viable, and commercially attractive. These technical advantages, combined with the cost advantages, have contributed to the rapid adoption of GM HT cultivars and the near-doubling of the NT soybean area in the United States. In both countries, GM HT soybean crops are estimated to account for 95 per cent of the NT soybean crop area. Substantial growth in NT production systems has also occurred in Canada, where the NT canola area increased from 0.8 to 2.6 million ha between 1996 and 2005. Similarly, the area planted to NT in the US cotton crop increased from 0.2 to 1 million ha over the same period. The increase in the NT cotton area has been substantial from a base of 200,000 ha to over 1.0 million ha between 1996 and 2005. The fuel savings resulting from changes in tillage systems are drawn from estimates from studies by Jasa and CTIC. The adoption of NT farming systems is estimated to reduce cultivation fuel usage by 32.52 L/ha compared with traditional conventional tillage and 14.7 L/ha compared with reduced tillage cultivation. In turn, this results in reductions in CO_2 emissions of 89.44 and 40.43 kg/ha, respectively.

ii) The use of reduced/no-tillage farming systems that utilise less plowing increase the amount of organic carbon in the form of crop residue that is stored or sequestered in the soil. This carbon sequestration reduces carbon dioxide emissions to the environment. Rates of carbon sequestration have been calculated for cropping systems using normal tillage and reduced tillage, and these were incorporated in our analysis on how GMcrop adoption has significantly facilitated the increase in carbon sequestration, ultimately reducing the release of CO_2 into the atmosphere. Of course, the amount of carbon sequestered varies by soil type, cropping system, and ecoregion. In North America, the International Panel on Climate Change estimates that the conversion from conventional tillage to no-tillage systems stores between 50 and 1300 kg C/ha annually. In the analysis presented below, a conservative savings of 300 kg C/ha per annum was applied to all no-tillage agriculture and 100 kg C/ha^{-1} $year^{-1}$ was applied to reduced-tillage agriculture. Where some countries aggregate their no/reduced-tillage data, the reduced-tillage saving value of 100 kg C/ha^{-1} $year^{-1}$ was used. One kilogram of carbon sequestered is equivalent to 3.67 kg of carbon dioxide. These assumptions were applied to the reduced pesticide spray applications data on GM IR crops, derived from the farm income literature review, and the GM HT crop areas using no/ reduced tillage.

The additional soil carbon sequestration gains resulting from reduced tillage with GM crops accounted for a reduction in 8.05 billion kg of CO_2 emissions in 2005. This is equivalent to removing nearly 3.6 million cars from the roads per year. In total, the carbon savings from reduced fuel use and soil carbon sequestration in 2005 were equal to removing 4 million cars from the road.

1.7. Market Value of Biotech Crops

Commercial production of biotech crop varieties has reached significant levels in four important commodities: corn, soybeans, cotton, and canola. As well, nearly all biotech production occurs in five countries: the United States, Argentina, Canada, Brazil, and China. There are other biotech countries and biotech crops, but these contribute only modestly to worldwide biotech production value.

Five countries had 67.5 million hectares planted to biotech varieties of maize, soybeans, cotton and canola in 2003/04 (Table 1. 7). They were the United States (42.8 million hectares); Argentina (13.9 million hectares); Canada (4.4 million hectares); Brazil (3.0 hectares); and China (2.8 million hectares). According to James these five countries and four crops constitute 98 percent of the total biotech cropland worldwide. Among the five countries 63 percent are planted in the United States, 21 percent in Argentina, 6 percent in Canada, and 4 percent each in Brazil and China. The rest of the world accounts for only 2 percent of the total area planted to biotech crop varieties.

In the United States, adoption of biotech varieties is led in area by corn and soybeans, followed by cotton and canola. In Argentina, corn, soybeans and cotton lead, while in Canada canola leads. Brazil's main biotech crop is soybeans. In China, the only reported acreage is in biotech cotton.

Using gross market values based on world prices we calculated in 2003/2004 the total value of biotech crops for the five leading countries and four leading crops at $43.9 billion

(Table 1.8). The United States accounted for $27.5 billion of this total, Argentina for $8.9 billion, China for $3.9 billion, Canada for $2.0 billion and Brazil for $1.6 billion.

Table 1.7. Global Biotech Crop Area: Leading Countries

Five leading counties	*Area in biotech crop production*	*Share of world biotech area*	*biotech crop varieties*
	67.5 million hectares	*98%*	
United States	42.8 million hectares	63%	maize, cotton, soy,
Argentina	13.9 million hectares	21%	canola soy, maize,
Canada	4.4 million hectares	6%	cotton canola,
Brazil	3.0 million hectares	4%	maize, soy cotton
China	2.8 million hectares	4%	

Table 1.8. Global Biotech Crop Value: Leading Countries

2003/04	*Biotech-related crop value*	*2003/04*	*Biotech-related crop value*
Five countries	*$43.9 billion*	*Four crops*	*$43.9 billion*
United States	$27.5 billion	Soybean	$23.5 billion
Argentina	$8.9 billion	Maize	$11.2 billion
China	$3.9 billion	Cotton	$7.8 billion
Canada	$2.0 billion	Canola	$1.4 billion
Brazil	$1.6 billion		

When broken out by crop, of the $43.9 billion total, biotech soybeans were valued at $23.5 billion, maize at $11.2 billion, cotton at $7.8 billion, and canola at $1.4 billion.

Soybeans were planted to 88 million hectares worldwide in 2003/04, with global production estimated at 190 million metric tons, and the world price averaging $250 per metric ton. The top five biotech countries represented 84 per cent of land area planted to soybeans and 90 per cent of production. More than half (54 percent) of soybean production in the top five biotech countries is from biotech varieties.

Total biotech soybean market value in 2003/04 was $23.5 billion - the highest of any biotech crop. The United States had the largest area in soybeans and highest biotech crop value ($13.3 billion). Brazil had the next largest biotech soy area, but due to a low (official) adoption rate, generated only $1.6 billion in biotech market value. Some reports suggest that the real biotech adoption rate in Brazil is as high as 30 per cent, which would more than double Brazil's biotech soybean production value. Argentina grew $8.3 billion in biotech soybeans in 2003/04. China grew 8.7 million hectares of conventional soybeans, but had no biotech production. The Canadian soybean area is just over a million hectares, and about half were biotech varieties.

Maize was grown on 140 million hectares worldwide in 2003-2004, producing 614 million metric tons, at an average world price of $100 per metric ton. The top five biotech countries represent 70 percent of worldwide maize production and 49 percent of the global maize land area. Biotech varieties are grown on 19 percent of maize production land in the top five biotech countries, which collectively produced $11.2 billion in biotech maize (Table 1.9).

Table 1.9. Global Biotech Maize Value: Leading Countries

Maize 2003/04 price = $100/MT	*Crop area M Ha*	*Production MMT*	*Biotech adoption rate*	*Biotech-related crop value*
Five countries	68.5	434.5	19%	$11.2 billion
United States	28.8	256.9	40%	$10.3 billion
China	23.5	114.0	–	–
Brazil	12.6	41.5	–	–
Argentina	2.1	12.5	40%	$500 million
Canada	1.2	9.6	40%	$384 million
Rest of the world	71.9	179.5	–	–

The United States is the leading biotech maize producer, with $10.3 billion in production market value. China's maize production area is nearly as large as the United States, but China does not grow biotech varieties commercially. Brazil has no reported biotech maize production, but significant maize production land area. Argentina has a modest area of land planted to maize, and an estimated $500 million in biotech soybean market value. Canada grew $384 million in biotech maize, based on a 40 percent adoption rate.

Cotton was planted to 32.6 million hectares worldwide in 2003/04. Production is estimated at 93.5 million bales of 480 pounds each. The adjusted world price averaged 59 cents per pound. Half of the world's cotton production takes place in the top five biotech countries, and 61 percent of that is from biotech varieties (Table 1. 10).

Table 1.10. Global Biotech Cotton Value: Leading Countries

Cotton 2003/04 price = $0.59/lb	*Crop area M Ha*	*Production M Bales*	*Biotech adoption rate*	*Biotech-related crop value*
Five countries:	11.2	46.7	61%	$7.8 billion
China	5.1	22.4	62%	$3.9 billion
United States	4.9	18.3	73%	$3.8 billion
Brazil	1.0	5.7	–	–
Argentina	0.3	0.4	60%	$75 million
Canada	–	–	–	–
Rest of the world	21.4	46.8	–	–

The global value of the biotech cotton in 2003/04 was $7.8 billion. China has the most area in cotton, the highest production and yields, and generates the most biotech cotton market value. The United States has almost as much area as China, higher adoption, lower yields, but essentially the same biotech production value. Argentina grew $75 million in biotech cotton on a relatively modest land area. This assumes a 60 per cent adoption rate, although some reports suggest it may be as low as 20 per cent. Brazil has more area in cotton production than Argentina, but no biotech adoption. No cotton is grown in Canada.

Canola was planted to 26 million hectares worldwide in 2003/04, with total production estimated at 39 million metric tons, and an average world price of $285 per metric ton. The top five biotech countries account for half the worldwide land area devoted to canola, and half the global production (Table 1.11).

Table 1.11. Global Biotech Canola Value: Leading Countries

Canola 2003/04 price = $285/MT	*Crop area M Ha*	*Production MMT*	*Biotech adoption rate*	*Biotech-related crop value*
Five countries	12.6	18.8	28%	$1.43 billion
China	7.5	11.4	–	–
Canada	4.7	6.7	68%	$1.29 billion
United States	.4	.7	73%	$138 million
Argentina	–	–	–	–
Brazil	–	–	–	–
Rest of the world	13.4	20.2	–	–

Among these top five countries, 28 percent of canola was a biotech variety. In 2003/04 the worldwide market value of the biotech canola crop was $1.4 billion. China grows the most canola among the five countries, but none in biotech varieties. Canada has the next largest land area planted to canola worldwide, but the majority of this crop is biotech, generating nearly $1.3 billion in biotech market value. The United States, by contrast, has modest canola production, but still produces $138 million in biotech canola value. Argentina and Brazil have no meaningful canola production and no biotech varieties in use.

Other countries grow biotech varieties of soybeans, cotton, and maize, apart from the five leading nations, and James identifies 13 countries with biotech crop production, with 8 of these at meaningful levels. Combined, these countries grew more than 600,000 hectares of biotech crops commercially in 2003/04, producing an additional $160 million in global biotech crop value. South Africa planted 400,000 hectares to a combination of maize, soybeans, and cotton, is the only country in Africa to produce a biotech crop, and could arguably be included as a top biotech country with its estimated $147 million in biotech crop value. In Latin America, Colombia grew 5,000 hectares of Bt cotton. Honduras planted less than 1,000 hectares of Bt maize, while Mexico also planted small areas of commercial biotech soybeans and Bt cotton. Uruguay grew 60,000 hectares of biotech soy, and produced its first crop of Bt maize. In the Asia-Pacific region, Australia planted 100,000 hectares or 59 percent of its cotton area in biotech varieties in 2003/04. The Philippines planted Bt corn for the first time on 20,000 hectares. India grew Bt cotton on 100,000 hectares, and Indonesia had unconfirmed reports of Bt cotton production. Several European countries grew biotech corn or soybeans in 2003/04. Romania grew 70,000 hectares of biotech soybeans. Spain put 32,000 hectares or 6 percent of its maize area in Bt varieties. Bulgaria had a few thousand acres of herbicide tolerant maize production, and Germany also had a small area of Bt maize.

Finally, there are other biotech food and fibber plants in commercial production around the globe. They have a relatively minor economic impact, compared with the four major crops, but are important to the areas and farmers that grow them. For example, biotech papaya varieties are grown in the United States that account for about half of the average $20 million papaya crop. Biotech tobacco is also grown in the United States, but the market penetration is too small to measure. Reports from other countries indicate market availability of biotech tomatoes and sweet peppers in China.

References

Beck, W.S., Liem, K.P., Simpson, G.G. *Life*. 3rd Edition. New York: Harper Collins. 1991.

Curtis, H. and Barnes, N. *Biology*. New York: Worth Publishers. 1989.

Dodds, John H. *Plant Genetic Engineering*. New York: Cambridge University Press. 1985.

Torrey, J.G. "The Development of Plant Biotechnology." *American Scientist*. 73:354-363. 1987.

Towle, Albert. *Modern Biology*. Austin, TX: Holt, Rinehart and Winston. 1988.

Walden, Richard. *Genetic Transformation in Plants*. Milton Keynes, England: Open University Press. 1988.

2

PLANT TISSUE CULTURE

Research in plant science often involves growing new plants in a controlled environment. These may be plants that we have genetically altered in some way or may be plants of which we need many copies all exactly alike. These things can be accomplished through tissue culture of small tissue pieces from the plant of interest. These small pieces may come from a single mother plant or they may be the result of genetic transformation of single plant cells which are then encouraged to grow and to ultimately develop into a whole plant. Tissue culture techniques are often used for commercial production of plants as well as for plant research.

Tissue culture involves the use of small pieces of plant tissue (explants) which are cultured in a nutrient medium under sterile conditions. Using the appropriate growing conditions for each explant type, plants can be induced to rapidly produce new shoots, and, with the addition of suitable hormones new roots. These plantlets can also be divided, usually at the shoot stage, to produce large numbers of new plantlets. The new plants can then be placed in soil and grown in the normal manner. Plant tissue culture is used widely in plant science; it also has a number of commercial applications.

2.1. Introduction to Cell and Tissue Culture

Tissue culture as a technique was first used almost 100 years ago to elucidate some of the most basic questions in developmental biology. Ross Harrison at the Rockefeller Institute, in an attempt to observe living, developing nerve fibers, cultured frog embryo tissues in plasma clots for 1 to 4 weeks. He was able to observe the development and outgrowth of nerve fibers in these cultures. In 1912, Alexis Carrel, also attempted to improve the state of the art of animal cell culture with experiments on the culture of chick embryo tissue. Carrel succeeded in expanding the possibilities of cell culture by keeping fragments of chick embryo heart alive and beating into the third month of culture and growing chick embryo connective tissue for over 3 months. Carrel reported growing chick embryo tissue for many years *in vitro,* and thus helped convince the scientific community that *in vitro* cultures were useful experimental systems.

The next important advance in the conceptualisation and technology of cell culture was the demonstration by Katherine Sanford and co-workers that single cells could be grown in culture. This, along with Harry Eagle's demonstration that the complex tissue extracts, clots, and so forth previously used to grow cells could be replaced by an arbitrary mixture of amino acids, vitamins, co-factors, carbohydrates, and salts, supplemented with a small amount of serum protein opened up a new area of cell culture. A vast range of manipulations that had not been possible previously could now be performed with cells, including production of genetically altered cell lines through mutagenesis and cloning, direct comparison of cells from normal and transformed tissues, the study of cellular physiology and metabolism, and the growth of normal and transformed human cells *in vitro*. Arising out of this work was the demonstration that cells in culture could be established as cell lines that maintained, at least in part, the differentiated functions characteristic of their cell type of origin. Thus, the creation of cell lines that maintained some functional properties of adrenal cells, pituitary cells, neurons, myocytes, and hepatocytes allowed the study not only of growth but of the response to hormones and other environmental factors and the production and secretion of hormones and other differentiated functions *in vitro.*

The demonstration that each cell type has an optimal mix of nutrients that supports its function has led to media derived to support specific cell types under specialised conditions. In parallel, the recognition that serum could be replaced by defined components such as attachment proteins, transport proteins, and hormones and growth factors once again opened up new possibilities for the maintenance of specialised cells and tissues in culture, and thus the ability to address important biological questions in new ways.

With the advent of recombinant expression in mammalian cells and the creation of antibody-producing hybridoma cell lines, along with the use of large-scale culture techniques for culturing mammalian cells, industrial cell culture has received widespread focus as a production system for recombinant proteins. The special considerations inherent in industrial production using large-scale cultures have further increased our understanding of the range of cell "behaviours," their inherent stability, the ability to genetically manipulate cell properties, and the technical challenges of growing mammalian cells in tanks large enough to have several atmospheres' difference in pressure from top to bottom.

Each of these observations and technical advances has brought new challenges and widened one's experience with that "microorganism", or "social organism," which is the mammalian cell *in vitro*.

2.2. Theory of Tissue Culture

Cell culture consists of a few basic concepts and techniques that can and should be mastered at the student or introductory level in a few weeks or months. These include sterile technique, subculture of cells, freezing and thawing cells, cloning cells, measuring cell growth and viability, and starting primary cultures. With these few techniques the scientist or student can usually successfully handle many of the experiments performed with established cell lines, especially those lines that are relatively hardy.

However, it is important for the scientist who makes tissue culture an important tool in his or her research to have a more complete understanding of the science and the years of experimentation behind these techniques. What does the medium do for the cells? How does the choice of incubator setting and medium interrelate? How can the environment be altered for optimal growth of cells at high density? How should the medium be changed for suspension culture? What does one do when the cells "just die"? There is an extensive body of information available that will help answer these questions; however, far too many scientists are content to ask someone else when they have a problem and consider the solution "magic." As flattering as it may be to be considered a magician, it is by far preferable that each person doing cell culture have a good basic understanding of the principles behind the subject.

2.2.1. Primary Culture

Several different types of culture are routinely performed. These can be roughly divided into "primary culture" and "culture of established cell lines." Primary culture can consist of the culture of a complex organ or tissue slice, a defined mixture of cells, or highly purified cells isolated directly from the organism, as illustrated in Figure 2.1 A, B, or C.

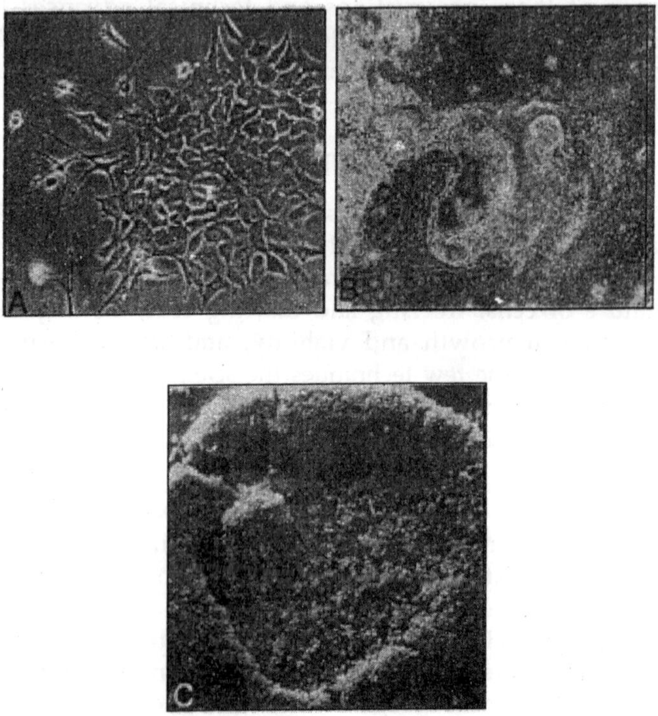

Figure 2.1. (A) Primary culture of isolated granulosa cells; (B) a coculture of several cell types from neonatal rat lung; and (C) a follicle derived from coculture of rat immature granulosa cells and oocytes that reassociate in vitro.

More commonly, techniques may be employed to purify the cell type of interest and start a primary culture consisting largely of that one cell type. Such cultures usually start at initial plating as containing 60°C 95 per cent of the cell type of interest, although this percentage may increase or decrease during the subsequent culture period. However, primary cell and organ cultures have an advantage in that they are recently removed from the *in vivo* situation and might therefore be expected to more closely resemble the function of that cell or tissue *in vivo*. The disadvantage is that these cultures are reacting to a constantly changing environment over the first days or weeks *in vitro*, including the damage sustained during the removal of cells from the animal and tissue and partial recovery from this damage, the change in environment from the animal to the *in vitro* culture, and the changing composition of the culture as some cells in the mixed cultures die and others proliferate and/or differentiate.

2.2.2. Established Cell Lines

The second type of cell culture is the culture of established or immortal cell lines. The vast majority of these are derived from tumours or from cells transformed *in vitro*, although some of the very earliest lines were established from normal embryonic tissue. There are also lines that have been widely used, such as WI-38, which are from normal human tissue and have a limited life span *in vitro*.

These cell lines have been the workhorses of cell culture, from their use in studying the control of the cell cycle to vaccine production and large-scale industrial production of recombinant proteins in 12,000 liter tanks. Not surprisingly, after many decades of growth in many laboratories they are both relatively tough and have altered from their original phenotype. Thus, cells having the same designation carried in different laboratories may vary considerably in their properties.

More recently, cell lines have been developed with the aim of maintaining a normal phenotype combined with the ability to grow the cell, or its precursor, indefinitely in culture. This can be accomplished using conditional transformation or by establishing the cell line from stem cell or precursor cells, which can then be induced to differentiate into a terminally differentiated cell type in culture.

2.2.3. Plasticity and Totipotency

Practically any plant transformation experiment relies at some point on tissue culture. There are some exceptions to this generalisation but the ability to regenerate plants from isolated cells or tissues in vitro underpins most plant transformation systems. Two concepts, plasticity and totipotency, are central to understanding plant cell culture and regeneration.

Plants, due to their sessile nature and long life span, have developed a greater ability to endure extreme conditions and predation than have animals. Many of the processes involved in plant growth and development adapt to environmental conditions. This plasticity allows plants to alter their metabolism, growth and development to best suit their environment. Particularly important aspects of this adaptation, as far as plant tissue culture and regeneration are concerned, are the abilities to initiate cell division from almost any tissue of the plant and to regenerate lost organs or undergo different developmental pathways in response to particular stimuli. When plant cells and tissues are cultured in vitro they generally exhibit a very high degree of plasticity, which allows one type of tissue or organ to be initiated from another type. In this way, whole plants can be subsequently regenerated. This regeneration of whole organisms depends upon the concept that all plant cells can, given the correct stimuli, express the total genetic potential of the parent plant. This maintenance of genetic potential is called 'totipotency'. Plant cell culture and regeneration do, in fact, provide the most compelling evidence for totipotency. In practical terms though, identifying the culture conditions and stimuli required to manifest this totipotency can be extremely difficult and it is still a largely empirical process.

2.3. Tissue Culture Laboratory

2.3.1. Space Requirements

The tissue culture laboratory should ideally be exclusive to tissue culture functions in order to minimise the introduction of potential contaminants. All people not cell culture should not enter the room. Avoid to entry and exit the culture room during the course of an experiment. Remove any equipment not needed for cell culture to another area of the laboratory.

The need to minimise the potential for contamination requires that the room be kept under positive pressure with high-efficiency particulate air (HEPA) filtered air flowing through it. The floors should be smooth and untextured. If a vinyl floor covering is used, it should be a continuous unseamed sheet. False ceilings are also a potential source of contamination and should not be used with a positive pressure air flow. If possible, a solid ceiling should be constructed. Minimally, new ceiling tiles should be installed every few years and the space above well cleaned, with any leaks from the outside or from condensation fixed immediately. Plumbing and all other "bulkhead" fittings and hardware should be well sealed where they pass through the wall or ceiling.

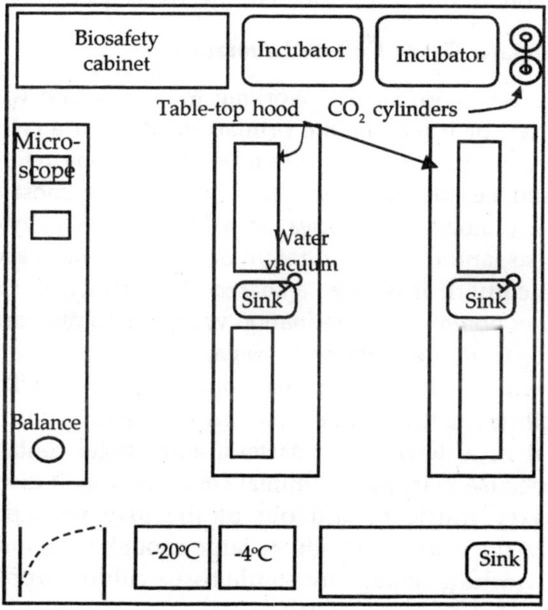

Figure 2.2. Suggested floor plan for a teaching laboratory uses bench top hoods and bench top or free-standing incubators.

Given that there are different needs, depending on the level of tissue culture being done, we will discuss the requirements for three types of tissue culture setups: the teaching laboratory, the standard research laboratory, and the optimal tissue culture laboratory. In all cases, however, the laboratory or designated area should be designed with minimal and optimal flow of traffic. This concept will affect placement of incubators, hoods, microscopes, freezers and refrigerators, and storage of sterile, disposable supplies, as well as positive and negative interaction of personnel. Figure 2.2. show floor plans for several different tissue culture laboratory configurations, depending on space limitations, funding, and usage level.

If you have the opportunity to design and construct or renovate your own tissue culture laboratory space, plan for plenty of "unused" space. This allows for easier cleaning, easier access to equipment, and the ability to add more equipment without costly new construction. There should be one biosafety hood for each person who is a full-time culture room user.

2.3.2. Standard Tissue Culture Laboratory

This may be a core facility, working in conjunction with other laboratories, or it may be the primary facility for a cell biology laboratory. All basic cell culture studies can be done in this facility that would be adequate for cell culture use for most cell and molecular biology laboratories, as well as use by physiologists, biochemists, and others who might occasionally need access to a culture facility. If it is to be a shared core facility, the need for minimising potential contamination vectors becomes paramount. This is critical not only with regard to bacterial or fungal contamination, but also with cross-contamination of cell cultures. Some facilities have an airlock that can serve as a buffer zone and somewhat of a deterrent to bacteria and fungal contaminants, largely because in its most minimal configuration, it discourages unnecessary traffic in and out of the primary laboratory. Optimally, the entryway might be large enough to include a sink for hand washing, storage for sterile tissue culture supplies, and even space for a freezer or cell counter.

When this is not possible, or is impractical, the laboratory should be designed so that cell cultures can be "compartmentalised," that is, primary cell cultures can be handled in a specifically designated hood(s), and kept in an incubator chamber separate from other cell lines being maintained in long-term culture. Cell cultures coming into the laboratory, as frozen vials or as viable cultures, primaries, or established cell lines, should be quarantined in an incubator chamber and handled in a designated hood until they are tested for mycoplasma.

Incubators are available as two-gas (CO_2 air) or three-gas models (CO_2, O_2, NO_2), this being largely determined by cost and specific needs of the investigators. It is always possible to augment a two-gas incubator chamber when this has not been incorporated into the original design. Insulation is maintained by either a water jacket or an air jacket, with corresponding advantages and disadvantages. The water jacketed incubator can maintain temperature over a longer period of time should there be a power outage, and this can be a critical feature for some installations. It is much heavier when filled, however, and the level must be maintained by periodic "topping off" and the jacket drained when the incubator has to be moved. The air-jacketed incubator is lighter, has more moving parts to fail, comes up to temperature faster, but will lose heat much faster when the fan goes off.

More importantly, the interior design and construction, materials used, and ease of assembly and disassembly can determine in part how well the cultures can be maintained free of contamination. Contaminating mold will grow on stainless steel, labelling tape, and even plastic, so shelving and the hardware securing it must be easy to remove and clean when necessary. Copper shelving and interior walls can inhibit the growth of such organisms but it is expensive, and unless all hardware components are of copper construction, one cannot completely inhibit the growth of mold on interior surfaces. Routine cleaning of stainless steel or aluminum shelves with a disinfectant and ethanol rinse will help to reduce these risks. Be careful to use a disinfectant recommended by the incubator supplier. Many excellent disinfectants are volatile and will kill cultured cells as well as contaminants. The chamber should be allowed to equilibrate overnight after a thorough cleaning prior to returning the cultures.

Incubator manufacturers have various proprietary methods of delivering and regulating gas flow into the chamber. Independent of this, the tissue culture laboratory needs a supply of CO_2 and any other gas that will be delivered to the chamber. In-house supplied gas can be a relatively inexpensive source; it need only be equipped with a miniature regulator to reduce the house gas to a flow rate optimal for the incubator and an in-line 0.2-μm filter to prevent introduction of mold and other potential contaminants that can accumulate on the inner wall of a gas hose. If O_2 experiments are to be carried out in an incubator that has such a provision, then house nitrogen should also be plumbed into the laboratory. Alternatively, when house gas is unavailable, commercial gas cylinders may be used, the critical point being to maintain an uninterrupted flow of gas. A manifold, connecting two or more supply cylinders with two-stage regulators, can be configured to supply the gas efficiently and economically. Reinforced silicon tubing that can be sterilised by autoclaving should be used to connect the gas source to the incubator.

It is important to know if the displays on the incubator control panel are reflecting in fact the actual conditions inside the chamber. For accurate temperature determination, a portable RTD thermometer is recommended. The appropriate thermometer-probe combination can provide accuracy to within 0.05°C. CO_2 can be monitored fairly accurately and inexpensively with a Fyrite unit.

The tissue culture hood can be as simple as an open, laminar flow unit, with air passing initially through a HEPA filter and moving parallel to the work surface, exiting at the front of the hood. However, current regulations may require the use of biosafety cabinets, in which HEPA filtered air is circulated within the hood and exhausted through appropriate filters and ductwork to the outside. These hoods are generally available in 4-ft and 6-ft lengths, the latter being somewhat more convenient in terms of work space. Two people can also work side by side in a 6-ft hood. This is convenient if the experimental protocol requires two people to work together. Larger hoods also may be necessary if large-scale tissue culture work is to be done where many large spinners or roller bottles are handled at the same time.

Regardless of the size being used, it is important that the interior of the hood be as free of obstruction as is practical to optimise airflow. Quite frequently one finds the hood being used as a repository for a variety of tissue culture supplies, vacuum units, personal belongings, and other sundry items, leaving little room for work space, thus minimising airflow and increasing the ever-present risk of contamination. Keep only the minimum necessary equipment in each hood, have one set of dedicated equipment and supplies for each hood, and restrict the use of that equipment to the hood.

No mouth pipetting should ever be done for tissue culture work. Currently available biosafety cabinets have duplex electrical outlets, convenient for plugging in pipetting aids and gas and vacuum valves, and are equipped with UV fixtures.

The only other item of major importance is a vacuum trap, consisting of two 1- or 2-liter Erlenmeyer flasks connected on one end to the vacuum source and on the other to a small hook attached to the hood (Figure 2.3). This apparatus should sit on the floor beneath the hood and be emptied regularly. Tubing, preferably silicon or latex, should have an inner diameter equivalent to the outer diameter of a Pasteur pipette and should be of sufficient length to facilitate aspirating medium and other reagents. The primary flask should contain disinfectant and the tubing should be flushed with disinfectant after the work in the hood is done. All hoods should have biohazard waste containers lined with autoclave bags.

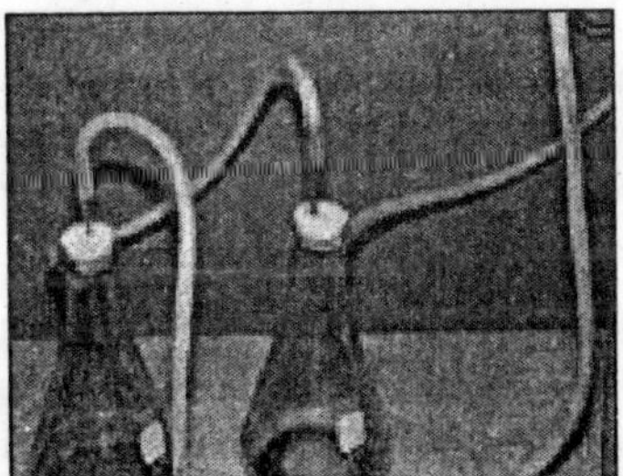

Figure 2.3. Vacuum Trap and Suction setup for Medium Disposal

The laboratory should have at least one inverted phase contrast microscope, equipped with 10x, 20x, and 40x objectives. If fluorescent microscopy will be needed, an epifluorescent attachment should be included. A 4x objective is useful for scanning large fields. For detecting mycoplasma, a 100x objective and fluorescence capability is necessary. A spring-loaded marker that screws into the nosepiece of the microscope is useful for marking areas of interest in culture dishes.

The standard tissue culture laboratory should have a reliable source of water for preparing medium. Considerable study has gone into water quality requirements for optimal cell growth, some cell types being far more sensitive to water quality than others. Nonetheless, all cells respond to water quality and it is important to be able to control this as much as possible. Ideally, water supplied by the city or county should first pass through a deionisation unit. Often, institutions will have a source of deionised water supplied to the laboratories. To this source, the investigator should connect a purification unit, usually in the form of several organic resin cartridges, a charcoal cartridge to remove organic compounds, and a final ultrafiltration cartridge, (Figure 2.4) with a 0.2- μm filter attached to the outlet. A still can also be used but this should be a double distillation with potassium permanganate in the first reservoir to remove organic material.

For teaching laboratories using tougher cells (and especially with serum-containing medium), bottled spring or distilled water can even be used. The water source should be tested, as some bottled waters are better than others for making media. This is not recommended, however, if a water purification system is available, as there are some cell types that will not grow if water quality is poor. Do not use tap water or water straight from a deionising column.

Liquid nitrogen freezers for long-term storage of cell cultures should be considered necessary for any standard tissue culture facility. These can be "portable" units, which can be moved when needed and hold up to 2,500 vials. This type of unit is filled with NO_2 manually. Alternatively, there are stationary freezers that can hold up to 10,000 vials or more and are filled automatically by a supplied NO_2 source tank. In either case, but particularly the former, it is absolutely necessary to regularly monitor the liquid nitrogen levels. Cells may be stored frozen at 80°C for a few months but lose viability rapidly at this temperature. Any facility that intends long-term storage of cultured cells should have liquid nitrogen storage capability, preferably a tank that stores cells in the vapour phase, since this minimises the possibility of cross-contamination of cells during storage.

Figure 2.4. Water Purification system that produced water suitable for medium preparation

For the student laboratory, cells can be obtained from a friendly research laboratory. Plate the cells at a relatively low density in 25-cm² flasks. Give a flask to each group of students. Prepare backup flasks, flush the flask with 5 per cent CO_2, and leave them sitting in a dark place at room temperature. Many cells will remain viable under these conditions for several weeks, although growth is extremely slow. If these need to be used, they should be placed in the incubator and grown to confluence before subculture.

2.3.3. Optimal Tissue Culture Laboratory

The optimal tissue culture laboratory is one that would serve a scientist whose main focus of research is cell biology. Thus, there would be multiple laboratory members using the facility and this use would go beyond merely growing cells to be used in experiments. The researchers might be studying many different aspects of cell behaviour and working out improvements in the technical aspects of cell culture. This setup would necessarily include all of the above-mentioned equipment as well as additional equipment for visualisation and maintenance of cell cultures. This could include multiple gas incubators, microscopes with upright/bright field capabilities, as well as a "dedicated" fluorescence microscope, upright or inverted, with objectives optimised for fluorescence. Such a laboratory also might need increased capability for observing cell cultures such as a facility for larger-scale cultures incorporating the use of roller bottles and spinner flasks. In addition, special microscopes for real-time three-dimensional observation and time-lapse videography can provide the investigator with powerful imaging tools. The ability to capture an image directly from the microscope is becoming increasingly important, and this may be done in a number of ways, such as videography of viable cultures over time or the utilisation of image capture boards to digitise, display, and analyse a single image.

Another consideration in the optimal setup is vibration isolation for microscopes. This becomes critical in situations where tables or benches used for the microscopes are situated adjacent to equipment with heavy and constant power drains: refrigerators, freezers, centrifuges, and so forth. Vibration isolation can be critically important in achieving success with a particular technique, microinjection, for example.

An electronic particle counter, such as the type made by Coulter Electronics, is extremely useful when conducting experiments that require numerous and regular cell counts, although alternative if somewhat indirect methods are becoming more widely available and may be more practical for experiments utilising multiwell plate formats.

2.3.4. Plasticware and Glassware

Plasticware, although expensive, is now preferred over glass for pipettes, tissue culture dishes, and so forth. If glassware, such as spinner flasks, is used for suspension cultures, it should be sialated to minimise the chance of cells sticking to the surface. Use of sialated tubes for media collection and storage before assay for secreted factors is also recommended. Do not sialate if cells are to be grown directly on glass surface or on matrixcoated surfaces, for example, coverslips. Glassware used for tissue culture should be autoclaved with moist heat for 30 min at a pressure of 15 lb/in^2. Empty bottles should have their screw caps loosely on and covered with foil. Partly filled bottles should have their caps tightened. The glass bottles should dry slowly after sterilisation to avoid condensation, which would otherwise be trapped inside. This condensation will dilute solutions placed in the bottle and may contaminate them with chemicals (e.g., boron) leached from the glass. Glass pipettes, tubes, and other small glass items can be sterilised in an oven with dry heat for 90 min at 160°C.

Tissue culture plasticware is available in a variety of formats, ranging from Terisaki plates to larger multiwell plates in 96-, 48, 24-, 12-, and 6-well configurations. In addition, chamber slides in different configurations are available, in which the upper chamber is removed so the slide can be used in immunocytochemistry or for viewing on an upright bright field microscope. Tissue culture flasks are available with "vented" caps that have filters incorporated in them for gas exchange without having to loosen the cap itself. Culture vessels useful for directly scaling up attached cultures are available. These include roller bottles, large stacked surfaces, hollow fibers, and so forth.

2.3.5. Maintaining the Laboratory

Routine maintenance is essential to having a functional cell culture laboratory. While one or more persons may be designated to oversee this task, all users must participate to some extent. This section will deal with those tasks necessary to keep a cell culture lab in good working order. Above all, the entire laboratory should be kept neat, clean, and uncluttered. Clutter provides a hiding place for contaminants and makes the space more difficult to use. Provide a designated space for all equipment and supplies and see that these items are returned to that space by all users. Optimally, when the tissue culture room is shared by several frequent users, each will have a designated space for their supplies and experiments in the refrigerator, a drawer, and incubator. Less frequent users might share a common space. It is generally more efficient for the hoods to be used of a "first-come-firstserved" basis, although a sign-up sheet may be necessary for large blocks of time or if there are a larger number of users.

2.3.5.1. *Daily Tasks*

The hoods should be wiped with a mild disinfectant such as 70 per cent ethanol before and after each use by each person using the hood. The trash receptacles and the media waste receptacles should be taken outside the tissue culture room and emptied at the end of the day, or more often if they are full. If any contaminated cultures have been discarded, they should be removed from the room immediately and autoclaved before disposal. In most areas, biosafety rules require that tissue culture waste be treated as a biohazard. Dry waste that has come in contact with cells can be discarded into biohazard bags and autoclaved. Liquid waste from the traps should be disinfected by the bleach or disinfectant in the trap. If there is excess medium, add more disinfectant before disposal. If any medium has been spilled in the hood, the trash, or on the floor, this should be cleaned up and the area disinfected immediately. Media are designed to grow cells, including prokaryotic cells, and do so just as well when spilled.

The hood sashes should be placed at the indicated level and the UV lights turned on in the hoods at the end of each use. Do not turn hoods off overnight, as continuous operation of the blowers and germicidal lamps will aid in maintaining aseptic conditions in the hood. Microscopes, cell counters, and other small equipment should be turned off. Visually check the temperature and CO_2 readouts on the incubators to confirm that they are operating within acceptable tolerances. All of the above should be rechecked at the end of the day by the last person leaving the room. Hoods, benches, microscopes, and water baths should be checked to make sure no reagents or cultures have been left out.

2.3.5.2. *Weekly Tasks*

The liquid nitrogen level should be checked in all storage tanks. The incubators should be checked for water. Sterile supply cabinets should be restocked. For serum-free culture work, it is imperative to have growth factors, hormones, and other required components on hand at all times. It may be necessary to maintain a list of these components on the outside of the refrigerator or freezer where the items to be ordered can be marked accordingly.

2.3.5.3. *Monthly Tasks*

These tasks may be performed less often if the facility is not heavily used, but all should be done at least annually. The room should be thoroughly cleaned, including the floors and bench tops. Ask the cleaning staff to use a new mop and fresh cleaning solution to clean the floor. The incubators should be disassembled, the removable parts autoclaved, and the rest wiped with disinfectant.

The CO_2 levels and temperature levels should be checked and calibrated. If a calibration service is available, have the incubators calibrated at least once a year and preferably every 6 months. In lieu of this service, check the temperature with an RTD thermometer and the CO_2 with a Fyrite unit on a regular basis. The hoods should also be cleared out, disassembled, and all surfaces wiped with a strong disinfectant, such as Betadine. All automatic pipettors should be checked for accuracy. Weighing the same nominal volume from several pipettors is a quick method to check accuracy. Coulter counters and other major equipment should be serviced.

Check the water purification unit and clean stills or replace cartridges on the filtration units. Clean out refrigerators and freezers. Throw out any expired solutions and check carefully that there is no mold growing on tubes, tape, or caps of containers in the refrigerator. Microscopes should be cleaned and the alignment checked. UV bulb life should be checked. Check to see that microscope bulbs and other supplies are well stocked.

Contamination and faulty equipment can destroy months of work overnight. The time and attention required for a well-maintained laboratory will repay the investigator with years of trouble-free service.

2.4. Physical Environment

The essential idea of the entire panoply of the chemical solution and mechanical apparatus surrounding the cell in *in vitro* culture is to recreate the physical, nutritional, and hormonal environment of the cell *in vivo*. This includes controlling the temperature, pH, osmolality, and gaseous environment; providing a supporting surface; and protecting the cell from chemical, physical, and mechanical stress. This section will deal with the control of the physical environment.

2.4.1. Temperature

Most mammalian cell cultures are grown in incubators that are set at 37°C. This was chosen because it is the core body temperature of *Homo sapiens.* One might therefore expect it to be the optimal temperature for growing many cell lines derived from human tissues. However, there is no reason to expect 37°C to be optimal for all types of human cells, and certainly not for all cells from all mammals, although most cells from warm-blooded animals will grow at this temperature. For example, the normal temperature of the human skin is well below the 37°C of the core of the body. Thus, it might be considered more "normal" to grow human epidermal cells at a lower temperature. Likewise, the testis of all mammalian species are maintained at a temperature several degrees lower than the core body temperature, and spermatogenesis will not proceed normally at core body temperature. It is therefore necessary to use a lower temperature if one wishes to create an *in vitro* environment appropriate for studying all aspects of the process of spermatogenesis. In contrast, many animals used to derive cell lines have a body temperature significantly different than that of humans. There is no *a priori* reason that cell lines from these animals' tissues should be grown at 37°C. However, if the line has been in common use for many years, the cells might be best adapted to this temperature. Insect, worm, and fish cells, on the other hand, cannot be grown at 37°C and require significantly lower temperatures.

There are several general statements about choosing the temperature set point for the incubator. Figure 2.5 shows packed cell volume and protein secretion by CHO cells grown at various temperatures. Growing cells at too high a temperature is more detrimental than too low a temperature. Even fairly short exposures to temperatures as little as 3°C above normal may lead to the death of the entire culture. Cells can tolerate temperature 10 to 20 degrees below normal for fairly long periods of time, but will grow slowly, if at all, at these low temperatures. If it is done correctly, cells can of course be frozen to extremely low temperatures of 80°C to 130°C and survive for years in cryostorage.

If studies are to be done comparing properties of cells grown at different temperatures, the two stock cultures should be grown at the two different temperatures for at least one subculture before setting up the comparison experiment. Abrupt changes of temperature cause many changes in the cells that may take some time to disappear. Likewise, when observing cells in the microscope, do not leave the plates out for long periods allowing them to cool to room temperature as this will affect the growth rate and other properties of the cells. Removing a few plates at a time for observation and returning them immediately after minimises cooling and the resulting condensation that also changes the properties of the medium.

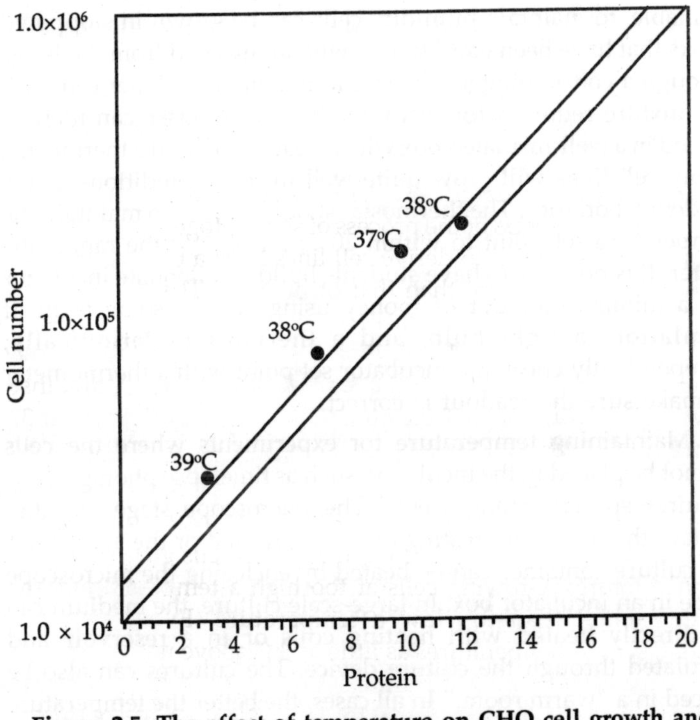

Figure 2.5. The effect of temperature on CHO cell growth and recombinant protein yield *in vitro*

The primary mode of maintaining temperature in cell culture is the incubator. While most cell culture incubators are set up to regulate humidity and the gaseous environment, as well as control temperature, temperature control is probably the most important part of this since there are alternative methods for controlling pH and humidity. Generally, incubators that maintain temperature by warming and circulating heated air are more responsive to temperature changes and will restore the temperature faster after the door of the incubator is opened and closed. However, in the event of a power failure or mechanical break-down, the water-jacketed incubators will maintain the temperature for several hours after power failure. The water also acts as insulation and the circulation of air only within the incubator makes maintaining a humidified atmosphere easier.

In student laboratories, if cell culture incubators are not available, to maintain humidity cells can be grown in stoppered flasks that have been equilibrated with air exhaled from the lungs through a cotton plugged pipette to maintain the 5 per cent CO_2 air mixture required for many media. These flasks can then be placed in a well-insulated box with a heat source and a thermostat. Many cell lines will grow quite well in these conditions in the student laboratory. The thermostat should be able to maintain the temperature set point to within 20°C; the tighter the range, the better. It is possible to have students build an adequate incubator for a minimal amount of money using acrylic, some form of insulation, a light bulb, and a thermostat. Periodically, independently check the incubator set point with a thermometer to make sure the readout is correct.

Maintaining temperature for experiments where the cells cannot be placed in the incubator, such as timelapse photography, requires special arrangements. The microscope stage or other surface the dishes are resting on can be heated or the air around the culture container can be heated by enclosing the microscope stage in an incubator box. In large-scale culture, the medium can be directly heated with heating coils or in a reservoir and circulated through the culture device. The cultures can also be placed in a "warm room." In all cases, the better the temperature control, the better and more reproducibly the cells will behave.

It is also important to be sure that the heating does not lead to undue evaporation of the medium. In an incubator this is done by maintaining a saturating humidity in the air in the incubator. Air blown over cultures to keep them warm should also have saturated humidity, and cultures kept on a heated surface will work best if the cells are in a sealed flask to minimise evaporation.

Since incubators, hoods, microscopes, freezers, and other equipment in the tissue culture room generate a good deal of heat, there should be good air turnover and a good cooling system for a culture room. If the temperature in the room exceeds 37°C by very much for any length of time, it is possible that all the cultures will be lost, since most tissue culture incubators do not cool. If non-mammalian cell lines, such as those from insects, fish, or amphibians, are being grown in a laboratory, an incubator that will cool the cultures is desirable, since these cell types must be grown at temperatures well below 37°C.

2.4.2. pH

Regulation of extracellular and intracellular pH is also essential for survival of individual mammalian cells and of animals as organisms. The pH is not only important for maintaining the appropriate ion balance, but also for maintaining the optimal function of cellular enzymes and for optimal binding of hormones and growth factors to cell surface receptors. Even transient changes in pH can alter cell metabolism and induce the production of heat-shock proteins, a process which can lead to apoptotic cell death. As with the organism, different compartments of the cells can have widely differing pHs, but the cell will maintain the appropriate pH of its subcellular compartments if the external environment is correct. This means not only providing the correct external pH but also the proper membrane components, ions, and ratio of ions, which allow the cells to maintain its internal pH through the integrity of its ion pumps and cell membranes.

Most media strive to achieve and maintain a pH between 7.0 and 7.4, with a median of 7.2. Different cell types may have an optimal pH slightly outside this range, and cells certainly differ widely in their ability to tolerate significant deviations from this level. As with temperature, slow changes in pH are tolerated better than rapid changes that cause apoptotic cell death. Most cells will tolerate a medium in the range of 6.5 to 7.8, but media much outside this range can be lethal. Even within this range, the growth rate and cellular functions vary widely as a function of pH. Cells will release cell-bound growth factors at a pH of 4 to 5 and remain intact but not viable. pH 2.0 is used to fix cells for some histochemical procedures. Cell membranes are solubilised at very basic pH, leaving basement membrane and cytoskeletal components on the culture vessel surface.

The regulation of pH is done through a variety of buffering systems. Most media use a bicarbonate CO_2 system as a major component of the buffer system. Other buffers included in most media formulations are the phosphate buffers. Media may be further supplemented with complex organic buffers or serum, which when present in levels of 5 20 per cent (vol/vol) of the medium provide significant buffering capacity.

An understanding of the metabolism of the cells is essential to be able to derive the best method to control cell pH using the media and buffering systems available. All cells produce lactic acid and CO_2 as a by-product of their energy metabolism. This is necessary for life. Generally, the faster-growing cells will produce more lactic acid than slow-growing or nongrowing cells, but there is great variability among different types of cells. In addition, there is an interdependence between medium composition and both the absolute and relative amounts of lactic acid produced by a given cell. Cells grown in low concentrations of glucose will convert a higher percentage of the glucose into macromolecules used in cell replication and less into lactic acid than the same cells grown in high concentrations of glucose. Lactic acid itself can be taken up and metabolised by some cells. Additionally, the CO_2 produced by the cells will affect the pH through the bicarbonate CO_2 buffering system, but will also be released into the atmosphere in the culture container at a rate determined by the culture vessel. Thus, a sealed flask will build up high levels of CO_2, while a fermenter with air sparging will strip much of the CO_2 produced by the cells out of the vessel. The actual pH in a culture at a given time is thus determined by the culture configuration, the medium the cells are grown in, the cell type, and the buffering capacity of the medium.

Cells grown in dishes in incubators are passively controlled by the type of dish, medium composition and depth, and the buffers, while cells grown in fermenters or instrumented spinners have active control of pH through the programmed addition of acid or base to maintain the pH at a given set point. This control comes at a cost, however. These added acids and bases result in an increase in salts in the medium and a resulting increase in osmolality, which eventually will become, of itself, limiting for cell growth.

The main buffering system in most media is the CO_2 bicarbonate system. The interaction of CO_2 derived from the cells or the atmosphere with water leads to a drop in pH described by the equation below:

$$H_2O + CO_2 = H_2CO_3 = H^+ + HCO^-_3$$

Increasing the bicarbonate concentration neutralises the effect of increased CO_2 due to the following:

$NaHCO_3 = Na^+ + HCO^{+3}$. The increased H drives the equation above to the left until equilibrium is reached at pH 7.4. The effect on cell growth of altering the pH by mismatched CO_2-bicarbonate concentrations is shown in Fig. 2.6.

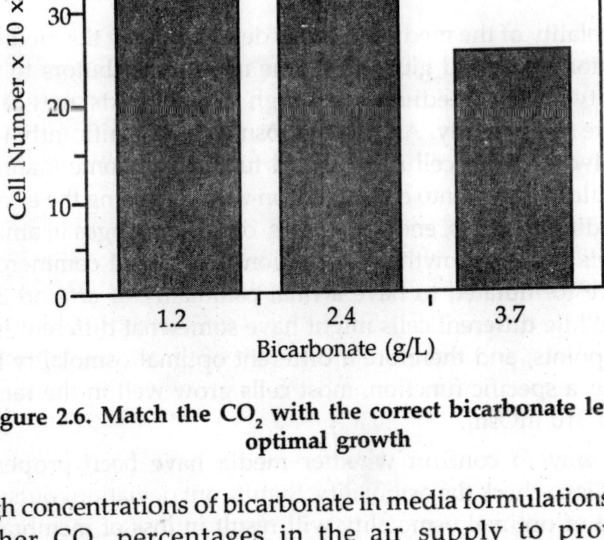

Figure 2.6. Match the CO_2 with the correct bicarbonate level for optimal growth

High concentrations of bicarbonate in media formulations require higher CO_2 percentages in the air supply to provide an appropriate pH in the medium but also to provide greater buffering capacity. The carbonate ions themselves also affect cell function to some extent, and some cell types require bicarbonate to grow, independent of its requirement as a buffer. The choice of buffering system will clearly affect both the pH and the osmolality of the final medium, as well as other aspects of cell physiology controlled by the ionic environment.

The bicarbonate buffering system, while inexpensive and well understood, may be supplemented with a variety of organic buffers. The most widely used zwitterionic buffer is HEPES (*N*-2-hydroxyethylpiperazine-*N*-2-ethane sulfonic acid). Since these organic buffers are insensitive to the CO_2 level, they provide a good backup system for rapidly metabolising cells that produce a lot of CO_2, and can stabilise pH swings that occur in CO_2 bicarbonate-buffered media when removing cells from the incubator for observation under the microscope. This is particularly important when eliminating serum that has a significant buffering capacity itself. Most cells tolerate HEPES buffer added in the range of 10 to 50 mM, or even higher. However, since this is a rather expensive medium additive, it is generally used at 10 to 20 mM.

2.4.3. Osmolality

The osmolality of the medium used is determined by the medium formulation. Salts and glucose are the major contributors to the osmolality of the medium, although amino acids may also contribute significantly. Altering the osmolality significantly will almost always affect cell growth and function in some manner. This should be taken into consideration when studying the effects of the addition of ions, energy sources, or large changes in amino acid levels on cell growth and function. Almost all commercial media are formulated to have a final osmolality of around 300 mOsm. While different cells might have somewhat different iso-osmotic points, and therefore a different optimal osmolality for growth or a specific function, most cells grow well in the range of 290 to 310 mOsm.

One way to confirm whether media have been properly prepared is to check the osmolality. Significant deviations outside the range of optimal osmolality will result in loss of membrane integrity, as the outside osmotic pressure becomes too much higher or lower than that which must be maintained inside the cell. In short, the cells explode or collapse. This phenomenon can be observed in minutes by removing medium from a dish of cells, replacing it with water, and observing under a microscope. If you wish to study the effects of osmolality on your cells, make up medium without sodium chloride. Add various concentrations of NaCl around the concentration that is normal for the medium you are using and determine the effect on the parameter to be studied.

Osmolality can be checked directly with an osmometer. One can also increase the osmolality of medium by adding sodium chloride. Adding 0.0292 g/liter of NaCl will raise the osmolality 1 mOsm. This is a rough estimate but useful if a quick experiment is to be performed to measure the effect of increasing osmolality. Remember, in media other components such as sugars, other ions, and amino acids also contribute to the final osmolarity of the medium. Also, be aware that changing osmolality by adding NaCl alters the Na:K ratio, and this may affect other aspects of cell function, including membrane transport.

2.4.4. CO_2, Oxygen, and other Gases

The CO_2 level is generally part of the buffering system of the medium and must be set accordingly. Oxygen levels in the incubator are provided by the 90 to 95 per cent air in the incubator mix. This means that the PO_2 in most incubators is at 18 per cent. To reduce the gaseous oxygen to less than this amount, special incubators must be used that mix CO_2, air, and nitrogen to obtain the desired PO_2. A cheaper but less precise method is to obtain a sealed plastic box with entry and exit ports and buy premixed gases with the appropriate PO_2. The cultures are placed in the box, the lid sealed, the box flushed with the compressed gas mixture through the ports, and the sealed box placed in the incubator to maintain temperature. Be sure to provide a source of moisture, such as an open plate of sterile water, since compressed air has an extremely low moisture content and can cause the medium in the dishes to evaporate, thus raising the osmolality. This method allows several different boxes with different gas mixes to be placed in the same incubator; however, a different tank will be needed for each mixture. Fermenters can be equipped directly with PO_2 sensors that can feed air or oxygen into a tank to maintain a set PO_2 level in the medium. This is probably the most accurate way to study the effects of minor variations in PO_2 on cell function.

However, most researchers do not deal with direct changes in the PO_2 level in the incubator atmosphere, but rather are concerned with changes in the PO_2 of the medium generated by the cells themselves and its effect on cell function. By far the most common problem is lack of adequate oxygen in cell culture. All mammalian cells require oxygen for their metabolism. This oxygen is obtained from the surrounding medium. In stationary cultures the oxygen enters the medium by diffusion from the medium atmosphere interface. If the diffusion rate is too low, the cells are too far from the medium surface or the cells are using the oxygen too fast, and oxygen deprivation, poor cell growth, and eventually death will result.

Oxygen generally will only diffuse freely through the first few millimeters from the surface. The amount of medium in a culture dish or flask will therefore be a balance between the cell's need for nutrients over the course of the experiment and the rate at which oxygen can diffuse from the surface to the cell monolayer. Too much medium will inhibit oxygen diffusion and slow cell growth; too little may be exhausted leading to cell death.

One constant threat to cell health is oxidative damage. Some cells are so sensitive to such damage that they cannot be maintained *in vitro* without antioxidants being added to the medium. One such cell type is the primary porcine Leydig cell, which will die between 72 and 90 hr of culture unless antioxidants such as vitamin E are added to the culture. This extreme sensitivity to an oxidative environment is found only in Leydig cells from the pig, although functional alterations may be present in other cell types when antioxidants are not present. In contrast, some cells do better in a reducing environment created by the addition of b-mercaptoethanol or other reducing agents. It should be recognised that reducing agents will inactivate some hormones, which are required for cell growth.

2.4.5. Surfaces and Cell Shape

A clear and obvious difference between the *in vitro* and *in vivo* environment is the surface to which cells attach and the shape they acquire as a result. While it seems intuitively obvious that cells in the organism do not grow attached to either glass or plastic, it is only in the last decade or two that researchers have demonstrated the profound effects that cell shape and the surface to which cells are attached can have on cell growth and differentiated function. This is due in part to the recent explosion in the discovery of the components that make up the extracellular matrix (ECM), which is part of basement membranes and connective tissue in all organs.

Many of these components are complex mixtures of components present in Engelbreth-Holm-Swarm and (EHS) tumour matrix, and culture dishes precoated with matrix proteins are now commercially available. Recent research has also expanded our understanding of the "receptors" that the cells use to attach to these ECM proteins and how this attachment affects cell function. An added complication is the fact that many cells *in vitro* can make their own attachment factors and can even secrete quite complex ECM, thus altering the surface they are plated on to one more amenable to that cell's function.

Serum also contains soluble attachment factors, such as fibronectin, which rapidly bind to tissue culture plastic and provide a modified surface for cell attachment. Cells may also add to or break down extracellular proteins that are provided for them.

2.4.5.1. *Adherent versus Nonadherent Cells*

Most cells in an animal grow attached to some structure in connective tissue, basement membrane, or mineral matrix such as bone. Cells in the blood, lymph, and other fluids are the only ones that normally grow "in suspension." Even so, many of these, such as lymphocytes, have the ability to move into solid tissue when required. Many of the cells which grow *in vivo* as attached cells cannot grow *in vitro* as a single cell suspension. If they are placed in culture in such a way that they cannot attach to a surface or another cell, they will die. Cells from tumours and those transformed *in vitro* frequently lose this need for attachment. In fact, the ability of cells to grow suspended in soft agar is considered one of the characteristics of transformed cells.

However, not all cells that grow in suspension are necessarily transformed. Lymphocytes readily grow in suspension and many cell types will survive and even function better when grown in conditions in which they do not attach and spread on a cell surface, but rather grow attached to each other as small aggregates. This highlights an important distinction between cell attachment and cell shape. Cells in a small cluster that are attached to each other are not growing as a single cell suspension but do have a more rounded shape than they would if they were attached to and spread on a surface. When interpreting the outcome of experimental manipulations, care must be taken to differentiate between the effects of changes in culture surface and the effects of changes in cell shape, which may result from altering the surface on which they are grown.

2.4.5.2. *Cell Shape*

The shape and size of cells *in vivo* will depend on their environment, what they are attached to, the hormonal signals they are receiving, and their function. Thus, epithelial cells may form a single cell layer attached to a basement membrane with thick, highly polarised cells. Peritubular cells may also be attached in a single layer but may be very thin, acting more as a physical barrier than a protein or mucopolysaccharide-secreting factory. Interstitial cells such as those in the ovary and testis may be almost round in cross-section and may occur in connected clumps in the tissue. In the artificial environment of *in vitro* culture systems, most cells will be rounded when grown in suspension and flattened when grown on a surface such as tissue culture plastic. Some cells, such as pancreatic b cells or testicular Leydig cells, may show increased function when they are allowed to maintain a rounded shape; other cell types may need to attach and spread to survive.

Plastic or glass surfaces can be coated with a purified attachment protein such as fibronectin, laminin, or collagen. Basement membrane can also be laid down on the culture dish by one cell type, the cells removed sterilely, and a second cell type cultured on this cell-produced ECM. This is less defined and more laborious than using purified protein components, but might well support a cell's growth or function better.

Cells can also be grown on a variety of artificial membranes. Generally speaking, there are several advantages to growing cells on membranes. The membranes can be used to separate the cell from the medium or one cell from another, and thus study cell communication and cell secretion. Cells will extend processes into the pores of some membranes or even migrate through membranes if the pores are large enough. Membranes can be chosen to have a specific molecular weight cutoff, and thus can pass or retain nutrients or cell-secreted products on the basis of size. In this case, the cells must be contained on one side of a continuous membrane and the other side will contain a separate medium reservoir. Most membranes are opaque, which makes direct observation of the cells growing on them impossible. However, several types of membranes have been developed for cell culture that are almost transparent when wet and allow observation of the cells in an inverted microscope. Cells grown on membranes can be fixed and processed for either scanning or electron microscopy if the membrane material is compatible with the solvents and fixatives used in these processes.

Membranes are also produced as hollow fibers within a sterile reservoir with inlets and outlets, as flat sheets or disks, or attached to plastic rings that can be inserted into a standard cell culture dish. While the use of membranes is expensive, one can obtain data in these configurations that cannot be obtained in other ways.

2.4.6. Stress

Another role of the supporting environment is to protect the cells in culture from stress that might kill them directly, set off natural genetic programs that may lead to their death (apoptosis), or cause mutations or other changes that would not reflect their natural functioning *in vivo* or their desired function *in vitro*. Stress encountered by the cell *in vitro* can be due to rapid changes in pH, temperature, or osmolality, which set off the transcription of "heat-shock" genes in response to this stress. Agents that damage DNA such as mutagenic chemicals or even light may also lead to cell death or mutation.

Agents that compromise membrane stability, such as detergent-like compounds, may lead to leakiness and the inability of the cells to maintain the pH and ion gradients necessary for function. Oxidative damage to cells is also an ever-present threat, given that oxygen is a necessary part of the cultural environment. Cells may be differentially sensitive to such damage, depending on their cytoplasmic enzymes, cellular composition, and ability to produce protective enzymes such as superoxide dismutase.

Finally, cells may be sensitive to contaminants and breakdown products found in medium, such as heavy metals contaminating medium components, organic compounds from insufficiently pure water, or ammonia produced as a by-product of cellular Again, it is important to keep in mind the difference between the goal of recreating an *in vitro* environment that allows the cells to function, to the greatest extent possible, the same way they do *in vivo* and other equally valid aims of culture. These model systems are invaluable for increasing our understanding of mammalian development, normal function, and disease states. However, there are many instances in which one might wish to use the flexibility and mutability of single cells *in vitro* to create a cell that is designed for a specific purpose, even if this might not reflect the *in vivo* situation. One such instance is the creation and optimisation of secretion of monoclonal antibodies by hybridomas.

If one wishes to use cells to provide a model for studying *in vivo* physiology, the cells should be protected from stress as much as possible from their initiation as a cell line, and certainly during the experimental procedures. Serum supplementation protects against many of these factors, since it is designed to protect cells in the organism from similar stress by maintaining homeostasis. Serum has significant buffering capacity, proteins that bind and detoxify heavy metals, and significant protease inhibitors to protect against damage from proteases released by living and lysed cells. Thus serum-free culture requires both more care and, to the greatest extent possible, replacement of protective properties by defined components such as organic buffers, protease inhibitors, and antioxidants. On the other hand, few cells exist in serum except in a wound. The presence of serum may therefore prime these cells to respond the way they would respond *in vivo* to this type of trauma, thus compromising the study of normal cell function.

If a cell line is to be engineered for a specific purpose, however, selecting for maximum resistance to stress may be desirable. This may be done by subjecting the cell line to sublethal stress and selecting for those cells most resistant. In fact, this has happened to most of the commonly used cell lines in the process of their being passed from laboratory to laboratory over the years.

Generally, the best idea is to protect cells from rapid changes in pH, temperature, and osmolality. These may shock the cells and lead to apoptotic death. Many researchers prefer not to remove the cells from the incubator at all, or even open the incubator, while a crucial experiment is in progress. This is obviously not possible in most laboratories or desirable where direct observation of cultures may add significant data. Incubators that contain several inner doors to the chambers may minimise drift in temperature and pH when the door is opened. In any case, all users should take care to keep incubators open for the minimum amount of time. Organic buffers, as discussed above, can be used to minimise pH swings.

Mechanical stress is generally not a concern to those growing cells in stationary culture such as tissue culture dishes or flasks. In fact, mechanical stress is a normal part of the environment of many cell types, including heart myocytes and other muscle cells that are stretched regularly, cells in the bloodstream that undergo a good deal of shear when travelling through small vessels and capillaries, and cells lining the lung airways that are stretched with each breath. Some experimenters recreate these stresses *in vitro* to study their effects. For example, myocytes can be grown on flexible substrates that can be stretched to study the effects of mechanical stress on these heart cells. In most cases, however, mechanical stress is best minimised *in vitro.*

Roller bottle culture represents minimal mechanical stress, but spinners, fermenters, and other methods of large-scale culture all present a certain amount of mechanical stress and the potential for mechanical damage to cells. These stresses include a significant change in pressure in large (e.g., 10,000 liter) tanks as the cells move from the top to the bottom of the tank, shear created by the mixing propellers and the bubbles created during oxygenation of the tank, and shear created when the cells are transferred through pumps from one vessel to another. Large-scale culture vessels and media are designed to minimise this mechanical stress.

Elements in the chemical environment of the cell such as heavy metals can also damage cells if they are not bound and sequestered. As mentioned above, many proteins in serum provide this protection, but serum can also contain toxic substances. Since these vary from serum to serum lot and the sensitivity varies from cell line to cell line, serum lots should be tested and compared using the cell lines you are planning to culture.

The best course to follow, in the absence or presence of serum, is to buy components of high purity, prepare tissue-culture-grade water in the laboratory for making media, and keep glassware and weighing devices for use exclusively in making media. In serum-free medium, transport proteins such as transferrin and ceruloplasmin can act not only to provide needed nutrient to the cells but to protect against damage caused by free iron, copper, and other metals that may also bind to these proteins.

2.4.7. Proteases

Proteases can be secreted by viable cells or released from dying cells into the medium. These may directly injure the cells, remove cell surface components such as receptors that are necessary for proper cell growth, or degrade growth factors added to the medium or proteins secreted by the cells. Serum contains protease inhibitors and proteases and substances such as plasminogen, which can be converted into active proteases by cell-secreted plasminogen activator. Thus, the proteolysis in serum-containing medium will reflect the cell type and amount of serum present. If cells are grown serum free, nonspecific protein such as bovine serum albumin or specific protease inhibitors can be added to protect from proteolytic degradation. The best-controlled physical environment still needs the addition of a well-thoughtout chemical and hormonal environment to produce the optimal growth and function of cells *in vitro*.

2.5. Tissue Culture Media

The nutrient mixtures used in tissue culture is termed as tissue culture media. In actual use, these will almost always be supplemented with serum or another complex biological fluid (milk, embryo extracts, and plasma), with a defined mixture of hormones and growth factors. All media are composed primarily of water. If media are prepared in the laboratory, the water source can make a critical difference in the quality of the media prepared. This is especially important for serum-free media where trace metals and organic compounds present in some distilled or deionised water can cause severe toxicity problems for some types of cells.

2.5.1. Basic Components of Culture Media

Culture media used for the in vitro cultivation of plant cells are composed of three basic components:

(1) Essential elements, or mineral ions, supplied as a complex mixture of salts;

(2) An organic supplement supplying vitamins and/or amino acids; and

(3) A source of fixed carbon; usually supplied as the sugar sucrose.

For practical purposes, the essential elements are further divided into the following categories:

(1) macroelements (or macronutrients);

(2) microelements (or micronutrients); and

(3) an iron source.

Complete, plant cell culture medium is usually made by combining several different components, as outlined in Table 2.2.

2.5.1.1. *Media components*

It is useful to briefly consider some of the individual components of the stock solutions.

2.5.1.1.1. *Macroelements*

As is implied by the name, the stock solution supplies those elements required in large amounts for plant growth and development. Nitrogen, phosphorus, potassium, magnesium, calcium and sulphur (and carbon, which is added separately) are usually regarded as macroelements. These elements usually comprise at least 0.1 per cent of the dry weight of plants.

Nitrogen is most commonly supplied as a mixture of nitrate ions (from the KNO_3) and ammonium ions (from the NH_4NO_3). Theoretically, there is an advantage in supplying nitrogen in the form of ammonium ions, as nitrogen must be in the reduced form to be incorporated into macromolecules. Nitrate ions therefore need to be reduced before incorporation. However, at high concentrations, ammonium ions can be toxic to plant cell cultures and uptake of ammonium ions from the medium causes acidification of the medium.

Table 2.1. Some of the elements important for plant nutrition and their physiological function. These elements have to supplied by the culture medium in order to support the growth of healthy cultures in vitro

Element	*Function*
Nitrogen	Component of proteins, nucleic acids and some coenzymes. Element required in greatest amount
Potassium	Regulates osmotic potential, principal inorganic cation
Calcium	Cell wall synthesis, membrane function, cell signalling
Magnesium	Enzyme cofactor, component of chlorophyll
Phosphorus	Component of nucleic acids, energy transfer, component of intermediates in respiration and photosynthesis
Sulphur	Component of some amino acids (methionine, cysteine) and some cofactors
Chlorine	Required for photosynthesis
Iron	Electron transfer as a component of cytochromes
Manganese	Enzyme cofactor
Cobalt	Component of some vitamins
Copper	Enzyme cofactor, electron-transfer reactions
Molybdenum	Enzyme cofactor, component of nitrate reductase
Zinc	Enzyme cofactor, chlorophyll biosynthesis

In order to use ammonium ions as the sole nitrogen source, the medium needs to be buffered. High concentrations of ammonium ions can also cause culture problems by increasing the frequency of vitrification (the culture appears pale and 'glassy' and is usually unsuitable for further culture). Using a mixture of nitrate and ammonium ions has the advantage of weakly buffering the medium as the uptake of nitrate ions causes OH- ions to be excreted. Phosphorus is usually supplied as the phosphate ion of ammonium, sodium or potassium salts. High concentrations of phosphate can lead to the precipitation of medium elements as insoluble phosphates.

Table 2.2. Composition of a typical plant culture medium. The medium described here is that of Murashige and Skoog (MS)a

Essential element	Concentration in stock solution (mg/l)	Concentration in medium (mg/l)
Macroelements		
NH_4NO_3	33000	1 650
KNO_3	38000	1 900
$CaCl_2.2H_2O$	8800	440
$MgSO_4.7H_2O$	7400	370
KH_2PO_4	3400	170
Microelements		
KI	166	0.83
H_3BO_3	1240	6.2
$MnSO_4.4H_2O$	4460	22.3
$ZnSO_4.7H_2O$	1720	8.6
$Na2MoO_4.2H_2O$	50	0.25
$CuSO_4.5H_2O$	5	0.025
$CoCl_2.6H_2O$	5	0.025
Iron sourcec		
$FeSO_4.7H_2O$	5560	27.8
$Na_2EDTA.2H_2O$	7460	37.3
Organic supplement		
Myoinositol	20000	100
Nicotinic acid	100	0.5
Pyridoxine-HCl	100	0.5
Thiamine-HCl	100	0.5
Glycine	400	2
Carbon source		
Sucrose	Added as solid	30 000

2.5.1.1.2. Microelements

These elements are required in trace amounts for plant growth and development, and have many and diverse roles. Manganese, iodine, copper, cobalt, boron, molybdenum, iron and zinc usually comprise the microelements, although other elements such as nickel and aluminium are frequently found in some formulations.

Iron is usually added as iron sulphate, although iron citrate can also be used. Ethylenediaminetetraacetic acid (EDTA) is usually used in conjunction with the iron sulphate. The EDTA complexes with the iron so as to allow the slow and continuous release of iron into the medium. Uncomplexed iron can precipitate out of the medium as ferric oxide.

2.5.1.1.3. Organic supplements

Only two vitamins, thiamine (vitamin B1) and myoinositol (considered a B vitamin) are considered essential for the culture of plant cells in vitro. However, other vitamins are often added to plant cell culture media for historical reasons.

Amino acids are also commonly included in the organic supplement. The most frequently used is glycine (arginine, asparagine, aspartic acid, alanine, glutamic acid, glutamine and proline are also used), but in many cases its inclusion is not essential. Amino acids provide a source of reduced nitrogen and, like ammonium ions, uptake causes acidification of the medium. Casein hydrolysate can be used as a relatively cheap source of a mix of amino acids.

2.5.1.1.4. Carbon source

Sucrose is cheap, easily available, readily assimilated and relatively stable and is therefore the most commonly used carbon source. Other carbohydrates (such as glucose, maltose, galactose and sorbitol) can also be used, and in specialised circumstances may prove superior to sucrose.

2.5.1.1.5. Gelling agents

Media for plant cell culture in vitro can be used in either liquid or 'solid' forms, depending on the type of culture being grown. For any culture types that require the plant cells or tissues to be grown on the surface of the medium, it must be solidified (more correctly termed 'gelled'). Agar, produced from seaweed, is the most common type of gelling agent, and is ideal for routine applications. However, because it is a natural product, the agar quality can vary from supplier to supplier and from batch to batch. For more demanding applications, a range of purer (and in some cases, considerably more expensive) gelling agents are available. Purified agar or agarose can be used, as can a variety of gellan gums.

2.5.1.2. *Plant Growth Regulators*

The essential point as far as plant cell culture is concerned is that, due to this plasticity and totipotency, specific media manipulations can be used to direct the development of plant cells in culture.

Plant growth regulators are the critical media components in determining the developmental pathway of the plant cells. The plant growth regulators used most commonly are plant hormones or their synthetic analogues.

There are five main classes of plant growth regulator used in plant cell culture, namely:

(1) auxins;
(2) cytokinins,
(3) gibberellins;
(4) abscisic acid;
(5) ethylene.

Each class of plant growth regulator will be briefly looked at.

2.5.1.2.1. *Auxins*

Auxins promote both cell division and cell growth The most important naturally occurring auxin is IAA (indole-3-acetic acid), but its use in plant cell culture media is limited because it is unstable to both heat and light. Occasionally, amino acid conjugates of IAA (such as indole-acetyl-L-alanine and indole-acetyl-L-glycine), which are more stable, are used to partially alleviate the problems associated with the use of IAA. It is more common, though, to use stable chemical analogues of IAA as a source of auxin in plant cell culture media. 2,4-Dichlorophenoxyacetic acid (2,4-D) is the most commonly used auxin and is extremely effective in most circumstances. Other auxins are available, and some may be more effective or 'potent' than 2,4-D in some instances.

2.5.1.2.2. Cytokinins

Cytokinins promote cell division. Naturally occurring cytokinins are a large group of structurally related (they are purine derivatives) compounds. Of the naturally occurring cytokinins, two have some use in plant tissue culture media. These are zeatin and 2iP (2-isopentyl adenine). Their use is not widespread as they are expensive (particularly zeatin) and relatively unstable. The synthetic analogues, kinetin and BAP (benzylaminopurine), are therefore used more frequently. Non-purine-based chemicals, such as substituted phenylureas, are also used as cytokinins in plant cell culture media. These substituted phenylureas can also substitute for auxin in some culture systems.

2.5.1.2.3. Gibberellins

There are numerous, naturally occurring, structurally related compounds termed 'gibberellins'. They are involved in regulating cell elongation, and are agronomically important in determining plant height and fruit-set. Only a few of the gibberellins are used in plant tissue culture media, GA3 being the most common.

2.5.1.2.4. Abscisic acid

Abscisic acid (ABA) inhibits cell division. It is most commonly used in plant tissue culture to promote distinct developmental pathways such as somatic embryogenesis.

2.5.1.2.5. Ethylene

Ethylene is a gaseous, naturally occurring, plant growth regulator most commonly associated with controlling fruit ripening in climacteric fruits, and its use in plant tissue culture is not widespread. It does, though, present a particular problem for plant tissue culture. Some plant cell cultures produce ethylene, which, if it builds up sufficiently, can inhibit the growth and development of the culture. The type of culture vessel used and its means of closure affect the gaseous exchange between the culture vessel and the outside atmosphere and thus the levels of ethylene present in the culture.

2.5.1.2.6. Plant growth regulators and tissue culture

Generalisations about plant growth regulators and their use in plant cell culture media have been developed from initial observations made in the 1950s. There is, however, some considerable difficulty in predicting the effects of plant growth regulators: this is because of the great differences in culture response between species, cultivars and even plants of the same cultivar grown under different conditions. However, some principles do hold true and have become the paradigm on which most plant tissue culture regimes are based.

Auxins and cytokinins are the most widely used plant growth regulators in plant tissue culture and are usually used together, the ratio of the auxin to the cytokinin determining the type of culture established or regenerated. A high auxin to cytokinin ratio generally favours root formation, whereas a high cytokinin to auxin ratio favours shoot formation. An intermediate ratio favours callus production.

2.5.2. Selecting an Appropriate Medium

If a new cell line is brought into the laboratory, determine what medium is recommended for its growth. This information can be obtained from the same source as the cells. If the recommended medium is incompatible with the CO_2 settings on the incubator used for other cells grown in the laboratory, or is not commonly prepared in the laboratory, one might wish to change the growth medium. It is best to initially grow the cells in their original medium and compare this with the more convenient medium after a passage or two in each. If the growth rate and morphology of the cells look the same, then a medium switch can be made.

However, keep in mind when trying to repeat published data that cells grown in a different medium may respond differently in some parameters measured. For example, estrogen-requiring cells previously grown in F12 medium and then switched to DME medium may have a diminished response to exogenously added estrogens in this medium because of the weak estrogenic activity of the phenol red, which is present at a much higher concentration in the DME medium. If one wishes to grow a primary cell in culture and no published data exist on growing that cell type *in vitro,* or one wishes to grow the cells in a different manner, it is best to screen several of the commercially available media before deciding on the one that is best for that particular use.

This can be done by obtaining five to ten candidate media powders from a supplier, preparing them all in the laboratory as described below using the same water and supplementary components, and doing a direct comparison of cell growth in the different conditions. If endpoints other than cell growth are important, measure these too in each of the media. Carry the cells in the selected medium for several passages and freeze them in this medium for future use. It is interesting to compare the formulations of the media tested and look for patterns of whether the cells of interest seem to prefer a low or high glucose, low or high calcium, low or high levels of amino acids, and so forth. This will be useful if you decide to further modify the medium, as described in the section below. It is important to know the composition of the test medium if conditioned medium is to be screened for biological activity. Since this is often done at 10 to 50 per cent conditioned medium, the test medium should minimally use the same bicarbonate buffer concentration as the assay medium and optimally be identical to the assay medium except for changes introduced by the conditioning cells.

2.5.3. Media Preparation

Media can be purchased as prepared liquid media, made up in the laboratory from dried powders containing most of the components of the nutrient mixtures, or prepared in the laboratory from individual stocks of the individual components or groups of the components. We do not recommend purchasing liquid medium, especially for serum-free culture work. Medium components deteriorate with time, and do so faster in solution. Some necessary components break down and are lost, others create toxic breakdown products or oxidise to toxic components. Some prepared liquid media can be frozen. Those that form a precipitate when thawed should not be frozen. In any case, it is always safe to store the prepared powdered medium and make liquid medium in the laboratory on a regular basis.

Powdered nutrient mixtures generally have a shelf life of a year or more if stored in moisture-proof, airtight containers in the dark. If the laboratory does not use large volumes of media, the 1-liter packages are convenient. Preparing medium in the laboratory from components immediately before use is obviously the best way to insure that the medium contains the desired components in the desired form. This is essential if the investigator wishes to study the role of the nutrients themselves or to optimise the nutrient portion of the medium as described below. However, most laboratories will find that preparation of medium from commercial powdered nutrient mixtures and a limited storage of the prepared media in a light-tight refrigerator will be adequate for their needs. This is also less costly than purchasing prepared media, especially when the cost of filters and so forth can be spread over large-volume use.

2.5.4. Serum Treatment

Serum is frequently spoken of as if it were a defined single substance. This is very far from the truth. Cell culture media can be supplemented with sera from any species of animal; bovine (fetal, newborn, or adult), horse, or human sera are the most frequently used. These are quite different in many ways and can have very different effects on the properties of cells grown in them. Additionally, serum varies from animal to animal, with changes in diet, and seasonally. Therefore, there is considerable variability from lot to lot of the commercially available sera. In addition to whole sera, which is allowed to clot and the clot removed, the blood can be collected with an anticlotting agent and the cellular portion spun out, resulting in plasma. Serum and plasma, even from the same animal, are quite different in composition and their effect on cells.

Sera can be treated before use in one or more ways: filtration, dialysis, diafiltration, heat treatment, or fractionation. These treatments can act as an added insurance against contamination, can remove or inactivate toxic components of the serum, can remove or inactivate growth-promoting or differentiating components of the serum, and specifically can remove low- or high-molecular-weight components of the serum or particular serum fractions. Clearly, this complex and undefined addition to medium must be treated with some care to insure consistent results. The only way to insure good results is to thoroughly test several lots of serum for their ability to support the desired cell characteristics and then buy a large quantity of the best lot, store it at 20 to 80°C, and use it for the next several years.

Most commercial sera come sterilely packaged. It is best to purchase serum that has been sterilely collected as well, as an added insurance against viruses or mycoplasma, which can go through some filters. When adding human serum to cultures, the entire culture and all waste should be treated as a biohazard. Human sera should be collected from known donors or blood banks that test for the common viruses such as human immunodeficiency virus (HIV) and hepatitis.

2.5.5. Quality Control

It is best to prepare media from commercially available powdered nutrient mixtures. It is important to keep good records and do quality control testing of reagents used in making the medium. We generally keep one set of glassware exclusively for medium making. This is rinsed well with distilled water but not washed with detergent between each use. This avoids the possibility of any detergent residue getting into the medium. The sodium bicarbonate and other reagents that are used in media should be used only for media. When weighing out these reagents, a disposable tongue depressor and weigh boat should be used. This avoids contaminating these reagents with other, potentially toxic chemicals that may be in use in the laboratory.

References

Bhojwani, S. S., Razdan M. K. *Plant tissue culture: theory and practice, Revised edition* Elsevier, 1996.

George, E. F. Hall, M. A. and G.-J. De Klerk (eds). *Plant propagation by tissue culture. Volume 1. The background. 3rd edition* Springer, Dordrecht, 2008.

Dodds, J. H. and Roberts, L. W. *Experiments in plant tissue culture.* Cambridge University Press, Cambridge.1995.

Gamborg, O. L. "Plant tissue culture. Biotechnology" . Milestones. *In vitro Cellular and Developmental Biology—Plant,* 38, 84–92. 2002.

Torbert, K. A., Rines, H. W. and Somers, D. A. "Transformation of oat using mature embryo-derived tissue cultures" . *Crop Science,* 38, 226–31.1998.

3

TYPES OF PLANT TISSUE CULTURES

Plant tissue culture techniques are essential to many types of academic inquiry, as well as to many applied aspects of plant science. In the past, plant tissue culture techniques have been used in academic investigations of totipotency and the roles of hormones in cytodifferentiation and organogenesis. Currently, tissue-cultured plants that have been genetically engineered provide insight into plant molecular biology and gene regulation. Plant tissue culture techniques are also central to innovative areas of applied plant science, including plant biotechnology and agriculture. For example, select plants can be cloned and cultured as suspended cells from which plant products can be harvested. In addition, the management of genetically engineered cells to form transgenic whole plants requires tissue culture procedures; tissue culture methods are also required in the formation of somatic haploid embryos from which homozygous plants can be generated. Thus, tissue culture techniques have been, and still are, prominent in academic and applied plant science.

3.1. Types of Cultures

Cultures are generally initiated from sterile pieces of a whole plant. These pieces are termed 'explants' and may consist of pieces of organs, such as leaves or roots, or may be specific cell types, such as pollen or endosperm. Many features of the explant are known to affect the efficiency of culture initiation. Generally, younger, more rapidly growing tissue is most effective. Several different culture types most commonly used in plant transformation studies will now be examined in more detail.

3.1.1. Callus

When explants are cultured on the appropriate medium, usually with both an auxin and a cytokinin, an unorganised, growing and dividing mass of cells called callus results. It is thought that any plant tissue can be used as an explant, if the correct conditions are found. In culture, this proliferation can be maintained more or less indefinitely, provided that the callus is subcultured on to fresh medium periodically. During callus formation there is some degree of dedifferentiation, both in morphology and metabolism.

One major consequence of this dedifferentiation is that most plant cultures lose the ability to photosynthesise. This has important consequences for the culture of callus tissue, as the metabolic profile will probably not match that of the donor plant. This necessitates the addition of other components such as vitamins and, most importantly, a carbon source to the culture medium, in addition to the usual mineral nutrients.

Callus culture is often performed in the dark as light can encourage differentiation of the callus. During long-term culture, the culture may lose the requirement for auxin and/or cytokinin. This process, known as 'habituation', is common in callus cultures from some plant species. Callus cultures are extremely important in plant biotechnology. Manipulation of the auxin to cytokinin ratio in the medium can lead to the development of shoots, roots or somatic embryos from which whole plants can subsequently be produced. Callus cultures can also be used to initiate cell suspensions, which are used in a variety of ways in plant transformation studies (Figure 3.1).

3.1.1.1. Cell-suspension cultures

Callus cultures can be broadly classified into compact or friable categories. In compact callus the cells are densely aggregated, whereas in friable callus the cells are only loosely associated with each other and the callus becomes soft and brittle. Friable callus provides the inoculum to form cell-suspension cultures. Explants from some plant species or particular cell types tend not to form friable callus, making cell-suspension initiation a difficult task. The friability of callus can sometimes be improved by controlling the medium components or by repeated subculturing or by culturing it on 'semi-solid' medium.

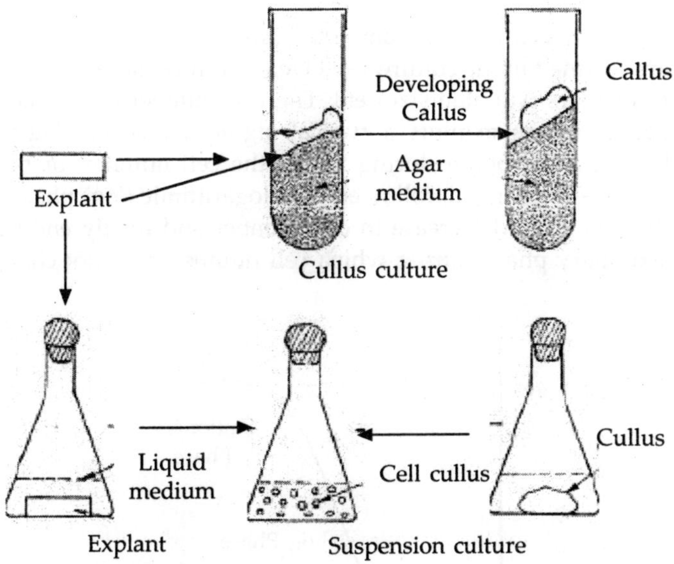

Figure 3.1. Initiation of Callus and Suspension Cultures.

When friable callus is placed into a liquid medium and then agitated, single cells and/or small clumps of cells are released into the medium. Under the correct conditions, these released cells continue to grow and divide, eventually producing a cell-suspension culture. A relatively large inoculum should be used when initiating cell suspensions so that the released cell numbers build up quickly. The inoculum should not be too large though, as toxic products released from damaged or stressed cells can build up to lethal levels. Large cell clumps can be removed during subculture of the cell suspension. Liquid cultures must be constantly agitated, generally by a gyratory shaker at 100-250 rpm, to facilitate aeration and dissociation of cell clumps into smaller pieces. Suspension cultures grow much faster than callus cultures, need to be subcultured about every week, allow a more accurate determination of the nutritional requirements of cells and are the only system amenable to scaling up for a large scale production of cells and even somatic embryos (SEs). The suspension cultures are broadly grouped as: (1) batch cultures, (2) continuous cultures, and (3) immobilised cell cultures.

3.1.1.1.1. Batch cultures

In a batch culture the same medium and all the cells produced are retained in the culture vessel, e.g., culture flasks (100-250 ml), fermenters (variable size) etc. The cell number or biomass of a batch culture exhibits a typical sigmoidal curve (Figure3.2), having a lag phase during which the cell number, or biomass remains unchanged, followed by a logarithmic (log) phase when there is a rapid increase in cell number and finally ending in a stationary phase during which cell number does not change.

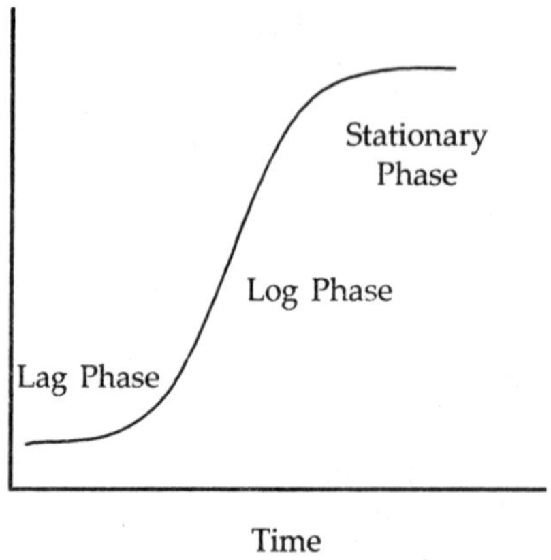

Figure 3. 2. A Model Curve for Cell Number in a Batch Culture

The lag phase duration depends mainly on inoculum size and growth phase of the culture from which inoculum is taken. The log phase lasts about 3-4 cell generations, and the duration of a cell generation may vary from 22-48 hr mainly depending on the plant species. The stationary phase is forced on the culture by a depletion of the nutrients and possibly due to an accumulation of cellular wastes. If the culture is kept in stationary phase for a prolonged period the cells may die. Batch cultures are maintained by subculturing. They are used for initiation of cell suspensions which may be used for cloning, cell selection or as seed cultures for scaling up or for continuous cultures. They are, however, unsuitable for studies on cell growth and metabolism because there is a constant change in cell density and nutritional status of the medium. But batch cultures are much more convenient than continuous cultures and as a result, are routinely used.

3.1.1.1.2. Continuous cultures

In a continuous culture, the cells are maintained in a steady state by regularly substituting a portion of the used or spent medium by fresh medium. Such culture systems are of either (1) closed or (2) open type. In a closed continuous culture, cells are detached from the used medium taken out for replacement, and added back to the culture/so that cell biomass keeps on increasing. In an open continuous culture both cells and the used medium are taken out and replaced by equal volume of fresh medium. The replacement volume is so adjusted that cultures remain at submaximal growth indefinitely.

The open cultures are of either turbidostat or chemostat types. In the former cells are allowed to grow up to a certain turbidity when a predetermined volume of the culture is replaced by fresh normal culture medium. In contrast, in a chemostat, a chosen nutrient is kept in a concentration so that it becomes depleted with limited growth, while other nutrients are still in concentrations higher than required. In such a situation, any addition of the growth-limiting nutrient is reflected in cell growth. Chemostats are ideal for the determination of effects of individual nutrients on cell growth and metabolism.

3.1.1.1.3. Immobilised cell cultures

Plant cells and cell groups may be enclosed in a suitable material, like agarose and calcium alginate gels, or entrapped in membranes or stainless steel screens. Cells receive liquid medium is continuously run through the column to provide them nutrients and aeration. Immobilisation of cells changes their cellular physiology in comparison to suspension culture cells thus offering several advantages for their use in biochemical production.

3.1.2. Subculture

After a period of time, it becomes necessary, chiefly due to nutrient depletion and medium drying, to transfer organs and tissues to fresh media. This is particularly true of tissue and cell cultures where a portion of tissue is used to inoculate new culture tubes or flasks; this is known as subculturing. In general, callus cultures are subcultured every 4-6 weeks, while suspension cultures need to be subcultured every 3-14 days. Plant cell and tissue cultures may be maintained indefinitely by serial subculturing. In case of suspension cultures, subculturing should be done about or somewhat prior to the time of their maximum growth. The inoculum volume should be 20-25 per cent of the fresh medium volume; in any case, the initial cell density of the fresh culture (just after inoculation) should be 5 x 104 cells /ml or higher otherwise the cells may fail to divide.

A suitable medium may be devised for a new system in several ways. A simple approach is to first test-several concentrations, e.g., 0, 0.5, 2.5, 5 and 10m mol 1-1 of an auxin and a cytokinin to identify a suitable combination of the two. Now, first different auxins and then different cytokinins available to the worker may be tested to identify the best of each. Using these GRs, the different standard recipes may be evaluated. One may then check the ½, full and even higher salt concentration of the selected medium as well as different (2-6 per cent) sucrose concentrations. A further refinement requires testing ½, full and x 2 concentrations of individual components of the selected recipe. Alternatively, the worker may evaluate the various combinations of low, medium and high concentrations of four solutions, i.e., minerals, organics, auxin and cytokinin, to arrive at a suitable recipe; De Fossard et al. have provided a broad spectrum experiment for this purpose.

3.1.3. Estimation of Growth

Cell number is the most informative measure of cell growth. This measurement is applicable to only suspension cultures, and even there cell aggregates must be treated, e.g., with pectinase, to dissociate them into single cells before counting the cell number in a haemocytometer. Therefore, cell number is estimated only where information obtained justifies the efforts. In contrast, packed cell volume of suspension cultures is easily determined by pipetting a known volume into a 15 ml graduated centrifuge tube, spinning at 200 x g for 5 min and reading the volume of cell pellet which is expressed as ml cells/l of culture.

Culture fresh and dry weights are the most commonly used measures of growth of both suspension and callus cultures. In case of callus cultures, the cell mass is placed on a preweighed dry filter paper or nylon filter and weighed to determine fresh weight. Cells from suspension cultures are filtered onto a filter paper or nylon filter, washed with distilled water, excess water removed under vacuum and weighed alongwith the filter; the filter is preweighed in wet condition. For dry weight determination the cells and the filter are dried in an oven at 60°C for 12 hr and weighed; the filter is pre-weighed in dry condition. Cell fresh and dry weights may either be expressed as per ml (suspension culture) or per culture.

3.1.4. Nuclear Cytology

In order to understand changes occurring at the chromosome and gene level nuclear cytology is done. Callus and suspension culture show both numerical (polyploidy and aneuploidy) and structural (deletions, translocations etc.) chromosome changes.

The cytogenetic status of cultured cells is influenced by several factors of the culture system, like GR concentrations and combination, culture age, liquid or agar medium, subculture interval, sucrose concentration, etc. Suspension cultures of many diploid species show a selection for diploid cells so that they remain predominantly diploid for long period.

3.2 Single Cell Cultures

Single cell cultures provide an excellent platform to investigate the properties and potentialities of plant cells help us to understand the interrelationships and complementary influences of cells in multicellular organisms. Free cells in cultures permit quick administration and withdrawal of diverse chemicals/ substances, thereby making them easy targets for mutant selection. Moreover, the individual cells within a mass of cultured cells invariably show cytogenetical and metabolic variations according to the stage of the growth cycle and culture conditions. Such variability, known as 'spatial heterogeneity', has evoked much interest since differences between cells in their karyotype and the ability to accumulate secondary metabolites would manifest during morphogenesis in the clones regenerated from single cells. Thereby the cell line selection technique can be usefully applied to produce high-yielding cultures as well as plants with superior agronomic traits.

3.2.1. Isolation of Single Cells

The leaf tissue is the most suitable material for the isolation of single cells since a more or less homogeneous mass of cells in the leaves raise defined and controlled large-scale cell cultures. From such healthy plant organs single cells can be isolated using mechanical or enzymatic methods.

1) *Mechanical method:* Successful experiments show how to isolate mesophyll cells, active in photosynthesis and respiration, from mature leaves of several species of dicots and monocots, including the grasses. Even metabolically active single cells from the bundle sheath of crab grass can be isolated using a similar procedure.
2) *Enzymatic method:* By treating tobacco leaf tissue with the enzyme pectinase can be obtained a large number of metabolically active cells. Isolating single cells by this method has been found it is possible to obtain high yields from preparations of spongy parenchyma with minimum damage or injury to the cells.

3.2.2. Cultured Tissues

The most prevalent method of obtaining a single cell system is from cultured tissues. Freshly cut pieces from surface sterilised plant organs are placed on a nutrient medium having a suitable proportion of auxins and cytokinins to initiate cultures. Explants on such a medium exhibit callusing at the cut ends, which gradually expand to the entire surface of the tissue. The callus is detached from an explant and transferred to a fresh medium of the same composition so as to build up a mass of tissue. Repeated subculture on an agar medium improves the friability of the callus, a prerequisite for raising a fine cell suspension in a liquid medium. The pieces of undifferentiated and friable callus are transferred in a continuously agitated liquid medium contained in autoclaved flasks or other suitable vials. Agitation of the culture medium exerts mild pressure on small pieces of tissue, breaking them into free cells and small cell aggregates, thus augmenting the gaseous exchange between the culture medium and the culture air, and also ensuring uniform distribution of cells and cell clumps in the medium.

A fine suspension of cells is usually obtained particularly when friable calli are used for the initiation of suspension cultures. Single cells can also be isolated from suspension cultures by suitably filtering out cell clumps and harvesting the cells by centrifugation.

3.2.3. Cloning

The process of obtaining a clone of cells consisting of all the cells derived through mitosis from a single cell is called cloning. Therefore, all the cells of clone are expected be identical with each other in their genotype and karyotype (chromosome constitution), and other attributes, except for the changes that are undergone during and after cloning. Cloning is based on single cells separated from tissues and cultured in a manner to allow separate recovery of the cell mass derived from them.

3.2.4. Culture of Single Cells

Single cells can be cultured by the following methods:

i) Filter paper raft nurse tissue technique
ii) Microchamber technique
iii) Microdrop method
iv) Bergman's plating technique
v) Thin layer liquid medium.

Since isolated single cells fail to divide in normal tissue culture media, either a nurse tissue or a conditioned medium in which plant cells have been grown for about 48 hr is used for their culture.

3.2.4.1. Filter paper raft-nurse tissue technique

Single cells are placed on small pieces of sterilised filter paper that are placed on top of established callus cultures several days in advance, allowing the filter papers to be wetted by the exudates from callus tissues. The single cells placed thus derive their nutrition from the callus exudates diffusing through the filters.

The cells divide and form macroscopic colonies on the filters; the colonies are isolated and cultured.

3.2.4.2. *Microchamber technique*

In a microchamber single cells are suspended in conditioned medium, and a drop of medium having a single cell is placed. The microchamber is then is covered by a coverslip. Microchambers allow microscopic observation, and they can be kept in a petricdish for incubation.

3.2.4.3. *Microdrop method*

Microdrops of 0.25-0.5 ml are distributed in the microwells known as the Cuprak dish. Which are specially designed, having a smaller outer chamber and a large chamber and the dish is sealed with parafilm. The cell density in the medium is adjusted to allow one cell per droplet. Which as success used for protoplasts and single cells as well.

3.2.4.4. *Bergmann's plating technique*

In this technique, the density of the cells suspended in a liquid medium is twice the desired density in the plate. Equal volumes of the liquid and agar media kept melted in a water bath are mixed thoroughly and quickly spread in a thick layer in a Petri dish. The cells remain embedded in the soft agar medium, are observable under a microscope, and when macroscopic colonies develop they are isolated and cultured separately.

3.2.5.Thin Layer Liquid Medium Culture

The technique plating cells in a thin layer of liquid medium adequate aeration is common for protoplast culture.

It is important to:

a) culture the single cells in dark and keep microscopic observations to the minimum since light has a detrimental effect on cell proliferation. In addition,

b) either a conditioned medium or a suitably enriched medium should be used since standard tissue culture media are unsuitable. Some highly enriched synthetic media permit the culture of as few as 25-50 cells/ml, while some media supplemented with casamino acids and coconut milk support cell division at 1-2 cells/ml. It is postulated that some biochemicals essential for cell division leach into the culture medium when they are subcultured; this produces the lag phase. Cells begin to divide only when an equilibrium is established for these metabolites between the medium and the cells; this happens much later at lower cell densities and below a critical cell density it may not be achieved. For this reason conditioned or specially enriched media are required for culture of cells at low densities.

3.3. Shoot and Meristem Culture

Most of the horticultural and forest crops are infected by systemic diseases caused by fungi, viruses, bacteria, mycoplasma and nematodes. Pathogen attack does not always lead to the death of the plant but very often the infection caused by pathogens considerably reduces the yield and quality of crops. While pathogens are nearly always transferred in plants through vegetative propagation, viral diseases occur in virtually all seed propagated as well as vegetatively propagated crop species.

Eradication of pathogens is highly desirable to optimise the yields and also to facilitate the movement of materials across the international boundaries. While plants infected with bacteria and fungi may respond to treatments with bactericidal and fungicidal compounds, there is no commercially available treatment to cure virus-infected plants. The therapeutic chemicals capable of eradicating virus from infected plants are not available though the use of virazole and vidarabine in the culture medium have resulted in the production of virus-free lily and apple plants.

To produce disease-free plants a healthy nucleus stock could be developed by selecting out one or more healthy plants and then multiplying them vegetatively, but where the entire population of a clone is infected the only way to obtain a pathogen-free plant is through tissue culture. Apical meristems in the infected plants are generally either free or carry a very low concentration of the viruses. However, the titre of the viruses increases in the older tissues corresponding to the increase in the distance from the meristern tips. Various reasons attributed to the escape of the meristems by virus invasion are:

a) viruses move readily in a plant body through the vascular system which in meristems is absent,

b) a high metabolite activity in the actively dividing meristematic cells does not allow virus replication and

c) a high endogenous auxin level in shoot apices may inhibit virus multiplication.

A knowledge of the gradient of virus distribution in shoot tips encouraged Holmes to obtain virus-free plants from infected individuals of Dahlia through shoot-tip cuttings. Morel and Martin applied tissue culture techniques for elimination of viral infection in plants. They cultured meristem tips excised from infected Dahlia and obtained disease-free plants. Since then the advances made in elimination of virus by tissue culture have been such that this approach has now become a popular horticultural practice.

The first application of meristem culture was to obtain virusfree plants of dahalias. In 1952, Morel and Martin isolated 100 um long meristems and cultured them to obtain virus-free shoots. Since then the technique of meristem culture has been greatly refined and used for obtaining plants free from viruses, viroids, mycoplasma and even fungi and bacteria in a range of crops. In India, some valuable clones of potato, sugarcane etc. have been freed from virus infections through meristem culture. Care must be taken to remove the apical meristem with as little surrounding tissue as possible to minimise the chances of virus particles being present in the explant. This application of meristem culture is of great value, particularly in the maintenance of breeding materials and germplasm exchange which are invaluable for any breeding programme.

Considerable expertise is required to dissect out the shoot apical meristem with only one or two leaf primordia. Care has to be taken to prevent desiccation, and contamination by the virus present in the surrounding tissue. Generally, growth regulators are added to the medium to support shoot growth from the cultured meristems. In general, the larger the meristem explant, the greater the chances of its survival and shoot development. But the risk of infection by the virus also increases with explant size (Table 3.1.). Therefore, a compromise has to be reached between these two opposing forces in deciding the explant size.

Table 3.1. The frequency of virus-contaminated plants of Cymbidium recovered by culturing shoot meristems of different lengths.

Length of Shoot Meristem	*Virus contaminated Plants*
(um)	*(%)*
600	100
270	87
120	52

It may be emphasized that many viruses have been shown to be present within the apical meristem, e.g., carnation mottle virus, chrysanthemum stunt virus etc. Further, all the plants derived from meristem culture are not necessarily virus-free; many of them may even contain latent bacterial and/or fungal infections. Therefore, it is essential that each tissue culture raised plant must be tested, using specific tests, for the presence of a given virus or pathogen before it is claimed that the plant is free from the concerned virus/pathogen. Further, freedom must be claimed only from the specific virus/pathogen for which it has been shown to be free.

3.3.1. Thermotherapy and Meristem Culture

Conventionally, viruses are eliminated by thermotherapy of whole plants in which plants are exposed to temperatures between 35-40°C for a few minutes to several weeks depending on the 'host-virus combination. Thermotherapy is based on the fact that most viruses are killed at temperatures much below those which kill their host plants. Thermotherapy is usually effective against isometric and thread-like viruses and against mycoplasmas, but is ineffective against many other viruses. Only a small proportion of plants survive heat treatment, and often these plants become susceptible to other diseases possibly due to inactivation of their defense mechanisms.

Thermotherapy is often combined profitably with meristem culture to obtain virus-free plants. In general, shoot tips are excised from heat-treated plants, but cultured meristems may also be given thermotherapy; the former is preferable since larger explants can be safely taken from heat-treated plants (this increases the survival rate of explants).

Potato virus S (PVS) and potato virus X (PVX) are not eliminated either by thermotherapy or by meristem culture alone. But, both these viruses have been eliminated by culturing meristems from heat-treated plants. But in case of some viruses, e.g., potato spindle tuber virus, plants recovered from thermotherapy plus meristem culture have to be subjected to a second cycle of heat-treatment coupled with shoot tip culture for effective virus elimination. The heat-treatment may either be continuous or a daily cycle of high and low temperatures may be applied; the latter is often preferable. The duration of treatment may vary from few to several days, e.g., 8-18 weeks for PVX, depending on the host-virus combination. This is perhaps the most successful approach of virus elimination.

3.3.2. Cryotherapy

More recently, prolonged exposure to a low temperature (5°C) followed by shoot tip culture has proved quite successful in virus elimination; this is often called cryotherapy. For example shoot-tip cultures from Chrysanthemum plants treated with 5°C for 4 months yielded 67 per cent plants free from chrysanthemum stunt virus (CSV), and 22 per cent plants free from chrysanthemum chlorotic mottle virus (CCMV). The frequency of virus-free plants increased to 73 per cent in case of CSV and to 49 per cent in case of CCMV after a 7.5 month exposure to 5°C. Cold treatment is preferable to heat-treatment as it is less injurious to the plants and often more effective in virus elimination.

3.3.3. Chemotherapy

Some chemicals, e.g., virazole (ribavirin), cycloheximide, actinomycin D etc., which interfere with virus multiplication may be added into the culture medium for curing the shoottips of viruses; this is known as chemotherapy. For example, virazole was shown to eradicate potato virus X (PVX) from shoots derived from tobacco protoplasts. Virazole is effective against a range of animal DNA and RNA viruses. It has also been used to eliminate ORSV (odontoglossum-ringspot virus) from Cymbidium.

3.3.4. Virus-Indexing

It may be emphasized that all the plants obtained through meristem culture, with or without chemo- or thermotherapy, are not virus-free. Therefore, such plants have to be tested for the presence of the concerned virus(es). This is called virus indexing.

i) The simplest method for virus-indexing is to score the plants for the presence of specific symptoms produced by the relevant virus. However, the development of visible symptoms may take considerable time.
ii) Alternatively, the saps from test plants may be used to inoculate highly sensitive and healthy indicator plants. An indicator plant for a virus is that plant species or variety which is highly susceptible to the virus and readily develops the symptoms.
iii) A highly sensitive and precise technique for virus-indexing employs antibodies specific to the concerned virus(es). This is known as serological test. ELISA is perhaps the most convenient, rapid and efficient test system especially when a large number of samples are to be handled.
iv) DNA and RNA probes can also be used for virus detection but they are time-taking and are not amenable to large scale application in a manner similar to ELISA.

3.3.5. Maintenance of Virus-Free Stocks

Plants freed from viruses are maintained in insect-free glasshouses or cages. This is necessary to prevent re-infection by the viruses since such plants are as susceptible to virus infection as the parent variety from which they were obtained. A large scale multiplication of such plants can be done in areas where and during seasons when the chances of infection by the virus are minimal. However, the best and cheapest means of maintaining virus-free plants is in *in vitro* cultures.

3.4. Organogenesis

Plasticity and totipotency are central to understanding cell culture and regeneration. Plants have a sessile nature and long lifespan, and hence developed a greater ability to endure extreme conditions and predation than have animals. Plasticity allows plants to alter their metabolism, growth and development to best suit their environment.

3.4.1. Totipotency

In vitro-cultured plant cells and tissues generally exhibit a very high degree of plasticity, allowing one type of tissue or organ to be initiated from another type thus regenrating whole plants. This depends upon all plant cells given the correct stimuli, being able to express the total genetic potential of the parent plant. This maintenance of genetic potential is called 'totipotency'. Plant cell culture and regeneration provide the most compelling evidence for totipotency, but identifying the culture conditions and stimuli required to manifest this totipotency can be extremely difficult and it is still a largely empirical process.

3.4.2. Regeneration of Plants

The outcome of the process of dedifferentiation followed by redifferentiation of cells is organogenic differentiation. Mostly, the whole plant regeneration from cultured cells may either occur through shoot-bud differentiation or somatic embryogenesis, thereby establishing the totipotency of somatic cells which. This, natural conditions, can also be observed to some extent during the vegetative reproduction of plant species. The cell(s) of stem, leaf and root cuttings of several plants are able to directly differentiate shoots and roots.

There are many factors which influence the process of regeneration, but it can be successfully achieved through proper selection of the inoculum, proper choice of the medium, a balanced combination of plant growth regulators and the control of physical environment.

The earliest reports on controlled organogenesis in vitro were by White 1939 who obtained shoots on callus of a tobacco hybrid and by Nobecourt 1939, who observed root formation in carrot callus. The finding of White was confirmed and extended by Skoog 1944, who showed that auxin could stimulate rooting and inhibit shoot formation. Further studies established that a balanced combination of auxin and cytokinin controls the root and shoot formation.

3.4.3. Source of Explant

Regenaration from any plant part can be achieved by proper combination of various factors. Stem segments and apices, root segments, leaf pieces, petioles, inflorescence sections, flower petals, ovular tissue, seedling parts (cotyledon, hypocotyl) and seed embryos have been used as an explant. All these explants can give rise to organs and embryos directly or indirectly via callus. A suitable explant is desirable for a given species for successful regeneration. Explants consisting of actively dividing cells have been useful in initiating the cultures and subsequent regeneration.

3.4.4. Regulation of Regenaration

A suitable explant and a proper medium callus or explants to form organs. Many growth active substances, plant growth regulators as well as other nutrients are included in the culture medium to regulate organogenesis in vitro.

Most of the plant species respond to a suitable auxin/cytokinin balance by forming shoots and roots. A variety of substituted purines, pyrimidines and ureas have been used as cytokinins to achieve organogenesis. Sometimes other compounds affecting organogenesis are amino acids, antiauxins, polyamines, oligosaccharides, etc.

3.4.5. Mechanism of Organ Development

Optimal conditions of medium constituents are important to achieve organogenesis. This is essential for each species or even each cultivar specifically for clonal propagation. The process of differentiation begins with the changes in individual cells in response to external stimuli, the cells producing an organised structure by cellular differentiation.

The morphogenetic signals generate the conditions suitable for competent.cells to undergo their internally controlled programme of differentiation. This comes about from selective gene action with the subsequent cellular processes of DNA replication and translation. As a result there are biochemical and biophysical changes in the target cells, including a change in the metabolic activity, followed by cytological and histological changes that lead to visible appearance of organs formation.

There is no doubt that the process of differentiation begins at single cell level but this does not mean that all organs are produced from a single cell. Once the stimulus is set in, centres of meristematic activity are formed surrounding the cell from which the process starts. The organs are developed from these multicellular meristematic centres. Clusters of meristematic cells, called nodules or meristemoids develop in areas that accumulate starch, which serves as an energy source for shoot bud differentiation. Meristemoids may develop vascular elements inside them, while their outside may be made of cambium like cells.

3.5. Somatic Embryogenisis

In somatic (asexual) embryogenesis, embryo-like structures, which can develop into whole plants in a way analogous to zygotic embryos, are formed from somatic tissues. These somatic embryos (SE) can be produced either directly or indirectly.

There are two ways in which somatic embryos are generated. In direct somatic embryogenesis, the embryo is formed directly from a cell or small group of cells without the production of an intervening callus. Though common from some tissues, direct somatic embryogenesis is generally rare in comparison with indirect somatic embryogenesis. In indirect somatic embryogenesis, callus is first produced from the explant. Embryos can then be produced from the callus tissue or from a cell suspension produced from that callus. Somatic embryogenesis from carrot is the classical example of indirect somatic embryogenesis.

3.5.1. Developmental Patterns of SEs

Somatic embryogenesis usually proceeds in two distinct stages. In the initial stage (embryo initiation), a high concentration of 2,4-D is used. In the second stage (embryo production) embryos are produced in a medium with no or very low levels of 2,4-D. In somatic (asexual) embryogenesis, embryo-like structures, which can develop into whole plants in a way analogous to zygotic embryos, are formed from somatic tissues. SEs generally originate from single cells which divide to form a group of meristematic cells. Usually, this multicellular group becomes isolated by breaking cytoplasmic connections with the other cells around it and subsequently by cutinisation of the outer walls of this differentiating cell mass. The cells of meristematic mass continue to divide to give rise to globular (round ball shaped), heart-shaped, torpedo and cotyledonary stages (Figure 3.3).

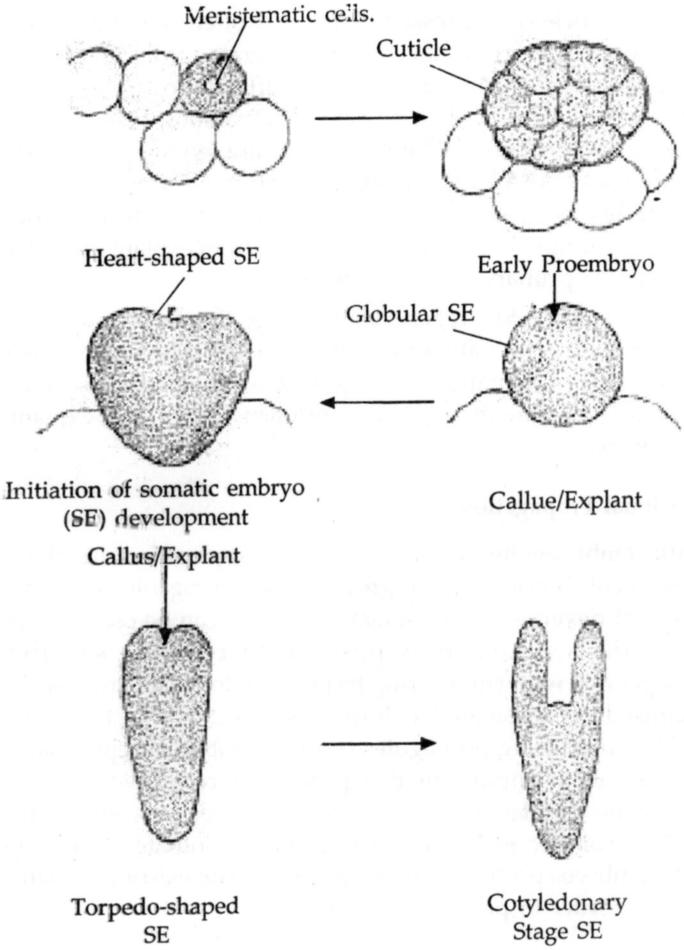

Figure 3.3. Development of a Somatic embryo from a single superficial cell of the explant.

In general, the essential features of SE development, especially after the globular stage, are comparable to those of zygotic embryos. Somatic embryos are bipolar structures in that they have a radicle and a plumule. The radicular end is always oriented towards the center of callus or cell mass, while the plumular end always sticks out from the cell mass. In contrast, a shoot bud is monopolar as it does not have a radicular end. In many SEs, radicle is suppressed so that they often do not produce roots; in such cases, roots have to be regenerated from the shoots produced by germinating SEs. SEs often show abnormal developmental features, e.g., 3 or more cotyledons, bell- shaped cotyledons, larger size etc. These problems are often overcome by the presence of ABA or a suitable concentration of mannitol. In some species, normal looking somatic embryos are formed but they fail to germinate. The SEs regenerating from explant or callus are termed as primary somatic embryos.

In many cases, SEs regenerate from the tissues of other SEs or the parts of germinating SEs; such SEs are called secondary somatic embryos. Somatic embryogenesis is influenced by several factors such as growth regulators, nirogen source and explant genotype etc.

3.5.2. Clonal Propagation

Somatic embryogenesis has a potential application in plant improvement. Since both the growth of embryogenic cells and subsequent development of somatic embryos can be carried out in a liquid medium, it is possible to combine somatic embryogenesis with engineering technology to create large-scale mechanised or automated culture systems. Such systems are capable of producing propagules (somatic embryos) repetitively with low labour inputs. In this process of repetitive somatic embryogenesis (also referred to as accessory, adventive, or secondary somatic embryogenesis) a cycle is initiated whereby somatic embryos proliferate from the previously existing somatic embryo in order to produce clones.

3.5.2.1. Cloning Zygotic Embryos

A wide range of soybean genotypes have been tested for their ability to undergo auxin-stimulated somatic embryogenesis during cloning of zygotic embryos. All of them are reported to form somatic embryos provided appropriate nutrients are provided in the medium. The role of genotypes in conferring regeneration capacity is further supported by studies on zygotic embryo cloning of wheat, rice and maize. Analysis of various cultivars demonstrated that the regeneration capacity of these crops was directly affected by non-additive, additive and cytoplasmic factors. However, the genotype which has the capacity to undergo repetitive somatic embryogenesis can be back-crossed to elite lines in order to transform the latter with capacity for high regeneration of somatic embryos. Such a type of transformation could play an important role in plant breeding since through in vitro techniques high quality somatic embryos have been produced in 80 species of tropical crops.

3.5.2.2. Raising Somaclonal Variants

Embryos formed directly from pre embryogenic cells appear to produce relatively uniform clonal material, whereas the indirect pathway generates a high frequency of somaclonal variants. Mutation during adventive embryogenesis may give rise to a mutant embryo which on germination would form a new strain of plant. Nucellar embryos, like shoot tips, are free of virus and can be used for raising virus-free clones, especially from some tree species where shoot tip culture has not been successful. For clonal propagation of tree species, somatic embryogenesis from nucellar cells may offer the only rapid means of obtaining juvenile plants equivalent to seedlings with parental genotype. Clonal propagation through somatic embryogenesis has been reported in 60 species of woody trees representing 25 families. Somatic embryos with genotype of a selected elite parent are potentially convenient organs for cryopreservation and germplasm storage.

3.5.3. Synthesis of Artificial Seeds

An artificial seed is a bead of gel containing a somatic embryo (or shoot bud), and the nutrients, growth regulators, pesticides, antibiotics etc. needed for the development of a complete plantlet from the enclosed SE/ shoot bud.

Artificial seeds, consisting of somatic embryos enclosed in a protective coating, have been proposed as a 'low cost – high volume' propagation system. The inherent advantages of artificial seeds are the production of many somatic embryos and the use of conventional seed.handling techniques for embryo delivery. The objective is to produce clonal 'seeds' at a cost comparable to true seeds. Two types of artificial seeds have been developed, namely, hydrated and desiccated.

Redenbergh et al. developed hydrated artificial seeds by mixing somatic embryos of alfalfa, celery and cauliflower with sodium alginate, followed by dropping into a solution of calcium chloride to form calcium-alginate beads. The beads become hardened as calcium alginate is formed. After about 20-30 minutes the artificial seeds are removed, washed with water and used for planting. Hydrated artificial seeds are sticky and difficult to handle on a large scale, dry rapidly in the open air. These problems can be resolved by providing a waxy coating over the beads. However, it is not possible to store, except at low temperatures and for limited periods, hydrated artificial seeds and they have to be planted soon after they are produced. Precision machines for large scale encapsulation of SEs have been devised.

Kim and Janick applied synthetic seed coats to clumps of carrot somatic embryos to develop desiccated artificial seeds. They mixed equal volumes of embryo suspension and 5 per cent solution of polyethylene oxide, a watersoluble resin, which subsequently dried to form polyembryonic desiccated wafers. The survival of encapsulated embryos was further achieved by embryo hardening' treatments with 12 per cent sucrose or 10-6M ABA, followed by chilling at high inoculum density.

Another delivery system for somatic embryos for obtaining transgenic plants is Fluid- drilling. Embryos are suspended in a viscous-carrier gel which extrudes into the soil. The primary problem in fluid-drilling is that the sucrose level necessary to permit conversion also promotes rapid growth of contaminating micro-organisms in a non-aseptic system. Gray found that somatic embryos of orchard grass became quiescent when desiccated in empty plastic petri dishes at 70 per cent relative humidity at 23°C which amounted to loss of 13 per cent water. However, after 21 days of storage, desiccated embryos when rehydrated in vitro germinated to produce viable plantlets though limited (4 per cent) in number. Senaratna et al., treated alfalfa somatic embryos with ABA at the torpedo to cotyledonary stages in order to increase their tolerance to desiccation. Over 60 per cent of such desiccated embryos germinated when placed on a moist filter paper or sown directly onto sterile soil and formed plantlets. Heat-shock treatments also induced a degree of desiccation tolerance comparable to that conferred by ABA treatment and had no detrimental effect on the subsequent growth of the plantlets.

In general, the gels contain inorganic nutrients, a carbon source, fungicides, bactericides and other growth promoting substances necessary for SE germination and seedling establishment in the field. When shoot buds are encapsulated, growth regulators needed for rooting and shoot growth are also included.

3.5.4. Source of Regenerable Protoplast System

Embryogenic callus, suspension cultures and somatic embryos are used as sources of protoplast isolation for a range of species. Cells or tissues, regenerating in cultures yield protoplasts that are capable of forming whole plants. Embryogenic cultures are a good source of regenerable protoplasts in the graminaceous, coniferous and citrus species. Embryogenic citrus suspension cultures also provide protoplasts that can be used in the production of interspecific and intergeneric somatic hybrid plants Among cereals, the development of a protoplast regeneration system for maize has been especially challenging.

3.5.5. Genetic Transformation

In seed embryogenesis, zygotic embryos, deeply embedded in the nucellar tissue live in a protected environment besides being genetically heterogeneous, while somatic embryos remain virtually unprotected and more or less give rise to genetically uniform plants. The leaf-disc transformation systems has made it possible tissues to regenerate via somatic embryogenesis. Here, isolated single cells can be transformed in cultures and grown on a selection medium to callus colonies which eventually form somatic embryos on removal of auxin from this medium. By using repetitive embryogenesis for Agrobacterium-transformed walnut cells, multiple crops of somatic embryos without employing the callus phase can be obtained.

Repetitive embryos can originate from single epidermal or subepidermal cells that can be readily exposed to agrobacterium. Thus, the transformation technique, when applied to a primary somatic embryo, instead of a zygotic embryo, give rise to totally transgenosic somatic embryos. Repetitive embryogenesis is also ideally suited to particle gun-mediated genetic transformation in which the particle gun literally shoots DNA that has been precipitated onto particles of a heavy metal, into the plant cells. The transformed cell lines can thus give rise to a number of transformed somatic embryos through repetitive embryogenesis.

3.5.6. Synthesis of Metabolites

The repetitive embryogenesis system is of potential use in the synthesis of metabolites such as pharmaceuticals and oils. Somatic embryos of borage containing high levels of ylinolenic acid, also produce this metabolite but through repetitive somatic embryogenesis a continuous supply of γ-linolenic acid is ensured, otherwise limited to the growing season in the zygotic embryo. The same principle can be applied for production *in vitro* of industrial lubricants from jojoba and leo-palmitostearin from Cacao.

3.6. Micropropagation

Micropropagation is the production of miniature planting material in large numbers by vegetative multiplication through regeneration.

3.6.1. Various Approaches used for Micropropagation

i) *Axillary budding:* Here shoots are developed from pre-existing meristems on nodal region ensuring genetic stability of the regenerants. Propagation of long (life) cycle plants such as forest and fruit trees is multiplied mainly through this method and also at the industrial level where high standards of homogeneity are to be maintained e.g., strawberry, rose, potato, artichoke, asparagus, pineapple, banana etc.

ii) *Adventitious budding: De novo* formation of adventitious buds may occur directly from the tissues of the explant, giving the required level of genetic stability in the regenerated plants except for chimeral plants. In some cases adventitious shoots form indirectly from which *calli* hence are unorganised structures, and hence prone to nuclear unstability during rapid cell divisions.

iii) *Somatic embryo development:* somatic embryos occur either directly on explants or more frequently in callus cultures. In indirect somatic embryogenesis where somatic embryos occur more frequently there is production of somaclonal variation in the regenerated plant population though this method does not ensure genetic stability. However, somatic embryogenesis in cell cultures is potentially useful for developing scale-up technology using bioreactor and in the production of synthetic seeds through encapsulation.

3.6.2. Advantages of Biotechnology

There are several advantages of plant tissue culture over conventional propagation methods, vegetative and sexual, for large-scale micropropagation.

i) Shoot multiplication can be achieved in flasks because hundred of miniature plantlets are produced.

ii) Propagation is carried out under sterile conditions eliminating damage from microbes insects and diseases.

iii) Using virus-free material ensures that a large number of virus-free plants are obtained.

iv) Plant tissue culture is carried out under defined conditions of environmental, nutritional and tissue system.

v) Seasonal variations do not affect production since uniform conditions are maintained.

vi) No care is required between two subcultures as compared to conventional vegetative propagation systems like watering, weeding.

vii) Just a small glass house space is required because of miniature size of plantlets.

viii) Mother plant or genotype of stock plant can be stored and maintained *in vitro* without damage to environmental factors and stock plants.

ix) Plants difficult to propagate vegetatively by conventional methods can be propagated biotechnically.

x) It is possible to mechanise the whole process of vegetative propagation for large-scale plantations.

xi) Miniature storage organs can be provided for genotype storage and subsequent plantations.

xii) Being sterile, transport across countries is permissible without difficulties.

3.6.3. Disadvantages of Biotechnology

The system also has certain disadvantages:

i) Capital-intensive expensive materials, like autoclave, laminar air flow bench, controlled culture rooms etc., make micropropagation methods through use of tissue culture-involve quite expensive.

ii) Being a technically skilled work, knowledge about material, techniques and decision making are required in the personnel.

iii) Contamination can severely damage material and make of production costly, affecting time schedule delivery of the material.

iv) Since specific conditions of micropropagation, rooting, and hardening may be required each material requires separate research methods.

v) Small delicate plantlets that are produced, take longer initial time to grow.

vi) Genetic stability is doubtful in certain methods.

vii) As this is a capital intensive industry, plants produced in small numbers cost too much. Cost is a major factor for the production and sale of tissue culture raised plants.

3.6.4. Micropropagation Methods

The information and knowledge about botany, reproductive biology, geographic distribution and growth seasons are required to obtain plant material in healthy state for optimal growth. The stages involved in the micropropagation of a given plant genotype are given in the following paragraphs. These stages are used irrespective of the approach used for multiplication of the species. The schematic layout of the method is given in Figure 3.4.

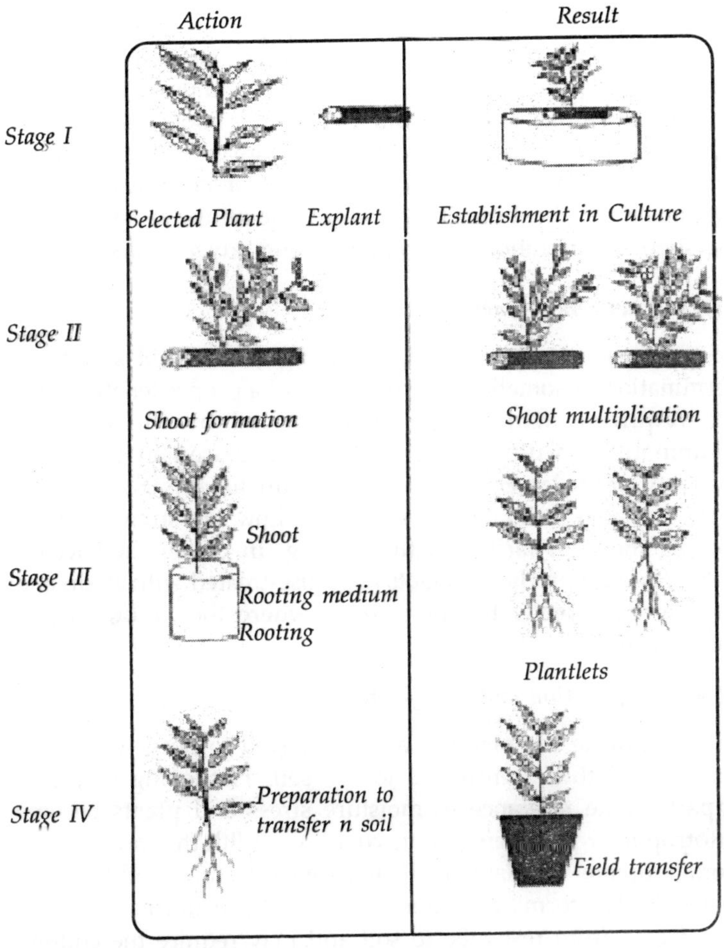

Figure 3.4. Micropropagation Method-stages to obtain in vitro multiplication

3.6.4.1. Selection and establishment of aseptic cultures

While selecting it is typical, healthy, disease free mother plants necessary to test for virus presence and if found positive it is eliminated. This is followed by preparation of explants, surface sterilisation and transfer to appropriate media. The medium should support the rapid proliferation of the explants.

3.6.4.2. *Multiplication of propagule*

To achieve a rapid multiplication of the regenerative system for obtaining large number of shoots. Medium and tissue factors are optimised empirically. *In vitro* produced shoots are used as explants to produce more shoots. Orchids are multiplied by proto corm formation. By using repetitive use of *in vitro* produced explants and subcultures of the propagule on the same medium the shoot proliferation gets enhanced and adventitious shoots.

3.6.4.3. *Plantlets regeneration*

To produce plantlets through rooting of isolated shoots or germination of somatic embryos shoots of a proper length or age are required, which may be produced by the medium combinations. Cultures for rooting are placed in low light intensity, or the lower portion of the culture tubes is covered with black paper to facilitate root induction. Low salt strength of the rooting medium facilitates the rooting. In certain herbaceous plants, *in vitro* produced shoots are pulse-treated with auxin and transferred directly to pot mixture where they root, e.g., in chrysanthemum.

3.6.4.4. *Preparation and transfer to field*

After culturing the plantlets are transferred to pots, when they harden and then transferred to the soil. Hardening of plants imparts some tolerance to moisture stress and plants become autotrophic from heterotrophic condition. With the formation of storage organs on plantlets, their establishment in soil becomes easier. Bulbs, corms and tubers may undergo dormancy when transferred from test tube to soil and may require the chilling treatment to germinate.

Plants are irrigated frequently and their growth and variation are monitored regularly. Plants may be kept up to flowering (flowering plants) or transferred to fields (plantation crops) after 4 to 6 weeks of acclimatisation.

3.6.5. Clonal Multiplication of Woody Species

Due to the long-term lifecycle of valuable crops, such as fruit and nut trees, plantation species (palms of various types, rubber, coffee, tea etc.), timbers (hardwood and softwood) and important trees of social forestry. Progress in the improvement of woody perennials through the application of conventional methods has been extremely slow. Rooting the cuttings from elite tree species has posed problems in the process of their clonal propagation. According to a researcher, vegetative propagation of many monocotyledonous palms and forest species is virtually impossible since forests are multiplied by the growth of natural seedlings and by collection of seeds from randomly pollinated elite trees. *Cryptomeria japonica* is an exception which has been clonally propagated for centuries in Japan. Recently, clones of Norway spruce on a large scale in the forests of Sweden and were developed Germany.

3.6.7. Hardwoods and Fruit Trees

Cloning of temperate fruit species and other trees, like populous, eucalyptus tectona, etc., through tissue culture has been successfully achieved. Tissue culture of nut trees also holds some promise. Recent success in tree tissue culture is the controlled flowering of *in vitro* propagated bamboo.

In order to establish cultures from adult explants it requires a lot of effort to produce a clonal multiplication of woody perennials in vitro on a commercial scale. Also, it requires effective treatments to induce efficient rooting of *in vitro* multiplied shoots and quality improvement of somatic embryos to achieve high frequency conversion to plantlets. Immature embryos of avocado and shoot tips from seedlings of citrus, jackfruit and plums have proved equally effective clonal materials in cultures. Since nucellar seedlings are naturally rejuvenated clones, they have been used as stock plants for micropropagation of several polyembryonic citrus cultivars and mangosteens.

Monoembryonic species of Citrus, varieties of mango, loquat and many tropical and subtropical fruit crops are reported to regenerate from somatic embryos induced on explants from mature trees. Shoot tips or nodal segments from tropical adult trees are reliable sources of explants. The conditioning of stock plants improves the culture response of shoot tips and nodal segments and so do warm as also and humid months. Sometimes horizontal placement of shoot-tip explants on the medium may prove superior to vertical orientation. Tropical fruit species such as banana, papaya and pineapple could be commercially propagated.

3.6.8. Softwoods

Clonal propagation of gymnosperms which account for most of the world's produce of timber (softwood) is desirable because of their long lifespan and open pollination. There are drawbacks in vegetative multiplication by cuttings as cuttings lack the vigour of seedlings and their growth declines with increasing age of the ortet. The *in vitro* techniques should help to overcome these problems and regenerate plants at faster rates.

Though a large number of adventitious buds can be induced on embryonal explants of gymnosperms. Yet only a few plants get finally established in soil because of losses during successive stages of development, rooting, or transplantation. Plagiotropic growth is seen in many species. Plantlets of these species have relatively less-field-survival than orthotropic plantlets. Plantlets recover faster in as in Douglas-fir, where soil and become orthotropic sooner than cuttings.

3.7. Anther and Pollen Culture

Haploid tissue can be cultured in vitro by using pollen or anthers as an explant. Pollen contains the male gametophyte, which is termed the 'microspore'. Both callus and embryos can be produced from pollen. Two main approaches can be taken to produce in vitro cultures from haploid tissue.

The first method depends on using the anther as the explant. Anthers can be cultured on solid medium. Pollen-derived embryos are subsequently produced via dehiscence of the mature anthers. The dehiscence of the anther depends both on its isolation at the correct stage and on the correct culture conditions. In some species, the reliance on natural dehiscence can be circumvented by cutting the wall of the anther, although this does, of course, take a considerable amount of time. Anthers can also be cultured in liquid medium, and pollen released from the anthers can be induced to form embryos, although the efficiency of plant regeneration is often very low. Immature pollen can also be extracted from developing anthers and cultured directly, although this is a very time-consuming process.

Both methods have advantages and disadvantages. Some beneficial effects to the culture are observed when anthers are used as the explant material. There is, however, the danger that some of the embryos produced from anther culture will originate from the somatic anther tissue rather than the haploid microspore cells. If isolated pollen is used there is no danger of mixed embryo formation, but the efficiency is low and the process is time-consuming. In microspore culture, the condition of the donor plant is of critical importance, as is the timing of isolation. Pretreatments, such as a cold treatment, are often found to increase the efficiency. These pretreatments can be applied before culture, or, in some species, after placing the anthers in culture.

Many of the cereals (rice, wheat, barley and maize) require medium supplemented with plant growth regulators for pollen/anther culture. Microspore explants can regenerate by direct embryogenesis, or via a callus stage and subsequent embryogenesis. Haploid tissue cultures can also be grown from the female gametophyte (the ovule). The ploidy of the plants obtained from haploid cultures may not be haploid due to chromosome doubling during the culture period. Chromosome doubling may be an advantage, as in many cases haploid plants are not desired from haploid tissues. Such plants are often referred to as 'di-haploids', because they contain two copies of the same haploid genome.

Haploid plants may be obtained from pollen grains by placing anthers or isolated pollen grains on a suitable culture medium, constituting anther and pollen culture respectively. Ideally, the anthers may be taken from plants grown under controlled temperature, light and humidity; the optimum condition may differ from species to species. Often, the effectiveness for haploid production declines with the age of donor plants.

At the appropriate developmental stage flower buds are collected, surface sterilised, and their anthers are excised and placed horizontally on culture medium or embed them in the culture medium. Flower buds with small anthers may themselves be cultured and, in some cases, the entire inflorescence is cultured. Anthers should be handled carefully, otherwise callus formation from anther walls may be induced. Alternatively, pollen grains may be separated from anthers and cultured on a suitable medium.

3.7.1. Pollen Culture

When isolated pollen grains are cultured *in vitro*, they produce haploid embryos or callus; this approach is called pollen culture.

Pollen may be isolated either by squeezing or float culturing the anthers. In squeezing culture about 50 anthers are placed in 20 ml of medium and squeezed with a glass rod. The solution is filtered through a nylon mesh of suitable pore size, and centrifuged at 500-800 rpm for 5 minutes. The pollen pellet is collected, washed twice and suspended at a final density of 103-104 pollen/ml.

In float culture, excised anthers are floated on a shallow liquid medium in Petri dishes. In a few days, the anthers start and continue to release their pollen grains into the medium so that their serial subculture yields pollen samples in different stages of androgenesis. Androgenesis can be markedly improved before pollen isolation by a preculture of anthers for 4-7 days. Successful pollen culture is reported for several species, e.g., tobacco, Datura, petunia, potato, barley, wheat, rice, maize, rapeseed etc. Pollen culture scores over anther culture due to the elimination of anther wall, e.g.,

i) studies on differentiation and development are easier and more precise.
ii) no callus formation can occur from wall tissue.
iii) products from different pollen grains ordinarily do not get mixed up.

3.8. Ovary/ Ovule Culture

Ovary/ ovule culture is another method used for the production of Haploid plants. Since, the haploids are important in plant breeding, methods have been devised to obtain them in cultures. This can be achieved by culture of unfertilised ovaries to obtain haploid plants from egg cell or other haploid cells of the embryo sac. This is called ovary culture, and the process is termed as gynogenesis. Success has been obtained with many species from gynogenesis, e.g., wheat, rice, maize, tobacco, petunia, gerbera, sunflower, sugarbeet, rubber, barley, etc. The rate of success varies considerably with species and significantly influenced by the genotype so that some cultivars do not respond at all. In rice, japonica genotypes show more response than indica cultivars.

While in most cases, the optimum stage for ovary culture is the nearly mature embryo sac, in rice ovaries at free nuclear embryo sac stage are the most responsive. Generally, culture of whole flowers, ovary and ovules appended to placenta respond better, but in gerbera and sunflower isolated ovules show better response. Cold pretreatment of the inflorescence before ovary culture angments gynogenesis.

Growth regulations which are crucial in gynogenesis, and at higher levels they may induce callusing of somatic tissues and even suppress gynogenesis.

Sucrose level also appears to be critical. Ovaries/ ovules are generally cultured in light, but in some species, like sunflower and rice, dark incubation favours gynogenesis and minimises somatic callusing. In rice, light may lead to degeneration of gynogenic proembryos.

3.8.1. Developmental Stages

Generally, the two stages of gynogenesis may each have distinct requirements. During induction, ovaries are floated on a low auxin liquid medium and kept out of light, while for regeneration they are transferred on to an agar medium with higher auxin concentration and incubated in light.

Haploid plants are generally produced from in egg cell in most of the species but in some species, e.g., rice, they originate chiefly from synergids. As in anther culture, gynogenesis may occur either via embryogenesis or through plantlet regeneration from callus. In rice a small amount of protocorm like callus are formed from which shoots and roots regenerate, while picloram promotes embryo regeneration. On the other hand, sugarbeet usually shows embryo development, while in sunflower embryos regenerate following a callus phase.

3.8.2. Limitations of Ovary Culture

Ovary culture has its two limitations. So far it has been success has been only in less than two dozen species, and the frequency of responding ovaries and the number of plantlets/ovary is quite low. Therefore, anther culture is preferred over ovary culture. Only in those cases where anther culture fails, e.g., sugarbeet, and for male sterile lines, ovary culture assumes significance.

3.8.3. Embryo Culture

In embryo culture an immature or mature embryo *in vitro* is isolated and grown with the goal of obtaining a viable plant. Embryo culture is now a well-established branch of plant tissue culture. There are two types of embryo culture: mature embryo culture and immature embryo culture.

Mature embryo cultures are derived from ripe seeds when embryos do not survive *in vivo* or become dormant for long periods of time or done to eliminate the inhibition of seed germination. Chemical inhibitors or acids, mechanical resistance present in the structures covering the embryo, rather than dormancy of the embryonic tissue, lead to seed dormancy.

The immature embryo culture is mainly used to avoid embryo abortion with the purpose of producing a viable plant. This process involves aseptic isolation of embryo and its transfer to a suitable medium for development under optimum culture conditions. At a suitable time florets are removed and either florets or ovaries are sterilised. Ovules are then be removed from the ovaries, with the tissue within the ovule, in which the embryo is embedded, already sterile. For mature embryo culture, the single mature seeds are disinfected or if the seeds are still unripe then the still encased fruit is sterilised. The embryos can then be aseptically removed from the ovules.

A different procedure is required for embryo culture to overcome seed dormancy. Seeds with hard coats are sterilised and soaked in water for few hours to few days. Sterile seeds are then split and the embryos excised. The selection of medium is necessary to sustain continued growth of the embryo. In most cases, a standard basal plant growth medium with major salts and trace elements may be used.

3.8.4. Technique

A basal salt medium with a carbon energy source such as sucrose can be used to grow mature embryos. Apart from the salt medium young embryos, require different vitamins, amino acids, and growth regulators, and in some cases natural endosperm extracts.

Embryo culture derives great benefits from reduced organic nitrogen as aspargine, glutamine or casein hydrolysate with the addition of malic acid. After one or two weeks when the embryo ceases to grow, it must be transferred, to a second medium with a normal sucrose concentration, low level of auxin and a moderate level of cytokinin. This allows for renewed embryo growth with direct shoot germination in many cases. Where the embryo does not show shoot formation directly, it can be transferred to a medium for callus induction followed by shoot induction. After the embryos have grown into plantlets *in vitro*, they are generally transferred to sterile soil and grown to maturity.

In crucifers, cereals and millets an embryo callus is reported to possess a high regenerative capacity compared to those derived from mature organs, such as the leaf, stem and root. The age of the embryo considerably affects the regenerative ability of maize calluses.

References

Bhojwani, S. S., Razdan, M. K. *Plant tissue culture: theory and practice, Revised edition* Elsevier. 1996.

George, E. F., Hall, M. A. and De Klerk G.J. (eds) 2008. *Plant propagation by tissue culture. Volume 1. The background.* Springer, Dordrecht.

Raven, Peter H., Evert, Ray F., & Eichhorn, Susan E. *Biology of Plants* (4th ed.). New York: Worth Publishers. 1986.

Morel, G. "Meristem culture techniques for the long-term storage of cultivated plants". In *Crops Genetic Resources for Today and Tomorrow.* Ed. 0. H Frankel, J. G. Hawkes. 1975.

Murashige, T. "Plant propagation through tissue culture". *Annual Reviews Plant Physiology* 25: 135-166. 1974.

Roca, W.M., Bryan, J.E., Roca, M.R. "Tissue culture for the international transfer of potato genetic resources". *Am. Pot.* J. 56: 1-11. 1979.

Wilkins, C.P., Dodds, J.H. "The application of tissue culture techniques to plant genetic conservation". *Science Progress* 68: 281-307. 1982.

4

TISSUE CULTURE APPLICATIONS

Plant tissue culture, the growth of plant cells outside an intact plant, is has direct commercial applications that include:

- Micropropagation used in forestry, and to conserve rare or endangered plant species.
- A plant breeder using tissue culture to screen cells rather than plants for advantageous characters, e.g. herbicide resistance/tolerance.
- Large-scale growth of plant cells in liquid culture inside bioreactors as a source of secondary products, like recombinant proteins used as biopharmaceuticals.
- Crossing distantly related species by protoplast fusion and regeneration of the novel hybrid.
- Cross-pollinating distantly related species and tissue culturing the resulting embryo that would otherwise normally die.
- Production of doubled monoploid (dihaploid) plants from haploid cultures to get homozygous lines more rapidly in breeding programmes, usually by treatment with colchicine that causes doubling of the chromosome number.
- As a tissue for transformation, followed by either short-term testing of genetic constructs or regeneration of transgenic plants.
- Employing certain techniques such as meristem tip culture that can be used to produce clean plant material from virused stock, such as potatoes and many species of soft fruit.

- Micropropagation using meristem and shoot culture to produce large numbers of identical individuals
- Screening programmes of cells, rather than plants for advantageous characters
- Large-scale growth of plant cells in liquid culture as a source of secondary products
- Crossing distantly related species by protoplast fusion and regeneration of the novel hybrid
- As a tissue for transformation, followed by either short-term testing of genetic constructs or regeneration of transgenic plants
- Removal of viruses by propagation from meristematic tissues.

4.1. Production of Haploids

Haploids can produced *in vitro* thus solving some problems in genetic studies since gene actions are readily manifested due to a single allelic dose present in chromosomes of an entire genome. This provides them suitable material for induction of mutations and various gene manipulations. Through rapid achievement of homozygous traits in double-haploids, pollen-derived haploid plants have been used in breeding and improvement of crop species.

4.1.1. Release New Varieties through F1 Double-haploid System

One cycle of meiotic recombination is usually insufficient for the improvement of agronomic quantitative traits. This is because the linkage between polygenes will not release all potential variations available in a cross. Combining anther-culture with sexual hybridisation among different genotypes of anther- derived plants has helped to overcome this problem. The anthers of the hybrid (F1) progeny are excellent breeding material for raising pollen-derived homozygous plants in which complementary parental characteristics are combined in one generation.

Studies related to inheritance of qualitative traits show that in breeding programmes the double-haploids derived from pollen cultures expressed genetic variability to an extent that new varieties have been synthesised in respect of barley, brassica, rice, maize, rye, potato, pepper and asparagus. By inducting and haploid then by chromosome doubling, exclusively male plants in dioecious species can be obtained.

4.1.2. Selection of Mutants Resistant to Disease

Screening mutants with resistance to disease is of prime importance in crop improvement. Haploids can be used in rapid selection of mutant that disease-resistant because they provide a relatively easier system for the induction of mutations. Tobacco mutants resistant to black shank disease and wheat lines resistant to scab are some examples of using anther culture technique in mutant selection successfully.

4.1.3. Developing Asexual Lines

A pollen-derived rubber tree that was taller by six metres multiplied by asexual propagation to raise several clones. Particularly interesting in the poplar tree species is that haploid seedlings selected for desired genotypes are naturally doubled as diploids after 7-8 years, this coinciding with their flowering and enabling them to be both asexually and sexually propagated. Pollen-derived haploid plantlets have also been produced in other perennial woody species too.

4.1.4. Transfer of Desired Alien Genes

The chromosomal instability in haploids attracts alien chromosomes or genes during wider crossing programmes. In rice, the breeding of high-yielding varieties for resistance to blast is conventionally achieved through back-cross which is a time-consuming process, taking probably more than twelve years to develop a cultivar with desired resistance. Hybridisation and anther culture can ensure that a standard pollen-plant breeding procedure can be developed in rice to introduce genes for high yield and resistance to blast in a short span of two years.

4.2. Somacional Variations

Genetic variations occur in undifferentiated cells, isolated protoplasts, calli, tissues and morphological traits of regenerated plants mostly due to changes in the chromosome number and structure. Cytological heterogeneity in cultures arises due to: The expression of chromosomal mosaicism or genetic disorders in cells of the initial explants, and new irregularities brought about by culture conditions.

Cell or tissue cultures undergo frequent genetic changes (polyploidy, aneuploidy, chromosomal breakage, deletion, translocation, gene amplifications and mutations) which are also expressed at biochemical or molecular levels. Plant cell and tissue cultures, therefore, provide increased genetic variability relatively quick and with a simple technology. The genetic variability in cultures expresses in the form of variant traits in regenerated plants which are then transmitted to the progeny through sexual or vegetative propagation. This genetic variability is called somaclonal variation or gametoclonal variation. Plants regenerated from tissue and cell cultures show heritable variation for both qualitative and quantitative traits. Somaclonal variation has been described in sugarcane, potato, tomato etc.

4.2.1. Isolation of Somaclonal Variants

It is far easier to isolation mutants for several traits from cell cultures than from whole plant populations since a large number of cells, can be easily and effectively screened for mutant traits which would otherwise be very difficult, ordinarily impossible. Mutants can be selectively chosen for disease resistance, improvement of nutritional quality, adaptation of plants to stress conditions, e.g., saline soils, low temperature, toxic metals (e.g., aluminium), resistance to herbicides and to enhanced the biosynthesis of plant products used for medicinal or industrial purposes. Isolation of somaclonal variants can be done by screening and cell selection.

i) *Screening.* This technique which involves the observation of a large number of cells or regenerated plants for the detection of variant individuals is the only feasible approach for the isolation of mutants for yield and yield traits. Screening has been profitable for the isolation of cell clones that produce higher quantities of certain biochemicals and for computer-based automated cell sorting devices.

ii) *Cell selection.* In this approach, a suitable selection pressure permits the preferential survival/growth of variant cells only, e.g., selection of cells resistant to various toxins, herbicides, high salt concentration etc. When the selection pressure allows only the mutant cells to survive or divide, it is called positive selection and where the wild type cells divide normally and therefore are killed by a counter selection agent it is called negative, selection. The mutant cells, unable to divide escape the counter selection agent are subsequently rescued by removal of the counter selection agent. The negative selection approach is utilised for the isolation of auxotrophic mutants.

The positive selection approach may be subdivided into four categories: (i) direct selection, (ii) rescue method, (iii) stepwise selection and (iv) double selection.

In direct selection, the selection-pressure-resistant cells survive and divide to form colonies while the wild type cells are killed by the selection agent. This common selection method is used for the isolation of cells resistant to toxins (produced by pathogens), herbicides, elevated salt concentration, antibiotics, amino acid analogues etc.

As in the direct selection in the rescue method too, the wild type cells are killed by the selection agent, and the variant cells remain alive but, usually, they do not divide due to the unfavourable environment. The selection agent is then eliminated to save the variant cells. This approach has been used to recover low temperature and aluminium-resistant variant cells.

In the selection pressure, e.g., salt concentration, which may be gradually increased from a relatively low level to the cytotoxic level, the resistant clones isolated at each stage are subjected to the higher selection pressure. Such a selection approach, called stepwise selection, may often favour gene amplification or mutations in the organelle DNA.

In certain cases, it may be viable to select for survival and/or growth and also some other feature reflecting resistance to the selection pressure; this is called double selection. An example of double selection is antibiotic streptomycin, which inhibits chlorophyll development in cultured cells. This approach has been used for the selection of cells resistant to the herbicide amitrole, tobacco mosaic virus (TMV) and aluminium.

4.2.3. Characterisation of Variants

Somaclonal variants show instability when they are isolated through cell selection. The frequency of stable variants may range from eight per cent to 62 per cent, perhaps depending on the species and the selection agent. During further screening or selection many selected clones fail to exhibit their resistance and are susceptible, misclassified as resistant.

Several clones, called unstable variants, lose their resistance to the selection agent after a period of growth in the absence of selection pressure. They may result from changes in gene expression and from gene amplification.

Some variant phenotypes, while quite stable during the cell culture phase, disappear when plants are regenerated from the variant cultures, or when the regenerated plants reproduce sexually, in case they are expressed in the regenerated plants. Such epigenetic changes are attributed to stable changes in gene expression e.g., hormone habituation of cell cultures.

The remaining variants that stably express the variant phenotypes during the cell culture and the regenerated plant phases, and exhibit the transmission of these phenotypes through the sexual reproduction cycle are called mutants. Only such a category of variants would find an application in crop improvement and may represent true gene mutations or some other types of changes.

4.2.4. Molecular Basis of Somaclonal Variation

Somaclonal variation may arise due to changes in chromosome number and/ or structure, gene mutation, plasmagene mutation, alteration in gene expression, gene amplification, mitotic crossing over, transposable element activation, and rearrangements in cytoplasmic genes.

Most mutants isolated from cell cultures involving single gene mutations, the mutant allele is either dominant or recessive. Gene amplification in some variants, are recovered though stepwise selection of plant cells *in vitro*. Sometimes, deamplification may also occur in somaclonal variants, e.g., for rRNA genes. Also, some previously silent mutator genes may get activated during tissue culture e.g., in maize. Clearly, transposable elements can be activated during *in vitro* culture. The breakage and fusion of chromosomes, which occur during culture, may activate some controlling elements. Mitotic crossing-over could also be responsible for some of the genetic variations detected in regenerated plants leading to the recovery of homozygous recessive single gene mutations in plants regenerated from cells *in vitro*.

4.2.5. Somaclonal Variations and Induced Mutations

In general, cell cultures are not given mutagenic treatments for the recovery of somaclonal variants. But where mutagenic treatments were used, usually an increase in the frequency of somaclonal variants was observed. While in some cases, mutagenesis is necessary for the recovery of the specific variant being isolated, it should be avoided as far as possible in view of the undesirable features associated with such treatments. Somaclonal variations are preferable to induced mutations for several reasons:

i) Chimareism is a major problem in induced mutations but not in somaclonal variations.

ii) Induced mutations are often linked with undesirable features like sterility etc.

iii) 'New' alleles and even 'new' mutations have been recovered through somaclonal mutations.

iv) The frequency of useful mutations is very high in the case of somaclonal variations.

v) A highly effective selection can be applied *in vitro* for several economically important characters which is virtually impossible in the case of mutation breeding,

vi) An astronomically large number of individuals can be effectively screened *in vitro*, etc. But, it may be stressed that the somaclonal variation applies to only those species where whole plant can be regenerated from cultured cells, while mutation breeding applies to all the species, and that the former is dependent on quite sophisticated facilities for tissue culture and of greenhouse.

4.3. Gametoclonal Variations

Haploid plants regenerated from callus cultures show

gametoclonal variation like in rice. Such a variation may be subjected to selection at the haploid level, and the chromosome number of the selected plants may be doubled to obtain homozygous plants.

Gametoclonal variation appears to be more desirable than somaclonal variation for the following reasons:

i) The mutant characteristic is expressed in the Ro plants.

ii) Cells with detrimental mutations may regenerate much less frequently in the case of haploid than in diploid cells. However, regeneration of haploid plants is feasible in fewer plant species than regeneration from somatic cells.

Somatic-derived somaclones and gametic-derived gametoclones can be distinguished in three ways:

i) Both dominant and recessive mutant genes induced by gametoclonal variation will express directly in haploid plants regenerated from microspores of diploid anthers.

ii) Recombinants recovered in gametoclones would be the result of meiotic crossing-over.

iii) The gametoclone can be used only after having stabilised by doubling its chromosome number.

The value of gametoclonal variation in crop improvement is evident from the development of double-haploid lines by anther culture of F1 hybrid plants of wheat and rice. Based on certain observations, it is suggested that factors causing meiotic recombination and mutation prior to initiating anther culture give rise to gametoclonal variation. Mutations may, however, occur even after or by the doubling process. Further, the genetic changes associated with the use of microspores for regeneration equally contribute to gametoclonal variation.

Variation in chromosome number of gametes or gametophytic tissue has a vital role in gametoclonal variation as evident from a range of aneuploids and mixoploids recovered from anther cultures of wheat, maize, and sexual hybrid of wheat and triticale. Also, variants uncovered after androgenesis are never recovered from somatic protoplast cultures. Further, it is observed that gene amplification contributes to reduction in the yield of gametoclonal variant plants. Thus, the mutation spectrum obtained by gametoclonal variation may differ from that obtained by somaclonal variation.

4.3.1. Applications in Plant Breeding

Somaclonal variation and gametoclonal variation are useful sources of introducing genetic variations valuable to the plant breeders.

Breeding of autotetraploid (4x) crops like potato, alfalfa etc. presents problems due to the complex segregation patterns at the autotetraploid level. Whereas breeding at the haploid level of crops like potato is much easier due to much simpler segregation patterns and smaller breeding populations involved. Haploids of tetraploid (4x) species are called dihaploid since they have two similar or dissimilar genomes. Potato dihaploids (2x) are fully fertile, and allow hybridisation and selection at this level. The selected dihaploid clones are given colchicine treatment to obtain tetraploid clones which are used as varieties. The process involving extraction of dihaploids from tetraploid species, breeding at the dihaploid level and then chromosome doubling of selected dihaploid lines to obtain tetraploid varieties is called analytical breeding which is becoming increasingly popular in crops like potato. Use of dihaploids in breeding of crops like potato offers the following advantages:

i) much simpler segregation ratios and, as a result, smaller segregating populations,
ii) a greater efficiency of selection, and
iii) much easier hybridisation with diploid wild species which offer many genes of great value.

There are several effective schemes for analytical breeding, including a scheme for extraction of monoploids (monoploids have a single genome) and then protoplast fusion of four selected monoploids to obtain autotetraploid varieties in crops like potato.

The following are the advantages:

i) Production of homozygous DH lines using haploids saves at least four years in comparison to selfing/close inbreeding.
ii) Selection among DH lines is at least 6-8 times as efficient as that among segregating populations thus reducing the size of breeding populations. In addition, the confusing effects of heterozygosity are absent in DH lines.

iii) Haploids may be useful for the isolation of desirable mutants whose chromosome number may be doubled to obtain homozygous mutant lines in a single generation.

iv) Pollen embryos are generally highly regenerative. Therefore they can be used for gene transfers mediated by agrobacterium or by a technique like particle gun.

But analytical breeding also has its limitations, such as:

i) In many crops, the application of this technique is not yet available, hence not possible for haploid production.

ii) In many other crops, since large numbers of haploid plants are not easily obtained a wider application of pollen culture in crop improvement depends primarily upon the development of techniques to produce more and quicker methods.

iii) High cost of obtaining haploids and doubled haploids is still a major problem.

iv) Sometimes deleterious mutations may be induced during the *in vitro* phase.

v) A non-random recovery of haploids may reduce the spectrum of variability recovered.

vi) A sophisticated tissue culture laboratory and a dependable greenhouse are essential for success.

vii) To carry out the various operations specialised skill is required.

viii) Plants other than haploids are often recovered from anther cultures/ interspecific hybrids; this necessitates cytological analyses.

ix) Occurrence of grametoclonal variation may limit the usefulness of pollen embryos for genetic transformation/ gene transfer.

x) The problem of high frequency of albinos produced in anther cultures of monocots, especially cereals needs to be resolved for a successful exploitation of haploids in breeding of such crops.

4.4. Germplasm Conservation

With population explosion, pressure on the forest and land

resources there has been a decline in the population of medicinal and economically important plant species, some of which are on the verge of vanishing. Such 'threatened' species are trying to be protected and preserved for future breeding programme.

In India, the Botanical Survey of India has brought out three volumes highlighting the threatened plants of the Indian subcontinent. Several other organisations, are engaged in the protection and preservation of such plants. Preserving the genetic material can be done in two ways:

i) *In situ preservation:* This aims at the preservation of the germplasm in their natural environment by establishing biosphere reserves, national parks, gene sanctuaries etc. This type of preservation there is the risk of declination of the preserved species due to environmental hazards.

ii) *Ex situ preservation:* Being the chief mode of preservation of germplasm, it provided a suitable condition in the gene bank to preserve the genetic materials in the form of seed or *in vitro* cultures. But this requires considerable knowledge of genetic structure as well as elements influencing them.

Conserving plant germplasm in the seed-propagated plants has certain limitations such as: Loss of seed viability with passage of time, seed destruction due to seed born pathogens, pest etc., and is confinement only to seed propagating plants.

In contrast the vegetatively propagated plants are preserved *in vitro* as shoots, meristems, embryos etc. The advantages of *in vitro* this over in situ preservation are that large amount of material can be preserved in small area as; it overcomes the destruction due to environmental hazards; and it provides large amounts of plant material for culturing.

4.4.1. Cryopreservation

In cryopreservation metabolic functions of biological material are reduced and subsequently arrested by imposition of ultra-low temperature. At the temperature of liquid nitrogen (–196°C) almost all metabolic activities of cells cease and the sample is then preserved in such a state for extended periods. However, only certain biological materials, in their natural state can be frozen to sub-freezing temperatures without adversely affecting the cell viability. Knowledge about the chemicals having cryopreservative properties such as glycerol and dimethyl sulfoxide has facilitated the development of effective cryopreservation technique.

4.4.1.1. Need for Cryopreservation

The main purpose and aim of developing cryopreservation technology is to preserve valuable genetic resources, especially of vegetatively propagated and also of those species which have short lived seeds. Therefore, it is essential to have efficient regeneration protocols through tissue culture of the species. Regeneration methods using apical meristem are advantageous as these are simple and there are no chances of genetic variation. It is important that tissue culture methods do not create genetic variability.

In contrast in cases where genetic variability has been induced in the cultures, this variability needs preservation for use in the future. Therefore, cryopreservation technique is equally good for preserving the genetic resources of existing genotypes and also of new variants.

4.4.1.2. Procedure of Cryoprotection and Pretreatment

Cryoprotection in some form is necessary for cryopreservation of plant materials unless they are naturally dehydrated, as in the case of dormant vegetative buds in the winter, or artificially cold acclimated. Several chemicals, such as dimethyl sulphoxide (DMSO), glycerol, various sugars and sugar alcohols, protect living cells against damage during freezing and thawing. The colligative properties of the cryoprotectants lower the temperature at which freezing first occurs and can alter the crystal habit of ice when it separates, thus minimising the harmful action of electrolyte concentration resulting from conversion of water into ice. Cryoprotectants having high solubility in aqueous phase and low toxicity to the cells are categorised as: Permeating, e.g. DMSO, methanol, glycerol; and non-permeating, e.g., sugars, sugar alcohols, high molecular weight polymers as dextran, polyvinyl pyrrolidone, hydroxy ethyl starch.

The cells require different pretreatment periods with different compounds for proper cryoprotection; e.g., DMSO enters more rapidly than glycerol, in permeating types, and therefore requires shorter period for treatment. Most of the cryoprotectants exhibit varying degrees of cytotoxicity at higher concentrations, with DMSO at 5 to 10 per cent and glycerol at 10 to 20 per cent. Sometimes, a mixture of cryoprotectants improves the efficacy, e.g., addition of osmotically active compounds in the culture medium such as mannitol, sorbitol, sucrose and proline increase the freezing resistance of the cells acting by their dehydration effect. Mannitol beneficial in reducing the mean cell volume of cells of *Acer pseudoplatanus* and *Capsicum annuum* and increase their post-freezing survival. Therefore, all such compounds also possess cryoprotective properties besides that they are osmotically active.

4.4.1.3. Freezing methods

Various plant materials can be cryopreserved by slow freezing, rapid freezing and droplet-freezing.

i) *Slow freezing*: When plant cells are cooled progressively, ice crystal formation begins extra-cellularly, presuming that plasma membrane acts as a barrier that prevents the ice crystal formation in the cytoplasm. In the absence of ice crystals, cytoplasm remains super cooled. On reducing temperature further, the concentration of extra-cellular liquid increases, as more water is converted to ice. The vapour pressure of the frozen solution being lower than the same concentration of the super cooled liquid, its of slowly cooled cells reach equilibrium with external ice by efflux of water thus, preventing the intra-cellular ice formation and consequently preventing freezing injury. It is believed that slow freezing increases the concentration of cytoplasm and increased dehydration increases the survival of cells. The development of efficient slow freezing method depends upon several factors like: Cooling rates; pretreatment and cryoprotection; type and physiological state of the material; and the temperature prior to immersion in liquid nitrogen. The most commonly used methods generally involves regulated slow cooling at a constant rate of 0.5 to 2°C/min. to terminal temperature between –3°C to -40°C followed by storage in liquid nitrogen. Slow cooling method used for different species is given in Table 4.1.

Table 4.1. Summarised conditions for optimal cryopreservation for a few species

Plant species	*Cryoprotectant*	*Cooling*	*Result*
Acer	DMSO 24%	Prefreezing	20-30%
pseudoplatanus	- 30°C-50°C,		
	Glucose 10%	Storage-196°C	reduction in survival
	Glycerol 10%	Prefreezing	cell survival
		1°C 28%	
	Storage –196°C		
Atropa belladonna	DMSO %	Prefreezing	40% cell survival
		2°C/min.	
		Storage –196°C	
Datura stramonium	DMSO 5%	Prefreezing 1-2 °C/min.	40% cell survival
		Till –100°C	
		Storage-196°C	
Daucus carota	DMSO 10% or	Prefreezing 2-4 °C/min	65% cell after 60 days.
	Glycerol 5%	Storage –196 °C	Storage, plant
			Regeneration.
Ipomoea sp.	DMSO 2.5%	Pre-freezing 2-4°C/min.	5.2% survival,
	Glycerol 2.5%	Storage-196°C	high growth in
	Sucrose 6.5%		few days.

ii) *Rapid freezing:* Rapid freezing, unsuitable for the cryopreservation of cell cultures, is employed to cryopreserve shoot tips of carnation, potato, strawberry and several others, and somatic emboyos of oil palm. Rapid freezing is done by direct immersion of the cryoprotectant-treated specimens in liquid nitrogen. At such high cooling rates, the intracellular fluids do not have sufficient time to equilibrate with the external ice with the possibility of intracellular ice formation, which is considered to be harmful for cells.

iii) *Droplet freezing:* In this method, the cryoprotectant treated meristems are dispensed in droplets of 2-3 ml on an aluminium foil in a petriplate being frozen by slow cooling (0.5°C/min.) to a subzero temperature between –20 to –40°C prior to immersion in liquid nitrogen.

Material can be kept stored in liquid nitrogen (–196°C) or in its vapour (–150°C), and rapidly thawed is to avoid the damaging ice recrystallisation which may occur during slow warming. In general, thawing is carried out by removing the sample by liquid nitrogen storage and transferring in a water bath (34–40°C) for about 1–2 minutes or until material is warmed up.

The most reliable and accurate estimate of viability is regrowth of cryopreserved specimens. The other technique used for vitality test of the cells such as fluoresce in diacetate stain, triphenyl tetrazolium chloride test etc. can give a quicker method of testing cell viability. The further growth and regeneration of the cryopreserved cells after thawing will be like normal cells which should be carefully handled, plated and while subculturing of such cells.

4.5. Production of Secondary Metabolities

Plant cell culture can be a potential means of producing useful plant products thereby doing away with conventional agriculture with all its attendant problems and variables like: environmental factors (drought, floods, etc.), disease, political and labour instabilities in the producing countries, uncontrollable variations in the crop quality, inability of authorities to prevent crop adulteration, losses in storage and handling, etc. Thus, the production of useful and valuable secondary metabolites in large bioreactors located in the consuming country leads to controlled production according to demand and a reduced time requirement.

Theoretically, that such large scale suspension cultures will be suitable for industrial production of useful plant chemicals such as pharmaceuticals and food additives, similar to that of microbial fermentation. Yet due to some significant differences between microbial and plant cell cultures consideration must be given attempting to apply plant cell cultures to the available technology.

Table 4.2 shows a comparison of some of the characteristics of plant and microbial cultures of relevance to fermentation. This table serves to demonstrate some of the problems that can be encountered with plant cell cultures. The sensitivity to shear is due both to the large size of the cells and to the relatively inflexible cellulose cell wall. Thus, with normal blade impellers the cells may twist which will inhibit mitoses and, for this reason, air- lift fermentors are recommended by some researchers. The large size of the plant cell contributes to its comparatively high doubling time, which thus prolongs the time required for a successful fermentation run.

Table 4.2. Characteristic of Microbial and Plant Cell Relevant to Fermentation

Characteristics	*Microorganism*	*Plant Cell*
Size	2 u	>10 u
Shear stress	Insensitive	Sensitive
Water content	75%	>90%
Duplication time	<1 hour	days
Aeration	1-2 vvm	0.3 vvm
Fermentation time	Days	Weeks
Product accumulation	Medium	Vacuole
Production phase	Uncoupled	Often growth-linked
Mutation	Possible	Requires haploids
Medium cost ($) (MS medium)	8-9/m	65-70/m

Since product secretion is uncommon in a vocuole, the high metabolite yields seen in microorganisms that secrete product cannot be expected. Membrane permeabilisation of plant cells may serve to relieve the constraints of product inhibition by facilitation of leakage into the extracellular medium, and by permitting recycling of the biomass it would help reduce production costs.

The low aeration requirement for plant cells is an advantage over microbial cultures in general. In addition, the high cost of running a fermentation vessel over several weeks should be considered, although media costs are much less than those of animal cell cultures.

In addition to the above problems concerned with fermentation technology, at the biochemical level, the two major problems concern poor expression of products and instability of cell lines.

Compared with intact plant cells cultured plant cells often produce reduced quantities and different profiles of secondary metabolites and these quantitative and qualitative features may change with time. The poor product expression may often be due to a lack of differentiation in cultures and in some cases of cultures there is over-production of metabolites compared with the whole plant.

Many cultured cells producing metabolites not observed in the plant, e.g., Lithospermum erythrorhison cultures synthesizing rosmarinic acid. It is clear that selecting original plant material having high yields of the desired phytochemical may be important in establishing high-yielding cultures. Furthermore, it is important to screen for high producing lines over and over again as also the nutritional.

4.5.1. Materials

any part obtained from any plant species can be employed to induce callus tissue, however the successful production of callus depends upon plant species and their qualities. Dicotyledons are rather amenable for callus tissue induction, as compared to monocotyledons; the callus of woody plants generally grow slowly. Stems, leaves, roots, flowers, seeds and any other parts of plants are used, but younger and fresh explants are preferable as explant materials. Explants obtained must be sterilised using ethanol, sodium hypochlorite and/or other chemicals to remove all microorganisms from the materials and a typical sterilisation procedure will be described later as an example.

4.5.2. Media

4.5.2.1. Inorganic Salts

Many media are designed induce a callus from an explant and to cultivate the callus and cells in suspension, agar or its substitutes into the media to prepare solid medium for callus induction. A most commonly useα media for plant tissue cultures is one that has very high concentration of nitrate, potassium and ammonia. In comparison, the levels of inorganic nutrients in the B5 medium are lower. Many other media have been developed and modified and nutrient however, it is not always necessary to test many kinds of basal media when a callus is induced, rather using only one or two kinds of basal media in combination of different kinds and concentrations of phytohormones. The most suitable medium composition should be optimised afterwards in order to obtain higher level of products and higher growth rate.

4.5.2.2. *Carbon Sources*

By adding suitable carbon sources such as sucrose or glucose, fructose, maltose and other sugars also support the growth of various plant cells is enhanced and the efficient production process of useful metabolites is established depending on plant species and products; therefore it is necessary to optimise the medium compositions including carbon sources in each case. From an economically also, the use of more inexpensive carbon sources is appropriate in industry and crude sugars such as molasses have been examined.

4.5.2.3. *Vitamins*

The basal media include myo-inositol, nicotinic acid, pyridoxine HCl and thiamine HCl thiamine being essential one for many plant cells, with other vitamins stimulating the growth of the cells in some cases.

4.5.2.4. *Phytohormones*

Phytohormones or growth regulators are required to induce callus tissues and to promote the growth of many cell lines. The concentration of auxins in the medium is generally between 0.1 and 50 μM. Since each plant species requires different kinds and levels of phytohormones for callus induction, its growth and metabolites production, it is important to select the most appropriate growth regulators and to determine their optimal concentrations. Gibberellic acid is also added to the medium if necessary.

4.5.2.5. *Organic Supplements*

In order to stimulate the growth of the cells, organic supplements like casamino acid, peptone, yeast extracts, malt extracts and coconut milk are sometimes added to the medium.

4.5.3. Large Scale Cultures

Commercial applications of tissue culture requires scaling up of culture systems; which can be achieved by (i) increasing the number or (ii) size of culture vessels, or (iii) a combination of both, depending chiefly on the physical condition of medium, i.e., agar or liquid, used for the cultures.

4.5.3.1. *Agar Cultures*

The most commonly used medium for experimental and some commercial activities like micropropagation, is agar-geiled. Generally during experimentation, scaling up can be achieved by increasing the number of culture vessels. Test tube baskets or stands containing culture tubes which are placed on culture racks. Half litre bottles are the most commonly used culture vessels in commercial micropropagation activities. While scaling up agar cultures each vessel has to be handled individually during subculture.

4.5.3.2. *Suspension Cultures*

Use of liquid medium in suspension cultures allows easy and extensive scaling up by employing bioreactors, although a limited scaling up can be achieved by using larger culture flasks.

A bioreactor is a culture vessel, generally of a large volume, e.g., 1L to over 1000L which has provisions for:

i) aeration,
ii) stirring to achieve medium and cell mixing,
iii) contamination control and
iv) replacement of used medium and/or used medium plus cells.

The bioreactors used for culture of plant cells may be of following four types: (a) batch bioreactors, (b) continuous bioreactors, (c) multistage bioreactors and (d) immobilised cell bioreactors; all bioreactors, except the last one are commonly called stirred tank reactors since the vessel has a device for stirring.

Batch bioreactors. In batch bioreactors, the medium and inoculum are loaded at the outset. The cells are allowed to grow and the entire cell mass is harvested at the end of incubation period. The characteristic features of such bioreactor systems are: continuous depletion of medium, accumulation of cellular wastes, alterations in growth rate, and continuous change in the composition of cells. The spin filter bioreactor can be used as a batch bioreactor by closing the inlet for medium and the outlets for medium/ medium plus cells.

The bioreactor has ports for (1) addition of fresh medium, (2) air space, (3) removal of only spent medium (minus cells; they are separated by the spin-filter) through the hollow central shaft, (4) entry of air to the sparger, and (5) removal of spent medium plus cells. When all the 5 ports are in operation, the bioreactor is of continuous flow type. But when only ports (2) and (4) (air outlet and inlet, respectively) are open with remaining ports of (1), (3) and (5) closed, the bioreactor becomes a batch type.

Continuous bioreactors. In continuous bioreactors, during the entire incubation period there is continuous inflow of fresh medium and outflow of used medium. A spin-filter bioreactor is a good example of continuous flow bioreactor.

The continuous bioreactor provides a highly versatile system for control on medium change rate and on cell density, being made possible due to the two routes for medium removal while only one of them allows the removal of cells. Using a continuous flow bioreactor are grown cells at a specified cell density in an active growth phase. Such cultures may either provide inocula for further culture or may serve as a continuous source of biomass yields.

Multistage bioreactors. Using two or more bioreactors in a specified sequence in such culture systems each carries out a specific step of the total production process.

A perfusion type bioreactor allows regular circulation or change of medium by regularly removing the used medium only through the spin-filter and adding fresh medium through the medium port. It also allows change of one type of medium by another as per the need of the culture system. The simplest situation would involve two bioreactors to produce a biochemical like shikonin, the first of the two bioreactor provides conditions for rapid cell proliferation and favours biomass production, while the other one has conditions conducive for shikonin biosysnthesis and accumulation. The cell biomass, is collected from the first stage bioreactor, is used as inoculum for the second stage reactor. Similarly, for large scale somatic embryo (SE) production, the first reactor may be in continuous mode, while the second may be of batch type. The continuous first stage reactor provides conducive conditions for rapid proliferation of embryogenic cells. The cell mass from this bioreactor serves as a continuous source of inoculum for the second stage batch type bioreactor which has suitable conditions for embryo development and maturation.

The use of continuous first stage bioreactor (i) avoids the time, labour and cost needed for cleaning etc. of a batch reactor between two runs, (ii) eliminates the lag phase of batch cultures, and (iii) provides a more homogeneous and actively growing cell population. However, due to their long run periods, continuous cultures have a greater contamination risk than batch bioreactors.

Immobilised cell bioreactors: These bioreactors are based on cells encased either in gels, such as, agarose, agar, chitosan, gelatine, gellan, polyacrylamide and calcium alginate to produce beads, or in a membrane or metal screen compartment or cylinder. The cell-containing membrane/screen cylinder is kept in a chamber through which the medium is circulated from a recycle chamber. The medium movement may be so adjusted as to flow across the screen compartment rather than parallel to it. The technology is being refined for commercialisation. Fresh medium is regularly added to and equivalent volume of used medium is withdrawn from the recycling chamber to maintain its nutrient status.

The technique cell immobilisation changing the physiology of cells as compared to that of cells in suspension is useful where the biochemical of interest is excreted by the cells into the medium. Product excretion may also be effected by immobilisation itself, or by certain treatments like altered pH, use of DMSO (dimethyl sulfoxide) as a permealising agent, changed ionic strength of medium, an elicitor etc. Immobilised cell reactors have no risk of cell wash out, have low contamination risk, protection of cells from liquid shear, better control on cell aggregate size, separation of growth phase (in a batch/continuous bioreactor) from production stage (in an immobilised cell bioreactor), cellular wastes regularly removed from the system, and cultures at high cell densities.

4.6. Enhancing the Production of Secondary Metabolites

Several products were found to be accumulated in cultured cells at a higher level than those in native plants through optimisation of cultural conditions. For example, ginsenosides by Panax ginseng, rosmarinic acid by *Colleus blumei*, shikonin by *Lithospermum erythrorhizon*, diosgenin by *Dioscorea*, ubiquinone-10 by *Nicotiana tabacum* were accumulated in much higher levels in cultured cells than in the intact plants.

However, many reports have described that yields of desired products were very low or sometimes not detectable in dedifferentiated cells such as callus tissues or suspension cultured cells. In order to obtain products in concentrations high enough for commercial manufacturing, therefore, many efforts have been made to stimulate or restore biosynthetic activities of cultured cells using various methods. The following are typical approaches that may increase productivity of cultured plant cells.

4.6.1. Optimisation of Cultural Conditions

4.6.1.1. Medium

A number of chemical and physical factors affecting cultivation. The most fundamental approach in plant cell culture technology is to have a number of chemical and physical factors affecting cultivation, such as media components, phytohormones, pH, temperature, aeration, agitation, light, etc., to be tested extensively with various plant cells. Auxins are commonly added to the medium for callus induction, but at a low concentration or omitted for production of metabolites.

4.6.1.2. Temperature, pH, Light and Oxygen

A suitable temperature for induction of callus tissues and growth of cultured cells 17–25°C. But, each plant species may favour a different temperature and lowering the cultivation temperature increased the total fatty acid content per cell in dry weight. The medium pH is usually adjusted to between 5 and 6 before autoclaving and extremes of pH are avoided. The optimum pH is determined and controlled using a small scale bioreactor or a jar fermentor with pH control equipment.

On an industrial scale stainless steel tanks are used for growth of plant cells thus eliminating the use of light. To avoid cases of light-stimulated secondary metabolites production, modification of fermentors with lighting facilities have also been carried out.

To increase the productivity of secondary metabolites, high cell density cultures have been investigated.

4.6.1.3. Selection of High-producing Strains

The physiological characteristics of individual plant cells vary most often. A number of researchers have used cell cloning methods as this is the most promising way of increasing the levels metabolites present.

Repeated cell cloning using cell aggregates of *Coptis japonica,* led to obtaining a strain which grew faster and produced a higher amount of berberine and cultivated the strain in a 14 L bioreactor. The selected cell line increased growth about six-fold in three weeks and the strain was very stable, producing a high level of berberine even after 27 generations.

Cultured, green *Lavendula vera* cells accumulated a high level of free biotin when grown in light and using pimelic acid, a precursor of biotin, as a selecting agent. The choice of a high ubiquinone-10 producing *Nicotiana tabacum* strain has produced excellent results.

High pyrethrin tissue cultures were produced from *Chrysanthemum cinerariaefolium* and analytical screening of the tissue lines enabled the selection of a few "high yielding" strains from high yielding plant selections. For a rapid assay method, crucial in the selection of a high yielding cell line, using fluorescence assay to select ajmalicine and other major heteroyohimbine alkaloids producing cells of *C. roseus* proved advantageous even over a radioimmunoassay procedure since almost unlimited numbers of colonies could be screened in this way. However, the cell cloning is especially compatible with selection for high pigment production as selection can be achieved visually, or with the use of simple spectrophotometric analysis.

Cell cloning, being a very useful technique to increase the level of secondary metabolites, should be applied as widely as possible. The lack of specific enzyme(s) represents the most important reaction for the inability of plant cell cultures to produce secondary metabolites. Comparing the highest cinnamoyl putrescine producing, p-flurophenylalanine resistant strain TX-4 or *N. tabacum L. CV Xanthi* with a low producing strain for five enzymes of the biosynthetic pathway, namely phenylalanine ammonia-lyase, trans-cinnamate-4-hydroxylase, 4-coumarate-CoA ligase, ornithine decarboxylase and arginine decarboxylase it was found to be 3 to 10 times higher in TX4 cells. Protoplasts were also used for the selection of high-shikonin producing cell lines of *L. erythroryzon* and thiophene producing *Tagetes patula* cell lines.

4.6.3. Addition of Precursors

Supplementing appropriate inexpensive precursors or related compounds to the culture media sometimes stimulates secondary metabolite production and hence advantageous. By adding amino acids to cell suspension culture media tropane alkaloids, indole alkaloids, and ephedorin are produced with some stimulative. It is true that some amino acids are precursors of various alkaloids, but generally the biosynthetic steps from amino acids to alkaloids are so complicated that there are doubts whether amino acids added are incorporated into the alkaloids directly in cell culture, affecting not only alkaloid biosynthesis directly as precursors, but also indirectly through other metabolic pathways in the cells.

Addition of phenylalanine to *Salvia officialis* suspension cultures stimulate the production of rosmarinic acid and shortens the production time as well. Addition of 500 mM tropic acid to the medium of Scopolia japonica increases the amount of alkaloids by up to 14 times. The level of an anticancer compound, tripdiolide, produced by *T. wilfordii* cultured cells increases by addition of 100 μg/L farnesol which is dephosphorylated farnesyl pyrophosphate, and an intermediate in the biosynthesis of terpenoids. Addition of phenylalanine into the agar medium of *Taxus cupsidata* cells stimulates the biosynthesis of an anticancer compound, taxol. However, biotransformation using intact cells or immobilised cells is an alternative way of producing a product by adding precursors into the culture media.

When cells cultivated in the presence of ajmalicine Rauwolfia serpentina form a new indole alkaloid, 6-hydroxytaumacline, in significant amounts. Arbutin, a skin depigmentation agent, is produced by biotransformation of hydroquinone using *C. roseus* cells and addition of it into the liquid culture medium of this cell line produces arbutin efficiently, possibly applicable for commercial application. Using *Mallotus japonicus* cells for the glucosylation of umebelliferone and salicylic acid it was found that salicylic acid was converted to its o-glucoside with a yield of 90-95 per cent. The glucoside showed as potent an analgesic activity as salicylic acid, while its effect was more rapid and more long-lived than that of salicylic acid in mice. It is interesting that plant cells are able to form prodrugs having commercial usefulness. Cultured *Salix matsudana* cells glucosylates salicylic acid to 2-o-salicylic acid-ß-D-glucoside. The cells also convert salicyl alcohol to both salicin (2-o-salicyl alcohol-ß-D-glucoside) and isosalicylicin (1-o-salicyl alcohol-ß-D-glucoside). However, salicylic aldehyde is converted to ß-D-glucosides of salicylic acid and salicyl alcohol instead of helicin.

The conversion of geraniol and nerol to neral and geranial by *Vitis vinifera* cell suspension cultures is also interesting. Using plant cell culture techniques, radioactive labelled compounds can be formed from appropriate substrates and is a useful application of the technology because of high value of the product.

4.6.4. Biotransformation

The approach of using a suitable substrate compound instead of the addition of a particular compound as a precursor into the culture medium of plant cells, biotransforming to a desired product using plant cells has been extensively applied in the fermentation industry using microorganisms and their enzymes. For example, L-aspartic acid and L-malic acid are being manufactured commercially from fumaric acid, respectively using microorganisms. And various steroids are also produced by microbial biotransformations. Biotransformation of ß-methyldigitoxin to ß-methyldigoxin using *D. lanata* cells is useful because digoxin has a large market as a cardiac glycoside.

The process for production of a very expensive antitumor drug, vinblastine, from catharanthine and vindoline using a biotransformation as well as a simple chemical synthesis is now being developed for commercialisation. Concerning vinblastine-related compounds production, a biotransformation of catharanthine and vindoline to anhydrovinblastine using horseradish peroxidase and glucose oxidase mediated coupling of vindoline and catharanthine is also a viable proposition.

Economically, biotransformation processes including addition of substrates into the cultures are one of the most commercially realistic approaches in plant tissue though the availability of inexpensive precursors is a key issue.

4.6.5. Elicitor Treatment

Specific secondary metabolites are often synthesised by microbial infections of intact plants. The best understood systems are those of fungal pathogens in which case the regulatory molecules have been identified as glucan polymers, glycoproteins and low molecular weight organic acids.

Researchers found that a cell line of *Papaver somniferm* synthesized and accumulated sanguinarine, a quaternary benzophenanthridine alkaloid when exposed to a homogenate of the fungus Botrytis. Sanguinarine extracted from the intact plants is being used for oral hygienic products.

By adding autoclaved culture homogenate of yeast, *Rhodotorula rubra,* into the suspension culture of *Ruta graveolens* S-adenosyl-L-methionine: anthranilic acid N-methyltransferase are elicited. A yeast polysaccharide preparation induced L-tyrosine decarboxylase in suspension cultures of *Thalictrum rugosum* and *Eschscholtzia californica.*

There is a noticeable increase in rosmarinic acid, a-0-caffeoyl-3, 4-dihydroxyphenyllactic acid, content in cultured cells of L.erythrorhizon after addition of yeast extract to the suspension cultures. In suspension cultured cells of parsley, chitosan elicited a rapid deposition of the 1,3-ß-glucan, callose.

Recent developments in phytochemical elicitation show that simple inorganic and organic molecules can induce product accumulation. Sodium orthovanadate and vanadyl sulphate induce the accumulation of isoflavone glucosides in *Vigna angularis* cultures and indole alkaloid accumulation in *Catharanthus roseus* cultures, respectively. Other substances found to stimulate alkaloid accumulation in *C. roseus* include sodium chloride, potassium chloride and sorbitol as well as abscisic acid. Such processes employing simple and cheap elicitors, have much promise in industrial scale plant cell cultures.

Addition of oxalate, a fungal elicitor, to the medium of *Gossypium hirsutum* suspension culture could reduce the amount of *Verticillium dahliae* elicitor to be employed to stimulate metabolite synthesis.

A combination of phosphate limitation and fungal elicitation synergistically increases production of secondary metabolites, e.g., *Rhizoctonia solani* with increased production of the sesquiterpene solavetivone by Agrobacterium rhizogenes-transformed hairy root cultures of *Hyoscyamus muticus*. However, when phosphate limitation is coupled with fungal elicitation, the productivity increase is considerably greater than that obtained with either method alone.

The stimulating activity of elicitors is quite significant if an appropriate one is chosen to stimulate synthesis of a particular product. However, the use of microbial elicitors may not be economical since an elicitor-producing microorganism should be cultivated in a fermentor separately from cultivation of plant cells using another fermentor. As such a simple and cheap compound should be employed as an elicitor.

4.6.6. Morphological Differentiation and Hairy Root Culture

4.6.6.1. Organ Culture

Using morphologically undifferentiated cells for the production of useful metabolites in ways similar to microorganisms show high productive ability of such cells compared with intact plants as already described. In fact, a callus of *Phytolacca americana* which was induced from a stem tissue more than a decade ago, still produces a high level of an alkaloid, betanine. However, there may be a putative link between differentiation and secondary metabolite accumulation.

Though a close correlation exists between the expression of secondary metabolism and morphological and cytological differentiation, it is not yet clear to what extent secondary metabolism depends on the development of specific structures or whether these two processes are genetically and/or physiologically linked.

The successful production of phytochemicals by cell cultures is dependent on the development of a certain level of differentiation in the literature demonstrating a relationship between differentiation and secondary metabolic accumulation. Hiraoka and Tabata successively transferred the callus of Datura meteloides from a medium with auxins to one without and then cultivated continuously.

In Digitalis purpurea cultures, organ redifferentiation in callus tissues stimulated digitalis cardenolides production. Similar was the case with rotenone formation using *Derris elliptica* and morphinane alkaloid production using *Papaver somniferum*. Using *P. somniferumi*, a green callus which differentiates epidermis or vascular bundles produced the alkaloids. A limited degree of tissue differentiation occurred and the cell contained codeine as a main alkaloid while the level of morphine increase as differentiation progressed.

Root cultures derived from suspension cultured cells of *P. bracteatum* produced the baine while axenic callus and shoot cultures of *Pyrethrum cinerarifolium* had an ability to produce pyrethrin. Root cultures of *Panax ginseng* have yielded high levels of ginsenosides.

4.6.6.2. Hairy Root Culture

A difficult but alternative possibility is to induce biochemical differentiation, but suppress morphological differentiation. For this the cells would have to be first selected for homogeneity in shape and metabolism. Using a range of species, has resulted in selection of regular green aggregates with a spherical shape and with cells of a regular morphology. Such aggregates have enabled high flow rates and easy medium removal, superior to those obtained with more dispersed cultures.

Regarding differentiated cell growth and phytochemical production, it has been observed that the insertion of a very fine platinum or titanium wire into a disorganised callus of *Helianthus tuberousus* brings about morphological differentiation, whereas control, untreated tissue remains in an undifferentiated state. If this phenomenon can be reproducible in other species it might well serve as a well-defined means of controlling differentiation and possibly secondary metabolism.

Another observation is that electric currents (1–2 μA) bring about shoot regeneration in callus culture at a rate five times greater than in controls. This has been patented as a process for the stimulation of growth and differentiation of plant tissue and for enhancing the accumulation of secondary metabolites.

In recent times there has been more focus on the use of Agrobacterium rhizogenes in secondary metabolism research. It inserts the Ri plasmid into wounded tissue, causing the growth of very fine adventitious roots, so-called "hairy-roots" which can be cultured in hormone-free medium.

4.7. Plant Cell Culture for Production of Chemicals

A large variety of biochemicals which are metabolites of both primary and secondary metabolism are derived from plants. But, secondary metabolites have impressive biological activities like antimicrobial, antibiotic, insecticidal, molluscidal, hormonal properties, and valuable pharmacological and pharmaceutical activities, and many are used as flavours, fragrances, colours, etc. Secondary metabolites include a variety of compounds e.g., alkaloids, terpenoids, phenyl propanoids, etc.

4.7.1. Pharmaceuticals

Plant secondary metabolites can be divided into several groups which are given below:

4.7.1.1. Alkaloids

A variety of alkaloids, most of which are plant metabolites, have been used as pharmaceuticals and most of them are plant metabolites.The typical tropane alkaloids, atropine, hyoscyamine, scopolamine and cocaine, were widely used as blockers of the parasympathetic nervous system, such as anodyne and antispasmodic. Various approaches to increase the productivity of alkaloids have been tried by many researchers Vinblastine, an antitumor alkaloid, is most likely to be produced commercially by a Japanese company using a combination process of plant cell culture and that of chemical synthesis.

4.7.1.2. Morphinan Alkaloids

Papaver somniferum L. (opium poppy) is a traditional commercial source of codeine, and morphine which can be converted to codeine an analgesic and cough-suppressing drug. Mature capsules of *P. bracteatum* accumulates up to 3.5 per cent of thebaine which also can be converted to codeine.

Researchers point out that morphogenetic differentiation from cultured cells of *P. bracteatum* is a prerequisite for producing higher levels of thebaine. Cells of *P. somniferum* produce norsanguinarine, sanguinarine, cryptopine and other various alkaloids, but not codeine and morphine.

Fungal mycelium of *Botrytis sp.* elicits production of sanguinarine by *P. somniferum* cells, increasing the level of this alkaloid 26 times in the presence of the elicitor, 29 per cent of the dry cell weight, compared to the medium without it. The alkaloid extracted from intact plants is added to toothpastes as an anti-plaque agent, but the production of sanguinarine by cell culture technology has not yet been commercialised.

Since no success was gained by the efficient production of thebaine and codeine using cell culture systems by de novo synthesis experimentation was done with the biotransformation of codeinone to codeine using immobilised cells of *P. somniferum*. The conversion yield was 70.4 per cent and about 88 per cent of codeine converted was excreted into the medium.

4.7.1.3. Berberine

Berberine, us used for intestinal disorders in eastern countries, is an isoquinoline alkaloid which is distributed in roots of *Coptis japonica* and cortex of *Phellondendron amurense*. It takes 5 to 6 years to produce Coptis roots as the raw material.

Researchers find that addition of 10-8 M gibberellic acid into the medium stimulates berberine productivity up to 1.66 g per L of the medium. Using a cell sorter and protoplasts of *C. japonica,* many higher alkaloid-producing cell lines are selected. A "high-density cell culture" process is used to produce berberine much more efficiently now. In order to achieve a cell mass of 70 g/L on a dry weight basis, stirring without damaging the cells, supply of sufficient amounts of oxygen, and that of appropriate nutrients are optimised. As a result, the yields have reached 70 g per L of cell mass and 0.45 g/day of berberine. The continuous culture with high cell density was also been conducted successfully.

Addition of a polyamine, spermidine stimulates the production of berberine by Thalictrum minus cell suspension cultures although other polyamines such as cadaverine, putrescine and spermine are ineffective. Spermidine effects an increase of ethylene generation which is associated with activation of berberine synthesis. The maximum stimulative effect is obtained by the addition of 2 mM spermidine.

4.7.1.4. Tropane Alkaloids

Scopolamine and hyoscyamine, occurring in the leaves of (*Solanaceae*) plants including *D. myoporoides* and *D. leichhardtii* are being used commercially as anesthetic and antispasmodic drugs. Scopolia, Atropa, Hyoscyamus and Datura also contain tropane alkaloids.

The concentrations of scopolamine and hyoscyamine in cultured cells are generally very low in spite of many efforts to increase the yield the plant cell culture has not yet been employed to manufacture these alkaloids.

Researchers report that roots differentiated from cultured cells accumulate scopolamine, hyoscyamine and/or nicotine, but that the alkaloids are not accumulated in leaves of the regenerated plantlets.

Since the alkaloids are synthesized in the root, cultivated hairy roots transformed with Agrobacterium rhizogenes are cultivated, producing hyoscyamine and other alkaloids at the similar levels to the normal roots.

4.7.2. Agricultural Drugs

4.7.2.1. Plant Virus Inhibitors

On examining the inhibitory effects of a large number of chemically synthesized compounds and natural molecules on plant viruses it is found that some of them have potent activity as protectors against virus infection.

In order to screen high producing cultured cells of plant viral inhibitors, a variety of callus extracts were examined for the inhibitory activity towards tobacco mosaic virus (TMV) infection. Among these extracts *Phytolacca americana* callus was selected as the most potent producer of the inhibitor. The level of the inhibitor accumulated in the suspension cultured cells which *P. americana* was homogenised and the supernatant was diluted up to 100 times with water. The diluted solution was found to inhibit TMV infection on tobacco and tomato plants significantly.

Using Agrostemma the growth of suspension cultured cells of *A. githago* was somewhat faster than *P. americana.* With Columnlite chromatography and electrofocusing Active principles of *P. americana* and *A. githago* were isolated. At least four basic proteins were obtained from *P. americana* cells whose molecular weights were $1.10X10^4$ to $3.15X10^4$. Among them the highest molecular weight component contained sugars in the molecule. On the other hand, only one basic protein was isolated as the principle from *A. githago* culture and its molecular weight was 2.5×10^4. The proteins obtained from *P.americana* have been widely investigated because of their activities to HIV.

After screening various plants *Mirabilis jalapa* was selected as a producer of an anti-plant virus protein, procuring the callus from leaves of *M. jalapa*. Its suspension cultured cells established were found to accumulate the protein intracellularly. Optimisation of the production and the cell growth as well as selection of high producing cell lines were conducted extensively. The molecular weight of one of the lines was 24 KD and the amino acid sequence was determined which had 24 per cent homology with a ribosome-inactivating protein, ricin D-A chain.

4.7.3. Food Additives

4.7.3.1. Pigments

In traditional dyeing shikonin and its derivatives such as acetyl shikonin and isobutyl shikonin accumulated in the roots of *Lithospermum erythrorhizon.* The reddish purple pigments used and as a herbal medicine. The plant has also been used as a herbal medicine. Mass cultivation of *L. erythrorhizon* cells produced shikonin compounds to offset the shortage of this plant. Its culture conditions were optimised extensively to increase the level of the products using flasks and various types of fermentors including a rotating cylindrical fermentor.

While *L. erythrorhizon* produced shikonins in White's medium the cell growth was poor in the same medium, whereas Linsmair-

Skoog's (LS) medium was recognised to support the growth but not shikonin production. Therefore, a two-stage culture for mass production of shikonin compounds was used. Namely, to proliferate the cells, LS medium was used at the first stage of the fermentation, and then the cells were transferred into White's medium for production of shikonin compounds. In order to improve the yields further, optimisation of components in both media was carried out extensively, and MG-5 and MG-9 media were established.

Most cultured cells in liquid and solid media occur as aggregates, and selection of high-producing cell lines from the aggregated cells is not effective and labour-intensive. Protoplasts from the cultured cells with appropriate enzymes and selected high shikonin compounds produced protoplasts using a cell sorter. The selected protoplasts were generated to cell lines and cultivated in suspension. From 48 cell lines, a cell line having 1.8 fold the productivity of the parent line was obtained. The cell line showed stable production of shikonin compounds.

To produce the compounds more efficiently, a high-cell density culture process in the second stage of the two-stage cultures. By feeding the nutrients into M-9 medium, the level of cell mass increased and that of the compounds produced increase twofold as much as without feeding. Shikonin and its derivatives are used for lipsticks.

The hairy root culture of *L. erythrorhizon* with Agrobacterium rhizogenes did not produce shikonin on solid MS medium but produced the pigment in the root culture medium and also secreted it into the medium. Addition of absorbents XAD-2, XAD-4, charcoal and so on increased the concentration of shikonin produced.

4.7.3.2. Anthocyanins

Anthocyanins, the water-soluble pigments responsible for the bright colours in flowers and fruit, are composed of an aglycone (anthocyanidin) and more than one sugar moiety. They normally change colour over the pH range due to the existence of four pH-dependent forms. Thus, at low pH they are red and at higher pH value (over 6) they turn blue. Soft drinks, sugar confectionary, jams and bakery toppings have anthocyanins as an ingredient to give them a red colour. The major source of anthocyanins for commercial purposes are grape pomaces and waste from juice and wine industries, but other potential sources have been investigated. Crude preparations of anthocyanins instead of pure ones are used extensively in the food industry as they are rather inexpensive.

Most of them seem to use an anthocyanin-producing cell line as a model system for secondary product production because of their colour which allows production to be easily visualised.

Accumulation of anthocyanins is enhanced by a high osmotic potential in *Vitis vinifera L.* (grape) cell suspension cultures. Sucrose or mannitol is added in the medium to increase the osmotic pressure and the level of anthocyanins accumulated.

4.7.3.3. *Safflower Yellow*

This yellow pigment obtained from the floret of the safflower plant (*Carthamus tinctorius L.*), also known as Mexican saffron or American saffron, has no relation to genuine saffron. The major pigment is carthamin, which exists in the flowers, and there is also a red pigment. Carthamin is the quinoid form of isocarthamin. Safflower yellow is stable to heat and light and is used in baked goods and beverages.

Carthamin is produced from safflower callus cultures. The callus, obtained from flower bud explants, and can also be put into suspension. Medium optimisation has been performed. The production of alpha-tocopherol has been described for safflower cultures. Selection with various media components and precursor feeding experiments have enhanced the production. Addition of polysaccharides, like cellulose, chitin or chitosan increased production of the red pigment. Addition of 1 mM Dphenylalanine and removal of Mg alone or both Mg and Ca from the culture medium also increased the production.

4.7.3.4. *Saffron*

Saffron made from the stamens of *Crocus sativus* are prized for their use as a flavouring and colourant and used in baked goods, soups, meat and curry products, cheese, confectionary and as a condiment for the rice of Spanish and Indian foods. The stigma of the plant contains crocin (yellow pigment), safranal (a fragrance) and picocrocin (bitter substance).

The water-soluble crocin, being a glycoside, is not soluble in oils and fats. Saffron is sensitive to pH changes and is unstable towards light and oxidative conditions, but it is moderately resistant to heat. It is used It is also reputed to have medicinal value for stomach ailments.

Since the life of the saffron flowers is very short, making harvesting difficult, it is an ideal target for plant tissue cultures. This problem was tackled through the propagation of saffron stigma-like structures *in vitro*. Further studies show that crocin and picrocrocin are present and, after heat treatment (as done with field-grown stigmas), safranal is produced. The composition of these phytochemicals correspond with that of similarly-treated young, intact stigmas.

4.8. Bioreactors Systems for Mass Cultivation of Plant Cells

Bioreactors, designed for large-scale cell, tissue or organ culture in liquid media, have demonstrated a number of important advantages over conventional semi-solid micropropagation, including several fold increase in multiplication rates, and reduction in space, energy and labour. These advantages have gained increased attention to the use of liquid systems generally.

The disadvantages in include contamination, lack of protocols and production procedures, increased hyperhydricity, and problems of foaming, shear stress and release of growth inhibiting compounds by the cultures. Contamination in liquid systems, and the lack of problem solving information compared to semi-solid growth media pose more risk and systems require greater skill. Thus, in many developing countries the potential of cost reduction will depend on the cost of labour, production capacity, crops being propagated, contamination rate, and energy and costs, hence not a viable proposition

4.8.1. Bioreactor Vessels

For micropropagation, "large-scale" bioreactors generally rang from 1.5 to 20 l, although even larger vessels have been used. These are in stark contrast to the standard plant tissue culture vessels, which typically contain less than 0.5l of semi-solid medium. The size of the vessels in bioreactors versus conventional micropropagation systems has important investment cost consequence. Standard semi-solid plant tissue culture vessels have a seal that permits limited gas exchange with minimal contamination. The light, temperature and gases in the conventional vessel are determined by the environment of the conventional growth room.

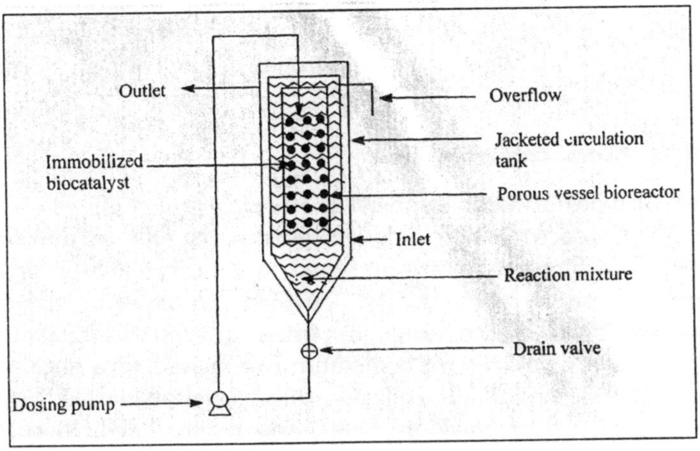

Figure 4.1. Porous vessel bioreactor

Some bioreactors are capable of growing 60-100,000 plants per cubic meter and providing more precise control of the plant growth gaseous exchange, illumination, medium agitation, temperature and pH. Being aseptically sealed, they can be operated most of the time in ordinary rooms. With only the handling of plant material performed under laminar flow. hence they are advantageous to developing countries for new or expanding plant culture facilities, in combination with a conventional laboratory. However, to be economical its use requires indexed plant cultures, and attention to aseptic procedures during the handling of plant material.

4.8.2. Bioreactor Models

The various bioreactor designs, developed for a wide range of crops, culture types and stages, can be divided into two broad types: those in which the cultures are immersed partially or temporarily in the medium, and those in which the cultures are continuously submerged. The latter are usually employed in high-density multiplication of cultures where submersion does not lead to abnormal plant development. Such cultures include protocorm-like bodies, embryogenic callus, somatic embryos, meristem clusters and nodules. Protocorm-like bodies are meristematic tissues which produce propagules. Meristem clusters and nodules are organogenic culture made up of compact masses of meristems without significant leaf, stem or root development. Submerged bioreactors can also be used for the growth of small bulbs or corms for field growing.

4.8.2.1. Partial Immersion Type

Several types of partial immersion bioreactors, include gaseous phase bioreactors, liquid layer bioreactors and temporary immersion bioreactors, are preferred to culture plant material sensitive to hyperhydricity. Cultures in gaseous phase bioreactors, are mechanically supported on a porous base and intermittently sprayed with medium or exposed to a nutrient mist. Excess medium is collected in a vessel fabricated from autoclavable clear glass or plastic, and re-circulated. In such bioreactors the and the environment of the growth room controls illumination and temperature, and such bioreactors can provide excellent growth and development for most tissue and organ cultures. However the mist generation systems available are complex to maintain.

In liquid layer bioreactors the 'LifeRaft' TM, a simple liquid layer system, consists of a disposable microporous membrane raft supported by a buoyant float at the surface of the liquid medium and only the base of cultures is exposed to the medium. The control of illumination, temperature and the gaseous environment is similar to that in standard tissue culture vessels. Stationary support systems for liquid layer bioreactors have been also developed from sealed clear plastic film with a wire frame. Instead of vessels, small bags made of plastic film are also being used in commercial laboratories.

In temporary immersion bioreactors, of the two vessels used one holds the liquid medium and the other the cultures. The cultures are immersed in the medium, for a preset duration at specified intervals. Due to their construction and operation being very simple, they are attractive low-cost alternatives. Another design includes a system where a single vessel with a reservoir on one side is mechanically tilted at preset intervals. In this manner, medium periodically bathes the cultures in the vessel and maintains the propagules in a vertical position.

Other temporary immersion bioreactors use different designed vessels and rotation, where, as the vessel turns, the culture is intermittently dipped in the medium. Since temporary immersion bioreactors are simple and cost-effective to run many commercial micropropagation laboratories have begun to adopt these systems into their production procedures.

4.8.2.2. Submerged Type

There are basically two types of submerged bioreactors: mechanically agitated and air-driven. Mechanically agitated bioreactors use devices such as propellers, turbines, impellers, paddles, and vibratory mixers to circulate the medium. Each device generates a somewhat different pattern of fluid flow in the vessel. For each type of culture and plant, the shear force and pattern of flow must be optimised. For example, it benefits to have sufficient shear force to maintain meristematic clusters as small aggregates which minimises stagnant regions in the vessel. Otherwise can cause cultures to stick to surfaces and form clogs that disrupt medium flow.

But high shear force can also produce excessive foaming, with a negative effect on culture growth. Air-driven bioreactors are the simplest submerged bioreactor in which filter-sterilised air is forced into the vessel's bottom, and aerates and agitates the medium, thereby lifting and circulating the culture. Such air-driven bioreactors, called simple aeration or bubble column bioreactors, depending on their volume, are usually fabricated from transparent or translucent materials, but are typically formed from autoclavable clear glass or plastic and fittings. Or else they are made of non-autoclavable clear flexible plastic, and sterilised with gamma radiation. The simple airlift or bubble column bioreactors are easy and cost efficient to run.

Submerged bioreactors can be integrated into automated or mechanised micropropagation systems which is a great soon to biotechnologists. In the automated systems, culture from the bioreactor is pumped to a bioprocessor for separating, sizing and distributing the culture to vessels for propagule development. In the mechanised system, an operator transfers the multiplied culture to a mechanical cutting device for separation and then to a vessel for propagule development.

4.8.3. Advantages and Disadvantages

Bioreactors have many advantages that include increased culture multiplication rate, faster culture growth, reduction in medium cost and reduction in energy, labour and laboratory space. The enhanced multiplication rate and growth primarily reflect the effect of liquid medium. The elimination of gelling agents (e.g. agar) reduces medium cost, and filter sterilisation of the medium saves the need for autoclaving. In bioreactors, the culture density in liquid media is much higher than in the conventional vessels with semisolid media. The use of bioreactors requires much smaller space in the growth room, fewer clean work stations, and less space for media preparation, vessel storage and washing than that used in the conventional micropropagation in which tissue culture vessels are typically placed on shelves with a large space between the shelves energy and economic costs make the smaller size of the laboratory and the lesser number of people reduce air-conditioning needs make bioreactors attractive to use. Reduced requirements for lighting the bioreactors and autoclaving, and less labour, simplification of medium preparation, washing of vessels, and simplified handling of the cultures, all lead to cost reduction.

But they also have their disadvantages in micropropagation, namely, contamination, lack of protocols and production procedures, increased hyperhydricity, and problems of foaming, shear stress and release of growth inhibiting compounds by the cultures. Unfortunately, culture contamination, is a major problem in conventional commercial micropropagation, is even more acute in bioreactors. In conventional micropropagation, the loss of a small number of the contaminated vessels is negligible when compared to that; in bioreactors, where even a single contaminated unit is a huge loss. However, despite these difficulties, a number of commercial laboratories have developed effective procedures to control contamination in bioreactors.

4.8.4. Control of Contamination

Contamination can be prevented in bioreactors by a proper handling of the plant material, of equipment during transfers, and cultures during production. Only the surface sterilised explants, multiplied in small vessels and indexed for freedom from diseases, are used to generate cultures in bioreactors. A bioreactor small is sterilised in an autoclavable plastic bag, sealed with a cotton wool plug, and opened only under the laminar flow cabinet. For bioreactor other methods to protect the vessel after autoclaving and during its transport to the sterile area must be provided. The vessel that is assembled under laminar flow, should be carefully guarded against non-sterile objects entering the sterile air stream in the vicinity of open ports. After inoculation, all the ports are sealed, and the bioreactor is brought to the growth room and connected to an air supply. For harvesting, the bioreactor is again taken to the sterile area, and the ports are opened so that the air stream cannot carry contamination from the outside. With proper procedures followed, the reactor can be reused without re-sterilisation.

In spite of safety norms followed in initiating cultures, bioreactors can become contaminated from the environment or from microbes, latent in the culture. The contamination can be checked with one or a combination of anti-microbial compounds, acidification of the media, and micro-filtration of the medium. Several commercial laboratories use anti-microbial compounds, sodium hypochlorite or sodium dichloroiso-cyanurate. Acidification of the medium (pH3) has controlled contamination in the multiplication of banana and in media for other crops. Contamination can also be controlled by circulation of the medium through a 0.2μ microporous filter. However, removing contaminants involves frequent filter changes, too expensive for use in a commercial laboratory. Chilling the filter retards filter clogging significantly.

4.8.5. Construction of a Simple Bioreactor

Laboratories with limited resources can use a simple temporary immersion bioreactor made of autoclavable glass or plastic vessels or bags with a wide mouth and a tight fitting rubber stopper two holes for glass tubing. One tube is inserted and attached to the stopper so that when the vessel is closed, the end of the tube is closer to the bottom of the vessel. A second tube is inserted and fixed in the stopper so that the end of this tube is near the top of the vessel when it is closed. A heat-resistant silicone sealer, can be used to fix the tubes. To add or remove cultures the vessel should have at least one port sealed with a rubber stopper or a metal closure with a rubber gasket. There are four ports, two for air-inlet and -outlet, and two for medium-inlet and -outlet. The air-inlet and -outlet ports are connected to hydrophobic Teflon submicron microporous air filters, which must be kept dry. The wet filters clog and may lead to contamination. Medium ports should be able to be opened or closed, connected or disconnted without leakage to prevent entry of contaminants. The bioreactors should fit under the laminar flow unit leaving enough space on the top and sides to allow easy access to the vessel for handling cultures and changing the medium.

The purpose-built bioreactors are either air-driven or temporary immersion type. A reservoir vessel similar to the culture vessel is prepared with the mouth of the vessel just wide enough to fix the two glass tubes. The tops of the tubes reaching the base of the culture vessel are connected to each other with silicone tubing.

For both the reservoir and culture vessel, the ends of the short tubes above the stopper are connected to sterilising filters with a short length of silicone tubing. The bioreactor, before being used, is assembled with the stoppers loosely fitted into the vessels and placed in an autoclavable bag with a cotton wool plug. The culture vessel can also contain an autoclavable plastic or stainless steel mesh screen to support the culture above the medium in the vessel.

After autoclaving the bioreactor is placed on the laminar flow bench. The culture is aseptically introduced into the vessel and sterile medium in the reservoir, the stoppers are tightly plugged into the vessels, and taped shut to prevent contamination. The bioreactor is then placed in the growth room, the two vessels are placed alongside each other, and the reservoir air filter is connected to a 6W aquarium pump with silicone tubing. When air is pumped into the reservoir vessel, the medium is forced into the culture vessel, and allowed to remain in the culture vessel for a short duration. The air pump is then connected to the culture vessel air filter, and the medium is pumped back into the reservoir. This process is repeated at preset intervals, and can be automated.

References

Conger, B.V. *Cloning agricultural plants via in vitro techniques.* CRC Press Inc. 1981.

Henshaw, G. "Technical aspect of tissue culture storage for genetic conservation". In: Frankel and Hawkes. *Crop Genetic Resources for Today and Tomorrow.* CU.P. 349-358. 1975.

Hu, C.Y., Wang, P.J. "Meristem, shoot tip, and bud cultures". In *handbook of plant cell culture* I. Ed. D. A. Evans, W. R. Sharp, P. V. Ammirato, and Y. Yamada. 1983.

Hussey, G. "In vitro propagation of horticultural and agricultural crops". In: *Plant Biotechnology* Ed. H. Smith. 1983.

Wilkins, C.P., Dodds, J.H. "The application of tissue culture techniques to plant genetic conservation". *Science Progress* 68: 281-307. 1982.

5

PLANT TRANSFORMATION TECHNOLOGY

In both basic and applied molecular biology DNA can be moved into an organism, thereby altering its genotype or genetic make-up. Genes derived from unrelated species and even other kingdoms, such as bacteria, fungi, plants, animals, that would otherwise be inaccessible to an organism, can be combined in the lab using genetic transformation techniques.

In a laboratory transformation involves *Escherichia coli* or *Saccharomyces cerivisiae* and a thermal or electrical stimulus. In the pharmaceutical industry, bacteria and yeast are transformed with selected human genes to produce therapeutic benefits to treat human disorders. Most of the insulin now supplied to diabetics is produced this way.

Conjugation takes place in nature, whereby many strains of bacteria exchange genetic information and the new genetic information is passed on to subsequent generations. The single-celled nature of the organisms enables the change of only one cell in order to integrate the new genetic information and allow for its transmission to the next generation. This natural exchange of genetic information allows the organism an increased capacity to adapt to its environment.

In multicellular organisms transformation is usually more difficult, particularly with some important plant species like cereals. The multicellular nature of most plants poses the problem of transforming each cell of the plant in order to fully integrate the new information. The transformation of an individual plant cell and then regenerating it into a whole organism calls upon the pluripotent nature of plant cells.

Bacteria help to genetically transform plants and other organisms. Viruses are also able to move DNA (or RNA) into an organism and cause immense changes. Soil bacteria, such as *Agrobacterium tumefaciens* and *Agrobacterium rhizogenes*, are great examples of natural transformation systems, causing crown gall disease and 'hairy root syndrome' respectively. *Agrobacterium* can transfer a section of its own DNA, known as the T-DNA or 'transfer DNA', into plant cells. For *A. tumefaciens* this normally occurs at wound sites and the transferred DNA causes the plant to develop crown gall disease. Apart from causing a gall, the *Agrobacterium* harnesses the plant's machinery to produce unique sugars called opines that the bacterium uses as a nutrient supply.

Scientists have used one of the most common methods for introduction of DNA into plant cells using 'disarmed' *Agrobacterium tumefaciens* – which the DNA of interest has replaced the disease genes. DNA of interest can be any valuable agricultural trait, such as genes that provide herbicide tolerance in canola and other crops. The particle bombardment or biolistics transformation method uses rapidly propelled tungsten microprojectiles coated with DNA that are 'fired' into plant cells. The DNA of interest is often incorporated at random into the plant DNA. Other methods, such as electroporation and microinjection, have also been employed to transform plant cells.

Cells that have been successfully transformed are selected and the DNA of interest is typically cloned adjacent to DNA for a selectable marker gene such as *nptII* (encoding kanamycin antibiotic resistance). The transformed plant cells are then cultured in the presence of the antibiotic and only cells or tissues expressing the new DNA survive. In order to eliminate or exclude the antibiotic resistance markers in transformation experiments, effective methods have been developed. Fluorescence genes such as GFP (green fluorescent protein) can also be used to report transformation events.

5.1. Genetic Transformation

The transfer and expression of foreign genes into plant cells has been one of the most recent and successful areas of plant cell and tissue culture developed for exploring potential benefits in agriculture. The transformation of bacterial cells by uptake of DNA from another bacterium, and subsequent incorporation of this foreign DNA into the genetic material of the host is well established. However, genetic modification of higher plants by introducing DNA into their cells is a highly complex process, involving a foreign nucleotide sequence that has to be introduced in such a way that allows expression of alien genetic information in the transformed cells.

Availability of restriction endonucleases, development of the DNA-DNA hybridisation technique and marker genes for selection of transformed cells have now enabled integration of exogenous DNA into cells of Monera, fungi, animals, and higher plants. The addition of new genetic information as a result of these processes in eukaryotic plants expresses at the cellular and later, at the whole organism level, and can also be transmitted in successive generations of transformed individuals. Such transgenic plants, since normally they originate following an exchange of genetic information between sexually incompatible species, are not known to exist in nature.

Using whole plant systems such as seeds and seedlings genetic modification of plants was done by introducing isolated DNA. Exogenously applied DNA is taken up by plants, and integrated into the host genome, based mainly on isopycnic centrifugation in cesium chloride (CsCl) gradients.

Integration and replication are two major types of experiments that were conducted. In the former radioactive labelled DNA of density different from the host DNA was administered to the seedlings, followed by isolating DNA from seedlings, or their excised parts, after a period of metabolism for analysis on CsCl gradients. The radioactive donor DNA was ascertained for integration by the occurrence of a radioactive band approximately intermediate in density between the host and donor DNA. The intermediate density band was further tested by sonication and denaturation. The replication experiments were also conducted in a similar manner and differed only in respect of the donor DNA, which was non-radioactive. These experiments also showed proof of bacterial plant DNA complexes in tomato, barley and Arabidopsis.

Trying transformation experiments using a similar donor DNA with the same host or different plant systems have not yielded the same results, showing that there was neither integration nor replication of the exogenous DNA inside the host cells. In fact, the donor DNA ultimately degraded in the host and the degraded products were used for endogenous DNA synthesis.

Genetic transformation of whole plant systems seems to have achieved success with cereals. There has been much interest in the reported rhizobial infection of non-legumes to form an effective nodular nitrogen-fixing symbiosis within the dicot plants belonging to the genus Parasponia so much as to extend nodular symbiotic association to monocots, especially the cereals. Cocking and co-workers reported induction of nodular structures on the roots of rice seedlings when these roots were treated with a cellulasepectolyase enzyme mixture followed by inoculation with either Rhizobium, or Bradyrhizobium, in the presence of PEG. This has opened new vistas in the genetic engineering of plants for possibly extending this technique for effective nodulation in respect of other cereals as well as non-legumes. It has to be seen whether nodular structures on transformed rice are effective in nitrogen fixation. Otherwise, other strains of genetically engineered Rhizobium, or its wild type, may be needed for this purpose.

5.1.1. Uptake of DNA by Pollen

In 1976, researchers incubated *Nicotiana glauca* pollen with DNA isolated from *Nicotiana langsdorffii* and then used this pollen to pollinate emasculated *N. glauca* flowers. This experiment led to the production of transformed plants with developed tumours at the wound site on the stem, characteristic feature exhibited by the sexual hybrids arising between these two species. Pollen from maize plant homozygous for aleurone and various endosperm marker colours were mixed with DNA extracted from a strain homozygous dominant for all these traits yielding a high pollen transformation efficiency. These observations suggest that pollen may act as vehicles for the transfer of foreign genes into plants.

Young floral tillers of rye plant were manually injected with a solution of plasmid DNA carrying a chimeric gene having T-DNA nopaline synthase gene flanking coding sequence of the Tn

5 neo gene (expressing for antibiotic aminoglycoside resistance). These Injected flowers were pollinated with normal pollen and progeny seeds plated on kanamycin medium. Of the 3,000 seeds, only two seedlings were isolated showing traits of kanamycin resistance and the nos-neo gene.

5.1.1.1. Transformation of Protoplasts

Genetic modification of cells has benefited by the isolated protoplast system which proved most rewarding. Since there is no wall around the plasma membrane, protoplasts are able to fuse and can also take up chloroplasts, nuclei, micro-organisms and isolated foreign DNA. This has induced researchers to investigate various modes of genetic modifications in cells using the protoplast system.

5.1.2. Uptake of Organelles

5.1.2.1. Chloroplast Transplantation

The uptake of chloroplasts by albino protoplasts isolated from petunia, tobacco and carrot plants conducted in experiments required a simple treatment exposing a mixture of albino protoplasts and chloroplasts to PEG in a manner similar to protoplast fusion. PEG caused fusion of the two external membranes of the plastids and while entering the protoplasts both the plastid and protoplast membranes fused, causing disruption of these organelles. Hence, focus was diverted from direct uptake to transfer of chloroplasts through the fusion of protoplasts, making it possible to restore photo-autotrophy in albino protoplasts by fusion with wild type (green) protoplasts in a number of species. A gamma-irradiated chloroplast-mutant of Nicotiana species was used for interspecific chloroplast transfer by protoplast fusion, producing cybrids, whereas protoplast fusion was suitable to rescue the chloroplast mutants of *N. plumbaginifolia* from undesirable nuclear backgrounds.

Cloroplast transformation via protoplast fusion in plant improvement can occur by transfer of genes for herbicide resistance between related species. For example, isolation of mutants by cell selection in populations of protoplast-derived colonies yield triazine-resistant Nicotiana plants. The resistant trait arises as a result of the mutation in the chloroplast gene psb-A. Similarly, researchers found that inhibitors of bacterial protein synthesis were used to induce chloroplast mutations in protoplast cultures of *N. plumbaginifolia* making it possible to isolate mutants resistant to antibiotics and photosynthesis-inhibiting herbicides. The bacterial APH (3) II gene from transposon Tn 5 and the cat gene from transposon Tn 9 are selectable marker or reporter genes that have been found suitable for transfer of mutant chloroplast genes.

Control of light-mediated proteins expressed by rbcs and cab gene families can be achieved by the interaction between chloroplast genes from weeds to sexually incompatible crop plants. This is another area that may have potential in future since it has played a useful role in understanding the nuclear-chloroplast interaction, physiology of isolated plastids and transfer of herbicide resistance.

5.1.2.2. Mitochondrial Transplantation

Though there is no evidence regarding the direct uptake of isolated mitochondria by protoplasts there are however reports on incorporation of mitochondrial DNA through protoplast fusion leading to the formation of cybrids. It may also be possible to transfer traits such as toxin sensitivity of Texas male sterile cytoplasm encoded by mt DNA in maize.

5.1.2.3. Nuclear Transplantation

The methods for isolation and uptake of nuclei by mesophyll protoplasts of petunia, tobacco and maize have been refined to facilitate expression of the transferred genome in the cells. Vicia hajastana nuclei in protoplasts of an auxotrophic cell line of Datura innoxia were transplanted and prototrophic cell colonies expressing for Vicia genomic DNA in cultures were isolated. However, these colonies on an appropriate regeneration medium developed shoots typical of wild type Datura.

The uptake of isolated chromosomes, instead of complete nuclei, Holds the potential for transfer of foreign genes into plant cells. Development of the flow cytometry technique permitted the introduction of isolated chromosomes into protoplasts of Haplopappus, Nicotiana and Petunia. Only metaphase chromosomes were found suitable. Isolated chromosomes can also be introduced into protoplasts by microinjunction. In most of these experiments transplanted chromosomes were eliminated due to incompatibility.

Experiments to transfer micronuclei into protoplasts were also conducted. Amiprophos-methyl (APM) treatment of potato and carrot cells encouraged accumulation of a large number of metaphases. The dispersed single or groups of chromosomes in these cells acquired a nuclear membrane and formed micronu-clei which could be separated by flow cytometry and then used for transplantation. This widens the application prospect of micronuclei for transferring specific intact chromosomes for the purpose of gene mapping.

5.1.2.4. Uptake of Microorganisms

There were trial to transfer whole bacteria (Rhizobium) into protoplasts for utilising the transformed protoplasts for nitrogen fixation but there yielded little success. Therefore, the use of viruses as vectors for genetic engineering studies gained focus. The mechanism of infection, nuclear-protein interaction, integration and expression of viral DNA in isolated protoplasts should be possible since in nature viruses do infect plants and multiply within their tissues. Earlier transducing phages were used to transfer bacterial genes into plant cell lines or callus tissues but there was no evidence of "transgenosis" in long term cultures.

Experiments on uptake of TMV by isolated protoplasts since 1969 have shown the incorporation of viral nucleic acids into protoplasts, using them as vectors in genetic transformation. Most plant viruses have encapsulated genomes, assuming that no specific virus receptor sites exist on the exterior of the plasma membrane. Viral based transforming agents may be of the integrating or non-integrating type.

The following factors are generally considered in selecting the progenitor for the transforming agent: (a) amenability to genetic engineering techniques, (b) packaging constraints in the particle and (c) development of a system in which expression can be studied. The most amenable and widely chosen progenitor for genetic engineering studies is the cauliflower mosaic virus

(CaMV). Moreover, the dsDNA genome in CaMV has appropriate restriction sites compatible with the methodology developed for *Agrobacterium tumefaciens* plasmid. A chimeral CaMV carrying methotrexate resistance was successfully introduced into turnip cells in such a way that the virus replicated normally in the transformed plants. Similarly, barley protoplasts infected with the ssRNA bromovirus (BMV), on having the chloramphenicol acetyltransferase (CAT) bacterial gene inserted into the RNA3 genome, transcribed for CAT activity. Gemini viruses may also be used for genetic transformations but their usefulness as vectors are still under study.

Agroinfection which involves a combination of viral and bacterial vectors is another approach being considered. Various strategies developed for uptake of viruses by plant protoplasts and their use as transforming agents. Use of fungal protoplasts as eukaryotic gene vectors for plant cell transformation has also been attempted. Fungal protoplasts have to be experimented upon since fungi and higher plants establish a range of nutritional relationships in nature. *Catharanthus roseus* protoplasts were fused with fungal protoplasts as an attempt to transfer the secondary metabolite synthesising ability of fungus into a higher plant protoplast system in cultures. Later, other researchers successfully mediated uptake of celery protoplasts by *Aspergillus nidulans* and *Fusarium oxysporium* protoplasts using a PEG solution, which was confirmed by ultrastructural studies and diamidino-2-phenylindole staining. Protoplasts of celery showed improvement in the level of viability after uptake by *A. nidulans,* whereas with *F. oxysporium* the level decreased. However, whether uptake of fungal protoplasts can be exploited for secondary metabolite synthesis in higher plant cell or protoplast system is yet to be seen.

5.2. Transformation Using Agrobacterium

In the past few years, there have been molecular studies of crown gall disease by Agrobacterium plasmids and the use of recombinant DNA technology, with further investigations using Agrobacterium for plant genetic manipulations.

Agrobacterium tumefaciens, a soil bacterium, can infect most dicotyledonous plants, usually at a wound site. Where the tissue develops a neoplastic growth known as a crown gall tumour. The tumour tissue excised from the plant is capable of growing on hormone-free medium independent of the bacterium. In cultures, the crown gall tissue produces a set of metabolites termed opines. Agrobacterium fosters the Ti plasmid, which is directly responsible for tumour induction. The transfer of small DNA segments from this plasmid and their integration into the genome of host cells generates tumour formation in the plants. The property of the Ti plasmid DNA segment (T-DNA) integrating with the genome of higher plants has boosted the use of Agrobacterium in genetic transformation techniques.

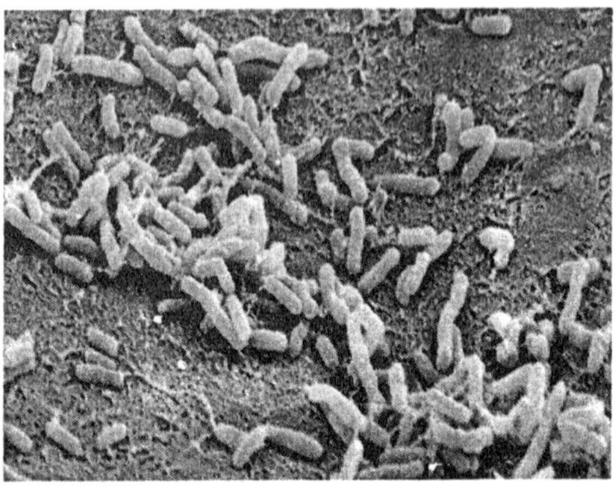

Figure 5.1. Agrobacterium tumefaciens

The infection by Agrobacterium rhizogenes another bacterium used in plant genetic manipulations, causes hairy roots which are of clonal origin. Unlike *A. tumefaciens* tumours, the tumour produced by *A. rhizogenes* is capable regenerating mature fertile plants. The tumour-inducing properties of rhizogenes strains are also carried on a large Ri plasmid DNA (Ri T-DNA). Two segments of this DNA (TL-DNA and TR- DNA) are transferred to plant cells during infection by *A. rhizogenes*.

5.2.1. Agrobacterium Mediated Gene Transfer

With the improvement in plant transformation vectors and methodologies the efficiency of plant transformation has increased and to achieve stable expression of transgenes in plants achieved. The simplicity of the transformation system and precise integration of transgenes make it viable for *Agrobacterium* Ti plasmid-based vectors to continue offering the best system for plant transformation. Binary vectors have been improved by the integration of supervirulent *vir* genes, matrix attachment regions (MAR) and the insertion of introns in marker genes and reporter genes. These improvements and the use of acetosyringone, have led to transformation of monocotyledonous plants using *Agrobacterium*. The green fluorescent protein (GFP) gene has been extensively modified for plant codon preference, ER/plastid targeting and for greater solubility thereby making it an adaptable vital reporter for transgenic plants. There is also development of transgenic plants devoid of antibiotic marker genes.

Marker genes in transcenic plants have been eliminated by cotransformation of multiple T-DNAs, site-specific recombination strategies and deployment of Ac/Ds-based transposition. Positive selection strategies using *ipt*, xylose isomerase and phosphomannose isomerase are shown to be useful in many crop plants. The development of BIBAC vectors, a demonstrated capability to transfer multiple genes of a pathway and successful T-DNA tagging in rice, are evidence to show that transformation technologies can be deployed for the study of 'functional genomics' in plants.

The particle bombardment system still has its use in organelle transformation and transformation of plants that lack efficient regeneration systems. This has led to the design of vectors that minimise transgene silencing while ensuring desired levels of transgene expression. The mechanism of T-DNA integration in plants has to be studied so that 'knock out' mutagenesis and homology-based gene replacements can be achieved in plants.

Plant transformation is performed using a wide range of tools such as *Agrobacterium* Ti plasmid vectors, microprojectile bombardment, microinjection, chemical (PEG) treatment of protoplasts and electroporation of protoplasts. Among all the methods employed transformation using *Agrobacterium* and microprojectile bombardment is currently the most extensively used method. Recent developments in the phenomenon of 'gene silencing' has come to centrestage after a large number of transgenic plants have been carefully evaluated for transgene expression in successive generations.

Agrobacterium tumefaciens have precisely transferred defined DNA sequences to plant cells been very effectively utilised in the design of a range of Ti plasmid-based vectors. Three genetic elements, *Agrobacterium* chromosomal virulence genes (*chv*), T-DNA delimited by a right border and a left border and Ti plasmid virulence genes (*vir*) constitute the T-DNA transfer machinery. The mechanisms governing the transfer of 'T-complex' via the conjugation channel and the roles of plant and *Agrobacterium* proteins in T-DNA integration are being intensely studied.

There are *Agrobacterium*-based DNA transfer system advantages in plant transformation which are:

i) The simplicity of *Agrobacterium* gene transfer makes it a 'poor man's vector.
ii) A precise transfer and integration of DNA sequences with defined ends.
iii) A linked transfer of genes of interest along with the transformation marker.
iv) The higher frequency of stable transformation with many single copy insertions.
v) Reasonably low incidence of transgene silencing.
vi) The ability to transfer long stretches of T-DNA (> 150 kb).

For long, *Agrobacterium* was unable to transfer DNA to monocotyledonous plants but with effective modifications in Ti plasmid vectors and finer modifications of transformation conditions, a number of monocotyledonous plants including rice, wheat, maize, sorghum and barley have now been transformed. In fact, *Agrobacterium* T-DNA transfer is now considered as 'universal' based on successful transformation of yeast, *Aspergillus*and human cells.

5.2.2. Ti Plasmid-based Vector System

5.2.2.1. *Agrobacterium vir Helper Strains*

Plant transformation use a typical binary vector system comprising an octopinetype *vir* helper strain, such as LBA4404, that harbours the disarmed Ach5 Ti plasmid and a binary vector such as pBin19. the nopaline-type MP90 and the L, L-succinamopine-type EHA101 have been added to the available range of *vir* helper strains. The bacterial kanamycin resistance gene in EHA101 was deleted to develop the *vir* helper strain EHA105. EHA101 and EHA105, due to harbouring the 'supervirulent' *vir* genes, exhibit show broader host-range and higher transformation efficiency. Many recalcitrant plants, such as rice, wheat and barley have been transformed using EHA101 and EHA105. A new *vir* helper strain pTiChry5 has been constructed from an *Agrobacterium* strain virulent on soyabean. This strain has good potential for legume transformation.

5.2.2.2. Binary Vectors

Since 1984, many modifications have been made in binary vectors to expand the range of their utility and to improve their transformation efficiency, one such being the construction of the superbinary vector pTOK233 by cloning the *vir*B, *vir*G and *vir*C genes of pTiBo542 in pGA472. This led to the successful transformation of *japonica* rice, *indica* rice, *javanica* rice, maize, *Sorghum* and *Allium cepa*were using the same strategy. A long-held notion that monocots are not amenable to *Agrobacterium*-mediated transformation has been proved wrong. Recently, pTOK233 was used to transform a pulse, moong bean.

Many finer improvements have been made in binary vectors: pBIN19 has been completely sequenced; its improved version, pBIN20, with many additional single restriction sites in the MCS was reported recently; a new series of pPZP vectors have been developed which are small in size and stable in *Agrobacterium*; the pPZP vector backbone was used to construct the pCAMBIA series of vectors with *npt*II, *hpt* or *bar* as selection markers and *gus* or *gfp* as reporters. Plant expression vectors of pRT100 series permit construction of gene cassettes with CaMV 35S promoter and its polyA signal, which can be excised and placed in the MCS of binary vectors.

The problem of availability of single restriction sites, particularly in the expression cassette with the gene of interest, was dealt, with by the construction of a set of plasmids, pART7 and pART27. The shuttle plasmid pART7 has a MCS placed between CaMV 35S promoter and *ocs* polyA signal. The expression cassette is flanked on either side by *Not*I sites. The coding sequence of a gene of interest is cloned in the MCS of pART7. The expression cassette is excised by using *Not*I and cloned in the *Not*I site of the binary vector, pART27 that is designed so as to have the *npt*II gene close to the left border to enables the selection marker is the last to be transferred into plants. This strategy ensures that all plants selected on kanamycin will have a complete TDNA inclusive of the gene of interest.

In functional genomics *Agrobacterium* Ti plasmid vectors have important operations. Random integration of TDNA in the plant genome causes mutations and simultaneously tags the gene into which insertion occurred. T-DNA tagging has cloned many *Arabidopsis* genes. A similar effort in rice resulted in the generation of 22,090 transgenic plants. Phenotypic analysis of these mutants is expected to lead to the discovery of many new gene functions.

Genome studies require binary vectors that can transfer T-DNAs several hundred kbs. Using binary bacterial artificial chromosome vectors (BIBAC) a T-DNA consisting of 150 kb human DNA was precisely transferred and integrated into tobacco genome. These vectors are expected to play a key role in positional cloning of plant genes and in functional genomics.

Engineering of plants with new pathways may require the transfer of multiple genes. The technical difficulty of placing multiple genes in one T-DNA encountered due to loss of single restriction sites in successive cloning steps was overcome in engineering provitamin A biosynthesis in rice. Daffodil *phytoene synthase* gene and *Erwinia phytoene desaturase* gene were placed in one T-DNA. The second T-DNA carried daffodil *lycopene-b-cyclase* and *E. coli hpt*. Interestingly, 12 out of 60 plants selected on hygromycin carried the genes of the second T-DNA as well. Therefore, 20 per cent of hygromycin-resistant plants comprised all the three genes of β-carotene biosynthetic pathway. Thus, multiple genes of a pathway can be transformed by cotransformation of multiple T-DNAs. Though particle bombardment permits cotransformation, *Agrobacterium* offers the additional advantage of precise integration of complete lengths of multiple T-DNAs.

Matrix attachment regions (MARs), the AT-rich DNA sequences that promote the attachment of DNA to the nuclear matrix or inner nuclear membrane, produce a series of loops. MARs have two functions: (a) Genomic compartmentalisation of coexpressed genes in individual loops and (b) Formation of the boundaries of transcriptionally active DNA loop domains, thereby blocking the inhibitory influence of neighbouring sequences. MARs have potential applications in transgenic plants. A transgene flanked on either side by MAR sequences may form an independently expressing loop domain thereby minimising position effects. Researchers constructed *gus* gene flanked by tobacco MAR sequences and transformed cell suspension cultures by particle bombardment, thereby increasing the expression of integrated *gus* gene by 140-fold. *Gus* and *gfp* genes were constructed with the tobacco Rb7 MAR sequences generating transgenic rice plants by particle bombardment. The presence of MAR sequences increased the expression of reporter genes in the range of 3.3-fold to 650-fold. The reduction in reporter gene expression, generally attributed to flanking sequences at the sites of integration, was also minimised.

5.2.3. Improvements in Marker Genes and Reporter Genes

Leaky expression of plant selection marker genes such as *npt*II and *hpt* in *Agrobacterium,* results in bacterial over growth during selection and in the emergence of false-positive plants. To avoid this, researchers placed the intron 2 of the potato ST-1 gene in the N-terminal part of the *npt*II gene. Lack of mRNA splicing in *Agrobacterium* completely eliminated leakiness in kanamycin selection. In another study, the castor bean catalase 1 intron was placed in the *hpt* gene to avoid leakiness in hygromycin selection. The binary vectors with intron-*hpt* significantly improved rice transformation efficiency by improving *hpt* expression and by limiting *Agrobacterium* overgrowth.

The histochemical analysis of expression of *gus* as a reporter gene help in Early detection of transformation events. A major limitation of this approach is the expression of this reporter gene in *Agrobacterium* despite the use of plant promoters. This problem was also tackled by inserting introns, which are processed in plants but not in *Agrobacterium*. The intron 2 of potato ST-LS1 gene was placed in the coding sequence of *gus* gene. In another study, the castor bean catalase gene intron was inserted in the Nterminal part of the coding sequence of *gus*. In both cases, *gus* activity was limited only to transformed tissues and was not detected in *Agrobacterium*.

Transgenic plants make use of green fluorescent protein gene (*gfp*) from jellyfish finds. GFP, unlike *gus*, is foreseen in living cells. Another reporter gene, luciferase (*luc*), requires an externally added substrate for detection. However, GFP fluorescence occurs with uv/blue light and oxygen without any externally added substrate. The first series of transformation experiments showed the following limitations of the application of wild type *gfp* gene in plants: (a) Codon preference was different between jellyfish and plants. (b) Presence of a cryptic intron in *gfp* was recognised and spliced in plants. (c) Low solubility of GFP in cytoplasm. (d) Interference of cytoplasm-accumulated GFP in the regeneration of transgenic shoots. Synthetic versions of *gfp* (e.g. modified *gfp*, *mgfp*) helped to avoid the cryptic intron and changed the codon preference to that of flowering plant genes. Thereby dramatically increasing the intensity of fluorescence in transgenic *Arabidopsis* and tobacco plants, and full length GFP being translated. Elevated expression of GFP unfortunately foiled the regeneration of transgenic shoots. However, fusion of a signal peptide-encoding sequence to *mgfp*4 and consequent targeting of mgfp4–ER protein to ER resulted in routine regeneration of highly fluorescent shoots.

The reduced solubility of GFP in cytoplasm was considered as an important reason for reduced fluorescence. Therefore the synthetic version, *mgfp*4 was subjected to triple sitedirected mutagenesis (F99S, M153T, V164A) to generate a soluble-modified *gfp* (*sm-gfp*). Expression of *sm-gfp*4 in *Arabidopsis* gave brighter fluorescence compared to *mgfp*4, possibly due to increased solubility of *sm-gfp*. Normally, GFP is excited at 396 nm or 475 nm and emits green fluorescence maximally at 508 nm. Experiments carried out with two site-directed mutagenesis experiments on *sm-gfp* generated spectral variants: a soluble-modified red-shifted version of GFP (smRSGFP) and a soluble-modified blue fluorescent version of GFP (sm-BFP).

Using the rbcSSU promoter and transit peptide (rbc-Tp-*sgfp*) plastid-targeted version of synthetic GFP (*sgfp*) was constructed. Transgenic rice plants generated using *Agrobacterium* with rbcS-Tp-*sgfp* showed 20-times higher fluorescence compared to a version without plastid targeting. Two groups placed intron-2 of potato ST-LS1 gene in the synthetic *gfp* versions. In the latter report, the intron was placed in *mgfp*4-ER, sm-*gfp*4, smRS-GFP and sm-BFP genes. Intron splicing occurred efficiently in tobacco and in maize cells.

5.2.4. Modifications in Cocultivation Conditions

Today, it is a common practice to have cocultivation of explants with *Agrobacterium* in the presence of acetosyringone, a *vir* gene inducer, in the transformation of recalcitrant crops such as rice, maize, barley and wheat. Preinduction of *Agrobacterium* with 400 mM acetosyringone prior to cocultivation is important in rice transformation. Finer changes such as supplementation of buffers to cocultivation medium and preincubation of explants that increase *vir* gene induction increase transformation efficiency of *Agrobacterium*.

Many physical factors influence transformation efficiency. Uniform wounding of tobacco leaves and sunflower apical meristem by particle bombardment, vacuum infiltration of *Arabidopsis* seedlings, rapid proliferation of rice calli at 30°C and cocultivation at a lower (22°C) temperature led to an increase in the efficiency of *Agrobacterium*-mediated transformation.

5.2.5. *'In planta'* transformation using *Agrobacterium*

Growth of *'in planta'* transformation by floral dips has come about due to the small size of *Arabidopsis* plant and improved efficiency of transformation by vacuum infiltration. This method obviates plant tissue culture and eliminates somaclonal variations. In a series of subsequent studies, it was found that ovules (female gametophytes) of immature flowers were more the often targets of floral dip transformation of *Arabidopsis*.

Medicago truncatula, an emerging model legume plant, has recently seen the *'In planta'* transformation. Infiltration of flowering plants was superior to infiltration of young seedlings. Transformation by vacuum infiltration used binary vectors with *bar* gene and 'BASTA' spray screened the TI progeny. This is a potentially useful approach in plants like pulses where regeneration protocols are either not available or regenerating parts of plants are not accessible to *Agrobacterium*-mediated transformation.

5.2.6. Marker-free Transgenic Plants

Antibiotic resistance markers (e.g. *npt*II, *hpt*) or herbicide resistance markers (e.g. *bar*) are required for selectively propagating transformed cells and tissues though their subsequent maintenance in transgenic plants is unnecessary. Markers have to be eliminated as the antibiotic resistance genes may be transferred to pathogenic bacteria or the herbicide resistance genes may be transferred to weeds. Besides removal of marker genes offers the following research advantages: It enables (i) successive rounds of transformation so that useful transgenes can be stacked without crossing, (ii) retention of promoters along with selection markers which will lead to the presence of multiple copies of a promoter, thereby activating signals for transcriptional gene silencing.

5.3.7. Marker Elimination by Site-specific Recombination

Site-specific recombination systems have proved beneficial for the elimination of selection markers. In the P1 bacteriophage recombinase (*Cre*) and its recognition site (*lox*P). Researchers cloned the luciferase (*luc*) gene and *lox*P/*hpt*/*lox*P cassette in a binary vector and generated transgenic tobacco plants. Then the transgenic plant with *luc*/*lox*P/*hpt*/ *lox*P was retransformed with a binary vector with *cre*/ *npt*II. In 10 of the 11 kanamycin-resistant plants obtained, excision of *hpt* had occurred. Chances were that *luc* of the first TDNA and *cre*/*npt*II of the second T-DNA were present in unlinked loci. Selfing of the T_0 transgenic plants was done and T_1 plants were analysed. In one case, 17 of 104 T_1 plants had luciferase (*luc*) activity but lacked kanamycin resistance. Thus, in a two-step exercise the marker genes (*hpt* and *npt*II) and the recombinase (*cre*) were eliminated. Instead of retransformation, crossing of *luc*/ *lox*P/*hpt*/*lox*P plants with *cre*/ *npt*II plants also yielded similar results.

In maize cells, the recombination system of the 2 mm plasmid of *Saccharomyces cereviseae* involving FLP recombinase and FRT recombination site has been successfully deployed for marker elimination. A combination of lambda attachment site (*att*P) and negative selection using *tms*2 and napthaleneacetamide (NAM) were used for marker elimination by intrachromosomal recombination.

The maize Ac/Ds system was tested for marker elimination in tomato using a T-DNA with Ds-*gus*-Ds/Ac/*npt*II for transformation. Transposition events in T_0 plants would place *gus* and *npt*II genes in unlinked loci thereby segregating them in T_1 plants. In a multi-auto-transformation vector (MAT) a combination of Ac functions and *ipt* gene were used thereby eliminating *ipt* without selfing and segregation.

A single-step *ipt*-MAT vector for rice incorporated the recombinase (R) and recombination sites (RS) of *Zygosaccharomyces rouxii* whereby the T-DNA constructed by them carried *nptII/gus/hpt*/RS/*ipt*/35SR/ *gfp*/RS sequences. *Agrobacterium*-mediated transformation of scutellum-derived callus was performed, with selection based on the ability to form normal shoots on cytokinin-minus medium. In 25 per cent of the infected calli, regenerated normal shoots carried the *gus* gene without *ipt*, showing that marker (*ipt*)-free rice plants could be generated in a very short span of one month.

5.2.7.1. *Elimination of Marker Genes by Cotransformation*

A simple approach for marker elimination uses cotransformation of two independent T-DNAs, one with selection marker and the other with the gene of interest. If the two T-DNAs integrate at different loci in the T0 plant, they will get segregated in the T_1 progeny leading to the generation of marker-free plants.

Based on the logical idea that conditions which support. Two separate T-DNAs were placed in superbinary vectors that harboured supervirulent *vir*B, *vir*G and *vir*C genes based on, the logical idea that conditions supporting high efficiency of transformation may lead to a high frequency of cotransformation. One T-DNA contained a drug resistance marker and the second T-DNA carried the *gus* gene. There was cotransformation of *gus* in more than 47 per cent of the drug-selected tobacco and rice plants. The non-selected *gus* gene segregated from *npt*II or *hpt* genes in T_1 progenies of more than 25 per cent of drug-selected plants. This simple two-step strategy is a convenient method to generate marker-free transgenic plants.

As an alternate strategy of cotransformation was to place two separate, compatible binary plasmids in one *Agrobacterium*. The selected T-DNA with *npt*II was placed in a RK2 replicon and the non-selected *gus* gene was placed in a pRiHR1 replicon. Cotransformation in T_0 plants and segregation of *npt*II and *gus* in the T_1 plants occurred at a high frequency. Two T-DNAs of variable lengths were placed on the binary plasmid with the *npt*II TDNA 7 kb long and the *hpt* T-DNA 3.2 kb long. The cotransformation frequency of the non-selected T-DNA was higher when the marker on the longer T-DNA (*npt*II) was used for selection. The selected marker could be genetically separated (segregated) in the T_1 plants. This shows that the T-DNA with the selection marker should be twice longer than the non-selected T-DNA to achieve high cotransformation in T_0 plants and to eliminate the marker by segregation in the T_1 plants.

5.2.7.2. *Positive Selection of Transgenic Plants*

While research to eliminate undesirable markers from transgenic plants was under way, scientists were making a parallel exploration of the use of positive selection markers. Positive selection markers unlike antibiotic and herbicide markers, which kill the untransformed cells, render metabolic/ developmental advantage to transformed cells and help in faster proliferation and regeneration. There is also the elimination of the problem of untransformed dead tissues being detrimental to the growth of transformed tissues. A glucuronide derivative of benzyladenine (BA). β-Glucuronidase, expressed in the transformed tissues, released BA and supported tissue proliferation and regeneration in a cytokinin-minus medium is a fine example of a positive selection agents. The transformation efficiency obtained thus was twice than that obtained with kanamycin. Another strategy for positive selection was based on the inability of plants to use xylose as carbon source unless it is converted to xylulose by xylose isomerase. *Thermoanaerobacterium thermosulfurogenes* xylose isomerase gene (*xyl*A) was successfully used as a positive selection marker in potato and tomato, the transformation efficiency of positive selection using xylose being comparable to that obtained with kanamycin.

With the positive selection method using *E. coli* phosphomannose isomerase gene (*pmi*) now well established, mannose is converted to mannose-6-phosphate by plant hexokinases. Plants lack phosphomannose isomerase. Mannose-6-phosphate that collects in cells, exerts severe growth restraint by inhibiting phosphoglucoisomerase and by depleting phosphate required for ATP synthesis. Mannose is not directly toxic and using is selection transformation of sugarbeet, maize, cassava and rice has been accomplished. Yeast 2-deoxyglucose-6-phosphate phosphatase conferring resistance for 2-deoxyglucose and *Catharanthus roseus* tryptophan decarboxylase that converts the toxic tryptophan analogue 4-methyl-tryptophan to non-toxic 4-methyltryptamine are other tested examples of positive transformation markers.

5.2.8. Problem of 'Long Transfer' of Plasmid Backbone

Agrobacterium T-DNA transfer has the unique advantage of accurately processing the T-DNA between the right and left borders and its precise transfer and integration to the plant genome. However, there are possibilities that non-T-DNA portions may also get transferred to the plant genome. T-DNA transfer initiates in low frequencies from the left borders as well. What is known as 'long-transfer' occurs when, at fairly high frequencies, TDNA transfer starting from the right border skips the left border and the entire binary plasmid is transferred to the plant genome. In both these processes, the vector backbone with bacterial antibiotic resistance markers is transferred into the plant genome. Which is a potential biohazard. Transgenic plants with such transformation events should be carefully identified and eliminated.

A simple modification in the binary vector was carried out to address the problem of 'long transfer' and to enrich typical transformation events from right to left T-DNA borders. A CaMV35SP-*barnase*-INT cassette was placed to the left of the left T-DNA border as a non-T-DNA lethal gene. Any long transfer event beyond the left T-DNA border will lead to the expression of *barnase* resulting in them death and elimination such transformed cells. By deploying this strategy, the chances of the transfer of binary vector backbone were totally eliminated.

5.2.9. Biolistic Gene Transfer

For transforming plants, physical DNA delivery methods are useful. Such methods are microinjection, PEG-mediated transformation of protoplasts, electroporation of protoplasts and tissues and microprojectile bombardment of tissues. Among these, the microprojectile bombardment is the most widely deployed method for genotype-independent plant transformation, first used to deliver DNA and RNA into the epidermal cells of *Allium cepa*. Since then, it has been used to transform yeast and filamentous fungi, algae, cereals and pulses. Chloroplasts of *Chlamydomonas* and mitochondria of yeast and *Chlamydomonas* have also been biolistically transformed.

Particle bombardment use three systems for with various modifications. Electric discharge particle acceleration device ACCE L utilises an instrument to accelerate DNA-coated gold particles with any desired velocity by varying the input voltage. The biolistic PDS 1000 He device is powered by a burst of helium gas that accelerates the macrocarrier, DNA-coated gold particles are uniformly bombarded over target cells upon the macrocarrier. The advantages of particle bombardment system are: (1) Plants not infected by *Agrobacterium* can be transformed. (2) DNA may be transferred without using specialised vectors. (3) Multiple DNA fragments/plasmids can be introduced by co-bombardment, thus eliminating the need to construct a single large plasmid containing multiple transforming sequences (4) False positives resulting from reporter gene expression in *Agrobacterium* are avoided. (5) Transformation protocols are applicable to plants which lack good regeneration systems. (6) Organelle transformation is achieved only by particle bombardment.

The biolistic gene delivery system has been successfully used to create transgenics, but certain drawbacks of the technique, high copy number and rearrangements of transgene(s), have been observed thus causing gene silencing or genomic rearrangements. However, many recent improvements have been made to offset problems relating to gene silencing and genetic integrity. The overall transformation efficiency of biolistics has been improved by osmotic conditioning of cells.

5.2.10. Gene Silencing in Transgenic Plants

One of the common problems in making transgenic plants with increased level of endogenous gene expression by the use of heterologous strong promoters or a foreign gene of some utility, is silencing of transgenes in a vast population of transgenic plants at the transcription level. This is referred to as transcriptional gene silencing (TGS); whereas silencing at post-transcriptional level is referred to as post-transcriptional gene silencing (PTGS). TGS involves restriction of transcription and association with methylation of promoter region whereas in PTGS, though the genes are transcribed, their mRNA is degraded. PTGS is associated with methylation of the coding region of the transgenes.

TGS is associated with multiple copies of transgenes and inverted repeats (IRs) in transgenes, which can act in *cis* as well as in *trans*, indicating a homology-dependent mechanism of gene silencing. Plants may have also evolved TGS as a device to safeguard themselves from the activities of endogenous transposable elements, which are exist in multiple copies. However, PTGS has been observed even when unrelated foreign genes have been transferred to plants. PTGS has been associated with high copy number of transgenes, strength of the promoter or the stability of the transcripts. Above the threshold level for accumulation of transcripts PTGS sets in. Viruses infect plants and accumulate their transcripts at very high level. PTGS may stop transcription of viral genes. Some plants use PTGS to recover from viral infection and to stay free from future infections. As organisms compete with each other for their survival, some plant viruses have evolved mechanisms to counter or suppress PTGS.

DNA methylation is involved in both TGS and PTGS, but it is not clear if this is the cause or the effect of gene silencing. Apart from the common sites of methylation CG and of CNG symmetric sequences, there are also some methylation at asymmetric sites. In order to determine the involvement of DNA methylation in gene silencing, a mutant form of 35S promoter that lacked all symmetrical methylation acceptor sites was synthesised and used to drive the *bar* gene. When resistant lines obtained were crossed with plants carrying the 271-silencer locus a significant reduction in the number of phosphinothricin (PPT) resistant progeny was observed.

Only asymmetric methylation acceptor sites could allow the silenced lines be methylated. When the 271-silencer locus was outcrossed, PPT resistance was restored. Other studies but using wild type 35S promoter showed that even after outcrossing of the silencer locus the silenced state was maintained, suggesting that for initiation of silencing, methylation at symmetric methylation sites is not required. But this is essential for maintenance of silencing through generations.

Mutation in *MET* gene, encoding a DNA methyltransferase, drastically reduces DNA methylation and results in the release of silencing at most loci; however some loci still remain silenced, probably because of the presence of other functional methyltransferases. Interestingly, the consequence the mutation in *DDM1* locus which encodes a plant homologue of SW12/SNF2 protein, a component of yeast chromatin remodelling complex, also results in loss of methylation and release of silencing, suggesting that normal chromatin structure is required for maintenance of normal methylation status as well as gene silencing. Studies with mammalian systems indicate that a methylated DNA-binding protein MeCP2, which can attach to even a single methylated CpG, recruits a repressor complex which has histone deacetylase. In a similar situation existing in plants, mutation in histone deacetylase (HADC) gene should be able to ease silencing without affecting methylation status. Transgenic inhibition of HADC in *Arabidopsis* is linked to pleiotropic effects similar to those observed in case of *MET/ DDMI* mutation, without any significant decrease in DNA methylation.

Mutation of *MOM* locus, which stands for 'morpheus molecule', also causes release of silencing without affecting DNA methylation. This gene could be another component of chromatin remodelling complex. TGS and PTGS operate differently but both involve DNA methylation, though of different regions. It has been proved that expressed RNA can trigger methylation of coding region as well as PTGS.

Upon infection, the potato spindle tuber viroid, which has a small RNA genome and does not code for any protein, can direct methylation of its transgene, integrated into plant genome. Production of double-stranded RNA complementary to coding sequence from a single transcript, which can pair to itself, can induce PTGS of genes having homologous sequence. Such double-stranded RNA can also be produced *in vivo* from promoter sequences having inverted repeats or other complex structures by read through transcription from nearby promoters. Recent experimental support this possibility.

Aberrant RNAs that are synthesized during PTGS are used by RNA-dependent RNA polymerase as substrates and synthesizes antisense RNAs. The pairing of both sense and antisense RNA leads to formation of double-stranded RNAs, which are targeted for degradation. Experimental evidence in tomato for the presence of RNA-dependent RNA polymerase has been obtained. The degradation products of about 25 bases to the silenced analogous ransgene have been observed during PTGS.

Transgene silencing is an unpredictable phenomenon, and hence plants showing silencing of the desired phenotype are discarded while plants which do not show silencing are used. Attempts to overexpress endogenous genes have often resulted in silencing of transgenes as well as endogenous genes, a phenomenon known as co-suppression, leading to PTGS, and attempts to overexpress genes of utility have suffered.

Attempts to overcome silencing by addressing the presence of duplicated sequences, high copy number of foreign genes, methylation acceptor sites, strength of promoter with silencing, are not well documented. Scientists have adopted the common occurrence of multiple copies of transgenes in case of biolistic method of transformation which has made *Agrobacterium*-mediated method of transformation more popular. However, there are reports of silencing of even single copy genes. Some attempts have been made to avoid silencing of transgenes.

Matrix attachment region (MAR) has been used in transformation vectors to prevent the influence of heterochromatin on the integrated genes. Insulation elements have proved beneficial in vertebrate systems to protect transgenes from effects of nearby *cis*-acting elements. Plant viruses have founds ways to avoid PTGS that plants use to avert viral infections. HC-protease (HC Pro) of potato virus Y (PVY) and 2b protein of cucumber mosaic virus (CMV) have been found to interfere with PTGS in tobacco plants infected by these viruses while HC Pro blocks the maintenance of gene silencing in tissues where silencing has already occurred, 2b protein prevents initiation of gene silencing at growing nodes of plants. Recently, a calmodulin-like protein, named rgs-CaM and identified from tobacco, has the ability to suppress PTGS and can be exploited to prevent gene silencing. However, whether such proteins would render these plants vulnerable to viral infections is yet to be investigated.

Depending upon the correlation of various features of vectors and gene silencing observed attempts should be made to avoid the use of potential methylation-acceptor sites. DNA repeats can be avoided by using different promoters and termination signals to drive expression of different genes of the vector. Vectors can be designed to incorporate MARs or insulators and even the genes for suppression of gene silencing could be incorporated.

5.2.11. Integration of Emerging Technologies of Transformation

Studying the developments in transformation, it is interesting to note that all improvements made in transformation systems were readily exploited for rice transformation. In the early stages, electroporation of protoplasts and PEG-mediated transformation of protoplasts were used to generate fertile transgenic rice plants. An electric discharge particle acceleration device was used to bombard immature rice embryos and generate fertile transgenic rice plants. Until 1994, it was strongly believed that rice is not amenable to *Agrobacterium*-mediated transformation. But with the construction and deployment of the superbinary vector pTOK233 for rice transformation, *Agrobacterium*-mediated rice transformation has become routine in many laboratories.

Agrobacterium-mediated cotransformation, a very simple and elegant approach for marker elimination, was simultaneously assessed in tobacco and rice. The cotransformation strategy was successfully utilised in rice for engineering provitamin A biosynthesis. A total of four genes contained in two separate T-DNA cassettes were transferred together.

Using tobacco MAR sequences, the usefulness of MARs in compartmentalising the expression of integrated transgenes was evident in the 140-fold increase in *gus* expression in rice. Immediately, a *hpt*-intron gene was successfully evaluated for rice transformation. As *gfp* was proving to be a useful, vital reporter of plants, a synthetic *gfp* version combined with *rbc*S promoter. Transit peptide was used for the first time to target the GFP protein to plastids to overcome its tendency to form insoluble complexes in cytoplasm. Isolated rice shoot apices were successfully used for *Agrobacterium mediated* transformation shoot apices were becoming attractive transformation targets. When an *ipt*-MAT strategy coupled to R/RS-based excision was used to obtain marker-free transgenic rice plants in one step, as demonstrated in rice.

Transformation technologies in rice are proving to be beneficial to meet the requirements of functional genomics. The clone with the *Xa21* gene was identified for bacterial blight resistance using a shot-gun approach with many rice genomic fragments. The clone new genes and to identify their functions, similarly generating 22,090 independent transgenic rice plants using *Agrobacterium*. It is estimated that as many as 25,700 TDNA tagged loci will be available for functional genomics analysis. Thus, with corresponding developments in transformation technologies and genome analysis, rice has become the first crop plant targeted for complete functional genomics analysis.

5.2.12. Deployment of Transformation Technologies in India

In India *Agrobacterium* Ti plasmid vectors and microprojectile bombardment have been successfully used for plant transformation. The early success of *Agrobacterium*-mediated transformation of *Brassica juncea* is now continued in the University of Delhi-South Campus which had also successfully disployed the *barnase/barstar*-based genetic engineering approach for introducing male sterility and its restoration in *B. juncea*.

Agrobacterium mediated transformation of *indica* rice by particle bombardment and Pusa Basmati 1 using both superbinary and binary vectors were achieved. Transgenic brinjal for salinity and drought tolerance by introducing *mannitol phosphodehydrogenase* gene were also developed. In order to develop abiotic stress resistance, *Arabidopsis* heat shock protein gene (*hsp*101) and rice pyruvate decarboxylase 1 (*pdc*1) was introduced into rice using *Agrobacterium* Ti plasmid vectors. *Rice tungro virus* genes cloned from RTBV and RTSV genomes are being introduced into rice to develop virus resistance and efficient permeabilisation of DNA into wheat zygotic embryos using membrane interactive agents, like saponin and toluene. Also currently being introduced into wheat are he *bar* gene for herbicide resistance, *pin*III for pest resistance and barley *HVA1* gene for abiotic stress resistance.

Researchers at IARI, New Delhi, have successfully integrated cryIAb in brinjal plants to develop resistance against the larvae of fruit borer. Transgenic tomato plants, engineered with cryIAc exhibited resistance to the larvae of *Helicoverpa armigera*. Researchers in ICGEB and IARI have successfully generated transgenic tobacco plants expressing cryIIa5, that conferred complete protection against *H. armigera*. A chloroplast transformation system in ICGEB is being developed to achieve a high level of expression of a wide range of useful proteins, and transformation of Basmati and IR64 cultivars of rice using both binary and superbinary vectors. Researchers at IARI are working towards developing tolerance for UV-B radiation by introducing the *cod*A gene from *Arthobacter* globiformis.

At Jawaharlal Nehru University, New Delhi, research is being carried on, desensitised a separate kinase for regulating amino acid biosynthesis in trans-genic plants. Also being raised are transgenic potato plants with increased nutritive value by engineering tuber-specific expression of *Amaranthus* albumin gene *AmA1*.

A significant achievement of the research is the generation of *CryIAc*-expressing IR64 transgenic rice that is highly toxic to the larvae of yellow stem borer. There is also transformation of embryo axis of chickpea by particle bombardment with cryIAc.

Various Indian laboratories are conducting plant genetic engineering work as per the guidelines evolved and implemented by the Review Committee on Genetic Manipulation (RCGM) of the Department of Biotechnology, Govt of India. The guidelines carry clear descriptions of greenhouse, net house and field evaluation of transgenic experiments.

Though there have been impressive progress in the transformation of crops like *Brassica,* rice, potato, tomato and brinjal, there are yet hurdles in transforming other important crops such as legumes, cotton and sugarcane, in which farmers face many problems that can be effectively solved by genetic engineering approaches. In both legumes and cotton non-availability of a good regeneration system is a limiting factor. Standardisation of conditions to regenerate plants via somatic embryogenesis can prove to be beneficial in achieving *Agrobacterium*-mediated transformation. Simultaneously, there should improved transformation efficiencies by using various approaches such as (i) .evaluation of different Ti plasmid backgrounds in *vir* helper strains, (ii) use of multiple copies of supervirulent *vir* genes, (iii) application of vacuum infiltration, (iv) use of marker genes with introns, (v) evaluating multiple markers including positive selection markers, (vi) performing co-cultivation at lower temperatures, and (vii) the use of matrix attachment regions in the T-DNA. *In planta* transformation and *bar* gene-based screening in T1 generation are attractive alternatives.

5.3. Direct Gene Transfer Methods

In 1984 the first transgenic plant was produced via Agrobacterium mediated modified transformation of *Nicotiana tabacum* protoplasts and thereafter several dozen plant species have been genetically engineered using different techniques.

Other techniques such as selectable markers were also developed that facilitated the development in genetic engineering for obtaining transformed plants. Since this technique is not suitable for monocotyledon plants as they are not natural host of Agrobacterium, other methods of direct gene transfer have been developed for use with monocots and other species, be categorised on the basis of the use of protoplasts or cell and tissue as the target materials. Freshly isolated protoplasts are used for genetic transformation using direct gene transfer methods.

a) *DNA Transfer in protoplasts:* This is accomplished by:
   i) Electroporation.
   ii) Chemically stimulated DNA uptake by protoplasts
   iii) Liposomes
   iv) Microinjection
   v) Sonication

b) *DNA Transfer in Plant Tissues:*
   i) Acceleration of DNA coated microparticles
   ii. Laser microbeam
   iii. Silicon carbide fibres

Direct uptake of DNA by isolated protoplasts is a genotype-dependent response, and regeneration from protoplasts is not common in all the species. Hence production of fertile transgenic plants has posed problems in most of the cereal species. However, transgenic fertile plants have been produced in *Sorghum vulgare, Oryza sativa* and *Hordeum vulgare.* Direct delivery of free DNA molecules into plant protoplasts by physical (electroporation and micro injection) and chemical (polyethylene glycol) methods have been developed to facilitate DNA delivery across the plasma membrane.

5.3.1. DNA Transfer in Protoplasts

5.3.1.1. Electroporation

Electroporation is based on the use of the short electrical pulses of high field strength, causing the uptake of DNA into protoplasts by temporary permeabilisation of the plasma membrane to macromolecules. Protoplasts and foreign DNA, placed in a buffer between two electrodes are treated to a high intensity electric current which damages membranes and creates pores in membranes. Immediately thereafter DNA diffuses through these pores, until the pores are resealed. The technique is optimised by using appropriate electric field strength. The optimum field strength that is dependent on the following:

i) The pulse length of electric current
ii) Composition and temperature of the buffer solution
iii) Concentration of foreign DNA in the suspension
iv) Protoplasts density, and
v) Size of the protoplasts.

The removal of pectin from the plant wall increases the amount of DNA which can be introduced by electroporation, tobacco mosaic virus being a case point. Electroporation has been used successfully for transient and stable transformation of protoplasts from a wide range of species. Plating efficiency of electroporated protoplasts grown on selection medium can be as high as 0.5 per cent, for tobacco being 0.2 per cent of electroporated leaf mesophyll protoplasts giving rise to transgenic calli. Low transformation efficiency is common in cereals, e.g. in rice 0.002 efficiency was recorded.

5.3.1.2. Chemically stimulated DNA uptake

Direct uptake of DNA by protoplasts is stimulated by

polyethylene glycol (PEG) which is the most widely used chemical for this purpose. Transformation involves stimulated by PEG mixing of freshly isolated protoplasts with DNA and immediately adding 15-20 per cent PEG dissolved in a buffer containing divalent cations. This mixture is incubated for 30 minutes, protoplasts are washed and then plated in petri plates for culture and growth. The optimisation of transformation frequencies by this method include factors that follows.

i) PEG concentration in the mixture
ii) Composition and concentration of salts used
iii) The pH of the solution
iv) Concentration of the foreign DNA
v) Size and form (linear, supercoiled) of the DNA molecules used
vi) Culture and selection techniques used for protoplasts.

There is preference for PEG mediated transformation over electroporation for stable transformation of monocot protoplasts due to relatively higher survival rates after treatment.

While PEG also stimulates the uptake of liposomes, improves the efficiency of electroporation, and causes precipitation of ionic macromolecules like DNA, it also stimulates their uptake by endocytosis. PEG mediated DNA uptake typically transforms 0.1 to 0.4 per cent of the total protoplasts treated. Production of transgenic plants depends upon the regeneration competence of the transformed protoplasts. In Petunia, 40 per cent transformed calli derived from mesophyll protoplasts could be induced to form fertile plants, equal to about 0.1 per cent transformation efficiency of the treated protoplasts. In different species different transformation efficiency has been observed, e.g., in embryogenic protoplasts suspension of rice was 0.0004 per cent and, soyabean and tobacco was 0.7-1 per cent.

5.3.1.3. *Liposomes*

Liposomes, small lipid sacs containing plasmids and prepared artificially have also been used as a carrier for the introduction of nucleic acid into plant protoplasts. PEG stimulates the fusion of liposomes with plant protoplasts. Liposomes mediated transformation includes positively charged agents such as cations in the transformation mixture or using the cationic liposome preparation.

For maize protoplast, other chemical agents like polycation polybrene or lipofectin have also been used for both transient and stable transformation. The new protoplasts transformation methods are cationic liposome and polycation mediated DNA delivery which are considered better than other methods of transformation. There are several advantages for the use of this technique.

i) Protection of DNA/RNA from nuclease digestion.

ii) They have low cell toxicity.

iii) Encapsulation of nucleic acids makes them more stable during storage.

iv) High degree of reproducibility.

v) This method is applicable to wide range of plant cell types.

5.3.1.4. *Microinjection*

Microinjecting nucleic acids into protoplasts or intact cells via is a labour-intensive procedure requiring special capillary needles, pumps, micromanipulators, inverted microscope and other equipment. However, nucleus or cytoplasm can be injected and cells can be cultured individually to produce callus or plants, thus avoiding transformants by drug resistance or marker genes. The worker has to skilfully insert the needle into the cytoplasm or in the nucleus. The basic technique is similar to that used for animal cell microinjection. In order to microinject protoplasts or other plant cells, the cells need to be immobilised which are done by:

i) The use of a holding pipette which holds the cells by vacuum.

ii) Attachment of cells to poly-L-lysine coated cover slips.

iii) Embedding the cells in agarose, agar or sodium alginate.

Glass micropipettes are with openings of about 0.3 uM in diameter, are inserted into plant cell cytoplasm and nuclei with the aid of a micro manipulator device. A syringe-like device is used for the controlled delivery of volume (10-11 - 10-4 ul) into the plant cell. Most plant cells are injected while keeping inside microdroplets (2-50 ul) of medium using a chamber which is sterile, vibration free and permits temperature and humidity regulation. Using this procedure, a maximum of 100-200 cells per hour can be microinjected.

The regeneration ability of the microinjected cells decides the recovery of transformants. Different methods have been used to grow injured single cells or protoplasts. Hanging droplets, covered under a thin layer of agar or agarose, and micro culture are being used. Attempts have been made to inject linear, or super coiled DNA, in cytoplasm or in nucleus. Nuclear injections are found better for transformations.

5.3.1.5. Sonication

In protoplasts of sugar beet and tobacco mild sonication facilitates the uptake and transient expression of a chloramphenicol acetyltransferase (CAT) gene. This is a better method than electroporation method used for the same material. Plating efficiency was also similar to untreated cells. However, transgenic plant production using this technique has not been reported so far.

5.3.2. Free DNA Transfer to Intact Tissue

5.3.2.1. Acceleration of DNA Loaded Microparticles

The latest technology to transfer DNA into intact tissues involves the transfer of micro-sized particles coated with DNA to penetrate the cells. Here micron size tungsten or gold particles are accelerated in a gun barrel to velocities sufficient for non-lethal penetration of cell walls and membranes. In 1987 the particle gun was first used to transfer chimeric DNA and viral RNA molecules into intact onion cells. Tungsten acted as a carrier of nucleic acids as it was available in micro-size balls, non-toxic to cells and was dense enough for rapid penetration of target material. Microprojectile mediated transformation, is a mechanical method of introducing DNA in to any plant species, can be successfully used where plasmids or protoplasts mediated transformation cannot be used. An acceleration device used to propel particles carrying plasmid DNA based on machine or technique used to accelerate the particles, are called 'particle gun technology', 'biolistic method', 'DNA bombardment', 'particle acceleration of

DNA method' and 'electric discharge particle acceleration method'.

This technique of stable transformation and testing a gene for cell and organelle specific expression has three components:

i) The basic equipment to generate particle acceleration.

ii) Metal particles coated with precipitated DNA (desired gene).

iii) Plant tissues to be used for particle penetration. The method for regeneration should be previously standardised and proper tissues be selected for bombardment.

In an electric discharge particle acceleration device a high voltage discharge (14 KV current) is delivered to a small water droplet which quickly vaporises and releases energy to propel DNA coated gold spheres into target cells. Similarly, a DNA carrier is attracted due to potential differences, stopped in-between by a screen, DNA coated particles cross the screen and fly towards target tissue and deliver the DNA into cells.

5.3.2.2. DNA Coating

This sophisticated technology requires precise preparation of DNA coated gold or tungsten particles, having the following properties:

i) High density (19 g/cm^3 or greater) to ensure proper acceleration and penetration through cell walls.

ii) Size (0.5-5 um) should match with size of the cells. Large sized spheres can be used with large cell size.

iii) Gold is costlier than tungsten, but it does not oxidise like tungsten.

DNA is precipitated onto the particles prior to bombardment. Most commonly, $CaCl_2$ and spermidine are used to precipitate plasmid DNA onto tungsten particles. Ethanol is also used for precipitation of the DNA on gold particles.

5.3.2.3. Plant Tissue

Competent generating plant tissue should be used for

transformation. Though mostly embryogenic tissue is an ideal material for transformation, yet leaf bombardment in tobacco, and stem bombardment in cranberry have resulted in transformed plants. Normally, reporter gene and selectable marker genes are used to isolate and select transformed cells/plantlets.

5.3.2.4. Laser Microbeam

Laser beam has been usefully employed for transformation of plant cells. To introduce DNA into plant cells and chloroplasts an ultraviolet (UV) laser. 343 nm beam is directed through an adjustable attenuator into the optical path of an inverted microscope. The focus of the laser beam is adjusted so that its focus is identical with that of the objective lens on a specimen in the microscope. This laser beam can then puncture holes in any part of cell which is in focus. Laser micropuncture of the cell wall and plasma membrane allows uptake of plasmid DNA into cells using Brassica napus cells and microspores. This technique has also been used to transfer genes into isolated chloroplasts and chloroplasts of intact protoplasts resulting in 20 per cent transformation achieved by this method but fertile plants are yet to be produced by this method.

5.3.2.5. Silicon Carbide Fibres

Transfer of DNA is done by microinjection and electroporation methods for using intact plant cells and tissues. Similarly, cells at low frequency are transformed vortexing plasmid DNA and plant cells with silicon carbide fibre. Under vortex, silicon fibres penetrate cells and create fine holes permitting entry of DNA. DNA uptake by inhibition in dried embryos of cereals and legume species has also been reported. This is a very simple method which can use dried somatic embryos. Only transient gene expression are observed and stable transformations are yet to be achieved by this method. Research is being carried out to introduce gene in pollen grains by microinjection in the anthers, which can be used for fertilisation to obtain transgenic plants.

References

Bajai, Y.P.S. (ed.). *Biotechnology in agriculture and forestry*, 17: 3-16. Berlin, Springer-Verlag. 1991.

Bhojwani, S.S. (ed.). *Plant tissue culture: applications and limitations.* Elsevier. Amsterdam.

Debergh Zimmerman, (ed.) *Micropropagation: technology and application.* Kluwer Academic Publishers. 1991.

Hanover J.W. & Keathley, D.E. (eds.). *Genetic manipulation of woody plants,* P. 413-432. Plenum Press. New York. 1988.

Haines, R.J. "Mass propagation of cuttings, biotechnologies and the capture of genetic gain". Paper presented at IUFRO meeting, Bordeaux, September 1992. pp. 15. 1992.

Hamdan, Y.Y. 1990. "Current status of plant biotechnologies in the Near East and North African". In Sasson & Constantini, eds. *Plant biotechnologies for developing countries,* proceedings. p. 131-139. Int. symposium organized by CTA and FAO, Luxembourg, June 1989.

Villalobos, V.M., Ferreira, P. & Mora, A. "The use of biotechnology in the conservation of tropical germplasm". *Biotech. Adv.,* 9: pp 197-215. 1991.

6

PLANT GENETIC ENGINEERING AND BIOTIC STRESS RESISTANCE

It is easier to engineer genetic plants than genetic animals, the reasons being: (a) there is a natural transformation system for plants (the bacterium *Agrobacterium tumefaciens*), (b) plant tissue can redifferentiate (a transformed piece of leaf may be regenerated to a whole plant), and (c) plant transformation and regeneration are relatively easy for a variety of plants.

The soil bacterium *Agrobacterium tumefaciens* can infect wounded plant tissue, thus transferring the Ti plasmid to the plant cell. Part of the Ti plasmid apparently randomly integrates into the chromosome of the plant. The integrated part of the plasmid contains genes for the synthesis of (a) food for the bacterium, and (b) plant hormones and are expressed at high levels in the plant. Overproduction of the plant hormones results in continuous growth of the transformed cells, causing plant tumours. Rapid, cancerous growth of the transformed plant tissue obviously is advantageous to the bacterium: more food gets produced.

The Ti plasmid has been genetically modified ("disarmed") whereby the genes involved in the production of bacterial food and of plant hormones get eliminated, and a gene that can be used as a selectable marker gets inserted. Selectable marker genes generally code for proteins involved in breakdown of antibiotics, such as kanamycin. Any selected gene can be inserted into the Ti plasmid as well. In principle, one can thus transform any plant tissue, and select transformants by screening for antibiotic resistance. Practically, however, there are some complications: (a) it has proven difficult to transform some monocots (grasses, etc.) by *Agrobacterium*, and (b) regeneration of plants from tissue culture or leaf discs is not always possible.

In a number of genetically engineered plant varieties traits that have been introduced by transformation include herbicide resistance, increased virus tolerance, or decreased sensitivity to insect or pathogen attack. Earlier genetically engineered plants like tobacco, petunia, or similar species had a relatively limited agricultural application. However, during the past decade it now has become possible to transform major staples such as corn and rice and to transform them to a fertile plant.

Increasingly, the transformation procedures used do not depend on *Agrobacterium tumefaciens*. Instead, μm-sized tungsten or gold bullets coated with the DNA can be integrated into cells. The bullets are fired from a device that works similar to a shotgun. The modernised device uses a sudden change in pressure of He gas to propel the particles, but the principle of "shooting" the DNA into the cell remains the same. This DNA-delivery device is nicknamed "gene gun", and has been shown to work for DNA delivery into chloroplasts as well.

Lately, the "gene gun" has become a very common method to transform plants, and is applicable to virtually all species investigated, for example, rice which is the most important stable crop for a large population world.

Another procedure to integrate foreign genes into cereals is by electroporation whereby: a jolt of electricity is used to puncture self-repairing holes in protoplasts (i.e., the cell without the cell wall), and DNA can penetrate these holes. However, it is often very difficult to regenerate fertile plants from protoplasts of cereals, there have been significant progress in overcoming these practical difficulties. Now even transgenic trees have been created, for example, introduction of the gene for a coat protein of the plum pox virus into apricot. The plum pox virus leads to the feared Sharka disease, for which there is no cure. The resulting transgenic tree shows a markedly decreased sensitivity to this virus.

With various new techniques to integrate foreign genes into plants, all major crop plants can be genetically engineered. In the

US there is little resistance to crops as long as the products are shown to be safe and advantageous but in other countries they are received poorly by the consumer. However, it should be remembered that essentially all agricultural products have been genetically modified by traditional breeding, so it may be difficult to define what is actually free of genetically modified organisms.

6.1. Applications of Plant Genetic Engineering

Apart from the production of highly valuable enzymes by plants and animals, production of human proteins in these organisms, other valuable proteins that are currently produced by microorganisms could very well be produced by higher organisms instead. Since eukaryotes glycosylate proteins animal and plant "bioreactors" in some respects may be superior to recombinant bacterial systems. Whereas the glycosylation pattern may be species-specific, appropriate glycosylation is often required for protein function compared to cell fermentation techniques Production through these organismal systems may also be cheaper.

Production of human serum albumin in transgenic tobacco and potato, and production of human insulin by tobacco are examples in case in which the produced protein appears to be fully effective in humans. Unfortunately, one cannot raise one's insulin level by eating transgenic tobacco leaves, as the protein in most cases get broken down to amino acids before reaching the bloodstream. Therefore, in such cases the practice of protein isolation and purification before transgenic leaves are converted into drugs is unavoidable.

The antibody's light and heavy chains earlier were produced in different plants, but a subsequent cross of these two varieties resulted in progeny carrying assembled and functional antibodies.

6.1.1. Reversible Male Sterility in Plants

Heterozygous individuals being often healthier and stronger than homozygous ones, the only way to guarantee heterozygoiscity in plants is to make sure self-pollination cannot occur. Earlier it was difficult for most crop plants to exclude selfing. To exclude self-pollination, it would be good to introduce male sterility in plants:

progeny from such plants are then expected to be 100 per cent heterozygous. A promoter was identified that was turned on exclusively in tapetum cells to introduce male sterility. This promoter then was linked up to a gene coding for a bacterial ribonuclease which selectively chops up ribonucleic acids. The promoter/ribonuclease construct was then generated into plants. Because the promoter allows expression only in tapetum cells, the gene construct disrupts only growth of the tapetal tissue and its end product, pollen. Plants transformed with this construct were male-sterile but otherwise normal.

Male-sterile plants are valued for hybrid seed production, but their value is limited when it comes to crop production. Crops, such as wheat, rice and tomato, in which the seed or fruit is the harvested product, must have the fertility restored. Fortunately, since the ribonuclease is inhibited very much by a simple protein, named barstar, the male-sterile plant can be crossed with a male-fertile variety in which the gene for barstar has been introduced, and the result is progeny with viable pollen and restored fertility.

6.1.2. Antisense RNA

Antisense RNA are nucleotide strands that are grown in a cell and that are complementary to a particular mRNA. Antisense RNA can be produced, for example, by inverting the coding region of a gene with respect to its promoter, and hybridised with its corresponding mRNA, making it double-stranded. The double-stranded mRNA no longer can be recognised by the protein-synthesizing machinery, and thus expression of this mRNA is suppressed or broken down quickly. Thus, one can inactivate specific genes while not interfering with others.

Antisense approaches protect plants from damage by plant viruses. For example, reversal of a gene from bean yellow mosaic virus (BYMV), and putting it into tobacco under a reasonably strong promoter, has resulted in a BYMV-resistant tobacco variety. A similar approach is used to transfer viral resistance to other plants which is significant, in that currently no effective, environmentally friendly methods exist to control many plant viruses.

The RNAi approach is very useful for both agriculture and medicine. The technology provides an excellent approach for reverse genetics in eukaryotes. With this method one can turn off genes, and see what the consequences are.

6.1.3. Agricultural Applications in Developing Countries

Two main reasons for the more rapid commercialisation of bioengineered crops in the developing world are: (a) less tight governmental approval mechanisms, and (b) hungrier populations. In developed countries biotechnological applications may be utilised to reduce production costs; in contrast, in the developing world the main factor is the production of more food. Indeed, genetic engineering applications seem to be pretty successful to cut down on pathogen-induced losses. For example, genetic modification of papaya plants protects very well against the very destructive ringspot virus.

6.2. Herbicide Resistance

Weeds compete with crops for water, nutrients and sunlight and can decrease crop yields from 35 per cent to 100 per cent. A number of options for minimising the impact of weeds on crop productivity are available to growers; one option is application of herbicides to the weeds.

Normally all crops have some degree of innate resistance to some herbicides. Today agriculturists minimise the negative impact of weeds by using herbicides and herbicide resistant varieties of most crops, including all of the major commodity crops, as well as small acreage horticultural crops, such as vegetables. An attendant benefit of herbicides is the impact their use has on soil erosion. Herbicides and herbicide- resistant crops allow growers to use minimal tillage or no-till techniques, which significantly reduces the impact of agriculture on soil erosion. Today more than 290 types of herbicide resistant weeds are present in agricultural fields around the world.

6.2.1. Developing New Herbicides and Herbicide-Resistant Crops

More than half a century ago, most herbicides used in agriculture were developed by generating many compounds in the lab and spraying them on weeds in the greenhouse to identify likely candidates for further development. Those that were herbicidal were then tested on crops to assess the crops' sensitivity. On obtaining promising candidates for new herbicidal compounds, tests were then conducted to determine whether the herbicide and herbicide resistant crop performed well under field conditions and to assess any impacts on the environment and human health.

The number of chemicals to be tested to develop a new herbicide began to increase exponentially from less than 1,000 in 1950 to approximately one million today. Consequently companies began to focus on developing new weed management options by modifying the crop, rather than finding a new herbicide. Beginning with herbicide already proven to be effective and then giving the crop plant the capacity to resist that herbicide is more efficient and allows herbicide companies to expand the applicability of their most effective and safest products.

Prior to the advent of modern biotechnology crop developers had to perforce use plants in the same or very closely related species, and most crop plants have very little genetic diversity for herbicide resistance. An exception was endogenous tolerance of soyabeans to metribuzin and 2,4-D, which varied among different cultivars. As a result, soyabean breeders were able to use conventional breeding and selection to increase soyabean's endogenous tolerance to these two herbicides, and to combine resistance to both in a cultivar that was resistant to one, but not the other. Crop developers have employed a series of techniques to create novel herbicide resistance traits. Screening and selection of plants with novel genetic variability led to the development of soyabeans tolerant to sulfonylurea herbicides.

Recent successes in plant cell and tissue culture linked with modern biotechnology enable researchers to screen cells and undifferentiated plant callus for herbicide resistance, which increases development efficiency. Treatment of corn tissue cultures with imidazolines followed by recurrent selection created imidazoline- resistant callus that could then be regenerated into corn plants and bred into a variety of cultivars.

Crop developers have also exploited the capacity of weeds to evolve resistance. For example, a breeding programme was initiated to move triazine resistance found in a weed, *Brassica rapa,* to canola, a crop closely related to the weed. Triazine-resistant canola was created, but unfortunately, the new resistance trait reduced crop productivity, which affected the cultivar's commercial value.

The tools of modern biotechnology have made it possible for a herbicide resistant gene present in a given organism to be incorporated into virtually any crop. This flexibility has revolutionised weed management by expanding the spectrum of herbicides that a crop can tolerate.

The first transgenic herbicide resistant crop, bromoxynil-resistant cotton, was commercialised in 1995, and glyphosate-resistant soyabeans came to market the following year. Glyphosate resistant soybeans, corn, cotton and canola are now grown on millions of hectares around the world.

6.2.2. Impact of Herbicide-Resistant Biotech Crops

The new herbicide resistant crops are a great boon to growers as the easy to use, low-cost and effective weed control programme does not harm their crops. Notwithstanding, the new herbicide resistant crops, developed with modern biotechnology, have significantly added to the growers' profits by decreasing their input costs, increasing yields, or both, in some cases.

These benefits have enhanced, the rate of adoption of biotech herbicide-resistant crops. Glyphosate-resistant soyabeans 60 per cent of the global soyabean acres. In 2005, 71 per cent (157 million acres) of the 222 million acres of biotech crops planted globally were herbicide-resistant varieties of corn, soyabean, canola and cotton. The majority of the corn, canola and cotton varieties were resistant to glyphosate and as a result, glyphosate-resistant weeds have begun appearing in fields with row crops like corn and soyabean where glyphosate has been used repeatedly. The effectiveness of glyphosate-resistant weed control deterred farmers from implementing diverse weed management practices, and lack of diversity in weed control measures encourages the evolution of resistance to a measure that is used repeatedly.

6.2.3. Future Prospects

All glyphosate-resistant biotech crops worldwide depend mainly on providing crops with a gene that encodes a glyphosate-insensitive form of the glyphosate target protein. Recently scientists have discovered novel sources of resistance to the sulfonylureas and imidazoline herbicides as well as glyphosate. A new biochemical mechanism of glyphosate resistance was recently discovered in a naturally-occurring microorganism which has a gene for an enzyme capable of deactivating glyphosate. Scientists optimised the activity of this enzyme. Once they transfer the gene for the optimised enzyme into crops, it will effectively protect the crops from glyphosate treatment.

Scientists have also found a new source of resistance to sulfonylureas and imidazolines in corn and soyabeans. When this mechanism is adequately expressed, the sulfonylureas, imidazolines and a number of other similar herbicides can be applied without harming the crop. These herbicides, like glyphosate, are highly advantageous to growers because they are effective against a broad spectrum of weeds.

These combinations of traits in one crop will provide growers with huge flexibility in weed control and a new strategy for controlling their most difficult weed problems.

6.3. Insect Resistance

Since development of insect-resistant plants in 1782 numerous insect-resistant plants have been developed by research centres through conventional method or biotechnology. For example, the International Rice Research Institute (IRRI) has developed and released numerous rice varieties resistant to brown planthopper, green leafhopper, rice stem borer, and rice gallmidge. Several transgenic insecticidal cultivars obtained through biotechnology approach have also been developed, such as Bt-corn, Bt-cotton, and Bt-rice.

Farmers all over the world are using many insect-resistant plants. Rice resistance to the rice brown planthopper and green rice leafhopper has been planted over 20 million hectares in Asia. Wheat resistant to Hessian fly are grown in the mid-western United States. In 1999, Bt-plants resistance to insect was planted over 8.9 million hectares, mostly in United Stated. The wide spreaduse of insect-resistant plants may be because these plants are the most economic, least complicated and environmentally soundest approach for protecting plants from insect damage.

New insect biotypes, such as brown planthopper biotypes and Hessian fly biotypes, have been developed. There has been development of new insect biotypes overcoming transgenic insecticidal plants derived from biotechnology approach which has not been recorded so far. Two strategies to delay the development of new insect bio-types have been reported. The first strategy relates to insect-plant resistant development and the second, to their application in the field in which insect-resistant plants are one of the components of integrated pest management approach.

6.3.1. Method to Develop Insect-Resistant Plants

In breeding programme for insectresistance plants, are first identified the parents or donors of resistance. These may be cultivated germplasm, landraces, weeds races, wild species, or different species. The breeding method is based on the following factors: the source of donor of resistance, the efficiency and certainty of selection of progenies. If the donor of resistance is of commercial varieties, they may be improved by pure line or mass selection or hybridised with elite germplasm. If they are land-races, weed races, or wild species, they have to be hybridised with the elite germplasm. In the case of the donor of resistance being of different species, biotechnology method may used, such as genetic transformation or protoplast fusion. If the donors are not available, biotechnology method may be used, such as somaclonal variation for insect-resistance. If the efficiency and certainty in the selection of desired traits are needed, the use of biotechnology method, such as marker assisted selection and anther culture, may be appropriate.

The method to develop insect-resistant plant can be grouped into conventional and biotechnology. The conventional approaches use the whole living organism and includes pure line selection, mass selection, hybridisation and while biotechnology approaches use cells or biomolecules to develop insect-resistant plants and includes protoplast fusion, somaclonal variation, molecular assisted selection, and genetic transformation. For anther culture and embryo culture, earlier they used organism of F1 plants to produce pollen grain and wild and cultivated species to produce embryo, but now they use cells culture, like the anther culture and embryo culture as a biotechnology approach.

6.3.1.1. Anther Culture

Using an artificial medium, anther or pollen grain can be cultured *in vitro*. On the artificial medium anther may form callus, shoot, root, and finally the entire plants. All the plants are haploid. This approach is helpful in speeding up the formation of homozygous population of insect-resistant plants.

6.3.1.2. Embryo Culture

Wide hybridisation involves transfer of genes conferring insect-resistant from wild species, that are often more resistant to insects to cultivated plants, but this will produce abnormal interspecific hybrid embryos. The embryo can be saved by culturing it on the nutrient medium to generate the entire plants. Several genes for insect resistance have been transferred from wild to cultivated germplasm (Table 6.1).

Table 6.1. Genes conferring insect resistance transferred from wild to cultivated species

Recipient	*Alien donor*	*Trait*
Bread wheat	*Secale cereale*	Resistance to greenbug
	S. cereale	Resistance to Hessian fly
	Aegilops squarrosa	Resistance to Hessian fly
	Rice Oryza officinalis	Resistance to BPH
	O. officinalis	Resistance to WBPH
	O. australiensis	Resistance to BPH
	O. minuta	Resistance to BPH
Peanut	*Arachis monticola*	Resistance to chewing insect
Lettuce	*Lactusa virusa*	Resistance to Aphids
Cotton	*Gossypium armourianum*	Boll weevil, leafworm, bollworm

6.3.1.3. Protoplast Fusion

Protoplasts are plant cells that could be isolated by digested the wall enzimatically. The traits of resistance to the pest may be present in the one of two species that cannot be hybridised sexually. The two species may be formed from hybrids through protoplasts fusion. The protoplasts may be cultured on artificial medium and some protoplasts will grow into entire plants which will carry the resistant traits.

6.3.1.4. Somaclonal Variation

Selecting insect-resistant variants can be done by somaclonal variation. Insect-resistant somaclones can be selected through the following steps (a) calli or cell suspension derived from high-yielding variety grown for several or long-term cycles, (b) the long-term cell lines regenerated into plants, and (c) the regenerated plants evaluated against target insects. About 2,000 plantlets of sugarcane were assessed for resistance to the sugarcane borer under artificial infestation and natural infestation in field plots. Some somaclones showed resistance to sugarcane borer.

6.3.1.5. Marker-assisted Selection

Nucleic acid probe, antibodies, or enzymes may be used as a marker-assisted selection in breeding programme to determine the genetic constitutions of plants or plant extracts, including the presence of resistant traits. The precise information from this method will increase the speed and certainty of selection progeny carrying resistance traits in conventional plant breeding.

6.3.1.6. Genetic Transformation

Genetic transformation integrates cloned DNA segments or genes from plants, bacteria, or animals into a new genetic background. These DNA segments or genes may confer resistance to insect. The genes may be delivered to the protoplast or cell by using one of the following method: Agrobacterium, biolistic, electroporation, PEG, vortex, and microinjection. Several resistant plants have been developed through genetic transformation (Table 6.2).

Table 6.2. Resistant plants develop through genetic transformation

Plants	*Traits*
Alfalfa	Insect resistance
Cabbage	Insect resistance
Cotton	Resistance to bollworms and budworm
Eggplant	Insect resistance
Maize/Corn	Resistance to corn borer
Potato	Resistance to Colorado potato beetle Resistance to Potato tuber moths

Soybean	Insect resistance
Sugar cane	Insect resistance
Sweet potato	Insect resistance
Tobacco	Insect resistance
Tomato	Insect resistance
Rice	Insect resistance (Bt toxin)

6.3.2. Strategies to Develop Insect-resistant Plants

To anticipate the breakdown of insect-resistant plants due to the development of new bio-types, various strategies should be adopted in developing those plants. One strategy is to prolong the useful life of insect-resistant plants by developing durable insect-resistant plants, governed by polygenes. Durable resistance is also referred to poligenic resistance or horizontal resistance. The other strategy is to develop varieties with different genes conferring insect-resistant which may be achieved by developing vertical resistant plants.

The vertical-resistant plant is governed by major genes. The more vertical resistant varieties with different genes is available, the easier it is for farmers to access new varieties to replace the previous varieties.

6.3.2.1. Durable Insect-resistant Plants

The level of resistance of durable insects-resistant plants is generally not very high the development of new biotypes is very slow. Polygenes resistance could be obtained from the landraces. Once crossing has been made between cultivated and landraces, there is the difficulty in selecting the desired segregants as not all the polygenes or QTL from landraces are transferred to cultivated, and the level of resistance is diluted. The drawback is that the screening techniques currently available is only applicable to identify segregants with high level of resistance, not segregants with low level of resistance.

In molecular-assisted selection, facilitating the development of durable insect-resistant plants, the first step should be to tag the QTL for insect resistance with molecular markers. Molecular marker-based selection will assist in the accumulation of polygenes. QTLs for resistance to rice brown planthopper have been tagged with RFLP markers.

6.3.2.2. Vertical Insect-resistant Plants

Vertical insect-resistant plants are more easier to develop since major genes are more easier to transfer from one variety to others. Numerous vertical insect-resistant plants with different genes should be developed through conventional or biotechnology in order to expect the new insect biotypes that overcome resistant plants. Numerous rice varieties with different genes resistance to BPH have been produced by IRRI.

Insect-resistant transgenic plants brought about by genetic transformation currently available is generally vertical resistance with the major genes, such as Bt-gene. Anticipating the development of new biotype overcoming Bt-plants, others resistant plants with different genes should be developed.

Pyramiding of major genes is a possible strategy to develop the longer useful life of resistance plants. This approach needs a reliable method to determine the presence of the two genes. If bioassay method will be adopted, several insect biotypes should be available for this purpose. The use of molecular markers is a very reliable method to diagnose the integration of the two genes in the plants. Pyramiding of major genes may be achieved through genetic transformation, for example, pyramiding the two Bt genes or combining genes encoding a toxin and a gene encoding a repellent.

6.3.2.3. Tissue-specific Expression

A great selection pressure that leads to the development of new insect biotypes cause the constitutive expression of resistance genes at all times and in the whole tissue. This problem can be dealt with by expressing the resistance traits in a specific tissue, or at certain growth stages, or only in response to insect feeding.

6.3. Disease Resistance

With transfer of genes from one organism to another becoming common place, it is possible to introduce genes conferring disease resistance into crop plants. Such gene transfers can be done by direct methods and or vector-mediated methods. Gene gun or biolistic method and *Agrobacterium*-mediated method are the best examples of the direct method and vector-mediated method, respectively.

6.3.1. Transgenic Plant Disease Management

Just as it was possible with coat protein-mediated plant viral resistance and with toxin-inactivating protein-mediated bacterial resistance disease resistance genes can be sourced from plant pathogens themselves. Host plants also contribute an enormous number of disease resistance genes, such as those encoding pathogenesis-related (PR) proteins, which have been used against fungal diseases.

6.3.1.1. Candidate Genes against Viral Pathogens

The best example of the use of transgenic resistance against plant diseases is one that was accomplished in the management of papaya ring spot virus (PRSV) in Hawaii. Traditional breeding in bringing about resistance against this disease posed crossability barriers. Hence coat-protein-mediated resistance using coat protein gene sourced from a Hawaiian strain of PRSV was attempted. One transgenic line was found to be completely resistant to PRSV.

Transgenic resistance against banana bunchytop resistance using a BBTV replicase gene is under study. Once it becomes routine, it will be easier to deliver human vaccines (cholera toxin vaccine and hepatitis B surface antigen) via transgenic banana fruits which form an excellent delivery material.

Recently, a gene silencing mechanism has been productive in containing rice yellow mottle virus. Similar attempts also have been made in containing multiple viral infections (tomato spotted wilt virus and turnip mosaic virus) in plants.

6.3.1.2. Candidate Genes against Bacterial Pathogens

Xa21, a wide-spectrum bacterial blight resistance gene sourced from an African rice, *Oryza longistaminata,* was backcrossed into a cultivated variety by scientists of the International Rice Research Institute, the Philippines (IRRI). The resistance gene was cloned using molecular means distributed to labs worldwide so that the gene could be put into rice cultivars of local importance.

Wildfire disease of tobacco caused by *Psuedomonas syringae* pv. *tabaci* is a serious disease wherein, a phytotoxin secreted by the pathogen, drastically modifies the amino acid metabolism of the plant with the eventual accumulation of ammonia in tobacco leaves, that causes extensive blighting. Interestingly, the pathogen that synthesises the phytotoxin remains unaffected by the toxin. Researchers carried out for a search of the candidate gene from the pathogen itself led to toxin-inactivating gene, which was named '*ttr*'. This was successfully isolated from the pathogen and the same was cloned into tobacco cultivars, which showed excellent wildfire resistance.

6.3.1.3. Candidate Genes for Fungal Resistance

PR protein genes that are a possible potential source for candidate genes for fungal resistance may play a direct role in defence by attacking and degrading the pathogen cell wall components. Typical candidate genes are that encoding chitinases and ß-1,3 glucanases. Increasing expression of individual and multiple PR-proteins in various crops have shown some success in enhancing disease resistance in particular pathogens. A chitinase gene from an anti-fungal biocontrol fungus species (*Trichoderma viride*) confers transgenic resistance against the rice sheath blight pathogen. A rice PR-5 protein gene in wheat delays the onset of symptoms caused by the wheat scab pathogen.

6.3.2. Biocontrol of Plant Pathogens

The worldwide environmental concern over pesticide use has initiated a large upsurge of biological disease control. Among the various antagonists used for the management of plant diseases, *Trichoderma* and *Pseudomonas* are most useful.

6.3.2.1. Trichoderma

Among the various isolates of *Trichoderma, T. viride, T. harzianum, T.virens* and *T. hamatum* are used to control various diseases of crop plants especially with the dreaded soil-borne pathogens. As a bio-control agent it has high rhizosphere competence, ability to synthesize polysaccharide–degrading enzymes, amenability for mass multiplication, broad spectrum of action against various pathogens and, above all, its environmental friendliness.

6.3.2.2. Fluorescent Pseudomonads in Induced Systemic Resistance (ISR) against Plant Diseases

Fluorescent pseudomonads inhibit the pathogens either directly through the production of various secondary metabolites or indirectly by inducing plant-mediated defence reactions. They have the ability to colonise the rhizosphere and the persistence throughout the growing season. They are root colonisers as they occur in the natural habitat of rhizosphere and when they are reintroduced to roots through seed or seed-piece inoculation, they colonise root surface profusely. They suppress the pathogens by antibiosis through the production of various antibiotic substances such as 2,4-diacetyl phloroglucinol, phenazine-1-carboxylic acid, oomycin A, oxychlororaphine, pyoluteorin, pyrrolnitrin and pyocyanine.

Siderophores that are extracellular, low molecular weight substances which selectively complex iron with high affinity are produced by fluorescent pseudomonads. These siderophores, such as pseudobactin and pyoverdine, chelate the iron available in the soil and make it unavailable to pathogen which due to deficiency of iron. In rice, seed treatment followed by root dipping and foliar spray with *Pseudomonas fluorescens* showed a higher induction of ISR against sheath blight pathogen, *R. solani*

6.3.3. Molecular Diagnostics of Plant Diseases

The foremost step in managing a plant disease is to correctly identify it. Some diseases can be diagnosed quickly by visual examination, and others require laboratory testing for diagnosis which may take days or even weeks to complete and are, in some cases, relatively insensitive. To prevent delays that can be frustrating appropriate disease control measures may be taken to prevent plant injury, especially when high value cash crops are involved. Fortunately, with advances in biotechnology, new products and techniques are becoming available that will complement or replace time-consuming laboratory procedures.

6.3.3.1. *ELISA Diagnostic Kits*

A number of disease detection kits, developed for use at the site where a disease is suspected, not requiring laboratory equipment, are especially useful to growers. Some tests only take five minutes to perform. The diagnostic kits uses a technique by which proteins called antibodies are used to detect disease causing organisms of plants (plant pathogens). The technique, called ELISA (enzyme-linked immunosorbent assay), is based on the ability of an antibody to recognise and bind to a specific antigen, a substance associated with a plant pathogen. The antibodies used in the diagnostic kits are highly purified proteins produced by injecting a warm-blooded animal (like a rabbit) with an antigen associated with one particular plant disease. The animal reacts to the antigen and produces antibodies that recognise and react only with the proteins associated with the causal agent of that plant disease. Colour changes on the unit's surface indicate a positive reaction. ELISA kits can detect bacterial canker of tomato and soyabean root rot.

6.3.3.2. *PCR*

A new technology, PCR (polymerase chain reaction), has great possibilities for enhancing the sensitivity of various assays that use nucleic acid probes. This technology is used to produce enormous numbers of copies of a specified nucleic acid sequence allows the detection of very small amounts of a pathogen in a sample by amplifying the pathogen sequences to a detectable level. PCR is especially useful in plant quarantine point of view owing to its fastness.

6.4. Virus Resistance

Many reports about the development of transgenic plants have become resistant to virus infections when they are transformed with sequences related to several gene functions.

Successes have been seen in areas related to sequences concerning viral capsid protein, viral movement protein, antisense RNA, antibody-mediated resistance, interferon-related genes and host genes involved in plant protection.

Biochemical complexes comprising an RNA or a DNA genome encased in a protein capsid which may or may not be surrounded by a membrane envelope are called viruses. The protein-coat-'covered genome is known as the nucleocapsid. The proteins on the surface of the capsid and envelope determine the interaction of the virus with the host and draw out the protective immune response against the virus. Some virus particles also contain enzymes required to facilitate the replication of the virus.

The tobacco mosaic virus (TMV) is an example of a virus with helical symmetry. Its capsomeres appear as projections that are assembled on the RNA genome into rods extending to the length of the genome. In contrast the capsomere arrange-ment in other viruses is cubical or icosahedral enclosing its nucleic acid component. Engineering resistance in plants involves either attacking the capsid properties or disrupting the virus replicating mechanisms.

References

British Medical Association. *The Impact of Genetic Modification on Agriculture, Food and Health.* BMJ Books. 1999.

Donnellan, Craig. *Genetic Modification (Issues).* Independence Educational Publishers. 2004.

Friedberg, E.C. *DNA Repair.* New York: WH Freeman and Company. 1985.

Morgan, Sally. *Superfoods: Genetic Modification of Foods (Science at the Edge).* Heinemann. 2003.

Morholt, E. and Brandwein, P.F. *Sourcebook for the Biological Sciences* 3rd Edition. San Diego: Harcourt Brace JOvanovich. 1986.

Smiley, Sophie. *Genetic Modification: Study Guide (Exploring the Issues).* Independence Educational Publishers. 2005.

Zaid, A; H.G. Hughes, E. Porceddu, F. Nicholas. *Glossary of Biotechnology for Food and Agriculture - A Revised and Augmented Edition of the Glossary of Biotechnology and Genetic Engineering. Available in English, French, Spanish, Arabic.* Rome, Italy: FAO.2001.

7

PLANT GENETIC ENGINEERING AND ABIOTIC STRESS RESISTANCE

Only a tenth of the world's 13 billion hectare is farmed other than urban areas much of the remaining 11.5 billion ha are lands are not suitable for any sort of agriculture. Moreover, almost all the land that is framable is under sub-optimal conditions, often to a considerable degree, for plant growth. Apart from losses due to pests and diseases, a further 70 per cent of yield potential is calculated to be lost due to unfavourable physiochemical environments, even in developed agricultures.

No doubt that in order to feed the eight billion mouths we expect by 2030, the world food production has to be doubled yet again. To do so, breeding of new varieties of foodcrops is necessary to improve yields on land presently being farmed on sub-optimal soils and extend productive agriculture into lands which are currently barren.

7.1. Abiotic Stress

7.1.1. Drought

Unpredictable drought is the worrisome catalyst affecting world food security and causing famines. Moreover, the world's water supply being fixed, increasing population pressures make successive droughts more severe added to which competition from industry will increasingly limit the water available for agriculture. Crops need plenty of water for example, for paddy rice 5000 litres of water is needed to produce 1 kilogram of grain. At present an unsustainable 70 per cent of the world's water is used for agriculture. By 2025 it is expected that most Asian countries too will face water shortages. Uncertainties over global warming raise yet further concerns.

7.1.2. Salt

Some 380 million ha, almost a third of the area farmed, saline with water logging and alkalinity. Over irrigation has raised the water table bringing underground salt, particularly NaCl, to the surface sixty million ha. Probably this agricultural salinisation now degrades as much land as is put under new irrigation each year. Pressures on water use will result in irrigated land becoming unproductive very soon and that secondary salinisation will become critical in Asia, Africa and South America.

Salt stress for rice is of particular significance for in Bangladesh, Orissa, Vietnam, Philippines, and inland salinity in the Indogangetic plain and Thailand. ICARDA secondary salinisation is a problem for all crops in Central Asia.

7.1.3. Acid and Degraded Soils

Some 40 per cent of the world's arable land, associated with acid soils, with pH less than 5, has high aluminium or manganese content that retards growth of crops. This is particularly important in South America where some 380 million ha are affected, including almost the whole of the Amazon basin. Other excess metal ion contents reduce the agricultural potential of other soils. For example, iron toxicity is a major problem affecting rice production in West Africa.

7.1.4. Low and High Temperatures

Temperature is also a limiting for the growth of crops, even at tropical latitudes. Occasional and unpredictable periods of low temperature can be harmful to yields. For example, in the Andes 70 per cent of land devoted to potato production is prone to cold stress.

In fact, abiotic stress tolerance, particularly drought, is the priority target trait for researchers dealing with crop plants. In the present economic and agricultural climate, with developed countries enjoying food surpluses, the private sector will continue to focus on protection from disease and improvements in aspects of quality. Even though abiotic stress is key factor for production in developed countries, it is doubtful if genetic solutions will be actively sought by commercial breeding companies. To overcome the problem at all, abiotic stress tolerance mechanisms and their genetic application in the crops of the developing world will need attention by the public sector working in the developing world.

7.1.5. Gene Mapping and Marker Development

Standard plant breeding genetic mapping as a prerequisite to genetic analysis. Rice, wheat, potatoes, etc., can use the well-developed public maps. Base maps for many of the 'orphan crops', which involve little international trade, are in use at centres of collaborations. Only a few very minor mandated species remain unmapped.

In the mapping of quantitative traits there is often little knowledge of the genetic control in advance of the analysis, as is usually the case with stress tolerance. It is usually carried out by 'QTL mapping' which requires a scan of the genome, with markers to identify those regions where segregation of the trait is concerned with segregation for the markers. Since only a subset of the available markers will segregate in any single population denser maps are needed. These locations establish a marker aided selection (MAS) breeding programme for tolerance and for eventual map-based cloning of the genes underlying the QTLs.

Breeders' markers, closely associated with the target gene, may be extracted straight from the base molecular map. Today the ideal marker system is micro satellites, also known as simple sequence repeats (SSRs). In the next few years single nucleotide polymorphisms (SNPs) that are more conducive to high throughput methods, will prevail as the ideal marker.

Rice, which will soon have the benefit of a full genome sequence will have never problems with markers. The sequence contains some 40,000 SSRs and SNPs, and base pair deletions or insertions indels are found in unique sequence at a rate of about 1 per cent, which works out at about 24 in every gene! However, apart from in the major cereals, good quality markers for all applications is still a supply problem for most mandate crops.

7.1.6. Stress Tolerance Control Mechanisms

The focus today is on physiological mechanisms underlying crop responses to stress and potential biochemical, physiological and architectural modifications that will allow crops to escape, avoid or tolerate stress. Two general approaches in relation to varietal improvement are 'empirical' and 'ideotype' breeding. The 'empirical' approach is from genotypic differences associated with the best sources of tolerance in the cultivated crop or its wild relatives. In this procedure sources of tolerance are identified and the underlying genetic control is investigated by QTL analysis in lines separated for high and low tolerance. The identification of a predominant causal physiological mechanism is beneficial, transfer of the improved trait to an already otherwise adapted variety can proceed simply for selecting for and accumulating 'beneficial' alleles. The second approach often described as 'ideotype' breeding, deals with specific morphologies or physiologies that might be expected to contribute to improved performance under stress. They are identified in diverse cultivated or wild germplasm and transferred to otherwise adapted varieties. The crossbreeding and, nowadays marker-aided pyramiding of the underlying alleles, progress in the same way in both approaches.

There is already considerable research being carried out at various centres to improve their mandate crops for stress tolerance. For example, drought, salt stress and cold temperature stress, all physiologically linked, result in limiting the crops' physiological access to water.

Thus many of the strategies for improved tolerance include osmotic adjustment in roots and leaves to retain water, erection of hydrophobic barriers in roots and leaves to retain water, and improvement of aquaporin efficiency to speed water movement in the plant. Although tolerance mechanisms might be expected to overlap, escape or avoidance mechanisms are more likely to be stress specific. For example, by reducing time to flowering late season drought may be avoided it will not help in a chronic saline situation. Deeper roots may be able to reach the last of the water in a drought but would only aggravate salt stress where the salt is being brought to the surface by a rising water table.

The situation prevailing now is that one would intuitively expect genetic control to be multigenic and complex, but to overlap somewhat in tolerance to the different stresses.

In aluminium tolerance, the selection screen is usually for improved growth of roots and shoots in Al supplemented nutrient solution at low pH, and the empirical approach to it reveals complex control. However, one locus dominates the network of genes which accounts for a major proportion of the genetic variation.

These results identify MAS breeding priorities and necessitate isolation of the key gene, either by cloning or the production of isogenic lines. Gene isolation and knowledge of the gene sequence are vital steps in a project to elucidate an understanding of the underlying mechanism.

Genetic analysis reveal that a few stress tolerances do usually appear to be under the control of major genes, submergence tolerance a being prime example. In SE Asia some 25 million ha, prone to flash flooding, can completely submerge the rice crop for several days. In this case a single locus, *Sub1*, is shown to provide substantial tolerance.

Wild species, where survival rather than yield is an inherent factor, are likely to retain useful variation that may have been bred out of the cultivated crop. There are many instances genes for tolerance being identified in wild relatives and being used to transfer useful variation to cultivated crops. In some cases the wild species themselves have been used directly to create new crop species.

Crop substitute is a possible alternative to varietal improvement which is probably a viable socio-economic strategy only in extreme stress environments. These differences are exemplified by investigations into the responses of tropical grasses to aluminium stress. In general, the genetic dissection of stress tolerance for developing countries is attracting considerable focus, especially in crops that already have advanced genetic maps.

Marker-aided pyramiding of several genes is probably the only possible course for the transfer of improved phenotypes that have been shown clearly to be under control of multiple loci. With these methods yet to impact stress tolerance breeding programmes, a few endogenous genes have been identified which are likely to have major beneficial effects when used as a transgene.

7.1.7. Genetic Engineering for Abiotic Stress Tolerance

Abiotic stresses refer to a number of abnormal environment parameters, such as drought, salinity, cold, freezing, high temperature, anoxia, high light intensity and nutrient imbalances etc. These stresses lead to dehydration or osmotic stress through scarcity of water for vital cellular functions and maintenance of turgour pressure. Stomata closure, reduced supply of CO_2 and slower rate of biochemical reactions during prolonged periods of dehydration, high light intensity, high and low temperatures lead to high production of Reactive Oxygen Intermediates (ROI) in the chloroplasts causing irreversible cellular damage and photo inhibition.

As a result of dehydration or osmotic stress, a series of compatible osmolytes get collected for osmotic adjustment, water retention and free radical scavenging. Likewise overexpression of certain enzymes, such as superoxide dismutase, ascorbate peroxidase and glutathione reductase, is involved in free radical detoxification and scavenging of free radicals under oxidative stress.

Table 7.1. Genetic engineering of plants for tolerance to abiotic stresses

Stress	*Gene/Enzyme*	*Source*
Osmotic	Delta-pyrroling-5-carboxylate synthetase (P5CS)	Mothbean (*V. aconitifolia*
Drought & Salinity	Mannitol-1-phosphate dehydrogenase (*mt1D*)	*E. coli*
Cold and Salt	Choline oxidase (*cod* A)	*Arthrobactor globiformis*
Salt	Choline dehydrogenase (*bet* A)	*E. coli*
Cold	Omega-3-fatty acid desaturase (*fad* 7)	*Arabidopsis*
Drought	Trehalose-6-phosphate synthase	Yeast
Drought	Levan sucrase (Sac B)	*Bacillus subtilis*

7.1.7.1. Complexed stresses by osmoticum, dehydration and salinity

A potent and compatible osmoprotectant is proline which accumulates in high concentrations in glycophytes and halophytes in response to osmotic stress, such as drought and high salinity. Two important enzymes for the biosynthesis of proline i.e. Δ^1 - pyrroline-5-carboxylate synthetase (P5CS) and Δ^1 - pyrroline-5-carboxylate reductase (P5CR), have been cloned from several plants and their expression studied under various abiotic stresses and ABA application.

Certain transgenic plants exhibiting a high level of P5CS mRNA also accumulate high level of P5CS protein, producing 10 to 18-fold more proline than the control plants. Under drought stress the proline content increased from about 80 mg/g fresh leaf (before stress) to about 3000 mg/g (after stress) in control and from 1000 mg/g to an average of 6500 mg/g in transgenic lines. Compared with the wild type (WT) control plants wilting in the transgenic plants was less severe and delayed by 2-3 days transgenic plants as compared. This shows that proline acts as an osmotic protectant and its increases production in the transgenic plants increased tolerance to drought and salt stress.

The compatible solute, glycinebetaine, a quaternary amine is widely distributed among plants and protects them on exposure to salt and cold stress. In plants, like spinach and barley, betaine is synthesized from choline by catalysed by choline mono-oxygenase oxidation of choline to betaine aldehyde. In the second step it is synthesized by nuclear coded gene for betaine aldehyde dehydrogenase. The transformed plants grew slowly at 200 mM NaCl whereas none of the WT plants grew. The collection of glycine betaine through genetic engineering in Arabidopsis enhanced its ability to tolerate salt and cold stress.

7.1.7.2. Anaerobiosis/Anoxia

Most plants are highly sensitive to anoxia during submergence. Avoidance of self are all metabolic changes that are an important aspect of the adaptation to oxygen limitation poisoning and cytoplasmic acidosis, maintenance of adequate supplies of energy and sugar. During anoxia, ATP and NAD^+ are generated via glycolysis and fermentation and not in the Krebs cycle and the respiratory chain via. A number of enzymes of the anaerobic pathways such as alcohol dehydrogenase and pyruvate decarboxylase induced during anoxia have been cloned and characterised.

7.1.7.3. Heavy Metal

Since soils have elevated levels of one or more inorganic ions such as sodium in saline soils; Al, and Mn in acidic soils and heavy metals Cu, Zn Pb, Ni, Cd etc. due to mining, industrial effluents and other human activities, optimum growth and productivity and even cultivation of most of the plants is severely restricted.

Table 7.2. Genetic engineering of plants for tolerance to heavy metal

Stress	*Gene/Enzyme*	*Source*
Cadmium	Metallothionein-I (MT-I)	Mouse
Copper	Metallothionein-like (PsMTA)	Pea
Aluminium	Citrate synthase (CSb)	P. *aeruginosa*

Some plants are genetically resistant to Al toxicity. In such cases the Al exclusion and uptake from root tips are found to be correlated to their increased capacity to release organic acids such as citric acid which chelates Al_3^+ outside the plasma membrane. Transgenic tobacco and papaya were found to overexpress a citrate synthase gene (CSb) from *Pseudomonas aeruginosa* in their cytoplasm. Tobacco lines expressing CSb had up to 10-fold higher level of citrate in their root tissues and one of the lines released 4-fold citrate extracellularly while in papaya there was only 2 to 3-fold increase of citric acid production which was shown to result in Al tolerance in both the species.

7.1.7.4. Heat and Cold

Temperate and subtropical plants are exposed to high temperature during early tillering, flower initiation, anthesis and grain filling stages leading to which results in substantial reduction in their productivity. In response to high temperature all organisms, including plants, blend a set of proteins called heat shock proteins (HSPs) which have been classified into several families based on their molecular masses. The induction of HSPs at permissive temperatures are linked with the acquisition of thermotolerance to tolerate short periods of an otherwise lethal temperature.

The chilling sensitivity of plants is closely associated with the degree of unsaturation of fatty acids in the phosphatidylglycerol of chloroplast membranes. Plants such as spinach and Arabidopsis having a high proportion of cis-unsaturated fatty acids, are resistant to chilling, whereas species like squash with only a small proportion are not. The chloroplast enzyme glycerol-3-phosphate acyltransferase is vital for assessing the level of phosphatidylglycerol fatty acid unsaturation. This shows that the level of fatty acid unsaturation of phosphatidylglycerol and the degree of chilling sensitivity of tobacco can be controlled by transformation with cDNAs for glycerol-3-phosphate acyltransferases from squash and Arabidopsis.

7.1.7.5. Shading

While optimum supply of nutrients and efficient photosynthesis contribute to biomass production, the allocation of assimilates within the developing plant determines the harvest index and economic yield. In pure stand canopy and mixed cropping, competition for light energy enforces shade avoidance syndrome embodied by rapid growth and extension of stem and petiole at the expense of leaves, storage and reproductive organs. This leads to predisposing plants to lodging, susceptibility to diseases and insect pests and a lower harvest index.

With the advent of semi-dwarf varieties of wheat and rice in the 1960s there was higher harvest index and grain yield by countering some of the defects of the tall genotypes. However, competition among plants for light energy continues to operate in canopies under intensive cultivation practices. The photosynthetic pigments in plants absorb the visible radiation (400-700 nm) and reflect and transmit far red (FR) radiation beyond 700 nm. The FR wave band between 700 and 800 nm salient in the dense plant stands have been involved in proximity perception for initiating shade avoidance syndrome.

Photoreceptors called phytochromes, possessing distinct photo sensory functions perceive the FR reflection signals. Phytochrome (phyA) mediating the inhibition of stem growth on etiolated plants in response to FR wave length 710-720 nm is rapidly degraded and down regulated in light grown plants. Transgenic tobacco lines expressing a high level of heterologous oat phyA apoprotein have been produced.

There is correlation between the level of growth inhibition of transgenic plants and the level of phyA production. Under field trials at various planting densities, from 20 to 100 cm, the transgenic plants were identical to the WT plants at the lowest plant density but became progressively shorter as the plant density increased. This phenomenon, termed as "proximity conditional dwarfing", led to a 15 to 20 per cent increase in harvest index in transgenic plants under high plant density

This shows the suppression of shade avoidance response under high level of phyA expression. Further studies on the molecular basis of interaction of various phytochromes among themselves and with R: FR rations in natural light environment may help to change crop plant structure to avoid shade stress and obtain maximum production under high plant density, mixed cropping and agroforestry.

7.1.7.6. UV-B

The high influxes and absorption of UV-B radiation affects terrestrial plants, damaging DNA directly or indirectly through formation of free radicals, membranes by peroxidation of unsaturated fatty acids, photosystemII, phytohormones and even symbiotic relationship of plants with micro-organisms. Many secondary metabolites such as flavonoids, tannins and lignins are increased at elevated levels of UV-B radiation which screen UV-B and protect the cellular components against the UV-B damage.

7.1.7.7. Oxidative Stress

Overproduction of reactive oxygen intermediates (ROI) including H_2O_2 results in a number of abiotic stresses such as extreme temperatures, high light intensity, osmotic stresses, heavy metals and a number of herbicides and toxins causing extensive cellular damage and inhibition of photosynthesis.

7.1.7.8. Strategies for Improving Tolerance

Within a decade of the molecular understanding of pathways

induced in response to one or more of the abiotic stresses the work on genetic engineering of tolerance to abiotic stresses began. In a majority of the cases the transgenes expressed faithfully but compared to the non-transformed wild type plants only a limited level of tolerance was provided under stress conditions. In many cases the transgenic plants had morphological abnormalities and slower growth under nonstressed environment. The very low level of many compatible osmolytes responsible for osmotic adjustment was not effective in providing the required water retention and osmotic adjustment.

High levels of tolerance for commercial exploitation can be achieved by the use of multiple tolerance mechanisms for one or more of the abiotic stresses through stepwise or co-transformation. The QTL mapping of stress tolerance in certain species, comparative mapping and map-based cloning in plants may be employed to screen genes which function under stress as well as those induced and expressed in response to stress.

Molecular understanding of the stress perception, signal transduction and transcriptional regulation of abiotic stress responsive genes may help in engineering tolerance for multiple stresses.

Studies on the molecular mechanism for providing protection against biotic and abiotic stresses may lead to a generalised master mechanism for stress tolerance. Optimum homeostasis is always a requisite to living organisms for adjusted environments. Thus, abiotic stress accompanying a number of biological phenomena must be precisely investigated by consideration of plant homeostasis.

7.2. Molecular Alphabets of Plant Abiotic Stress Responses

All plants have a latent ability to adjust to circadian and seasonal environmental variables which are often decisive factors in controlling certain physiological attributes. Apart from those variables, there may be certain other rapid and unpredicted disturbances in the environment resulting in stressful conditions.

For instance, water shortage for long periods due to lack of irrigation, infrequent rains or lowering of water table causes drought stress; excess water through rain, cyclones or frequent irrigation results in flooding or submergence or anaerobic stress; cultivating plants on saline soils or frequent irrigation with ground water leads to salinity stress; and sudden atmospheric heating or cooling due to transient changes in wind patterns, cloud formation or excessive sunlight causes temperature stress. Most crop plants, not selected for meeting exigencies caused by such abiotic stress factors, are unable to adjust to such conditions.

Abiotic stress affects survival, biomass production and accumulation, and grain yield of most crops. Different crop ecosystems are affected by different abiotic stress factors, and to a differential extent, with varying degrees of susceptibility. There is also some level of variation associated with specific developmental stages of the plant. Today, how plants face stress in terms of molecular alterations is the key issue in plant stress molecular biology studies.

The best studied environmental factor in plant research with respect to molecular details is both the quality and quantity of light affecting photosynthesis and growth of plants. Light is perceived through several different photoreceptors. A battery of molecular components is involved in transduction of light signal to the nuclei where it regulates transcription of selective genes, in both positive as well as negative manner. Also, light-responsive elements (LREs) present in the promoters of the lightregulated genes and the transcription factors associated with light-regulated promoters interact to produce regulated gene expression. However, as against light, there is scarcity of information on how changes in temperature, water and salt levels are perceived and translated into cellular events.

The period of 1980s onwards has been the 'Phase of Recombinant DNA Technology' with Genetic engineering as the most revolutionary tool to impact agricultural research in recent years. In the last few decades techniques of protein analysis, identification, isolation, cloning and characterisation of genes, promoter analysis, genetic transformation and new research on genomics and proteomics have contributed significantly in to plant molecular biology. This progress has led to production of a range of transgenics for varied traits. The study of genetically engineered abiotic stress-tolerant crops for high level stress tolerance has gained a great deal of attention too. There has been a lot of information on stress proteins that are specifically induced in response to abiotic stresses. Gene libraries enriched for stress-related cDNA clones have been constructed for several plant species. Availability of the genomic clones of stress genes has helped to identify the stress-related promoter sequences. Due to a higher level of sophistication achieved in the isolation of cDNA clones that are present in minute amounts in the gene libraries, identification of genes encoding transcription factors and for proteins which mediate stress signalling has become possible in selected instances.

Both plant molecular biologists and plant biotechnologists feel the need to further unravel the fundamentals of the plant stress responses.

7.2.1. Macro- and Micro-level Stress Effects

It is generally believed that the response of plants to stress agents is particularly of adaptive nature when the stress is sublethal. In contrast, response shown may be biased towards senescence or cell death if the stress is lethal. Application of sublethal stress regimes in experimentation has become an effective approach for unveiling the fundamentals of stress responses.

Using tools of physiological, biochemical and molecular relevance, there are intensive efforts to unveil both the constitutive as well as inducible mechanisms connected with the survival of plant cells under stress conditions. While some of the stress effects are common amongst different abiotic stresses, others are unique to a particular stress type.

At the whole plant level, common effects caused by the stress factors mentioned above include reduced seed germination and seedling establishment, poor seedling vigour, decrease in the root length, leaf rolling, reduced pollen viability, leaf senescence, incomplete grain filling and reduction in grain yield. At the cellular and sub-cellular level, stress-induced unique changes include increased unsaturation of the membrane lipids in response to low temperature stress, increased levels of different osmolytes in response to osmotic factors (such as dehydration, salinity and low temperature stresses), general repression of protein biosynthesis in response to water and high temperature stresses, selective changes in K^+/Na^+ levels in response to salt stress and finally, and upregulation of glycolytic and enzymes required for alcoholic fermentation by anaerobic stress.

These cellular changes have been dealt with mostly specific probes, with a parallel development in stress studies and is the application of shot-gun approach which aims at analysing what happens in the system under sublethal stress regimes. This approach relies on more sophisticated molecular tools and techniques, such as examining stress-related changes in protein profiles and differential screening of gene libraries. This development has led to an unprecedented set of discoveries of stress proteins and stress genes which are described below.

7.2.2. Stress Proteins and Stress Genes

One minute of increased temperature bringing about an altered puffing pattern of the polytene chromosome in *Drosophila* has led to more research on the molecular basis of the stress responses. This showed that the heat shock (HS) conditions result in an altered protein profile in *Drosophila* cells. Soon thereafter, it was reported that HS induces comparable alterations in the protein profiles of plants too. Subsequently heat shock proteins (HSPs) have been detected and characterised in a number of plant species.

Heat shock protein (hsp) genes are the most extensively characterised of the stress genes so far, particularly in *Drosophila, Saccharomyces cerevisiae* and *E. coli.* HSPs are mostly encoded by nuclear genes, but these proteins are localised in different cell compartments, including cytoplasm, mitochondria, chloroplast and endoplasmic reticulum. Depending upon the mode of expression, broadly, hsp genes fall into the following two categories: (i) those that are constitutively expressed, often referred to as heat shock cognates (hsc), and (ii) those that are strongly induced under heat stress (hsp).

Several plant hsp genes have been cloned and sequenced since the beginning of 1980s, the most thoroughly characterised plant hsp genes being those encoding low molecular weight HSPs and HSP70. The nucleotide sequence and the structural features of the hsp are remarkably conserved. Apart from hsp genes, a large number of genes induced in response to low temperature, water, salt and anaerobic stress have been cloned and characterised in recent years.

It has been a difficult task to identify the precise physiological roles of most stress genes/proteins. The best relevant details are perhaps available for 'Anaerobic Proteins' as most of these characterise enzymes of the glycolytic and alcohol fermentation pathways. This analysis shows that respiratory pathway is affected in a major way in response to anaerobic stress. Turning to high temperature stress, a great success was achieved when biochemical and genetic tools enabled two groups to suggest that HSP70 is involved in the role of chaperoning. Thereafter, HSP60 and HSP90 have also been found important for chaperoning activities. HSP100 is shown to be critically required for resolubilising protein aggregates formed due to action of stress.

Regarding other stress responses, 'Water Stress Proteins' (WSPs) have been used in several important metabolic functions. The 'Late Embryogenesis Abundant' or LEA proteins and osmotin may protect macromolecules and membranes, and chaperones and proteases are involved in protein turnover and protein translocation. The detoxification enzymes, such as glutathione S-transferase, catalases, superoxide dismutase and ascorbate peroxidases, are useful in protection from reactive singlet oxygen species and finally proteins involved in regulatory functions and in signal transduction, including various protein kinases and transcriptional factors have a broader role in managing stress responses. With precise information on the biochemical role(s) of stress genes/proteins being made available it should be possible to replace the operational stress gene/protein terms with more specific functional nomenclature.

Though there a lack of information on the functioning of most stress proteins/genes, work employing stress proteins of known functions as well as on those with relatively unknown functions has largely contributed in planning strategies for improving stress tolerance through transgenic technology. This stresses the fact that there is a constant need for identification, isolation and characterisation of increasing number of stress proteins as well as for unravelling the functions of those stress proteins which have not been assigned any biochemical role thus far.

7.3. Abiotic Stress Tolerance

With increasing global scarcity of water resources and the

increased salinisation of soil and water, abiotic stress is already a major restrictive factor in plant growth and will soon worsen as desertification covers more and more of the world's terrestrial area. Drought and salinity, already widespread in many regions, are expected to cause serious salinisation of more than half of all arable lands by the year 2050. With population growth exceeding food supply, agricultural and plant biotechnologies aimed at overcoming severe environmental stresses need to be fully implemented.

Plant adaptation to environmental stresses is controlled by molecular networks which activate stress-responsive mechanisms to re-establish homeostasis, and to protect and repair damaged proteins and membranes. In contrast to plant resistance to biotic stresses, which is mostly dependent on monogenic traits, the genetically complex responses to abiotic stresses are multigenic, and thus more difficult to manage and engineer. Plant engineering strategies for abiotic stress tolerance depend on the expression of genes that are involved in signalling and regulatory pathways or genes.

These genes encode proteins conferring stress tolerance or enzymes present in pathways leading to the synthesis of functional and structural metabolites. Current efforts to improve plant stress tolerance by genetic transformation have led to several important achievements; however, the genetically complex mechanisms of abiotic stress tolerance makes the task extremely arduous.

Therefore, biotechnology should be fully integrated with classical physiology and breeding. Several comprehensive reviews have been published on molecular mechanisms of abiotic stress tolerance and on engineering tolerance to stress, mainly profiling a large number of stress associated genes and signal-transduction systems. Several successful approaches to achieving tolerance through the genetic engineering of specific genes have been assessed and some pitfalls also highlighted.

7.3.1. Stress-Associated Genes and Proteins

7.3.1.1. Signalling Cascades and Transcriptional Control

There have been extensive studies on genes involved in signalling cascades and in transcriptional control, such as mitogen-activated protein (MAP) and salt overly sensitive (SOS) kinases, phospholipases and transcription factors. However, most of these studies deal with short-term experiments, which cannot provide conclusions about the actual stress tolerance of crops. This depends on longer term plant performance with respect to biomass, yield data and the degree of recovery from stress. It has been recently proved that constitutive expression of the tobacco mitogen-activated protein kinase kinase kinase/ Nicotinia protein kinase 1 in maize activates an oxidative signal cascade and leads to cold, heat, and salinity tolerance in the transgenic plants. Tne transgenic maize maintained significantly higher photosynthetic rates, indicating that NPK1 induces a mechanism that protects the photosynthetic machinery. Most current genomic research is done at the DNA and RNA levels, assuming that gene up- or down-regulation will explain abiotic stress processes. However, there has been an increasing number of demonstrations of translational and post-translational regulation of stress-associated proteins in response to environmental stress. AnnAt1 is a member of a multigene family of Ca_2^+-dependent membrane-binding osmotic stress-responsive proteins. Interestingly, its gene transcript was constitutively expressed, while protein expression was upregulated in response to stress and the protein translocated from the cytosol to the membrane. The mutant plants annAt1 and annAt4 were seen to be hypersensitive to salt and abscisic acid (ABA) during seed germination and early seedling growth.

7.3.1.2. *Heat-shock Proteins and Chaperones*

Plant abiotic stress tolerance involves heat-shock proteins (Hsps) and molecular chaperones, as well as late embryogenesis abundant (LEA) protein families. High temperature, salinity and drought stress can cause denaturation and dysfunction of many proteins. We thus expect that Hsps and LEA proteins help to safeguard against stress by controlling the proper folding and conformation of both structural and functional proteins. The increase in Hsp expression under conditions of abiotic stress was researched extensively by functional genomics and proteomics in different plant species. A significant osmoprotective effect was got in *Escherichia coli* transformed with the cytosolic chaperonin CCP-1a from Bruguiera sexangula. Studies on plant genetic transformation with Hsp genes have focused mostly on heat stress and thermotolerance. Overexpression of HSP101 from Arabidopsis in rice plants culminated in a significant improvement of growth performance during recovery from heat stress. Overexpression of LEA proteins was associated in several cases with desiccation tolerance, although the actual function of these proteins is still unknown. Recently, overexpression of HVA1, a group 3 LEA protein isolated from barley (*Hordeum vulgare* L.), conferring dehydration tolerance to transgenic rice plants was reported.

7.3.2. Stress-associated Changes in Metabolites and Metabolomics

Severe osmotic stress brings about adverse changes in cellular components. A wide range of metabolites that can prevent these harmful changes have been identified, including amino acids (e.g. proline), quaternary and other amines, and a variety of sugars and sugar alcohols (e.g. mannitol and trehalose). Two general plans for the metabolic engineering of abiotic stress tolerance have been put forth: increased production of specific desired compounds or reduction in the levels of needless (toxic) compounds. However, modulation of a single enzymatic step is usually monitored by the tendency of cell systems to restore homeostasis, thus limiting the potential of this approach. Targeting multiple steps in the same pathway could help to manages metabolic fluxes in a more predictable manner.

7.3.2.1. Amino Acids

Accumulation of proline was associated with better plant performance under salt stress. Proline-level increments can be achieved in plants by overexpressing Δ^1-pyrroline-5-carboxylate synthetase (P5CS), as found, for example, in tobacco. This approach also brought about the upregulation of proline dehydrogenase, which reduces proline levels. Indeed, Arabidopsis transformation with proline dehydrogenase antisense or a knockout of this enzyme produced increased free proline accumulation and better growth performance under salt stress. To check whether plant growth reduction in response to osmotic stress might actually result from osmolyte accumulation, proline levels were manipulated by expressing mutated derivatives of P5CS from tomato in *Saccharomyces cerevisiae* The levels of proline accumulation and cell growth were inversely correlated in cells grown under normal osmotic conditions. Alternatively, proline might confer a protective effect by generating stress-protective proteins. Exogenously applied proline and/or salt stress in *Pancratium maritimum* were found to induce the expression of ubiquitin, antioxidative enzymes, and dehydrins.

7.3.2.2. Amines

The accumulation of glycine-betaine a widely studied osmoprotectant, has been researched with respect to modifications of several metabolic steps. Betaine aldehyde decarboxylase from the halophyte *Suaeda liaotungensis* was integrated into tobacco plants and the *in vitro* plantlets were significantly resistant to salt conditions. Similar outcomes were obtained by transforming rice with the choline dehydrogenase gene (codA) from *Arthrobacter globiformis*; the gene product of codA catalyses the oxidation of choline to glycine betaine via betaine aldehyde as intermediate transgenic rice plants recovered from salt stress and set seeds, in contrast to wild-type plants.

Tomato plants transformed with a bacterial codA gene targeted to the chloroplast were highly tolerant to chilling and oxidative stress, showing an enhancement in photosynthetic rate, plant survival, flower retention and fruit set. Co-targeting multiple steps in the same pathway was a successful strategy for overexpressing glycine-betaines in plants and bacteria. Indeed, stress tolerance was increased by the genetic engineering of *E. coli* and tobacco plants with the betaine aldehyde–choline dehydrogenase fusion protein. Plant polyamines have earlier been shown to be involved in plant response to salinity. More recently, genetic engineering for increased biosynthesis of several specific polyamines resulted, in several cases, in stress-tolerant plants.

Overexpression of arginine decarboxylase (ADC), ornithine decarboxylase and S-adenosylmethionine decarboxylase produced a significant increase in putrescine levels and a small increase in spermidine and spermine levels. Transgenic rice plants expressing *Datura stramonium* ADC under the control of the monocot Ubi-1 promoter mugenerated ch higher levels of putrescine under drought stress only, promoting spermidine and spermine synthesis and ultimately protecting the plants from drought. Overexpression of spermidine synthase cDNA from *Cucurbita ficifolia* in *Arabidopsis thaliana* significantly increased spermidine levels, and consequently enhanced tolerance to various stresses.

7.4. Drought

Drought, a meteorological event, results from absence of rainfall for a period of time to cause moisture-depletion in soil and waterdeficit with a decrease of water potential in plant tissues. But from the agricultural aspect, its working definition would be lack of water availability, including precipitation and soil-moisture storage capacity, in quantity and distribution during the lifecycle of a crop plant. This restricts the expression of full genetic potential of the plant which acts as a serious limiting factor in agricultural production by preventing a crop from reaching the genetically determined theoretical maximum yield. The effect of drought on crop production and overall economy is evident in the most affected states in India which are Rajasthan, parts of Gujarat, Haryana and Andhra Pradesh. Most of the crops are sensitive to water deficits, particularly during flowering to seed development stage. Even crops grown in arid and semi-arid regions, such as pearl millet, sorghum and pigeon pea, are also affected by drought at the reproductive stage.

In agriculture, drought resistance is associated with the ability of a crop plant to produce its yield with minimum loss in a water-deficit environment relative to the water-constraint-free management. An understanding of genetic basis of drought resistance in crop plants is a prerequisite for a geneticist to sudevelop a perior genotype through either conventional breeding methodology or biotechnological approach.

7.4.1. Mechanisms of Drought Resistance

Genetically the mechanisms of drought resistance can be grouped into drought escape, drought avoidance and drought tolerance. However, crop plants use more than one mechanism at a time to resist drought. Drought escape is referred to as the ability of a plant to complete its lifecycle before serious soil and plant water deficits set in. This mechanism involves rapid phenological development, developmental plasticity and remobilisation of preanthesis assimilates to grain.

Drought avoidance is a plant's ability to maintain relatively high tissue water potential regardless of a shortage of soil-moisture, whereas drought tolerance is the ability to tolerate water-deficit with low tissue water potential. Improving water uptake, storing in plant cell and reducing water loss prevent drought situations. The responses of plants to tissue water-deficit determine their level of drought tolerance. Drought can be avoided by maintenance of turgour through increased rooting depth, efficient root system and increased hydraulic conductance and by reduction of water loss through reduced epidermal (stomatal and lenticular) conductance, reduced absorption of radiation by leaf rolling or folding and reduced evaporation surface (leaf area). Plants under drought condition survive by balancing between maintenance of turgour and reduction of water loss. The mechanisms of drought tolerance are maintenance of turgour through osmotic adjustment (a process which induces solute accumulation in cell), increase in elasticity in cell and decrease in cell size, and desiccation tolerance by protoplasmic resistance.

Unfortunately, adapting to drought has its own disadvantages. A genotype of short duration usually yields less compared to that of normal duration. The consequence of mechanisms that confer drought resistance by reducing water loss (such as stomatal closure and reduced leaf area) is reduced assimilation of carbon dioxide. Osmotic adjustment enhances drought resistance by maintaining plant turgour, but the increased solute concentration responsible for osmotic adjustment may have harmful effect in addition to energy requirement for osmotic adjustment. Consequently, crop adaptation must show a balance among escape, avoidance and tolerance while maintaining adequate productivity.

7.4.2. Genetics

Drought resistance being complex its expression depends on the action and interaction of different morphological (earliness, reduced leaf area, leaf rolling, wax content, efficient rooting system, awn, stability in yield and reduced tillering), physiological (reduced transpiration, high water-use efficiency, stomatal closure and osmotic adjustment) and biochemical (accumulation of proline, polyamine, trehalose, etc., increased nitrate reductase activity and increased storage of carbohydrate) characters. Not much is known about the genetic mechanisms that quantify these characters. The identification of genes responsible for morphological and physiological characteristics and their location on chromosome have proved futile, but their inheritance pattern and nature of gene action have been recorded. Polygenic inheritance of root characters has also been reported. The long root and high root numbers are controlled by dominant alleles, and thick root tip by recessive alleles. However, leaf rolling and osmotic adjustment have shown monogenic inheritance. Researchers have reported a drought resistance gene, *Drt1* in rice, which is related to genes for plant height, pigmentation, hull colour and awn, and has pleiotropic effect on the root system. Similarly, in cowpea drought resistance is said to be controlled by a single dominant gene.

Some more reports for other traits are available, but there is need for further investigation to better understand genetic control of morphological and physiological traits contributing to drought resistance.

In addition to there changes, biochemical change involving induction of compatible solute biosynthesis is one way to bring about drought. Under drought, plants try to maintain water content by accumulating various solutes that are non-toxic and do not interfere with plant processes. Therefore, they called compatible solutes. Some of them are fructan, trehalose, polyols, glycine betaine, proline and polyamines. The various genes responsible for different enzymes involved in biosynthesis of these solutes have been identified and cloned from different organisms (bacteria, yeast, human and plant).

7.4.3. Breeding Approach

Three breeding approaches for drought resistance have been developed. The first is to breed for high yield under optimum (water-stress-free) condition. The basic philosophy of this is that, as the maximum genetic potential of yield is expected to be realised under optimum condition and a high positive correlation exists between performance in optimum and stress conditions, a genotype superior under optimum level will also yield relatively well under drought condition.

However, the concept of expression of maximum genetic potential in optimum condition questioned is as genotype environment interaction may limit the high-yielding genotype to perform well under drought. Thus, the second approach, i.e. to breed under actual drought condition, has been suggested.

In the second approach the intensity of drought is highly variable from year to year and so environmental selection pressure on breeding materials changes drastically from generation to generation. This problem compounded with low heritability of yield makes the breeding programme complicated and slow.

Counter to the above two approaches would be to improve drought resistance in high-yielding genotypes through absorption of morphological and physiological mechanisms of drought resistance. However, transferring drought resistance in high-yielding genotypes is problematic as there is lack of understanding of the physiological and genetic basis of adaptation in drought condition. In contrast, improving the yield potential of an already resistant material may be a more paying proposition, provided there is genetic variation within such a material. Simultaneous selection in non-stress environment for yield and in drought condition for stability may be done to achieve the desired goal of generating drought-resistant genotype with high yield. As such, the breeding methodology to be applied for drought resistance is the same as that applied for other purposes. In general, pedigree and bulk method could be adopted for self-pollinated crops and recurrent selection for cross-pollinated crops. However, if transfer of a few traits relating to drought resistance to a high-yielding genotype is the goal, then backcross is the appropriate technique. On the other hand, biparental mating (half sib and full sib) preserves the broad genetic base as well as provides the scope to generate the desired genotype of drought resistance. The success of any breeding programme depends on the availability of the screening technique, especially for drought resistance.

7.4.3.1. Screening Techniques

Genetic improvement in drought resistance utilising the existing genetic variability can be achieved by an efficient screening technique, which should be fast and capable of assessing plant performance at the critical developmental stages and screening a large population with only a small sample of plant material. A combination of different traits of direct relevance, rather than a single trait, should be preferred as selection criteria. Researchers provided a framework for estimating how combination of traits influences plant water status and growth. This may usefully bridge physiology and breeding into the integrated programme of plant improvement.

The different screening techniques used so far are as follows:

i) Use of infrared thermometry for screening efficient water uptake.

ii) Banding herbicide metribusin at a certain depth of soil and use of iodine-131 and hydroponic culture under stress of 15 bar for screening root growth.

iii) Psychrometric procedure for evaluating osmotic adjustment.

iv) Diffusion porometry for leaf water conductance.

v) Use of the mini-rhizotron technique for root penetration, distribution and density in the field with minimum disturbance.

vi) Infrared aerial photography for dehydration postponement.

vii) Use of carbon isotope discrimination for selecting increased water-use efficiency.

viii) As loss of yield is the main concern for the crop plant from the agricultural aspect, plant breeders stress on yield performance under moisture-stress condition.

A drought index which gives a measure of drought based on loss of yield under drought-condition in comparison to moist condition has been employed for screening drought-resistant genotype. An artificially created water-stress environment is used to provide the condition in selecting superior genotype out of a large population. Visual scoring or measurement for maturity, leaf rolling, leaf length, angle, root morphology and other morphological characters of direct relevance to drought resistance are also considered.

7.4.4. Biotechnological Approach

The techniques for gene transformation of crop plants have been used to identify genes responsible for drought resistance and their transfer. Mainly two approaches, namely targeted and shotgun approaches, make it possible for genetic engineering to obtain transgenic plants conferring drought resistance.

7.4.4.1. Targeted Approach

Metabolic pathways that synthesize different metabolites such as polyamine, carbohydrate, proline, glycine betaine and trehalose are associated with drought resistance. This approach depends on relevant information on biochemical reaction available for synthesis of these metabolites, and uses the related genes to transfer them from different sources to crop plants. This approach being more precise and methodical, has a higher probability of success when compared to the shotgun approach.

Lately, introducing drought-induced genes involved in different biochemical pathways from different sources to sensitive plants has emerged as one of the promising methods. The gene *TPS1* found in yeast encodes for trehalose-6-phosphate synthetase and is involved in biosynthesis of trehalose. Tobacco has been transformed with the yeast *TPS1* gene. By determining the water loss of detached leaves or the effect of withholding irrigation on the death and damage of leaf, it is evident that the transgenic plants have increased drought resistance.

The gene *P5CS* encodes for pyrroline-5- carboxylate synthetase, involved in proline synthesis, and the over-production of proline provides drought resistance. Transgenic tobacco over-expressing *P5CS* gene transferred from mothbean showed a high level of enzyme and produced 10–18-fold more proline than control plant. The over-production of proline encouraged root biomass and flower development under drought condition.

The bacterial gene *SacB*, found in *Bacillus subtilis*, encodes for levan sucrase, which has a role in fructan synthesis. When this gene was transferred to tobacco, the transgenic plant produced fructan and performed better in comparison to control under PEG-mediated drought condition.

The genes *betA* encoding for choline dehydrogenase and *betB* encoding for betaine aldehyde dehydrogenase are implicated in the biosynthesis of glycine betaine, and accumulation of glycine betaine provides drought resistance. Holmstrom *et al.* transferred *betB* gene from *Escherichia coli* to tobacco.

With the availability of genes involved in polyamine biosynthesis such as *ADC* (encodes for arginine decarboxylase), *ODC* (encodes for ornithine decarboxylase) and *SAMDC* (encodes for S-adenosyl-methionine decarboxylase), it is now easier to manage polyamine content using sense and antisense constructs of these genes in transgenic plants. Transgenic tobacco plants with *ODC* gene from yeast and mouse, *ADC* gene from oat and *SAMDC* gene from human have been recorded, yet more needs to be done to observe whether the transgenics show any tolerance to drought. Only the expression level of polyamines in plants has been studied. However, *SOD* (superoxide dismutase) gene from pea has been transferred to tobacco and transgenics were found to be drought resistant.

References

Gao, Ji-Ping, et al. "Understanding Abiotic Stress Tolerance Mechanisms: Recent Studies on Stress Response in Rice." *Journal of Integrative Plant Biology* 49 (6): 742"750, 2007.

Mittler, Ron. "Abiotic stress, the field environment and stress combination." *Trends in Plant Science* 11(1): 15–19, 2006.

Palta, Jiwan P. and Farag, Karim. "Methods for enhancing plant health, protecting plants from biotic and abiotic stress related injuries and enhancing the recovery of plants injured as a result of such stresses." *United States Patent* 7101828, September 2006.

Roelofs, D. et al. "Functional ecological genomics to demonstrate general and specific responses to abiotic stress." *Functional Ecology* 22: 8–18, 2008.

Wang, W., Vinocur, B. and Altman, A. "Plant responses to drought, salinity and extreme temperatures towards genetic engineering for stress tolerance." *Planta* 218: 1-14, 2007.

8

MOLECULAR FARMING AND INDUSTRIAL PRODUCTS

With the advent of biotechnology, agricultural crops and livestock can be utilised for entirely new purposes. Just as genetically engineered bacteria are routinely being used for economical, efficient production of a wide range of medicinal proteins and industrial enzymes, plants and animals may be engineered to produce a variety of valuable biological molecules ranging from medicinals such as vaccines to polymers such as biodegradable plastics.

Molecular farming has many advantages, one of which is that biologically derived substances are most efficiently produced through biological means, utilising natural and renewable resources and ensuring optimal environmental compatibility. This will lead to the creation of new markets for farmers which complement their existing business.

The "proof of concept" exists in the form of numerous examples of transgenic plants comprising a variety of new products in their leaves or seeds, and transgenic animals producing therapeutic proteins in their milk. Some changes in production methods may be necessary, as with the espousal of any new plant or animal as an agricultural commodity; for example, tobacco producing new materials in the leaf might be machine-harvested green before extraction and purification of the product. However, it is expected that molecular farming will relate directly to traditional agricultural skills and resources. There has recently been transition from "experimental" status to profitable opportunities for farmers.

8.1. Molecular Farming Process

Until recently, improvement in an agricultural enterprise depended largely on breeding programmes to enhance the productive capacity of crops and livestock. But now genetic engineering has introduce entirely new characteristics very efficiently, by introducing the gene(s) coding for the desired characteristic directly into the genetic makeup of a plant or animal. This technology uses natural resources and is both specific and precise only a defined minimal set of genes is introduced into the host. Genetic engineering is thus far more efficient than traditional breeding. Additionally, the ability to introduce non-native genes allows plants and animals to produce entirely novel materials for innovative applications.

Plants, in particular, for long have natural chemicals and materials, such as flavourings, fragrances, medicines, dyes, rubber and oils. By adding or deleting genes from these plants, it is possible to alter the traits of the products or create entirely new ones. For example, rapeseed (canola) oil, normally used for cooking and margarine production, has been improved upon not only to be healthier for cooking but also for use in manufacturing detergents and novel lubricants. Soyabeans have also been modified for novel bio-based lubricants that are renewable and fully biodegradable, having fewer risks for humans and the environment in shipping, handling, and production. In some certain applications, these lubricants may be competitive with petrochemical-derived lubricants. In another example, tobacco modified to manufacture several products including antibiotics, novel polymers, a dental treatment, and anti-cancer drugs. Dozens of compounds have been produced in a whole multitude of crop plants ranging from traditional commodity crops like corn, soyabeans and tobacco to special crops like spinach, potatoes and beets. Several pharmaceuticals made in plants thus are in late-stage clinical trials and are soon to hit the market. The possibilities of plant molecular farming for these types of pharmaceutical and industrial proteins are considered huge because of the ease of boosting up their production from experimental levels to field production and their estimated low production costs. Some highly valuable pharmaceuticals that are required only in small doses will likely be manufactured on a very small scale (e.g., in a greenhouse). Other products such as biodegradable plastics, which are presently at an initial stage of development, may eventually require thousands or even millions of acres for production.

With considerable interest in livestock-based production systems, goats, pigs, cows, sheep, and chickens have all been engineered to produce a variety of new proteins, mostly for the medical industry. Advantages accruing from these systems are their relatively low operating costs and unlimited scale-up. Another benefit of the animal-based systems is that the desired protein can be produced in the milk of mammals or the eggs of thereby allowing easier extraction and purification. As an example, one recent development in animal molecular farming is the scale-up of a herd of goats engineered to produce a spider's web protein in their milk. This "biosteel" will revolutionise the materials industry, as it is both stronger and more flexible than steel, and offers a lightweight alternative to carbon fibre. Potential uses include applications for which strength and lightness are essential, such as aircraft, racing vehicles and bulletproof clothing, and for artificial tendons, ligaments and limbs. In addition, 40 to 50 pharmaceuticals are already being developed for production in animal systems; at least two reached the late-stage clinical trials in the summer of 2000.

8.2. Economic Opportunities

The economic viability of molecular farming systems cannot be generalised as each application is different, with variation in farming practices, efficiency of the gene-expression system, ease with which the product can be extracted and purified, and prospects for obtaining useful by-products. Though the plant systems are different from animal systems, there are extreme contrasts even among the plants and among the animals. For example, tobacco is a very different crop compared to corn, soyabeans or canola, as it has the advantage of producing large quantities of green leaf material per acre, and convenient to work with from a biotechnology aspect. Hence it may be the ideal "factory" for products produced in green leaves. In cases where the products need to be produced in seeds, a better option may be corn, soyabeans, or canola because tobacco seeds are extremely small. All crops have different genetic composition, different characteristics and different production methods, making each a unique "vehicle" for molecular farming.

These differences make it impossible to establish the "cost" of generating a product using molecular farming. However, certain principles are quite clear: for both plants and animals, profitability requires high expression of the introduced gene, maintenance of product integrity, and the ability to boost up, harvest, recover, purify, and store the target product as economically as possible. For products that require high purity, the processing and purification costs are expected to be very high.

Today, in the commercial scale-up of molecular farming several suitable products have been identified, especially in the medical field, and the technology is being perfected for large-scale implementation. The focus first is on high-value pharmaceuticals molecular-farmed owing to their high profit potential. Several such products are under development and both large and small companies. For success in the competitive marketplace, cost cutting will be crucial, especially in the area of purification and product recovery. Ultimately, if the products and processes of molecular farming are to be commercially successful, they must have a competitive edge over existing, alternative products and processes (e.g., pharmaceuticals produced through microbial fermentation). Or, if the products are entirely new (e.g., biodegradable plastics or "biosteel"), there must be a parallel new market for them. With this technology continuing to develop and molecular farming production becoming more effective larger-scale, lower-value products such as industrial chemicals or biodegradable plastics should lead to larger-scale opportunities for agricultural producers.

Ensuring environmental and process regulations being developed for this industry, it will be necessary to set a regulatory regime that balances safeguarding the public and the environment and sustaining the innovative process. Though the commercial progress is advancing, molecular farming is still very much a nascent industry. Contract production is likely, and profit margins may differ considerably depending on the value of the target and the increased production requirements. But what is unique about molecular farming is that this opening should provide a captive new market for agricultural products, one that can increase over time and one that provides the pioneer producers with a clear advantage.

8.3. Host Systems and Expression Technology

In the recent past plant-based expression systems have mushroomed as a serious competitive force in the large-scale production of recombinant proteins. The first plant-derived technical proteins have already hit the market, and detailed economic estimates have shown their viability against established market sectors. Now, several plant-derived biopharmaceutical proteins are nearing the penultimate stages of commercial development. For these products that include antibodies, vaccines, human blood products, hormones and growth regulators, plants offer practical and safety advantages as well as lower production costs compared with traditional systems based on microbial or animal cells, or transgenic animals. With several products in development, molecular farming in plants is gaining importance. In this, recent technological developments in molecular farming focus on plant host systems and expression technology.

8.3.1. Transgenic Plants: Production of Recombinant Proteins

8.3.1.1. Production Costs

The comparatively low cost of large-scale production a major advantage makes transgenic plants for molecular farming a great asset. Both capital and running costs are markedly lower than those of cell-based production systems since fermenters or the skilled personnel to run them are eliminated. It is estimated that recombinant proteins can be produced in plants at 2–10 per cent of the cost of microbial fermentation systems and at 0.1 per cent of the cost of mammalian cell cultures, depending on the product yield. For high-yielding proteins, the economic advantages of plant production systems are clear. One bushel of maize producing recombinant avidin at 20 per cent total soluble seed protein has the same total yield as one tonne of chicken eggs – the natural source of avidin – but at 0.5 per cent of the cost. Not many proteins can be produced at this level, but yields of 0.1–1.0 per cent total soluble protein (TSP), the typical levels necessary for the production of pharmaceutical proteins, such as recombinant antibodies, are sufficiently competitive with other expression systems to make plants economically feasible. Where such yields cannot be gained, molecular farming in plants might not be viable particularly if the product has a low market value.

For any given expression system, scalability is a consequential commercial advantage. Fermentation systems and transgenic animals have definite potential in this respect, whereas in plant-based production the scale can be modulated quickly in response to market demand simply by using more or less land as per demand. The speed of scale-up is also important as it can take several years to achieve tenfold boost in a host of transgenic sheep using natural breeding cycles, but a field of transgenic plants can be scaled-up more than 1000-fold in a single generation due to the fertile seed output. Where scale-down is required, surplus transgenic animals have to be forsaken or maintained at a loss, whereas the acreage of land planted to a specific transgenic crop can be adjusted as needed, and transgenic plant lines can be stored indefinitely and inexpensively as seed.

The costs of production of a recombinant protein also depend to a large extent on the required purity because.85 per cent of expenditure show downstream processing rather than production as such. The same methods are used for protein purification despite of the expression system, and the costs of processing are similar across the board when high purity is needed. However, plants are beneficial as several types of recombinant protein can be used in unprocessed or partially processed material, therefore eliminating many of the downstream costs. For example, recombinant subunit vaccines procured from plants can be dispensed by the consumption of raw or part-processed fruits and vegetables, and antibodies for passive immunisation can be given as topically applied pastes following minimal purification. Similarly, industrial enzymes such as glucanase and phytase can be integrated into the industrial process either as part-processed plant material or expressed directly in the plant that needs to be processed.

For purifying pharmaceutical proteins before use, several techniques have been developed to bring down downstream processing costs. As the cost of processing is inversely proportionate to the concentration of product in the starting material, the yield of protein per unit of plant biomass is imperative, depending on both the expression system and the product itself. This needs to be evaluated on an empirical basis. One of the benefits of recombinant protein expression in the seeds of transgenic cereal plants is that high levels of the product can collect in a small volume, which reduces the costs related to processing. Where conventional extraction from seeds is too costly, further strategies to aid in purification can be used. An example is the oleosin-fusion platform in which the target recombinant protein is expressed in oilseed crops as a blending with oleosin. The fusion protein can be recovered from oil bodies using a simple extraction procedure, and the recombinant protein is isolated from its fusion partner by endoprotease digestion. Similarly, a strategy is devised in which recombinant proteins are expressed as fusion constructs comprising an integral membrane-spanning domain procured from the human T-cell receptor. The recombinant protein accumulates at the plasma membrane and can be extracted in a small volume using appropriate buffers and detergents.

Exploiting the ability of vegetative tissues (e.g. leaves or roots) is another system to secrete recombinant proteins in their exudates, enabling proteins to be collected continuously. Such an approach for the production of human secreted alkaline phosphatase is used for the synthesis of recombinant antibodies, but is suitable only for small-scale production in high-containment facilities. This is because field deployment would result in large amounts of recombinant protein leaching into the soil and groundwater.

8.3.1.2. Development Timescale

The time required for the gene-to-protein process for transgenic plants includes the preparation of expression constructs, transformation, regeneration and the production and experimentation of several generations of plants. The testing phase is essential to ensure the soundness of transgene and expression, and the biochemical activity of the product, as well as the absence of adverse phenotypic changes in the host plant. These processes may take up to two years depending on the plant species, although milligram amounts of protein might be available after several months for initial testing. It is more beneficial for microbial expression systems and animal cells here as they can produce the first batches of recombinant protein more quickly than plants, but this might be outbalanced by their limitations in terms of overall scalability and cost. In addition many plant-based transient expression systems are available for the quick production and testing of recombinant proteins on a small scale, and these can be used to assess expression constructs and product quality before relegating to the expense of transgenic plants. Plant-cell-suspension cultures also produce recombinant proteins more rapidly than transgenic plants because the development and testing schedule is much shorter. Suspension cells are therefore being used for molecular farming, especially where high containment is an advantage.

8.3.1.3. Product Authenticity

In the production of pharmaceutical proteins, plants are viewed as being much safer than both microbes and animals as they generally lack human pathogens, oncogenic DNA sequences and endotoxins. However, it is important to weigh the intrinsic safety of plant-derived human proteins because the structural authenticity of such proteins affects their behaviour *in vivo*. The protein synthesis pathway is conserved between plants and animals, so plants appear to fold and assemble recombinant human proteins efficiently, a great advantage over bacterial expression systems, in which many proteins fail to fold properly and are therefore, degraded (resulting in low yields) or accumulate as insoluble inclusion bodies. The efficiency of plants to fold and assemble complex proteins correctly is proved by their capacity to produce functional serum antibodies, which contain four polypeptide chains covalently joined by disulphide bonds. More remarkably, plants produce functional secretory antibodies, which are dimers of the typical serum immunoglobulins and have two additional polypeptide components, making ten separate polypeptides in total. Two different cell types are needed to assemble such antibodies in mammals.

Although plants and animals have the same the protein synthesis pathway in they have some differences in posttranslational modifications, particularly with respect to glycan-chain structure. Thus, plant-derived recombinant human proteins tend to have the carbohydrate groups $\beta(1 \rightarrow 2)$-xylose and α(1 → 3)-fucose, which are absent in mammals, but do not have the terminal galactose and sialic acid residues that are found on many native human glycoproteins. Compared with other production systems it is clear that even minor differences in glycan structures can change the distribution, activity or longevity of recombinant proteins in contrast with their native counterparts, and could render such proteins immunogenic when administered to humans. Several modifications in the glycosylation pathway are needed to produce proteins with typical human glycan structures in plants. For this, the strategies to 'humanise' the glycan structures of recombinant human glycoproteins include: (i) the use of purified human β(1,4)-galactosyltransferase and sialyltransferase enzymes for the *in vitro* modification of plant-derived recombinant proteins; and (ii) expression of human b(1,4)-galactosyltransferase in transgenic plants to produce recombinant antibodies with galactose-extended glycans. In the case of the latter, 30 per cent of the antibody was galactosylated, which is similar to the proportion found in hybridoma cells. *In vivo* sialylation will be more difficult to achieve because plants lack the precursors and metabolic capability to produce this carbohydrate group. Although these methods involve the addition of carbohydrate groups missing in plants, there is also the difficulty of removing plant-specific carbohydrates. Inhibition of the enzymes fucosyltransferase and xylyltransferase using antibodies, ribozymes orRNA interference can be used to achieve this, and gene targeting by homologous recombination has produced recombinant proteins lacking plant-specific glycans in the moss *Physcomitrella patens*.

Although achieving authenticity should be the target in molecular farming, it is worth noting that studies in which mice were administered a recombinant antibody containing plant-specific glycans showed no proof of an immune reaction.

8.3.2. Maximising Yields in Transgenic Plants

Establishing the commercial viability of molecular farming is one of the most important factors in the achievement of adequate recombinant protein yields. Absolute yields depend on the species used for production but the selection of production crop depends on several factors in addition to yield potential.

8.3.2.1. High-level Transgene Expression

Expression-construct design can help in procuring high yields in transgenic plants by maximising the rates of transcription and translation. For dicotyledonous species, the strong and constitutive cauliflower mosaic virus 35S (CaMV 35S) promoter is often selected to get transgene expression. In cereals, where the CaMV35S promoter has a lower activity, themaizeubiquitin-1 (ubi-1) promoter is preferred. The rate of transcription in cereals is also enhanced by the addition of an intron; this phenomenon is known as intron-mediated enhancement.

Regulated promoters have both practical and biosafety advantages and hence can be used in preference to constitutive promoters. For example, although constitutive promoters enable high-level accumulation of recombinant proteins in seeds, the proteins are also expressed in leaves, pollen and roots. This might affect the growth and development of vegetative parts of the plant negatively and could expose herbivores, pollinating insects and microbes in the rhizosphere to the effects of the recombinant protein. Restriction of protein accumulation to seeds helps to offset these risks. Inducible promoters can be availed of to limit recombinant protein expression to just before or following harvest, as has been shown with recombinant glucocerebrosidase expressed in tobacco.

The rate of translation can be increased by establishing that any mRNA instability sequences are removed from the transgene construct, and that the translational startsite such the Kozak consensus for plants. Modification might be need to codon usage in some transgenes, not only to maximise the rate of protein synthesis, but also to freeze out cryptic introns and instability sequences.

There can be maximisation of both transcription and translation can be maximised by taking precautions against transgene silencing, a group of epigenetic phenomena that can restrain the expression of even structurally intact transgenes at either the transcriptional or post-transcriptional level. Strong triggers for silencing encompass prokaryotic DNA sequences and sequences that have the potential to form hairpin secondary structures at the DNA level, or double-stranded RNA when expressed. Useful techniques to minimise the frequency of silencing include the use of clean DNA procedures to escape vector backbone integration.

8.3.2.2. Protein Targeting

Targeting of recombinant proteins to oil bodies or the plasma membrane is a useful strategy to facilitate isolation and purification. However, subcellular targeting can be a general approach to increase the yield of recombinant proteins as the compartment in which a recombinant protein collect strongly affects the interrelated processes of folding, assembly and post-translational modification. All of these contribute to protein stability and, hence, help to determine the final yield.

It is more advantageous for high-level protein accumulation comparative targeting experiments with full-size immunoglobulins and single-chain Fv fragments which have shown that the secretory pathway is a more suitable compartment for folding and assembly than the cytosol. Since many plant-derived recombinant proteins under development are human proteins that normally pass through the endomembrane system, this principle can be applied not only to antibodies but also more generally. Antibodies directed to the secretory pathway using either plant or animal N-terminal signal peptides usually collect to levels that are extremely greater than those of antibodies expressed in the cytosol. Rare exceptions to this generalisation indicate that intrinsic features of each antibody might also contribute to overall stability.

The endoplasmic reticulum (ER) contributes an oxidising environment and a profusion of molecular chaperones, while there are few proteases. These are probably the most important factors influencing protein folding and assembly. Recently research, shows that antibodies targeted to the secretory pathway in transgenic plants interact specifically with the molecular chaperone BiP (binding protein).

Due to lack of further targeting information, the expressed protein is secreted to the apoplast. Depending on its size, the protein can be kept therein or might leach from the cell, with important implications for production systems based on cell-suspension cultures. The stability of antibodies in the apoplast is lower than in the lumen of the ER. Therefore, antibody expression levels can be enhanced even more if the protein is regenerated to the ER lumen using an H/KDEL C-terminal tetrapeptide tag. Accumulation levels are generally 2–10-fold greater compared with an identical protein lacking the H/KDEL signal. As an added advantage antibodies recovered in this way are not modified in the Golgi apparatus, suggesting they have high-mannose glycans but not plant-specific xylose and fucose residues. Again, although the principles of ER-retention in molecular farming have been determined using antibodies, it is likely that they will also apply to many other proteins. High levels of protein expression have also been shown when transgenes are introduced into the chloroplast rather than the nuclear genome. Several examples of chloroplast-based molecular farming in tobacco have been reported. Here the technology for chloroplast gene transfer that is most advanced include: (i) production of human growth hormone at levels approaching 8 per cent TSP; (ii) production of human serum albumin at levels exceeding 11 per cent TSP; (iii) production of cholera and tetanus toxin fragments at up to 25 per cent TSP; and (iv) production of a thermostable xylanase at 6 per cent TSP, with 85 per cent recovery.

The high turnover with the chloroplast transgenic system is advantageous, and chloroplasts also offer biosafety advantages in terms of transgene containment, due to maternal inheritance. However, the use of this system as a general approach in molecular farming is restrained by the inability of chloroplasts to carry out many post-translational modifications, including glycosylation, and biosafety concerns have been raised by the recent demonstration of horizontal gene transfer from the chloroplasts of transplastomic plants tobacteria, under laboratoryconditions. An alternative approach is to express the protein from the nuclear genome but initiate a chloroplast targeting sequence. In an interesting recent report, researchers expressed a camelid heavy-chain antibody in potatoes, and succeeded in targeting the protein to the chloroplast where it suppressed the starch-branching enzyme. The aim of the study was metabolic engineering rather than molecular farming, but that camelid heavy-chain antibodies are extremely stable makes these molecules useful candidates for both medical and industrial applications.

8.4. Developments in Molecular Farming

Plant molecular farming using GM techniques is, for the most part, still in its nascent stage, with only a small number of plant-made research and analytical-grade proteins ready for commercialisation of the product development pipeline.

8.4.1. Pharmaceutical Applications

Pharmaceutical therapeutics are which have medical applications, can be classified into three broad categories — antibodies, vaccines and 'other therapeutics'.

8.4.1.1. Antibodies

While antibodies are proteins produced by the immune system in response to the presence of a specific antigen, an antigen is any substance, for example, a bacterial or viral protein, which is capable of causing an immune response. Monoclonal antibodies (mAbs) are specific antibodies that have been produced in a laboratory, purified and which can recognise single antigens. Certain mAbs for human tare useful herapeutic purposes. Traditionally, such mAbs were derived from mice and cell cultures, but this system has several disadvantages, including the high set-up and running costs, the limited opportunities to scale up production and the potential for contamination with human and animal pathogens. Furthermore, mice-derived mAbs may result in an immunogenic reaction in humans; a reaction which may be avoided by using mAbs derived from plants.

Plants have the capability to churn out large amounts of mAbs, with low production costs, the ability to be rapidly scaled up to meet market demand and reduced risk of contamination with human and animal pathogens. Systems for expressing mAbs have been extensively tested to permits in tobacco, but cereal grains are now being used because protein accumulation in dry seeds allows long-term storage at ambient temperatures with little degradation or loss of activity. Other crop plants that are used to produce antibodies include potatoes, alfalfa and rice.

8.4.1.2. Vaccines

A vaccine is an antigenic substance — either a protein, peptide, attenuated (weakened) living organism or a dead organism —

which is able to draw forth an immune response that safeguards against development of an infectious disease. Vaccines are designed to 'educate' the immune system, so that upon exposure to the infectious agent it will already have the antibodies that will enable it to combat the infection.

Considerable research into the use of transgenic plants as an alternative to conventional vaccine production systems. PMV production appears to offer advantages in protein production and storage; and potential advantages in the distribution of vaccines to developing countries.

8.4.1.3. Other Therapeutics

Apart from antibodies and PMVs, scientists have also examined the potential for using plants to produce a wide assortment of other therapeutic agents, including hormones, enzymes, interleukins, interferons and human serum albumin. Proteins found in breast milk, such as lactoferrin, lysozyme and α-casein, have been expressed in plants. Addition of these therapeutic proteins to fortify other food products could aid in improving infant health.

8.4.2. Industrial Products

Plant molecular farming can also be beneficial making products with industrial applications including biofuels (biodiesel and ethanol), industrial oils, natural sweeteners, taste-modifying proteins, bioplastics, spider silk and elastic proteins.

8.4.2.1. Biofuels

Since biofuels currently cost more to produce than petroleum fuels, situation has generated considerable interest in developing new technologies to reduce the production costs of biofuels. Plant molecular farming is seen by many researchers to be one way of reducing these costs.

8.4.2.2. Sweeteners

Artificial sweeteners with a low calorific value (e.g. saccharin) have been developed as a substitute for sucrose and other natural sugars, especially for individuals who ail from diseases related to the consumption of sugar, such as diabetes, hyperlipemia, dental caries and obesity. However, some artificial sweeteners have unwelcome baking properties and aftertaste, which has led to research into alternatives. Potential alternatives to current artificial sweeteners that can be produced in plants include palatinose, brazzein, miraculin, isomaltulose and sorbitol.

8.4.2.3. Bioplastics

Bioplastics are considered an environmentally beneficial alternative to synthetic plastics due to their biodegradability and potential to decrease our reliance on petrochemical resources. The focus of current research into biodegradable plastics has largely been on polyhydroxyalkanoates (PHAs) —polyesters of hydroxy acids—that are produced by more than 100 different genera of bacteria. The monomers found in PHAs are highly diverse, suggesting that these plastics have a wide spectrum of physical properties and could be used as substitutes for polyethylene, polystyrene, polypropylene and PET. PHAs are also biocompatible, breaking down into molecules that are naturally inherent in animals, so they may have medical applications, for example, as implants, gauzes and suture filaments.

The cost of synthesizing of PHAs in GM or non-GM bacteria is high, with bacterial PHA currently priced five to ten times higher than petroleum-derived plastics. This constrains the use of bacterial PHAs to high-value products such as medical applications. There have been attempts to synthesise PHA in other organisms, including agricultural crop plants, to develop large scale production of biodegradable plastics economically. It has been calculated that bioplastic concentrations in plants would have to reach 15 per cent dry weight for commercial production to be economically feasible.

The polymer poly(3-hydroxybutyrate) (PHB) bacterium *Ralstonia metallidurans* from glucose is the most extensively studied PHA naturally synthesised by the. It has similar properties to polypropylene, although it is more brittle, making it less stress resistant for industrial applications.

It is naturally present at low levels in the cell walls of some plants. Higher levels of it were first produced in plants in 1992 in the model plant *Arabidopsis thaliana*. Since then researchers have attempted to produce PHB in a number of different plants with varying results; including oilseed rape, maize, sugarcane, flax, cotton, soyabean, palm oil, tobacco, switch grass, sugar beet, potato and alfalfa.

8.4.2.4. Spider Silk

Spiders spin silks with different properties for constructing webs, cocoons and draglines. For example, 'dragline' silk, used for the frames of a spider's web and for safety lines, is sturdier than high-tensile steel, but weighs less than one-tenth as much. Its tensile strength compares to that of the synthetic fibre, Kevlar, but it is more elastic. 'Capture spiral' silk has a lower tensile strength but can be extended to more than twice its length before breaking. These exceptional material properties have encouraged research into developing spider silks for industrial applications. Creating GM plants that can synthesise spider silk may allow large-scale production of these fibres.

8.4.2.5. Resilin

Resilin, an elastic protein found in specialised regions of the cuticle of most insects has low stiffness, high strain and efficient energy storage, and hence facilitates flight and jumping in insects. Its resilience and durability of could have both industrial and medical applications; for example, in spinal disc implants, vascular prostheses and high-efficiency rubber. There is an ongoing research on the possibility of using GM plants as a production system for resilin.

8.4.3. Broadacre Grain Crops and Oilseeds

8.4.3.1. Antibodies

The antibody 'T84.66' which recognises a marker for colorectal, lung, breast and pancreatic cancers, has been tested in Germany in both cancer imaging and therapy and has been produced in several different plant systems, including wheat, rice, tobacco and pea. An anti-Herpes Simplex Virus-2 (HSV-2) antibody, developed and expressed in soyabeans in the United States, has been shown to prevent vaginal HSV-2 transmission in mice. If a similar efficacy is shown in human clinical trials, then this antibody could represent an inexpensive protective measure for this disease.

8.4.3.2. Vaccines

A vaccine for the *E. coli* Heat-Labile Toxin protein diarrhoea caused by this toxin has been expressed in maize and potato in the United States to combat. The safety of the vaccine from each source is currently being tested in two independent phase I clinical trials.

8.4.3.3. Treatment of Allergic Reactions

Two proteins separated from Japanese cedar pollen (Cryj1 and Cryj2) have been expressed at a high level in GM rice in Japan. Initial testing in mice has shown that the consumption of this rice successfully treated allergic reactions to Japanese cedar pollen by causing immune system tolerance to develop. If this concept proves to work in people, treatments for a range of allergic reactions could be devised.

8.4.3.4. Research-grade Proteins

Aprotinin extracted from bovine lung tissue or derived from cows is an intravenously administered protein that helps prevent bleeding following cardiac surgery by blocking the action of certain enzymes in the bloodstream that dissolve blood clotsTP PT. It also has some other related research and manufacturing applications through maintaining protein stability. It has been successfully expressed in maize and tobacco plants.

Avidin is a protein naturally found in egg whites and has applications in research and diagnostics. The chicken avidin gene has been successfully expressed in maize. Avidin from GM maize was commercialised in the United States in 1997 and

β-glucuronidase derived from bacteria is an easily visualised reporter gene that is used in plant research and diagnostics has been expressed in many plant species, including maize. Lactoferrin, an iron-binding protein found in milk and some white blood cells, can be used to treat gastrointestinal infections in addition to topical infections and inflammations. The lactoferrin gene from humans has been successfully expressed in rice and maize. Lactoferrin derived from rice for research purposes has been commercialised in the United States.

Lysozymes are enzymes that can destroy the cell walls of certain bacteria and have practical application as antimicrobial agents. A human version of lysozyme has been successfully integrated in GM rice plants. Lysozyme isolated from this GM rice has been commercialised in the United States.

Trypsin is an enzyme that cleaves peptides and proteins at specific places. The trypsin gene from cows has been successfully expressed in GM maize plants.

8.4.3.5. Treatment for Cystic Fibrosis

Gastric lipase may be of used to treat cystic fibrosis and pancreatitis. Researchers from the French company Meristem Therapeutics have successfully expressed gastric lipase in GM maize plants. The resulting drug is currently undergoing 'formulation optimisation' to optimise the delivery and activity of Merispase®.

8.4.3.6. Cost-effective Biofuels

GM maize seeds are produced with increased sugar production. The modified seeds contain an enzyme that turns maize starch into sugar for ethanol production. This heat-stable enzyme should make ethanol easier to produce and reduce costs by eliminating the need for mills to add liquid enzymes.

8.4.3.7. Alternative Sweeteners

Brazzein with an intrinsic sweetness 500–2000 times that of sucrose is a protein. It can be isolated from the fruit of the African plant *Pentadiplandra brazzeana,* but it is uneconomical to produce on a commercial scale due to the limited availability of the fruit and complications associated with large-scale production of the native plant. Widespread commercial production of brazzein is, therefore, only likely to occur if the protein is expressed in an existing crop plant. GM maize plants expressing high levels of brazzein of up to 4 per cent of total soluble protein in maize seed were developed in the United States by ProdiGene, who reported brazzein accumulation. Tests indicated that maize germ flour containing brazzein could be directly used in food sweetening applications.

8.4.3.8. Production of Bioplastics

Moderate accumulation of PHB in commercial crop plants, with concentrations of 5.7 per cent dry weight and 7.7 per cent mature seed dry weight have been achieved for maize and rapeseed, respectively. Leaf chlorosis developed at higher levels of PHB accumulation in maize. No changes in size, appearance and germination frequency of GM rapeseed seeds were observed.

8.4.4. Tobacco

Tobacco produces abundant biomass, making it a useful platform crop for many plant molecular farming applications.

8.4.4.1. Antibodies

The antibody 'CaroRx™' binds specifically to *Streptococcus mutans* preventing the bacteria from stricking to the teeth. CaroRx™ produced in tobacco has reached phase II clinical trials in the United States.

A mAb recognising human chorionic gonadotropin is being expressed in tobacco in Germany. This antibody could potentially be used for pregnancy detection, emergency contraception and diagnosis or therapy of tumours that produce the pregnancy hormone human chorionic gonadotropin.

Antibodies have been developed that could be used to treat non-Hodgkin's Lymphoma. These have been expressed in a GM tobacco virus, which is used to infect tobacco plants, which then express the antibodies in the infected plant tissue. Researchers in Cuba have developed GM tobacco plants that make a mAb that is used for purifying a component of a vaccine for Hepatitis B.

8.4.4.2. Vaccines

An oral PMV for hepatitis B in humans has been successfully produced in tobacco plants, potato and lettuce.

A PMV to combat the infectious agent of bubonic plague has been developed in the United States. Researchers have modified tobacco plants to generate high levels of different antigens, from the plague bacterium *Yersinia pestis*. Early animal trials have shown that the vaccine elicits a protective immune response. Genetically modified tomato plants have also been used for plague vaccine proof-of-concept studies.

8.4.4.3. *Alternative Sweeteners*

Palatinose, a low calorie alternative to sucrose, has been produced industrially by using bacteria such as *Erwinia rhapontici* to convert sucrose into palatinose. Since this method is costly and the scale is limited, the widespread use of palatinose as an alternative sweetener is limited. The enzyme that converts sucrose to palatinose has been successfully cloned from *E. rhapontici* and expressed in both tobacco and potatoes in Germany. Tobacco though is not the ideal plant for this application whereas GM potatoes converted almost all their sucrose into palatinose, without affecting growth or development.

8.4.4.4. *Spider Silk*

GM plants that can synthesise spider silk have to be produced to allow large-scale production of these fibres. Successful expression of spider silk proteins has been achieved in tobacco and potato in Germany, with an accumulation of up to 2 per cent of total soluble proteins in leaves. Spider silk proteins have also been synthesised in *A. thaliana*. However, silk proteins isolated from these GM plants have not yet been successfully spun into a fibre. Given that silk properties are dependent on the structure of the fibre, which is influenced by the way the proteins are spun, advances in artificial spinning technologies are required before spider silk fibres will be commercially available.

8.4.5. Horticultural Crops

8.4.5.1. *Vaccines*

Oral PMVs for bacterial and Norwalk Virus-induced diarrhoea have been produced in GM potatoes, though it is not yet at the appropriate level for the next stage of product development to continue. A vaccine against a glycoprotein from rabies has been expressed in spinach, wherein a GM plant virus expressing the rabies vaccine has been used to infect spinach plants. The vaccine is then expressed in the infected plant tissue. This vaccine is yet being researched.

8.4.5.2. *Alternative Sweeteners*

Miraculin, a taste-modifying protein, does not have a sweet taste by itself, but can change a sour taste into a sweet taste. The maximum sweetness was estimated to be approximately 400, 000 times that of sucrose on a molar basis; which means the same sweetening effect can be achieved with only minute amounts of miraculin. As with brazzein, commercial production of miraculin is currently not economically viable, as the protein comes from a tropical plant that is difficult to cultivate outside its natural environment. The expression of miraculin in GM plants could provide an attractive alternative production method. To date, the protein has been successfully expressed in GM lettuce in Japan and displayed strong sweetness-inducing activity.

References

Sijmons, Peter J. "Production of Correctly Processed Human Serum Albumin in Transgenic Plants". Mogen International NV, Einsteinweg 97, 2333 CB Leiden, The Netherlands.

Twyman, Richard M. "Molecular farming in plants: host systems and expression technology". *Science Direct*. 28 October 2003.

Christou; P, Ma JK, Drake PM. "Genetic modification: The production of recombinant pharmaceutical proteins in plants". *Nature Reviews Genetics*. October 2003.

Mandel, Charles. "Molecular Farming Under Fire". wired. http://www.wired.com/medtech/health/news/2001/11/48108.

9

METABOLIC ENGINEERING

The aim of metabolic engineering in plants, involving the modification of endogenous pathways to increase flux towards particular desirable molecules, is to increase the production of a natural product, whereas in others it is to synthesize a novel compound or macromolecule. Since a few years, there have been appreciable advances in metabolic engineering through the application of genomics and proteomics technologies to elucidate and characterise metabolic pathways in a holisticmanner, rather than on a step-by-step basis, supported by direct studies at the level of the metabolome. Further work has been carried out on a genomic scale on the modelling of metabolic pathways. Such studies reveal that metabolic pathways are managed at multiple levels and any form of interference can have wide-ranging effects at the whole system level.

As a consequence the manipulation of single genes is of only limited value in metabolic engineering, and focus has moved towards more complex and sophisticated strategies in which several steps in a given pathway are modified simultaneously to achieve optimal flux.

9.1. Strategies for Metabolic Engineering

The basic goals of metabolic engineering in plants include the production of more of a specific desired compound, the production of less of a specific unwanted compound, and the production of a novel compound (i.e. a molecule that is produced in nature, but not usually in the host plant, or a completely novel compound). The goals need to be achieved by the engineering of single steps in a pathway to increase or decrease metabolic flux to target compounds, blocking competitive pathways or to introduce short cuts that divert metabolic flux in a particular way. However, the effects of modulating single enzymatic steps are often absorbed by the system in an attempt to restore homeostasis thus limiting the application. Targeting multiple steps in the same pathway could help to control metabolic flux in a more predictable manner involving upregulating several consecutive enzymes in a pathway, upregulating enzymes in one pathway while suppressing those in another competing pathway, or using regulatory genes to establish multipoint control over one or more pathways in the cell.

9.1.1. Engineering Primary Metabolic Pathways

9.1.1.1. Carbohydrate Metabolism

Carbohydrate metabolism in plants involves the reversible conversion of sugars, the direct products of photosynthesis, into storage and structural carbohydrates such as starch and cellulose. Starch is a storage carbohydrate that collects fleetingly in leaves and stably in seeds, tubers and roots. Apart from being a staple source of dietary carbohydrate for animals it also has a large number of industrial uses. Metabolic engineering increases starch yields and modifies the properties of starch by changing the relative proportions of its two components, amylose and amylopectin. There has also been some efforts in producing novel starches using bacterial enzymes to change the nature and frequency of branching. Recent examples of metabolic engineering to enrich the quantity of starch include overexpressing the potato sucrose transporter in transgenic pea seeds to accelerate sugar uptake and boosting the adenylate pool in potato tubers, resulting in a 60 per cent increase in the level of starch in combination with an increased tuber yield. Researchers have shown that wheat and rice plants transformed with a modified form of the maize gene for ADP-glucose pyrophosphorylase utilise their spare capacity for starch biosynthesis and produce 40 per cent and 20 per cent more seed biomass, respectively, than wild-type controls.

Several enzymes that control starch synthesis and branching have been modulated in potatoes to yield high-amylose starch and high-amylopectin starch. More recently, with similar experiments in other starch-producing crops, a low-amylose rice grain has been produced by expressing an antisense waxy gene, and rice grains with modified amylopectin have been produced by antisense inhibition of the gene encoding isoamylase. There is an ongoing research on several novel strategies for starch metabolic engineering and the use of starch-binding domains to target recombinant enzymes to starch granules during starch biosynthesis.

Since cellulose is valuable as pulp, fibre and as a starting point for the synthesis of commercially important polymers the cellulose biosynthetic pathway is also a useful target for metabolic engineering. Although the cellulose biosynthesis pathway is ye to be fully understood, several enzymes required for cellulose synthesis have been identified, including sucrose synthase, the CesA family of cellulose synthases, and Korrigan cellulases. The inhibition of sucrose synthase in cotton is shown to stunt fibre formation, whereas antisense and dominant negative experiments involving CesA genes in Arabidopsis demonstrate roles for several of the CesA proteins in primary and secondary cell-wall formation.

The fructans, another group of storage carbohydrates that are important targets for metabolic engineering, have a similar texture to fats and can substitute for them in many foods, yet contain far fewer calories. Presently, the only commercially available fructan is inulin from chicory, but bacterial and plant enzymes for fructan biosynthesis have been integrated into several crops to increase fructan levels to levels suitable for extraction. Fructans are also seen to protect plants from abiotic stress. This application is shared with sugars such as trehalose, as well as other small-molecule compounds, like glycine betaine, mannitol and sorbitol, all of which have been subject to metabolic engineering to produce improved plant varieties.

9.1.1.2. Amino Acid and Polyamine Metabolism

Metabolic engineering targets amino acid synthesis pathways predominantly to increase the abundance of essential amino acids (lysine, threonine, methionine and tryptophan) in food and feed crops. Other amino acids, such as proline, are essential in plant stress responses and their modulation provides a way to engineer stress tolerance in transgenic plants. Recent instance of amino acid pathway engineering show the impact of RNA interference (RNAi) on metabolic engineering and the importance of multiple point intervention. A new version of the maize *opaque2* mutant has been generated by RNAi, using a construct targeting the 22 kDa zein genes. Like the original mutant, the new version of *opaque2* has an enhanced lysine content, but unlike the original it behaves in a dominant rather than recessive manner. The lysine content of *Arabidopsis* seeds can be increased either by expressing a bacterial, feedback-insensitive dihydrodipicolinate synthase (DHPS) transgene or by knocking out the lysine catabolism pathway, leading to twelvefold or fivefold gains in lysine, respectively. However, where both the transgene and knockout are merged in the same plant line, an 80-fold increase over wild-type levels can be achieved.

Polyamines, small aliphatic amines, are extracted from the amino acids ornithine and arginine by decarboxylation. They are important in plant physiology and development. The three main polyamines found in plants are putrescine, spermidine and spermine. Recent analysis of the polyamine biosynthesis pathway has centred on tobacco and rice, and many enzymatic steps being studied by overexpression or antisense suppression. Rice plants expressing S-adenosylmethionine decarboxylase (SAMDC) from carrot indicated increased levels of the higher polyamines (spermine and spermidine) in seeds, while only spermine levels were affected in leaves. When a yeast SAMDC gene was expressed in tomato, higher levels of spermine, spermidine and lycopene were detected in fruits, increasing the nutrient value of the tomato juice. Inhibition of the arginine decarboxylase (ADC) gene in rice using an oat antisense ADC construct lowers the levels of putrescine and spermine in transgenic plants, but does not effect downstream genes in the pathway.

9.1.1.3. Lipid Metabolism

The modification of lipid metabolism to alter the quantity and quality of fatty acids in plants is important in the food industry, and in the production of detergents, fuels, lubricants, paints and plastics. Since extensive modifications do not affect the growth and development of the vegetative parts of the plant the metabolic engineering of seed oils is attractive. Oilseed rape (*Brassica napus*) was transformed with multiple cDNAs encoding enzymes for the desaturation and functional modification of long-chain fatty acids. Researcher expressed 18:1 Δ12 desaturase either alone or in combination with 18:2 Δ6 desaturase. With the 18:1 Δ12 desaturase alone, the seeds produced up to 46 per cent linoleic acid, whereas with both enzymes the seeds produced up to 43 per cent γ-linolenic acid. Researcher expressed a combination of β-ketoacyl-CoA synthase and 22:1 acyl-CoA:lysophosphatidic acid acyltransferase in oilseed rape, yielding 60 per cent erucic acid. More recently, the a-linolenic acid content of rice grains has been increased tenfold by expressing the soybean FAD3 gene encoding Ω3 fatty acid desaturase. Researcher expressing a cytochrome P450 isolated from seeds of *Euphorbia lagascae*, have produced up to 7 per cent epoxy-fatty acids without inhibiting oleate desaturation.

9.1.2. Engineering Secondary Metabolic Pathways

9.1.2.1. Alkaloids

The largest group of secondary products synthesized by plants are alkaloids. The pathways, enzymes and regulatory genes involved in their synthesis have been extensively researched. Several distinct types of alkaloids can be distinguished, including the isoquinoline, indole and tropane classes. Recent studies shows have recently demonstrated antisense suppression of the berberine bridge enzyme in poppy cells, reducing the amount of benzophenanthridine alkaloids but increasing the levels of several amino acids. There have been great efforts to modulate the indole alkaloid biosynthetic pathway, which produces important pharmaceutical compounds, such as the anti-cancer drugs vinblastine and vincristine.

9.1.2.2. Terpenoids

The structurally diverse polymers, terpenoids (isoprenoids) are constructed from the basic building block isoprene. They are classified according to the number of isoprene units they contain, which increase by fives. Thus, hemiterpenes comprise five isoprene units, monoterpenes contain 10, sesquiterpenes contain 15 and diterpenes contain 20. Carotenoids, another important group of plant secondary metabolites, are terpenoids with 40 isoprene units. The largest polymers comprise thousands of units and include natural products such as rubber. Regardless of their diversity, all terpenoids are synthesized from the common precursors dimethylallyl pyrophosphate and isopentyl pyrophosphate. This takes place through two distinct pathways, the mevalonate-independent pathway and the mevalonate pathway, both of which have been targets for metabolic engineering.

So far, there have been few examples of monoterpenoid metabolic engineering. In one study, petunia was transformed with the (S)-limonene synthase gene from *Clarkia breweri* the outcome of which was the production of linalool that repels aphids. In another, the same gene was inserted into tomato, under the control of a fruit-specific promoter, resulting in the collection of linalool in tomato fruits.

More focus has been on the carotenoid metabolic pathway, which produces pigments with importance in light harvesting and photoprotection, and which form vital components of human and animal diets. In plants, the first step in carotenoid biosynthesis is the pairing of two geranyl phosphate molecules to yield geranylgeranyl pyrophosphate (GGPP). This is converted into phytoene by the enzyme phytoene synthase and then to ζ-carotene by the enzyme phytoene desaturase. Further desaturation produces formation of lycopene, which is converted either into β-carotene (provitamin A) by the enzyme β-cyclase, or into δ-carotene and then α-carotene by consecutive enzymatic steps catalysed by ε-cyclase and β-cyclase.

The best example of carotenoid metabolic engineering in plants is of course the synthesis of b-carotene in rice endosperm (Golden Rice), which normally collects GGPP but is deficient in the subsequent enzymes in the pathway. More 'Golden Rice' lines have been generated lately, containing only two foreign enzymes. Studies show the production of indica Golden Rice varieties that are amenable suitable to deregulation and also ensure freedom to operate in developing countries due to the removal of constraints brought about by intellectual property and regulatory issues.

Transgenic tomatoes are also shown expressing phytoene desaturase or phytoene synthase. In the former case, there was no overall increase in carotenoid content, although b-carotene levels were three times higher than normal; in the latter case, the total carotenoid content was up to fourfold normal levels.

9.1.2.3. Flavonoids

The flavonoids, which are phenolic compounds derived ultimately from phenylalanine, share a common metabolic source with lignins and lignans. The first committed step in flavonoid biosynthesis is the conversion of the precursor 4-coumaroyl-CoA into chalcone by the enzyme chalcone synthase. Chalcone is then procured in a series of enzymatic steps to produce a variety of molecules that act as pigments, defence chemicals (phytoalexins) and regulatory molecules. Flavonoids have been a favourite useful for metabolic engineering since the early 1990s, because the result of different modifications can be determined through alterations in flower colour. As for other pathways, there are now more experiments involving multiple gene transfer strategies. A recent example is the stacking of dihydroflavonol 4-reductase and anthocyanidin synthase transgenes in forsythia by sequential transformation. Increments in the levels of flavonoids in food plants can provide health benefits, as these molecules often have antioxidant activity, as shown in the case of tomatoes by overexpressing the petunia gene for chalcone isomerase, leading to an 80-fold increase in the flavonoid content of the tomato peel and a corresponding 20-fold increase in the flavonoid level in tomato paste. Also, chalcone synthase and flavonol synthase transgenes were found to act synergistically to upregulate flavonol biosynthesis significantly in the flesh of tomato fruits.

9.1.2.4. Lignins

Lignins are complex polymers composed of caffeic, ferulic and sinapic acid monomers joined together and to the cell wall. Like flavonoids, they are derived from phenylalanine and can be classed as phenylpropanoid compounds. One interesting example is the use of the LIM regulatory protein from tobacco to control several of the metabolic enzymes used in the lignification pathway coordinately. Similarly, two Myb-like transcription factors have been isolated from Eucalyptus gunni, which are expressed in the xylem. These recognise sites in the hydroxycinnamoyl-CoA reductase (CCR) and cinnamyl alcohol dehydrogenase (CAD) genes which encode sequential enzymes in the synthesis of p-coumaryl alcohol. Recent reports also show the viability of multiple gene transfer to modify several lignin characteristics simultaneously in trees and the simultaneous suppression of multiple lignin biosynthetic genes in tobacco.

9.1.2.5. Quinones and other Benzoic Acid Derivatives

Quinones derived from benzoic acid are cyclic compounds and they often act as electron transporters in plants. So far, three major metabolic pathways that begin with benzoic acid have been modified in plants — those leading to salicylic acid, shikonin and α-tocopherol (vitamin E) production. Vitamin E synthesis in plants has obvious roles in the nutritional value of foods, and recent work has aimed to increase the levels of this molecule in the model plant *Arabidopsis thaliana*. For example, researchers overexpressed *A. thaliana* homogentisate prenyltransferase producing twice the level of vitamin E found in normal seeds, and *A. thaliana* VTE3 and VTE4 genes, encoding 2-methyl-6-phytylbenzoquinol methyltransferase and γ-tocopherol methyltransferase, in transgenic soyabean. The transgenic seeds manifested a greater than eightfold increase in α-tocopherol and a fivefold increase in vitamin E levels.

9.1.3. Engineering Novel Metabolic Pathways

Existing pathways in the host plant care often modified or extended using metabolic engineering. In some cases multigene engineering has been used to initiate completely novel pathways and thus to produce completely alien products. In one such experiment two multifunctional cytochrome P450 enzymes and a uridine diphosphate glucose (UDPG)-glucosyltransferase were transferred from sorghum (*Sorghum bicolor*) into *A. thaliana,* producing cyanogenic glucoside dhurrin (which confers resistance to the flea beetle *Phyllotreta nemorum*). Transgenic plants carrying one or both of the cytochrome P450 enzyme were also generated, but they lacked UDPG-glucosyltransferase, and demonstrated the accumulation of appropriate metabolic intermediates.

Employing this general approach, another example is the production of polyhydroxyalkanoate (PHA) polymers in transgenic plants. PHAs are polyesters of hydroxyacids synthesized by a range of bacteria, which can be used as biodegradable plastics. The best-studied PHA, poly(3-hydroxybutyrate) (PHB), can be made to collect in the plastids of transgenic *A. thaliana* plants at levels exceeding 40 per cent dry shoot weight. This is done through the introduction of three bacterial genes, encoding the enzymes 3-ketothiolase, acetacetyl-CoA reductase and PHA synthase. PHB is a brittle plastic with few potential uses. The copolymer poly(3-hydroxybutyrate-co-3-hydroxyvalerate), more beneficial produced in *A. thaliana* and rapeseed through the coordinated expression of four bacterial genes: the three mentioned above plus *Escherichia coli* threonine deaminase. Although high-level accumulation of these plastics generally had adverse effects on plant growth, both plastics have recently been produced at relatively high levels in alfalfa with no harmful effects.

9.2. Metabolic Engineering for Plant Secondary Metabolites

Over the past few years plant cell biotechnology has emerged as a potential area within the field of biotechnology, focusing on the production of plant secondary metabolites. For most compounds of interest, e.g. morphine, quinine, vinblastine, atropine, scopolamine and digoxin, there is to yet be a commercially feasible process. In contrast with the production of antibiotics by micro-organisms, the plant is an already existing, although not always a reliable source of these compounds. This has constrained the amount of research put into developing alternative biotechnological production for each of the products mentioned. Half a century of extensive research on the production of penicilline has yielded 50 g/l and more. In plant cell biotechnology only some products have been able to considerably increase the production levels, e.g. shikonin (3.5 gA) and berberine (7 gA). Presently research particularly focuses on the viability to apply metabolic engineering to improve yields to commercially interesting levels.

9.2.1. Technological and Economical Feasibility

Since plant cells are extremely sensitive for shear forces, there is a need to the use special low-shear bioreactors, e.g. air-lift bioreactors. In industry such bioreactors are not common, as most processes are run in stirred-tanks. Hence such a bioreactor is preferable for plant cell cultures, for it is the lowest cost process-unit. More recent research on the shear sensitivity of plant cells, among others, have shown that in fact plant cells in general are quite shear-stress tolerant. This is endorsed by the fact that a series of large scale processes have been reported with plant cell cultures, e.g. shikonin production. Plant cells have even been cultured in a 60 m^3 stirred tank.

Regarding its economy assuming a yearly production of 3000 kg/year of a compound produced by a cell culture at a level of 0.3 g/l, resulted in a calculated price of 1500 US $/kg (1 1,12), an increase of productivity with a factor 10 (i.e. 3 g/l) results in a price of 430 $/kg. In both cases a fed-batch type of process is applied. These prices are high, but several natural products have even much higher prices. However, most of the high-value speciality chemicals are produced at too low levels in the plant cell cultures.

9.2.2. Improving Production

In order to improve yields of secondary metabolites in plant cell cultures many strategies are being employed. First of all the screening and selection of high producing cell lines and the optimisation of growth and production media with success in shikonin and berberine with and in many others with limited success.

New approaches on the past years have been developed: the culturing of differentiated cells, induction by elicitors and metabolic engineering. With the culture of differentiated cells mostly production of the desired compounds in levels has been achieved comparable to that of the plant. However the culture of such differentiated tissues on a large-scale in bioreactors is a major constraint. For studies of the biosynthesis, such systems are very useful.

The second approach the use of elicitors, has been successful in several cases but limited to a certain type of compound for each plant, compounds which most likely act as phytoalexins in these plants.

Therefore attention is more and more focused on metabolic engineering to increase yields. Several possibilities are:

- increase activity of enzymes which are limiting in a path-vay;
- induce expression of regulatory genes;
- block competitive pathways;
- block catabolism.

The first two possibilities need the expression of genes producing active enzymes with the latter two approaches blocking of genes by antisense genes. In all cases the respective biosynthetic pathway has to be understood on the level of products, enzymes and genes, as well as the regulation on all these levels, including aspects as compartmentation and transport. In studies of the production of terpenoid indole alkaloids cloned genes coding for tryptophan decarboxylase (TDC), smctosidine synthase (SSS) and NADPH:cytochrome P-450 reductase. The latter enzyme serves all cytochrome P-450 enzymes in the plant, among others geraniol-10-hydroxylase (GlOH), a crucial enzyme in the terpenoid part of the alkaloid biosynthesis. The gene encoding for GlOH has not yet been cloned, despite the fact that the enzyme has been purified to homogeneity. This is because the plant contains many, very similar, P-450 genes. By means of PCR, 18 closely related genes coding for P-450 enzymes were detected in *Carharunthus roseus*.

The *fdc*-gene, expressed in tobacco plants, resulted in an active enzyme. These plants produce upto 1 per cent of DW in tryptamine, i.e., similar levels as found for the tryptamine-derived indole alkaloids in plants producing such compounds. The activity of anthranilate synthase, the first committed enzyme of the tryptophan-pathway from chorismate, was not enhanced in the tryptamine producing transgenic tobacco plants. Plants can apparently make aoout 1 per cent of its DW in tryptophan-derived compounds with its normal primary metabolic machinary. Integration of the rdc-gene into C. *roseus* showed callus cultures increasing up to 10-fold levels of TDC activity and a concomminant increase of tryptamine levels, but no significant increase of strictosidine or other alkaloids. This authenticates other observations that the terpenoid part of the pathway is also a limiting factor. The *tdc*- and sss-genes are both single copy genes, repressed by auxins and induced by elicitation. Due to this similar regulation, they might be controlled by one transacting factor. Identification of this factor and the encoding gene generates great interest, as such a regulatory gene might also control further steps in the biosynthetic pathway. The cloning of such a regulatory gene would eliminate the need to clone a large number of genes coding for the individual steps in the biosynthetic pathway.

In C. *roseus* a 25-50 per cent increase is observed in the activity of AS, TDC and S S S upon elicitation, whereas the activities of PAL decreases and chalcone synthase is not affected. However, there is a significant change in the content of phenolics, 2,3-dihydroxybenzoic acid (DHBA) being the major compound formed. Simultaneously a strong induction is seen of the enzyme isochorismate synthase, which convert chorismate into isochorismate. In bacteria this pathway is known to result in DHBA. This compound has antimicrobial activity, particularly in combination with UV-light, and has also antifeedant properties, thus likely to be a major defence compound. Phytoalexin-biosynthesis in *Cinchona* cell cultures (anthraquinones) also use chorismate as precursor as well as mevalonic acid. In *Tabernaemontana* species mevalonate is used for the mterpene biosynthesis after elicitation, siphoning of this precursor from the alkaloid pathway. Blocking such pathways by means of antisense genes might increase the availability of these precursors for the alkaloid biosynthesis.

Catabolism is another key aspect in the collection of the alkaloids. Studies feeding ^{15}N or ^{14}C labelled alkaloids to *Tabernaemntana divaricata* and *C. roseus* cell cultures showed that at a certain point of the growth phase the rate of catabolism equals the *de-novo* biosynthesis. Identification of the enzymes involved and the cloning of the encoding genes are thus of interest to eventually use antisense gene technology to block catabolism and thus increase the production.

9.3. Metabolic Engineering of Plant Volatiles

Secondary or specialised metabolites an amazing diversity of low molecular weight organic compounds. More than 1 per cent of these metabolites are lipophilic molecules with low boiling points and high vapour pressures at ambient temperature, and mainly represented by terpenoids, phenylpropanoids/ benzenoids, fatty acid derivatives and amino acid derivatives. These volatile compounds are released from leaves, flowers and fruits into the atmosphere and from roots into the soil. The main functions of airborne volatiles are to safeguard plants against herbivores and pathogens, to attract pollinators, seed dispersers, and other beneficial animals and microorganisms, and to serve as signals in plant–plant interaction. The contribution of volatiles to plant survival and overall reproductive success in natural ecosystems, and their effect on agronomic and other commercial traits, including yield and food quality, indicate that modification of volatile production via genetic engineering has the potential to improve cultivated plant species.

Metabolic engineering involves biochemical pathways and the identification of the genes and enzymes involved in the synthesis of volatile compounds. In the recent past a renewed interest in these questions combined with technical advances have resulted in both a large increase in the number of plant volatiles identified and remarkable progress in discovering the genes and enzymes of volatile biosynthesis. Various attempts have been made to modulate volatile profiles in plants via metabolic engineering to enhance direct and indirect plant defence and to improve scent and aroma quality of flowers and fruits. There have been success in achieving the desired goals, but many other attempts show meagre enhancement of volatiles or in other unpredicted metabolic consequences such as further metabolism of the intended end products or deleterious effects on plant growth and development.

Recently in response to herbivore attack plants emit diverse volatile blends that may be comprise more than 200 different compounds. These emitted volatiles can directly intoxicate, repel or deter herbivorous insects, or they may attract natural predators and parasitoids of the offending herbivores, thus indirectly protecting the signalling plant from further harm.

Several reports on the involvement of volatiles in plant defence suggest that plant protection in agricultural and forest ecosystems can be enhanced by the modulation of the volatile spectrum via metabolic engineering, thereby creating an alternative pest-management strategy based on biological control.

The herbivore enemies being efficient in controlling the herbivore population must be present in the area where the crop is grown. Second, the introduced or enhanced plant volatile blends must provide key elements to attract herbivore enemies, and volatile release must be synchronised with herbivore activity. Finally, the released volatiles should not enhance the attractiveness of the plant to non-specific herbivores. Apart from direct and indirect induced defences, herbivore-induced volatiles can act as airborne signals that warn neighbouring plants about the pathogen attack and encourage them to respond more strongly against future insect attack or serve as signals within a plant and prime systemic defences.

Although the molecular mechanisms involving priming are not known, priming prepares the plant or its undamaged parts for rapid defence but delays the response until the actual herbivore attack. Since the defence network remains mostly inactive until actual herbivore attack priming-related expenses are much lower than those of the induced direct defence, and the benefits of priming outweigh its costs when disease occurs. Priming crops by planting a few transgenic plants that constantly produce defence volatiles in the field may offer an efficient form of plant protection and provide an advantage to non-transgenic receiver plants. However, the full benefits of this approach will only become possible with a clear understanding of the molecular mechanisms of volatile-induced priming, the determination of the major volatile signal components that trigger it and their species specificity, and the identification of reliable molecular markers for the primed state.

Probably the array of volatiles produced from a given species emerged as adaptations to specific challenges and, it is not surprising that some plant enemies have in turn emerged to take utilise such emissions to locate their plant 'prey' for food, egg deposition, and the raising of their young. In addition, recently studies show that volatiles released from plants can also provide chemical cues to parasitic weeds, the most damaging agricultural pests, for host location and discrimination.

A study of the parasite–host interactions and, in particular, the role of volatiles in these plant–plant interactions, will help in the development of new techniques for the non-herbicidal control of weed populations via the metabolic alteration of the hosts' volatile spectrum. So far very little is known about the attractive and repelling properties of specific plant volatiles, the main compounds involved in plant–insect and plant–plant interactions, and the molecular mechanisms of their action. The modulation of the volatile spectrum has already proved a useful strategy for increasing volatile-based plant defences. The overexpression of strawberry linalool/nerolidol synthase *(FaNES1)* targeted to chloroplasts showed transgenic Arabidopsis producing high levels of linalool which repelled the aphid *Mysus persicae* in dual-choice assays.

The ectopic expression of the same *FaNES1* gene in transgenic potato enhanced the linalool level and affected tritrophic interactions, producing transgenic plants that were more attractive to predatory mites than the uninfested wild-type plants. Improvement of Arabidopsis indirect defence was also attained by overexpressing *FaNES1* in mitochondria, which has the sesquiterpene precursor farnesyl diphosphate (FPP). This manipulation led to the synthesis and emission of (3S)-(E)-nerolidol as well as the C_{11} homoterpene 4,8-dimethyl-1,3(E),7-nonatriene [(E)-DMNT], believed to be a degradation product of nerolidol, and made the plants attractive to the carnivorous predatory mites *Phytoseiulus persimilis,* natural enemies of spider mites.

A similar successful example involved the overexpression of a maize terpene synthase gene (TPS10) in Arabidopsis that produced transgenic plants with strong emission of several sesquiterpenes that are typically released (in maize) after herbivory by lepidopteran larvae. These transgenic plants were more appealing to the female parasitic wasp *Cotesia marginiventris,* which had had a previous oviposition experience with larvae of the potential host. Moreover, production of the volatile patchoulol and 13 additional sesquiterpene products in transgenic tobacco overexpressing patchoulol synthase (PTS) inhibited tobacco hornworms, a majority of which had drifted away from leaves of the transgenic plants to the leaves of wild-type plants and devoured 20–50 per cent more of the wild-type plants within six hours.

Metabolic engineering of volatile signals involved in direct and indirect defences are not confined to terpenoids. An increase in (Z)-3-hexenal, a major green leaf volatile, was obtained in transgenic tobacco plants overexpressing either the yeast acyl-CoA Δ9 desaturase or the insect acyl-CoA Δ11 desaturase. The expression of these transgenes elevated the levels of 16:1 fatty acids and increased 13-lipoxygenase activity, which catalyses the first step to hexenal production from α-linolenic acid. While the effect of elevated levels of (Z)-3-hexenal on insect behaviour is yet to be researched, the negative effect of this compound on aphid performance was proved in transgenic potato plants with reduced levels of the hydroperoxide lyase enzyme, which is responsible for the splitting of fatty acid hydroperoxides to C_6 aldehydes.

The potential of genetic engineering for the improvement of plant defence as well as the role of some volatile compounds in plant–insect interactions have been proved by researchers. Despite this development there has also been the effect of genetic perturbations on plant growth and development, and some challenges to achieving efficient production of the desired volatile terpenoid compounds have been revealed. In fact, the transfer of carbon to linalool production in Arabisopsis via *FaNES1* overexpression, while not effecting the levels of plastidderived isoprenoids such as chlorophylls, lutein and bcarotene, resulted in a growth-retardation phenotype that was inherited through several generations of transgenic plants. Emission of linalool in transgenic potato showed a more severe phenotype; apart from growth retardation, plants had bleached leaves after their transfer from *in vitro* to the greenhouse. Leaf chlorosis, vein clearing, and reduced stature were also seen in transgenic tobacco producing high levels of patchoulol as a consequence of the expression of PTS coupled with FPP synthase, both targeted to the plastids. These observed phenotypes could be the effects of the exhaustion of isoprenoid precursors for other metabolites essential for plant growth and development, or possibly the toxic effects of the newly introduced terpenoids in plant cells.

For metabolic engineering of the volatile spectrum to be successful, it is essential to produce and emit sufficient amounts of the desired compounds. However, the metabolic consequence of newly synthesized compounds will be determined by the entire biochemical repertoire of the plant used. There is a complete lack of understanding of the biochemical repertoire in any plant species, and hence, it is difficult to predict how much of the desired compound will actually sustain in the desired form. For example, the newly synthesized compounds may be affected by enzymes that are normally present in the cell and have broad substrate specificity, such as dehydrogenases, glucosyl transferases, and others. Presently there is lack of knowledge of such enzymes in general and their specific distribution in different plant species.

In the beginning there were efforts to increase terpenoid production in transgenic plants which showed that metabolic engineering of sesquiterpenes is not as simple as the generation of monoterpenes, which are formed exclusively or at least predominantly via the methylerythritol phosphate (MEP) pathway in the plastids. In several cases, FPP, which is expected to be produced in relatively large amounts for sterol biosynthesis, is not readily available for catalysis by introduced sesquiterpene synthases. Also the contribution of the cytosolic mevalonic acid (MVA) and plastidic MEP pathways to sesquiterpene formation and thus movement of isoprenoid intermediates between organelles, depends on the plant species, tissue and physiological state of the plant. So far the overexpression of TPS10 and PTS in Arabidopsis and tobacco, respectively, represent the two most successful attempts at generating high levels of volatile sesquiterpenes by enzymes targeted to the cytosol. However, to achieve emission of (3S)-(E)-nerolidol and (E)-DMNT in Arabidopsis, *FaNES1* had to be targeted to the mitochondria. In addition, targeting PTS along with FPP synthase to the plastids increased the amount of produced patchoulol up to 100 times compared with its cytosolic formation.

Of late there is realisation that plants emit volatile compounds from their roots into the rhizosphere which may help the plant attract beneficial microorganisms and deter harmful ones. They may also be useful in competition between plant species. However, it has been shown that some parasitic plants use underground volatile compounds to locate their hosts. As of now there are no proofs of genetic engineering changing root volatile emission in order to improve plant fitness; this is clearly a very fertile area for future work.

9.3.1. Metabolic Engineering of Floral Volatiles

As against metabolic engineering of vegetative volatiles where the result of altered emission profiles on insect behaviour was investigated, the impact of changes in floral scent on insect attraction is yet to be studied. Metabolic engineering of floral volatiles was considered successful when the transformation in scent profiles were significant enough for human detection. The olfactorily detectable enhancement of volatiles emitted from flowers and leaves was achieved in transgenic tobacco through the introduction of three citrus monoterpene synthases. In another experiment, the redirection of the metabolic flux from the anthocyanin pathway towards benzoic acid in transgenic carnations effected an increase of methylbenzoate production which was sufficient for olfactory detection by humans. However, much more effort to modify the scent bouquet is needed to in the absence of suitable substrates for the introduced reaction, modification of the scent compound into a non-volatile form, insufficient levels of emitted volatiles for olfactory detection by humans, or masking of introduced compound(s) by other volatiles.

Expelling some volatile compounds from the floral bouquet is another approach for scent modifications. Transgenic petunias lacking methylbenzoate, phenylacetaldehyde, benzylbenzoate and phenylethylbenzoate, and isoeugenol were obtained via RNAi-mediated post-transcriptional gene silencing. Improvement of aroma quality of fruits, vegetables and herbs volatiles are important determinants of the overall aroma properties and taste of fruits. In nature, volatiles contribute to seed dispersion by increasing fruit appeal. Volatiles released from vegetative parts of plants may also appeal to some animals or insects as foodstuff, even though they are unpalatable to most other herbivores.

The presence of volatiles in fruits, vegetables and herbs has significant effect on the cultivation of many plant species. Extensive breeding programmes to maximise certain attributes of foodstuff – for example, overall yield (i.e. size), total solids, sugar content, or pigmentation–involves less attention to enhancing or even maintaining volatile production. As a consequence many current cultivars of domesticated plant species produce less volatiles than their wild relatives or earlier cultivars.

Classical breeding will be helpful in reintroducing aroma volatiles as was done in tomato by crossing it with its relative L. *peruvianum*. However, this is a laborious and time-consuming process, requiring the monitoring of a complex trait. For example, volatile collections and analyses must initially be done with expensive gas chromatography–mass spectrometry (GC–MS) instruments and thereafter human evaluations must also be made by subjective test panels as it is not yet possible to predict how humans will react to a given mixture of volatile compounds. Hence, there is lack of emphasis in most breeding programmes on the aroma of produce.

Genetic engineering can ease some drawbacks of classical plant breeding and enhance aroma of fruits, one advantage being that it is less complex –introducing a single trait at a time. Another is that it allows the introduction of genes whose coding information may not be present in the cultivar. There are general problems and pitfalls of genetic engineering for biochemical traits. The addition of a single gene is unlikely to result in a substantial production of the desired volatile if the formation of this compound is the end result of a long metabolic pathway. Even more important is the fact that a single new volatile is generally unlikely to change the consumers' overall perception regarding the flavour and aroma quality of produce. The overexpression of a yeast Δ9-desaturase and a non-specific alcohol dehydrogenease (ADH) in tomato fruit were attempted to circumvent these problems. In these cases, the concentrations of various aroma compounds derived from fatty acids such as (Z)-3-hexenol, (Z)-3-hexenal and/or the ratios of the aldehydes to alcohols changed. The higher levels of alcohols in transgenic fruits were related to more intense ripe flavour by taste panelists. However, neither of these manipulations introduced new aroma compounds.

The introduction of the *Clarkia breweri* linalool synthase (LIS) gene into tomato under the control of the fruitspecific E8 resulted in the accumulation in the fruit of small amounts of linalool and its oxidation product, 8-hydroxylinalool, which were detectable by both GC–MS and the human nose. This was accomplished because linalool is produced from geranyl diphosphate (GPP) by LIS in a single step, and GPP is an intermediate in the synthesis of carotenoids, a pathway that is highly active in ripening tomato fruits. The synthesis of linalool and 8-hydroxylinalool did not affect the total amounts of carotenoids produced by the fruit; however, the amounts of these monoterpenes were also not enough to significantly change the overall flavour perception by humans.

A much stronger effect on flavour perception was recently achieved by researchers by expressing geraniol synthase (GES) in tomato under the polygalacturonase promoter, another fruit ripening-specific promoter. Geraniol is also an acyclic monoterpene alcohol that is synthesized in one step from GPP. Unlike linalool, which is a tertiary alcohol whose hydroxyl group cannot be further oxidised, geraniol is a primary alcohol that can easily be oxidised to geranial by non-specific alcohol dehydrogenases. GES-transgenic tomato fruits synthesized large amounts of geraniol, leading to a noticeable reduction in pigmentation. Moreover, transgenic fruits further metabolised geraniol to geranial, which underwent spontaneous tautomerisation to neral. Neral and geranial combine to make a mixture called citral, which imparts a strong lemon flavour. They were also further metabolised to geranic and neric acids, respectively. Additional modifications in them resulted in the formation of nerol, citronellol, citronellal, citronelic acid, citronellyl acetate, and rose oxide. When these transgenic fruits were assessed there were indications that the fruits had stronger aroma, and transgenic fruits were preferred over the non-transgenic ones.

Recent work in peppermint on boosting the output of terpenes favoured by humans (e.g. menthol) and decreasing the synthesis of unfavoured compounds were partially successful. Using antisense technology, transgenic plants were obtained that had a 50 per cent reduction in menthofuran concentration, while other transgenic plants showed some increase in the concentrations of limonene, a cyclic monoterpene. However, an assessment of consumer responses to the aroma properties of these transgenic mint plants is yet to be recorded.

References

Domach, M. M. *Introduction to biomedical engineering*. Upper Saddle River: Pearson Prentice Hall. 2004.

Goodman, D.C. *From Farming to Biotechnology: A Theory of Agro-industrial Development*. Oxford: Blackwell. 1987.

Stephanopoulos, G. N., Aristidou, A. A., Nielsen, J. *Metabolic Engineering: Principles and Methodologies*. San Diego: Academic Press. 1998.

Vemuri, G.N. and Aristidou, A. A. Metabolic Engineering in the -omics Era: Elucidating and Modulating Regulatory Networks. Microbiol. *Mol. Biol. Rev*. 69(2):197-216. 2005.

Windass, J. D. et.al., Improved conversion of methanol to single-cell protein by *Methylophilus methylotrophus*. In *Nature, 287*, 396 – 401. 1980.

Bibliography

Adelberg, Edward A. (ed.) *Papers on Bacterial Genetics*. Boston: Little, Brown and Company. 1960.

Ausubel, F.M. et al. *Current Protocols in Molecular Biology*. New York: John Wiley and Sons. 1989.

Bajai, Y.P.S. (ed.). *Biotechnology in agriculture and forestry*, 17: 3-16. Berlin, Springer-Verlag. 1991.

Beck, W.S., Liem, K.P., Simpson, G.G. *Life*. 3rd Edition. New York: Harper Collins. 1991.

Bhojwani, S. S., Razdan M. K. *Plant tissue culture: theory and practice, Revised edition* Elsevier, 1996.

British Medical Association. *The Impact of Genetic Modification on Agriculture, Food and Health*. BMJ Books. 1999.

Christou; P, Ma JK, Drake PM. "Genetic modification: The production of recombinant pharmaceutical proteins in plants". *Nature Reviews Genetics*. October 2003.

Conger, B.V. *Cloning agricultural plants via in vitro techniques*. CRC Press Inc. 1981.

Curtis, H. and Barnes, N. *Biology*. New York: Worth Publishers. 1989.

Debergh Zimmerman, (ed.) *Micropropagation: technology and application*. Kluwer Academic Publishers. 1991.

Dodds, J. H. and Roberts, L. W. *Experiments in plant tissue culture*. Cambridge University Press, Cambridge.1995.

Domach, M. M. *Introduction to biomedical engineering*. Upper Saddle River: Pearson Prentice Hall. 2004.

Donnellan, Craig. *Genetic Modification (Issues)*. Independence Educational Publishers. 2004.

Friedberg, E.C. *DNA Repair*. New York: WH Freeman and Company. 1985.

Gamborg, O. L. "Plant tissue culture. Biotechnology" . Milestones. *In vitro Cellular and Developmental Biology—Plant*, 38, 84–92. 2002.

Gao, Ji-Ping, et al. "Understanding Abiotic Stress Tolerance Mechanisms: Recent Studies on Stress Response in Rice." *Journal of Integrative Plant Biology* 49 (6): 742"750, 2007.

George, E. F. Hall, M. A. and G.-J. De Klerk (eds). *Plant propagation by tissue culture. Volume 1. The background. 3rd edition* Springer, Dordrecht, 2008.

Goodman, D.C. *From Farming to Biotechnology: A Theory of Agro-industrial Development.* Oxford: Blackwell. 1987.

Haines, R.J. "Mass propagation of cuttings, biotechnologies and the capture of genetic gain". Paper presented at IUFRO meeting, Bordeaux, September 1992. pp. 15. 1992.

Hamdan, Y.Y. 1990. "Current status of plant biotechnologies in the Near East and North African". In Sasson & Constantini, eds. *Plant biotechnologies for developing countries,* proceedings. p. 131-139. Int. symposium organized by CTA and FAO, Luxembourg, June 1989.

Hanover J.W. & Keathley, D.E. (eds.). *Genetic manipulation of woody plants,* P. 413-432. Plenum Press. New York. 1988.

Henshaw, G. "Technical aspect of tissue culture storage for genetic conservation". In: Frankel and Hawkes. *Crop Genetic Resources for Today and Tomorrow.* CU.P. 349-358. 1975.

Hu, C.Y., Wang, P.J. "Meristem, shoot tip, and bud cultures". In *handbook of plant cell culture* I. Ed. D. A. Evans, W. R. Sharp, P. V. Ammirato, and Y. Yamada. 1983.

Hussey, G. "In vitro propagation of horticultural and agricultural crops". In: *Plant Biotechnology* Ed. H. Smith. 1983.

Mandel, Charles. "Molecular Farming Under Fire". wired. http://www.wired.com/medtech/health/news/2001/11/48108.

Mittler, Ron. "Abiotic stress, the field environment and stress combination." *Trends in Plant Science* 11(1): 15–19, 2006.

Morel, G. "Meristem culture techniques for the long-term storage of cultivated plants". In *Crops Genetic Resources for Today and Tomorrow.* Ed. 0. H. Frankel, J. G. Hawkes. 1975.

Morgan, Sally. *Superfoods: Genetic Modification of Foods (Science at the Edge).* Heinemann. 2003.

Morholt, E. and Brandwein, P.F. *Sourcebook for the Biological Sciences* 3rd Edition. San Diego: Harcourt Brace JOvanovich. 1986.

Murashige, T. "Plant propagation through tissue culture". *Annual Reviews Plant Physiology* 25: 135-166. 1974.

Palta, Jiwan P. and Farag, Karim. "Methods for enhancing plant health, protecting plants from biotic and abiotic stress related injuries and enhancing the recovery of plants injured as a result of such stresses." *United States Patent* 7101828, September 2006.

Raven, Peter H., Evert, Ray F., & Eichhorn, Susan E. *Biology of Plants* (4th ed.). New York: Worth Publishers. 1986.

Roca, W.M., Bryan, J.E., Roca, M.R. "Tissue culture for the international transfer of potato genetic resources". *Am. Pot.* J. 56: 1-11. 1979.

Roelofs, D. et al. "Functional ecological genomics to demonstrate general and specific responses to abiotic stress." *Functional Ecology* 22: 8–18, 2008.

Sijmons, Peter J. "Production of Correctly Processed Human Serum Albumin in Transgenic Plants". Mogen International NV, Einsteinweg 97, 2333 CB Leiden, The Netherlands.

Smiley, Sophie. *Genetic Modification: Study Guide (Exploring the Issues).* Independence Educational Publishers. 2005.

Stephanopoulos, G. N., Aristidou, A. A., Nielsen, J. *Metabolic Engineering: Principles and Methodologies*. San Diego: Academic Press. 1998.

Torbert, K. A., Rines, H. W. and Somers, D. A. "Transformation of oat using mature embryo-derived tissue cultures" . *Crop Science,* 38, 226–31.1998.

Torrey, J.G. "The Development of Plant Biotechnology." *American Scientist*. 73:354-363. 1987.

Towle, Albert. *Modern Biology*. Austin, TX: Holt, Rinehart and Winston. 1988.

Twyman, Richard M. "Molecular farming in plants: host systems and expression technology". *Science Direct*. 28 October 2003.

Vemuri, G.N. and Aristidou, A. A. Metabolic Engineering in the -omics Era: Elucidating and Modulating Regulatory Networks. Microbiol. *Mol. Biol. Rev*. 69(2):197-216. 2005.

Villalobos, V.M., Ferreira, P. & Mora, A. "The use of biotechnology in the conservation of tropical germplasm". *Biotech. Adv.,* 9: pp 197-215. 1991.

Walden, Richard. *Genetic Transformation in Plants*. Milton Keynes, England: Open University Press. 1988.

Wang, W., Vinocur, B. and Altman, A. "Plant responses to drought, salinity and extreme temperatures towards genetic engineering for stress tolerance." *Planta* 218: 1-14, 2007.

Wilkins, C.P., Dodds, J.H. "The application of tissue culture techniques to plant genetic conservation". *Science Progress* 68: 281-307. 1982.

Windass, J. D. et.al., Improved conversion of methanol to single-cell protein by *Methylophilus methylotrophus*. In *Nature, 287,* 396 – 401. 1980.

Zaid, A; H.G. Hughes, E. Porceddu, F. Nicholas. *Glossary of Biotechnology for Food and Agriculture - A Revised and Augmented Edition of the Glossary of Biotechnology and Genetic Engineering. Available in English, French, Spanish, Arabic*. Rome, Italy: FAO.2001.

Index

Measurement Techniques in Industrial Instrumentation

S.K.Sen
Professor of Instrumentation Engineering
Department of Applied Physics
University of Calcutta
West Bengal

Publishing Globally
NEW AGE INTERNATIONAL (P) LIMITED, PUBLISHERS
New Delhi • Bangalore • Chennai • Cochin • Guwahati
Hyderabad • Kolkata • Lucknow • Mumbai
Visit us at www.newagepublishers.com

Published by New Age International (P) Ltd., Publishers
First Edition: 2012

BRANCHES

- **Bangalore** 37/10, 8th Cross (Near Hanuman Temple), Azad Nagar, Chamarajpet, Bangalore-560 018
  Tel.: (080) 26756823, Telefax: 26756820, **E-mail: bangalore@newagepublishers.com**
- **Chennai** 26, Damodaran Street, T. Nagar, Chennai-600 017
  Tel.: (044) 24353401, Telefax: 24351463, **E-mail: chennai@newagepublishers.com**
- **Cochin** CC-39/1016, Carrier Station Road, Ernakulam South, Cochin-682 016
  Tel.: (0484) 2377004, Telefax: 4051303, **E-mail: cochin@newagepublishers.com**
- **Guwahati** Hemsen Complex, Mohd. Shah Road, Paltan Bazar, Near Starline Hotel, Guwahati-781 008
  Tel.: (0361) 2513881, Telefax: 2543669, **E-mail: guwahati@newagepublishers.com**
- **Hyderabad** 105, 1st Floor, Madhiray Kaveri Tower, 3-2-19, Azam Jahi Road, Nimboliadda, Hyderabad-500 027
  Tel.: (040) 24652456, Telefax: 24652457, **E-mail: hyderabad@newagepublishers.com**
- **Kolkata** RDB Chambers (Formerly Lotus Cinema) 106A, 1st Floor, S.N. Banerjee Road, Kolkata-700 014
  Tel.: (033) 22273773, Telefax: 22275247, **E-mail: kolkata@newagepublishers.com**
- **Lucknow** 16-A, Jopling Road, Lucknow-226 001
  Tel.: (0522) 2209578, 4045297, Telefax: 2204098, **E-mail: lucknow@newagepublishers.com**
- **Mumbai** 142C, Victor House, Ground Floor, N.M. Joshi Marg, Lower Parel, Mumbai-400 013
  Tel.: (022) 24927869, Telefax: 24915415, **E-mail: mumbai@newagepublishers.com**
- **New Delhi** 22, Golden House, Daryaganj, New Delhi-110 002
  Tel.: (011) 23262368, 23262370, Telefax: 43551305, **E-mail: sales@newagepublishers.com**

ISBN: 978-81-224-3338-8

₹ 299.00

C-12-04-6321

Printed in India at Parmanand Offset, Delhi.

Typeset at A.V. Compositors, Delhi.

PUBLISHING FOR ONE WORLD

NEW AGE INTERNATIONAL (P) LIMITED, PUBLISHERS
4835/24, Ansari Road, Daryaganj, New Delhi-110002
Visit us at **www.newagepublishers.com**

PREFACE

The field of instrumentation has witnessed rapid strides over the last couple of decades. The intricacies involved in measurement and its subsequent control in an effective and precise manner makes the study in this field to be an extremely important one. The advent in the field of instrumentation has become possible because of the ability to measure the state, characteristics and conditions of the physical processes in a quantitative manner with a fair degree of accuracy.

The book *Measurement Techniques in Industrial Instrumentation* contains ten chapters. It is intended to serve as a textbook for undergraduate students of engineering who pursue a course in instrumentation and control engineering. A brief introductory course is covered in chapter 1 while chapters 2 to 4 cover static characteristics, statistical error analysis and dynamic characteristics. A new chapter on PFDs and P & IDs is included in the book for completeness since an instrumentation engineer must understand the above stated diagrams fully. This chapter is not covered in other standard textbooks available in the market. A clear understanding and basic knowledge of the diagrams is a must for today's process engineers.

Since it is literally impossible to include the vast measurement areas in a single book, as such only the more important ones have been taken up for elaborate discussion. Numerous review questions, worked out and sample problems have been included in all the chapters for better understanding of the study material.

I am especially thankful to New Age International (P) Limited, Publishers, as well as the editorial team for their effort to complete this project within a short span of time with nice get up.

S.K. Sen

CONTENTS

CHAPTER 1

INTRODUCTION TO MEASUREMENTS

1.1 INTRODUCTION

Man is, in general, curious in nature and this trait, coupled with his imagination, skill and intuition has seen rapid strides in the field of measurement over the last few decades. During the early days of human civilisation, barter trade system was in vogue as measurement and it was used as a means to quantify the exchange of goods. Although very crude and unscientific as the system was, human beings had to do with this system because of non-availability of scientific and reliable forms of measurements.

As human civilisation continued to evolve and progress, man tried to improve upon the then existing systems in order to have more meaningful exchange of trade. Man's continued stride for betterment in measurement systems gave rise to industrial revolution in the last century, followed by electronic revolution a few decades later. These developments, followed by the introduction of microprocessors in 70s and digital computers, have led to hitherto unseen and unimaginable measurement techniques which are very accurate, reliable and at the same time cheap.

Measurement has become an all pervading one, be it chemical, aeronautical, automobiles or manufacturing systems of all sorts. As consumers are demanding higher and higher limits of accuracy coupled with cost control, measurement techniques are reaching the sky to achieve the above. The success on the above fronts have been made possible because of the ability and a quest for betterment in measuring the state, condition and the characteristics of physical systems with sufficient and dependable degree of accuracy. Techniques and technologies employed nowadays in manufacturing/process industries have undergone a sea change with the introduction of new technologies like robotics, continuous condition based monitoring (CBM), remote control, etc.

1.2 ELEMENTS OF A MEASURING SYSTEM

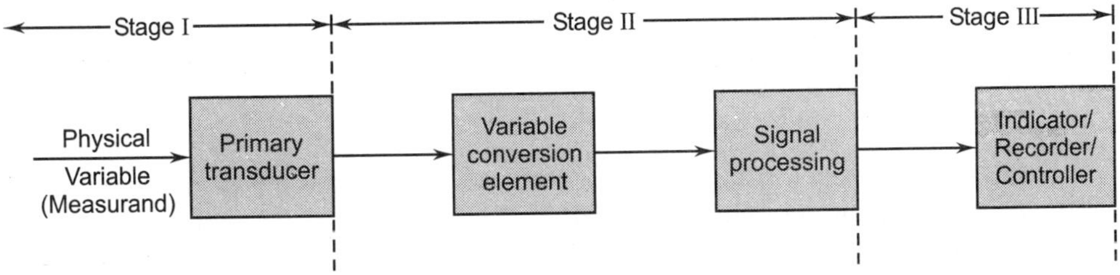

Fig. 1.1: Basic elements of a measuring system

The basic elements of a measuring system consists of three stages—I, II and III—shown in figure 1.1. Stage I consists of sensor/transducer part, stage II the signal processing part and the last or final stage may be indicator, recorder or it may include controller also.

Stage I gives an output which is a function of the measurand or input parameter. An ideal transducer should give an output which is proportional to the input parameter variations, but should be insensitive to every other possible input. Thermocouples, strain gauges, liquid-in-glass thermometers are examples of primary transducer.

Measurands, which are normally measured are temperature, pressure, flow, level, velocity, acceleration, pH, humidity, force, torque, etc.

Stage II is the intermediate stage and may include variable conversion element and signal processing circuitry. For force and displacement measurements, a strain gauge is used which gives an output in the form of a varying resistance. This is then converted into a voltage change by employing a bridge circuit. The signal processing circuit increases the amplitude or power of the signal (transducer's output) or both; it all depends on the transducer output. This enhances the sensitivity and resolution of measurement. This stage also has several other functions like filtering, integration, differentiation and sometimes telemetering also.

Final or stage III provides an indication, record or utilised for control of the process. The measurand may be presented as an indicator (as in a pressure gauge) or a digital output (as in pulse counting).

1.3 APPLICATION AREAS OF MEASURING INSTRUMENTS

Varied application areas exist for measuring instruments. These are:

- System parameter informations
- Perform mathematical manipulations
- Design studies
- Monitoring
- Simulation of system conditions
- Verification of physical phenomena/scientific formulae
- Testing of materials, product specifications and adherence to standards
- Quality control
- Automatic control of a process/operation

System Parameter Informations

System parameter informations: By determining various system parameters, healthiness of a process can be ascertained. In fact, condition-based monitoring leads to a continuous assessment and subsequent corrective actions can be taken, *e.g.*, the health of a patient.

Perform Mathematical Manipulations

A simple pocket calculator performs the functions of addition, subtraction, multiplication, division, integration, differentiation, logarithmic calculations, etc. In addition, several instruments are designed to perform functions like signal sampling, linearisation, averaging, ratio control, etc.

Design Studies

Before launching a product in the market, extensive design studies are made on the drawing board. Thus, before formally launching a car/aircraft, its prototype is developed taking the help of design data. It is then tested under working conditions and various operating parameters are noted. Any deviations from expected parameter values are corrected at the design stage before marketing the product. Thus, experiments supplement design data in the intermediate stage to ensure acceptability of the final product by the discerning customers.

Monitoring

Weather forecasting is made possible by employing instruments like barometers, anemometers, thermometers, etc. Again, by measuring flow, temperature, pressure, etc. in a process plant, an operator would be able to take corrective actions. Temperature of a greenhouse can be taken care of by a thermometer.

Simulation of System Conditions

A prototype of an aircraft is tested in controlled air streams generated in a wind tunnel that simulates the flow, turbulence conditions that the aircraft would actually face in its flights. Such simulation testing helps improve the design parameters of different parts of an aircraft. Again, a passenger car is tested under simulated road conditions to determine its worthiness/acceptability by customers.

Verification of Physical Phenomena/Scientific Formulae

Experimental studies are at times undertaken where substantial theoretical backgrounds are unavailable. Postulates proposed by scientists are verified experimentally to ascertain their veracity. Thus, Coulomb's postulate that friction between two dry surfaces is proportional to the normal reaction and independent of area of contact was experimentally verified and since then is known as Coulomb's law.

Testing of Materials, Product Specifications and Adherence to Standards

To improve longevity, reliability of product parts and a product as a whole, standard organisations specify material and product standards which the manufacturers of such parts/products must adhere to. Again, as accuracy in measurement goes on improving, the specifications/standards are also changed from time to time. These ensure less downtime for parts/products.

Quality Control

Each and every part of a machine is extensively put under quality control tests so that the defective parts are outright rejected. This ensures that the end product is flawless and reliable over a considerable length of time. The machine parts normally tested are for blow holes, cracks, etc, apart from some long duration tests, such as creep, fatigue, etc.

Automatic Control of a Process/Operation

The block diagram of a closed loop control system is shown in figure 1.2. This is one of the most important application areas of a measuring instrument. The deviation (called the error signal) between the reference input and the measured value is fed to a controller, the output of which drives a servo valve and an actuator. This closed loop system thus helps in maintaining the value of the measurand to a value in close proximity to the reference input value.

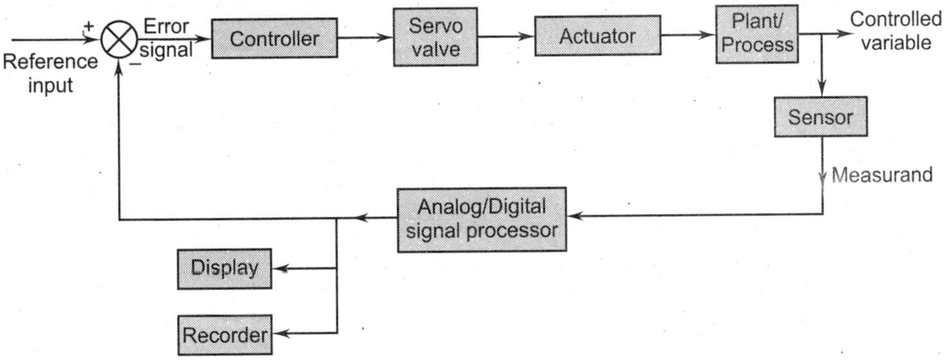

Fig. 1.2: Block diagram of a closed loop control system (for monitoring and control)

The automatic control system has widespread use in process industries like oil refineries, fertiliser plants, power plants, textile industries as also in sophisticated systems like radar tracking system, missile guidance, automatic aircraft landing, etc.

1.4 CLASSIFICATION OF INSTRUMENTS

It is based on method of energy conversion, nature of output signal and mode of operation. Different types of instruments are:

- Manual/automatic type
- Active/passive type
- Null/deflection type
- Analog/digital type
- Contact/non-contact type

They are discussed as under:

1.4.1 Manual/Automatic Type

An instrument, which requires human intervention and operation for recording/displaying the value of the input (measurand), is manual in nature. Thermocouples and RTDs employed for measurement of temperature employ null type potentiometers and are examples of manual type. Employing an automatic self-balancing feedback device in the above will lead to an automatic type of instrument.

1.4.2 Active/Passive Type

Active or power operated instruments require an auxiliary power source, such as electrical supply, compressed air, etc, for their operation. Examples of such instruments are linear variable differential transformer (LVDT), a float-type water tank level indicator, etc. When an LVDT is used for displacement measurements, its output changes, but it needs an auxiliary power across its primary for it to be operational. Similarly, in a water tank level indicator, the position of the wiper along the potentiometer changes when the water level changes. The output is taken from the wiper as shown in figure 1.3.

In passive type instruments, the force/energy required for driving their indicator/output are derived from the measurand. Examples are a Bourdon gauge, a mercury-in-glass thermometer, a Pitot tube, etc.

For active instruments, resolution can be varied considerably while the same for passive ones can be varied in a limited manner. For example, in the case of a Bourdon type pressure gauge (a passive instrument), resolution can be increased to some extent by making the pointer longer such that the pointer tip can move through a longer arc. But obviously, there is a practical limit to an increased pointer length. In

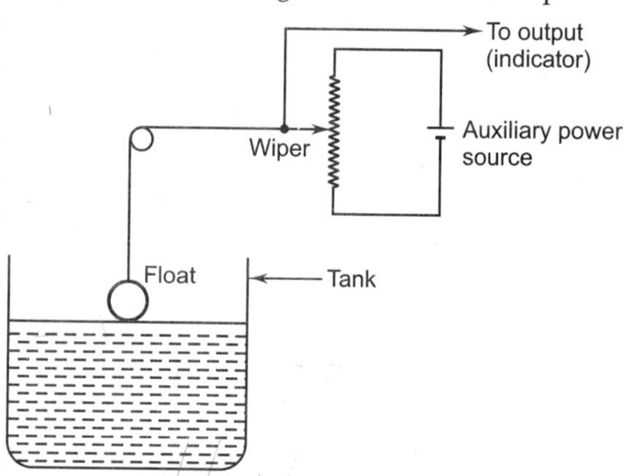

Fig. 1.3: An example of an active instrument

the case of an active instrument, resolution can be increased by increasing say, the supply voltage. But in such a case, heat dissipation is a restraining factor.

In general, passive instruments are generally cheaper and simpler in construction than their active counterparts.

1.4.3 Null/Deflection Type

Null type instruments can be manual or automatic type. A dead weight tester is a null type instrument where the applied external pressure is balanced by adjusting weights placed on a pan. Another example of a null type instrument is that the thermo-emf developed in a thermocouple is balanced (nulled) by a potentiometer.

A common Bourdon type pressure gauge is an example of a deflection type instrument. In this type of instrument, the physical effect produced by the measurand (pressure, temperature, etc.) is opposed by the moving parts of the instrument, the displacement of which is thus a measure of the measurand.

Deflection type instruments are simple in construction and have good dynamic response. However, they interfere with the measurand and hence accuracy is low. Null type instruments are slow in response and hence their dynamic response is poor. But their accuracy is much more and they are normally used for calibration purposes.

1.4.4 Analog/Digital Type

An analog instrument gives an output which is continuous or stepless in nature. Thus, the output can have an infinite number of values within the range of the instrument's span. Examples of analog instruments are a Bourdon type pressure gauge, a mercury-in-glass thermometer, a U-tube manometer, etc. Although theoretically an analog instrument can assume an infinite number of values, practically it is dependent on the scale length and its graduations.

In digital instruments, the measurand (physical parameter or variable) is expressed digitally, *i.e.*, by numbers. Thus, the display is discrete and it varies in steps. The measurand is sensed first and quantised (*i.e.*, assigned a particular analog value) and then it is digitalised, say by means of an ADC (Analog to digital converter). It is then electronically processed, calibrated and the measurand is displayed digitally.

Digital instruments have a number of advantages like ease in processing, noise immunity and direct compatibility with microprocessors/microcomputers, apart from data coding, error detection, error correction capability, if needed. Digital instruments are more reliable, requires less calibration but is a little more costly.

1.4.5 Contact/Non-contact Type

A contact type instrument is kept in the measuring medium. A clinical thermometer is an example of a contact type instrument.

Non-contact type instruments are placed at a distance from the measuring medium. An example is an optical pyrometer which measures very high temperature from a considerable distance. A variable reluctance tachometer, which measures the rpm of a rotating device, is also an example of a non-contact type of instrument.

1.5 MEASUREMENT STANDARDS

During the early days of civilisation, very crude methods for units of measurements were used. For example, human torso like foot or hand were used as standard for measurement of length. As human civilisation evolved over the years, better and accurate methods for different standards were tried out and improved upon. In 1960, a standard meter was defined in terms of 1.65076373×10^6 wavelengths of the radiation from Krypton-86 in vacuum. Later on, in 1983, it was redefined as the length of path travelled by light in an interval of 1/299792458 seconds. Similar improvements were also effected for other kinds of measurement units.

For different types of applications and functions, different standards are specified. These are: International Standards, Primary Standards, Secondary Standards and Working Standards.

1.5.1 International Standards

These are standards conforming to highest possible accuracy achievable using very advanced measurement technologies. These are maintained by International Bureau of Weights and Measures at Se'vre's in France. These standards are maintained under recommended environment conditions and are not available to the ordinary user. International standards for Kilogram (M), length (L) and time (T) are prototype kilogram, wavelength of Krypton-86 orange-red lamp in vacuum and cesium clock respectively.

1.5.2 Primary Standards

These are maintained by national laboratories/standard organisations. These are calibrated independently by absolute measurements and are used to calibrate and check secondary standards. Primary standards are also not available to the ordinary user for use.

1.5.3 Secondary Standards

These are maintained by industrial units to act as basic reference standards. These are periodically checked and calibrated against primary standards. Secondary standards are available to the ordinary user for calibration and checking of their instruments.

1.5.4 Working Standards

Working standards are commercially available in the market after their certification by primary or secondary standards. As for example, the industry working standard for length is the precision gauge block made of steel of specified compositions. This has a accuracy tolerance of 0.25 to 0.5 micron range. As other working standards, standard cells and standard resistors are available conforming to environmental specifications. These standards are very widely used for calibration of laboratory/field instruments, for checking products, etc.

Table 1.1 shows definitions of standard units, while table 1.2 gives the details of fundamental, supplementary and derived quantities, their units and symbols.

Table 1.1: Definition of Standard Units

Physical Quantity	*Standard Unit*	*Definition*
Length	metre	The length of path travelled by light in an interval of 1/299 792 458 seconds
Mass	kilogram	The mass of a platinum-iridium cylinder kept in the International Bureau of Weights and Measures, Sévres, Paris
Time	second	$9.192\ 631\ 770 \times 10^9$ cycles of radiation from vaporized caesium-133 (an accuracy of 1 in 10^{12} or 1 second in 36000 years)
Temperature	kelvin	The temperature difference between absolute zero and the triple point of water is defined as 273.16° kelvin
Current	ampere	One ampere is the current flowing through two infinitely long parallel conductors of negligible cross-section placed 1 metre apart in vacuum and producing a force of 2×10^{-7} Newtons per metre length of conductor.
Luminous intensity	candela	One candela is the luminous intensity in a given direction from a source emitting monochromatic radiation at a frequency of 540 terahertz (Hz $\times 10^{12}$) and with a radiant density in that direction of 1.4641 mW/steradian. (1 steradian is the solid angle which, having its vertex at the centre of a sphere, cuts off an area of the sphere surface equal to that of a square with sides of length equal to the sphere radius)
Matter	mole	The number of atoms in a 0.012 kg mass of carbon-12

Table 1.2: Fundamental, Supplementary and Derived Units

(*a*) Fundamental units

Quantity	*Standard unit*	*Symbol*	
Length	metre	m	
Mass	kilogram	kg	
Time	second	s	
Electric current	ampere	A	
Temperature	kelvin	K	
Luminous intensity	candela	cd	
Matter	mole	mol	

(*b*) Supplementary fundamental units

Quantity	*Standard unit*	*Symbol*	
Plane angle	radian	rad	
Solid angle	steradian	sr	

(c) Derived units

Quantity	*Standard unit*	*Symbol*	*Derivation formula*
Area	square metre	m^2	
Volume	cube metre	m^3	
Velocity	metre per second	m/s	
Acceleration	metre per second squared	m/s^2	
Angular velocity	radian per second	ras/s	
Angular acceleration	radian per second squared	rad/s^2	
Density	kilogram per cubic metre	kg/m^3	
Specific volume	cubic metre per kilogram	m^3/kg	
Mass flow rate	kilogram per second	kg/s	
Volume flow rate	cubic metre per second	m^3/s	
Force	newton	N	$kg\ m/s^2$
Pressure	newton per square metre	N/m^2	
Torque	newton metre	Nm	
Momentum	kilogram metre per second	kg m/s	
Moment of inertia	kilogram metre squared	$kg\ m^2$	
Kinematic viscosity	square metre per second	m^2/s	
Dynamic viscosity	newton second per sq. metre	$N\ s/m^2$	
Work, energy, heat	joule	J	N m
Specific energy	joule per cubic metre	J/m^3	
Power	watt	W	J/s
Thermal conductivity	watt per metre kelvin	W/m K	
Electric charge	coulomb	C	A s
Voltage, e.m.f., pot. diff.	volt	V	W/A
Electric field strength	volt per metre	V/m	
Electric resistance	ohm	Ω	V/A
Electric capacitance	farad	F	A s/V
Electric inductance	henry	H	V s/A
Electric conductance	siemen	S	A/V
Resistivity	ohm metre	Ω m	
Permittivity	farad per metre	F/m	
Permeability	henry per metre	H/m	
Current density	ampere per square metre	A/m^2	
Magnetic flux	weber	Wb	V s
Magnetic flux density	tesla	T	Wb/m^2
Magnetic field strength	ampere per metre	A/m	
Frequency	hertz	Hz	s^{-1}
Luminous flux	lumen	lm	cd sr
Luminance	candela per square metre	cd/m^2	
Illumination	lux	lx	lm/m^2
Molar volume	cubic metre per mole	m^3/mol	
Molarity	mole per kilogram	mol/kg	
Molar energy	joule per mole	J/mol	

REVIEW QUESTIONS

1. Draw the basic elements of a measuring system and discuss the individual blocks.
2. Mention the different application areas of a measuring instrument. Discuss any two of them.
3. State the basis on which instruments are classified and mention their types. Discuss the active and passive types of instruments.
4. What are the different standards of measurements? Discuss them.

Fill in the blanks:

1. During the early part of civilisation, human hand was used as a measurement standard for ....
2. The elements of a measuring system, in general, consist of .... stages.
3. A closed loop automatic system maintains the value of the measurand very close to the .... value.
4. Prototype of aircrafts are tested under .... conditions.
5. A clinical type thermometer is an example of .... type instrument.
6. LVDT is an example of .... type instrument.

Tick the correct answer:

1. Signal processing circuitry enhances the sensitivity/resolution/both of measurement. ☐
2. An active instrument requires/does not require an auxiliary power source for their operation. ☐
3. A Bourdon type pressure gauge is an example of a passive/active instrument. ☐
4. In the case of an active instrument, resolution can be increased by increasing/decreasing the supply voltage. ☐
5. Null type instruments are fast/slow in response. ☐
6. A clinical thermometer is a contact/non-contact type instrument. ☐
7. International standard of measurement is more/less accurate than working standard. ☐

CHAPTER 2

STATIC CHARACTERISTICS OF INSTRUMENTS

2.1 INTRODUCTION

Performance characteristics of instruments are of great importance which determine their suitability for a given application. The characteristics of instruments are of two types: static and dynamic.

Static characteristics refer to static measurement of quantities which either remain time invariant or varies very slowly with time. Examples are emf of a cell or the resistance value of a resistor, both at constant temperature or the height of a man, all refer to static measurements.

A static characteristic tree is shown below in Figure 2.1. The different static performance parameters will now be discussed

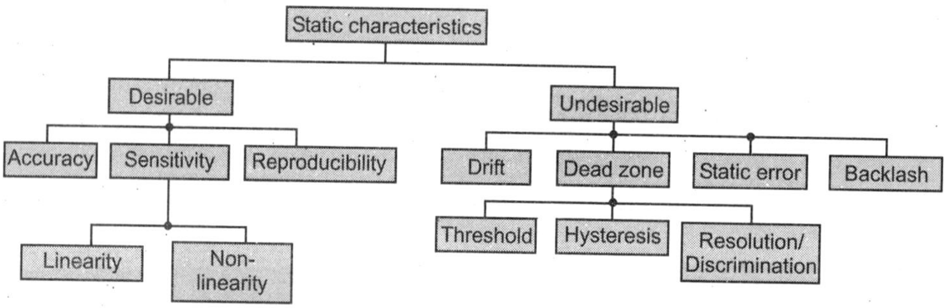

Fig. 2.1: The static characteristic tree

2.2 ACCURACY

Accuracy is defined as the nearness (or closeness) of the indicated value to the true value of the quantity being measured.

Accuracy, in terms of percentage, is expressed as:

1. $$\text{Accuracy (\%)} = \frac{\text{Indicated (measured) value} - \text{Actual (True) value}}{\text{True value}} \times 100$$
(in terms of % of true value)

2. $$\text{Accuracy (\%)} = \frac{\text{Indicated (measured) value} - \text{Actual (True) value}}{\text{Maximum scale value}} \times 100$$
(in terms of full scale division or FSD)

Definition (2) for accuracy is less accurate than definition (1).

Generally, the measured value of a quantity differs from the true value due to the effects of temperature, humidity as also the performance characteristics of the measuring system. Because of these disturbing effects, accuracy is sometimes expressed in terms of uncertainty (or inaccuracy) with specified probability limits.

For an instrument, accuracy indicates the maximum likely departure of the indicated (recorded) value from the true value while for a transducer it indicates quantitatively the likely deviation (departure) from the characteristic curve (of the transducer) due to the disturbing inputs.

2.2.1 Precision

It is defined as the ability of an instrument to reproduce a certain fixed input within a given accuracy. It is tacitly assumed in the above definition that the measurement is done by the same observer for the same fixed input quantity using identical type of measurement and under the same operating conditions, but over a short time span. If, following the above definition, the repeated readings for a fixed input lie within ± 0.5%, then the precision of the instrument is ± 0.5%.

Repeatability is a characteristic of a precision instrument. For measuring transducers, repeatability must be very high from considerations of consistency in final product quality. Non-repeatable performance of transducers/instruments is caused by random or accidental errors. Precision can be quantified by the statistical analysis of a set of measured data.

Precision has two components—conformity and number of significant figures. A resistance, whose actual value is 1,734,589 Ω, is measured by an ohmmeter which consistently gives a value of 1.8 MΩ. Repeated measurements are giving the value 1.8 MΩ. Here, there is a limitation of the scale reading giving rise to precision error. This example indicates that conformity is a necessary but not sufficient condition for precision. Low precision in this case is because of lack of significant figures obtained. Likewise precision is a necessary but not sufficient condition for accuracy.

2.2.2 Accuracy Vs. Precision

Here is the difference between the two:

Accuracy	*Precision*
1. It represents the correctness of the measured value (measurand) w.r.t. the true value.	1. It represents the degree of repeatability of several independent measurements of the same input under identical measurement conditions.
2. It depends on systematic and random errors.	2. It depends on systematic and random errors.
3. Accuracy is determined by proper calibration of instrument.	3. Precision is evaluated by performing statistical analysis of a set of measured data.

It must be noted at this point that high precision does not mean high accuracy and vice versa.

2.2.3 Significant Figures

Number of significant figures is an indication of precision in a measurement. It conveys the actual information regarding the magnitude and measurement precision when a quantity is measured The more the number of significant figures, the greater the precision in measurement.

A standard resistance of value 47 Ω means that the resistance value is closer to 47 Ω than to 46 Ω or 48 Ω. If it is specified as 47.0 Ω, then its value is closer to 47.0 Ω than it is to

46.9 Ω or 47.1 Ω. When it is specified as 47 Ω, the number of significant figures is 2 while in 47.0 Ω, it is 3. Any value with more number of significant figures expresses greater precision.

Simple rules exist when mathematical manipulation of data are involved. In simple terms, they are:

Addition and Subtraction:

After completion of the respective operation, the result is rounded off to the same number of decimal places as that of the least accurate figure.

Multiplication and Division:

After completion of the respective operation, the result is rounded off to the same number of significant figures as that of the least accurate number.

2.3 SENSITIVITY

Sensitivity (also called gain or scale factor) is defined as the absolute ratio of the change in the output signal to a change in the input signal (or measurand).

Sensitivity (or static sensitivity, as defined here) is expressed as follows:

$$\text{Static Sensitivity } K = \frac{\text{Change in output signal}}{\text{Change in input signal}}$$

$$= \frac{\varnothing q_o}{\varnothing q_i}$$

where, Δq_i and Δq_o are the changes in the input signal and the corresponding change in the value of the output signal respectively.

Unit of sensitivity of a thermocouple is $\mu V/°c$ while that for a spring under tension/compression is usually expressed in N/m. A highly sensitive system means a readable or detectable change in output signal for a very small change in input quantity.

Reciprocal of sensitivity is sometimes used and is called inverse sensitivity or deflection factor.

2.3.1 Linearity and Non-linearity

If the change in the output quantity to the corresponding change in the input quantity is constant throughout the range of an instrument, then it is called a linear instrument. For such an instrument, sensitivity is constant while that for a non-linear instrument it varies. Figure 2.2 (*a*) shows the linear characteristics of an instrument while figure 2.2 (*b*) is representative of a non-linear type. A perfectly linear instrument throughout its whole range is practically not possible. The deviations from

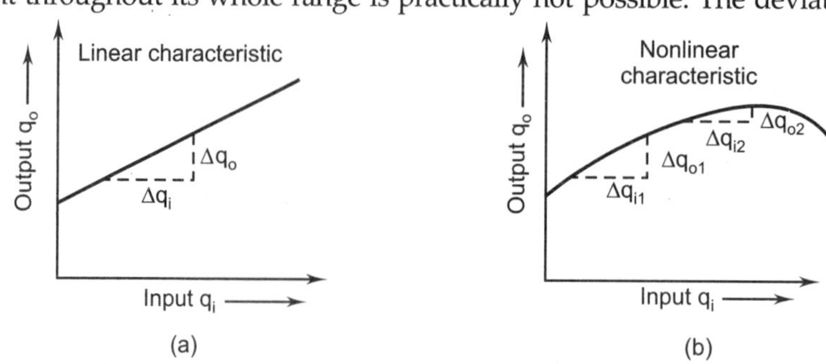

Fig. 2.2: Sensitivity of instruments: (*a*) Linear characteristics (*b*) Nonlinear characteristics

2.8.3.1.3 Gross Error

It is classified into three types: personal, faulty components and improper applications. These are now detailed below.

2.8.3.1.3.1 Personal Error

It is related to the operator's ability to record correct data. An individual may record the observed readings either high or low and thus would introduce an error.

2.8.3.1.3.2 Faulty Components

Sometimes, there may be misalignment between moving parts, loosening of coupling, leakage of magnetic flux, etc. This would lead to zero shift and zero drift of the recorder.

2.8.3.1.3.3 Improper Applications

In some cases, the operating environment may differ widely from design/calibration environment. For example, under actual operating conditions, electrical noise, vibration, mechanical shock may be present. The instrument reading would thus be erroneous.

2.8.3.2 Random Error

Some incidents, such as changes in geomagnetism, earth tremors, cosmic ray showers, etc., lead to random errors. Sometimes these errors may be positive and sometimes negative and thus may compensate. Presence of such errors is detected by an apparent lack of consistency in the measured values. An error estimation arising out of presence of random error may be obtained from a statistical treatment of data.

2.8.3.3 Systematic Error

It is divided into two: environmental and instrumental. These are now discussed.

2.8.3.3.1 Environmental Error

Environmental condition variations, such as atmospheric temperature, pressure, humidity may lead to environmental errors. A metal scale is affected by temperature while a plastic scale is affected by humidity.

Such errors can be remedied by taking the measurements under controlled conditions.

2.8.3.3.2 Instrumental Error

This is subdivided into four types. These are: inherent shortcomings, interferences, multiple earths and loading effects. These are now going to be discussed.

2.8.3.3.2.1 Inherent Shortcomings

Malfunctioning of an instrument may occur because of ageing, spring tension becoming weak, etc. These lead to change in calibration. Thus, calibration of the instrument should be checked from time to time to overcome inherent shortcomings.

2.8.3.3.2.2 Interferences

In this, some kind of interference in the form of voltage or current sources are superposed. It is of two types—series mode interference and common mode interference.

In series mode interference, the interfering source may be voltage/current type. For an interfering voltage source, it is in series with the actual voltage source. For maximum voltage to be transferred to the load, load impedance must be much higher than the source and line impedances. Again, for a high S/N ratio to be obtained, the voltage of the interfering source must be very less compared to actual voltage source. For such an analysis, a Thevenin model is employed.

Again, for a series mode interference, if the interfering source is of current type, Norton model is applied. In this case, the load impedance must be much less than the internal impedance of the current source.

For a common mode interference, if the common mode interfering voltage is applied on either side of the signal circuit, it will not affect the voltage across the load.

2.8.3.3.2.3 Multiple Earths

When a circuit is grounded at more than one point, a potential difference may exist between these points. This will lead to series and common mode interference. As a consequence, measurement error will be there.

When a source and a receiver (*i.e.* measuring instrument) are earthed, their leakage impedances will come into play. Since these two impedance values are not same, a *p.d.* would exist between these two earth points. To eliminate such effects, these two leakage impedances must be very very high.

2.8.3.3.2.4 Loading Effects

Any measurement system would draw some energy, however small it might be, to measure the value of the measurand (*i.e.*, the unknown quantity being measured). Thus, ideal measurement of the measurand is never possible. The term 'loading' refers to the change in the value of the measurand when it is connected to a measuring system. The loading error in measurements can be reduced, by suitable design, but can never be made zero.

Mercury in a thermometer or thermocouple bead extract some heat energy from the measurand, whose temperature is being measured. Thus, temperature recorded would be less. These are examples of loading the measurand by the measuring systems (here, thermometer, thermocouple etc).

It is most appropriate at this point to discuss input impedance and output impedance. Input impedance relates to an instrument, which say, measures the current or voltage. While output impedance refers to power supplies, batteries, oscillators, amplifiers and active transducers.

We assume that a voltage source (AC or DC) is connected across the input terminals of an instrument. The ratio of the voltage across the input terminals (V_{in}) of the instrument to the current flowing (I_{in}) into it is termed as input impedance of the instrument. Thus, the input impedance (of the instrument) is

$$Z_{in} = \frac{V_{in}}{I_{in}}$$

Let V_{out} is the voltage appearing across the open-circuited output terminals of the source and I_{out} is the calculated current that would flow if the output terminals of the source were shorted. Then, the output impedance of the source is

$$Z_{out} = \frac{V_{out}}{I_{out}}$$

However, if I_{out} is attempted to be measured in the above referred manner, it may result in burning out of the source. An example of the output impedance is the internal resistance of a battery.

Figure 2.7 shows how the variation of input impedance affects measurement. Due to 'loading' of the source by the instrument, a measurement error results. A 100 V source is in series with a 2 K resistance.

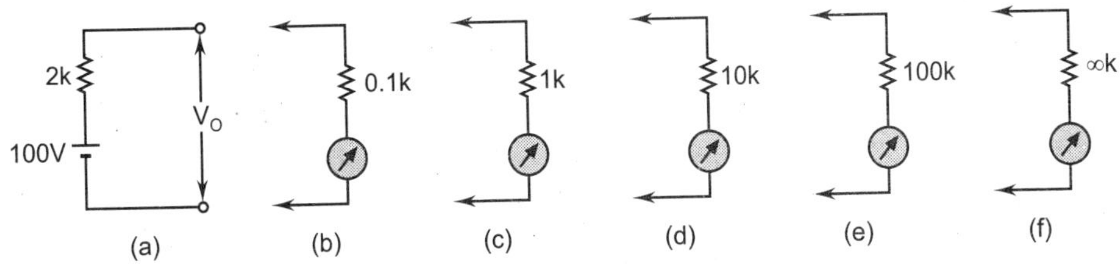

V_o(V)	*Load resistance (kΩ)*	$V_{measured}$ (V)	*Per cent error*
100	0.1	4.7619	95.238
100	1.0	33.3333	66.6667
100	10	83.3333	16.6667
100	100	98.0392	1.9608
100	∞	100.0	0

Fig. 2.7: Measurement error caused by meter input impedance

This voltage is measured by a voltmeter with various input impedances as shown in Figs. (*b*) to (*f*), as the input impedance is lowered, percent error in measurement increases. This error becomes zero when the input impedance is infinity. If the voltmeter does not have a high input impedance, it would draw a significant current from the test circuit. The measuring device (here voltmeter) is said to be loading down the test circuit.

For the voltage of a voltage source being measured by a measuring system, the output (source) impedance of the source should be low and the input impedance of the instrument (measuring system) must be very high. To contain the percent error to within 1%, the thumb rule suggests that the load impedance be 100 times the source (output) impedance. If the signal source is a current source, minimum loading demands the reversal of the ratio of impedances.

2.8.4 Limiting Errors

It is also known by the name guarantee errors. In the majority of indicating instruments, accuracy is guaranteed to a certain percentage of full-scale reading. Passive components like resistors, capacitors etc are guaranteed within a certain percentage of their rated value. Limiting error is the deviation from the specified values. As an example, if a certain resistance is given as 1000 Ω ± 10%, then the resistance variation is limited between 900 Ω to 1100 Ω.

2.8.5 Some Important Terms

2.8.5.1 Responsiveness

It is the smallest change in the quantity under measurement that causes a motion of the indicating portion of the instrument.

2.8.5.2 Stability

It is the ability of a measuring system to maintain its level of performance over a considerable period of time. A highly stable instrument/transducer needs infrequent calibration for its operation.

2.8.5.3 Span and Range

For a 0 – 10 A ammeter, the range is from 0 A to 10 A and the span is 10 A. For a 10 V – 500 V voltmeter, the range is from 10 V to 500 V while the span is 490 V.

2.8.5.4 Live Zero

Output of some transducers have a range 4 – 20 mA. It means that when the input to the transducer is zero, the output is a 4 mA instead of 0 mA. This is called live zero while if the transducer goes wrong, the reading would be 0 mA and is known as dead zero.

2.8.5.5 Scale Readability

It is the ability of the reader to record a reading as precisely as possible. It should be borne in mind that a precise reading does not necessarily mean that the reading is accurate.

SOLVED EXAMPLES

1. Find the (*a*) average current (*b*) range of error when a current measurement yields the following readings:

1.71 A, 1.61 A, 1.65 A and 1.77 A.

Ans. (*a*) Average current $= I_{av} = \frac{1.71 + 1.61 + 1.65 + 1.77}{4}$ A

$= 1.685$ A

(*b*) $I_{max} - I_{av} = (1.77 - 1.685)\text{ A} = 0.085\text{ A}$

$I_{av} - I_{min} = (1.685 - 1.61)\text{A} = 0.075\text{ A}$

Thus, range of error $= \frac{0.085 + 0.075}{2}\text{ A} = 0.08\text{ A}$

2. Values of two resistances are 29.7 Ω and 5.764 Ω. Find the total resistance to the appropriate number of significant figures when they are in series.

Ans. R_1 = 29.7 Ω (three significant figures)

R_2 = 5.764 Ω (four significant figures)

∴ Total resistance = $R_1 + R_2$ = 35.464 Ω (five significant figures)

= 35.4 Ω (three significant figures)

Thus, the value of total resistance is 35.4 Ω and is expressed in three significant figures, since one of the resistances, viz., R_1 is expressed in three significant figures.

3. A current 2.12 A flows through a resistance of value 41.517 Ω. Express the drop across the resistor to the appropriate number of significant figures.

Ans. The drop is = (2.12 × 41.517)V

= 88.01604 V

= 88.0 V

The drop is expressed to three significant figures–as that of the current.

4. Sensitivity of a voltmeter is 5000 Ω/V and it reads 100 V on a 150 V scale when it is connected across an unknown resistance. Current through the resistance is 10 mA, measured by a millimeter and connected in series with the resistance. Find (*a*) apparent resistance of the resistor (*b*) actual resistance of the resistor (*c*) error percentage due to loading effect of the voltmeter.

Ans. (*a*) The apparent resistance is $= \frac{100\text{ V}}{10\text{ mA}} = 10\text{ k}\Omega$

(here, we neglect the value of resistance of the milliammeter) (*b*) Voltmeter resistance for the range selected

$$= 5000\frac{\Omega}{V} \times 150\text{ V} = 750\text{ k}\Omega$$

Now, the voltmeter is in parallel with the unknown resistance. If R_a = actual value of the unknown resistor, then,

$$10 = \frac{750 \times R_a}{750 + R_a} \Rightarrow R_a = 10.135\text{ k}\Omega$$

(*c*)

$$\text{Percentage error} = \frac{\text{actual value} - \text{apparent value}}{\text{actual value}} \times 100$$

$$= \frac{10.135 - 10}{10.135} \times 100 = 1.33\%$$

5. A 0 – 200 V voltmeter has a guaranteed accuracy of 1% of f.s.d. A voltage measured by this voltmeter reads 79 V. Calculate the limiting error in per cent as also the per cent error at this reading.

Ans. The magnitude of the limiting error = 0.01 × 200 V = 2 V.

The percentage error when the meter reading is 79 V is

$$= \frac{2}{79} \times 100 = 2.53\text{ per cent}$$

Note: (*i*) when the meter reads 60 V, the per cent error is

$$= \frac{2}{60} \times 100 = 3.3\ \%\text{ per cent}$$

(*ii*) when the meter reads 40 V, the per cent error is

$$= \frac{2}{40} \times 100 = 5\ \%\text{ per cent}$$

Thus, per cent error increases with decreasing meter reading. Thus, it is very important to take measurements as close to full scale as possible.

6. A voltmeter reads 99 V when the true value is 100 V. Determine the absolute error and the relative percentage error.

Ans. Absolute error = 100 V – 99 V = 1 V

$$\text{Relative percentage error} = \frac{1\text{ V}}{100\text{ V}} \times 100$$

$$= 1\%$$

REVIEW QUESTIONS

1. Draw the static characteristics tree. What are meant by precision and accuracy? List their differences.
2. Discuss the following terms: sensitivity, drift, backlash, dead band and hysteresis.
3. Draw the static error tree. Explain the difference between systematic and random errors. What are the typical sources for these two types of errors? How are these errors minimised?
4. Explain linearity and non-linearity. How is sensitivity defined for the above two cases?
5. What is meant by loading effects? Establish that meter input impedance is a significant factor in measurement errors.

Fill in the blanks:

1. In static characteristics, the measurand remains time ...., or varies .... with time.
2. Accuracy is defined as the .... of the indicated value to the true value.
3. .... is a characteristics of a precision instrument.
4. Sensitivity is the ratio of the magnitude of .... to the ....signal.
5. Significant figure is an indication of the .... with which a measurement is made.
6. Unit of sensitivity for a thermocouple is ....
7. Mostly, transducer outputs are .... in nature.
8. When the measurand remains constant, the change in instrument reading with time is termed as ....
9. Drift is classified into .... categories.
10. For a transducer, hysteresis effect should be as .... as possible.
11. Threshold is the .... value of input that is needed to cause a detectable change in the output signal from zero position.
12. Human error is subdivided into .... types.
13. Parallax error is a type of .... error.
14. Changes in geomagnetism, earth tremors, cosmic ray showers lead to .... error.
15. If the input impedance of a voltmeter is very high, error in measurement of a voltage will be ....
16. Span for a 100 V – 1000 V voltmeter is .... V.
17. A spurious signal which causes a change in the value of the instrument reading is termed as ....
18. Error is the actual difference between the .... value and .... value.

Tick the correct answer:

1. While making a measurement, the errors committed by a human being are:
   (*i*) gross errors (*ii*) random errors (*iii*) instrumental errors (*iv*) environmental errors. ☐
2. In an instrument, the smallest measurable quantity is known as:
   (*i*) dead zone (*ii*) resolution (*iii*) threshold ☐
3. A resistance measurement is recorded as 200.01 Ω. The reading has:
   (*i*) 4 (*ii*) 5 (*iii*) 6 significant figures. ☐
4. Precision and accuracy are (*i*) same (*ii*) different. ☐
5. Measurement error is dependent/ not dependent on meter input impedance. ☐
6. Errors caused by changes in geomagnetism, earth tremors or cosmic ray showers are (*i*) random errors (*ii*) systematic errors. ☐
7. A scale has 1 mm as least count. Its resolution is (*i*) 0.5 mm, (*ii*) 1 mm ☐
8. In the absence of drift, reproducibility becomes (*i*) maximum, (*ii*) minimum. ☐
9. A common practice of reducing hysteresis error in the output for a given value of input is to maintain (*a*) high rate of change of input (*b*) low rate of change of input (*c*) taking observations in ascending or in the descending order (*d*) taking observations both in ascending and descending orders and take average value of the output. [GATE 2002] ☐

10. A zero error in a vernier caliper is termed as: □

(*a*) accidental error (*b*) systematic error (*c*) interference error (*d*) random error [GATE 2004]

REVIEW PROBLEMS

1. Indicate the number of significant figures in (*a*) 100 (*b*) 100.0. [**Ans.** (*a*) 3 (*b*) 4]

2. A 0 – 100 V voltmeter has 400 divisions. With eye estimation, 1/2 division can be read. Determine the resolution of the meter in volts. [**Ans.** 0.125 V]

3. 1.85 A of current is passing through a resistor, the voltage across which is 130.5 V. Find the power dissipated in the resistor. The answer should take into consideration four significant figures. [**Ans.** 241.4 W]

4. A voltmeter has a sensitivity of 10 kΩ/V and reads 80 V on a 100 V scale when connected across a certain resistor. If the current through the resistor is 2 mA, calculate the (*a*) apparent resistance of the resistor (*b*) actual resistance of the resistor (*c*) error per cent due to loading effect of voltmeter. [**Ans.** (*a*) 40 kΩ (*b*) 41.67 kΩ (*c*) 4%]

5. Find the voltmeter reading when it is connected between the paints A and B when the voltmeter scale range is adjusted to (*a*) 40 V (*b*) 50 V. In both the cases, the voltmeter sensitivity is 20 kΩ/V. [**Ans.** (*a*) 16.84 V (*b*) 17.39 V]

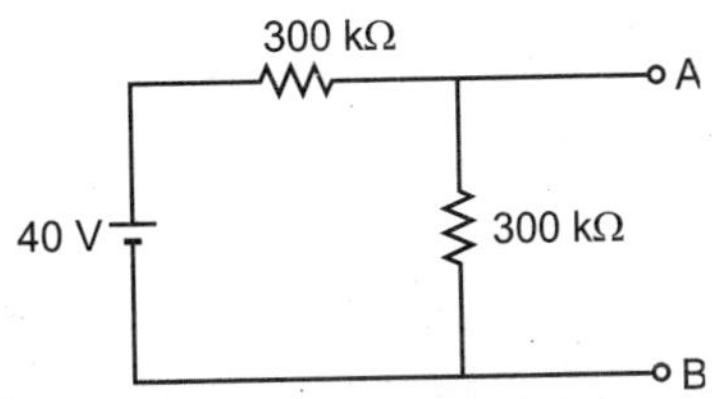

6. A temperature sensor having a range of 0°–100°C has a worst case accuracy specification of ± 10% of true value or ± 3% of full scale, whichever is higher. The sensitivity of the sensor is experimentally obtained by taking its readings at true temperatures 25°C and 75°C. The maximum error in sensitivity, expressed as a percentage of its ideal value is

(*a*) ± 0.105% (*b*) ± 0.21% (*c*) ± 1.05% (*d*) ± 2.1% [GATE 2005] [**Ans.** (*c*)]

7. A thermometer with time constant τ, initially at ambient temperature, is used to measure the temperature of a liquid in a bath. The excess temperature of the thermometer and the liquid over the ambient are $\theta(t)$ and $\theta_1(t)$, respectively, where t denotes the time. If $\theta_1(t) = kt$, where k is a constant, the steady state error, defined as $\lim_{t \to \infty} [\theta(t) - \theta_1(t)]$, is:

(*a*) ∞ (*b*) 0 (*c*) $-k$ (*d*) $-k\tau$ [GATE 2007] [**Ans.** (*d*)]

8. A batch of 11 resistors of 100 Ω nominal value, when measured gave the following readings:

1	2	3	4	5	6	7
104.000	96.000	95.000	101.414	105.000	98.586	102.000
8	9	10	11			
96.000	105.000	95.000	102.000			

Determine the % tolerance of the batch. 120 Ω resistors are made using these 100 Ω resistors having a tolerance of 3%. Determine the tolerance of the 120 Ω resistors. [GATE 2001] [**Ans.** 3.52% ; 15%]

Chapter 2

9. A $3\frac{1}{2}$ digit multimeter has an accuracy specification of ± 0.5% of reading ± 5 counts. If the meter reads 2.00 mA on a full scale of 20 mA, the worst case error in the reading is:

(*a*) 0.5% (*b*) 2.5% (*c*) 3.0% (*d*) 5.5% [GATE 2003] [**Ans.** (*c*)]

10. For a proving ring used in a press, the value of force and its uncertainty are given as 12355 ± 2354 N without any rounding off. If the error should be rounded off at the maximum to two significant figures only, the value of the reading is rounded off as:

(*a*) 12300 ± 2300 N (*b*) 12360 ± 2350 N (*c*) 12400 ± 2300 N (*d*) 12400 ± 2400 N

[GATE 2004] [**Ans.** (*d*)]

11. An iron-constantan thermocouple is to be used to measure temperature between 0° and 300°C. With reference junction at 0°C, the emf induced at 100°C and 300°C are

$$E_{100} = 5268\ \mu V, \quad E_{300} = 16325\ \mu V$$

Then the non-linearity at 100°C as a percentage of full scale is:

(*a*) + 3.23% (*b*) – 1.06% (*c*) – 0.96% (*d*) + 3.23% [GATE 2004] [**Ans.** (*a*)]

CHAPTER 3

STATISTICAL ERROR ANALYSIS

3.1 INTRODUCTION

Now-a-days a statistical analysis of measurement data is undertaken for error evaluation and consequent correction. It permits the best value of a measured quantity to be estimated from repeated, independent determinations. It also allows an analytical estimation in the uncertainty of the final result. A quantitative error estimation is also possible. Statistical analysis can also estimate and predict the value of a quantity which has not been studied explicitly.

For the statistical estimation and prediction to be meaningful, a large number of measurements are usually required. Also, systematic errors must be small compared to random or residual errors–for statistical data analysis; it cannot remove a fixed bias contained in the measurements.

3.2 VARIATE

Let v_1, v_2, ... v_n be the values of a quantity when it is measured independently. Then each v_i (i = 1, 2, ... n) of the set is called a variate. The group consisting of n variates constitutes the raw data.

3.3 MEAN, MODE, MEDIAN

The 'mean' of a set of variates is defined as:

$$\bar{v} = \frac{1}{n}(v_1 + v_2 + \cdots + v_n) = \frac{1}{n}\sum_{n=1}^{n} v_i \qquad ...(3.1)$$

'mean' thus represents arithmetic average or arithmetic mean.

The variates are sometimes associated with certain weighing factors w_1, w_2, ...w_n respectively. In such a case, the weighted arithmetic mean is

$$\overline{V} = \frac{w_1 v_1 + w_2 v_2 + \cdots + w_n v_n}{w_1 + w_2 + \cdots + w_n} = \frac{\sum_{i=1}^{n} w_i v_i}{\sum_{i=1}^{n} w_i} \qquad ...(3.2)$$

The 'mode' of a set of variates are those variates that occur more frequently than others. A set of variates may have more than one mode. When the set contains a single mode, it is called 'unimodal'.

'Median' of a set of variates is that particular variate for which there are equal numbers of variates greater and less than that particular variate.

Thus, when the variates are written either in ascending or descending order, the median lies at the middle of the list.

A best estimate for true value of data:

Central tendency of data is best described by mode, median and arithmetic mean. For moderately asymmetrical (or skewed) curves which are unimodal in nature, these three parameters are related by:

$$\text{Mean} - \text{Mode} = 3[\text{Mean} - \text{Median}]$$

Figure 3.1 shows the relative positions of mode, median and mean for frequency curves skewed to the right as well as to the left. For curves which are symmetrical, these three coincide with each other.

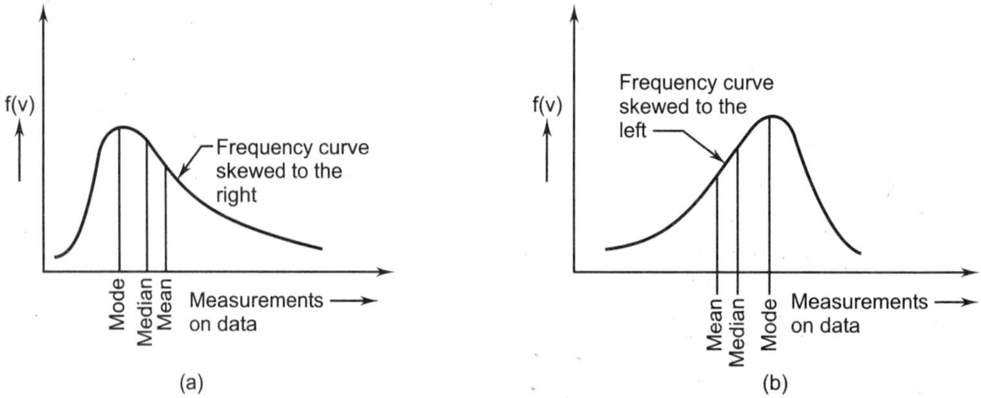

Fig. 3.1: Relative positions of mean, median and mode of moderately skewed (asymmetrical) frequency curves

3.4 DEVIATION

Deviation of a variate v_i belonging to the set $v_1, v_2, \ldots v_n$ is defined as

$$y_i = v_i - v$$

where, v is an arbitrary number. The relation suggests that deviation may either be positive or negative.

The sum of the deviations is

$$\sum_{i=1}^{n} y_i = y_1 + y_2 + \cdots + y_n$$

$$= (v_1 + v_2 + \cdots + v_n) - nv. \qquad \ldots(3.3)$$

If the sum of the deviations is zero, then

$$v = \frac{1}{n}(v_1 + v_2 + \cdots + v_n) = \bar{v} \qquad \ldots(3.4)$$

The above gives another definition for mean—mean is that number for which the sum of the deviations is zero.

The deviation from the mean of a variate v_i is designated as

$$x_i = v_i - \bar{v}$$

Obviously, $$\sum_{i=1}^{n} x_i = 0$$

Thus, the magnitude of sum of the deviations of variates less than the mean is always equal to the magnitude of the sum of the deviations of variates greater than the mean.

3.5 SPREAD OF DATA : VARIABILITY

The original raw data is lost when the variates are expressed by a single number, such as mean. One way to express the variates is in the form of $\bar{v} \pm \delta$ when δ carries the information regarding the spread (or dispersion) of a variate from the mean. There are four ways of representing δ. They are: limit of error, L; probable error, P; average deviation, A and standard deviation σ.

3.5.1 Limit of Error

The limit of error, symbolised by L, is defined as the value of $\delta = L$ so that all the variates belonging to the set lie between $\bar{v} - L$ and $\bar{v} + L$. When the lower and upper limits differ, it is the larger of the two which is used for both. Magnitude of L encompasses all the variates and thus it is representative of the maximum spread or dispersion of data about the mean.

3.5.2 Probable Error

The probable error, symbolised by P, is defined as the value of $\delta = P$ so that half the variates lie between $\bar{v} - P$ and $\bar{v} + P$. As in the case of limit of error, it is quite likely that the number of variates residing below and above the limit would vary. This is because of the likelihood of the asymmetrical spread of data below and above the mean. It is quite obvious that the value of P is always less than that of L.

3.5.3 Average Deviation

The average deviation, symbolised by A, is defined as the value of $\delta = A$ so that:

$$A = \frac{1}{n}(|x_1| + |x_2| + \cdots + |x_n|) = \frac{1}{n}\sum_{n=1}^{n}|x_i| \quad \text{...(3.5)}$$

where, $x_i = v_i - \bar{v}$ = deviation from the mean

Thus,

$$A = \frac{1}{n}\sum_{n=1}^{n}(|v_i - \bar{v}|) \quad \text{...(3.6)}$$

Average deviation, A, can never exceed the value of limit of error, L but may be greater than, equal to or less than the probable error, P.

If v_i's (v_1, v_2, ... v_n) occur with frequencies f_i (f_1, f_2, ... f_n), then,

$$A = \frac{\sum_{i=1}^{n} f_i|(v_i - \bar{v})|}{n} \quad \text{...(3.7)}$$

3.5.4 Standard Deviation

The standard deviation, symbolised by σ, is defined as the value of $\delta = \sigma$ so that,

$$\sigma = \sqrt{\frac{1}{(n-1)}(x_1^2 + x_2^2 + \cdots + x_n^2)} = \sqrt{\frac{1}{(n-1)}\sum_{i=1}^{n}x_i^2} \quad (\text{for } n \le 20) \quad \text{...(3.8)}$$

and

$$\sigma = \sqrt{\frac{1}{n}\sum_{i=1}^{n}x_i^2} \quad (\text{for } n > 20) \quad \text{...(3.9)}$$

Chapter 3

Since the expression involves square of the deviations, negative deviations are automatically eliminated. Again, larger deviations are more emphasized than smaller deviations. Thus, standard deviation is usually larger than average deviation A and the probable error, P.

The square of the standard deviation is known as variance, V. Variances are additive in nature and thus very convenient to use, while standard deviation has the same unit as the variable and hence magnitude comparison is easy. Scientific results are mostly expressed in terms of standard deviation.

3.6 UNCERTAINTY

Systematic and random errors are responsible for deviations from the true value of a variable when it is measured during an experiment. These two errors give rise to gross error. Since systematic errors are deterministic in nature, hence their effects are nullified by introducing a suitable correction factor. Thus, random errors, in a major way, contribute to the uncertainties in measurement. It is evaluated from the dispersion or scatter of data in an experimental measurement. Random errors follow a definite mathematical law, known as normal or Gaussian distribution, although the magnitude or sign of the same may vary. Thus, with the help of the above distribution, the random errors can be quantified with a reasonable degree of accuracy.

Now, uncertainties are of two types: internal and external in nature.

3.6.1 Internal Estimate of Uncertainty

Represented by U_I, it occurs because an instrument may record a slightly different value each time a given input is fed to it. The internal estimate of uncertainty is estimated as follows:

1. A considerable number of data of a particular measurement is obtained, known as population.
2. The population is divided into several portions, known as samples. Samples are selected so as to have an equitable distribution.
3. The 'means' and 'standard deviations' of each of the samples are evaluated. These represent estimated true value and uncertainty for the sample respectively.
4. The population mean is calculated by taking the mean of means.
5. The standard deviation of serial no. 4 gives the estimate of uncertainty.

3.6.2 External Estimate of Uncertainty

Represented by U_E, the external estimate of uncertainty is determined from the instrument manufacturer's literature, the limitation of the apparatus, etc. As a thumb rule, the resolution of an instrument is generally considered to be its external estimate of uncertainty.

3.6.3 Propagation of Uncertainties in Compound Quantities

Measurement of force involves the measurement of both mass and acceleration. Likewise, measurement of DC power involves measurement of both current and voltage. Obviously, measurement of uncertainty of force is due to the uncertainties involved in the measurement of mass and acceleration. To find out the uncertainty of a compound quantity, help is taken of the general equation:

$$y = f(x_1, x_2, ..., x_n) \qquad ...(3.10)$$

In the present case, $y = f(x_1, x_2)$ with y representative of force and x_1, x_2 are mass and acceleration respectively.

Thus,
$$dy = \frac{\partial y}{\partial x_1}dx_1 + \frac{\partial y}{\partial x_2}dx_2 + \cdots + \frac{\partial y}{\partial x_n}dx_n. \qquad ...(3.11)$$

The deviations in $y, x_1, x_2 \ldots x_n$ are $dy, dx_1, dx_2 \ldots dx_n$. These are the uncertainties and expressed as $U_y, U_{x_1}, U_{x_2}, \ldots U_{x_n}$.

Hence,
$$U_y = \frac{\partial y}{\partial x_1}U_{x_1} + \frac{\partial y}{\partial x_2}U_{x_2} + \cdots + \frac{\partial y}{\partial x_n}U_{x_n} \qquad ...(3.12)$$

The maximum of U_y is $\left.U_y\right)_{\max}$ and it would be so when all the uncertainties have the same sign.

$$\left.U_y\right)_{\max} = \left|\frac{\partial y}{\partial x_1}U_{x_1}\right| + \left|\frac{\partial y}{\partial x_2}U_{x_2}\right| + \cdots + \left|\frac{\partial y}{\partial x_n}.U_{x_n}\right| \qquad ...(3.13)$$

The probability that all the uncertainties have the same sign is very small. A practical approach is to square both sides and hence,

$$U_y^2 = \left[\left(\frac{\partial y}{\partial x_1}\right)^2 (U_{x_1})^2 + \left(\frac{\partial y}{\partial x_2}\right)^2 (U_{x_2})^2 + \cdots + \left(\frac{\partial y}{\partial x_n}\right)^n (U_{x_n})^2\right] + \left[\left(\frac{\partial y}{\partial x_1}\right)\left(\frac{\partial y}{\partial x_2}\right)(U_{x_1})(U_{x_2}) + \cdots + \left(\frac{\partial y}{\partial x_i}\right)\left(\frac{\partial y}{\partial x_j}\right)(U_{x_i})(U_{x_j}) + \cdots\right]$$

Simplifying

$$U_y = \pm\left[\left(\frac{\partial y}{\partial x_1}\right)^2 (U_{x_1})^2 + \left(\frac{\partial y}{\partial x_2}\right)^2 (U_{x_2})^2 + \cdots + \left(\frac{\partial y}{\partial x_n}\right)^2 (U_{x_n})^2\right]^{1/2} \qquad ...(3.14)$$

Expressing in a general way

$$U_y = \left[\sum_{i=1}^{n}\left(\frac{\partial y}{\partial x_i}\right)^2 (U_{x_i})^2\right]^{1/2} \qquad ...(3.15)$$

3.7 DATA GROUPING

Instead of a single set encompassing all the variates, sometimes they are grouped. Data grouping gives rise to the following two advantages: (1) the amount of computation to determine $\bar{v}$ and σ is reduced significantly, especially when n is large and (2) it is easier to pictorially present the grouped data from which their salient features may be extracted.

The procedure that is followed for data grouping is as follows:

- The total interval over which the variates lie is divided into smaller intervals. Each of these smaller intervals is called a range. Usually, the ranges are equal in length and are specified by their midpoints.
- All variates lying in a given range are assigned a value equal to the range midpoint.
- For a given set, number of ranges is determined by the desired accuracy on the one hand and the computational effort on the other.
- No variate should coincide with the end points of the ranges. If some variates coincide with the range end points, half of them are assigned the higher range and the other half the lower range.

Now, for the set consisting a total of n variates, let n be the number of variates in the rth range of width Δv_r. n_r is called the frequency of the variates in that range and is designated by $f(v_r)$. If there are a total of R ranges in the span,

Chapter 3

then,
$$\sum_{r=1}^{R} n_r = \sum_{r=1}^{R} f(v_r) = n \qquad ...(3.16)$$

All the variates in the rth range are assigned a value v_r without introducing a considerable error. Thus, the sum of all variates is approximately given by

$$\sum_{i=1}^{n} v_i = \sum_{r=1}^{R} v_r n_r = \sum_{r=1}^{R} v_r f(v_r)$$

Thus, the mean is expressed by

$$\bar{v} = \frac{1}{n}\sum_{r=1}^{R} v_r f(v_r) \qquad ...(3.17)$$

3.7.1 Histogram

If a large number of measurements are carried out on a measurand, and the frequency of occurrence of a particular value is plotted against the corresponding values, the resulting plot is called a histogram or a frequency-distribution curve. Figure 3.2 shows a histogram consisting of twenty variates (n = 20), which have been grouped in ascending order. The range midpoints are whole numbers (100, 101 ... etc.) varying from 100 to 110. All ranges are indentical with Δv_r = 1.0. The ordinate indicates the frequency of occurrence of variates in a given range. For example, four variables lie in the interval 104 ± 0.5.

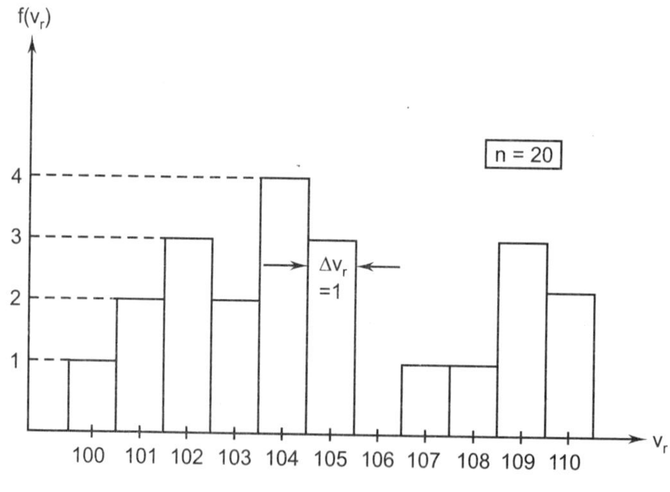

Fig. 3.2: Histogram of grouped data

An histogram is called a frequency distribution graph and potrays visually the distribution of variates in different intervals. The frequency distribution of the variates have maximum value in the range 104 ± 0.5. It is to be noted that no variate lies in the range of 106 ± 0.5.

3.7.2 Normalization of Histogram Ordinates

It is expected that with increasing n, the bar heights will increase. Thus, if two histograms are drawn with different values of n, comparison will be difficult. For higher values of n, the bar heights will, in general, increase.

A normalisation procedure is adopted for comparison of histograms with different values of n. In this, each bar height for a given histogram is divided by its n. Figure 3.3 shows two normalised histograms. It will lead to identical bar height scales and the comparison will then be meaningful. The normalised bar height is obtained by dividing $f(v_r)$ by n. The ratio $f(v_r)/n$ is the probability of occurrence of a variate in a given range Δv_r. The normalised histogram, so obtained, is also called probability histogram and $f(v_r)/n$ is known by probability distribution.

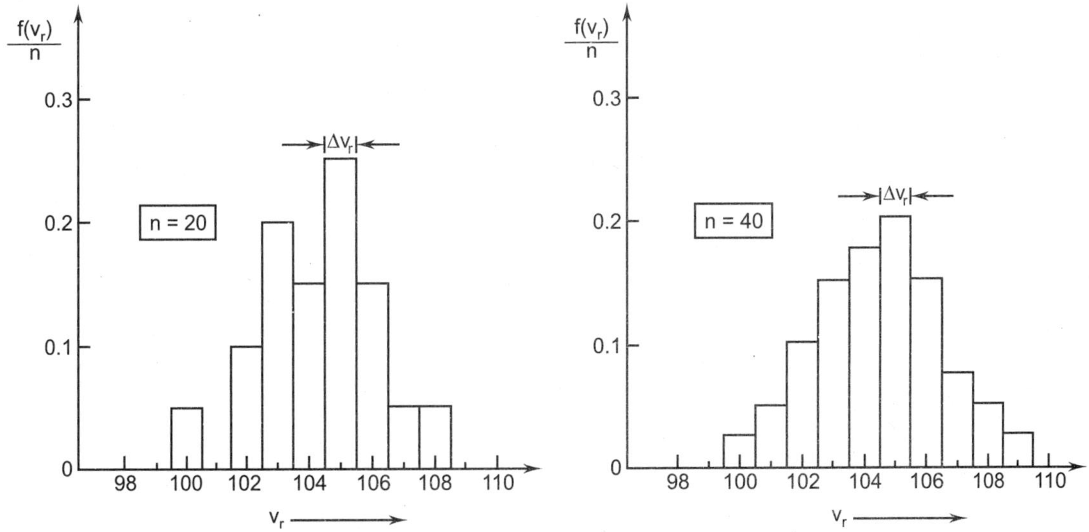

Fig. 3.3: The normalised histograms

3.8 PROBABILITY

It is defined in terms of occurrence of events. Probability is applied to any measurable entity and variates are also measurable. In discussions for proability, events are assumed to be equally likely and independent of each other. For equally likely events, it cannot be speculated which one of all possible events will occur at a given instance. Again, independent events are those for which the occurrence of one event does not influence the occurrence of any other event.

Let, in a particular case, there are E number of equally likely independent events in which E_i is an event. Probability of E_i to occur is then,

$$p(E_i) = \frac{E_i}{E} \quad ...(3.18)$$

Here, $p(E_i)$ is called the 'probability of occurrence of E_i' or simply 'probability of E_i'. Thus,

$$E = E_1 + E_2 + \cdots + E_k = \sum_{i=1}^{k} E_i \quad ...(3.19)$$

Here, it is assumed that E consists of k separate portions.

Since, sum of all the probabilities can never exceed 1,

$$\sum_{i=1}^{k} p(E_i) = \sum_{i=1}^{k} \frac{E_i}{E} = 1.$$

Applying probability to grouped data,

$$p(v_r) = \frac{f(v_r)}{n}$$

and thus,

$$\sum_{r=1}^{R} p(v_r) = \sum_{r=1}^{R} \frac{f(v_r)}{n} = \frac{n}{n} = 1. \qquad \text{...(3.20)}$$

3.9 ERROR ESTIMATION FROM GAUSSIAN (NORMAL) DISTRIBUTION

Figure 3.4 shows the Gaussian or normal distribution of height of several persons in a gathering.

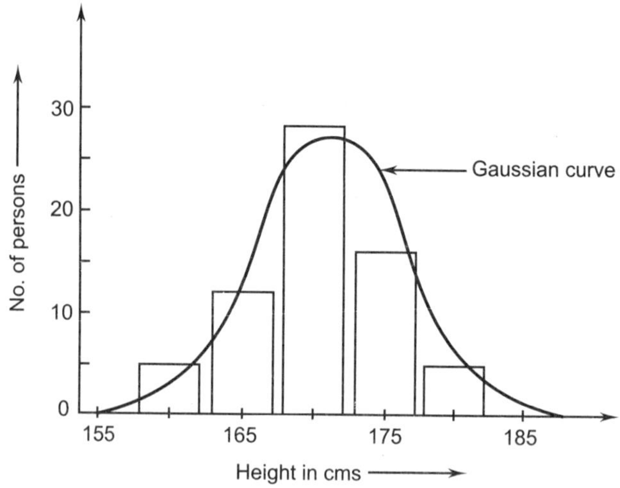

Fig. 3.4: Height vs. number of persons in a gathering exhibiting Gaussian distribution

This distribution can be expressed by the following equation:

$$y = \frac{1}{\sigma\sqrt{2\pi}} \exp\left[-\frac{1}{2}(v-\bar{v})^2\right] \qquad \text{...(3.21)}$$

where, v is the variate (or measurand), $\bar{v}$ is arithmetic mean (or mean)$_{\pm}$ and σ is the standard deviation.

The total area bounded by the normal distribution curve is 1 and is bell shaped. Hence,

$$\int_{-\infty}^{+\infty} y dx = \frac{1}{\sigma\sqrt{2\pi}} \int_{-\infty}^{+\infty} \exp\left[-\frac{1}{2}(v-\bar{v})^2\right] dx$$

$$= \frac{1}{\sqrt{\pi}} \int_{-\infty}^{+\infty} \exp(-g^2) dg, \; g = \frac{v-\bar{v}}{\sigma\sqrt{2}}$$

$$= \frac{\Gamma\left(\frac{1}{2}\right)}{\sqrt{\pi}} = 1. \qquad \text{...(3.22)}$$

The area under the curve between any two arbitrary points $v = c$ and $v = d$ $(c < d)$ represents the probability that v will lie in between c and d. This is expressed as $P_r\{c < v < d\}$.

If, $z = \frac{v - \bar{v}}{\sigma}$, then

$$y = \frac{1}{\sqrt{2\pi}} \exp\left[-\frac{1}{2}z^2\right] \quad ...(3.23)$$

The equation expresses the standard form of the normal distribution and is shown in figure 3.5.

In general, a certain range of permissible values of scatter or dispersion from the mean value is specified of a data and its probability to remain in the specified range is found out. This probability can be evaluated by calculating the area under the Gaussian (normal) curve for the specified range. When this probability is specified in percentage, it is known as confidence level.

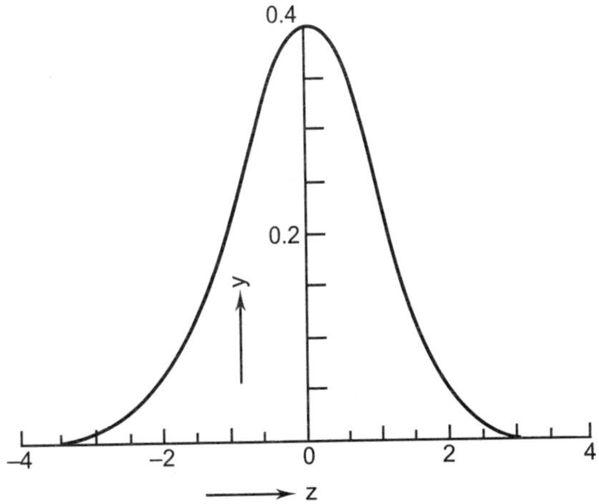

Fig. 3.5: Standard normal curve

3.10 CURVE FITTING : METHOD OF LEAST SQUARES

There is, sometimes, a scatter of data when the values of a variate is measured experimentally over its measurement range. In such cases, an average curve is drawn by eye estimation. Apart from conventional graphical method, there are three methods of curve fitting. These are: (*a*) method of sequential differences (MSD) (*b*) method of extended differences (MED) and (*c*) method of least squares (MLS).

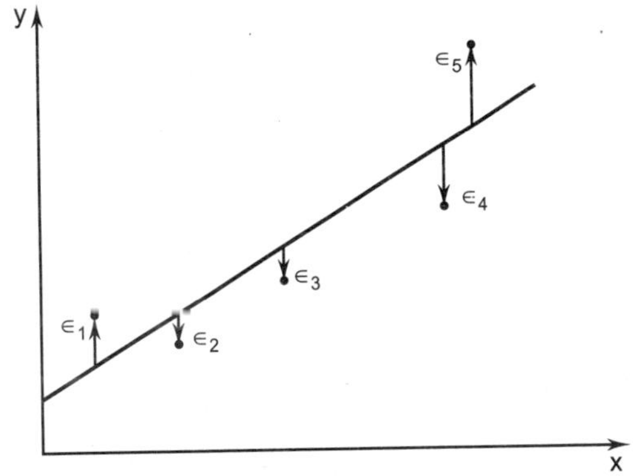

Fig. 3.6: Least squares method showing scatter of points for plotting best curves through the points

In the MLS or least squares method, a series of test points are found out which are distributed normally and independently. A curve is then drawn through the series of test points in such a manner that the sum of the squares of the deviation from the line would be a minimum. This holds best when the curve so drawn is a straight line. Figure 3.6 shows the best curve drawn through the scatter of points by the least squares method.

When least squares method is applied for a simple linear model $y = a_0 + a_1x$, the constants a_0 and a_1 are determined so that the sum of the squares of the vertical distance between the n datum points and the straight line is a minimum.

Mathematically,

$$\sum_{i=1}^{n} \epsilon_i^2 = \text{minimum}$$

This is so when,

$$\frac{\partial \sum_{i=1}^{n} \in_i^2}{\partial a_0} = 0$$

and,
$$\frac{\partial \sum_{i=1}^{n} \in_i^2}{\partial a_1} = 0$$

The sum may be expressed as per the following:

$$\in_i = y_i - (a_0 + a_1 x_i)$$

Thus,

$$\frac{\partial \sum_{i=1}^{n} \in_i^2}{\partial a_0} = \frac{\partial \sum_{i=1}^{n} [y_i - (a_0 + a_1 x_i)]^2}{\partial a_0} = 0 \qquad ...(3.24)$$

and,
$$\frac{\partial \sum_{i=1}^{n} \in_i^2}{\partial a_1} = \frac{\partial \sum_{i=1}^{n} [y_i - (a_0 + a_1 x_i)]^2}{\partial a_1} = 0 \qquad ...(3.25)$$

Simplification of the above yields:

$$\sum_{i=1}^{n} y_i - \sum_{i=1}^{n} a_1 x - n a_0 = 0 \qquad ...(3.26)$$

and
$$\sum_{i=1}^{n} x_i y_i - a_1 \sum_{i=1}^{n} x_i^2 - a_0 \sum_{i=1}^{n} x_i = 0 \qquad ...(3.27)$$

Parameters a_0 and a_1 are solved from these two linear non-homogeneous equation to yield

$$a_0 = \frac{\Sigma x \Sigma xy - \Sigma y \Sigma x^2}{(\Sigma x)^2 - n\Sigma x^2}$$

and,
$$a_1 = \frac{\Sigma x \Sigma y - n\Sigma xy}{(\Sigma x)^2 - n\Sigma x^2}$$

Thus,

$$\Sigma y = n a_0 + a_1 \Sigma x \qquad ...(3.28)$$

and,
$$\Sigma xy = a_0 \Sigma x + a_1 \Sigma x^2 \qquad ...(3.29)$$

The least squares method requires elaborate calculations. This method is therefore is used not too often and is applied where precise curve-fitting is mandatory for high accuracy demand from the system.

3.11 CHI-SQUARE TEST FOR NORMAL DISTRIBUTION OF DATA

Chi-square test is a method to verify whether the distribution of experimentally obtained data is normal or not and follow the normal or Gaussian distribution. This test helps in the evaluation of error.

When a set of measurements is made, it is assumed that the said measurements are a part of some known theoretical distribution. For the purpose of comparison, the whole distribution spectrum is divided into several classes. Observed frequency in each class is then determined. The expected frequency in each class is also estimated from normal distribution.

It is assumed that the whole distribution is divided into n classes and the expected and observed frequencies in various classes are $f_{e1}, f_{e2}, f_{e3} \ldots f_{en}$ and $f_{o1}, f_{o2}, f_{o3} \ldots f_{on}$. The parameter χ^2 is defined as:

$$\chi^2_{(n-m)} = \sum_{i=1}^{n} \frac{(f_{oi} - f_{ei})^2}{f_{ei}}$$

where, m is the number of constants used in the calculation of f_{ei} and $(n - m)$ is the degrees of freedom.

The numerator of the above expression is positive and represents the deviations between the observed and expected frequencies in each individual class. These values are normalised by dividing them by the respective expected frequency of each class. If it is assumed that the numerator of the above expression is the same for each class, the contribution to the value of χ^2 parameter from the two tail portions of the normalised curve would be more. This is because the value of expected frequency at the two tails are much less than at the middle of the curve and that f_{ei} appears in the denominator. In order to overcome this, the regrouping in each class is so made that the expected frequency in each class is made not less than 5.0.

Chapter 3

When the continuous distribution is replaced by discrete data, a correction, known as Yate's correction, is applied for cases when the degree of freedom, *i.e.*, $(n - m)$ is of the order of 1. This is written in the following form:

$$\chi^2_{(n-m)\simeq 1} = \sum_{i=1}^{n} \frac{[|(f_{oi} - f_{ei})| - 0.5]^2}{f_{ei}} \quad \ldots(3.30)$$

In case the experimental and theoretical distributions coincide with each other fully, $\chi^2 = 0$, which is mostly unlikely. As the value of χ^2 increases, the two distributions start deviating from each other.

χ^2 parameter is very useful in statistical analysis of data.

3.11.1 Criteria of Chi-square Test for Goodness of Fit

For the practical and theoretical distributions to reasonably match each other, *i.e.*, goodness of fit, the following criteria are assumed:

1. If the probability value in the χ^2 test lies in between 0.1 and 0.9, the two distributions, more or less, match each other. In some cases, the lower probability limit (also called significance level or simply level) is put at 0.05.
2. If probability goes below the lower limit, there is wide divergence between the two distributions, and in such cases, χ^2 assumes a considerable value.

3. If probability goes above 0.9, χ^2 parameter is nearly zero. But, in practice, such high probability cases are seldom encountered in practice. In these cases, data obtained is termed as suspiciously good.

SOLVED EXAMPLES

1. Three resistors have the values: $R_1 = 100 \pm 10\%$, $R_2 = 500 \pm 5\%$ and $R_3 = 200 \pm 10\%$. Find the magnitude of the resultant resistances and the limiting errors if they are connected in (*a*) series and (*b*) parallel.

Ans. (*a*) $R = R_1 + R_2 + R_3 = 100 + 500 + 200 = 800\ \Omega$

We have: $$\delta R = \delta R_1 + \delta R_2 + \delta R_3$$

or, $$\frac{\delta R}{R} = \frac{\delta R_1}{R}\cdot\frac{R_1}{R_1} + \frac{\delta R_2}{R}\cdot\frac{R_2}{R_2} + \frac{\delta R_3}{R}\cdot\frac{R_3}{R_3}$$

$$= \frac{R_1}{R}\cdot\frac{\delta R_1}{R_1} + \frac{R_2}{R}\cdot\frac{\delta R_2}{R_2} + \frac{R_3}{R}\cdot\frac{\delta R_3}{R_3}$$

now, $\frac{\delta R_1}{R_1} = 0.1$; $\frac{\delta R_2}{R_2} = 0.05$; $\frac{\delta R_3}{R_3} = 0.1$

$\therefore$ $$\frac{\delta R}{R} = \frac{100}{800}\times 0.1 + \frac{500}{800}\times 0.05 + \frac{200}{800}\times 0.1 = 0.069$$

Hence, $$R = 800 \pm 6.9\%$$

(*b*) Here, $$\frac{1}{R} = \frac{1}{R_1} + \frac{1}{R_2} + \frac{1}{R_3} = \frac{1}{100} + \frac{1}{500} + \frac{1}{200} \quad \text{or,} \quad R = 58.8\ \Omega$$

We have: $$\frac{\delta R}{R^2} = \frac{\delta R_1}{R_1^2} + \frac{\delta R_2}{R_2^2} + \frac{\delta R_3}{R_3^2}$$

or, $$\frac{\delta R}{R} = \frac{R}{R_1}\cdot\frac{\delta R_1}{R_1} + \frac{R}{R_2}\cdot\frac{\delta R_2}{R_2} + \frac{R}{R_3}\cdot\frac{\delta R_3}{R_3} = 0.094$$

$\therefore$ $$R = 58.8 \pm 9.4\%$$

2. Find the percentage error in the measurement of momentum of a body if the percentage errors in the measurement of mass and velocity are 2% and 4% respectively.

Ans. Momentum is given by: $p = mv$

or, $$\frac{\delta p}{p} = \frac{\delta m}{m} + \frac{\delta v}{v} = 0.02 + 0.04 = 0.06.$$

$\therefore$ Thus, the percentage error in the measurement of momentum = 6%

3. A student scored 90, 80 and 70 marks in papers I, II and III respectively. If weights of 5, 3 and 4 are assigned to these papers, find the weighted average of the marks.

Ans. The weighted average is given by:

$$= \frac{5\times 90 + 3\times 80 + 4\times 70}{5+3+4} = 80.8$$

4. Calculate the (*a*) arithmetic mean and (*b*) average deviation for 50 students whose heights are given below:

Height (cm)	160 – 162	163 – 165	166 – 168	169 – 171	172 – 174
No. of students	5	8	14	19	4

Ans. (*a*) The arithmetic mean is given by

$$\bar{v} = \frac{161 \times 5 + 164 \times 8 + 167 \times 14 + 170 \times 19 + 173 \times 4}{5 + 8 + 14 + 19 + 4}$$

$$= 167.54 \text{ cms}$$

(*b*)

Height (cms)	f_i	$\lvert v_i - \bar{v} \rvert$	$f_i \lvert v_i - \bar{v} \rvert$
160 – 162	5	6.54	32.7
163 – 165	8	3.54	28.32
166 – 168	14	0.54	7.56
169 – 171	19	2.46	46.74
172 – 174	4	5.46	21.84
	Σ = 50		Σ = 137.16

$\therefore$ average deviation $= \dfrac{137.16}{50} = 2.74$ cms

5. Ten measurements were made of a rod and the readings were: 2.597, 2.585, 2.587, 2.579, 2.591, 2.581, 2.583, 2.584, 2.578 and 2.589 cms. Determine the standard deviation.

Ans.

v_i	$\bar{v} = \Sigma v_i/n$	$x_i^2 = (v_i - \bar{v})^2$	Σx_i^2	$\sigma = \dfrac{\Sigma x_i^2}{n}$
2.597		0.000144		
2.585		0		
2.587		0.000004		
2.579		0.000036		
2.591	2.585	0.000036	0.000306	0.0000306
2.581		0.000016		
2.583		0.000004		
2.584		0.000001		
2.578		0.000049		
2.589		0.000016		

Thus, the standard deviation value is = 0.0000306

6. Apply the method of least squares to draw the most probable straight line from the following data: $x = 1, 3, 5, 7, 10$ and corresponding y values are $y = 1, 4, 5, 6, 9$.

Ans.

x	y	x^2	xy
1	1	1	1
3	4	9	12
5	5	25	25
7	6	49	42
10	9	100	90
26	25	184	170

Thus,

$$a_0 = \frac{\Sigma x\Sigma xy - \Sigma y\Sigma x^2}{(\Sigma x)^2 - n\Sigma x^2} = \frac{26\times 170 - 25\times 184}{(26)^2 - 5\times 184} = 0.7378$$

$$a_1 = \frac{\Sigma x\Sigma y - n\Sigma xy}{(\Sigma x)^2 - n\Sigma x^2} = \frac{26\times 25 - 5\times 170}{(26)^2 - 5\times 184} = 0.8196$$

$\therefore$

$$y = a_0 + a_1 x = 0.7378 + 0.8196x$$

N.B.: A plot between x and y can now be drawn.

7. Two resistors are connected in series and in parallel. The values of resistances are $R_1 = 100.0\ \Omega \pm 0.1\%$, $R_2 = 50\ \Omega \pm 0.06\%$. Calculate the uncertainty in the combined resistance of both (*a*) series and (2) parallel arrangements. *[WBUT (EI 402) 2004]*

Ans. (*a*) In the series connection case:

$$R = R_1 + R_2;\ \frac{\partial R}{\partial R_1} = 1,\ \frac{\partial R}{\partial R_2} = 1$$

$$U_{R_1} = 0.1\ ;\ U_{R_2} = 50\times\frac{0.06}{100} = 0.03$$

$$U_R = \left[\left(\frac{\partial R}{\partial R_1}\right)^2 U_{R_1}^2 + \left(\frac{\partial R}{\partial R_2}\right).U_{R_2}\right]^{1/2} = [1^2 + 0.1^2 \times 1^1 \times 0.03^2]^{1/2} = 0.1044$$

Thus, uncertainty = $\pm$ 0.1

(*b*) In the parallel connection case

$$R = \frac{R_1 R_2}{R_1 + R_2}$$

$$\frac{\partial R}{\partial R_1} = \frac{R_2}{R_1 + R_2} - \frac{R_1 R_2}{(R_1 + R_2)^2} = 0.111$$

$$\frac{\partial R}{\partial R_2} = \frac{R_1}{R_1 + R_2} - \frac{R_1 R_2}{(R_1 + R_2)^2} = 0.444$$

As before, $U_R = [0.111^2 \times 0.1^2 + 0.444^2 \times 0.03^2]^{1/2} = 0.01733$

Thus, uncertainty = $\pm$ 0.017

REVIEW QUESTIONS

1. Discuss mean, mode and median.
2. What is meant by limit of error, probable error, average deviation and standard deviation?
3. How is the internal estimate of uncertainty estimated?
4. Find an expression for total uncertainty in a compound quantity.
5. What is a histogram? Discuss the same and also the advantages obtained by normalising the histogram ordinate.

6. Discuss the estimation of error from Gaussian distribution.
7. Discuss the method of least squares for curve fitting.
8. Discuss the chi-square test for normal distribution of data and also mention the criteria of chi-square test for goodness of fit.

Fill in the blanks:

1. For statistical data analysis to be meaningful, ... errors must be small compared to .... or .... errors.
2. The term 'mean' is also known by arithmetic .... or arithmetic ....
3. When a set of data contains a single mode, it is called ....
4. When a set of variates is written either in ascending or descending order, the median lies at the ....
5. Standard deviation is always ....
6. The normalised histogram is also called .... histogram.
7. There are .... standard methods of curve fitting.
8. Chi-square test quantitatively measures the closeness of the experimentally observed distribution of a set of data with .... distribution.
9. In experimental analysis, ... is considered the best estimate of the true value.
10. Several averages are used in statistics, they are ...., ... and ....

Tick the correct answer:

1. Systematic errors can/cannot remove a fixed bias in instruments. ☐
2. When a set of variates contains a single mode, it is called unimodal/multimodal. ☐
3. Mean of a set of variates is that number for which the sum of deviations is zero/non-zero. ☐
4. Usually, the value of standard deviation is larger/smaller than average deviation. ☐
5. Random/gross error follow the Gaussian distribution. ☐
6. Amount of computation to determine mean/standard deviation/mean and standard deviation is significantly reduced when data grouping is done. ☐
7. A normalised histogram leads to an identical/non-identical bar height scales. ☐

REVIEW PROBLEMS

1. A student scored 89, 79 and 85 in three papers. Find the arithmetic mean of the marks obtained by the student. If the papers have 2, 3, 4 as the weighing factors, find the weighted arithmetic mean.

   [**Ans.** 84–33, 83–88]

2. A 0–10 A ammeter has a guaranteed accuracy of 1% of FSD. The limiting error while reading 2.5 A is

   (*i*) 1% (*ii*) 2% (*iii*) 4% (*iv*) none of these. *[WBUT (EI 402) 2003]*

   [**Ans.** (*iii*)]

3. What is the percentage error in the measurement of kinetic energy of a body if percentage errors in the measurement of mass and speed are 2% and 3% respectively. *[WBUT (EI 402) 2004]*

[Ans. 4%]

4. A 0–10 A ammeter has an accuracy of 1.5% of FSD. The current indicated by the ammeter is 2·5 A. Calculate the limiting values of current and percentage limiting error.

[Ans. 2.35 A, 2.65 A, 6%]

5. During a test, current was measured as 20 A. It flows in a resistor of value 0.2 Ω. The ammeter reads 1.5% low while the resistance reading was high by 0.5%. Find the true power express it as a percentage of power that was originally calculated. **[Ans.** 82W, 102.5%]

6. A 125 mA range ammeter reads 70 mA and a 50 V range voltmeter reads 30 V in an experimental set up. Both the meters have a guaranteed accuracy of ± 2% FSD. Find the limiting error in measuring power. **[Ans.** 6.9 %]

7. For the following given data, calculate (*a*) arithmetic mean (*b*) average deviation (*c*) algebraic sum of deviations (*d*) standard deviation and (*e*) variance. Data : $x_1 = 99.7$, $x_2 = 100.1$, $x_3 = 100.2$, $x_4 = 99.6$, $x_5 = 99.7$. *[WBUT (EI 402) 2003]*

[Ans. (*a*) 99.86, (*b*) 0.232 (*c*) 0 (*d*) 0.2702 (*e*) 0.073]

8. A large number of 230 Ω resistors are obtained by combining 120 Ω resistors with a standard deviation of 4.0 Ω and 110.0 Ω resistors with a standard deviation of 3.0 Ω. The standard deviation of the 230 Ω resistors thus formed will be

   (*a*) 3.5 Ω (*b*) 5.0 Ω (*c*) 7.0 Ω (*d*) 12.0 Ω *[GATE 2003]* **[Ans.** (*b*)]

9. A mass of 10 kg is measured with an instrument and the readings are normally distributed with respect to the mean of 10 kg. Given that $\frac{2}{\sqrt{2\pi}} \int_0^{0.84} \exp\left(-\eta^2/2\right) d\eta = 0.6$ and that 60% of the readings are found to be within 0.05 kg from the mean, the standard deviation of the data is:

   (*a*) 0.02 (*b*) 0.04 (*c*) 0.06 (*d*) 0.08 *[GATE 2005]* **[Ans.** (*c*)]

10. The measurements of a source voltage are 5.9 V, 5.7 V and 6.1 V. The sample standard deriation of the readings is:

    (*a*) 0.013 (*b*) 0.04 (*c*) 0.115 (*d*) 0.2 *[GATE 2005]* **[Ans.** (*d*)]

CHAPTER 4

DYNAMIC CHARACTERISTICS OF INSTRUMENTS

4.1 INTRODUCTION

In dynamic characteristics, the behaviour/performance of a system is studied when the measurand (input) is a function of time. The dynamic response depends both on the nature and complexity of the input function (measurand) as also on the parameters of the instrument under study.

Response to changes in the input signal to a measuring device/instrument is not instantaneous. This is because of the sluggishness of the instrument. Some time is needed after which the response (output signal) bears a fixed relationship with the input signal. Hence, before this time, response consists of two components: transient and steady state. With time, the transient component dies out and the system response becomes steady.

Mathematical modelling is sought to study the behaviour of a measuring system under dynamic conditions. The mathematical models are linear differential equations having constant coefficients and the measuring system is classified according to the order of the differential equation. Most of the measuring systems are either first or second order type. The behaviour of the measuring systems to time-varying inputs are based on certain distinguishable parameters, such as: (*a*) dynamic error (*b*) fidelity (*c*) bandwidth and (*d*) speed of response.

4.1.1 Dynamic Error

It is the algebraic difference between the true value and recorded value of the measurand at any instant, assuming that the measurand is time dependent. The error is zero for an ideal measuring system. But for higher order measuring systems, the dynamic error consists of two portions: transient error and the steady state error.

For a step-input, the dynamic error consists only of transient error. This error becomes zero when the system assumes steady state. Under steady state, the ratio of output signal to that of the input (measurand) is a constant and that this ratio represents the static sensitivity of the system.

For ramp or sinusoidal type input functions, dynamic error has both transient and steady-state errors. The steady state error is a constant for ramp type input while it is time dependent for sinusoidal type input. Dynamic sensitivity is defined as the ratio of the peak amplitude of the output signal to that of the input signal and that it is frequency dependent. This is known as gain-frequency characteristics. A phase error also exists which is the difference in phase between the output and input signals. The phase error is also frequency dependent. The variation of phase difference with frequency is known as phase-frequency characteristics.

4.1.2 Fidelity

It refers to the ability of a measuring system/device to reproduce at the output the variations in the measurand (input) with time. A zero order system has a very high fidelity and reproduces the time-varying variations of the input at the output end in a very truthful manner. Systems of higher order are incapable of reproducing at the output, these time varying variations of the input.

Fidelity performance is evaluated under steady state conditions for sinusoidal inputs. Under such conditions, for a high fidelity system, the output signal should follow the variations in the input at all instants of time and there would neither be amplitude or phase error. A high fidelity system has a flat amplitude-frequency characteristics over a wide range of frequencies starting from zero. Systems which do not respond to DC inputs, the flatness is assessed from dynamic sensitivity at some mid-frequency.

For different application areas, fidelity requirements vary. Thus, moving iron type instruments, RMS indicating meters, should have same sensitivity from DC to 50 Hz AC. Waveform recorders, should have an excellent fidelity over a wide frequency range with no amplitude or phase distortion. As a thumb rule, the amplitude frequency characteristics should remain flat within ±2% with no phase error.

For entertainment, fidelity requirements are relaxed, because human sensory systems are poor detectors for minor variations at the output end. For such cases, the dynamic sensitivity at zero frequency (or mid frequency, as the case may be) is allowed to fall down up to 70.7% over the frequency range of operation.

4.1.3 Bandwidth

For measuring devices and systems, dynamic sensitivity should remain within ±2% of its static sensitivity. The bandwidth is stated to cover this range for which the above criterion is fulfilled.

For filters and amplifiers, the bandwidth is the range of frequencies for which the dynamic sensitivity is 70.7% of that at either zero or mid frequency.

4.1.4 Speed of Response

It is related to the ability of a system to respond to sudden changes in the amplitude of the input signal. It is measured as the time taken by the system to come very close to steady state conditions after a sudden change in the input. Time constant, settling time, measurement lag, dead time, settling time and dynamic range are the other parameters related to speed of response.

Time constant (T) is defined as the time taken by the system to reach 63.2% of its final output value. A system with a large time constant is sluggish in nature and has a slow speed of response. Time constant governs the behaviour of a first order system.

Changes in the input to a system take some definite time to be reflected at the output. This time difference is called measurement lag. It is a characteristic of the system.

Irrespective of the nature of the input signal, some systems do not respond until after some time elapses. This time is known as the dead time of the system. Figure 4.1 shows the dead time for different types of inputs.

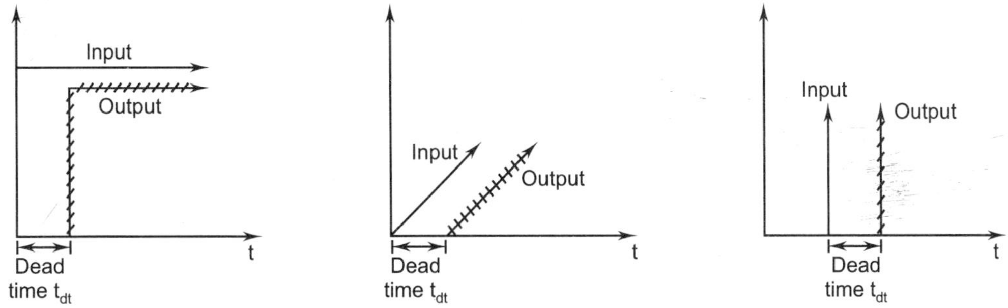

Fig. 4.1: Dead time for different input types

Settling time for an indicating instrument is defined as the time taken by the pointer to reach a value within ± 1% of its final value. Settling time depends on the parameters of the system and is affected by the conditions under which the system works.

Dynamic range is the range of input signals which a measuring system would be able to reproduce faithfully under dynamic conditions. It is expressed in dB and is the ratio of maximum to minimum input signal which the measuring system can handle faithfully.

4.2 GENERALISED PERFORMANCE OF INSTRUMENTS

Most physical systems can be modelled into ordinary linear differential equations, with constant coefficients and can be expressed as:

$$a_n \frac{d^n y_0(t)}{dt^n} + a_{n-1} \frac{d^{n-1} y_0(t)}{dt^{n-1}} + \cdots + a_1 \frac{dy_0(t)}{dt} + a_0 y_0(t)$$

$$= b_m \frac{d^m x_i(t)}{dt^m} + b_{m-1} \frac{d^{m-1} x_i(t)}{dt^{m-1}} \quad \text{...(4.1)}$$

In the above, a's and b's are constants representing the physical parameters of the system while $y_o(t)$ and $x_i(t)$ are the output and input variables respectively. The order of the system is n.

The transfer function, in Laplace domain, is given by:

$$G(s) = \frac{Y_o(s)}{X_i(s)} = \frac{P(s)}{Q(s)} = \frac{b_m s^m + b_{m-1} s^{m-1} + \cdots + b_1 s + b_0}{a_n s^n + a_{n-1} s^{n-1} + \cdots + a_1 s + a_0} \quad \text{...(4.2)}$$

where, s is a complex quantity and $s = \sigma + j\omega$. In the above, it has been assumed that all initial conditions are zero at $t = 0$.

The time domain output $y_0(t)$ is expressed as:

$$y_0(t) = y_0(t)|_{\text{I}} + y_0(t)|_{\text{II}} \quad \text{...(4.3)}$$

The first part, *i.e.*, $y_0(t)|_{\text{I}}$ is known as the complementary function and is obtained by making $Q(s) = 0$. $Q(s)$ is also called characteristic equation and is used to find out the transient part of the solution. After the transient time elapses, the output is $y_0(t)|_{\text{II}}$ which is also called particular integral. This represents the steady state behaviour of the system due to forcing function (or input function) such as step, ramp or sinusoid type.

System response when $x_i(t)$ is sinusoid in nature can be had by finding out the relationship between output and input in terms of the ratio of their amplitudes and phase difference between them. It is assumed that the system is linear in nature and steady state response is obtained by replacing $j\omega$ for s and varying the frequency from zero (DC) to infinity. In this case, thus,

$$\frac{Y_o(j\omega)}{X_i(j\omega)} = |M| \angle \phi \qquad ...(4.4)$$

$|M|$ represents the ratio of the amplitudes and $\angle\phi$ is the difference in phase between the input and output variables. It should be noted that both, $|M|$ and $\angle\phi$ are functions of the frequency of the input functions.

4.3 STANDARD INPUT TEST SIGNALS

Dynamic behaviour of measurement systems are studied by subjecting them to some standard input signals. These are: step input, ramp input, sinusoidal input and impulse input. These are discussed below and shown in figure 4.2.

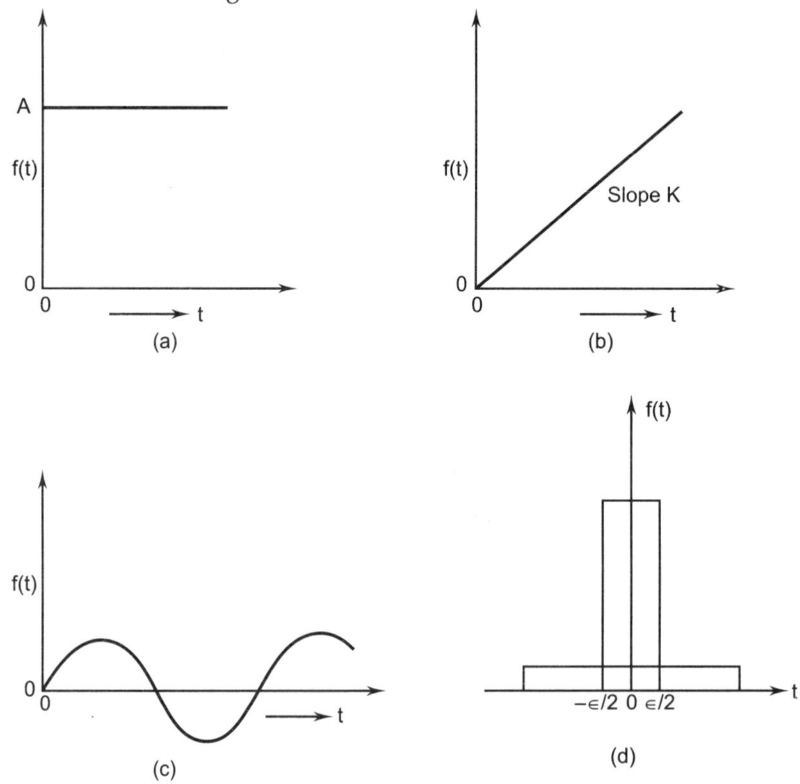

Fig. 4.2: Standard input test signals: (*a*) step-input; (*b*) ramp-input; (*c*) sinusoidal input; (*d*) impulse input

For the step input function,

$$f(t) = Au(t) \qquad ...(4.5)$$

and,

$$u(t) = 1 \text{ for } t > 0$$
$$= 0 \text{ for } t < 0$$

A being the height of the amplitude, as shown. If $A = 1$, then the function is called a unit step function.

The Laplace transform of the step input function is

$$F(s) = A/s \quad ...(4.6)$$

The ramp function is expressed as

$$f(t) = Kt \text{ for } t > 0 \quad ...(4.7)$$
$$= 0 \quad \text{for } t < 0$$

K being the slope of the ramp.

Its Laplace transform is

$$F(s) = K/s^2 \quad ...(4.8)$$

For a unit ramp type, $F(s) = 1/s^2$

The sinusoidal function is expressed as:

$$f(t) = A \sin(2\pi ft) \quad \text{for } t > 0$$
$$= A \sin \omega t \quad ...(4.9)$$

Its Laplace transform is

$$F(s) = \frac{A\omega}{s^2 + \omega^2}, \; s > 0 \quad ...(4.10)$$

Now, consider a rectangular function (a pulse signal) defined by:

$$f(t) = \left(\frac{1}{\in}\right)\left[\sigma\left(t + \frac{\in}{2}\right) - \sigma\left(t - \frac{\in}{2}\right)\right] \quad ...(4.11)$$

Let, $f(t)$ is characterised by having unit area always, irrespective of the parameter $\in$.

Thus, area of the above pulse is

$$= \int_{-\infty}^{+\infty} f(t)\, dt = 1$$

Now, when the parameter $\in$ tends to zero, the pulse contracts but retains its unit area characteristies.

Thus the height of the pulse increases without bound. The limit of a sequence of such functions as $\in \to 0$ is termed the unit impulse function, the delta function or the Dirac delta function. Thus,

$$\delta(t) = \lim_{\in \to 0} f(t, \in) \quad ...(4.12)$$

Mathematically, the derivative of a step function is an impulse function and thus, the Laplace transform of a delta function is

$$F(s) = 1 \quad ...(4.13)$$

Laplace transform of some common time functions $f(t)$ are shown in table 4.1.

Table 4.1: Laplace Transforms of Some Common Time Functions $f(t)$: $f(t)$ is finite for ≥ 0

$$\alpha[f(t)] = F(s) = \int_0^{\infty} e^{-st} f(t)\, dt$$

Function	Symbol	Graph	Transform
1st Derivative	$\frac{d}{dt} f(t)$		$sF(s) - f(0-)$
Integration	$\int_0^t f(t)\, dt$		$\frac{1}{s} F(s)$

Contd.

Chapter 4

Function	*Symbol*	*Graph*	*Transform*
Unit Impulse	$\delta(t)$	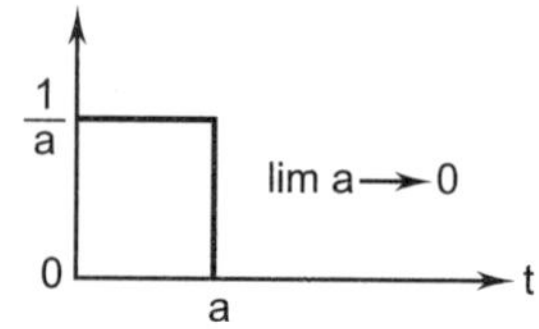	1
Unit Step	$u(t)$	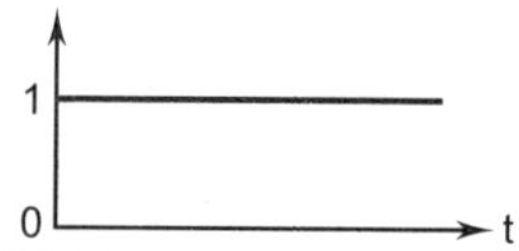	$\frac{1}{s}$
Exponential decay	$\exp(-\alpha t)$	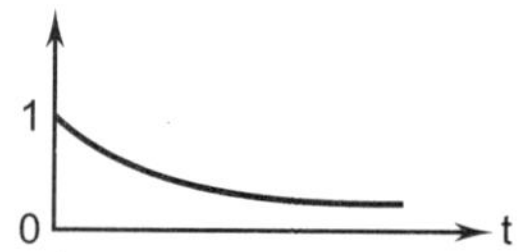	$\frac{1}{s+\alpha}$
Exponential growth	$1 - \exp(-\alpha t)$	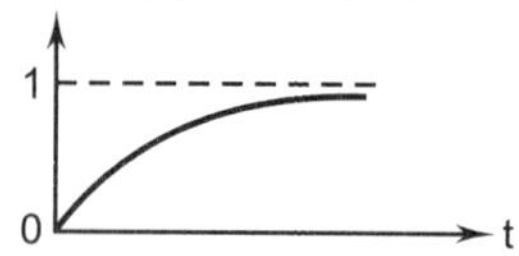	$\frac{\alpha}{s(s+\alpha)}$
Sine wave	$\sin \omega t$	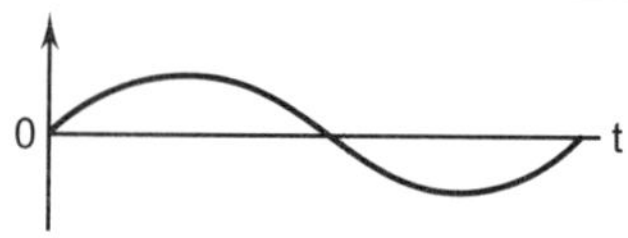	$\frac{\omega}{s^2+\omega^2}$
Phase shifted sine wave	$\sin(\omega t + \phi)$	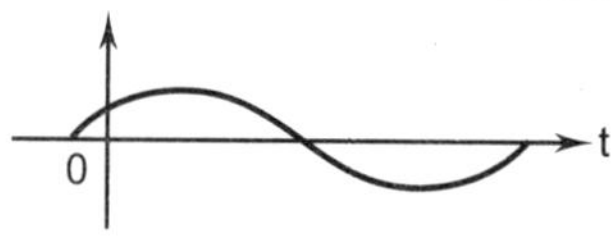	$\frac{\omega \cos \phi + s \sin \phi}{s^2+\omega^2}$
Exponentially damped sine wave	$\exp(-\alpha t) \sin \omega t$	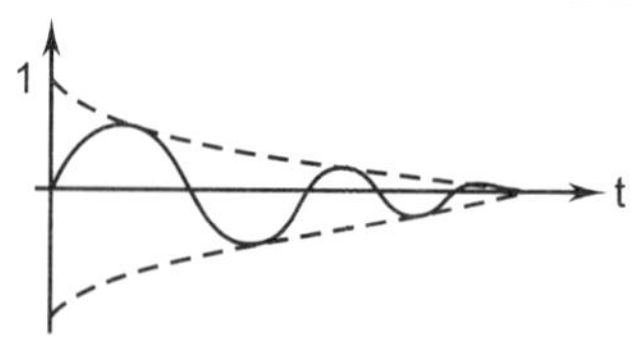	$\frac{\omega}{(s+\alpha)^2+\omega^2}$
Ramp with exponential decay	$t \exp(-\alpha t)$	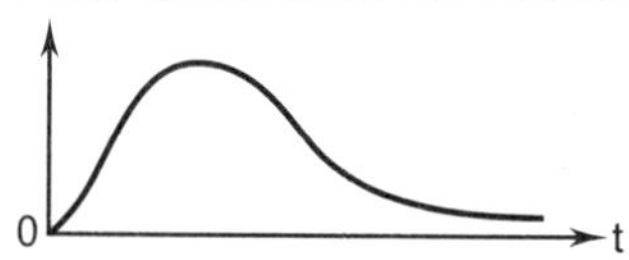	$\frac{1}{(s+\alpha)^2}$

4.4 DIFFERENT ORDERED SYSTEMS

4.4.1 Zero Order System

Such a physical system is represented as:

$$a_0 y_0(t) = b_0 x_i(t) \quad ...(4.14)$$

Thus,

$$y_0(t) = \frac{b_0}{a_0} x_i(t) = K x_i(t)$$

$$\therefore \qquad \frac{Y_o(s)}{X_i(s)} = K \qquad \ldots(4.15)$$

where, $\quad K = \dfrac{b_0}{a_0}$ = static sensitivity

Systems which possess a constant value for their transfer function is called a zero order system. The constant value represents the static sensitivity of the system and its block diagram representation is shown in figure 4.3.

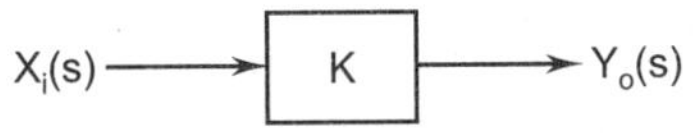

Fig. 4.3: Block diagram representation of a zero order system

The output of a zero order system follows the input changes very closely without any time or phase lag. Such a system does not have any dynamic error. It is very difficult to get a physical system which belongs to zero order although potentiometer used for measuring displacement may be thought of as a zero order system.

4.4.2 First Order System

Examples of a first order system are a mercury-in-glass thermometer, a thermocouple, a thermistor or an *RC* circuit.

A first order system is of the form

$$a_1 \frac{dy_o(t)}{dt} + a_0 y_0(t) = b_o x_i(t) \qquad \ldots(4.16)$$

So that the transfer function is

$$G(s) = \frac{Y_o(s)}{X_i(s)} = \frac{b_0}{a_1 s + a_0} = \frac{b_0/a_0}{1 + \dfrac{a_1}{a_0} s}$$

$$= \frac{K}{1 + s\tau} \qquad \ldots(4.17)$$

where, $K = \dfrac{b_0}{a_0}$ represents the static sensitivity and

$\tau = \dfrac{a_1}{a_0}$ is the time constant, seconds.

The block diagrammatic representation of a first order system is shown in figure 4.4.

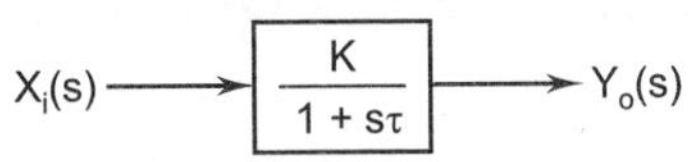

Fig. 4.4: Block diagram representation of a first order system

4.4.3 Second Order System

A second order system is of the form:

$$a_2 \frac{d^2 y_0(t)}{dt^2} + a_1 \frac{dy_0(t)}{dt} + a_0 y_0(t) = b_0 x_i(t) \qquad \ldots(4.18)$$

which yields the transfer function as

$$G(s) = \frac{Y_o(s)}{X_i(s)} = \frac{b_0}{a_2 s^2 + a_1 s + a_0} = \frac{b_0/a_0}{s^2\left(\frac{a_2}{a_0}\right) + s\left(\frac{a_1}{a_0}\right) + 1}$$

$$= \frac{K}{s^2\left(\frac{a_2}{a_0}\right) + s\left(\frac{a_1}{a_0}\right) + 1} \quad \text{...(4.19)}$$

where,

K = static sensitivity = b_0/a_0

ω_n = undamped natural frequency

$$= \left(\frac{a_0}{a_2}\right)^{1/2}$$

a_1/a_0 = damping condition of the system

ξ = damping factor (or damping ratio)

$$= \frac{\text{damping coefficient under existing damping condition}}{\text{damping coefficient under critical condition}}$$

$$= \frac{a_1}{2\sqrt{a_0 a_2}}$$

Thus,
$$\frac{Y_o(s)}{X_i(s)} = \frac{K}{\left(\frac{s^2}{\omega_n^2} + \frac{2\xi s}{\omega_n} + 1\right)} \quad \text{...(4.20)}$$

Different standard input conditions are applied in the expression for $Y_o(s)/X_i(s)$ for the second order systems to find out the value of $y_0(t)$.

The block diagram representation of a second order system is shown in figure 4.5.

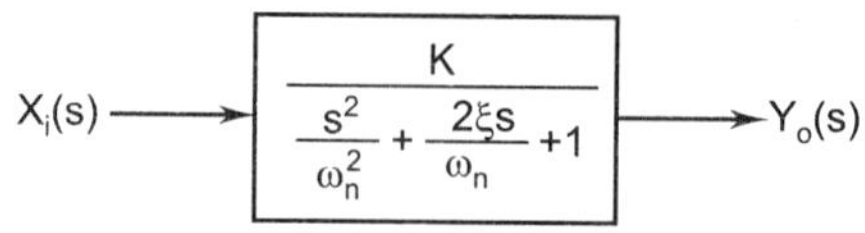

Fig. 4.5: Block diagram representation of a second order system

For first and second order systems, some assumptions are made. These are:

- Damping is viscous in nature.
- The restoring elements are linear.
- The system is a single degree of freedom type.

4.5 DYNAMIC RESPONSE OF FIRST ORDER SYSTEMS

4.5.1 Unit Step Function

Here a unit step function acts as an input to the first order system.

The transfer function of a first order system is

$$G(s) = \frac{Y_o(s)}{X_i(s)} = \frac{K}{1+s\tau} \quad \text{...(4.21)}$$

Now, for a unit step input, $X_i(s) = \frac{1}{s}$

Thus, $Y_o(s) = s \cdot \frac{K}{1+s\tau}$

Taking inverse Laplace transform,

$$y_0(t) = K(1 - e^{-t/\tau}) \qquad ...(4.22)$$

The fastness with which output $y_0(t)$ reaches the final value K depends on the time constant τ of the system. For low τ values, speed of response is faster. This is a desirable feature of any instrument. A useful feature that is adopted for any instrument is the time that it takes to reach ± 5% of the final value, shown in figure 4.6. The output reaches 5% and 1% of final value (response) in approximately 3τ and 5τ time respectively. A normalised response curve can be drawn with $y_0(t)/k$ as the Y-axis and t/τ as the X-axis and the unit step function switched off at $t = T$.

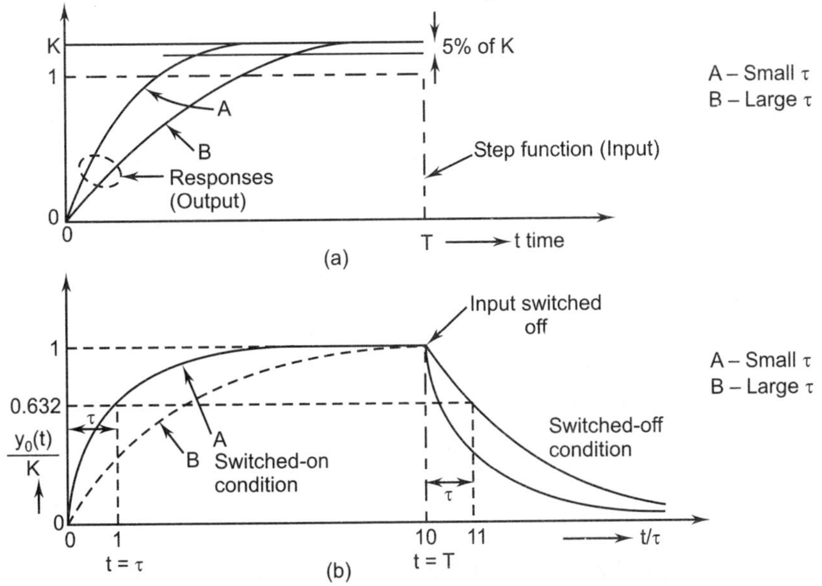

Fig. 4.6: (*a*) Response of a first-order system for a step-input function; (*b*) normalised response under switched on and switched-off conditions

4.5.2 Unit Ramp Function

Here a unit ramp function acts as an input to the first order system.

For a unit ramp input, $X_i(s) = \frac{1}{s^2}$ and thus,

$$Y_o(s) = \frac{1}{s^2} \cdot \frac{K}{1 + s\tau} \qquad ...(4.23)$$

Inverse Laplace transform of the above yields.

$$y_0(t) = K(\tau e^{-t/\tau} + t - \tau) \qquad ...(4.24)$$

The system response is shown in figure 4.7. If the response is ideal, the output value will be kt. Hence, the dynamic measurement error is

$$= \underset{\substack{\text{Transient} \\ \text{error}}}{K(\tau e^{-t/\tau})} - \underset{\substack{\text{Steady state} \\ \text{error}}}{k\tau} \qquad ...(4.25)$$

or, $$e_d = A - B$$

The transient error dies out with time, leaving only the steady state error. From the steady state error it is seen that the instrument reading always lags behind the actual value and the instrument, at any instant of time, shows a value what the input was τ seconds ago. Thus, a small τ value for the system is always desirable.

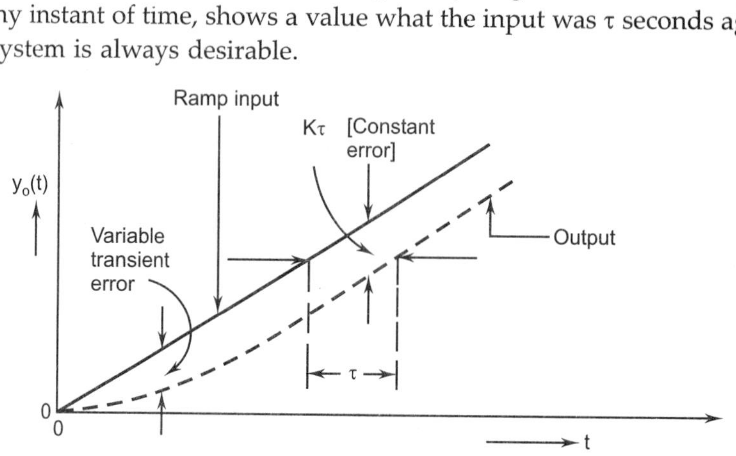

Fig. 4.7: Response of a first-order system for a ramp-input

The error plot with ramp input is shown in figure 4.8, with X-axis representing t/τ and Y-axis $ed/k\tau$.

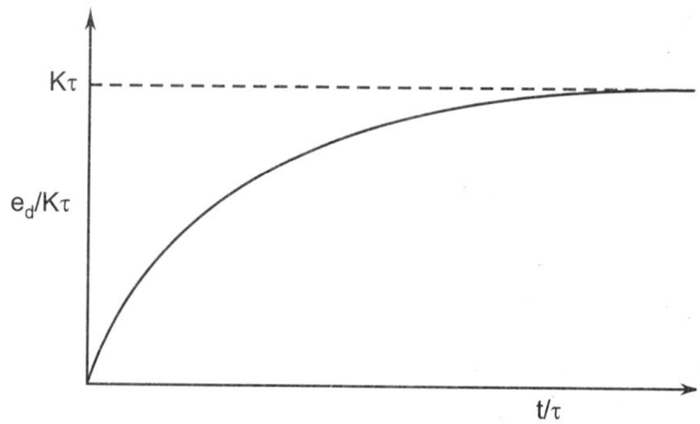

Fig. 4.8: Error plot for a ramp input

4.5.3 Unit Impulse Function

Here a unit impulse function acts as an input to the first order system. For a unit impulse function, transfer function = 1.

Thus, for a first order system,

$$Y_0(s) = \frac{K}{1+s\tau} \qquad \text{...(4.26)}$$

$$\therefore \qquad y_0(t) = \frac{K}{\tau} \cdot e^{-t/\tau} \qquad \text{...(4.27)}$$

It is practically impossible to apply an ideal impulse input (i.e. infinite magnitude existing for zero time duration) to a first order system. Figure 4.9(*a*) shows the response of a first order system to an impulse of finite strength *A* and duration $\in$ while figure 4.9(*b*) shows the response for an ideal impulse function.

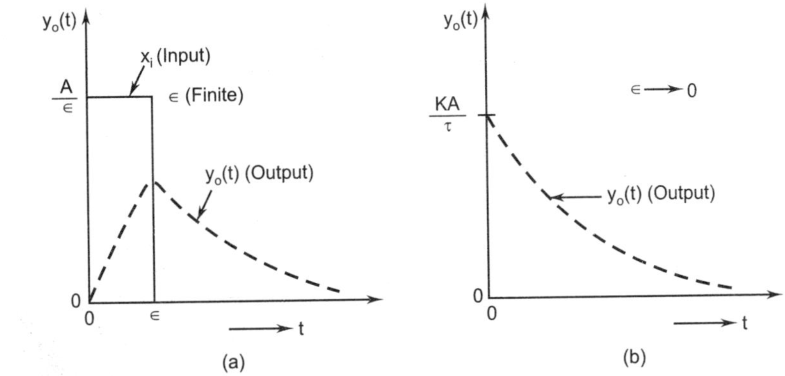

Fig. 4.9: Response of *a* first-order system for (*a*) prolonged impulse-input; (*b*) ideal impulse input

4.5.4 Sinusoidal Function

Here a sinusoidal function acts as an input to the first order system.

Frequency response for sinusoidal input is given by

$$G(j\omega) = \frac{Y_o(j\omega)}{X_i(j\omega)} = \frac{K}{1 + j\omega\tau}$$

$$= \frac{K}{\sqrt{1+\omega^2\tau^2}} \angle \tan^{-1}(-\omega t) \quad \text{...(4.28)}$$

Thus, the amplitude ratio $= \frac{Y_o}{X_i} = \frac{K}{\sqrt{1+\omega^2\tau^2}} = M$

and, $\phi = \tan^{-1}(\omega\tau)$

The plots of amplitude ratio and phase angle with frequency is shown in figure 4.10.

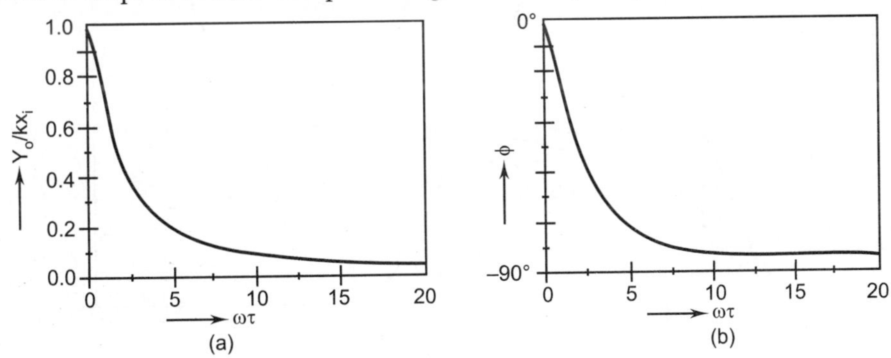

Fig. 4.10: Plots for a first-order system: (*a*) amplitude ratio vs. frequency, and (*b*) phase angle vs. frequency

4.6 DYNAMIC RESPONSE OF SECOND ORDER SYSTEMS

4.6.1 Unit Step Function

Here a units step function acts as an input to the second order system.

For a second order system, $$G(s) = \frac{Y_o(s)}{X_i(s)} = \frac{K}{\left(\frac{s^2}{\omega_n^2} + \frac{2\xi s}{\omega_n} + 1\right)} \quad ...(4.29)$$

If the input is a unit step function, then $x_i(s) = \frac{1}{s}$

$$\therefore \quad Y_o(s) = \frac{K}{s\left(\frac{s^2}{\omega_n^2} + \frac{2\xi s}{\omega_n} + 1\right)} \quad ...(4.30)$$

Three different outputs would result depending on the damping condition. These are given by:

$$\frac{y_0(t)}{K} = \left[1 - \frac{\exp(-\omega_n \xi t)}{\sqrt{1-\xi^2}} \sin\left\{\omega_n \sqrt{1-\xi^2}\, t + \sin^{-1}\sqrt{1-\xi^2}\right\}\right]$$

for $\xi < 1$ (underdamped condition) ...(4.31)

$$\frac{y_0(t)}{K} = [1 - (1 + \omega_n t)\exp(-\omega_n t)]$$

for $\xi = 1$ (critically-damped condition) ...(4.32)

and $$\frac{y_0(t)}{K} = \left[1 - \frac{\exp(-\omega_n \xi t)}{\sqrt{\xi^2 - 1}} \sinh\left\{\omega_n \sqrt{\xi^2 - 1}\, t + \sinh^{-1}\sqrt{\xi^2 - 1}\right\}\right]$$

for $\xi > 1$ (overdamped condition) ...(4.33)

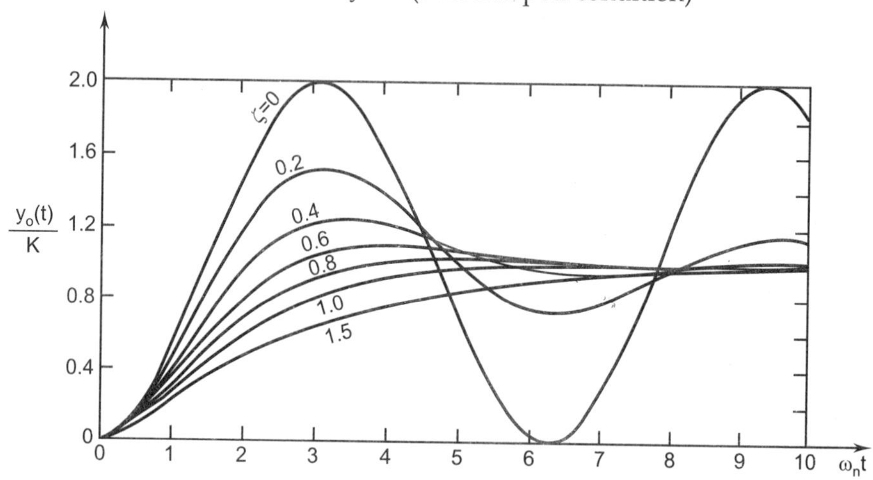

Fig. 4.11: Response of a second-order system for a step-input

A plot of $y_0(t)/k$ is shown in figure 4.11. The curves are shown for different ξ values. The manner in which $y_0(t)$ reaches its final value is ξ dependent. For $\xi = 0$, the system is totally undamped, $y_0(t)/k$ becomes equal to $(1 - \cos \omega_n t)$ and the system continues to oscillate around its final value of unity, having its peaks in between 0 and 2. As damping factor ξ increases from 0, the amplitude of oscillations decreases progressively with time. For values of ξ less than 1, the system is underdamped. As ξ increases from 0, the setling time (*i.e.*, the time to reach final value) continues to decrease. At $\xi = 1$, the system is said to be critically-damped, reaches the final value in shortest possible time and no oscillations take place. When $\xi > 1$, the system is overdamped and settling time also increases. Second order indicating instruments are designed to work under damping factors a bit less than unity, normally a value of 0.7 to 0.8 for ξ is taken. This ensures that the pointer does not get stuck up midway through, although some overshooting (beyond the final value) alongwith oscillations would be there.

4.6.1.1 Different Input Parameters

Figure 4.12 shows the response of a second-order underdamped system subjected to a unit step input. For such a case, different time-dependent quantities are specified to evaluate the performance. These are now discussed.

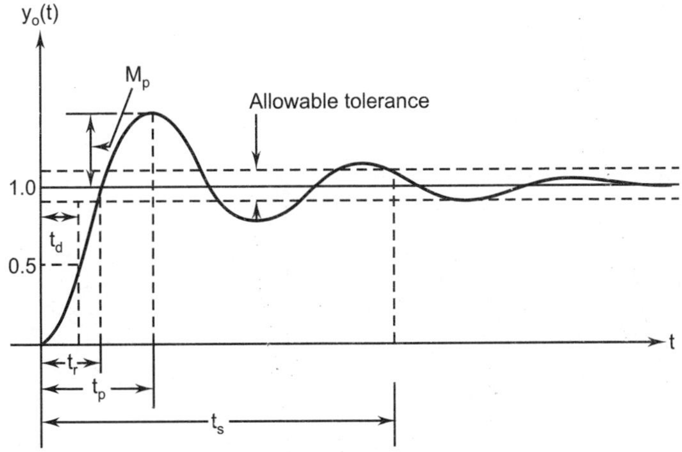

Fig. 4.12: Response of a second-order underdamped system for a unit step function

Delay time (t_d)

It is the time required for the response to reach 50% of the final value for the first time.

Rise time (t_r)

It is the time required for the response to attain the steady state value for the first time. This is given by:

$$t_r = \frac{\pi - \cos^{-1}\xi}{\omega_n\sqrt{1-\xi^2}}$$

Settling time (t_s)

It is the time required for the response to attain and remain within 5% of the final value for the first time. Settling time is given by:

$$t_s = \frac{4}{\xi\omega_n}$$

Peak time (t_p)

It is the time required for the response to reach the first peak value of the overshoot. It is given by:

$$t_p = \frac{\pi}{\omega_n\sqrt{1-\xi^2}}$$

Maximum or Peak Overshoot (M_p)

It is the maximum (peak) value of the response measured from unity. It is given by:

$$M_p = \exp\left[-\frac{\pi\xi}{\sqrt{1-\xi^2}}\right]$$

Chapter 4

4.6.2 Unit Ramp Function

Here a unit ramp function acts as an input to the second order system.

For a unit ramp input, $X_0(s) = \frac{1}{s}$ and hence the system response is

$$Y_0(s) = \frac{K}{s^2\left(\frac{s^2}{\omega_n^2} + \frac{2\xi s}{\omega_n} + 1\right)} \quad ...(4.34)$$

Assuming underdamped condition, the time-domain response is

$$\frac{y_0(t)}{K} = \left[t - \frac{2\xi}{\omega_n} + \frac{\exp(-\omega_n \xi t)}{\omega_n\sqrt{1-\xi^2}} \sin(\omega_n\sqrt{1-\xi^2}\, t + \phi)\right] \quad ...(4.35)$$

where, $$\tan\phi = \frac{2\xi\sqrt{1-\xi^2}}{2\xi^2 - 1}$$

The nature of the response is shown in figure 4.13. It is seen that the steady error is $2\xi k/\omega_n$. Obviously, the steady state error decreases with increasing ω_n. True response value, under steady state conditions, is obtained after a time gap of $2\xi/\omega_n$. If damping factor ξ is reduced, oscillations would persist for a longer time, but, at the same time, both steady state time lag and steady state error would decrease.

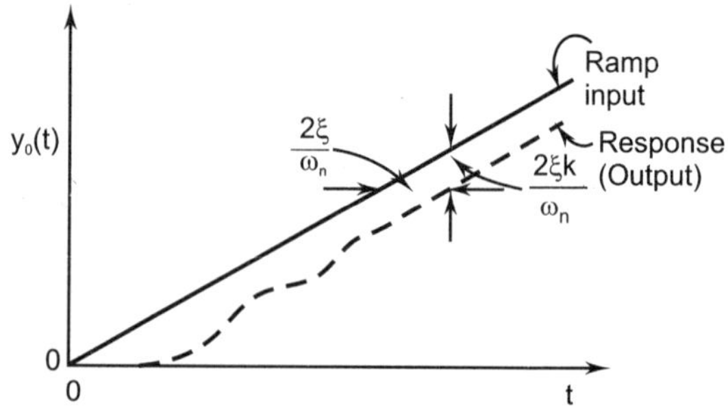

Fig. 4.13: Response of second-order system for a ramp-input

Some instruments, designed to measure the process parameters, like temperature or pressure, experience a change in input in the form of a ramp followed by a constant value. For such a terminated ramp function the response is shown in figure 4.14 for the case of a highly damped system.

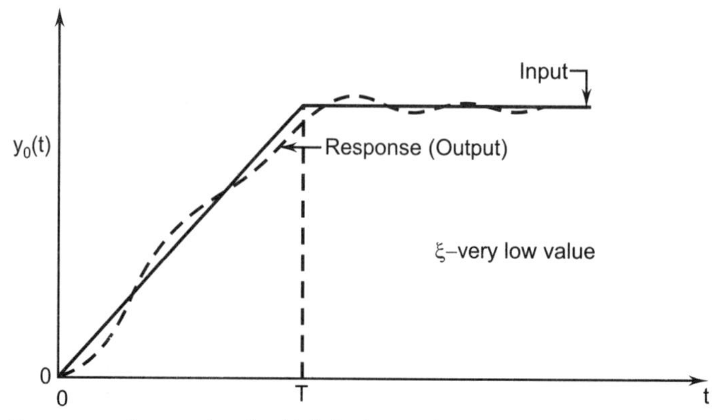

Fig. 4.14: Response of second-order hightly damped system for a terminated ramp-input

4.6.3 Frequency Response Characteristics for Sinusoidal Input

The frequency response of a second-order system subjected to a sinusoidal input is given by:

$$\frac{Y_o(j\omega)}{X_i(j\omega)} = \frac{K}{\left[-\left(\frac{\omega}{\omega_n}\right)^2 + \frac{2j\omega\xi}{\omega_n} + 1\right]} \quad ...(4.36)$$

If, $$\eta = \frac{\text{frequency of the input}}{\text{natural frequency}} = \frac{\omega}{\omega_n}$$

Thus, $$\frac{Y_o(j\omega)}{X_i(j\omega)} = \frac{K}{[(1-\eta^2)^2 + 4\xi^2\eta^2]^{1/2}} \angle\phi \qquad ...(4.37)$$

where, $$\phi = \tan^{-1}\frac{2\xi\eta}{\sqrt{1-\eta^2}} = |M| \angle\phi$$

The frequency response curves showing the variations of $|M|$ and $\angle\phi$ with ξ are shown in figure 4.15 under steady state conditions.

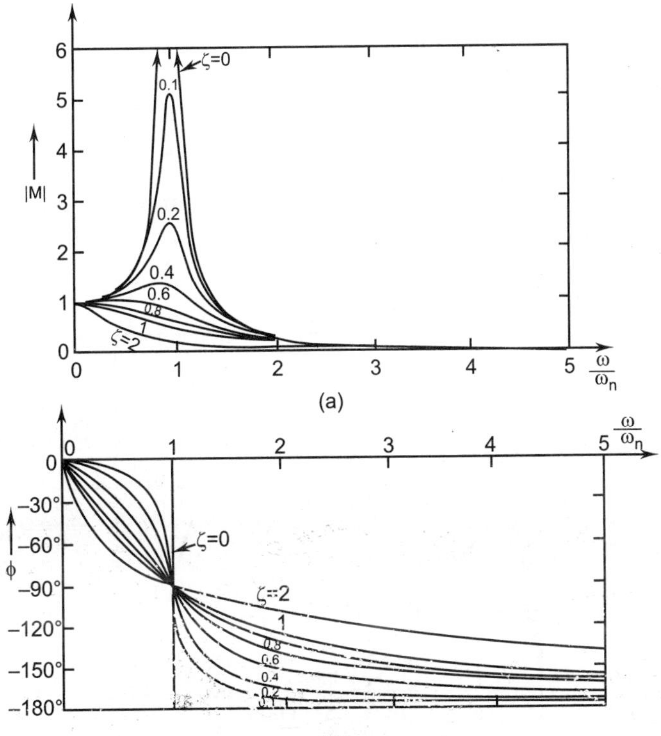

Fig. 4.15: Frequency response characteristics of a second-order system for (*a*) amplitude; (*b*) phase

If the damping factor is maintained at 0.707, the magnitude of $|M|$ will be K for frequencies upto 0.4 ω_n and the bandwidth becomes maximum. The above design considerations are adhered to for a recording instrument. Such instruments record the magnitude and phase of the input signal truthfully with no error, provided ϕ and η maintain a linear relationship between themselves for frequencies upto 0.4 ω_n.

4.6.4 Impulse Function

Here a impulse function acts as an input to a second order system.

The impulse response of second-order systems is dependent on the strength of the input impulse function. Sometimes, pulses of very short duration are also used for which the system response begins after the pulse dies. If the input is an unit impulse function, then the system response is

$$Y_0(s) = \frac{1}{\left(\frac{s^2}{\omega_n^2} + \frac{2\xi s}{\omega_n} + 1\right)} \qquad ...(4.38)$$

From which the normalized response $y_0(t)$ is obtained as:

$$\frac{y_0(t)}{KA\omega_n^2} = \frac{1}{\omega_n\sqrt{1-\xi^2}} \exp(-\xi\omega_n t)\sin\left(\omega_n\sqrt{1-\xi^2}\,t\right), \qquad ...(4.39)$$

when $\xi < 1$, *i.e.*, underdamped condition

$$= t\exp(-\omega_n t)$$

when $\xi = 1$, *i.e.*, critically damped condition

$$= \frac{1}{\omega_n\sqrt{\xi^2-1}} \exp(-\omega_n \xi t)\sinh\left(\omega_n\sqrt{\xi^2-1}\,t\right) \qquad ...(4.40)$$

when $\xi > 1$, *i.e.*, overdamped condition.

The normalised system response for a second-order system for different ξ values are shown in figure 4.16. It is seen that smaller the value of ξ, the larger is the value of the first peak.

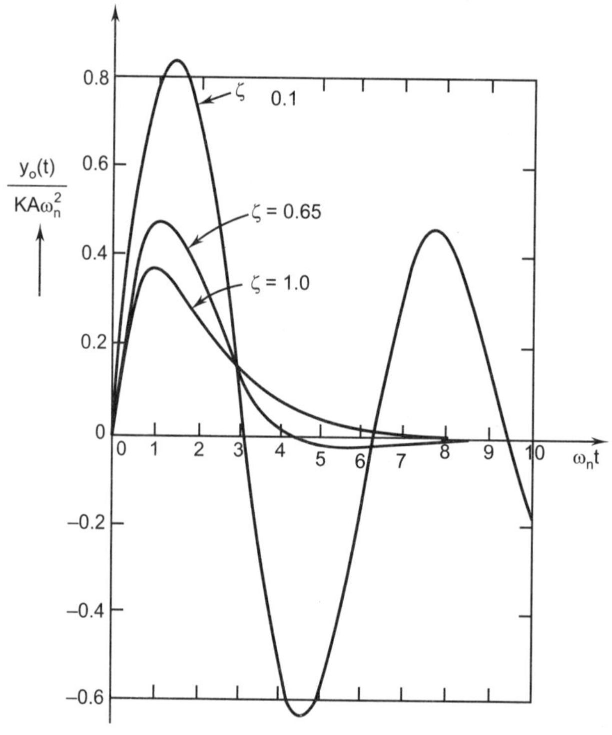

Fig. 4.16: Impulse response of a second-order system

4.7 COMPENSATION OF FIRST AND SECOND ORDER SYSTEMS

Compensation of first and second order systems are effected to reduce time constant and increase the damping ratio ξ respectively. These are now discussed.

4.7.1 Compensation of a First Order System

First order systems, like mercury-in-glass thermometer or a thermocouple, have the following governing relation:

$$(1 + \tau D)v_0 = K\theta_i(t) \qquad \text{...(4.41)}$$

τ is the time constant, D is the time derivative operator, *i.e.*, $D = d/dt$; v_0 is the output voltage, K is the static sensitivity and $\theta_i\ (t)$ is the temperature of the measurand (input) which is time dependent.

To reduce the time constant τ of this system, it is applied to a RC combination as shown in figure 4.17.

We have,

$$v_0' = i_0 R = (i_1 + i_2)R$$

now,

$$i_1 = \frac{v_0 - v_0'}{R_0} \text{ and } i_2 = C_0 D(v_0 - v_0')$$

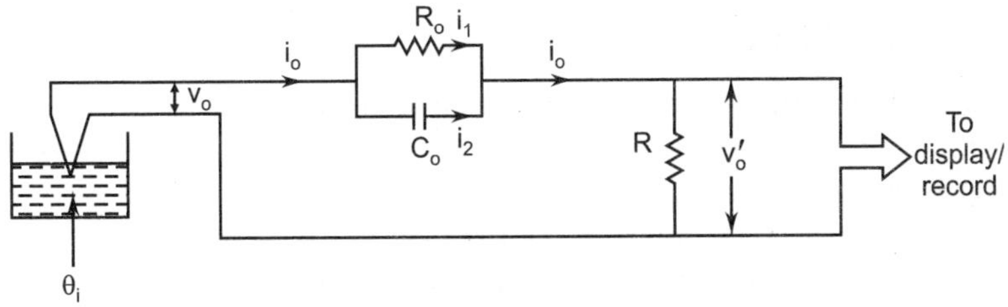

Fig. 4.17: A first order system with compensation applied

Thus,

$$v_0' = (i_1 + i_2)R$$

$$= \left[\frac{v_0 - v_0'}{R_0} + C_0 D(v_0 - v_0')\right]R$$

It can be rewritten as:

$$\frac{v_0'}{v_0} = \beta\left[\frac{1 + \tau_0 D}{1 + \beta\tau_0 D}\right] \qquad \text{...(4.42)}$$

where,

$$\tau_0 = R_0 C_0$$

and

$$\beta = \frac{R}{R + R_0}$$

If $\tau_0 = \tau$ (by adjusting R_0 and C_0), then,

$$\frac{v_0'}{v_0} = \frac{\beta(1 + \tau D)}{(1 + \beta\tau D)}$$

or

$$(1 + \beta\tau D)v_0' = v_0 \cdot \beta(1 + \tau D) = k\beta o_i \qquad \text{...(4.43)}$$

This is the expression for the compensated system of the first order. It is seen from the expression that the effective time constant becomes $\beta\tau$, which is less than the time constant τ for the uncompensated system, since $\beta < 1$.

Static sensitivity in this case becomes $K\beta$, which is less than the static sensitivity K for the uncompensated system. (since, $\beta < 1$).

Thus, reduced time constant for the compensated system is realised at the expense of reduced sensitivity.

For a sinusoidal input, $v_i = V_i \sin \omega t$, let the output voltages be v_0 and v_0' for a first order system, without and with compensation respectively. Their frequency response characteristics for both uncompensated and compensated systems are shown in figure 4.18.

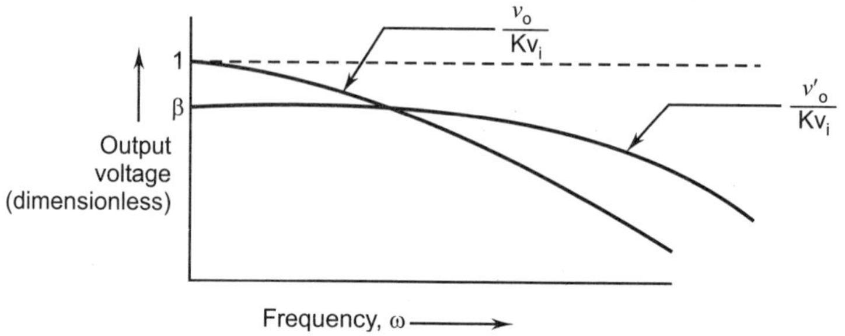

Fig. 4.18: Frequency response of uncompensated and compensated systems

4.7.2 Compensation of a Second Order System

Like the first order system, a second order system can be compensated to improve upon its performance. With compensation, effective damping is increased.

The compensated circuit is shown in figure 4.19. v_0 is the output of the uncompensated system. This is acting as the input to the compensating network, whose output is v_0'.

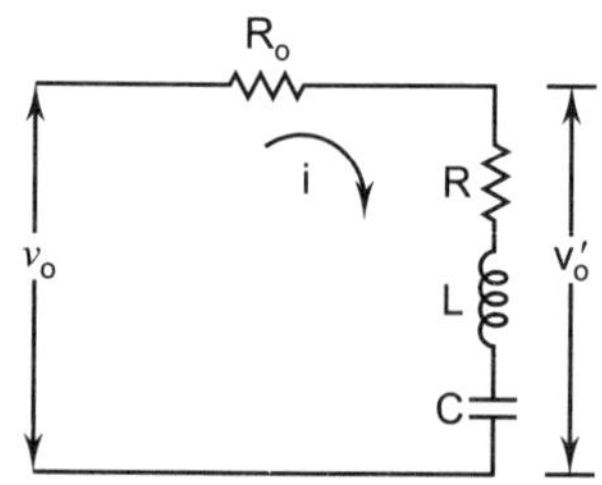

Fig. 4.19: A second-order system with compensation

We have

$$v_0 = i\left(R_0 + R + LD + \frac{1}{CD}\right)$$

and
$$v_0' = i\left(R + LD + \frac{1}{CD}\right)$$

$$\frac{v_0'}{v_0} = \frac{LCD^2 + R\,CD + 1}{LCD^2 + (R_0 + R)\,CD + 1} = \frac{\frac{D^2}{\omega_1^2} + \frac{2\xi_1}{\omega_1}D + 1}{\frac{D^2}{\omega_1^2} + \frac{2\xi_2}{\omega_1}D + 1} \qquad \text{...(4.44)}$$

where,
$$\omega_1 = \left(\frac{1}{LC}\right)^{1/2};\ \xi_1 = \frac{R}{2\sqrt{L/C}} \text{ and } \xi_2 = \frac{(R_0 + R)}{2\sqrt{L/C}}$$

For an uncompensated system:

$$\left(\frac{D^2}{\omega_n^2} + \frac{2\xi}{\omega_n}D + 1\right)v_0 = Kv_i \qquad \text{...(4.45)}$$

v_i is the input to the uncompensated system, ω_n the undamped natural frequency, K the static sensitivity and ξ the damping ratio.

If the components of the compensating network are so designed such that,

$$\omega_1 = \omega_n \text{ and } \xi_1 = \xi,$$

then,
$$\left(\frac{D^2}{\omega_n^2} + \frac{2\xi_2}{\omega_n}D + 1\right)v_0' = Kv_0 \qquad \text{...(4.46)}$$

Thus, it is seen that while the damping ratio for the uncompensated system is ξ, the same for the compensated system is ξ_2. Again, $\xi_2 > \xi_1$ and $\xi_1 = \xi$ (assumed), hence $\xi_2 > \xi_0$. Hence, employing compensation for a second order system effectively increases its damping.

SOLVED PROBLEMS

1. An amplitude inaccuracy of 5% is allowed for a first order instrument, the maximum frequency being 150 Hz. What is the maximum time constant of the system? Find the phase shift at 100 Hz.

Ans. We have, $$\frac{A_0}{KA_i} = \frac{1}{\sqrt{1+\omega^2\tau^2}} = 0.95$$

$$\omega = 2\pi f = 2 \times 3.14 \times 150 = 942 \text{ rad/sec.}$$

$$\therefore \quad \tau = 0.32 \times 10^{-3} \text{ sec.}$$

Phase shift at 100 Hz would be:

$$= -\tan^{-1}(\omega\tau)$$

$$= -\tan^{-1}(2 \times 3.14 \times 100 \times 0.32 \times 10^{-3})$$

$$= -11.36°.$$

2. A thermometer, initially at 50°C, is suddenly dipped in a liquid bath at 400°C. The thermometer reaches 300°C after 3 seconds. After what time the thermometer is expected to reach within 5% of its final value?

Ans. We can write, as per the data supplied:

$$300 = 350[1 - \exp(-3/\tau)] + 50$$

$$\tau = 1.9583 \text{ sec.}$$

A value within 5% of final value is reached in approximately 3τ time.

$$\therefore \quad \text{Required time} = 3\tau = 3 \times 1.9583$$

$$= 5.8749 \text{ sec.}$$

3. The natural frequency and the damping ratio of a second order system are 20 rps and 0.5 respectively. Find (*a*) damped natural frequency, (*b*) time constant, (*c*) maximum overshoot, (*d*) peak time and (*e*) settling time.

Ans. (*a*) Damped natural frequency $= \omega_d = \omega_n\sqrt{1-\xi^2}$

$$= 20\sqrt{1-0.5^2} = 17.32 \text{ rps}$$

(*b*) Time constant $= \tau = \dfrac{1}{\xi\omega_n} = \dfrac{1}{0.5 \times 20} = 0.1$ sec.

(*c*) Maximum overshoot

$$M_p = \exp\left[\xi\pi/\sqrt{1-\xi^2}\right] = \exp\left[0.5\pi/\sqrt{1-0.5^2}\right]$$

$$= 0.21$$

(*d*) Peak time

$$= \frac{\pi}{\omega_d} = \frac{\pi}{17.32} = 0.1813 \text{ sec.}$$

Chapter 4

(e) Settling time $= 4\tau = 4 \times 0.1 = 0.4$ sec.

4. A second order instrument has an undamped natural frequency of 4 Hz and damping ratio 0.5. It is subjected to a sinusoidal input. Find the amplitude and phase ratio for an input of 3 Hz .

Ans. We have,

$$Y_o = \frac{KX_i}{\sqrt{(1-\eta^2)^2 + (2\xi\eta)^2}}$$

and

$$\phi = -\left(\tan^{-1}\frac{2\xi\eta}{1-\eta^2}\right), \quad \eta = \frac{\omega}{\omega_n}$$

$$\omega_n = 2\pi \times 4 = 8\pi = 25.12 \text{ rad/sec.}$$

$$\omega = 2\pi \times 3 = 18.85 \text{ rad/sec.}$$

$$\eta = \omega/\omega_n = 0.75$$

$$\xi = 0.5$$

It is assumed that, K = static sensitivity = 1

$$\therefore \quad \frac{Y_o}{X_i} = \frac{1}{\sqrt{(1-0.75^2)^2 + (2 \times 0.5 \times 0.75)^2}} = 1.152$$

$$\phi = -\left(\tan^{-1}\frac{2\xi\eta}{1-\eta^2}\right) = -\left(\tan^{-1}\frac{2\times 0.5 \times 0.75}{1-0.75^2}\right)$$

$$\therefore \quad = -59.74°.$$

REVIEW QUESTIONS

1. What are the assumptions made of first and second-order systems?
2. Discuss the different types of standard input signals employed to evaluate the dynamic characteristic of instruments.
3. Deduce the transfer function of a generalized physical system. Discuss the time domain response.
4. Find the time domain response of a first order system subjected to a unit step function. How the time constant affects its performance? Draw the normalised response curve and discuss the same.
5. Find the time domain response of a first order system subjected to a unit impulse function. Draw the response curves when the impulse is of (a) finite strength A and time duration E (b) ideal in nature.
6. Find the outputs of a second-order system subjected to a unit step function for different damping factors. Draw the curves and discuss them.
7. Mention the different important input parameters when a second order underdamped system is subjected to a unit step input function.

Fill in the blanks:

1. The dynamic response depends on the .... of an instrument.
2. Response to an input signal change to an instrument is not instantaneous because of .... of the instrument.

3. A potentiometer is considered as a .... order instrument.
4. The ability of an instrument to reproduce at the output the instantaneous changes in the input is called ....
5. A zero order instrument has a very high ....
6. A high fidelity system has a .... amplitude-frequency characteristics over a wide range of frequencies.
7. Irrespective of the nature of the input signal, some systems do not respond until after some time elapses. This time is known as the .... time.
8. Dynamic range is expressed in ....
9. Systems which possess a constant value for their transfer function is called a .... order instrument.
10. Depending on the .... condition, three different outputs would result for an input change to an instrument for a second-order system.
11. Delay time is the time required for the instrument to reach .... percent of the final value for the first time.
12. Maximum or peak overshoot is the maximum value of the response measured from ....

Tick the correct answer:

1. The dynamic response of a system depends/does not depend on the nature and complexity of the input function. ☐
2. If an instrument becomes more sluggish, the transient response time increases/decreases. ☐
3. Most of the measuring systems are of zero order/first or second order type. ☐
4. For a step input, dynamic error of a system consists only of transient error/steady state error. ☐
5. A mercury-in-glass thermometer is a first/second order instrument. ☐
6. Dynamic sensitivity is frequency dependent/independent. ☐
7. Settling time is defined as the time for the output to reach a value within ±1%/±5% of the final value. ☐
8. Particular integral in the expression of time domain system response represents transient/steady state behaviour of the system. ☐
9. A system which possesses a constant value for its transfer function is a zero/first order system. ☐
10. The steady state error for a first order system subjected to a unit ramp function can be reduced by increasing/decreasing the value of the time constant τ. ☐
11. For a temperature measuring system, if the time constant is increased/decreased, the frequency response would improve. ☐
12. Compensation of an instrument leads to the improvement/degradation of its dynamic characteristics. ☐
13. Dynamic response of a second order instrument depends/does not depend upon damping factor. ☐
14. For a second order system, the final value is reached when the system is underdamped/critically damped. ☐
15. For a first order instrument, a 5% settling time is equal to (*a*) three times the time constant (*b*) two times the time constant (*c*) the time constant (*d*) time required for the output signal to reach 5% of the final value. [*GATE 2004*] ☐
16. Probability density function $p(x)$ of a random variable x is as shown. The value of α is: ☐
[*GATE 2006*]
(*a*) $2/c$ (*b*) $1/c$ (*c*) $2/(b + c)$ (*d*) $1/(b + c)$

Chapter 4

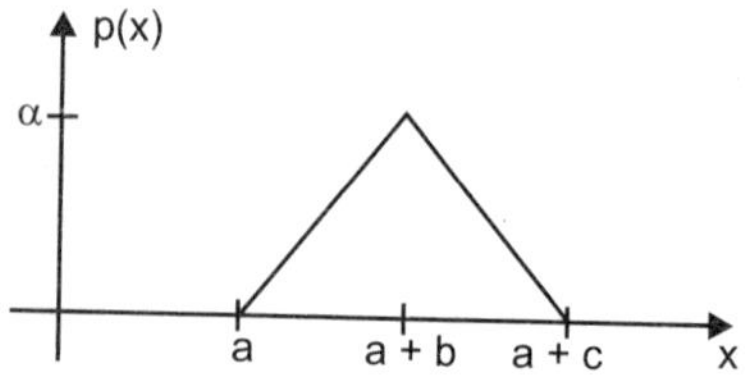

REVIEW PROBLEMS

1. A temperature measuring device, having a time constant of 3 sec., measures the temperature of a medium whose temperature changes sinusoidally between a maximum and minimum of 250°C and 200°C, the period being 20 sec. Determine the indicated maximum and minimum values of temperatures as also the time lag between the input and output signals.

   [**Ans.** 243.2°C, 206.8°C, 2.41 sec.]

2. Pressure is abruptly changed from 5 bar to 30 bar at $t = 0$. The transducer (being of first order) indicates a value of 20 bar after 30 seconds. Determine the time required to reach the pressure to 95% of the final value.

   [*WBUT (EI 402) 2004*] [**Ans.** 92.11 Sec.]

3. A temperature probe having a first order response with a time constant 1 sec. is given a step input of 100°C from 0°C. Calculate the indicated temperature every 0.5 sec. interval upto 2 sec.

   [*UOK 1996*] [**Ans.**

t(Sec)	0.5	1.0	1.5	2.0
Y_0(°C)	39.3	63.2	77.7	86.5

   ]

4. A thermocouple temperature indicator with reference to junction at room temperature has a time constant of 1 sec. It is dipped in hot bath of 120°C. If the room temperature is 20°C, the thermocouple type temperature indicator will read after 1 sec.

   (*i*) 120°C (*ii*) 100°C (*iii*) 63.2°C (*iv*) 140°C

   [*ET-23 W 1996*] [**Ans.** (*iii*)]

5. A first order temperature sensor has a time constant 5 sec. It is subjected to a harmonically varying temperature of 25°–250°C with a period of 15 seconds Find (*i*) amplitude ratio and (*ii*) phase lag.

   [**Ans.** (*i*) 0.43 (*ii*) 64.46°]

6. A thermocouple can be considered as a single capacity system with a time constant of 100 ms and unity gain. It is immersed in a heat bath whose temperature is varying at 1°C/sec from 0°C. When the actual temperature of the heat bath is 70°C, the reading of the thermocouple will be:

   (*a*) 70.1°C (*b*) 71°C (*c*) 69.9°C (*d*) 69°C [*GATE 2004*] [**Ans.** (*c*)]

7. A thermometer at a temperature 30°C is dipped suddenly into a bath of boiling water at 100°C. It takes 30 seconds to reach 96.5°C. The time required to reach a temperature of 98°C is:

   (*a*) 32.5 s (*b*) 34.6 s (*c*) 35.6 s (*d*) 38.6 s [*GATE 2005*] [**Ans.** (*b*)]

CHAPTER 5

INSTRUMENT SYMBOLS, PFDs AND P & IDs

5.1 INTRODUCTION

Measurement, monitoring and control of an industrial process involve the depiction and identification of analog/digital instruments, instrumentation systems and application softwares. Graphic symbols, identification codes (both alpha and numeric), functional diagramming techniques, abbreviations, function blocks and standard connecting lines are employed to facilitate the above.

ISA–The Instrumentation, Systems and Automation Society plays a pivotal part in formulating the various standards that are in vogue. ANSI/ISA's 5.1 – 1984 (*R* 1992) has been updated as 5.01.01 and is designated as ANSI/ISA–5.01.01 (for details of ISA standards, refer to: www. isa. org). This standard is very suitable for control instrumentation, measurement, control device functions, software applications for the purpose of identification and symbolisation in the fields of process industries and discrete parts manufacturing industries.

Several specialised industries, such as navigation, medicine, astronomy, etc. require specialised instruments and ISA does not have any specific standards to meet the requirements in these fields.

5.2 LOCATION WISE INSTRUMENT CLASSES

In general, there are four graphical elements—discrete, shared display and control, computer function and programmable logic controller (PLC). These are grouped into three location categories: primary location, auxiliary location and field mounted. This is shown in Fig. 5.1.

	Primary location accessible to operator	*Field mounted*	*Auxiliary location accessible to operator*
Discrete instruments	1 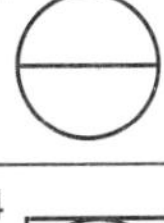	2 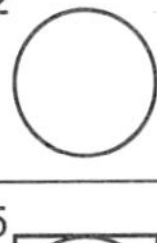	3 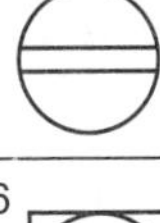
Shared display, shared control	4 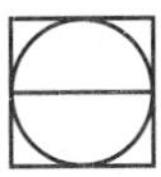	5 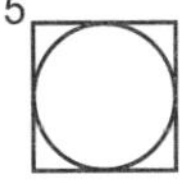	6 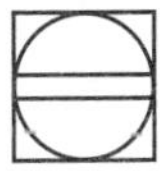
Computer function	7 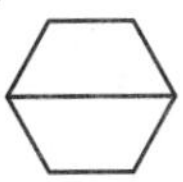	8 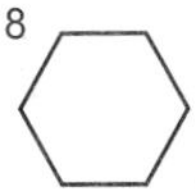	9 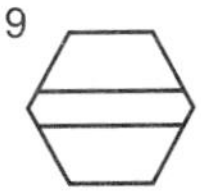

Contd.

	Primary location accessible to operator	*Field mounted*	*Auxiliary location accessible to operator*
Programmable logic control	10 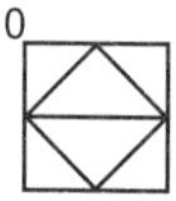	11 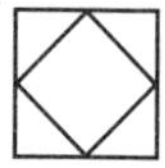	12 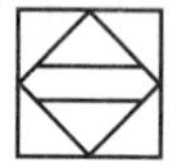

1. Symbol size may vary according to the user's needs and the type of document.
2. Abbreviations of the user's choice may be used when necessary to specify location.
3. Inaccessible (behind the panel) devices may be depicted using the same symbol but with a dashed horizontal bar.

Source: *Control Engineering* with data from ISA S5.1 standard

Fig. 5.1: General Instrument or Function Symbols

5.3 PROCESS FLOW DIAGRAMS (PFDs)

Process diagrams are generally categorised into two types — process flow diagrams (PFDs) and process and instrument diagrams (P & IDs) – also called piping and instrumentation diagrams. A PFD is a simple depiction that uses process symbols to describe the primary flow path of a process till the final end product is obtained. PFDs provide an overview of the process and includes all primary equipments and flow. It does not include secondary/minor flow, complex control loops, piping details, designations, etc. PFDs include the following:

- Process piping.
- Major equipment symbols, names and identification numbers.
- Direction of flow.
- Interconnection to other systems.
- Fluid compositions.
- System ratings and operational values as minimum, maximum flow, temperature and pressure.
- Major bypass and recirculation lines.
- Control loops that affect operation of the system.

PFDs do not include the following:

- Pipe classes/pipe line numbers.
- Sensors ands final elements.
- Minor bypass lines.
- Isolation and shut off valves.
- Maintenance vents and drains.
- Relief and safety valves.
- Flanges.

5.3.1 PFD : An Example

An example of a process flow diagram is shown in figure 5.2.

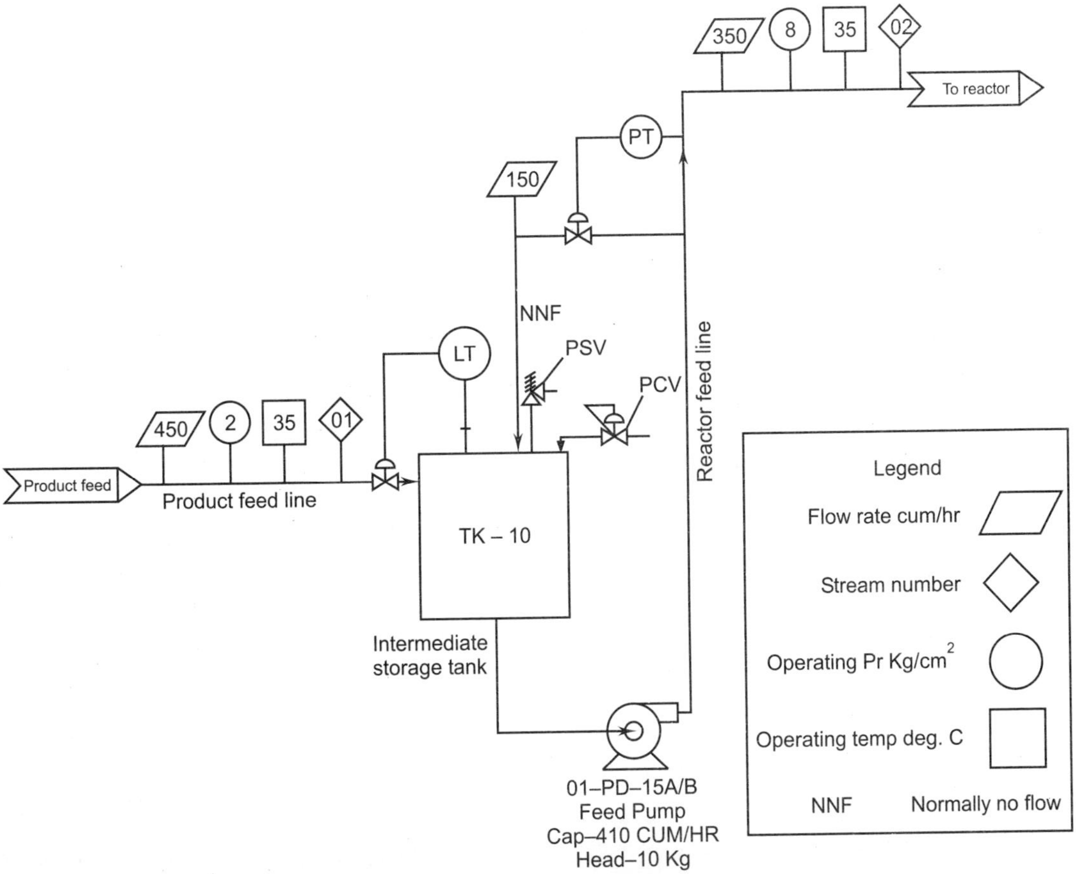

STREAM NO.	01	02
FLUID DESCRIPTION	Naptha	Naptha
PHASE	Liquid	Liquid
DENSITY Kg/M3/Mol Wt	795	795
VISCOSITY cp.	2.4	2.4
COMPRESSIBILITY	–	–

Fig. 5.2: An example of a PFD

The PFD for an intermediate pressurised storage naptha tank is shown in figure 5.2. Process data (variable) at different streams are indicated in the PFD. Naptha (shown as 'product feed' in the figure) is fed into the tank TK-10, which is subsequently fed into the reactor with the help of feed pump. Normally, two feed pumps are employed: one working and one standby (1W + 1S).

Temperature, pressure and flow rate of the incoming naptha are measured. The level in the tank is monitored by a level transmitter (LT) and subsequently controlled by a control valve.

Pressure of the reactor feed line is monitored by a pressure transmitter (PT). Pressure in this line is maintained constant by a control valve attached to the PT.

Since naptha is highly flammable, it is made blanket inside the tank with the help of nitrogen through an automatic pressure control valve. If any high pressure is developed in the tank, the pressure safety valve (PSV) opens and thus the tank is saved. The exhaust is flared into the atmosphere.

5.4 PIPING AND INSTRUMENTATION DIAGRAMS (P & IDs)

P & IDs symbolically represent a process plant with all its associated equipments, piping and instrumentation. It is also known as mechanical flow diagram (MFD).

The different equipments are connected through the piping and the process streams flow through it. Pipes come in different sizes and materials, although carbon steel is the most common. Thermal insulation of pipes, if required, is also specified. Piping encompasses tees, elbows, flanges, valves, etc.

The main process variables are: temperature, pressure, flow and level. Instrumentation includes the monitoring and control of these process variables. Instrumentation details also include the different types of valves. Different signals like pneumatic, electric, hydraulic, electromagnetic, etc. have different standardised representations.

P & IDs include the following:

- Total instrumentation and their corresponding designations.
- Process piping, their sizes, identification by number, etc.
- Different types of valves and their identification.
- Mechanical equipments with their names and numbers.
- Direction of flow.
- Control inputs and outputs.
- References for interconnections.
- Start-up and flush lines.
- Miscellaneous– vents, drains, reducers, increasers, swagers, fittings, sampling lines, etc.
- Class change interfaces.
- Vendor/contractor interfaces.
- Identification of components/subsystems delivered by other vendors.

P & IDs do not include the following:

- Equipment rating/capacity.
- Data for pressure, temperature, flow, etc.
- Explanatory notes.

5.4.1 Classification of P & IDs

P & IDs are classified as per the following:

- Process P & IDs
  - On-plot process unit design
  - Off-plot tankage and shipping systems
- Utility plant P & IDs — includes cooling towers, boilers, air compressors, water treatment plants, drain collection systems, and fire water systems.
- Utility Distribution P & ID – shows how the utilities are distributed for a given process. Piping and valve instrumentation are shown on main headers upto and including branch root valves.

- Interconnecting (Rack) P & ID– it shows the connecting link between individual processes, utility plant and utility distribution P & IDs.
- Vendor P & ID: It is prepared for systems that support major equipment packages.
- Legends and symbol diagrams/standard details.

5.4.2 Interconnecting P & IDs

A moderate to big process plant is spread over a substantial area and includes equipments/utilities both for process and associated utilities, relevant to the process plant. This is where the different P & IDs are interconnected and it is based as follows:

- Above the ground level process lines.
- Above the ground level utility lines.
- Below the ground level process and utility lines.

5.4.3 P & ID : An Example

An example of a piping and instrumentation diagram is shown in figure 5.3. The P & ID for an intermediate pressurised storage naptha tank is shown. Naptha is fed into the tank TK-10 via HV-13, which is a solenoid operated on-off valve. ZSO and ZSC are the sensors for providing the open close feedback while ZAH and ZAL are the open close feedback for the control room. LT-13 is the level transmitter to which are connected a level indicator (LI-13) along with alarm annunciation level high (LAH-13) and level low (LAL-13). These alarms are software generated. The valve HV-13 can also be manually operated through a soft switch situated at the operator station in the control room.

The level switch (LSLL-15) provides the level alarm low (LALL-15) in the control room which is interlocked with the feed pump for dry run protection.

Two feed pumps are employed: one working and one standby (1W + 1S). These pumps are interlocked with the pressure transmitter (PT-11) which provides pressure alarm low (PAL-11) that indicates reactor needs naptha from the tank and pressure alarm high (PAH-11) that indicates that the reactor side has some problem and hence the feed should be stopped immediately. Thus, these two pressure alarms are interlocked with feed pumps for starting/stopping of the pump.

Individual local panel for each pump has a start (HS-15B1), stop (HS-15B2) and run (YL-15B) indication which is located near to the pump. There is provision for remote manual start (HS-15A1), stop (HS-15A2) from control room and run (YL-15A) indication at control room. Feed pump discharge pressure is indicated in pressure gauge (PG-XX).

The pressure control valve (PV-11) is provided in the return line of the tank to maintain constant pressure in the reactor feed line where constant pressure is monitored by pressure transmitter (PT-11). Every valve has a bypass line for valve maintenance without interrupting production. The flow transmitter (FT-12) monitors the reactor feed flow as well as for recording weekly/monthly consumption of naptha by reactor. A stand pipe mounted level indicator (LG-10) indicates the naptha level inside the tank. PCV 10 and PSV-10 are identical to PCV and PSV in the PFD.

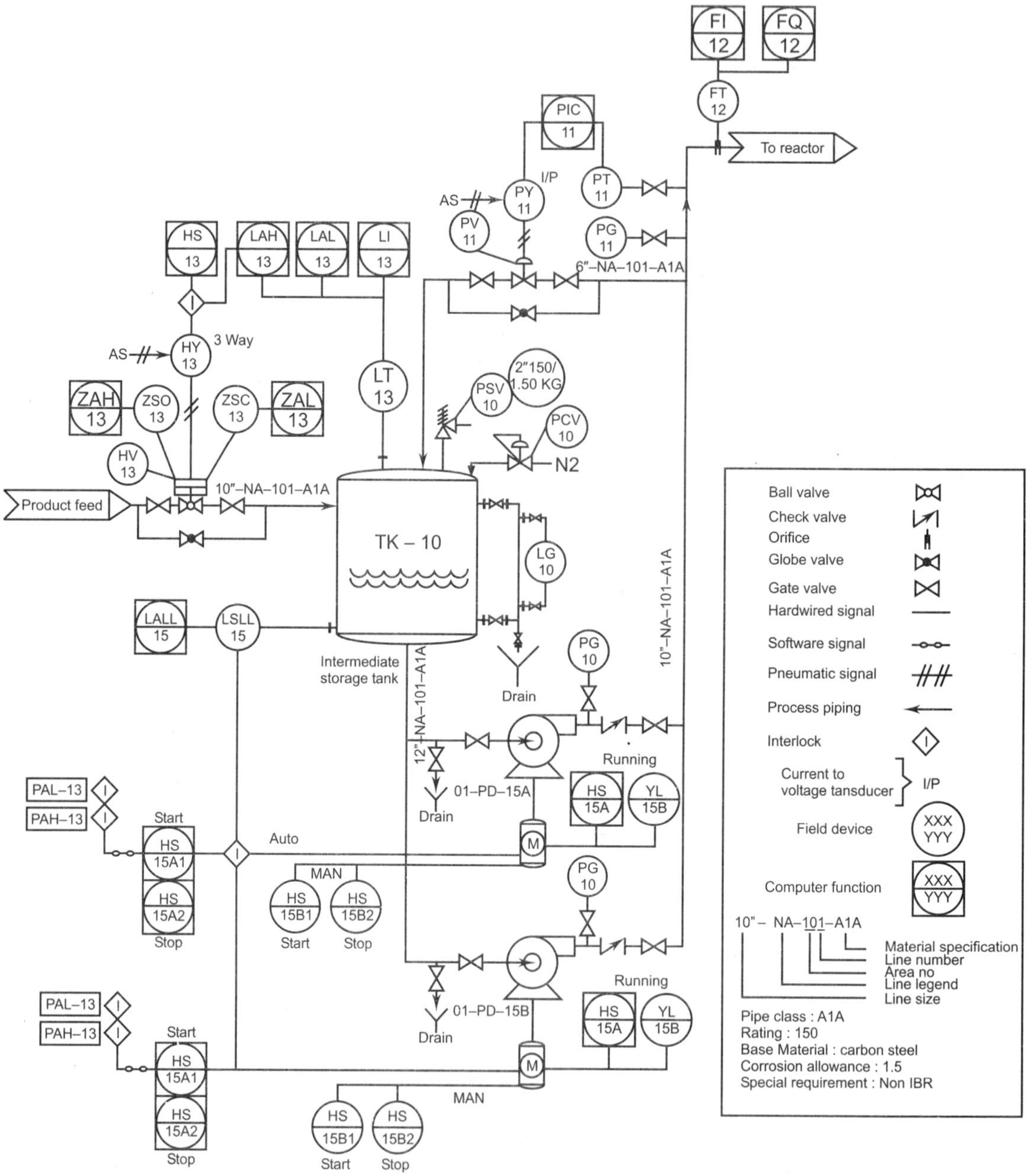

Fig. 5.3: An example of a P & ID

5.5 GRAPHIC SYMBOLS FOR INSTRUMENTATION SYSTEMS

ISA–The Instrumentation, Systems and Automation Society has revised the standard S 5.1 and which is now known as ANSI/ISA – 5.01.01. The graphic symbols is used in sketches and drawings and when used with identification letters and numbers, it would identify the functionality of each device.

The graphic symbols for instrumentation systems help in formulating schematic loop diagramming, functional diagramming and electrical schematic diagramming for any process instrumentation system that necessitates measurement, control, indication and a host of other functions.

5.5.1 Instrument Line Symbols

This is shown in Table 5.1. The lines should be fine relative to process piping and equipments, and as short as possible.

The different symbols represent (*a*) instrument/device connection at process measurement points (*b*) instrument power supplies (*c*) instrument air supplies (*d*) numerous signals between measurement instruments and control instruments.

Table 5.1: Instrument Line Symbols

No.	*Symbols*	*Application*
01		1. Instrument impulse line from process 2. Instrument impulse line from equipment 3. Analyzer sample line from process 4. Functional instrument diagram signal lines
02	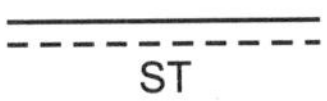	1. Heat [cool] traced instrument impulse line from process 2. Heat [cool] traced instrument impulse line from equipment 3. Heat [cool] traced analyzer sample line from process 4. Type of tracing may be indicated as ET = electrical, RT = refrigerated. ST = steam.
03	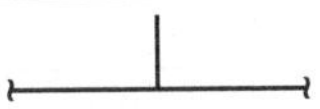	1. Generic instrument impulse line connected to process line 2. Generic instrument impulse line connected to equipment
04	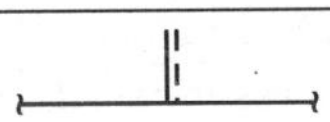	1. Heat [cool] traced generic instrument impulse line connected to process line 2. Heat [cool] traced generic instrument impulse line connected to equipment 3. Process line or equipment may or may not be traced
05	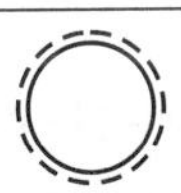	1. Heat [cool] traced instrument connected to process impulse line 2. Instrument impulse line may or may not be traced
06	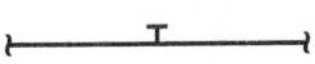	1. Flanged instrument connection to process line 2. Flanged instrument connection to equipment
07		1. Threaded instrument connection to process line 2. Threaded instrument connection to equipment
08		1. Socket welded instrument connection to process line 2. Socket welded instrument connection to equipment
09		1. Welded instrument connection to process line 2. Welded instrument connection to equipment
10		1. Instrument air supply 2. Indicate supply pressure as required, e.g., AS-60psig. AS-400kPa. etc. 3. IA [instrument air] or PA [plant air] may be used for AS 4. Use as required

Contd.

No.	Symbols	Application
11		1. Instrument electric power supply
	ES ———	2. Indicate voltage and type as required, e.g., ES-24VDC. ES-120VAC, etc. 3. Use as required
12		1. Undefined signal 2. Use of PFDs 3. Use for discussions or diagrams where type of signal, pneumatic or electronic is not of concern
13		1. Pneumatic signal
14		1. Electric signal 2. Electronic signal 3. Functional instrument diagram signal lines
15		1. Hydraulic signal
16		1. Filled thermal element capillary tube
17		1. Guided electromagnetic signal 2. Fiber-optic cable 3. Guided sonic signal
18		1. Unguided electromagnetic signal 2. Unguided sonic signal 3. Alternate radio communication link; see symbol 22
19		1. Communication link or system bus. between devices and functions of a microprocessor-based system 2. System internal software link.
20		1. Shared communication link or bus (not system bus) between two or more independent microprocessor-based systems 2. Shared data link from/between field located microprocessor-based devices or functions
21		1. Dedicated communication link or bus (not systems bus) between two or more independent microprocessor-based systems 2. Dedicated data link from a field-located microprocessor-based device and/or function
22		1. Dedicated radio communication link (not systems bus) between radio transmitting and receiving devices or systems 2. Unguided radio signal 3. Alternate unguided electromagnetic signal: see symbol 18
23	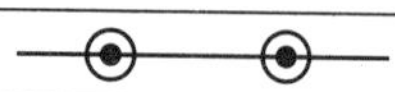	1. Mechanical link or connection
24	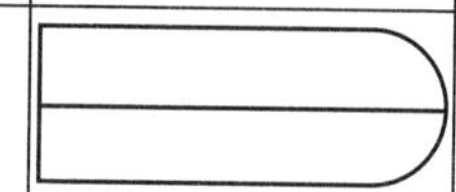	1. Signal connector 2. Drawing-to-drawing signal connector 3. Internal signal connector used to avoid long signal lines
25	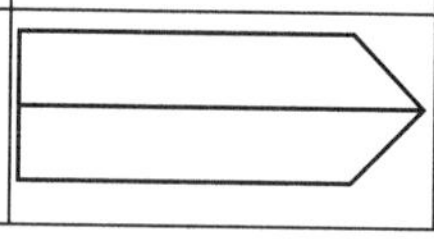	1. Signal connector 2. Internal signal connector used to avoid long signal lines 3. Drawing-to-drawing signal connector

(Courtesy: ISA ANSI/ISA – 5.01.01)

5.5.2 Discrete Devices and/or Functions

This is shown in Table 5.2. The symbols represent discrete devices that perform continuous and/or on-off functions. The devices do not share control or display functions for measurement, indication or control.

Table 5.2: Discrete Devices and/or Functions

No.	*Symbols*	*Location and accessibility*
01	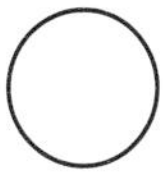	1. Field or locally mounted 2. Not panel or cabinet mounted 3. Normally accessible to an operator
02	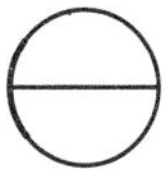	1. Central or main control room 2. Front of main panel mounted 3. Normally accessible to an operator
03	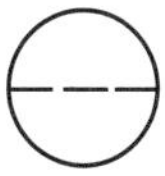	1. Central or main control room 2. Rear of main panel mounted 3. Not normally accessible to an operator
04	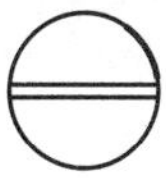	1. Secondary or local control room 2. Field or local control panel 3. Front of secondary or local panel mounted 4. Normally accessible to an operator
05	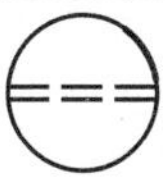	1. Secondary or local control room 2. Field or local control panel 3. Rear or secondary or local panel or cabinet mounted 4. Not normally accessible to an operator
06		1. Signal processor identifier located in upper right or left quadrant of symbols above 2. Signal processor identifier attached to symbols where affected signals are connected

(Courtesy: ISA ANSI/ISA–5.01.01)

5.5.3 Shared Continuous Devices and/or Functions

This is shown in Table 5.3. It pertains to analog, digital, discrete shared control devices and functions for continuous indication, control and calculations which are microprocessor based and are configurable.

Table 5.3: Shared Continuous Devices and/or Functions

No.	*Symbol*	*Location and accessibility*
01	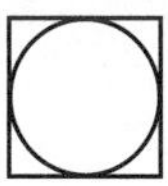	1. Dedicated single function device 2. Field or locally mounted 3. Not panel or cabinet mounted 4. Normally accessible to an operator device
02	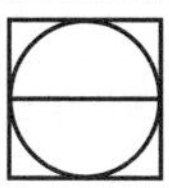	1. Central or main console 2. Visible on video display 3. Normally accessible to an operator console
03	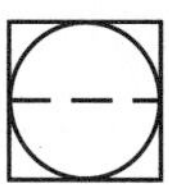	1. Central or main console 2. Not variable on video display 3. Not normally accessible to an operator console
04	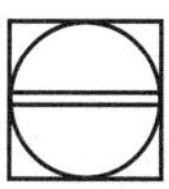	1. Secondary or local console 2. Field or local control panel 3. Visible on video display 4. Normally accessible to an operator console

Contd.

No.	Symbol	Location and accessibility
05	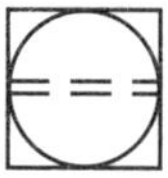	1. Secondary or local console 2. Field or local control panel 3. Not visible on video display 4. Not normally accessible to an operator console
06		1. Mathematical function located in upper or left quadrant of symbols above 2. Mathematical function attached to system where affected signals are connected

(Courtesy: ISA ANSI/ISA–5.01.01)

5.5.4 Shared On-Off Devices and/or Functions

This is shown in Table 5.4. It pertains to analog, digital, discrete control devices or functions for on-off or binary control, indication and calculations which are microprocessor based and are configurable.

Table 5.4: Shared On-Off Devices and/or Functions

No.	Symbols	Location and accessibility
01		1. Field or locally mounted 2. Not panel or cabinet mounted 3. Normally accessible to an operator at device
02		1. Central or main console 2. Visible on video display 3. Normally accessible to an operator at console
03		1. Central or main console 2. Not visible on video display 3. Not normally accessible to an operator at console
04	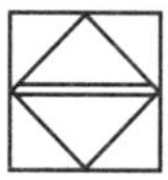	1. Secondary or local console 2. Field or local control panel 3. Visible on video display 4. Accessible to an operator at console
05	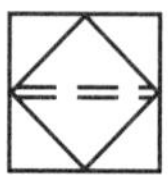	1. Secondary or local console 2. Field or local control panel 3. Not visible on video display 4. Not normally accessible to an operator at console
06		1. Mathematical function located in upper right or left quadrant of symbols above 2. Mathematical function attached to symbols where affected signals are connected

(Courtesy: ISA ANSI/ISA–5.01.01)

5.5.5 Computer Devices and/or Functions

This is shown in Table 5.5. In this case, the process plant operation is computer based with advanced analog/digital/discrete control.

Table 5.5: Computer Devices and/or Functions

No.	*Symbols*	*Location and accessibility*
01		1. Undefined location 2. Undefined visibility 3. Undefined accessibility
02	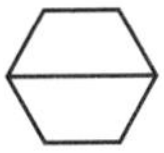	1. Central or main computer 2. Visible on video display 3. Normally accessible to an operator at console or computer terminal
03	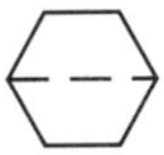	1. Central or main computer 2. Not visible on video display 3. Not normally accessible to an operator at console or computer terminal
04	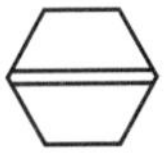	1. Secondary or local computer 2. Visible on video display 3. Normally accessible to an operator at console or computer terminal
05	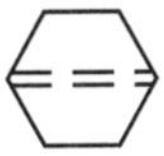	1. Secondary or local computer 2. Not visible on video display 3. Not normally accessible to an operator at console or computer terminal

(Courtesy: ISA ANSI/ISA–5.01.01)

5.5.6 Instruments and Mathematical Functions

This is shown in Table 5.6. These symbols are not normally used on P & IDs, but are used at the hardware/functional level for configurational purposes.

Table 5.6: Instruments and Mathematical Functions

No.	*Symbols*	*Description*
01	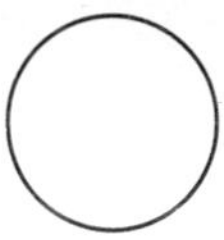	1. Measuring device 2. Input device 3. Readout device 4. Output device 5. Symbols from Tables 5.2 to 5.5 may be used
02		1. Automatic controller 2. Single mode controller 3. Discrete device driver 4. Insert function symbols, as required, to define controller algorithm from Table 5.7 5. Use for vertical diagramming
03	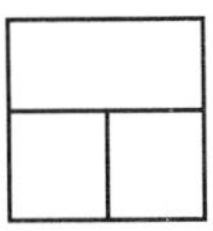	1. Automatic controller 2. Two-mode controller 3. Insert function symbols, as required to define controller algorithm, from Table 5.7 4. Use for vertical diagramming
04	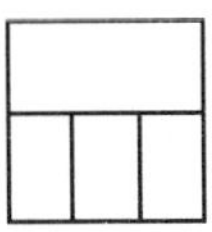	1. Automatic controller 2. Three-mode controller 3. Insert function symbols, as required to define controller algorithm, from Table 5.7 4. Use for vertical diagramming

Contd.

No.	*Symbols*	*Description*
05	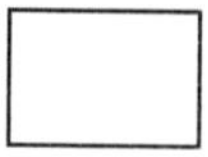	1. Automatic signal processor 2. Insert function symbol from Table 5.7 3. Use for vertical diagramming
06	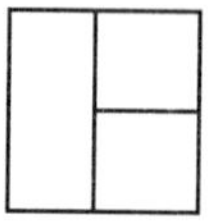	1. Automatic controller 2. Two-mode controller 3. Insert function symbols, as required to define controller algorithm, from Table 5.7 4. Use for horizontal diagramming
07	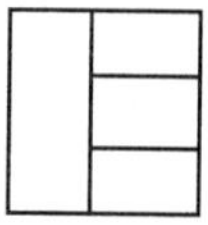	1. Automatic controller 2. Two-mode controller 3. Insert function symbols, as required to define controller algorithm, from Table 5.7 4. Use for horizontal diagramming
08	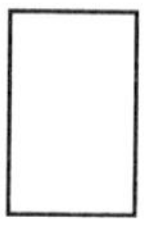	1. Automatic signal processor 2. Insert function symbol from Table 5.7 3. Use for horizontal diagramming 4. May be rotated 90°
09	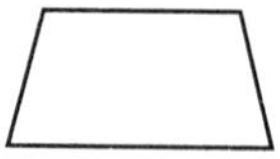	1. Final control element 2. Control valve 3. Insert function symbol identifier from Table 5.7 (No. 14)
10	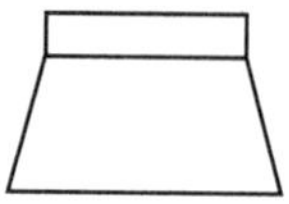	1. Final control element with positioner 2. Control valve with positioner 3. Insert function symbol identifier from Table 5.7 (No. 14)
11		1. Manual signal processor 2. (=) = A, adjustable signal generator (=) = T, signal transfer
12		1. Manual/auto station

(Courtesy: ISA ANSI/ISA–5.01.01)

5.5.7 Mathematical Function Block Symbols

This is shown in Table 5.7.

Table 5.7: Mathematical Function Block Symbols

No.	*Symbol/ Function*	*Equation/Graph*	*Definition*
01	Σ Summation	$M = X_1 + X_2 \ldots + X_n$ 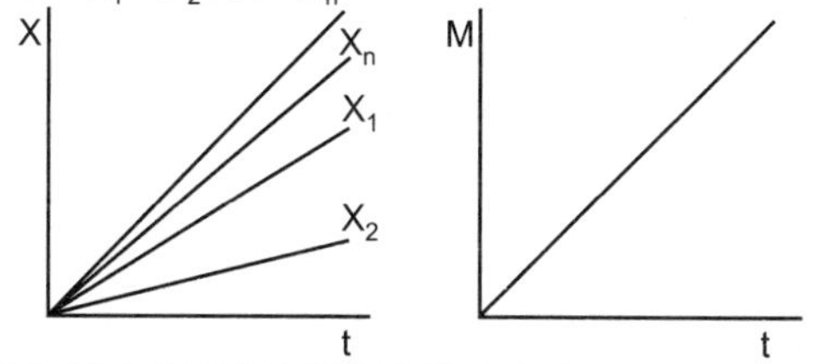	1. Output equals algebraic sum of inputs
02	Σ/n Average	$M = X_1 + X_2 \ldots + X_n/n$ 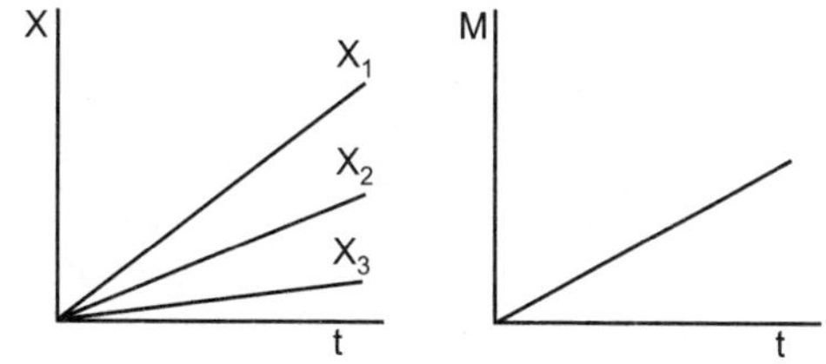	1. Output equals algebraic sum of inputs divided by number of inputs
03	Δ Difference	$M = X_1 - X_2$ 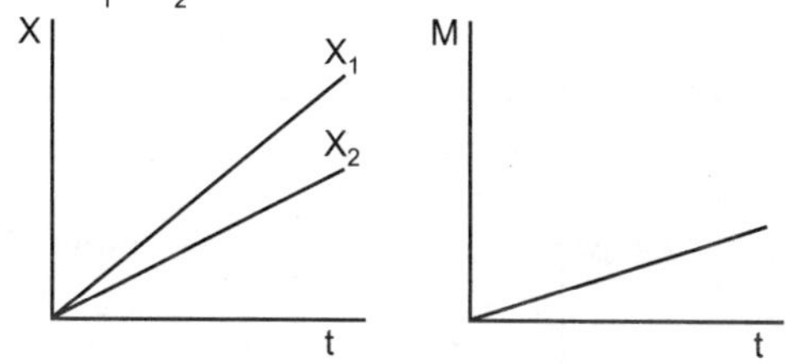	1. Output equals difference of two inputs
04	X Multiplication	$M = X_1X_2$ 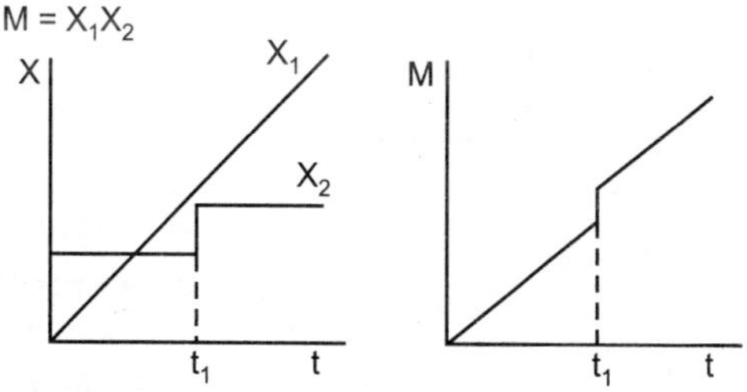	1. Output equals product of two inputs
05	$\div$ Division	$M = X_1/X_2$ 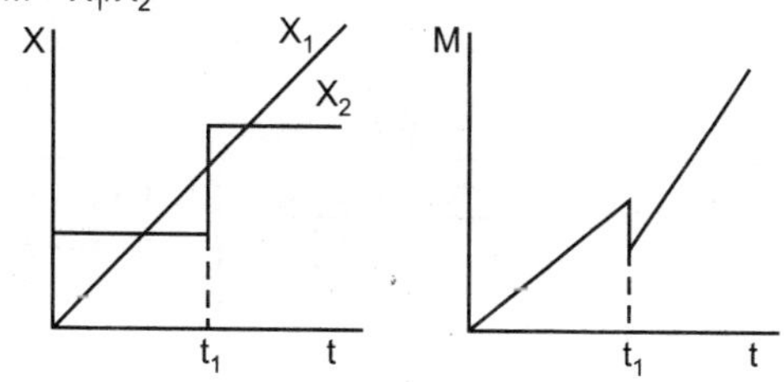	1. Output equals quotient of two inputs

Contd.

No.	Symbol/ Function	Equation/Graph	Definition
06	X^2 Square	$M = X^2$ 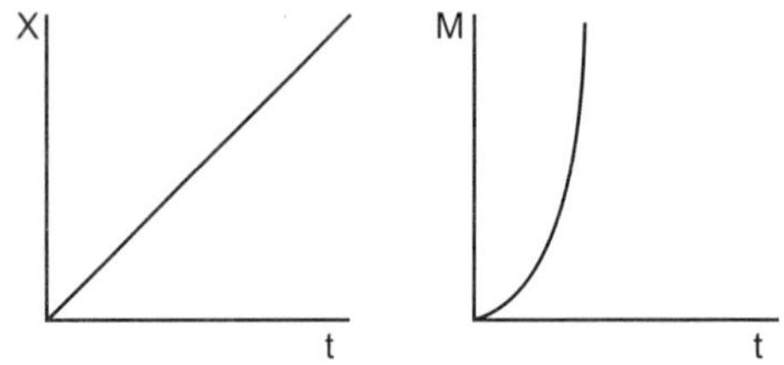	1. Output equal to square of input
07	X^n Exponential	$M = X^n$ 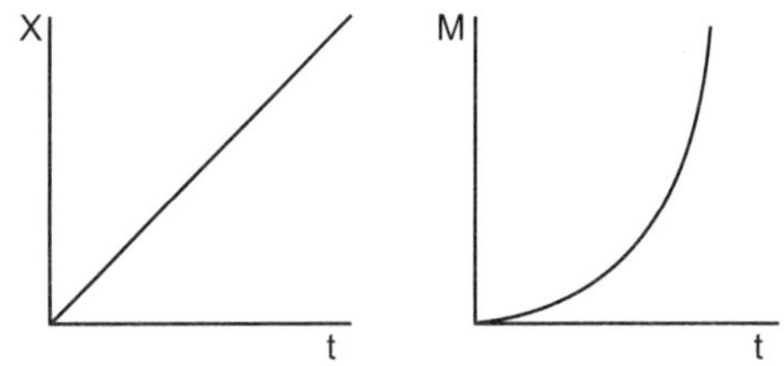	1. Output equal to *n*th power of input
08	√ Square Root	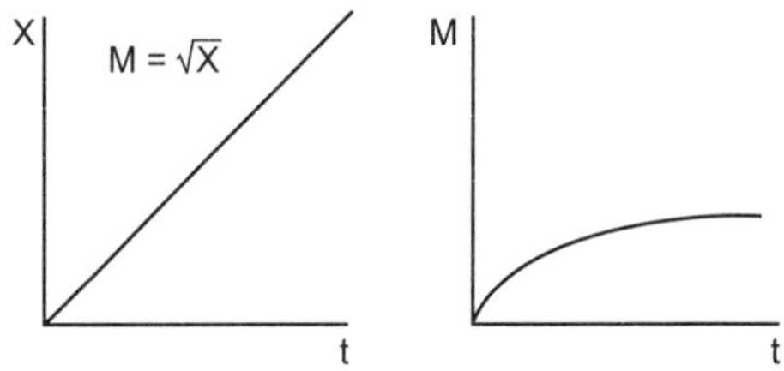	1. Output equal to square root of input
09	$\sqrt[n]{\ }$ *n*th Root	$M = \sqrt[n]{X}$ 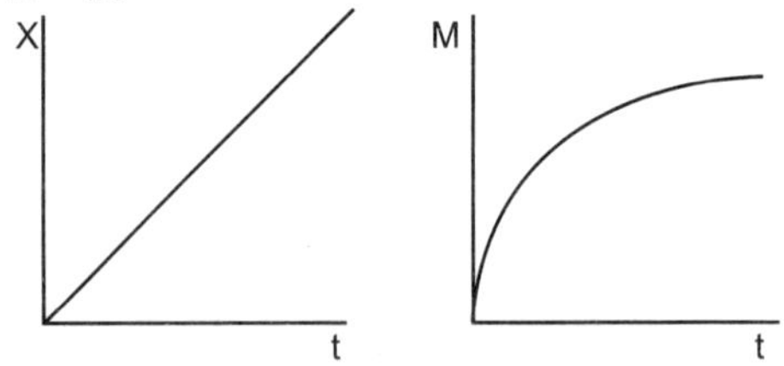	1. Output equal to *n*th root of input
10	K or P Proportional	$M = KX$ 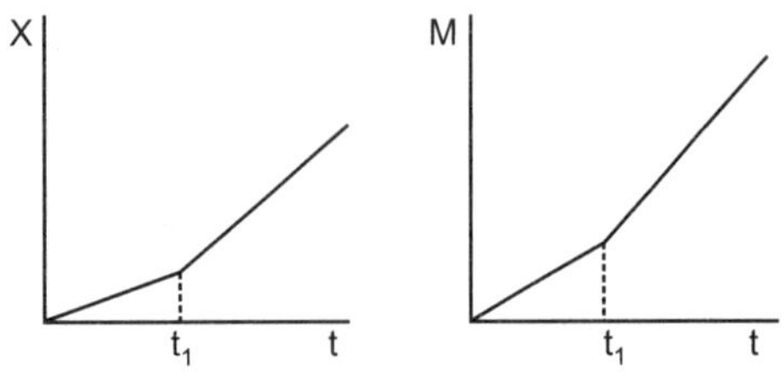	1. Output proportional to input 2. Replace "K" with "1:1" for volume boosters 3. Replace "K" with "2:1," "3:1," etc. for integer gains.
11	– K or – P Reverse Proportional	$M = -KX$ 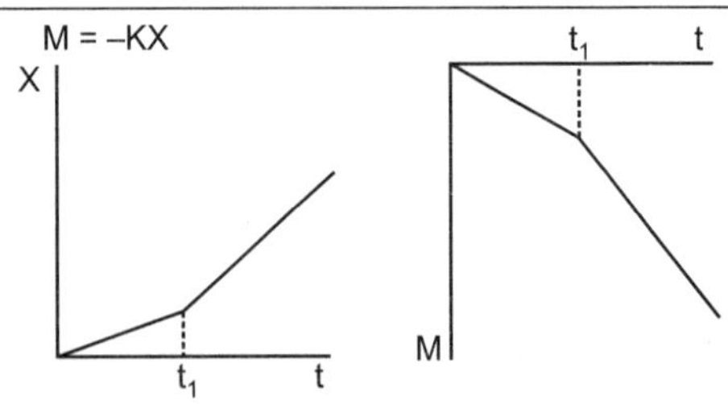	1. Output inversely proportional to input

Contd.

No.	*Symbol/ Function*	*Equation/Graph*	*Definition*
12	Integral	$M = (1/T_1) \int X\,dt$ 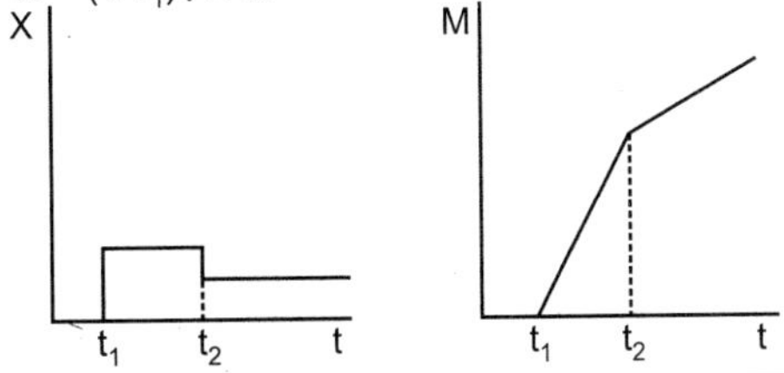	1. Output varies with magnitude and time duration of input 2. Output proportional to time integral of input 3. T_I integral time constant
13	*d/dt* or D Derivative	$M = T_D\,(dX/dt)$ 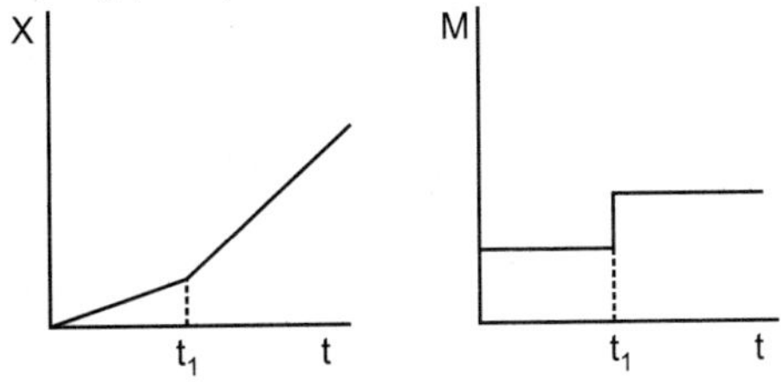	1. Output proportional to time rate of change of input 2. T_D derivative time constant
14	*f*(X) Unspecified function	$M = f(X)$ 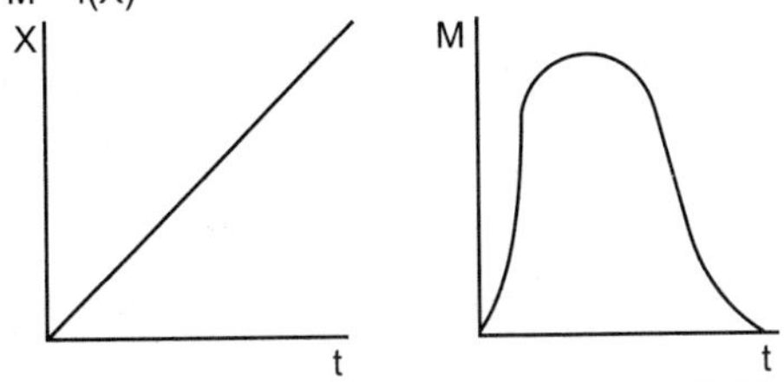	1. Output is a nonlinear or unspecified function of the input 2. Function defined in note or other text
15	*f*(*t*) Time Function	$M = Xf(t)$ 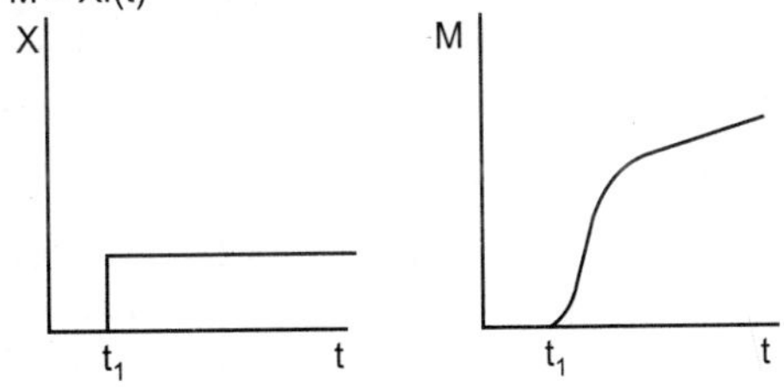	1. Output equals a nonlinear or unspecified time function times the input 2. Output is a nonlinear or unspecified time function
16	> High Select	$M = X_1$ for $X_1 \geq X_2$, $M = X_2$ for $X_1 \leq X_2$ 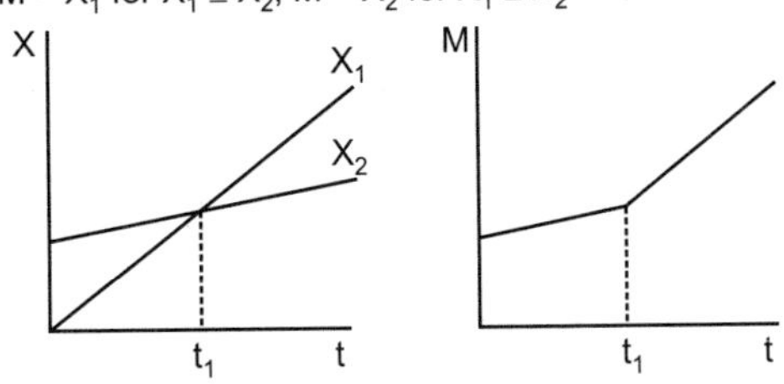	1. Output equals greater of two or more inputs.

Contd.

No.	*Symbol/ Function*	*Equation/Graph*	*Definition*
17	< Low Select	$M = X_1$ for $X_1 \le X_2$, $M = X_2$ for $X_1 \ge X_2$ 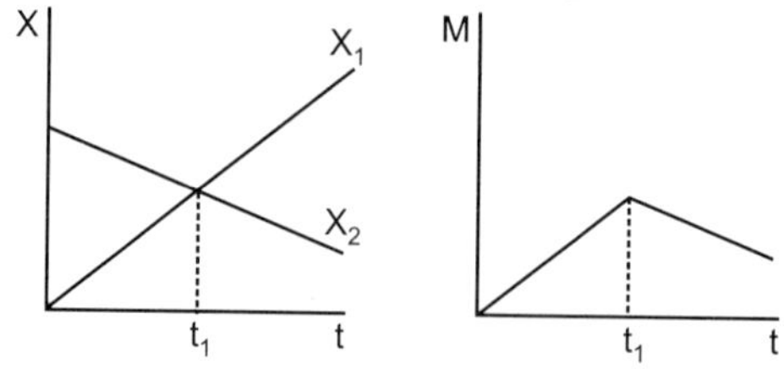	1. Output equals lesser of two or more inputs
18	≯ High Limit	$M = X_1$ for $X_1 \le H$, $M = X_2$ for $X_1 \ge H$ 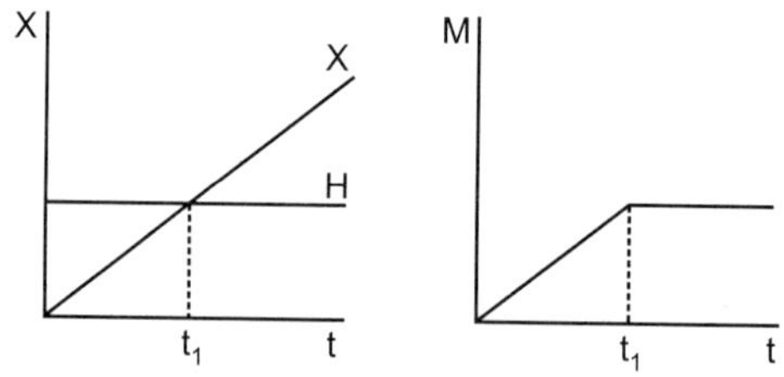	1. Output equals the lower of the input or high limit values
19	≮ Low Limit	$M = X_1$ for $X_1 \ge L$, $M = X_2$ for $X_1 \le L$ 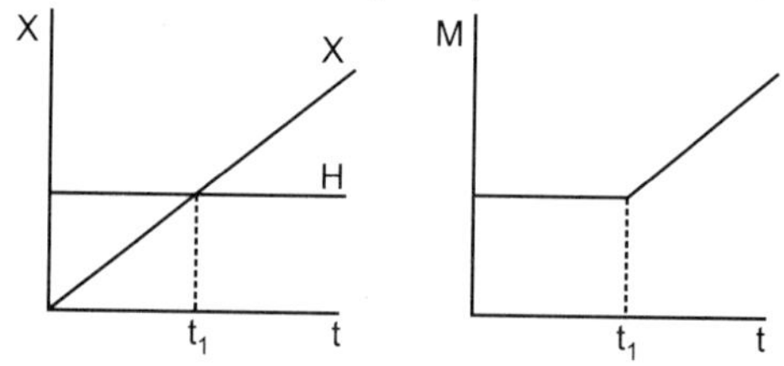	1. Output equals the higher of the input or low limit values
20	∀ Velocity Limiter	$dM/dt = dX/dt$ ($dX/dt \le H$, $M = X$) $dM/dt = H$ ($dX/dt \ge H$, $M \ne X$) 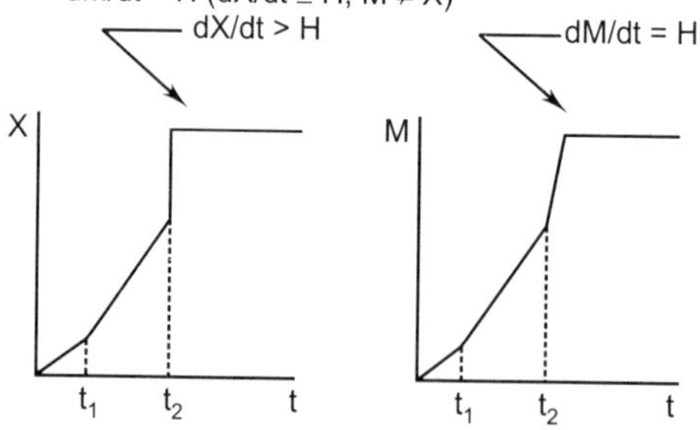	1. Output equals input as long as the input rate of change does not exceed the limit value that establishes the ouput rate of change until the output again equals the input

Contd.

No.	*Symbol/ Function*	*Equation/Graph*	*Definition*
21	+ Positive Bias	$M = X + b$ 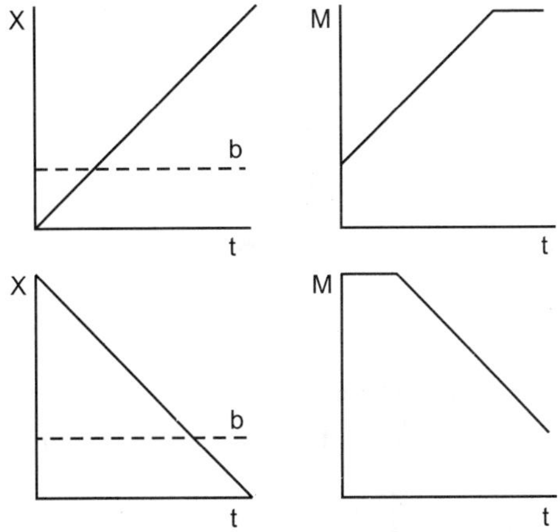	1. Output equal to input plus an arbitrary value
22	– Negative Bias	$M = X - b$ 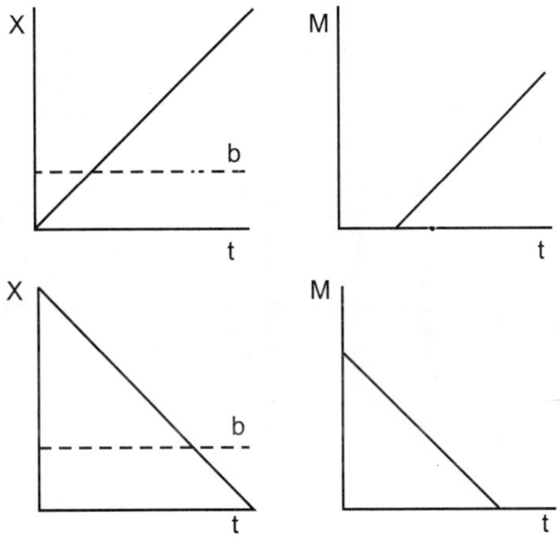	1. Output equal to input minus an arbitrary value
23	*/* Conversion	I = P, P = I, etc. 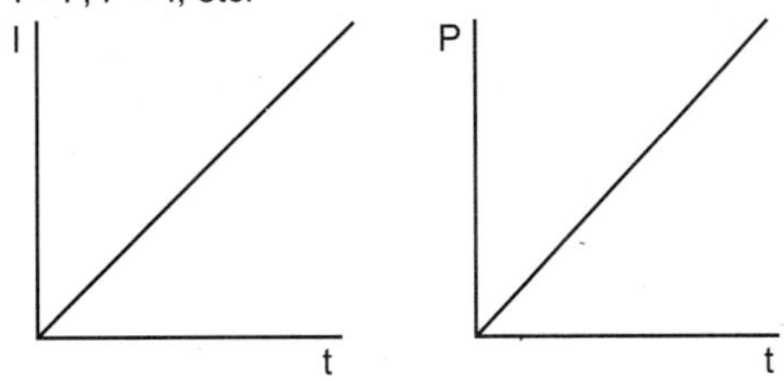	1. Output signal type different from that of input signal 2. * is equal to: E = Voltage A = Analog I = Current B = Binary P = Pneumatic D = Digital R = Resistance H = Hydraulic O = Electromagnetic, sonic

Contd.

No.	Symbol/ Function	Equation/Graph	Definition
24	* H High Signal Monitor	(State 1) M = 0 @ X < H (State 2) M = 1 @ X ≥ H 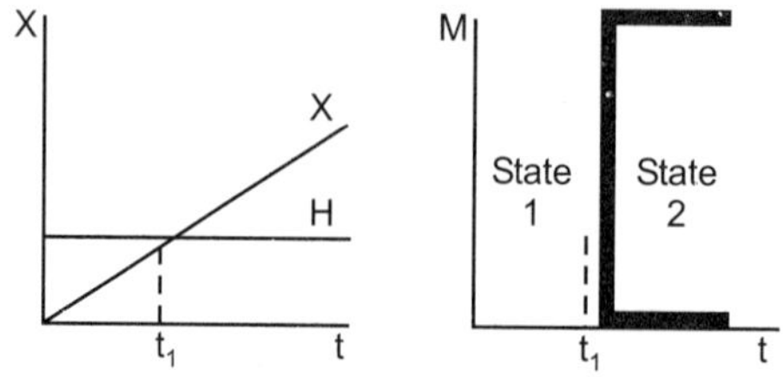	1. Output state is dependent on value of input 2. Output changes state when input is equal to or higher than an arbitrary high limit
25	*L Low Signal Monitor	(State 1) M = 0 @ X ≤ H (State 2) M = 1 @ X > H 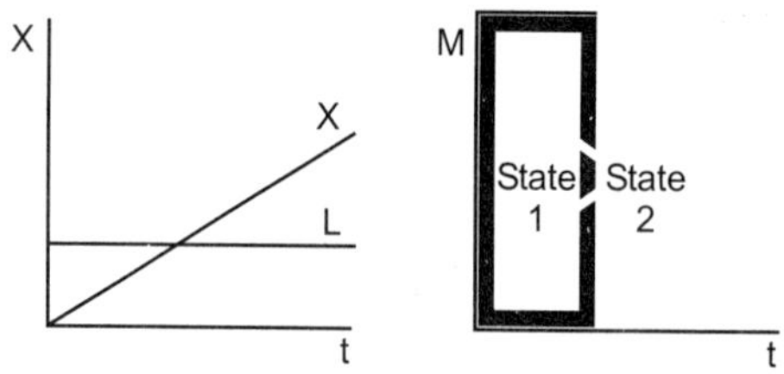	1. Output state is dependent on value of input 2. Output changes state when input is equal to or lower than an arbitrary low limit
26	*HL High/ Low Signal Monitor	(State 1) M = 1 @ X ≤ L (State 2) M = 0 @ L < X < H (State 3) M = 1 @ X ≥ H 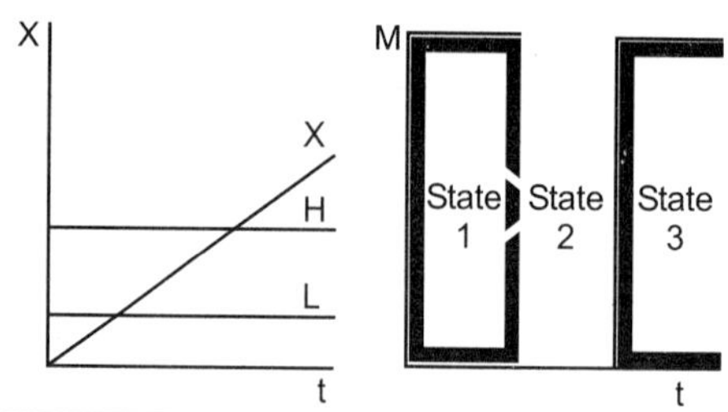	1. Output states are dependent on value of input 2. Output changes state when input is equal to or lower than an arbitrary low limit or equal to or higher than an arbitrary high limit
27	T Transfer	(State 1) M = X_1 (State 2) M = X_2 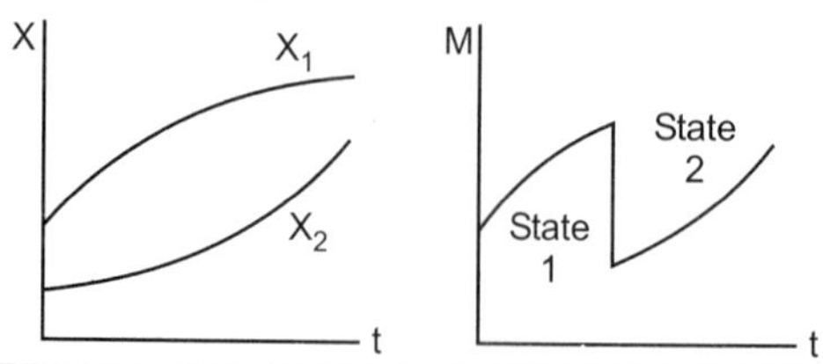	1. Output equals input that is selected by transfer 2. Transfer actuated by external signal.

(Courtesy: ISA ANSI/IS–5.01.01)

5.6 SOME IMPORTANT TERMINOLOGIES

Accessible

It is a feature associated with a discrete device function/interactive shared system function or a feature that can be used (and seen) for performing control operations like changing set point, auto-manual transfer or on-off operations.

Analog

A signal or device that does not have discrete positions (states). Changes value as per input changes. An analog signal implies a continuously varying quantity.

Application Software

It is an user specific software and is configurable. It contains logic sequences, control algorithms, codes, etc., to control input, output, calculations and make decisions.

Assignable

It is a system feature that channels/directs a signal from one device to another. This can be effected without any change in wiring by means of patching, switching or keyboard commands.

Auto-Manual Station

It operates in auto/manual mode to control operations of a control loop.

Binary

A signal/device having only two discrete positions/states. A binary signal is also known by 'on-off' or 'high-low' states.

Bubble

It is a preferred term for the circular symbol used to denote/identify the purpose/function of an instrument that may contain a tag number.

Communication Link

It is a wire/cable/transmitter network/bus system that connects microprocessor or computer-based systems which share a common data base and communicate amongst them in a hierarchical and or peer-to-peer relationship.

The wires/cables may be twisted pair, co-axial, fiber optic type while transmitters may be radio, telephone or microwave devices.

Computer Control System

It is a system in which all control action takes place within a control computer (main frame/mini computer) which may be single/redundant.

Computing Device

A device that performs calculations and/or logical operations and transmits one/more output signal.

Computing Function

It is a hardware/software function that performs one or more calculations or logic operations or both and transmits one or more resultant output signal.

Configurable

It is a term used for devices/systems whose functional/communication characteristics can be selected/altered by means of application software, program switches, fill-in-the-blank forms, 'pull-down' menus, entered values or text or by other methods – other than rewiring as a means of altering the configuration.

Controller

A device whose output can be varied to regulate a controlled variable in a specified manner. It may be a self-contained analog or digital instrument or may be an equivalent of either of the two in a shared-control system.

A controller may be an integral part of a control loop. A manual controller varies its output in response to manual adjustments—it is independent of the measured process variable.

An automatic controller varies its output automatically in response to a direct/indirect input of a measured process variable.

Control Valve

A device (separate from a hand actuated process block valve or self-actuated check valve) that manipulates the flow of one or more process fluid streams.

Converter

A device that receives signal in one form and outputs a signal in another form. Examples are V to I (voltage to current) or I to P (current to pneumatic pressure) converter.

Data Link

A wire/cable/transmitter network/bus system that connects field devices with dedicated (earmarked) microprocessors so that they share a common data base and communicate, as per established protocol, in a hierarchical or peer-to-peer relationship.

Detector

It is a device that detects the presence of flammable or toxic gases or some discrete part.

Device

It is an instrument that performs a specific action or function- *e.g.,* transmitter, controller, indicator, annunciator or a control valve, etc.

Digital

A signal or device that uses/generates a binary signal.

Discrete

It is a term that is used to describe a signal that is distinct/discontinuous/defined values. As example is a binary signal.

Distributed Control System

An instrumentation system having input/output devices, control devices, operator interface devices that performs the tasks of measurement, indication and control functions and capable of transmitting measurement, control and operation informations to or from a single or multiple user-specific locations which are connected by a single or multiple communication links.

Field Instrument

An instrument that is placed in the vicinity of its primary element or final control element and not in the control room or panel or console.

Final Control Element

A device, like a control valve, which directly controls the value of the manipulated variable of a control loop.

Function

The purpose of, or the action performed by, a device or application software.

Instrument

A device which is used for direct/indirect measurement, monitoring or control of a variable.

Instrumentation

An assortment of instruments or functions or their application for measurement/monitoring/ control of a process.

Local Instrument

An instrument placed in the vicinity of its primary element or final control element but not mounted on a panel or console of in the control room.

Loop

A combination of two or more instruments or control functions arranged in such a way that the signals pass from one to another for the purpose of measurement, indication or control of a process variable.

Measurement

It is the determination of the magnitude of a process variable.

Monitor

It is a general term for an instrument or instrument system to sense the status or measure the magnitude of one or more variables in order to derive useful information.

Panel

It is a structure having one or more sections, cubicles, consoles or desks that has groups of instruments mounted on it. It is used as an operator-process interface.

Primary Element

An external or internal instrument or system element that quantitatively converts the measured variable into a form suitable for measurement.

Process

Any operation or sequence of operations involving a change of energy, state, composition, dimension or other properties that may be defined with respect to zero or some other defined initial value.

Process Variable

Any measurable property of a process-like temperature, pressure of the same, other than instrument signals between devices in a loop.

Program

It is a repeatable sequence of actions which defines the output states having a fixed relationship to input states.

Programmable Logic Controller (PLC)

A PLC usually has multiple inputs/outputs and program which can be changed as per the needs of the process. A PLC is used to provide

- continuous control functions
- binary and/or discrete logic or sequencing functions.

Relay

It passes an information in an unchanged/changed form. It is applied to electric/pneumatic/ hydraulic switching.

Scan

This is sampling of a variable, in a predetermined manner, either periodically or intermittently.

Sensor

It senses the value of a process variable. It is either a separate or integral part of function of a loop or an instrument.

Set Point

It is an input variable that sets the desired value of the controlled variable manually, automatically or by means of a program in the same units as the controlled variable.

Software

Programs, algorithm, code, procedure, rules, etc., which are required for the operation or maintenance of a microprocessor or computer-based system.

Software Link

The interconnection/integration of different system components via communication networks or functions by means of software or keyboard instructions.

Test Point

It is a process connection to which no instrument is permanently connected but which is intended for the temporary connection of an instrument.

Transducer

It is a general term for a device, such as a primary element, transmitter, converter or other device which receives information in the form of one or more physical quantities, modifies the information or its form if so required and produces an output signal.

Transmitter

It is a device that senses a process variable via a sensor and produces an output value which is a function of the value of the process variable.

5.7 SOME BASIC AND SPECIALISED SYMBOLS

Given below are some basic and specialised symbols as used in varied industries with different types of requirements and applications.

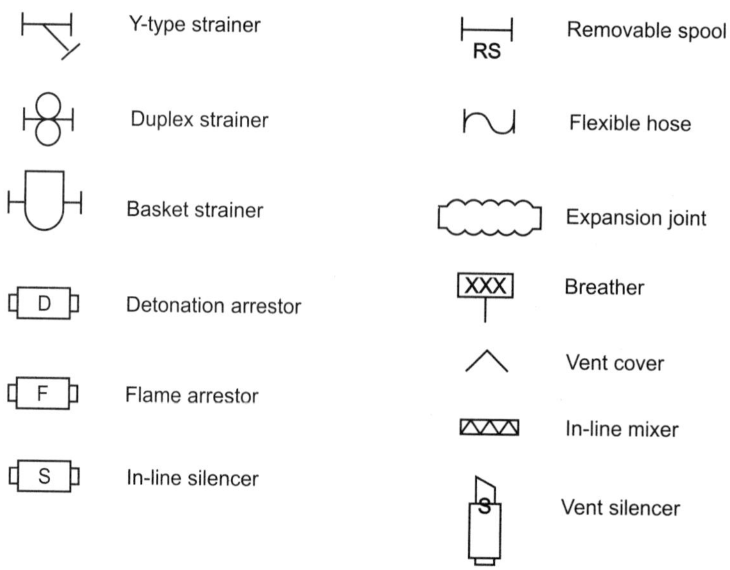

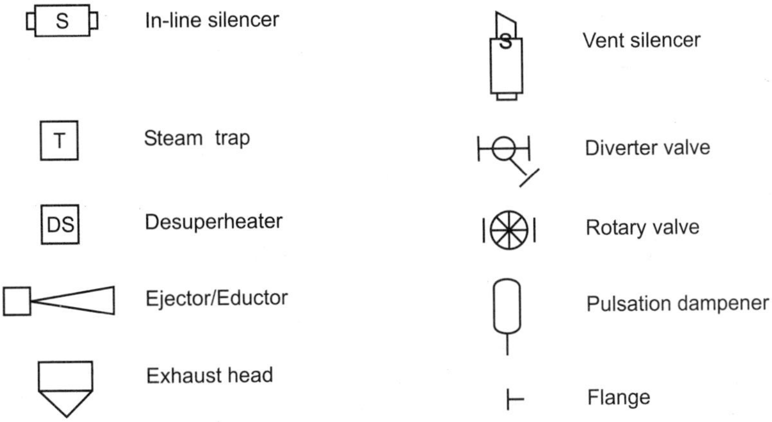

Fig. 5.4: Piping symbols

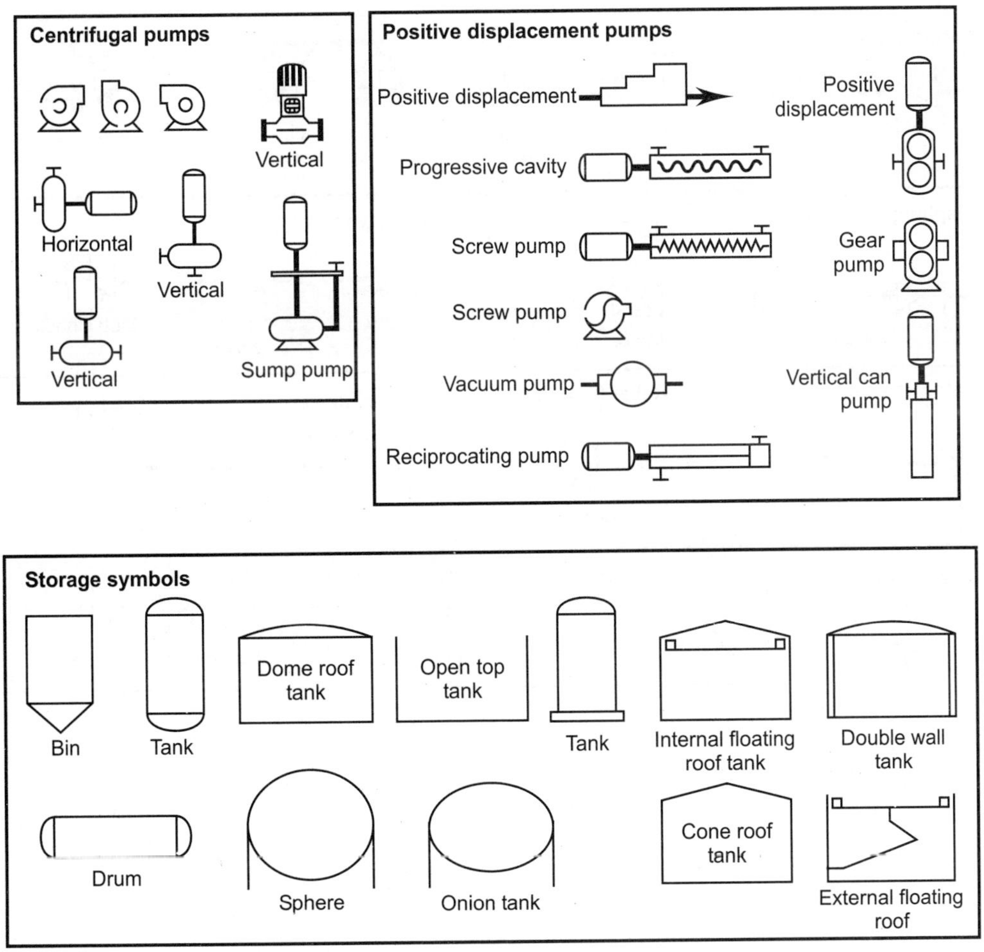

Fig. 5.5: Different types of pumps and tank

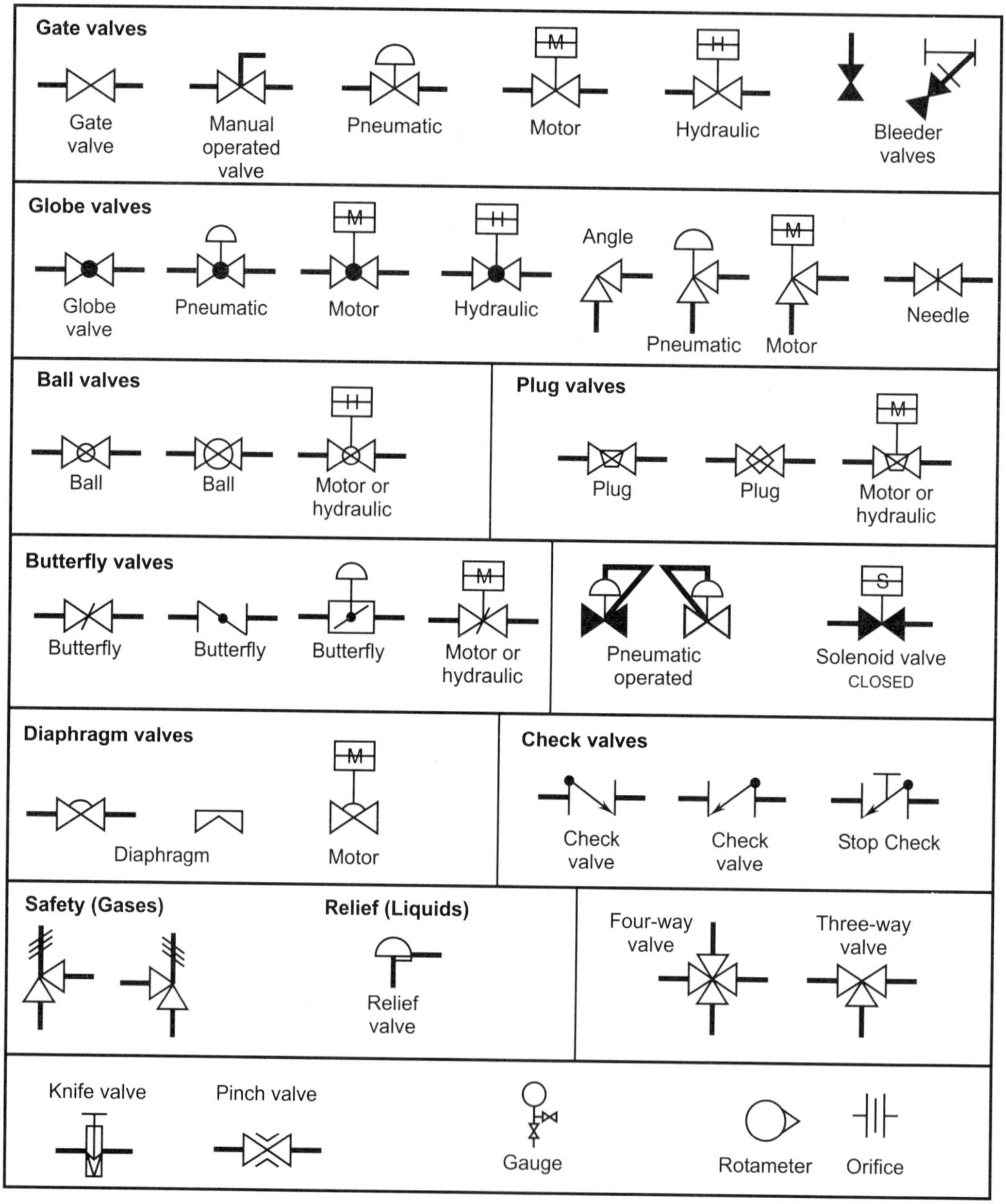

Fig. 5.6: Different types of valves

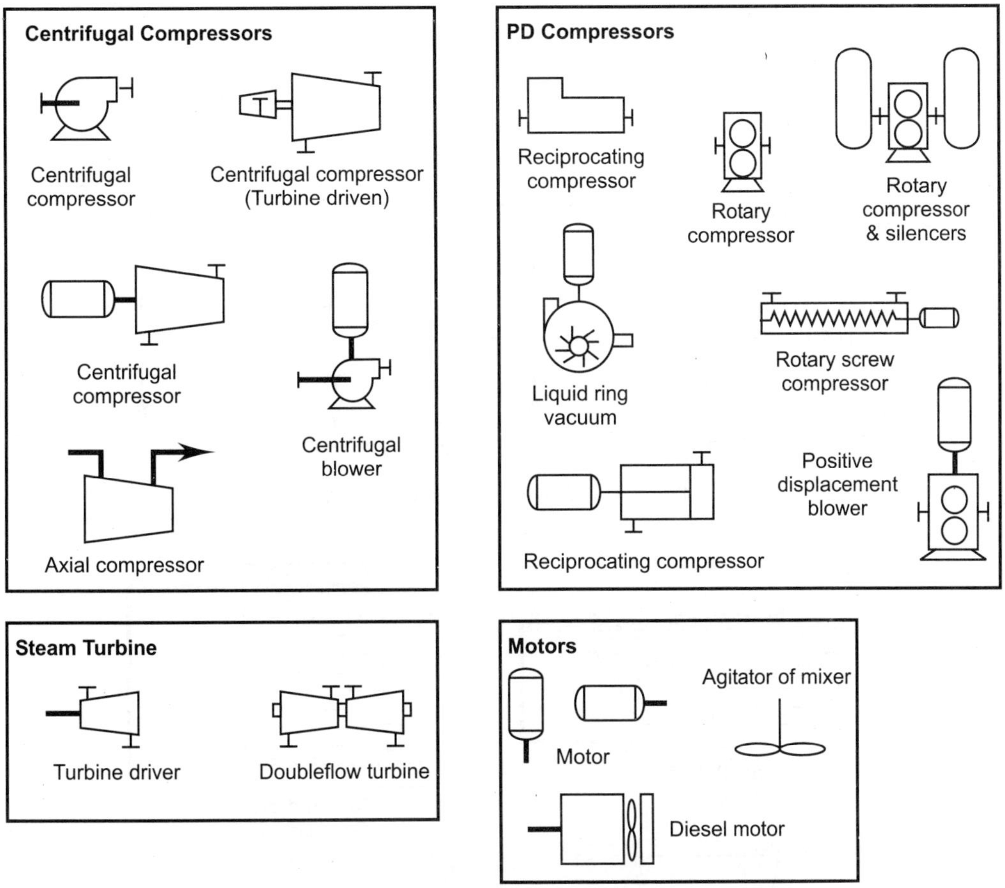

Fig. 5.7: Different types of compressors, turbines and motors

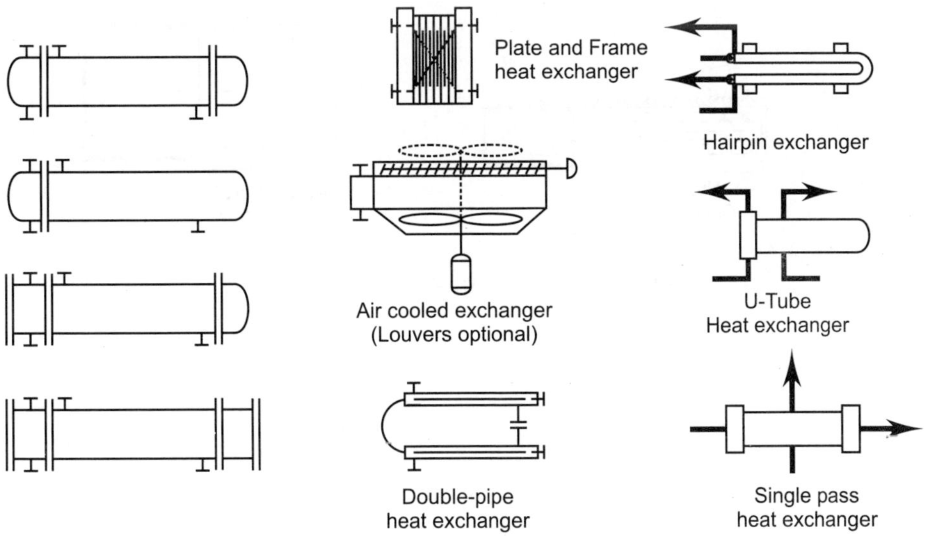

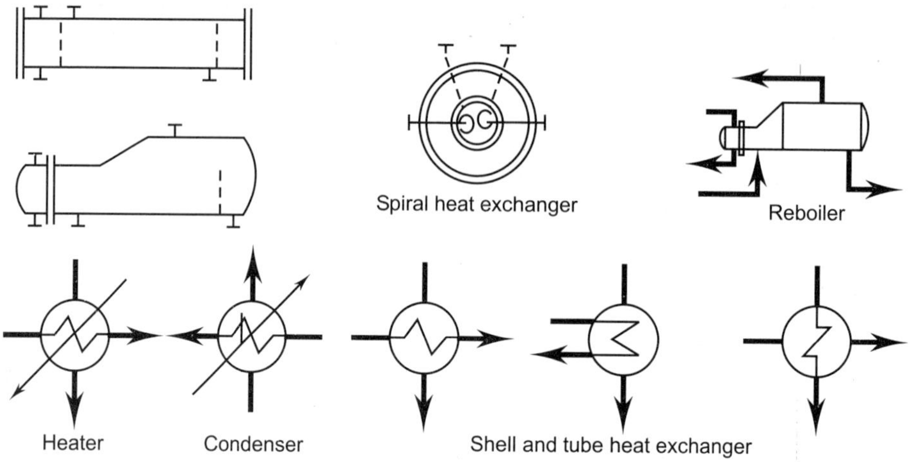

Fig. 5.8: Different types of exchangers

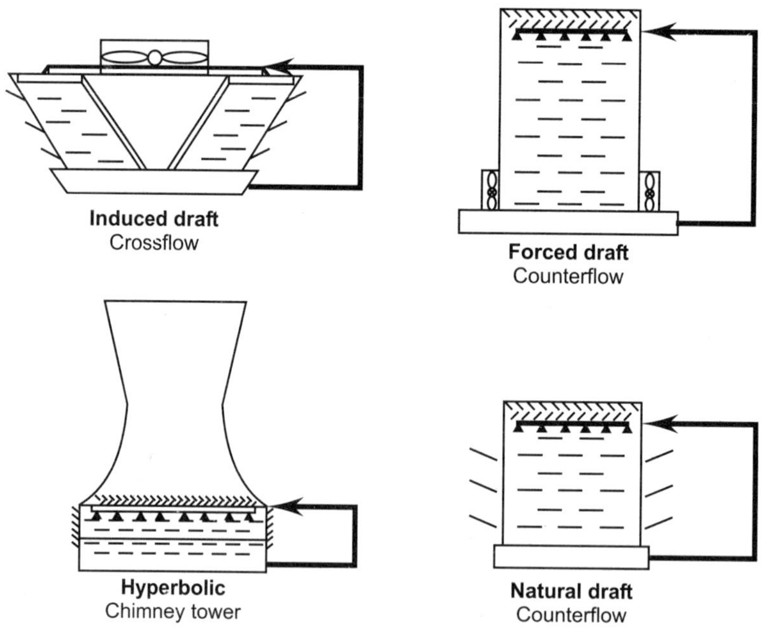

Fig. 5.9: Different types of cooling towers

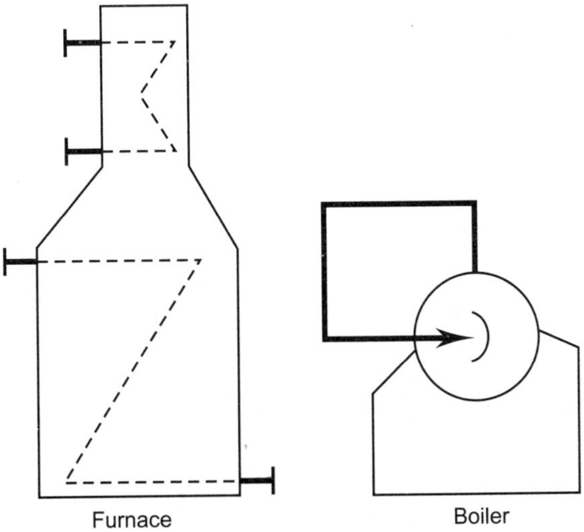

Fig. 5.10: Different types of furnaces and boilers

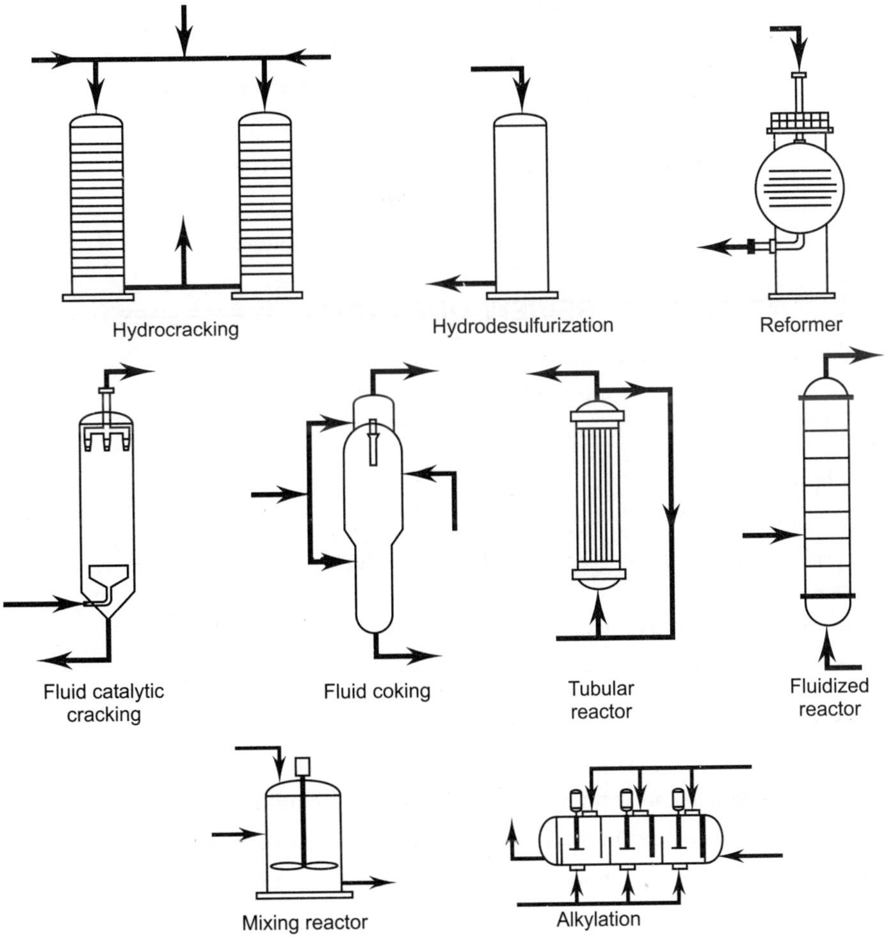

Fig. 5.11: Different types of reactors

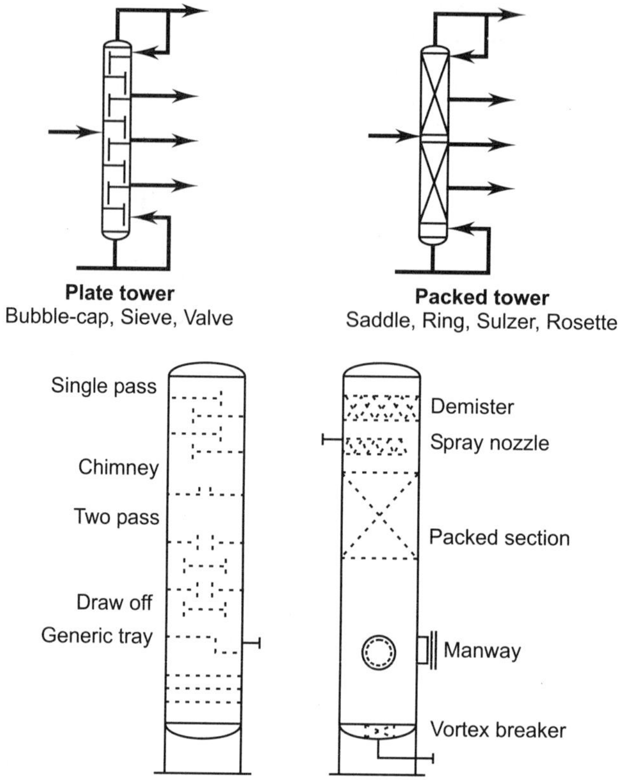

Fig. 5.12: Different types of cooling towers

REVIEW QUESTIONS

1. What details are included in a process flow diagram?
2. What details are included in a piping and instrumentation diagram?
3. Draw a PFD and describe the same.

Fill in the blanks:

1. ISA stands for ....
2. In general, there are four graphical elements which are grouped into .... location categories.
3. P & IDs are also known as ....
4. P & IDs are classified into .... categories.

Tick the correct answer:

1. Equipment rating and capacity are included/not included in a P & ID. ☐
2. Cooling towers, boilers, air compressors belong to process/utility plant P & ID. ☐
3. A field instrument is placed in the vicinity of its primary element/control room. ☐
4. A PLC is used to provide continuous/discontinuous control functions. ☐

CHAPTER 6

TEMPERATURE MEASUREMENT

6.1 INTRODUCTION

Measurement of temperature plays a very important and pivotal role and contributes in a major way to the quality of end product in many process industries. Depending on the type of application, measurement of temperature needs to be very precise in nature. In steel plants, the accuracy of measurement of molten steel in steel melting shop (SMS) is very vital to ensure the quality of end product. In chemical processes, a particular temperature is to be maintained for specific chemical reactions to take place effectively. In essence, temperature is directly related to the quality of end product.

Temperature measurement is based on the basic fact that two bodies in intimate contact will share heat in such a way that ultimately both attain the same temperature. The principle is based on the second law of thermodynamics.

There are many ways for measuring temperature. However, selection of a particular type of temperature sensing element depends primarily on various factors, namely speed of response, environment in which the measurement is to be carried out, range of temperature, accuracy of measurement, repeatability, compatibility with control elements, distance between the point of measurements and its reading, contact/non-contact type, installation pertaining to the sensor and lastly, the economics involved.

Temperature measurement is broadly divided into two main classes, *viz.*,

- Non-electrical methods and
- Electrical methods.

The non-electrical methods include:

- Change in physical dimensions of a solid (expansion/contraction of a solid with increase/decrease of temperature),
- Change in volume of a liquid (Charle's law),
- Change in pressure of a gas (Pressure law) and
- Change in vapour pressure.

Unlike mass, length etc., temperature is a derived quantity. Thus, measurement of temperature involves the measurement of certain properties of matter which undergo changes with heat. Such changes are:

- Changes in physical dimensions,
- Changes in physical state,

- Changes in chemical state,
- Changes in electrical properties, and
- Changes affected by radiation.

Figure 6.1 shows the temperature measurement range of different measuring instruments/sensors.

6.2 DIFFERENT TEMPERATURE SCALES AND FIXED POINTS

Table 6.1 shows the comparison of Kelvin and Celsius temperature scales.

Table 6.1: Comparison of Kelvin and Celsius Temperature Scales

	K	°C
Absolute zero	0	-273.15
Boiling point of O_2	90.19	-182.96
Zero Farenheit	255.37	-17.78
Ice point	273.15	0
Steam point	373.15	100
Freezing point of silver	1235.08	961.93

The Normal Hydrogen Scale (NHS) was proposed by chappnis in 1888. It used hydrogen gas in a gas thermometer as the thermometic material which covered a temperature range of –25° to 100C°. Thus, NHS had a very small temperature range within which it was used as a primary standard in fundamental laboratory measurements.

To circumvent this drawback, an International Temperature Scale of 1927 (ITS-27) was defined in 1927 having six defining or fixed points. In ITS-27, platinum resistance thermometers (PRTs) with defined properties were acknowledged as precision thermometers in the range –190° to 660°C. The scale was interpolated to 1063°C using platinum-rhodium thermocouples having specified compositions and properties, and beyond 1063°C, Wein's law was used.

The drawbacks of ITS-27 gave rise to International Temperature Scale of 1948 (ITS-48) having the same number of fixed points as in ITS-27. But in ITS-48, the melting point of silver was changed to 960.8°C instead of 960.5°C as in ITS–27. In ITS-48, the lower PRT interpolating point was raised to – 183°C from –190°C as in ITS-27. Above 1063°C, Planck's law was applied replacing Wein's law. In 1960, a revision of ITS-48 was effected, giving rise to International Practical Temperature Scale of 1948 (60) or IPTS-48(60). In this, the triple point of water was recommended at 273.16 K, giving rise to Kelvin Thermodynamic Scale.

ITS-48 provides the following six basic (or primary) fixed points on the Celsius scale, at one atmospheric pressure.

- Oxygen Point – Boiling point of liquid oxygen (– 182.97°C)
- Ice Point – Melting point of pure ice (0°C) [a fundamental point]
- Steam Point – Boiling point of pure water (100°C) [a fundamental point]
- Sulphur Point – Boiling point of liquid sulphur (444.60°C)
- Silver Point – Melting point of silver (960.8°C)
- Gold Point – Melting point of gold (1063.0°C)

On the same IPTS-48 scale, some of the secondary fixed points are:

- Temperature at which solid CO_2 changes to gaseous state at normal atmospheric pressure (–78.50°C)
- Freezing point of mercury (–38.87°C)
- Freezing point of antimony (+630.5°C)
- Freezing point of palladium (+1552°C)
- Melting point of tungsten (+3380°C)

Another revision was made in 1968, known as International Practical Temperature Scale or IPTS-68. It has 13 fixed points. IPTS-68 involved the extension of temperature on the lower side and employed improved measurement techniques. IPTS-68 was modified in 1975 and is known as IPTS-68(75). An Extended Practical Scale of 1976 or EPT-76 extended still the lower temperature side with the addition of 11 fixed points in the cryogenic range. This range is from 0.519 K (superconducting transition point of cadmium) to 27.102 K (boiling point of neon).

All the earlier temperature scales have lastly been superseded by the International Temperature Scale of 1990 (ITS-90). In this scale, a distinction exists between the International Kelvin Temperature (T) and the International Celsius Temperature (t) and given by

$$t_{90}\ (°C) = T_{90}\ (K) - 273.15$$

Temperature fixed points as provided by ITS-90 is shown in Table 6.2.

Table 6.2: The Temperature Fixed Points of ITS-90

Equilibrium state	*Scale*	
	T_{90}	t_{90} °C
Vapour-pressure point of helium	3 to 5	-270.15 to -268.19
Triple point of equilibrium hydrogen	13.8033	-259.3467
Boiling point of hydrogen at a pressure 33330.6 Pa	17	-256.15
Boiling point of equilibrium hydrogen	20.3	-252.85
Triple point of neon	24.5561	-248.5939
Triple point of oxygen	54.3584	-218.7916
Triple point of argon	83.8058	-189.3442
Triple point of mercury	234.3156	-38.8344
Triple point of water	273.16	0.01
Melting point of gallium	302.9146	29.7646
Freezing point of indium	429.7485	156.5985
Freezing point of tin	505.078	231.928
Freezing point of zinc	692.677	419.527
Freezing point of aluminium	933.473	660.323
Freezing point of silver	1234.93	961.78
Freezing point of gold	1337.33	1064.18
Freezing point of copper	1357.77	1084.62

The values of the temperature fixed points with the exception of the triple points are given at pressure, P_0 = 101 325 *Pa*

Chapter 6

6.3 EXPANSION THERMOMETERS

Temperature measuring expansion thermometers fall under the following categories:

- Thermometers based on expansion of solids,
- Thermometers based on expansion of liquids (variable volume thermometers) and
- Thermometers based on expansions of gases (variable pressure thermometers).

6.3.1 Thermometers Based on Expansions of Solids

There are three types of thermometers falling under this category. These are: Dilation thermometer, thermostat and bimetallic thermometer. The first two works are on the principle of differential expansion of materials, and the third depends on their bending properties.

6.3.1.1 Dilation Thermometer

Its principle of operation depends on the difference in expansion of two metals when exposed to some hot bodies. The difference is amplified by mechanical means as the difference in expansion is too small to be read directly in a correct manner. This type of thermometer can measure upto 1000°C with an accuracy limit of ±1 to ±2% of their temperature ranges. These are soldom used for industrial applications.

6.3.1.2 Thermostat

It is generally used to maintain a constant temperature of an electrical appliance. For example, whenever the temperature exceeds a preset value in an electric heater, the thermostat cuts off the electric supply. It is also used to monitor/regulate the supply of gas to gas ovens.

The principle of operation is based on the differential expansion of two materials. The materials normally used are invar (an iron-nickel alloy) and brass. The coefficient of linear expansion of invar is about $\frac{1}{20}$th of that of brass. In a thermostat, the invar rod is covered by a brass tube, one end of which is firmly soldered. To the other end, a microswitch and a micrometer are connected.

When the thermostat is put in a hot bath, differential expansion at the free end causes the microswitch to cut off the electric supply to the heater. The micrometer is used to preset the cut off temperature, *i.e.*, it is needed for initial adjustment of temperature.

6.3.1.3 Bimetallic Thermometer

A bimetallic thermometer consists of two metal sheets (usually of same thickness) of

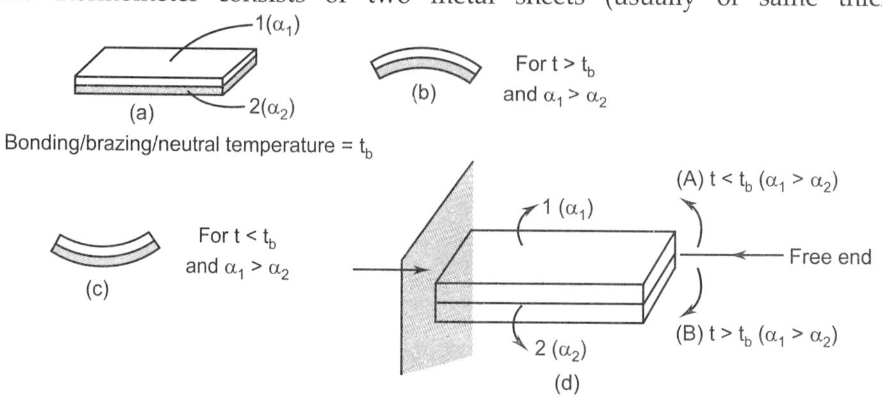

Fig. 6.1: Bimetallic thermometer (*a*) at neutral temperature t_b (*b*) temperature $t > t_b$ and $\alpha_1 > \alpha_2$ (*c*) temperature $t < t_b$ and $\alpha_1 > \alpha_2$ (*d*) temperature $t \gtrless t_b$ and $\alpha_1 > \alpha_2$.

different coefficients of linear expansions. The two sheets are brazed together. The temperature at which this is done is known as brazing/bonding temperature t_b and the combination remains flat at this temperature. The two sheets, 1 and 2, have different coefficients of linear expansions α_1 and α_2 ($\alpha_1 > \alpha_2$ assumed). But they will assume the shapes as shown in figure 6.1 for different temperatures. Thus, it (*i.e.*, the two pieces taken together) will bend in the directions shown for the two cases, *viz.*, $t > t_b$ or $t < t_b$ (figure 6.1(*b*) and (*c*) respectively) where t is the temperature to be measured. The metal having higher α value is called active metal and the one having lower α is called passive metal. Figure 6.1(*d*) shows the bending of the bimetal either way for $t \gtrless t_b$.

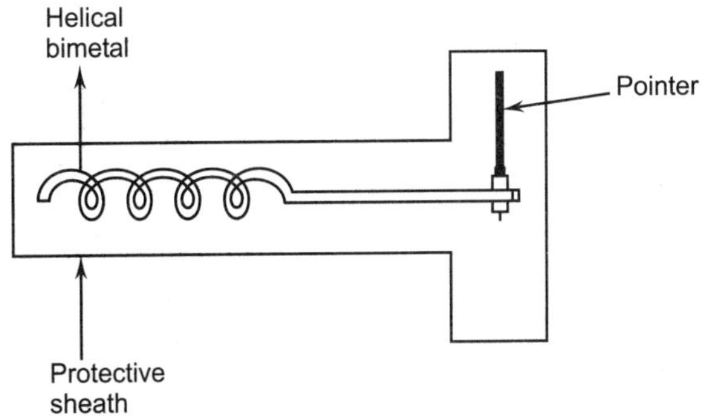

Fig. 6.2: A helical bimetal thermometer with pointer

Apart from the sheet type, different shapes of bimetal thermometers are available, such as spiral, cantilever, helix type. A pointer is attached to the free end of an industrial bimetallic thermometer. The pointer sweeps over a circular scale to indicate the temperature. This is shown in figure 6.2. Sometimes a protective sheath (thermal well) is used with the thermometer for protection against mechanical damage or corrosion. The well material may be of steel, brass, etc. A helical thermometer is used for higher accuracy.

Industrial bimetallic thermometers are used in the range of –40° to 500°C with errors between ±1% to ±2% of full scale. The upper and lower limits of temperatures are due to (*a*) creep of the thermometer materials at high temperatures and (*b*) stress on the thermometer materials at the two limits of temperature.

The speed of response of this type of thermometer depends on whether it is bare or having a thermal well for its protection. This type of thermometer is an accurate and simple device and at the same time rugged in nature.

The radius curvature (r) of the thermometer is given by the following simplified relationship, on the assumption that the Young's modulus of both the elements are almost equal.

$$r \simeq \frac{2S}{3(\alpha_1 - \alpha_2)(t - t_b)}$$

where, S = combined thickness of the thermometer

t = unknown temperature to be measured.

Bimetallic thermometers find applications in measurement of temperatures of live electrical equipments, temperatures of liquids and gases in containers, boilers, temperature of oil in transformers and, in some cases, surface temperatures also.

6.3.2 Thermometers Based on Expansion of Liquids (Variable Volume Thermometers)

They are of two types: Mercury-in-glass and mercury-in-steel thermometers.

6.3.2.1 Mercury-in-Glass Thermometer

Its principle of operation is based on the volumetric expansion of mercury with temperature for indication of unknown temperature. It has widespread use both in laboratories and industries.

Figure 6.3 shows a typical industrial mercury-in-glass thermometer. It consists of a bulb containing mercury. The bulb is enclosed in a metal well. When it is exposed to a high

Chapter 6

temperature, heat is transferred to the bulb via the well. As a consequence, mercury in the bulb expands driving mercury in the capillary. A scale adjoining the capillary is calibrated in terms of the unknown temperature.

The scale of a mercury thermometer is not exactly linear due to the following reasons:

- The coefficient of volume expansion of mercury is not constant over the entire scale range.
- The bulb expands/contracts on being heated/cooled. Such variation in volume of the bulb is to be taken into consideration while calibrating the thermometer.

  The space above the mercury is sometimes filled with dry pressurised nitrogen. This is done to increase the boiling point of mercury. Thus, when higher temperature is being measured, mercury column in the capillary rises and nitrogen gas pressure increases. It causes an expansion of the upper part of the capillary and compression of mercury volume.

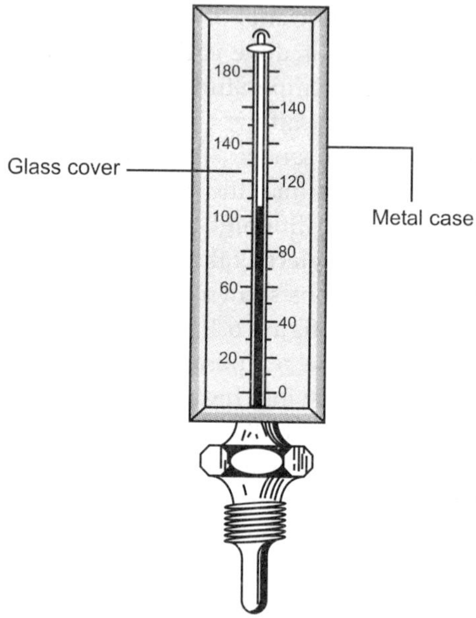

Fig. 6.3: An Industrial mercury-in-glass thermometer

- Aging effect causes a gradual change in volume of the thermometer bulb.

The thermal well, covering the glass bulb, is used to (1) prevent breakage of the glass bulb (2) provides a means for sealing at the place of installation.

The thermal well is made of steel, brass or cast iron depending on the type of exposure of the temperature measurement point. Sometimes for measurement of temperature of ovens or air ducts, perforated wells are used.

For overrange protection (means that by mistake if the thermometer is subjected to a higher temperature than its recommended maximum value), a small reservoir is kept at the top end of the capillary. In case of overshoot, mercury fills in the reservoir thereby preventing breakage.

The normal range of temperature measurement of this type of thermometer is from –39° to 510°C. The achievable accuracy of an industrial thermometer is about ±1 per cent of the span. This specified range can be extended from –40° to 540°C. Lower limit is achieved by employing eutectic mixture of mercury amalgam and higher limit by employing mercury at higher pressure to raise its boiling point.

The speed of response of an industrial thermometer depends on the following factors:

- The size of the bulb as also the size of the thermal well, and
- Characteristics of the fluid medium, *i.e.*, whether it is static or flowing.

To get a better speed of response, the space between bulb and the well is sometimes filled with a conducting liquid (normally mercury). This obviously facilitates heat transfer to the bulb from outside

An error, known as 'immersion error' (sometimes also called 'conduction error') may come into play with this kind of thermometers. Conduction of heat takes place along the bulb and the thermal well. It thereby reduces the temperature of bulb and consequently the temperature recorded by the thermometer would be less than the actual value. Thus, the thermometer bulb and the well should be so immersed that all parts of the bulb and the well are maintained at the same temperature.

This type of industrial thermometers are recommended for 'in situ' measurement of temperatures of liquids in tanks, air ducts, steam pipe and molten metal baths. But they are not a good choice for rapidly fluctuating temperatures.

There are three grades—*A*, *B* and *C*—in which mercury-in-glass thermometers are specified. While grades *A* and *B* are specified by National Physical Laboratory, grade *C* corresponds to commercial grade.

A thermometer should be calibrated under the condition of its actual use. That is, if in a particular case, a thermal well is to be used then the calibration of the thermometer should be done with the thermal well in place. A thermometer, calibrated without thermal well, can as well be used with the well so long as the consistencies of the reading so obtained is permitted for a given situation; in such a case a constant error will always be there and the reading shown by the thermometer is less than what it actually would have been.

Sometimes liquids other than mercury are used as a thermometric liquid for temperature ranges which cannot be covered by a mercury thermometer. As an example, liquid Pentance has a temperature range of –200° to 30°C.

A memory-in-glass thermometer can be used as a very simple on-off temperature controller. For such an on-off controller to be effective, two metallic wires are taken out—one from the lower end of the stem and the other at the point of temperature control. When the desired temperature is reached, the two metallic wires are shorted via the mercury column of the stem which can now be effectively used for switching off the heater circuit.

6.3.2.2 Mercury-in-Steel Thermometer

For industrial purposes, this type of thermometer has mostly replaced the mercury-in-glass thermometers because of two basic drawbacks of the latter—mainly (1) fragility of glass leads to its susceptibility to breakage, (2) it basically gives an 'in-situ' type of measurement.

These drawbacks are removed in mercury-in-steel thermometer by replacing the glass bulb and capillary by steel and measurements can be made at a considerable distance away by extension steel capillary provided with a protective armour.

Its basic principle is the same as that of a mercury-in-glass thermometer. The steel bulb and the capillary are totally filled with mercury, usually at a high pressure and connected to a Bourdon tube. To the other free end of the Bourdon tube is connected a pointer or a pen arm. With increase in temperature, more amount of mercury will be forced into the Bourdon tube, the free end of which will then uncurl, the pointer will move along a scale calibrated in terms of temperature.

The bulb, the capillary and the Bourdon tube are all hermetically sealed.

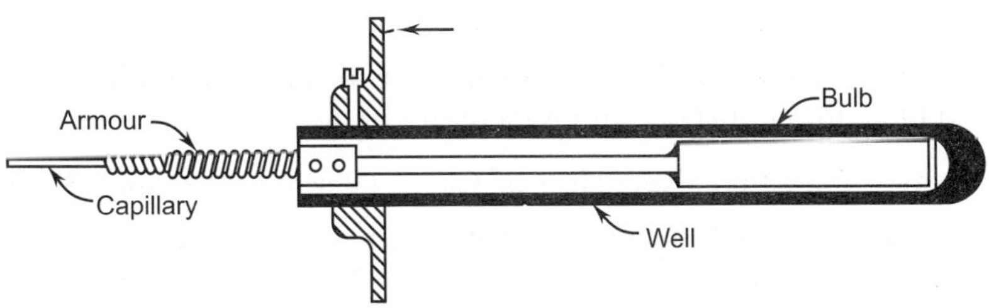

Fig. 6.4: Thermometer bulb and well

Figure 6.4 shows the thermometer bulb and the well while figure 6.5 below shows the mercury-in-steel thermometer installed in an industrial environment.

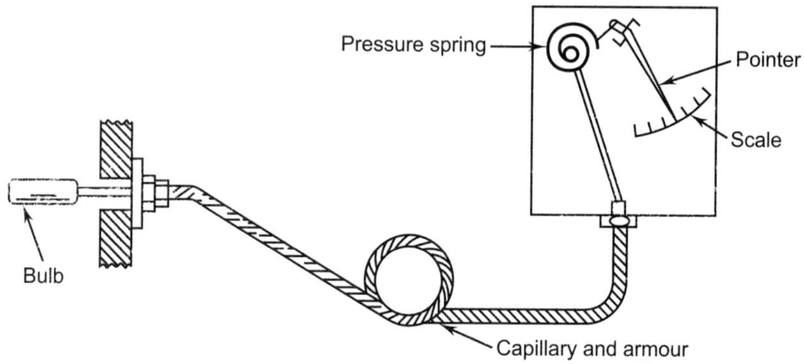

Fig. 6.5: Industrial pressure-spring thermometer mounted in a low pressure line

In mercury-in-steel thermometers, an effect, known as 'dip effect', exists. When such a thermometer is inserted into a hot process, a sudden expansion of the thermometer bulb occurs ahead of the expansion of memory. Thus, temporarily bulb volume increases and consequently the instrument records a fall in temperature as shown in figure 6.6. The dip effect is much less prominent for gas thermometers or vapour actuated pressure thermometers.

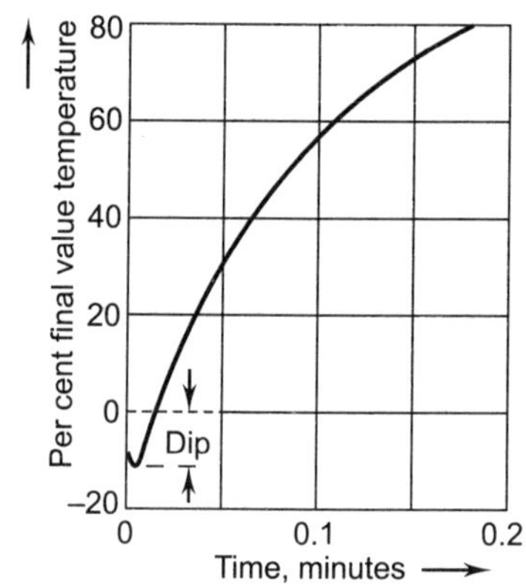

Fig. 6.6: Dip effect in mercury thermometer

Depending on the requirement of temperature measurement desired, the thermometer bulb may take different shapes like cylindrical, *U* type or wound in the form of an helix. For example, for average gas temperature measurement, the bulb may be in the form of a tube of considerable length having small diameter. This is because such an arrangement presents a considerable surface area to the gas. In cases where point temperature measurements are required, a bulb having a very small exposed area is employed. Apart from the above, the main consideration should be to maintain the ratio of surface area to volume of the bulb to a maximum so that the speed of thermal response becomes very fast and consequently the thermal lag is reduced.

For housing the bulb properly, flanges of varied forms are utilised. The type of attachment for the flange depends on the kind of application. For gas temperature measurements, obviously gas-tight flanges are employed.

The sensitivity of the thermometer is given by: $\frac{\Delta v}{\Delta t} = V_b(\beta - 3\alpha)$, where V_b = bulb volume, α and β are coefficients of linear and cubical expansions of bulb material and liquid. ΔV is the increase in bulb volume for an increase in temperature by Δt.

6.3.3 Thermometers Based on Expansion of Gases (Variable Pressure Thermometers)

They are of two types: Gas Thermometers and Vapour pressure Thermometers.

6.3.3.1 Glass Thermometer

In a gas expansion thermometer, the volume remains essentially constant. At constant volume, the relation between pressure and temperature is given by

$$P = P_o (1 + \gamma_v t)$$

where, P and P_0 are the pressures at t°C and 0°C respectively, and γ_v is the pressure coefficient at constant volume.

Thus, if the temperature increases from $t_1^o C$ and $t_2^o C$, then the corresponding pressure increase is given by:

$$P_2 - P_1 = P = P_0\gamma_v(t_2 - t_1)$$

Thus, the pressure increase (or decrease) bears a constant proportional relationship with temperature increase (or decrease).

The above discussion holds true for ideal gases, and a slight deviation takes place for real gases. In most cases, nitrogen is used as a filling gas in the thermometers, having a temperature range from –120° to 450°C. At the higher end of the temperature scale, the thermal system becomes somewhat porous and the nitrogen gas might leak out through this. Then the calibration of the temperature scale will become invalid.

A gas filled thermometer contains the gas in a bulb, a capillary and a Bourdon tube. At constant volume, the pressure of a gas increases by a constant factor 1/273 of its pressure at 0°C for every degree rise of temperature. Thus, the scale would be a linear one—relating the pressure and temperature of the gas, assuming, of course, that the volume increase of the capillary and the Bourdon tube with temperature are neglected. For this the volume of the bulb must be made many times larger than the combined volume of the capillary and the Bourdon tube. Viewing from this aspect, since the bulb volume is appreciably high, it obviously reduces the dynamic response of the thermometer. Looking from another perspective, since the gas has a low thermal capacity compared to an identical amount of liquid, the speed of response of the thermometer to temperature changing would be faster.

Accurate ambient temperature compensation, as was realised in the case of liqiud filled thermometers by inserting invar wire into the capillary (described earlier), is not possible in this case of gas-filled thermometers because invar wire would occupy an appreciable portion of the capillary bore volume and would not exactly compensate for very small changes of the same due to temperature changes.

Nitrogen is used mostly in gas thermometers because of its following advantages: (*a*) It is abundantly available—both in quality and quantity, (*b*) covers a wide range of temperature, (*c*) scale of a nitrogen filled gas thermometer is almost linear.

Although hydrogen has a low viscosity, its specific heat is many times more than other gases. Again, hydrogen can diffuse through the thermal system, particularly at higher temperatures. These points rule out the possibility of using hydrogen as a filling liquid.

Nitrogen gas, which is used as a filling gas in the bulbs, is normally filled between 150 to 500 psi at room temperature. This high pressure is maintained in the bulb to ensure that even for small temperature changes, the change of pressure developed would be great enough to overcome the frictional effect of the device and thus the system will be able to record the new temperature correctly.

To measure the temperature of an extended surface, an 'averaging bulb' is used, *i.e.*, in this case the bulb will have to a considerable sized with the most exposed surface area.

6.3.3.2 Vapour Pressure Thermometer

A vapour pressure thermometer employs the principle of variation (or dependence) of saturated vapour pressure of a liquid with temperature for its record. Figure 6.7 below shows the variation of saturated vapour pressure with temperature for a liquid. The system consists of a bulb, a capillary and a Bourdon tube-as in the case of liquid filled thermometers.

The pressure due to vapour of a liquid depends on the temperature existing at the free surface of liquid. Thus, the free surface must always remain in the bulb for the entire range of intended temperature measurement.

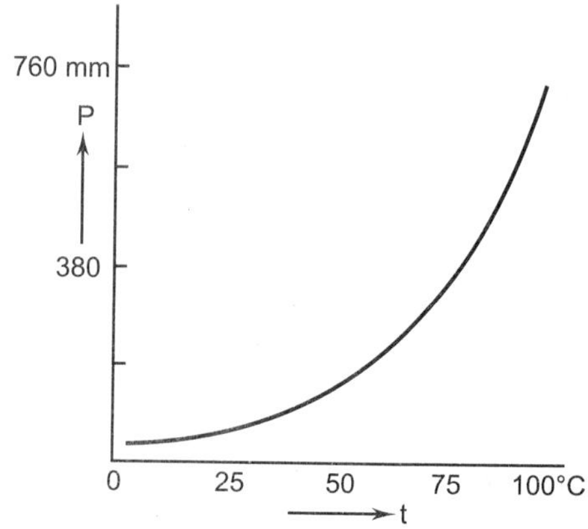

Fig. 6.7: Variation of vapour pressure of water with temperature

Some serious considerations must be given both for the bulb size and the amount of liquid in the bulb. When the bulb temperature falls below the capillary and Bourdon tube temperature, the bulb must be in a position to hold all the liquid. Again when it is the other way round, *i.e.*, bulb temperature is more than the temperatures of both the capillary and the Bourdon tube, the liquid volume should be adequate enough to fill in completely the capillary and the Bourdon tube. The volume of liquid should be so taken that at the highest temperature measurement point not all the liquid is vapourised and at the lowest temperature measurement point not all the condensed liquid fill up the entire thermal system.

The construction of a vapour pressure thermometer is shown, in figure 6.8. The non-volatile liquid fills the capillary, the Bourdon tube and a part of the bulb while the rest part of the bulb is filled up with a volatile liquid along with its vapour. As the bulb is exposed to unknown high temperature, the vapour pressure of the volatile liquid increases. The pressure is transmitted to the scale and pointer mechanism via the non-volatile liquid, the capillary and the Bourdon tube. The range of a vapour pressure thermometer is dependent on the fluid used—the most common ones being toluene, propane, butane, hexane, methyl chloride, etc. The scale of such a thermometer can be made linear by employing special linearising linkages. The material for the bulb are metals and their alloys, which are sometimes plated with nickel, chromium or silver.

A vapour pressure thermometer is not influenced by ambient temperature variations but slight temporary oscillations may occur when the unknown temperature and the ambient temperature are in proximity of each other. To explain, let the unknown temperature, as recorded by the thermometer, is slightly higher than the ambient temperature around the capillary and the Bourdon tube. Assume for the present that the said ambient temperature overshoots the unknown temperature suddenly. This will result in an increase in the vapour pressure inside and the temperature record will rise. But condensation will take place till equilibrium is reached so that finally the pointer will indicate the unknown temperature correctly. Similarly, a temporary downswing of temperature record will be there if suddenly the ambient temperature around the capillary and the Bourdon tube comes below the unknown temperature from a higher one. This effect is known as 'cross-ambient effect'. Thus, a vapour pressure thermometer should not be used in cross ambient ranges if a high speed of response is desired.

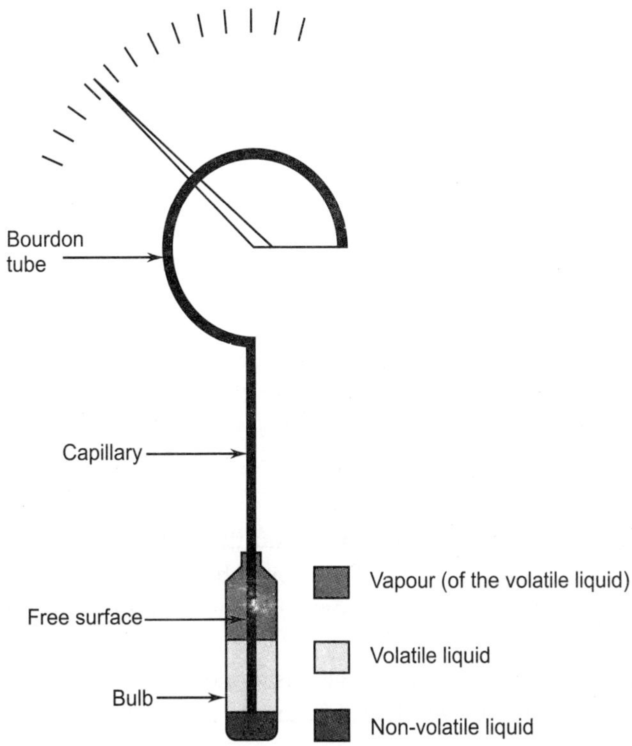

Fig. 6.8: Schematic of a vapour pressure thermometer

Several advantages can be cited for the widespread use of vapour pressure thermometers in industry. These are:

- Smaller bulb size than liquid or gas filled thermometers,
- Less expensive,
- Faster dynamic response than liquid thermometers,
- Do not suffer from ambient temperature effects, and
- Ease of maintenance.

The disadvantages associated with this kind of thermometers are:

- Scale is non-linear and cramped at the lower end of the temperature scale,
- Temperature range is limited by 'critical point' of the liquid, and
- Organic vapours affect the performance severely at higher end of the temperature scale.

A point to be noted at this stage is that the liquid and gas thermometers as also the vapour pressure thermometers are clustered under a special class and are known as "filled in sytems" in industries.

Vapour pressure thermometers are applied in the range of –50° to 350°C with accuracies of the order of ±1% to ±2% of full scale. A maximum capillary length of 50 m is permissible with saturated vapour pressure thermometers usually lying in the range of 0.7 MPa to 2.5 MPa.

6.3.4 Errors in "Filled-in" Systems and Their Remedies

The "filled-in" system are subject to the following types of errors:

- Ambient temperature effect,
- Barometric effect,
- Head (or elevation) effect,
- Immersion effect, and
- Radiation effect.

6.3.4.1 Ambient Temperature Effect

In case of liquid expansion thermometers, changes in ambient temperature at the capillary and at the receiving elements (*i.e.*, Bourdon tube) will cause the instrument temperature reading to be erroneous. The ambient temperature effect is best manifested by the following relationship:

$$t_{\text{error}} = \frac{v_c t_c + v_b t_b}{v_B}$$

where v_b, v_c are the volumes of the Bourdon tube and the capillary respectively.

t_b, t_c are the temperature changes at the Bourdon tube and the capillary.

V_B is the volume of the bulb and t_{error} represents the total error in measurement owing to ambient temperature variations.

Calibration of a liquid-in-metal thermometer is done at a particular ambient temperature. With deviation in temperature from the temperature at which the thermometer was calibrated, an error-called ambient temperature error comes into play. The situation gets aggravated as the length of the capillary is increased more and more.

With changes in temperature, the bore of the capillary changes, thereby affecting calibration. Thus this calls for offsetting this drawback. This can be achieved in several ways:

(*a*) By making the bulb volume much larger compared to the volume of the capillary, such that the effect due to capillary volume change is minimised. But this calls for a large bulb size—which increases its thermal capacity and thus would absorb greater amount of heat from the known high temperature surface/metal—and in consequence would reduce its temperature. Thus, it would introduce an error. Also, in such a case, the instrument response would become sluggish.

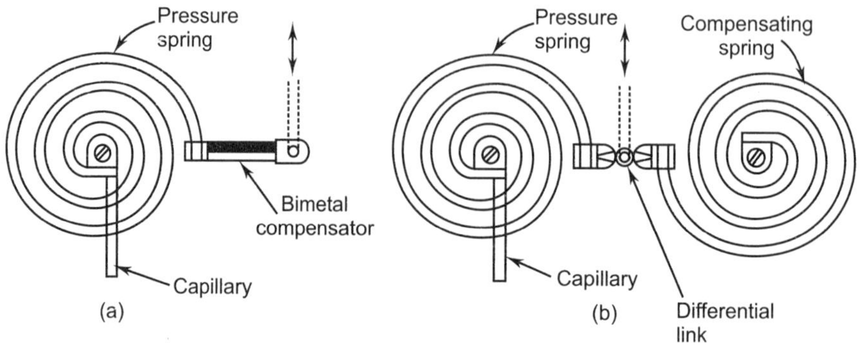

Fig. 6.9: Ambient temperature compensation technique at the receiving element

(*b*) Change in the ambient temperature at the receiving elements can be taken care of (compensated) by the methods shown in figure 6.9.
Figure 6.9 (*a*) employs a bimetal compensator while figure 6.9 (*b*) employs a second receiving element (called compensating spring.)

(*c*) Changes in the ambient temperature at the capillary and at the receiving elements can be compensated by employing the methods shown in figure 6.10.
In figure 6.10 (*a*), the compensating system consists of the compensating spring and a second capillary running adjacent to the measuring system capillary. The capillary belonging

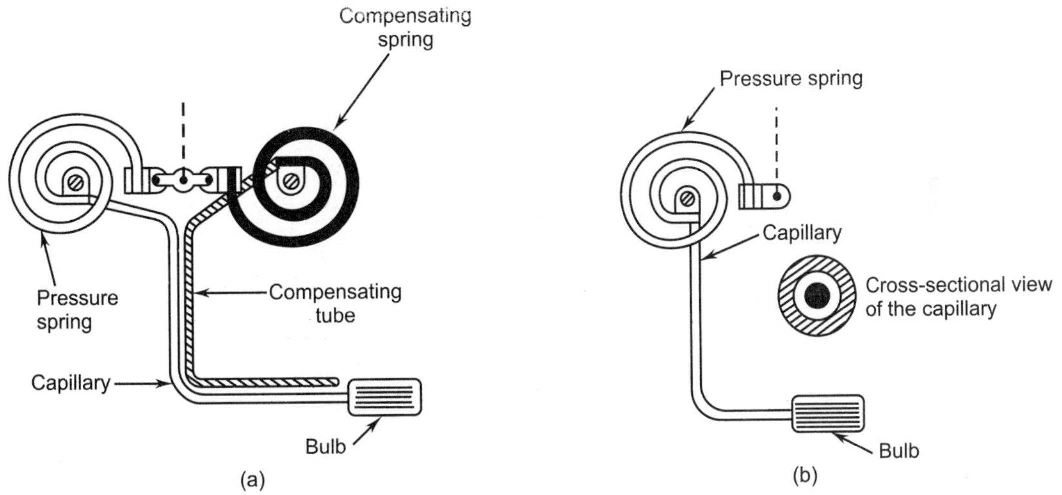

Fig. 6.10: Ambient temperature compensation along the capillary and at the receiving elements

to the compensating system is not connected to the bulb. The temperature compensation at the capillary and at the receiving element is mechanically compensated by connecting the pressure spring and the compensating spring in opposition.

Figure 6.10 (*a*) shows the ambient temperature compensation at the capillary and at the receiving element by introducing invar (an alloy having a very low coefficient of volume expansion) into the bore of the capillary. Mercury has a cubical coefficient of expansion which is around six times that of stainless steel. Thus, 5/6 of the bore volume of the capillary is occupied by invar and the the rest 1/6 by mercury. This arrangement, therefore, exactly compensates for the increase in capillary volume with temperature increase so that the pointer indication becomes a function of temperature variations at the bulb only.

The vapour pressure thermometer is not subject to ambient temperature correction because vapour pressure is dependent on the liquid surface temperature only. This liquid surface always remains in the bulb which is inmmersed in the process whose temperature is to be measured.

6.3.4.2 Barometric Effect

The pressure spring deflection is proportional to the difference between the internal pressure and the external barometric pressure. The instruments are calibrated for standard barometric pressure and hence when the barometric pressure changes, the instrument calibration does not remain valid. In the case of liquid and gas expansion thermometers, the internal pressure is considerably higher than the barometric pressure. Thus, any small change occurring in the barometric pressure would affect the calibration in a very small way and hence can be neglected. The vapour pressure thermometers, on the other hand, operate at much lower internal pressure and thus their calibrations get affected by changes in the barometric pressure. In some cases this error may not altogether be neglected. But compensation of this error is intricate and costly at the same time and thus not undertaken except in special cases.

6.3.4.3 Head or Elevation Effect

If there is any difference in level between the bulb and the pressure spring (the receiving element), an error, known as head or elevation error, will come into play as shown in figure 6.11. The recorded temperature will then be erroneous.

If the bulb is at a higher place than the pressure spring, this head of liquid column will add up to instrument reading and hence it will show higher temperature. On the other hand, if the bulb is at a lower point compared to the receiving element, the reading of the temperature indicator will be less than the true value of temperature.

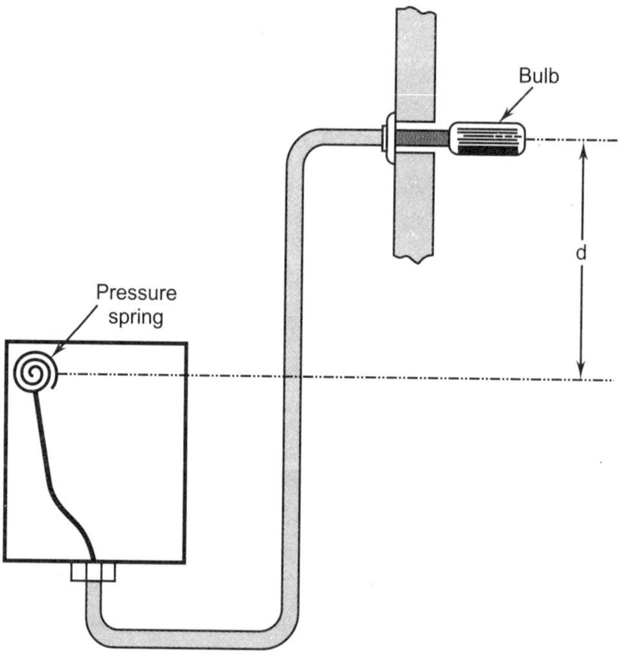

Fig. 6.11: Vapor-actuated thermometer installation and the head effect

In mercury-filled thermometers, an internal pressure of around 1000 psi is maintained. Thus, the error due to head effect is inconsequential and can be neglected. The head effect in gas expansion thermometers is also negligible because of two reasons—firstly the pressure of gas due to the head effect (*i.e.*, pressure created by '*h*' height of gas column) is inconsequential and secondly, the gas (normally nitrogen) in gas thermometers are normally filled at several hundred psi. On the other hand, the head effect in vapour pressure thermometers can sometimes becomes appreciable owing to the fact that internal pressure is not very high. To remedy the situation, calibration is done only after the bulb is installed in place, so that the head effect is eliminated altogether.

6.3.4.4 Immersion Effect

In case of mercury (*i.e.,* liquid) thermometer or gas expansion thermometer, an error due to immersion effect will cause the thermometer reading to be erroneous. It can be caused by:

(*a*) If the bulb is not fully immersed, fluid will expand for that portion of the bulb which is exposed to the unknown temperature. In this case, the thermometer will record a lower temperature and the error is roughly proportional to the unexposed part of the bulb.

(*b*) The bulb is fully immersed but not properly insulated. Then a part of the heat of the bulb is lost due to conduction through the extension neck and the thermal well.

So, proper care should be taken to overcome the immersion effect in liquid and gas thermometers.

The immersion effect does not affect a vapour pressure thermometer in a big way if the liquid surface within the bulb remains inside the process whose temperature is to be measured.

6.3.4.5 Radiation Effect

Error due to radiation effect occours in the measurement of temperature if there is a temperature difference between the thermometer bulb and the solid surrounding bodies. A radiation shield is housed around the bulb to minimise this effect.

6.3.5 Reproducibility

When ambient temperature compensation is taken care of in thermometers, the reproducibility or precision of the instrument becomes dependent on calibration drift. Some of the factors on which this drift depends are:

- Mechanical/thermal stress in the materials,
- Leakage from the thermal system,
- At higher temperatures leakage becomes more, and
- Fatigue and creep of metals.

6.3.6 Dead Zone

In industrial pressure thermometers, a minimum force is needed to overcome the (*a*) starting friction (*b*) lost motion between linkages. This minimum force would correspond to a minimum value of temperature change for this force to be developed. This is called the 'dead zone'. The mercury thermometer has a smaller dead zone compared to gas or vapour actulated thermometer because the former develops a larger force due to liquid expansion, unlike the other two.

6.3.7 Response of Thermometers

Thermometers are generally sluggish in their response. This response is governed by the heat transfer characteristics of the thermometer bulb, the well and also on the characteristics of the fluid in which the bulb is housed. The filling liquid, the bulb and the thermal well should be such that the thermometer can respond in a faster manner. Their main considerations are thermal capacitance, thermal conductivity and the surface area per unit mass.

The fluid surrounding the bulb also affects the speed of response, the main characteristics being thermal capacitance, thermal conductivity, film coefficient of heat transfer and the mass velocity.

For a given volume of the bulb, it should have a large area, a small mass, a small specific heat and a high thermal conductivity.

Although a thermal well protects a thermometer from possible mechanical damage, it, on the other hand, introduces a time lag so that the thermometer response becomes a bit sluggish. Figure 6.12 shows the response curve for a particular thermometer with and without a thermal well. The well is separated from the bulb by a space which offers a high thermal resistance. This is the cause of poor response characteristics of a thermal system that consists of a well. Again, as the process fluid velocity increases, the time lag decreases, as shown in figure 6.13.

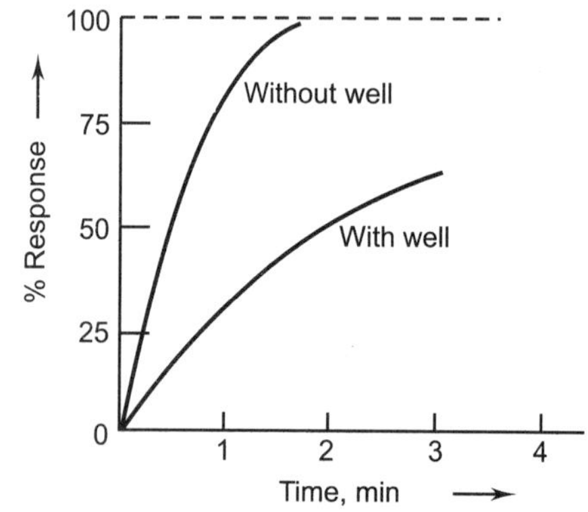

Fig. 6.12: Response characteristics of a filled-in system with and without well

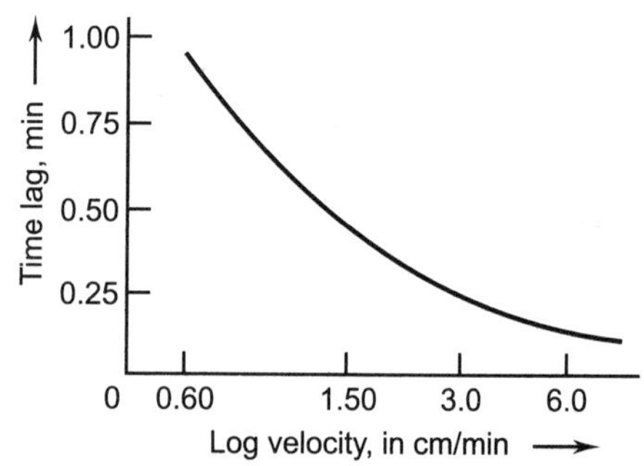

Fig. 6.13: Time lag vs. velocity curve in log scale

6.4 THERMOCOUPLES

6.4.1 Introduction

A thermocouple is an active electrical transducer consisting of two dissimilar conductors, which may be metals or alloys of metals or non-metals and shown in figure 6.14. The junction, exposed to the unknown temperature to be measured, is termed as 'sensing' or 'measuring' or 'hot' junction. The other junction is known as reference or cold junction whose temperature is maintained at either 0°C or at any other reference temperature. When a temperature difference is maintained between the hot and cold junctions, an emf is produced between them, causing a current, also known as thermoelectric current, to flow through the circuit.

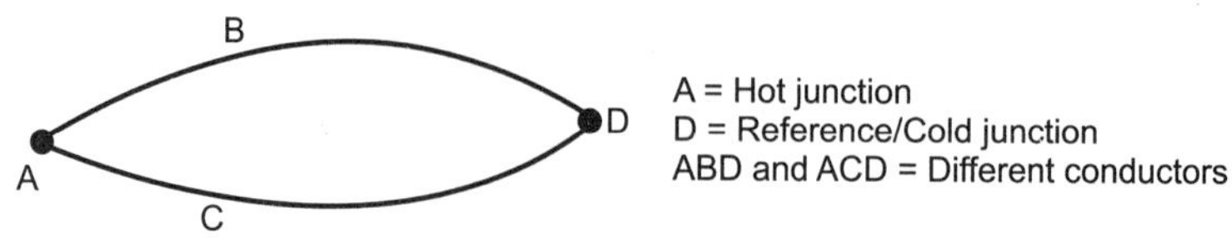

Fig. 6.14: A thermocouple

This current can be measured by placing a suitable instrument. The meter reading will then be proportional to the differential emf. of the two junctions and hence to the temperature difference between the hot and cold junctions. If the cold or reference junction is maintained at a particular value, the meter reading can be calibrated in terms of the temperature of the hot junction.

Since the emf developed between the two junctions is directly proportional to the temperature difference between them, hence measurements based on thermocouples are essentially 'differential' in nature.

The magnitude of emf (or thermal emf) is dependent on the (*a*) temperature difference between the two junctions as also (*b*) on the type of wire materials employed. The emf is given by: $e = \alpha (T_2 - T_1)$ where α = a constant, V/K, T_1, T_2 = cold and hot junction temperatures, K.

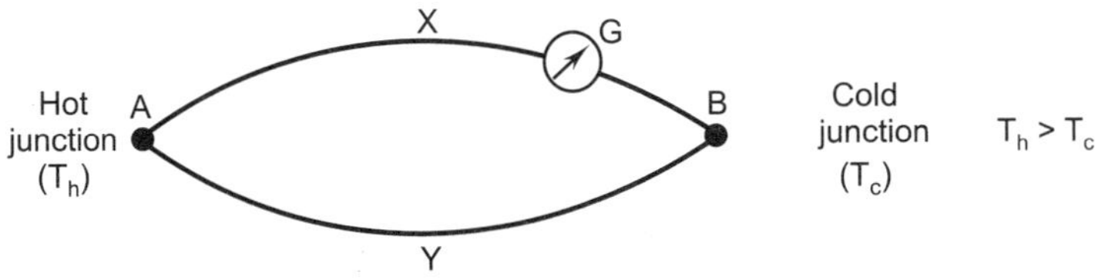

Fig. 6.15: A simple thermocouple circuit

Figure 6.15 shows a very simple thermocouple circuit consisting of two dissimilar metals *X* and *Y* forming two junctions—junction A is the hot junction and *B* the cold junction, maintained at temperatures T_h and T_c . *A* sensitive galvanometer *G* placed anywhere in the circuit will record a thermoelectric current proportional to the temperature difference ($T_h - T_c$). To note at this point that emf developed across the two junctions maintained at different temperatures is not affected by temperature gradients along either of the wires.

Thermocouples are economical and rugged at the same time. Because of their small size, the time constant is small and hence speed of response is fast. Thus, thermocouples are a good choice where speed of response is a major criterion for selection of a particular type of transducer.

Industrial thermocouples cover a very wide range as diverse from cryogenic temperature on the one hand to jet-engine exhaust temperature on the other. Temperature spanning from –270° to 2500°C can be measured by thermocouples.

Thermocouples are available with standard calibration charts and hence are easily replaceable. Another advantage of using thermocouples is their easy availability.

As the voltage available from thermocouples is in the *mV* range, care should be taken so that common mode signals must not interfere with the differential signal. Also noisy electric and magnetic fields must be prevented from affecting the electronic signal conditioning circuitry, the input to which is the output *mV* of the thermocouple. Electronic field radiated by a voltage source present in the vicinity of a thermocouple may be capacitive coupled to it giving rise to static noise. In such a case, capacitive grounded shield would eliminate such a static noise.

6.4.2 Characteristics and Properties of Industrial Thermocouples

An industrial thermocouple should have the following properties:

- A large thermo-emf for a given temperature range,
- Linearity between emf and temperature relationship,
- Precision of calibration,
- Immunity to corrosion and oxidation and atmospheric influences,
- Longevity,
- Repeatability, *i.e.*, properties must remain stable with time, *i.e.*, time-invariant behaviour,
- A high melting point,
- A low value of restivity,
- A low value of thermal coefficient of resistance,
- Properties must be repeatable during manufacture, and
- Stability at high temperatures.

Chapter 6

6.4.3 Laws of Thermoelectric Circuits

(*i*) Law of intermediate temperatures

The emf of a thermocouple between any two temperatures t_1 and t_3 is the sum of the emfs of the same thermocouple between t_1 and any intermediate temperature t_2 and between t_2 and t_3. In symbolic form:

$$e\Big|_{t_1}^{t_3} = e\Big|_{t_1}^{t_2} + e\Big|_{t_2}^{t_3}$$

As an example, the emf of a copper-iron thermocouple between 0° and 100°C is 1.21 mV and that between 100° and 200°C is 0.84 mV. Thus, the emf of the same thermocouple between the temperatures 0° and 200°C will be, as per the above:

$$e\Big|_{0°C}^{200°C} = e\Big|_{0°C}^{100°C} + e\Big|_{100°C}^{200°C} = (1.21 + 0.84)\ \text{mV} = 2.05\ \text{mV}$$

(*ii*) Law of intermediate metals

The emf of a thermocouple, consisting of two metals, remains unaltered if one junction is opened and either one metal or several different metals are inserted in between, provided that all the new junctions are maintained at the temperature of the junction which is opened.

For example, figure 6.16 (*a*) shows two metals *A*, *B* while figure 6.16 (*b*) shows three metals *A*, *B*, *C* and the junctions are maintained at temperatures as shown. Then from this law it can be concluded that both the figures would yield the same thermo emf.

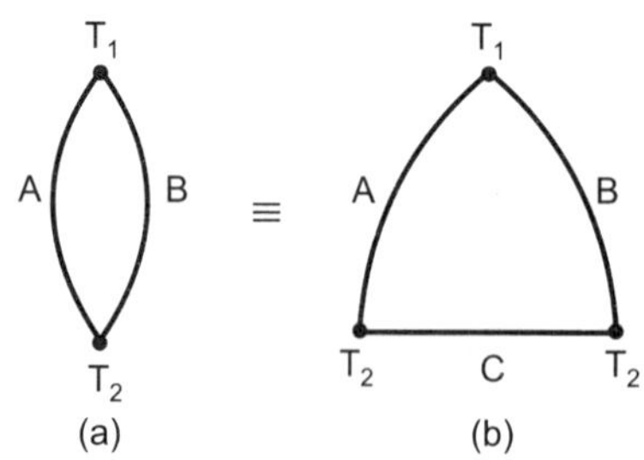

Fig. 6.16: Law of intermediate metals

6.4.4 Sensitivity of a Thermocouple

The slope of the thermo-emf vs. temperature curve gives the sensitivity figure of the given thermocouple. Thus, it is defined as:

$$\alpha = \frac{de}{dt}$$

and the cold junction is referenced at 0°C.

A high value of α is desirable since it indicates a large thermo emf for a small change of temperature. Again, since the emf vs. temperature relationship for any thermocouple is not linear in nature, the value of α is not constant for a couple for its entire range of operation.

Sensitivity for Type *J* and Type *S* thermocouples are roughly 0.05 mV/°C and 0.006 mV/°C respectively.

6.4.5 Temperature Measurement by Thermocouple

Every pair of dissimilar metals in contact forms a thermocouple. Now, a minimum of two contacts in series are required to have an electrical circuit. Thus, whenever temperature measurements are to be made using thermocouple circuits, care must be taken to offset/minimise the effects of such undesired/incidental thermocouples.

Figure 6.17 shows a circuit by which an unknown temperature can be measured using thermocouples. T_r represents the reference junction temperature and in some cases is maintained at ice (0°C) point. T_m is the temperature of the measuring junction. The voltage developed at the two temperature points T_m and T_r are V_{Tm} and V_{Tr} respectively. So, the output voltage is

$$V_{out} = V_{Tm} - V_{Tr}$$

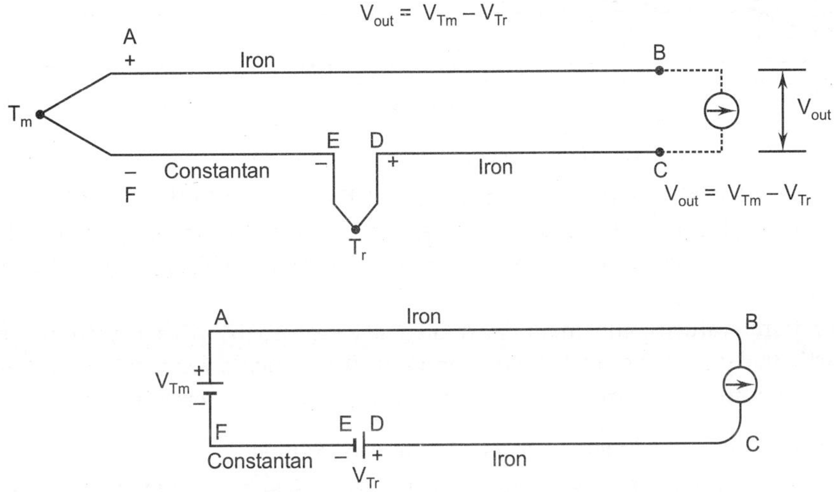

Fig. 6.17: Temperature measurement by thermocouple

Now, if the temperature of the hot (measuring) junction does not vary, V_{Tm} will remain constant, but V_{Tr} will vary if the reference junction temperature is not maintained constant. In such a case V_{om}, *i.e.*, output voltage available across the two points B and C will vary. In case the ambient temperature increases, V_{Tr} will increase, making V_{out} to decrease even if T_m remains constant.

On the other hand, if the reference junction temperature (*i.e.*, T_r) is maintained at a fixed value, V_{Tr} will remain fixed and in that case V_{out} will be solely dependent on V_{Tm}. Thus, it is obvious that V_{out} can be calibrated in terms of unknown temperature T_m.

6.4.6 Characteristic Curves of Different Thermocouples

Figure 6.18 shows the outputs of several thermocouples (in mV) with temperature (°C), the reference junction maintained at 0°C.

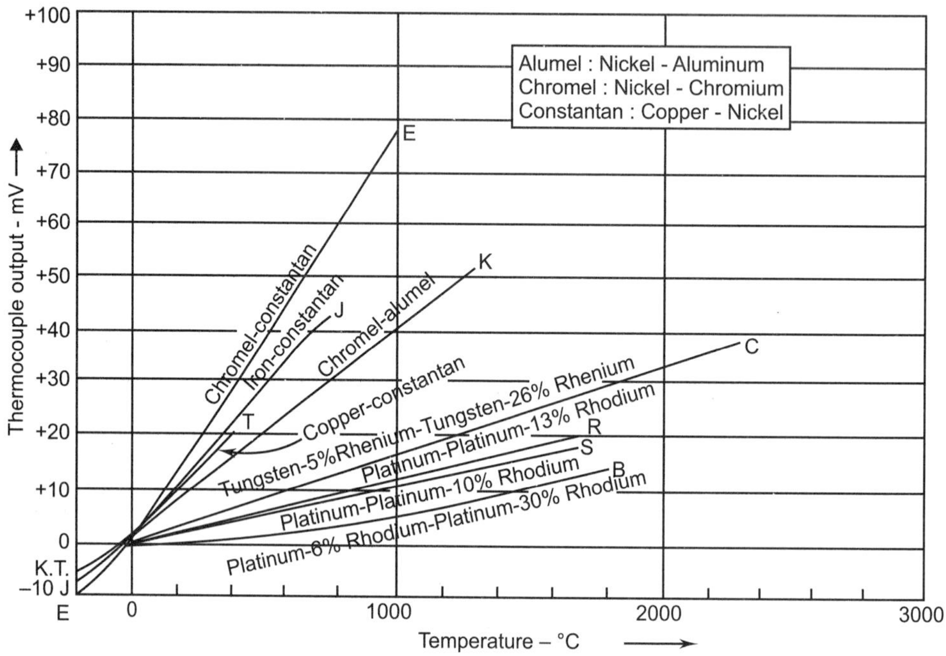

Fig. 6.18: Output characteristics of thermocouples

It should be noted at this point that the type of thermocouple that is to be employed for a particular situation depends on sevaral factors, *viz.*, temperature range, availability, cost, type of the environment in which the thermocouple is to be used (oxidising/reducing), accuracy requirement, etc.

At higher temperatures, the major part of heat received by a thermocouple is by way of radiation, particularly so if it is in still environment. If for such a furnace, the different parts are at different temperatures, thermocouple will indicate an average temperature.

6.4.7 ANSI and IEC 584 Designations of Thermocouples

Table 6.3 shows a number of commonly used thermocouples, their ANSI designations, typical useful temperature range and their voltage swing over the total temperature range.

Table 6.3: Thermocouple emf Vs. Temperature Range of Different Thermocouples

Junction material	*ANSI designation*	*Typical useful temperature range (°C)*	*Voltage swing over range (mV)*
Chromel/constantan	E	0 to 982	75
Iron/constantan	J	-184 to 760	50
Chromel/Alumel	K	-184 to 1260	56
Pt-Pt 13%/Rhodium	R	0 to 1593	18.7
Pt-Pt 10%/Rhodium	S	0 to 1538	16
Copper/Constantan	T	-184 to 400	26
Pt 6%—Rh/Pt 30%—Rh	B	38 to 1800	13.6
Tungsten 5%—Rhenium/ Tungsten 26%—Rhenium	C	0 to 2300	37

The following thermocouples, with compositions as shown, are covered by the international standard IEC 584 of 1995 and shown in table 6.4.

Table 6.4: Thermocouples and their Codes as per IEC 584 of 1995.

Thermocouples	*Codes*
Pt 10%—Rhodium/Pt	S
Pt 13%—Rhodium/Pt	R
Pt 30%—Rhodium/Pt	B
Iron/Copper-Nickel	J
Copper/Copper-Nickel	T
Nickel-Chromium/Copper-Nickel	E
Nickel-Chromium/Nickel-Aluminium	K
Nickel-Chromium-Silicon/Nickel-Silicon	N

6.4.8 Construction of Thermocouples

Depending on the type and requirements of use, several varieties of thermocouples are available in different mechanical constructions. These are:

- Straight plain wire thermocouple
- Surface Contact thermocouple
- Mineral insulated thermocouple
- Sheathed thermocouple
- Hot metal thermocouple
- Liquid metal thermocouple and
- Conical well thermocouple.

6.4.8.1 Straight Plain Wire Thermocouple

It is used when fastest response is desired but are susceptible to chemical corrosion and mechanical damage. Thus, this type of thermocouple is used in laboratories and protective environments.

6.4.8.2 Surface Contact Thermocouple

It is used for measurement of surface temperature like that of a pipe or tube or any particular point of interest in a plant. Since contact is very less, measurement is not very accurate. For higher accuracies, this type of thermocouple is embedded in a metal plate and the whole arrangement is clamped to the place of temperature measurement.

6.4.8.3 Mineral Insulated Thermocouple

Thermocouple wires are mineral insulated which are finely powdered and are compact ceramic or magnesium oxide etc.

The different types of MI thermocouples are shown in Figure 6.19.

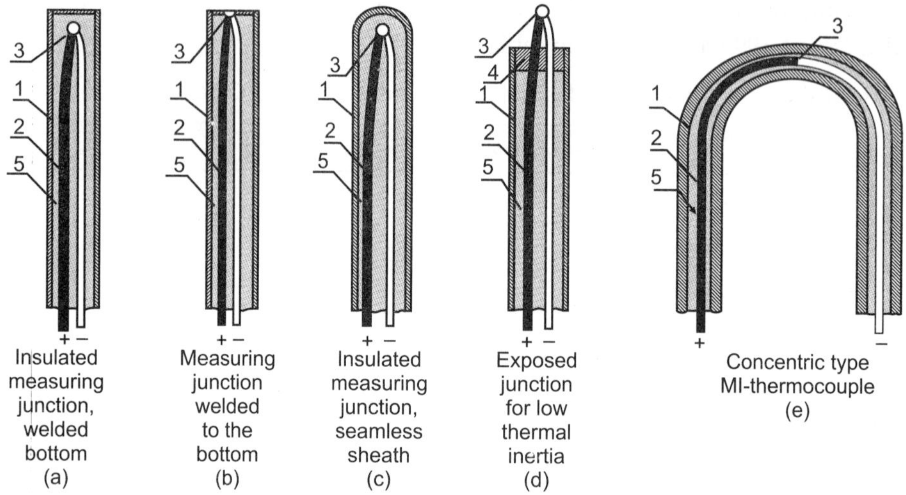

Fig. 6.19: Types of MI thermocouples: 1-metal sheath, 2-conductor, 3-measuring junction, 4-refractory cement and 5-mineral insulant

The advantages of MI thermocouples over the normal types are that they can be made as per requirements at the site and that the cables can be rolled and transported to the site very easily. This is of particular importance when replacements are intermittently needed.

Other advantages are: (*a*) low thermal inertia resulting in faster response (*b*) long working life (*c*) high electrical resistance between sheath and conductor (*d*) high resistance to corrosion (*e*) two measuring junctions in a single thermocouple ensures measurement of small temperature differences.

6.4.8.4 Sheathed Thermocouple

These are employed where the thermocouples can be replaced in practical plant situations. The sheaths are made of steel, inconel, ceramic etc., depending on the requirement. The response of sheathed thermocouples is obviously slower than the plain wire thermocouples. But its use is obviously an advantage in furances where temperatures do not change fast.

6.4.8.5 Hot Metal Thermocouple

These are used for measurement of hot metal temperature. The two thermocouple wires are connected to the two opposite ends of the solid hot metal slab.

Sufficient time is needed for this kind of temperature measurement but the advantage lies in the fact that the method is simple. Also, replacement of this kind of thermocouple is very very simple.

6.4.8.6 Liquid Metal Thermocouple

For measurements of hot liquid metal temperatures, like the temperature of molten steel in steel melting shop (temperature ~ 1640°C), liquid metal thermocouples are used. Another name of this kind of thermocouple is 'throw away thermocouple' *i.e.*, it can be used only once and then to be thrown off. To the expendable probe, the Pt – Rh/Pt thermocouple tip is attached. When dipped into the liquid steel for several seconds, it develops an emf which is consequently measured. After the measurement, the thermocouple is to be thrown off to be replaced by a fresh one for another round of temperature measurement. These are also called right-angle (corner) thermocouple. Their shape ensures that the junction head of the thermocouple does not 'see' the high temperature of the hot metal.

6.4.8.7 Conical Well Thermocouple

These are machined from solid bars of different varieties of stainless steel metals, etc. and are used for measurement of temperature in high pressure environment.

6.4.9 Industrial Thermocouples

The measuring junction of an industrial thermocouple must have high mechanical strength and low electrical resistance and may be formed or prepared by brazing, soft or hard soldering, arc, spot or flame welding.

The characteristic requirements of an industrial thermocouple have already been mentioned. A given thermocouple should conform to the shandard emf-temperature relationship over a period of time without any appreciable drift. Industrial thermocouples have very good repeatability and thus drift is really not a major area of concern. Thus, a damaged/broken thermocouple can be replaced with another one of the same type without going in for recalibration of the new one.

In order that a thermocouple has a long life, it should be immune to corrosion, contamination and oxidation. Chromel alumel thermocouple can be employed at even higher temperatures than the maximum range indicated in the table, under excess oxygen presence while the same is true of iron-constantan thermocouples in the absence of oxygen. Platinum thermocouples are contaminated at higher temperatures by hydrogen, silcon, arsenic, some metal vapours. Thus, it is recommended and sometimes used with gas tight thermal wells in which an oxidising atmosphere is maintained. Industrial thermocouples are available with different degrees of accuracy. Platinum, Platinum-Rhodium thermocouples offer the best accuracy (± 1/4 per cent), followed by iron-constantan (± 1/2 per cent) and lastly chromel—Alumel (±3/8 per cent).

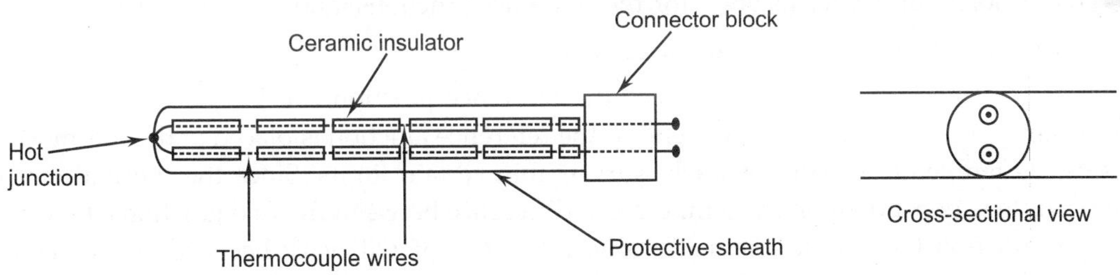

Fig. 6.20: An industrial thermocouple

Figure 6.20 shows the form of an industrial thermocouple in which the hot junction is open. The hot or measuring junction may be twisted, weld type or butt weld type. In the former type the two wires are twisted for several turns and then welded. This type is used for larger sized thermocouples that gives it a mechanical strength. An example is the measurement and control of blast furnace temperature using chromel-Alumel thermocouples. In the butt weld type the two ends are fused and are normally used for small sized thermocouples.

Industrial thermocouple wires are available in the market with different diameters. For example, chromel-Alumel wires are available in 1 mm, 2 mm, 3.2 mm diameter sizes. A diameter size of 0.45 mm for platinum thermocouple is used.

Normally, thermocouple wires are of considerable lengths, so that there is every chance of the two thermocouple wires to touch each other or with the sheath. To prevent this, twin-hole ceramic insulators are used through which the thermocouple wires are made to pass. At the end of the thermocouple, a fixing insulating head is there to which the two wires are terminated by means of fixing screws. Some convention is followed at the fixing head so that from the shape of the fixing screws, the positive and negative wires can be identified.

For temperature measurements in the range of 1500°C to beyond 2000°C, special type thermocouples, namely, Ir–W, W–(W–Re), Ir–(Ir–Rh) are used. For a thermocouple being used at the higher end of its ranges, it is recommended to use thicker wires, particularly so if the thermocouple is used for continuous temperature measurements. But thicker wires has a bearing on the speed of response—the thicker the wires the poorer the speed of response. This is because of obvious increase of thermal mass.

The physical characteristics of the two wires of a thermocouple may change in adverse industrial environments. In such cases, it is recommended to change the wires at regular intervals, the duration depends on the specific use in which the measurement is done.

6.4.10 Compensating Cables/Lead Wires

The thermocouples are installed at the site, whereas measurement of temperature is done in the control room. In most cases, considerable distance is involved in between these two places, sometimes to the tune of several hundred feet. Utmost care must be taken to connect the ends of the thermocouple wires (at the site) to the instrument (in the control room). These connecting leads are called 'lead wires' or 'compensating cables'.

There are two considerations which must be borne in mind while this connection is established with lead wires. These are:

- the problem of proper location for the reference junctions, and
- errors, if any, introduced by the lead wires.

There are obviously two places where the reference junction can be placed—either at the thermocouple end or at the instrument end. If the reference junction is placed at the thermocouple end side, its temperature will obviously vary within certain limits. Since the thermal emf of a thermocouple is dependent on the temperature difference between the two junctions, hence even if the hot junction temperature remains constant, this emf will vary because of the reference junction temperature variations. Thus, the reference junction is always placed at the instrument end side within the control room, where temperature variation problem is much less severe. If the control room is air-conditioned or maintained at a particular temperature, then the reference junction temperature can be maintained at a constant value.

The lead wires to be used for a particular type of thermocouple must have indentical emf-temperature characteristics over the range of use of the employed thermocouple. The lead wires are normally different from the thermocouple wires. Thus, obviously an error will be introduced on this count and hence utmost care should be taken in selecting the lead wires for a particular use so that the emf-temperature characteristics of the thermocouple and the lead wires match each other as closely as possible.

Normally the lead wires are stranded for flexibility. Four specific types of insulation and protective covering are used for lead wires, the types being (*a*) general purpose (*b*) high temperature use (*c*) moisture resistive type and (*d*) metal clad, corrosion resistive type. Of these, type (*d*) is most expensive.

Table 6.5: Lead Wires used for Different Thermocouples

Type of Thermocouple	*Type of lead wires*	*Temperature range in °C*	*Limit of error in °C*
'+' '–'	'+' '–'		
Copper-Constantan	Copper-Constantan	-60 to 95	± 0.83
Chromel-Constantan	Chromel-Constantan	-20 to 200	± 2.22
Iron-Constantan	Iron-Constantan	-20 to 200	±2.22
Chromel-Alumel	Chromel-Alumel	-20 to 200	± 3.33
	Copper-Constantan	25 to 95	± 3.33
	Iron—(Copper-Nickel alloy)	25 to 200	± 3.33
Pt — (Rh - Pt)	Copper—(Copper-Nickel alloy)	25 to 200	±6.66

Polarity consideration is an important aspect while connecting the lead wires to the thermocouple—wrong polarity connections would lead to very large error.

Table 6.5 shows the type of lead wires to be used with corresponding thermocouples, the temperature range of use of the lead wires and the limits of error.

6.4.11 Thermocouple Installation

A major area of concern is the proper place in the process at which a thermocouple should be installed so that the temperature indicated by the instrument gives a true and representative idea of temperature of the point of measurement.

Two considerations are there for placement of a thermocouple for fluid temperature measurement. If the fluid is static, the themocouple can be put at any place within the fluid. But for a flowing fluid, care should be observed not to place the measuring junction of the thermocouple at places of low velocity or where the fluid is stagnant in nature. This will ensure a true representative temperature of the fluid to be obtained and secondly, transfer lag will be reduced.

For a furnace of appreciable size, the inside temperature at different places within varies. In this case, the best location is selected after securing the temperature distribution at different places.

When hot and cold fluids are mixed in a pipe and the temperature of the mixed fluid is desired, care should be taken to place the thermocouple downstream at which complete mixing is ensured.

It may so happen that in some cases, the point of measurement of temperature at which the thermocouple is placed, is surrounded by hotter temperatures, such as higher flame temperatures, etc. In such cases, care should be taken to install the thermocouple in such a way that it does not receive heat from these higher temperature points.

Depth of immersion of a thermocouple is another area of concern and must be given proper thought. If the protecting tube is not inserted sufficiently enough into the process (whose temperature is desired) there will be heat transfer along both the tube as well as the conductors. This will thus lead to erroneous temperature measurement. The general industry rule is that the depth of immersion should be roughly 10 times the outside diameter of the tube to minimise this heat loss. When the hot junction is welded to the tube or else touches it, the depth of immersion could be reduced to 5–6 times the outside diameter of the tube. In some cases, this immersion depth is based on practical experiences.

6.4.12 Thermal Well (Thermowell) Considerations

Thermal wells are a protective cover for bare thermocouples to increase their longevity. A corrosive and oxidising atmosphere reduces the life of a thermocouple. Thus, a thermo well is used to reduce the effect of the above. Also, it provides protection against thermal and mechanical shock, etc. The thermal well should not contaminate the thermocouple and should be impervious to gases.

The exact shape of a thermal well may vary depending on the specific need, but normally it is made in the form of a tube with a closed end. Sometimes it is used with an open end which provides protection against mechanical shocks only. The total assembly consists of the thermocouple, the thermal well, head, a mounting flange and a connector for connecting the lead wires.

Sometimes two thermal wells are used—the secondary thermal well covering the primary one. Two Thermal wells are used when (*a*) corrosion is very severe (*b*) temperature is very high, exceeding, 1100°C (*c*) costly thermocouples are used—like [(Pt, Rh)–Pt]. Depending on the requirement, the secondary well may be made of metal or ceramic. The purpose of the secondary well is: (*a*) it prevents the sagging of the primary thermowell assembly, (*b*) protects the primary well to remain gas tight.

The properties which a thermal well should have are:

- high mechanical strength
- resistive to thermal and mechanical shock
- resistance to corrosion and oxidation
- resistance to gas leakage

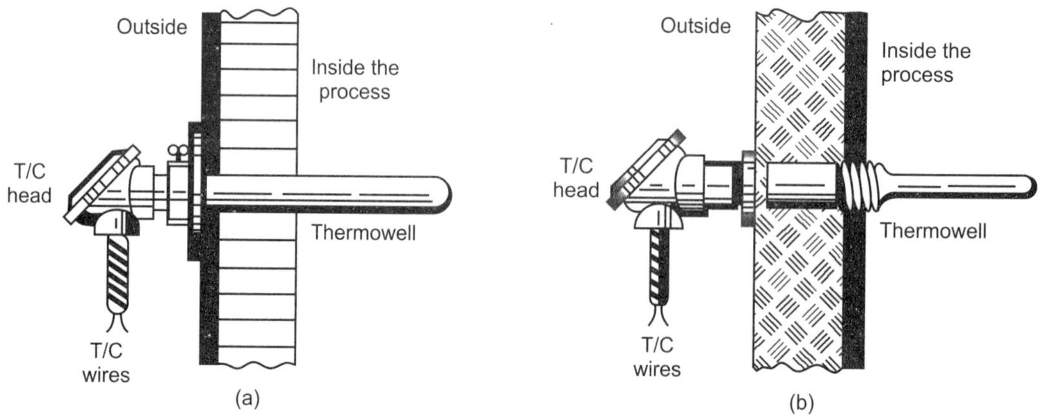

Fig. 6.21: Thermocouple installations with a single thermowell

Figures 6.21 and 6.22 show some of the techniques of how industrial thermocouples are used alongwith thermal wells. Figure 6.21 (*a*) is employed for measurement of furance gas temperatures, while the system shown in figure 6.21 (*b*) is employed for situations to measure temperature of pressure vessels and pipelines which are under pressure.

The assembly shown in figure 6.22 (*a*) (it is an open end thermal well) is employed for furnaces at lower temperatures. It protects the thermocouple from mechanical damage. Figure 6.22 (*b*) employs two thermal wells. Time constant of the system increases in this case.

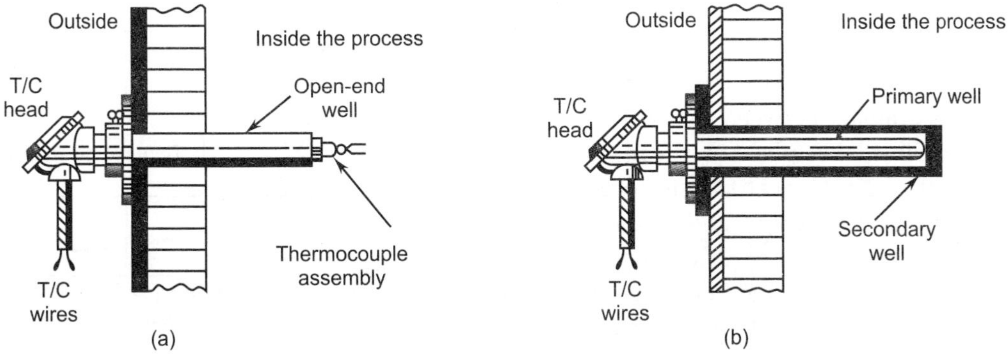

Fig. 6.22: Thermocouple installations (*a*) open-end thermowell, (*b*) Two thermowells, primary and secondary

6.4.13 Multiple Junction Thermocouples (Thermopiles)

Figure 6.23 (*a*) shows several thermocouples in series. It is used to increase measurement sensitivity when all the measuring point temperatures (T_M) are equal. Such an arrangement finds applications in IR pyrometers. Total emf available from the circuit is 3*e*.

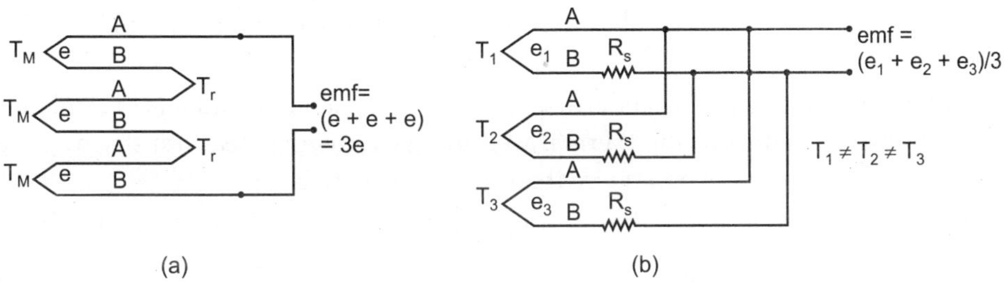

Fig. 6.23: (*a*) Series (*b*) Parallel thermocouples

The parallel combination of thermocouples, shown in figure 6.23 (*b*), is normally used to measure average temperature of a place. The different points at which the thermocouples are inserted are assumed to differ in temperatures slightly. This parallel combination is reduced to a single thermocouple when $T_1 = T_2 = T_3$.

For measurement of average temperature, a 'swamping' resistor R_S is put in series with each thermocouple to minimise the effects of unlike resistances in individual thermocouples and their lead resistances upto the point of parallel connection. R_S should have a large value compared to overall circuit resistance. Insertion of R_S prevents current flow between thermocouples when they are at different temperatures.

6.4.14 Speed of Response of Thermocouples

Thermocouples take a finite time to record the final voltage; this depends on whether a bare thermocouple is used or is used in conjunction with a thermowell. Secondly, it depends upon the conduction, convection and radiation heat transfer between the thermocouple and the surrounding medium in which the thermocouple is placed. Thirdly, it is dependent on the condition of the fluid - "still" or agitated condition, as in a flowing fluid the final record of temperature will obviously be much faster in which heat transfer by means of conduction plays a significant part. Figure 6.24 shows a typical thermocouple lag at various temperatures.

At lower temperatures (200° – 250°C), a major part of heat transfer between the thermocouple and the surrounding takes place by means of conduction between 250° – 330°C, heat transfer occurs by means of conduction, convection and radiation, whereas above this range most of the heat transfer takes place by radiation. At higher temperatures, thermocouple lag reduces because radiation heat transfer is much faster than the other two forms.

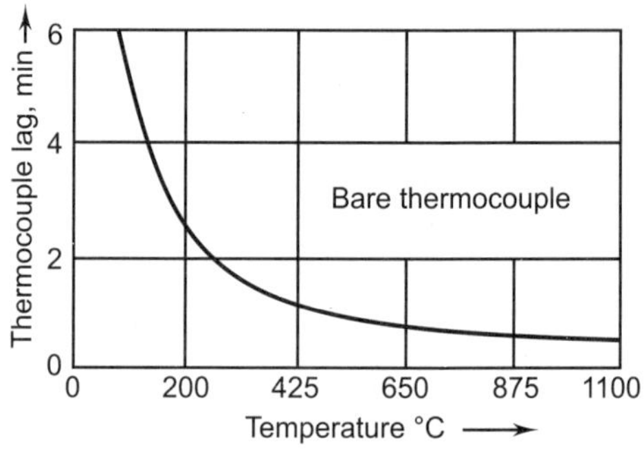

Fig. 6.24: Thermocouple lag at various temperatures for a bare T/C

The speed of response of a thermocouple is influenced by

- Thermal characteristics of the well, and
- The size of the thermocouple and the well

At lower temperatures, thermal capacity plays an important part since the film coefficient of resistance is high compared to the thermal resistance of the well material. Thermal capacity of ceramic is less than those of metals although the thermal resistance of ceramics is higher than that of metals.

At higher temperatures, heat transfer is mostly effected by radiation and it is the emissivity factor of the thermocouple as well as of the thermal well that governs this heat transfer by radiation. In general, oxidised metals have less emissivity than ceramic materials.

Thus, the speed of response of a thermocouple assembly depends on the above factors either at low or high temperatures.

For some translucent thermal well materials like glass, quartz or vycor (which is a type of glass), the speed of response is mostly dependent on the thermocouple materials because the above mentioned translucent materials transmit the heat instead of absorbing the radiation.

Figure 6.25 shows the speed of response of a 14 gauge Chromel-Alumel thermocouple when it is (*a*) bare (*b*) with a vycor thermal well and (*c*) with calorised iron thermal well. It is seen that the speed of response of the bare thermocouple is much faster than the other two.

Figure 6.26 gives a comparative idea about the response of a bare thermocouple, but of different SWGs.

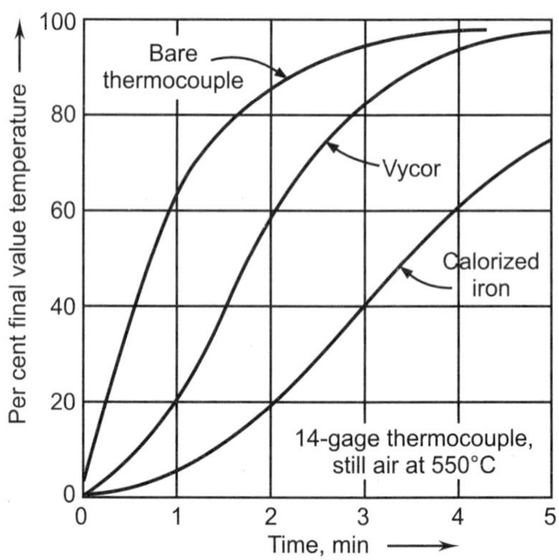

Fig. 6.25: Effect of thermowell on response of thermocouples

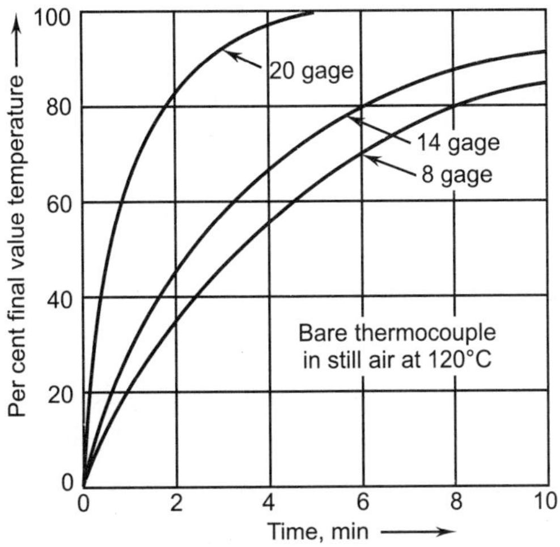

Fig. 6.26: Response of a bare thermocouple of different sizes

6.4.15 Radiation Error

Heat transfer by radiation becomes prominent at temperatures of 350°C or more. Radiation error occurs while trying to measure high air or gas temperature employing thermocouples. For a furnace whose different inside points are at different temperatures, it is likely that the temperature as recorded by the thermocouple would correspond to the percentage radiation of heat received by the different places which are at different temperatures. Thus, it is most likely that the temperature record would be anything in between the maximum and minimum temperatures of the furnace.

6.4.16 Time Constant of a Thermocouple

The time constant of a thermocouple is given by:

$$t = \frac{mc}{hA},$$

where, m = mass of the thermocouple junction
c = specific heat of the thermocouple material
h = heat transfer coefficient
and A = area along which heat transfer takes place.

The protection sheath introduces a second time constant and the thermal well a third. Thus the thermocouple time constant increases, delaying its response.

6.4.17 Measurement of Thermocouple Output

The mV output obtained from the thermocouple is to be measured and calibrated so that the scale calibration is a representation of the unknown temperature. There are several methods available to measure the output. These are:

- Millivoltmeter method,
- Potentiometer method, and
- Electronic method.

6.4.17.1 Millivoltmeter Method

Figure 6.27 shows a millivoltmeter arrangement for measurement of unknown temperature.

R_{th} = resistance of the thermocouple wires.

R_{ld} = resistance of the lead wires.

R_c = resistance of the millivoltmeter coil.

R_s = resistance in series with R_c

e_{th} = emf available from thermocouple.

e_m = voltage as measured by the millivoltmeter

T_h = temperature of the hot junction.

T_{amb} = temperature of the ambient.

It is obvious that:

$$e_m = \frac{e_{th}}{(R_{th} + R_{ld} + R_s + R_c)} \times (R_s + R_c)$$

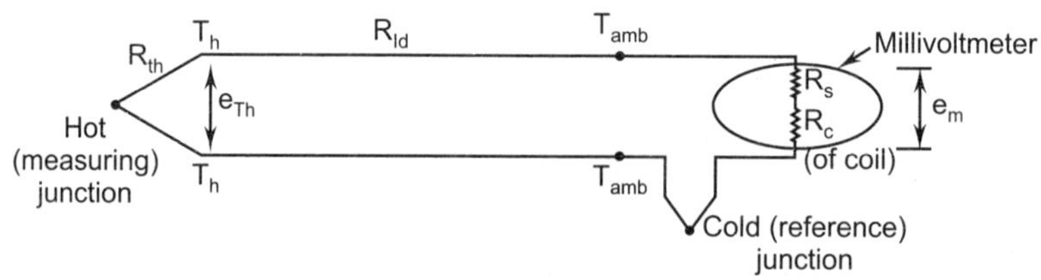

Fig. 6.27: Unknown temperature measurement by millivoltmeter method

Thus, e_m is less than e_{Th}. But the scale of the meter can still be calibrated in terms of the unknown temperature, for a given $(R_{th} + R_{ld})$.

$R_s + R_c$ should be as large as possible, to reduce the effect of any possible change in $(R_{th} + R_{ld})$.

In order to reduce the error caused by changes in external resistance (mostly R_{ld}), the resistance of the millivoltmeter should be as high as possible. Normally $(R_s + R_c) \sim 600\ \Omega$. Again normally, the meter is always calibrated with a fixed value of external resistance. The value of this resistance is 2Ω. Under such condition, the meter reading, which is in terms of unknown temperatures, becomes totally independent of any variation of lead resistance variation due to temperature change.

The meter calibration is affected due to a change in R_c. Since R_c is made of copper, its value changes with ambient temperature. This error is minimised by making R_S to be made of manganin. Another way out to eliminate this error is to choose R_S to have a negative temperature coefficient of resistance so that it offsets the positive temperature coefficient of the coil.

Reference junction compensation can also be achieved in the millivoltmeter by arranging a bimetallic strip. The bimetal strip is so arranged that the torque developed in the meter remains the same for a given hot junction temperature, even if the ambient temperature T_{amb}, varies.

6.4.17.2 Potentiometer Method

The potentiometer method of temperature measurement involving thermocouples is based on the well known method of comparing the unknown thermo-emf with a known emf. Under balanced condition, no current flows and hence this method is very accurate.

The circuit arrangement is as shown in the figure 6.28. A current passes through the resistors *X*, *Y* and the slide wire *S*, depending on the battery emf. The drop across X and the position of the slider in the slide wire is being balanced by the unknown thermo emf. When these are unequal, a current will flow through the galvanometer *G*, showing a deflection.

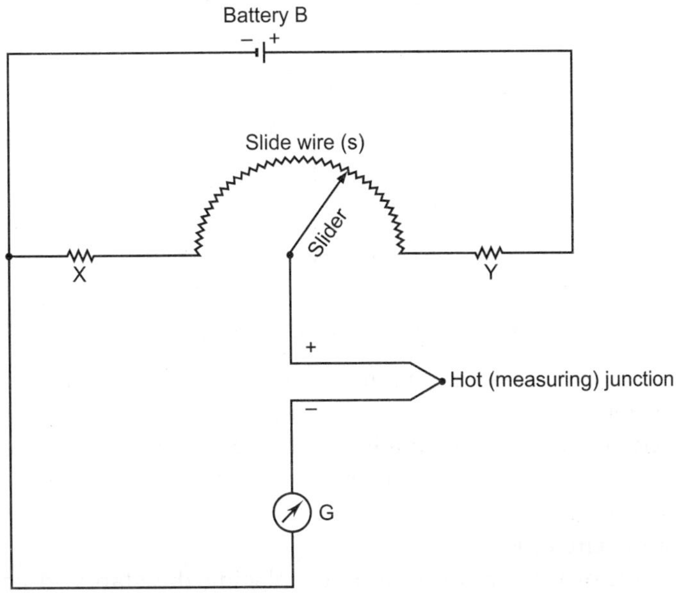

Fig. 6.28: Potentiometer method

This deflection in G can be either way, depending on whether the thermo emf is greater or less than the drop across *X* and up to the point of the slider in the slide wire.

A scale is attached to the slide wire S. The position of the slider on the scale can be calibrated for the emf of the thermocouple. Thus, for a thermocouple having a known emf-temperature relationship, the scale can be straightway calibrated in terms of temperature. Obviously, the scale has to be recalibrated if the thermocouple type is changed. Resistance *X* is used for calibration of the 'zero' on the slide-wire whereas *Y* is used to adjust the range of the instrument.

The drawbacks associated with this circuit are:

- The battery voltage, if varies, will lead to erroneous result.
- No provision to nullify the effect of reference junction temperature variation.
- Resistance changes with change of temperature has not been taken into account.

6.4.17.3 Improved Potentiometer Circuit

The problems associated with the earlier circuit are removed here, as follows, shown in figure 6.29:

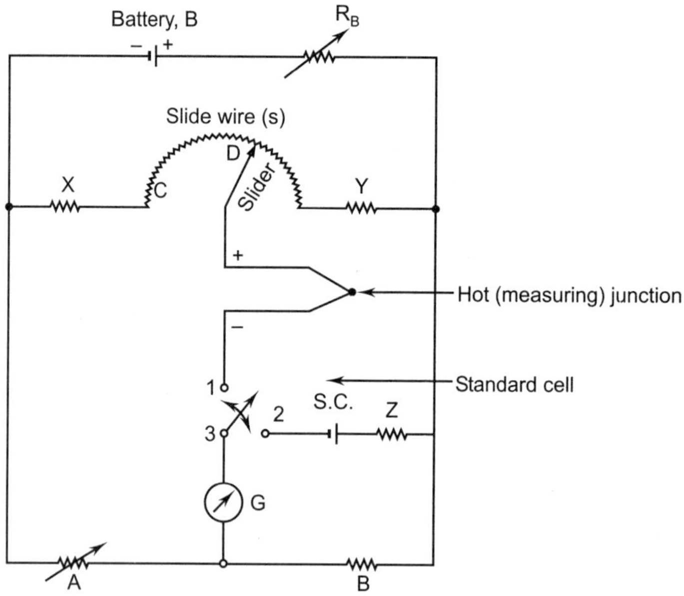

Fig. 6.29: Improved potentiometer circuit

- The emf of the battery *B*, even if it varies, does not lead to any problem. It is calibrated by means of the standard cell (S.C.) by varying R_B.
- Reference junction temperature variation is compensated for by the resistor A.
- All resistors employed are made of manganin—the temperature coefficient of resistance of the same is very small.

The operation of the circuit is as follows:

The standardisation of the circuit is done by placing the standard cell in the circuit by connecting the points 2 and 3. *Z* is used for limiting the standard cell current in such a way that unnecessary current drain through the circuit is avoided. The current through the resistors *X*, *Y* and *S* must be maintained at a fixed value when the standard cell is in the circuit with the galvanometer showing no deflection. This is achieved by adjusting R_B.

Once standardisation is over, the standard cell is isolated from the circuit by disconnecting it. Connection is now established between points 1 and 3. The scale attached to the slide wire is now calibrated for a particular type of thermocouple and a particular ambient temperature.

Reference junction temperature compensation is achieved by varying the resistor *A* either manually or automatically. The slider position on the slide wire *S* must remain at the same position even if ambient temperature varies. It is seen from the figure that the sum of the drops across *X* and CD (of the slide wire) must equal the sum of the drops across A and the thermo-emf of the thermocouple for null condition of G. Let the ambient temperature increases. Then the thermo-emf will decrease slightly. Hence, presently the sum of thermo-emf and the drop across A becomes less than the sum of the drops across *X* and the slide wire CD. Since slider position (here position D) must not vary for ambient temperature variation, hence A is increased a bit until the galvanometer again shows no deflection. Once this is done, since position of D is left untouched, thus it will show the same hot junction temperature despite the fact that the ambient temperature has changed (increased in this case).

6.4.17.4 *Electronic Method (with reference junction compensation)*

An electronic circuit for measurement of unknown temperature with automatic reference junction compensation is shown in the figure 6.30. IC 1 operates as a current source, whereas IC 2 and IC 3 operate as a summer and an inverting amplifier respectively.

If the ambient temperature increases, more current flows through **AD 590**, at the rate of 1 $\mu A/K$ and, at the same time, the thermo-emf decreases. Because of more current flow through **AD 590**, drop across R_1 increases. R_1 is so adjusted that the circuit output (at the point X) remains invariant with changes in the ambient temperature. Thus, the output voltage can now be processed and calibrated (at a particular temperature) in terms of unknown temperature for a given thermocouple.

6.4.18 Linearisation Technique of Thermocouples

The general form of emf-temperature relationship for a thermocouple is like this:

$$e = a + bt + ct^2 + dt^3$$

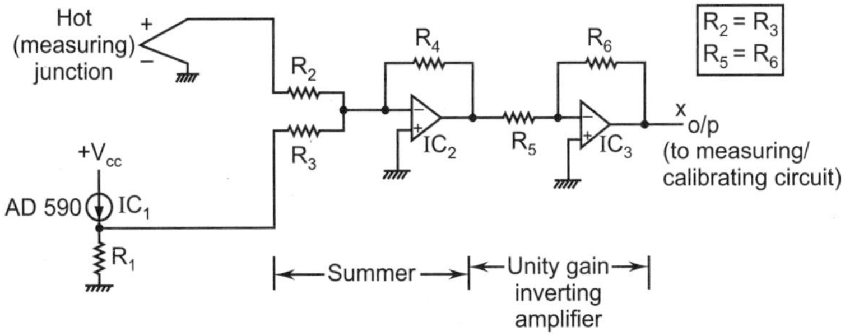

Fig. 6.30: Electronic method

where, e = thermocouple emf, volts

t = temperature of the measurement/hot junction

a, b, c, d = constants for a given thermocouple

As can be seen from the above, the emf-temperature relationship for a thermocouple is non-linear in nature. For thermocouples which are used over a small temperature range (of its total span), a single value of slope for that thermocouple in that temperature range was calculated and the relationship was utilised to be a linear one.

But for thermocouples which are used for a wide temperature range (of its total span), it is better to calibrate temperature indicator/recorder taking into consideration the above non-linear relationship.

There are several methods by which this non-linearity problem can be overcome.

These are:

(*a*) **Piece-wise (segmented) linearisation technique:** The non-linear emf-temperature curve is pieced into several parts, all having different slopes, but each part (portion) is assumed to have a single value of slope. For a given thermocouple the emf obtained for an unknown temperature to be measured, is fitted into the particular portion of the emf-temperature curve, with a particular value of slope. Then using op-amps, the unknown temperature is found out.

(*b*) **Look up table technique:** The second method involves a look-up table which stores the emfs and their corresponding temperatures for a given type of thermocouple. Then having a reference to this look-up table, the unknown temperature is noted from the emf obtained from a process.

(*c*) **Algorithm technique:** The third method involves an algorithm which involves the above non-linear relationship. This algorithm will then solve for the unknown temperature for a measured emf obtained from a process.

In both (*b*) and (*c*), the computer is to be fed with the knowledge about the type of thermocouple that is being used, before referring to either the look-up table or the algorithm.

6.4.19 Thermocouple Protection Against Open-Circuit

During operation, thermocouples often break due to natural wear and tear or for other causes. It will result in a low-temperature indication and consequently the controller will open the valve fully.

A simple thermocouple break protection circuit shown in figure 6.31 can prevent such a dangerous situation. Under normal condition, the break protection loop causes a very negligible voltage to be developed across the thermocouple junction. When the thermocouple breaks due to any reason whatsover, a voltage appears across the junction. This voltage is of such polarity which will drive the temperature indicator to a high setting.

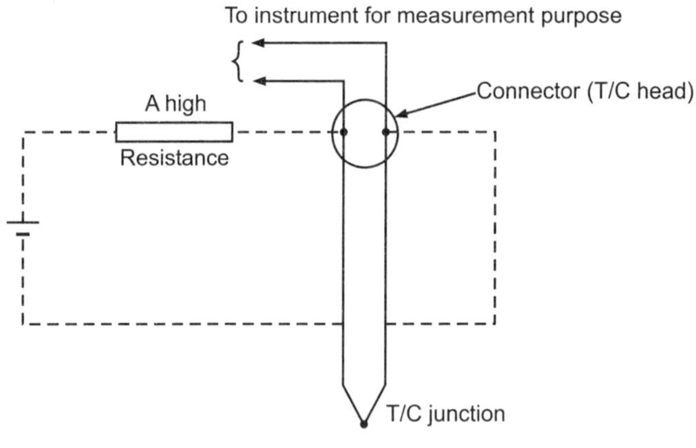

Fig. 6.31: Thermocouple break protection circuit

6.4.20 Special Thermocouple Probes

To suit particular needs of specialised industrial applications, special probes have been devised. Several such specialised probes are:

- molten metal probes — for hot metal temperature measurements
- hyperdermic probes — for insertion into rubber, etc.
- chisel probes — for hammering into frozen food, etc.
- pipe-clamp probes — for encircling pipes
- fast-acting probes for air-temperature measurement. These are surrounded by blackened spheres to measure radiation. But in such a case the time constant would be considerably increased.

6.4.21 Thermocouples for Measurement of High Temperatures

Different varieties of thermocouples cover different temperature ranges with Pt-Rh/Pt thermocouples able to cover temperatures well above 1000°C. In fact, Pt-Rh/Pt thermocouples are employed in Steel Melting Shop (SMS) where the temperature is in the vicinity of 1640°C.

A few non-standardised metallic or non-metallic thermocouples are employed for temperature measurements above 1000°C. Their emf vs. temperature characteristics are shown graphically in figure 6.32.

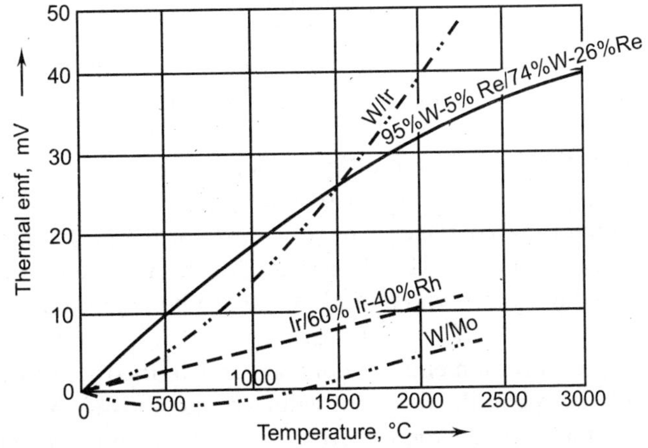

Fig. 6.32: Thermal emf E of high temperature thermocouples as a function of temperature

W/Mo and W/MoW thermocouples can measure temperature up to 2400°C in reducing atmospheres. Their disadvantages are low mechanical strength and an unstable emf values. For better mechanical strengths, W/WRe and WRe/Re thermocouples are employed, with different compositions of W and Re. A particular variety of the above can measure temperatures up to 2500°C. The W/WRe thermocouples are employed for high temperature measurements of exhaust gases in gas turbines, nuclear reactor cores, etc.

Graphite/W thermocouples can measure temperatures up to 1800°C in vacuum and 2500°C in reducing atmospheres. Graphite/SiC thermocouples can measure temperatures up to 1800°C in reducing atmospheres. Its major advantage is that it gives the highest emf output for a given temperature than any other thermocouple. Graphite/Graphite with 0.1% to 0.2% Be thermocouples can continuously measure temperatures up to 2000°C and 2600°C if used intermittently. Another thermocouple variety, namely, Ir/IrRh with different Rh proportions can measure temperatures up to 2000°C, but its drawbacks are an unstable emf vs. temperature characteristics, short lifetime and poor mechanical shock absorbing capability.

6.4.22 Conclusion

In conclusion, it can be mentioned that thermocouples are simple in their construction, their characteristics consistent, interchangeable and reliable. Also, standard output tables for various thermocouples are there and different thermocouples are available to fit into different temperature ranges.

But their low sensitivity, low output level and non-linear response contribute to the difficulties for applications of thermocouples for temperature measurement purposes.

6.5 RESISTANCE THERMOMETERS

6.5.1 Introduction

A very effective method of measuring temperature is to employ a resistance element. The principle involved is that the resistance of metals change with temperature and this change of resistance is calibrated in terms of temperature. Not only temperatures can be measured with great accuracy by employing RTDs (Resistance Temperature Detectors) but also very small incremental changes in temperatures (to the tune of 0.017°C) can also be discerned. Platinum RTDs are very stable in their performance and conform very closely to a standard curve (within 0.02°C–0.2°C).

In industries platinum, copper or nickel are employed as RTDs to measure the unknown temperature.

The relation between resistance and temperature is given by:

$$R_t = R_o(1 + \alpha t + \beta t^2 + \gamma t^3 + \ldots\ldots)$$

where, α, β and γ are temperature coefficients of resistance for the particular resistance material. Their values are positive for a metal RTD. R_t and R_o are the resistance values at $t^o C$ and 0°C respectively. β and γ are normally very small and hence can be neglected. Thus:

$$R_t = R_o\,(1 + \alpha t)$$

The coefficients α, β, γ, etc., for a metal are not constants in the true sense but depends on its purity. In fact, for a particular metal the different values of the above is a measure of the amount of impurity content in the metal. Thus, the material of the RTDs must be very very pure so that it corresponds to the actual standard/recommended temperature coefficient of resistance values. Thus, it should be ensured that the RTDs are not exposed to oxidation or contamination which, in its turn will alter the coefficient values.

When the temperature of hot body changes from t_1 to t_2, resistance changes from R_1 to R_2 and we have $R_2 = R_1 + R_o\,\alpha(t_2 - t_1)$ from which $t_2 = t_1 + \dfrac{R_2 - R_1}{\alpha R_o}$.

Although copper has a much lower value of resistivity compared to nickel or platinum, its stability and a highly linear nature of resistance vs. temperature curve makes copper RTDs to be a very strong contender for temperature measurement purposes at or near ambient.

6.5.2 Characteristic Properties of Industrial RTDs

Industrial RTDs should have the following characteristic properties:

- High resistance to oxidation and corrosion,
- Adequate ductility,
- Linearity of resistance—temperature relationship; this reduces circuit complexities to measure/record temperature.
- A high value of temperature coefficient of resistance: It will ensure small incremental changes in temperature values to be measured. It also ensures high sensitivity.
- Electrical characteristics of RTDs must not vary with time: This ensures repeatability and no recalibration of the measured instrument is necessary. In fact, the RTDs are properly aged/ annealed before being put to use.
- Sensitivity around 0.004/°C for platinum, 0.005/°C for nickel.

- A high value of resistivity. We know that $R = e.\frac{l}{A}$. Thus, if e is high, it means that for a given length (l) and cross-section (A) of the wire of the RTD, a higher value of resistance will be obtained. Thus, the physical size of the RTD will be small.
- Must possess enough mechanical strength. This will ensure a very fine wire RTD can be made. Thus, for a given length, the resistance value will be high since cross-sectional area can be made very small.
- Secondly, because of reduced mass of RTD, time constant will be low and faster response obtained. Although RTDs can be used up to a temperature of 1000°C, for industrial use it is limited to a much lesser temperature of 400°C. This is because at higher temperatures, distortion or strain of the resistance wire on its support may take place, thereby affecting measurements.
- Stability in physical properties.
- Resistance vs. temperature relationship should remain constant with no hysteresis.
- Easy reproducibility of the metal with identical properties.

Again, for measuring very small temperature differences, balanced bridge type of measurements are preferred over deflection type measurements because the former type provides more precision than the latter.

6.5.3 Comparison of Different Industrial RTDs

Table 6.6 shows the relative merits/demerits of several RTDs which are used industrially.

Table 6.6: Comparison of different RTD Materials

Material	*Temp. coeff. of resistance, α (Ω/Ω/°C) × 10^3*	*Resistivity e (μΩ cm)*	*Minimum wire diameter (cm)*	*Melting point (°C)*	*Density (gm/cm³)*	*Tensile strength (psi)*	*Range of use °C (approximate)*
Platinum	3.92	9.83	0.00508	1775.5	21.48	18,000	-250° to + 700° C
Copper	4.30	1.56	0.00508	1083	8.9	30, 000	- 200° to +250° C
Nickel	6.70	6.38	0.00508	1455	8.9	1,20,000	-100° to +350° C

Figure 6.33 indicates the variation of resistance with temperature for a few commonly used RTD materials. The graphs show that platinum is most linear in its resistance vs. temperature characteristics, while for copper and silver the resistance variation with temperature is less linear. For nickel the curve is in two parts and its characteristics is decidedly non-linear in nature.

Although very low temperatures are not normally encountered in industrial operations, RTDs can be used at temperatures as low as –200°C. The resistance vs. temperature curves are very much altered at very low temperatures because of superconductivity effects.

Resistance - temperature relationship for platinum RTDs is given by the well-known Callender equation:

$$t = \left(\frac{R_t - R_o}{R_{100} - R_0}\right) \times 100 + \delta\left(\frac{t}{100} - 1\right)\frac{t}{100}$$

where R_t, R_{100} and R_o are resistances at $t°$, 100° and 0° celsius respectively; δ is a characteristic constant of platinum.

The accuracy of the Callender equation pertains to the fact that it is in very good agreement with the thermodynamic scale with no correction needed between 0° – 1000°C.

The following points give an idea for the wide acceptance of RTDs as a means of temperature measuring device within their application range:

- Low to moderate cost,
- Easy availability,
- Easily replaceable,
- Simplicity of operation,
- Accuracy is high,
- Repeatability is very good, and
- Very small changes in temperature deviations can be measured.

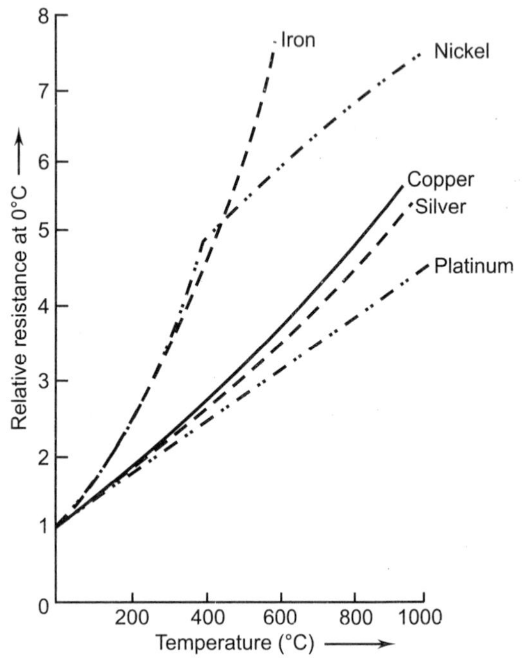

Fig. 6.33: Relative resistance (R_t/R_{ref}) versus temperature for some pure metals

6.5.4 Considerations for Industrial RTDs

Various forms of industrial resistance temperature detectors are available in the market. Considerations which go into the making of an industrial RTD are as follows:

- The coil must be wound non-inductively.
- Since electrical resistance changes with strain, utmost care must be taken so that as the wire expands or contracts on account of temperature increase/decrease, it is not strained. Only then the changes in resistance would be a true representative of changes of temperature.
- Coil should be so formed that a very high rate of heat transfer takes place. This enhances process response, and consequently time constant decreases. The resistance wire must be aged and annealed so that repeatability over a considerable period of time is ensured.
- The resistance wire must be pure—It will ensure a unique value of temperature coefficient of resistance. Secondly, even if all parts of the resistance bulb are not at the same temperature, no localised emf would develop.

6.5.5 Construction of RTDs

Typical construction of an RTD should be such that it is small in size and consequently improve the thermal conductivity. Almost always a protective sheath (well) is used to prevent mechanical shock, but delayed response is obtained. Bare RTDs are used when the medium is dry air. Figure 6.34 shows different constructional varieties of RTDs.

For general purpose temperature measurements, a perforated mica frame is used on which the wire is wound so that somewhat it takes the shape of a round coil. This is shown in figure (*a*). A rectangular shaped frame, shown in figure (*b*), is used in which the wire is clamped between two mica plates, making it very compact. Sometimes a metal spring frame is used alongwith it to improve heat transfer. Thus, this is a very fast response type. The spring acts as a shock absorber and as such it is used in vibratory environments. Still, a third type, shown in figure (*c*) is used for surface temperature measurements and goes by the name strain gauge (mesh) type. The RTD element is in shape of woven-wire mesh cloth. Figure (*d*) show a mineral insulated RTD (MIRTD). Its connecting leads are embedded in compact powder of Mg in a steel sheath. These are normally used in chemical and power engineering.

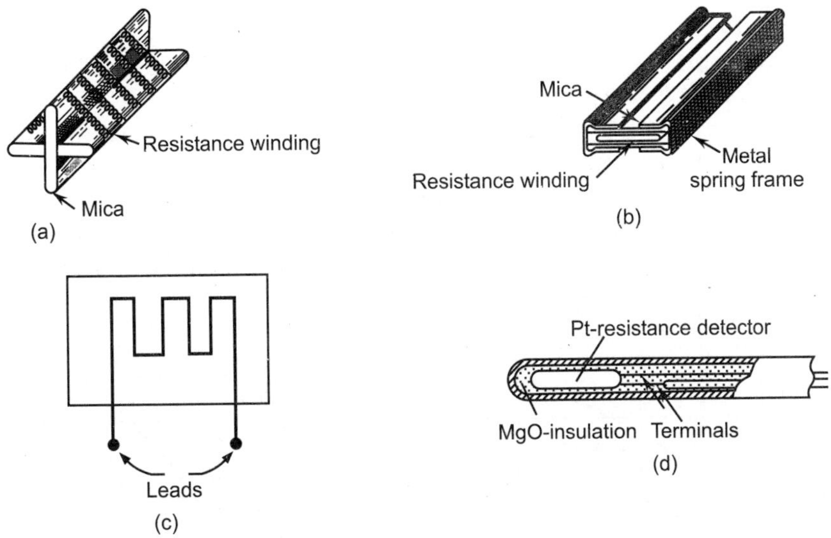

Fig. 6.34: Different RTD types: (*a*) General purpose type, (*b*) Rectangular type, (*c*) Strain gauge (mesh) type, (*d*) Mineral insulated type

Resistance thermometers may be tip type or stem type. Tip type RTDs have less thermal mass so that it response time is low compared to the other type. For stem type RTDs, heat transfer takes place along the stem on which the wire is round. Obviously, its response time is more because of involvement of more thermal mass.

Industrial resistance thermometers have values ranging from a fraction of an ohm to several hundred ohms. 46 Ω and 100 Ω (at 0°C) platinum RTDs and 53 Ω (at 0°C) Cu RTDs are common in the market.

One area concern is the connection of the RTD element to the rest of the circuit. This should be done to avoid (*a*) contact resistance and (*b*) thermoelectric effects. If the resistance value of the RTD employed is of considerable magnitude, the error that may creep into because of the above two mentioned effects are greatly reduced.

6.5.6 Comparison between Copper and Platinum RTDs

While Comparing the relative merits and demerits of industrial copper and platinum RTDs, the following gives a basis based on which a particular RTD (*i.e.*, either copper or platinum) may be suitable in a given situation.

- Copper is relatively much less costly than platinum.
- Copper has poor resistance to corrosion.
- Resistance-temperature curve for copper is more linear than platinum.
- Temperature range over which a copper RTD can be applied is somewhat more limited compared to platinum RTD.

- Copper has a low value of resistivity, compared to platinum. Thus, a copper RTD will be bigger in size than its platinum counterpart, given the same value of resistance. The bigger size of copper RTD will lead to slower response.

6.5.7 Response of RTDs

The speed of response of industrial RTDs is not a fixed quantity. It depends on many factors like whether the RTD is a bare one or has a well around it for protection purposes. Again, it is a function of the surrounding fluid medium – air or water. Again, response time of RTDs varies for static air or air having some velocity.

Figures 6.35 and 6.36 show the response curves of RTDs for the cases as indicated on the curves themselves. 'Dead' air causes a significant delay because the thermal resistance of the air space in between the RTD and the well is quite appreciable.

Precision, repeatability and accuracy of RTDs are better than thermocouples. One of the major contributory reasons for this is the absence of 'reference' junction in the case of RTDs.

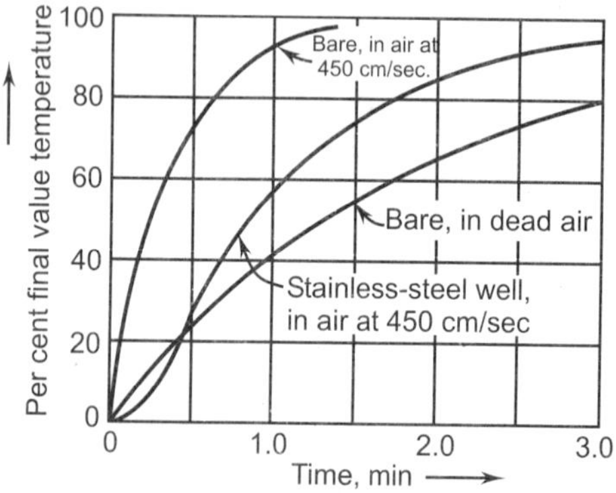

Fig. 6.35: Response of resistance thermometer in air at 90°C

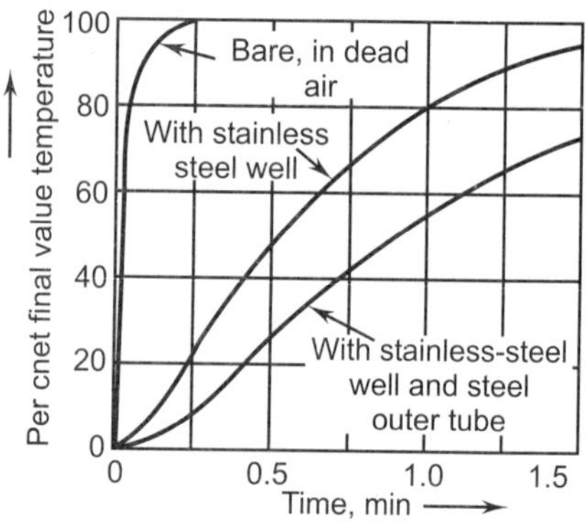

Fig. 6.36: Response of resistance thermometer in agitated water at 65°C

6.5.8 Measurement of Temperature by RTDs

6.5.8.1 Introduction

Resistance temperature detectors (RTDs) use some form of bridge configurations to measure unknown temperature. Two common methods are used to measure the bridge unbalance (which arises due to temperature variations): (*a*) the deflection and (*b*) the null method. In the first method, deflection is calibrated in terms of the unknown temperature. Null method employs potentiometric arrangements and obviously a better one than the deflection method of temperature measurement. Sometimes, there is considerable distance involved in between the point of measurement (where the RTD is mounted) and the place where the temperature is recorded. These two ends are connected mostly by copper conductors. These leads have some resistance values and should be of very small value compared to the resistance of the RTD. Also, ambient temperature variations affect the resistance values of the lead wires.

Thus, methods should be adopted so that both the lead resistance and its variation with temperature are nullified in such a way that they do not affect measurement accuracies. This is achieved by using extra lead wires placed in the opposite sides of the bridge.

Another aspect which should be accorded due importance is that the current through the RTD should be kept to a small value [around 2 to 20 mA] so that self-heating (I^2R loss) of the same is avoided to the extent possible. The error arising out of self-heating of the RTD can be minimised by (*a*) pulse excitation of the bridge or (*b*) taking a higher value of resistance for the employed RTD. In such cases, the resistance variation of the RTD can be claimed to be due to the variation of the unknown temperature only, which is to be measured. However, it should be noted at this stage that self-heating effect of RTDs is employed in certain applications, such as flow meters, anemometers and psychrometers, etc.

6.5.8.2 A Simple Wheatstone's Bridge Circuit

Figure 6.37 shows the simplest form of temperature measurement by RTD employing a wheatstone's bridge circuit. The battery provides the supply to the bridge and the galvanometer is employed to achieve null under balance condition. A and B are two fixed resistors. S is an adjustable slide wire. The resistance of the RTD is R (varies with temperature) and may be made of platinum, nickel or copper, depending on the application. Apart from RTD, all other resistances are made of manganin so that these resistances are least affected by temperature changes.

With variation in temperature of the place in which the RTD is housed, the null point is disturbed. It is restored by varying the slider C on the slider-wire S. The position of the slide wire is calibrated in terms of the unknown temperature.

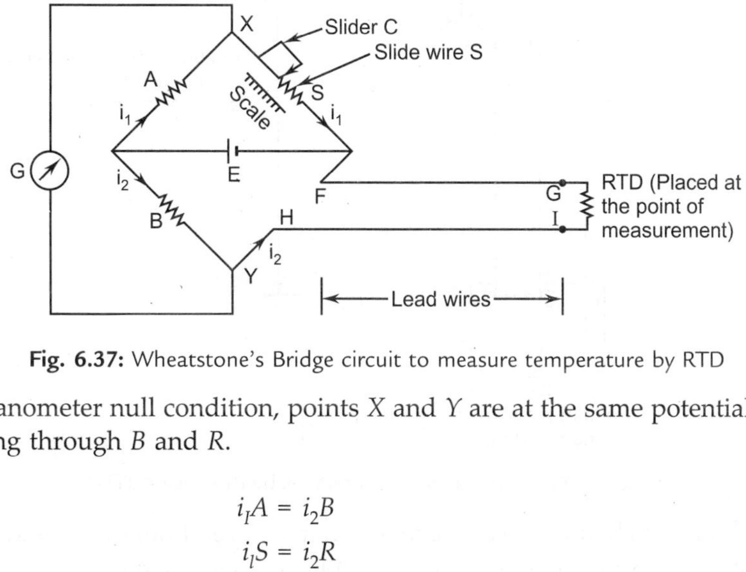

Fig. 6.37: Wheatstone's Bridge circuit to measure temperature by RTD

Under galvanometer null condition, points X and Y are at the same potential. Then a current (say i_2) is passing through B and R.

Then,

$$i_1 A = i_2 B$$

and,

$$i_1 S = i_2 R$$

Thus,

$$\frac{B}{A} = \frac{R}{S} \quad \text{or,} \quad \boxed{R = S.\frac{B}{A}}$$

Adjusting the slider C on the slide wire S gives the balance of the bridge. The scale can be calibrated in terms of temperature.

The drawbacks associated with this circuit are as follows:

The RTD is connected to the bridge by means of wires here, (FG and HI), which may be of considerable length and sometimes running into a few hundred metres. These lead wires are in series with the RTD. Also with temperature variation along the lead wires their resistances vary. This problem therefore affects bridge balance.

Contact resistance between slider and the slidewire remains in one arm of the bridge. Depending on wear and tear, dust, etc., this contact resistance value may be significant.

Joule heating (because of current flow) in the RTD increases its temperature. Hence, the temperature record shown on the scale will be higher than the actual one.

These problems are eliminated in three and four wire RTD configurations.

6.5.8.3 Siemens 3-lead RTD

Procedure: The circuit connections is shown in figure 6.38 and is due to Sir W. Siemens who designed it in 1871. The RTD is housed at the place of temperature measurement while the bridge circuitry is in the control room. It is a three lead RTD having three leads EF, BG and HG, all having same length l and cross-section. This l is the distance between the point of measurement

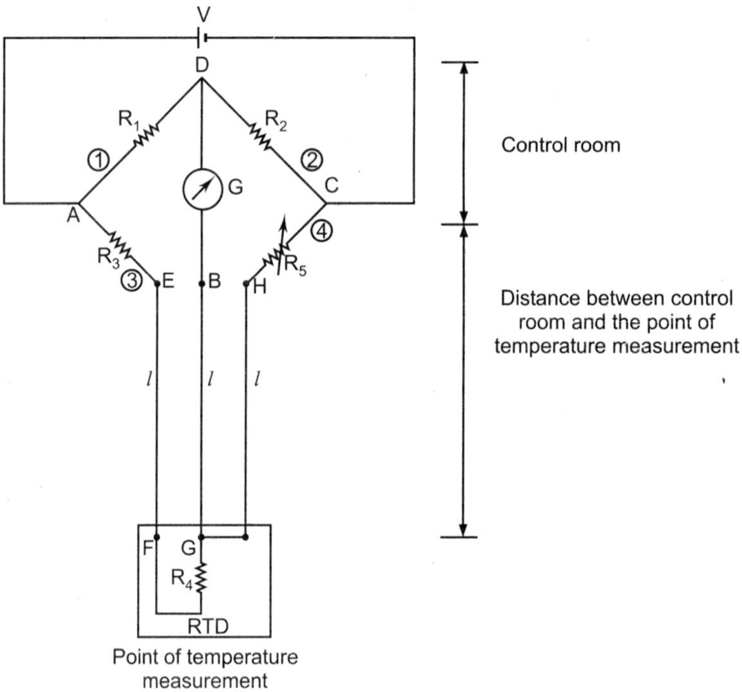

Fig. 6.38: Temperature measurement by three lead RTD

and the control room which, in some cases, may run into several hundred metres. The RTD is in the third arm [(3)] of the bridge with which one l (=EF) is in series alongwith R_3. Another l (= GH) is in the fourth arm [(4)]. Since (3) and (4) are in opposite arms of the Wheatstone's bridge, the effect of these two l's (*viz,* EF & GH) cancel out because any temperature variations along these two leads affect (3) and (4) identically and thus their effect is nullified. This leaves the third l (= GB) which is in series with the detector, and hence its resistance variation due to temperature change does not affect/influence the temperature measurements.

In practice, R_1 is made equal to R_2 and the ratio of $(R_3 + R_4)$ to R_5 is made equal to that of R_1/R_2 at a particular temperature. Any variation in RTD (*i.e.* R_4) results in unbalance and R_5 can calibrated in terms of temperature.

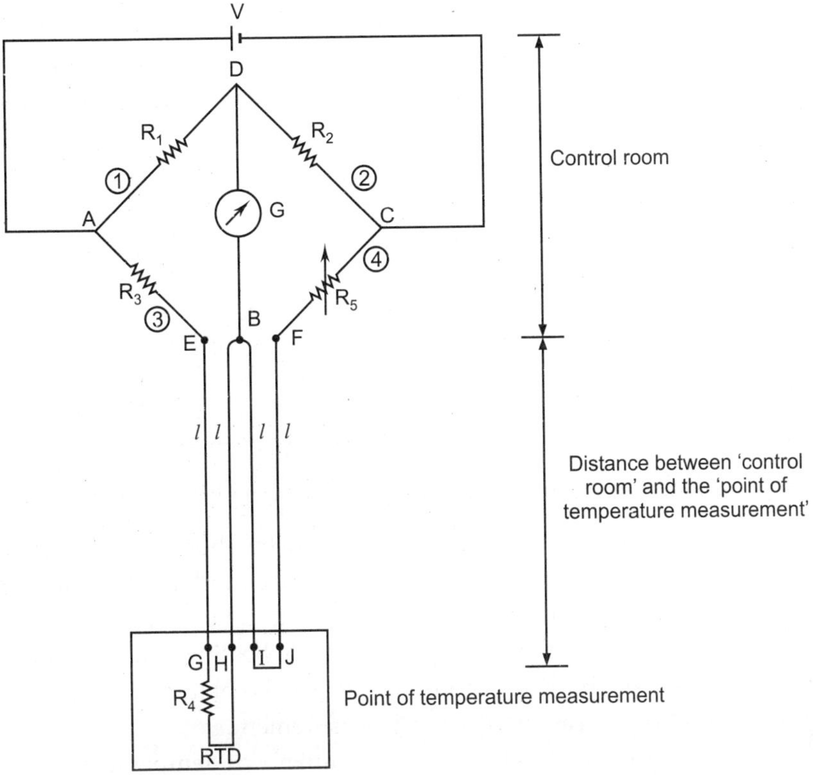

Fig. 6.39: Callender's four lead method of temperature measurement

Since the lead resistance BG lies in the galvanometer arm, the resistance of this arm is increased. This reduces the sensitivity of measurement. Also, as environment temperature varies, the resistance BG of this lead wire varies. This also affects sensitivity.

6.5.8.4 Callender's 4-lead RTD

It was introduced by Callender in 1886 and shown in figure 6.39. This is a 4-lead RTD, the leads being EG, BH, BI and FJ. All the leads are l in length and having the same cross-section. The first two leads, *i.e.* EG and BH are in the third arm [(3)], while leads BI and FJ are in the fourth [(4)] arm. Since the arms (3) and (4) are ratio arms, thus any temperature variation along the four leads affect both these two arms identically (in the form of resistance variation) and hence lead resistance variation due to temperature changes is nullified. Here lead wire compensation is achieved, but at the cost of employing one additional lead wire compared to 3-lead RTD.

In this method, none of the leads are in the galvanometer arm. Thus, in this case, galvanometer sensitivity is not impaired as it is in the case of 3-lead RTDs.

Use: Callender's method is very useful for differential temperature measurements in which two RTDs are used, one in arm (3) and the other in arm (4). Referring to the figure, one RTD is placed across the points G, H [*i.e.* in arm (3)] and the second RTD is placed across the points I, J [*i.e.* in arm (4)] and both the RTDs are two terminal in nature.

6.5.8.5 Pulse Excitation Technique

A pulse excitation technique, shown in figure 6.40, is an improved method to measure the temperature of several points in a process. It gives less error for the same current being passed through the RTDs. In the figure, three RTDs must have the same value so that a single calibration remains valid for all the points.

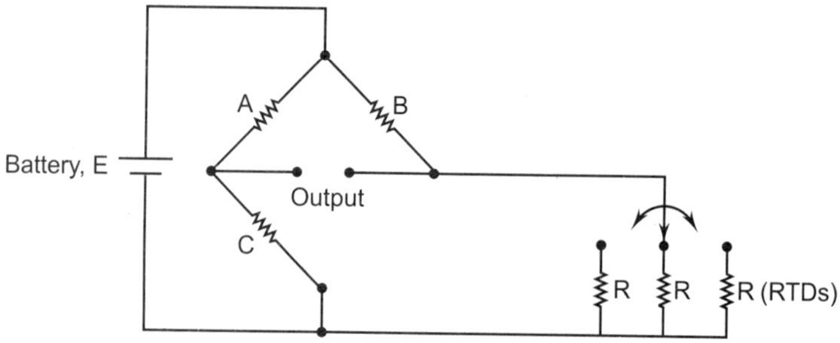

Fig. 6.40: Pulse excitation method to measure temperature at multi points

In the pulse excitation method, the RTDs are excited by short duration pulses, instead of constant excitation. The less the heat developed in a RTD due to Joule heating (I^2Rt), the better. If the time duration during which an RTD is excited is reduced by a factor of 16, then the current through it can be increased by four times, keeping the Joule heating at a constant level. For more current being passed through the RTD, more output will be obtained for the same unknown temperature. This, thus, will increase accuracy of measurement.

Thus, it is a time-sharing method to measure unknown temperature of several points employing only a bridge and offering more accuracy at the same time.

6.5.8.6 Average Temperature Measurement

To measure the average temperature of a furnace or of a surface, the method shown in the figure 6.41 is employed. The number of RTDs to be employed depends on volume of the furnace or the area of the surface and also on the accuracy with which the averaging is required. The more the number of RTDs employed the more will be the unbalance for a given temperature, thereby increasing accuracy.

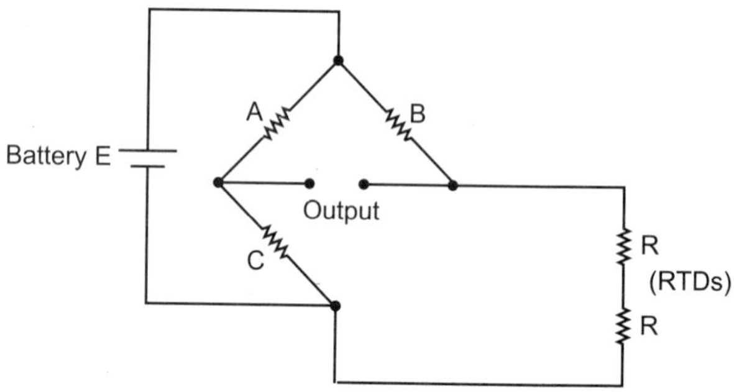

Fig. 6.41: Average temperature measurements

6.6 THERMISTORS

6.6.1 Introduction

Thermistors (also known as thermal resistors) are resistance type temperature transducers formed of solid semiconducting materials. Two types of thermistors are available — NTC (negative temperature coefficient) and PTC (positive temperature coefficient), although the NTC type is the one mostly used.

A thermistor is a resistor (variable) whose resistance value changes with temperature in an exponential manner. Commercially available thermistor sensors can cover a temperature range of – 100° to 450°C but sometimes the upper limit can reach upto 1000°C – but at such high temperatures their stability and accuracy are reduced. The upper operating limit of temperature for a thermistor is dependent on the physical changes in the material of the thermistor while the lower operating limit is determined by the resistance of the thermistor reaching such a high value that it cannot be measured by standard methods. The temperature range indicated above, of course, cannot be covered by a single thermistor — at least 4 to 5 different thermistor versions are needed, obviously for reasons of non-linear temperature vs. resistance relationship. The variation of resistance with temperature is much higher than in case of metals. Thus, it leads to very high sensitivity and they are the best choice in high resolution measurement and control applications. It is possible to control temperature within very close limits. A temperature change, as low as 0.001°C, is detectable with the right choice (*a*) of device and (*b*) portion of the non-linear temperature vs. resistance characteristics. Sometimes, the trade-off of using a thermistor for a particular application lies in between high resolution and non-linearity.

A thermistor is highly temperature sensitive—a resistance change of 10% per °C is common and in certain cases it may be even higher.

6.6.2 Different Characteristic Curves of Thermistors

There are three important characteristic curves for a thermistor. They are:

(*i*) Resistance-temperature characteristics.
(*ii*) Voltage-current characteristics.
(*iii*) Current-time characteristics.

Figure 6.42 shows the resistance-temperature characteristics curves of two typical thermistors, alongwith platinum. It is seen that within the temperature range of –100° to 400°C shown on the characteristics curve, the resistance of type A thermistor changes from 10^7 to 1 Ω-cm while that of platinum varies by a factor of 10 only.

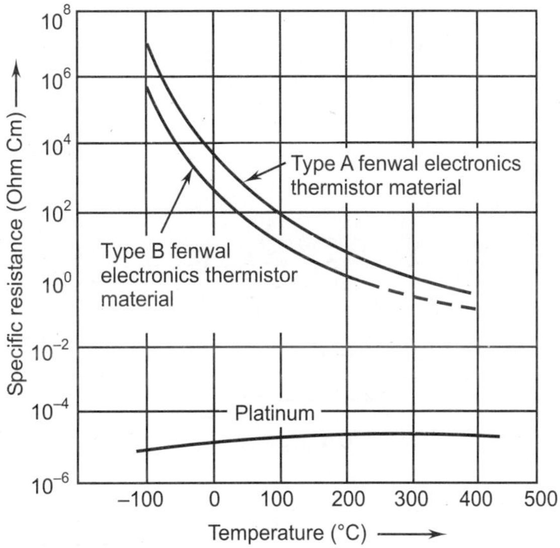

Fig. 6.42: Resistance-temperature characteristic curves of two thermistors, alongwith platinum

6.6.3 Advantages/Disadvantages of Thermistors

A thermistor temperature sensor has the following advantages and disadvantages:

Advantages:

- Very high temperature sensitivity.
- Thermistor dimension is very small; thus low thermal inertia, faster response and localised temperature measurement possible.

- High value of sensor resistance; thus, less affected by connecting wire (lead resistances) resistances.
- Small temperature difference measurements can be done.

Disadvantages:

- Highly non-linear resistance vs. temperature relationship, resulting in non-linear temperature scale.
- Temperature measurement range not high.
- At higher temperatures, susceptibility to permanent decalibration.

6.6.4 Construction of Thermistors

A thermistor is composed of sintered mixture of metallic oxides, such as manganese, nickel, cobalt, copper, uranium and iron. The sinter is compressed into desired forms and shapes and heat treated. The electrical connection is made in two ways: (*a*) the lead wires are embedded firmly into the material before heat treating is done (*b*) wires are attached to the plating of the thermistor. The various forms available are bead, probe, rod, disc or chip type and are shown in figure 6.43.

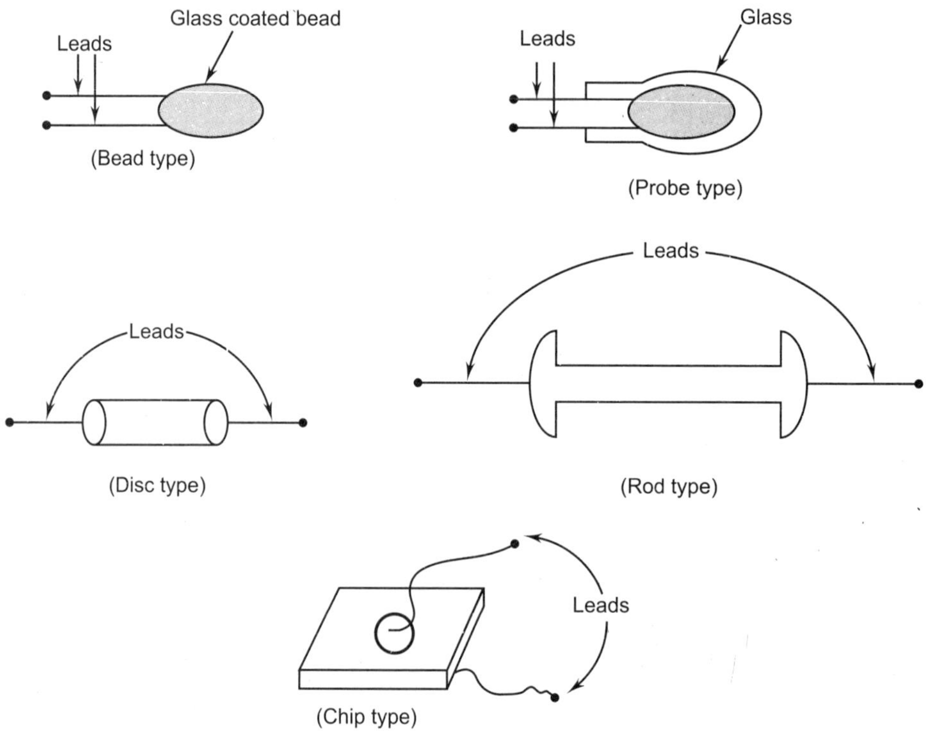

Fig. 6.43: Construction of different types of thermistors

Bead type is the smallest in size with diameters in the range of 0.15 to 1.25 mm, disc type diameters in the range of 5 to 25 mm, rod type diameters in the range of 0.8 to 6.0 mm with a length of around 50 mm. Amongst the many types of thermistors available, the one that is best suited for a particular application is determined by factors such as speed of response, nature of curve (which decides both accuracy and sensitivity) and also power dissipation.

Thermistor resistance values are mostly specified at 20° or 25°C. The following empirical relationship governs the resistance vs. temperature variation of a thermistor:

$$R = R_o \exp\left[\beta\left(\frac{1}{T} - \frac{1}{T_o}\right)\right]$$

where, R is the resistance of the thermistor at T K,

R_o is the resistance of the thermistor at reference temperature T_0 K,

β is a characteristic temperature constant for the thermistor.

The temperature coefficient of resistance is given by:

$$\alpha = \frac{dR/dT}{R} = -\frac{\beta}{T^2}$$

with $\beta \simeq 4000$ K, $T = 298$ K, $\alpha = -0.045$ K

The value of α for platinum is the region of 0.0036 per K. This obviously points to more sensitivity (about 10 times) for thermistors than platinum.

A thermistor can also be closely approximated through the Steinhart - Hart equation:

$$\frac{1}{T} = A + B \log_e R + C (\log_e R)^3$$

where T is the temperature in K, R is the thermistor resistor; A, B and C are curve fitting constants. These three constants can be found out from available data curves and solving three simultaneous equations.

6.6.5 Time Constant of Thermistors

This may vary from less than a second to a few minutes, depending on the size and the material of the thermistor. Time constant is given by:

$$t = \frac{ms}{hA}$$

For semiconductor materials, specific heat (s) is low, giving rise to a low value of time constant, *i.e.* the system is fast responding. Also, a low value of specific heat means that it takes little heat away from the process whose temperature is to be determined. Thus, it leads to precise temperature measurement possible.

6.6.6 Measurement of Temperature by Thermistors

Thermistors are used for detection of small temperature differences, measurement and control of temperature, etc. These are now discussed below.

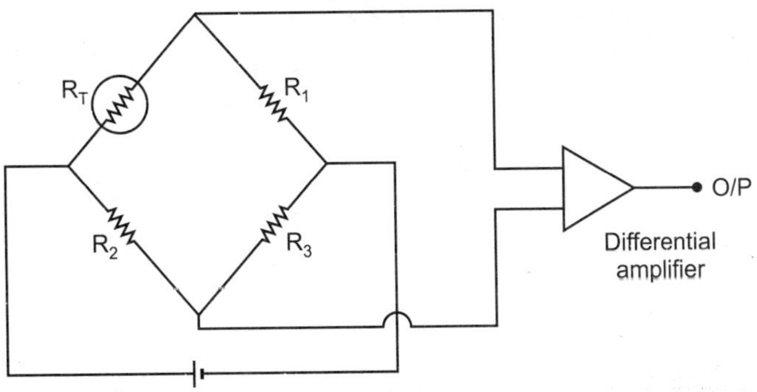

Fig. 6.44: Detection of small temperature difference by thermistors

6.6.6.1 *Detection of Small Temperature Difference*

The circuit shown in figure 6.44 can detect very small changes of temperature, about some reference temperature. R_T represents the resistance of the thermistor, while the other three resistances are made of manganin. The bridge is balanced if

$$\frac{R_T}{R_2} = \frac{R_1}{R_3}$$

The resistance change of the thermistor with temperature can be very safely assumed to be linear since very small change in temperature measurements is sought.

6.6.6.2 *Microammeter Circuit*

It is a simple series circuit in which the thermistor is in series with the microammeter and shown in figure 6.45. The μA is first calibrated in terms of known temperatures so that the scale of the μA becomes a direct reading one for measurement of unknown temperatures.

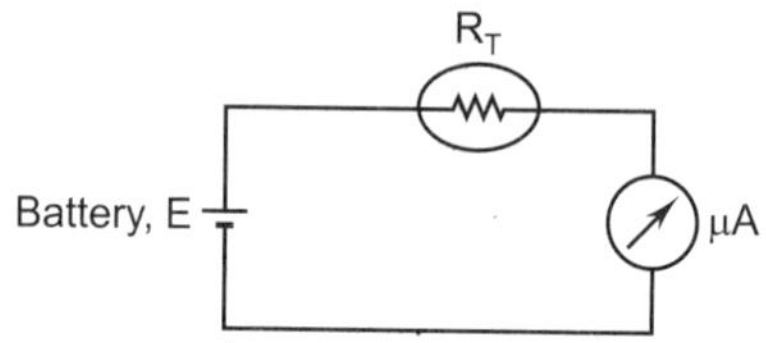

Fig. 6.45: Microammeter circuit to measure temperature

6.6.6.2.1 *Temperature Compensator*

The microammeter of the above circuit, which is calibrated in terms of unknown temperature, has a copper coil inside it. As current is passed through it, its resistance will increase due to Joule heating (I^2Rt loss) or else due to ambient temperature increase. Thus, it will lead to erroneous reading. To compensate for this, a temperature compensator circuit is used as shown in figure 6.46. The compensator has a thermistor paralleled by a shunt. This combination (called a compensator) is put in series with the microammeter (shown by the copper conductor in the figure). The compensator is placed very near to the copper coil so as to be subjected to identical temperature variations as that of the copper coil. What is ensured is that the negative temperature coefficient of the compensator is made equal to the positive temperature coefficient of the coil so that when curves *A* and *B* are added up, we get curve (C) — the compensated copper coil resistance. It is seen from the figure that compensated copper (*i.e.*, the copper conductor along with the compensating network) resistance almost remains constant with temperature variations.

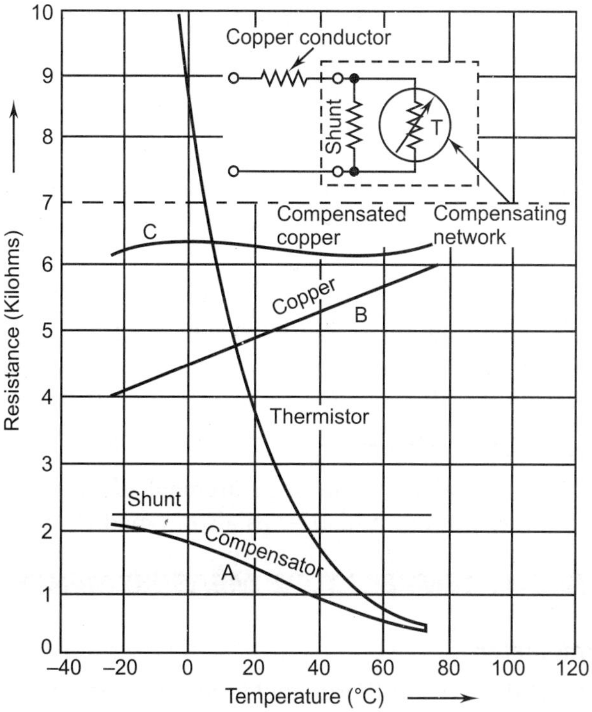

Fig. 6.46: Temperature compensation of a copper conductor by means of a thermistor network

6.6.6.3 A Simple Temperature Control Circuit

This Wheatstone's bridge circuit is ac excited and shown in figure 6.47. At a particular predetermined temperature, R_T (resistance of the thermistor) is equal to R_2. Then bridge output

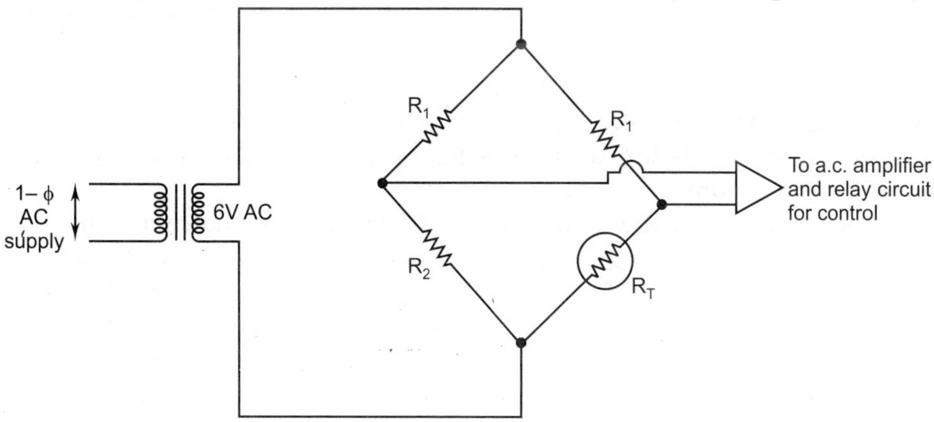

Fig. 6.47: A temperature control circuit

is zero, such that the relay circuit remains off. The relay contacts are used to control the current in the circuit that generates the heat. Temperature control to a precision of 0.00005°C can be achieved by this circuit. The three resistors, named R_1 an R_2 are made of manganin.

6.6.7 Comparison of RTD and NTC Thermistor

NTC thermistors have the following advantages compared to RTD.

- Smaller temperature difference measurements possible .
- A higher temperature sensitivity.
- Smaller in size
- Lesser thermal inertia, *i.e.* lesser value of time constant compared to a RTD.
- Higher detector resistance – hence readings are less affected by the connecting lead wires.

While the disadvantages of a NTC thermistor are:

- More susceptible to decalibration at higher temperatures.
- Non-standard characteristics.
- Less measuring temperature range.

6.6.8 Application Areas of Thermistors

Thermistors find extensive applications in the measurement of temperature of machine parts, electronic components, food processing industries as also in medicine industries.

6.7 MISCELLANEOUS TEMPERATURE MEASUREMENTS

6.7.1 A Quartz Thermometer

The principle of operation of this type of thermometer lies in the fact that the resonant frequency of a quartz crystal is dependent on its temperature.

In practice, the quartz crystal is put inside a sheath electrically connected as a part of the resonant element of an oscillator. The oscillator frequency varies with the variation of temperature of the quartz crystal. This frequency is calibrated in terms of the unknown temperature of the place in which the quartz crystal is placed.

Quartz thermometers operate within a temperature range of –40° to 230°C and having a good resolution of about 0.1°C and linearty of the order of ± 0.05%. The advantage lies in the linear temperature vs. frequency relationship although it is fairly costly because of complex circuitries involved to measure the frequency. It offers better stability and repeatability over RTD, thermocouples but its use is only confined to laboratory environments.

6.7.2 An Ultrasonic Thermometer

Its principle of operation is based on the fact that velocity of sound waves in a medium is dependant on its temperature.

The schematic of arrangement for an ultrasonic thermometer is shown in figure 6.48.

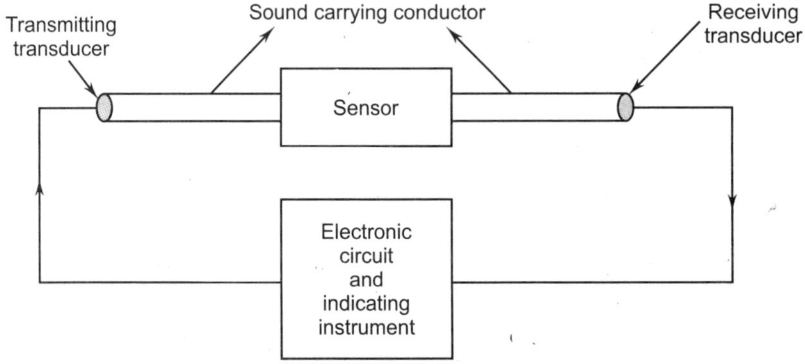

Fig. 6.48: Schematic of an ultrasonic thermometer.

It consists of an oscillator generating a frequency in the range of 0.1 to 3 MHz in the electronic circuit. It is then converted into sound waves by means of a piezoelectric or magnetostrictive transmitting transduer. This sound wave is then carried into the medium, whose temperature is to be measured, by means of a sound carrying conductor. The medium contains a sensor. The receiving transducer receives the output sonic signal after a time lag-dependent on the temperature of the medium. The electronic circuit measures this time lag and it is calibrated in terms of temperature of the medium.

Normally, rods and wires are used in sensors. In some cases, the medium itself also acts as sensor so that no separate sensor is necessary.

Typical application areas of ultrasonic thermometers, which can measure temperature in the range of 1500° – 3000°C, are nuclear reactors, furnaces, inside solid bodies, etc.

This type of thermometers are not mass produced and applied only in well specified problems. They cannot make point temperature measurements and are also very costly. Ultrasonic sensors have high accuracy. Some ultrasonic sensors can measure temperatures from the cryogenic range to plasma levels (20,000°C).

6.7.3 Temperature Indicators

These are used for an estimation of temperature—not applicable for precise temperature measurements. Some of the temperature indicators are pyrometric cones, thermochromic paints, temperature indicating crayons, self-adhesive indicators, liquid crystals, etc.

6.7.3.1 Pyrometric Cones

The pyrometric cones are commonly known as Seger cones or Orton cones. Variation in composition of the cones give them varied melting points. Each such cone is distinguished from the other by its distinguishing number.

Cones, with increasing numbers, bend at increasing temperatures. The bending temperature is also dependent on the rate of rise of temperature of the cone.

Pyrometric cones find widespread applications in ceramic industries to determine firing temperature of ceramic cones. Bar and ring type cones are also available.

Measurement tolerance for industrial and laboratory cones are around ±15°C and ±10°C respectively. Depending on the composition of the cones, a range of temperature between 600° to 2000°C can be covered and defined in BS 1041: Part 7.

6.7.3.2 Thermochromic Paints

Change of colour of these types of paints is a function of temperature and also the colour transition temperature is a function of time. Some of the paints change colours once, twice or even four times also at well defined temperatures. Some paints are reversible in nature; they regain their colour when cooled down while some other paints are not so—they are irreversible by nature.

Paints which change colour in a continuous fashion are manufactured in the range of 120°C to 400°C while discontinuous ones are manufactured from 40° to 1350°C. Manufacturers recommend a nominal paint temperature as that at which colour transition takes places after 30 minutes heating time.

Thermochromic paints find applications in estimating temperatures of ventilator fan blades, rotating wheels, pipelines, electrical motors, high voltage apparatus. They are also used as warning indicators while transporting costly chemicals, films, etc.

6.7.3.3 *Temperature Indicating Crayons*

They cover a range of 65° to 670°C, in steps of 10°C to 100°C. These are used for rapid temperature determination within 1 to 2 seconds.

6.7.3.4 *Self-adhesive Indicators*

A typical indicator covers a temperature range from 30° to 280°C in 3° to 10°C steps. They are very small in size and are used to determine the temperature of electronic components like transistors, ICs, etc.

6.7.3.5 *Liquid Crystals*

They are compounds and are formed by mixing esters and acids—the mixtures depend on the temperature range. Their reflection factors change with temperature, *i.e.* colour of liquid crystals is a function of temperature.

A temperature range of –40° to 283°C can be covered with resolution between 0.01°C to 10°C.

Liquid crystals find applications in labelling function, room temperature and body temperature measurements and have a precision of about ± 1°C.

6.7.4 Johnson Noise Thermometer

It is also known as thermal noise thermometer. The principle of operation to this thermometer depends on the fact that resistors provide a voltage which is straightway related to thermal noise. Sometimes rhenium sensors are also used.

6.7.5 Magnetic Thermometry

It involves some temperature dependence property of paramagnetic salts, like magnetic susceptibility to measure unknown temperature. Temperatures very close to absolute zero (–272°C) is measured by this method with an accuracy of the order of 0.00056°C obtained.

The paramagnetic salt is made to undergo isothermal magnetisation followed by adiabatic demagnetisation at the above referred low temperature. The entropy of the salt drastically decreases because of heat outflow with the magnetic ions becoming parallel to the field.

Measurement technique involves placing sphere of isotropic paramagnetic salt in the homogeneous portion of the magnetic field of a coil of a self or mutual inductance.The inductance of the coil is a function of temperature which can be measured by an ac bridge.

Inductance and susceptibility of the salt bear a linear relationship for a spherical or ellipsoidal paramagnetic salt.

Proper shielding of the bridge is a must. The balance is independent of frequency. A sensitive galvanometer may be used for detection which can be calibrated in terms of unknown low temperature.

6.7.6 Fluidic Sensors

Such a sensor converts gas temperature into gas pressure and the heart of such a sensor is a beat-frequency detector system.

The temperature to pressure fluidic converter is shown in figure 6.49. The input gas (at an unknown temperature) is split by a knife-edge inside a two chamber resonator oscillator. An oscillation is set into this chamber whose frequency is proportional to the square root of the absolute temperature.

There are two inputs to the beat frequency detector— one the output from the two chamber resonator oscillator and the other is a reference signal obtained from a temperature sensitive resonant oscillator. The beat frequency detector is a function of the reference signal and the output of the two chamber resonator oscillator. A steady output pressure (which is proportional to beat frequency and ultimately to unknown temperature) is obtained from the frequency converter.

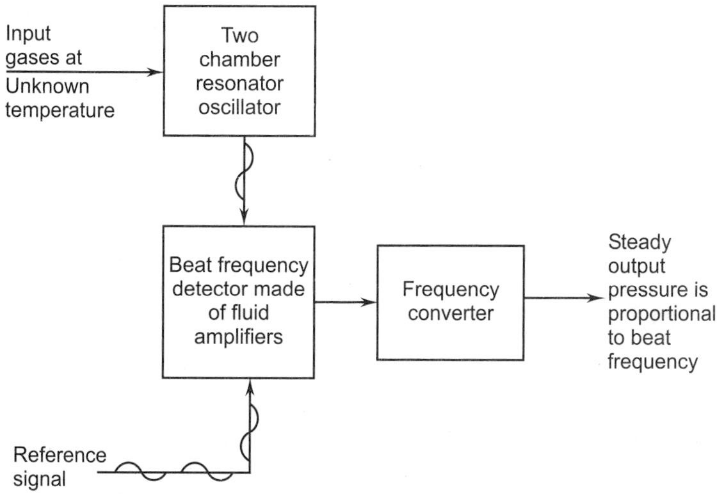

Fig. 6.49: A temperature to pressure fluidic converter

The maximum range of the instrument is approximately 1100°C and the accuracy limit is around 2 per cent.

6.7.7 Solid State Temperature Measurements

6.7.7.1 Diode Thermometers

Both silicon and germanium diodes are used as temperature sensors. Silicon diode covers a temperature range of –50° to 150°C while for germanium it is –251° to 43°C. Within these ranges, the forward bias voltages bears an almost linear relationship with temperature.

Temperature measuring range of diode thermometers is limited by the following: maximum permissible junction temperature and the range of linearity of the output signal.

Control of diode current is a must for proper operation of the diode thermometer. If the diode current is too low, the voltage-temperature characteristic becomes non-linear, while considerable diode current would result in self-heating of the diode - thereby leading to appreciable errors in measurements. A silicon diode is destroyed if it is heated beyond 200°C.

A typical circuit for temperature measurement using diode is shown in figure 6.50. A high input impedance analog voltmeter can be used which can be calibrated in terms of temperature.

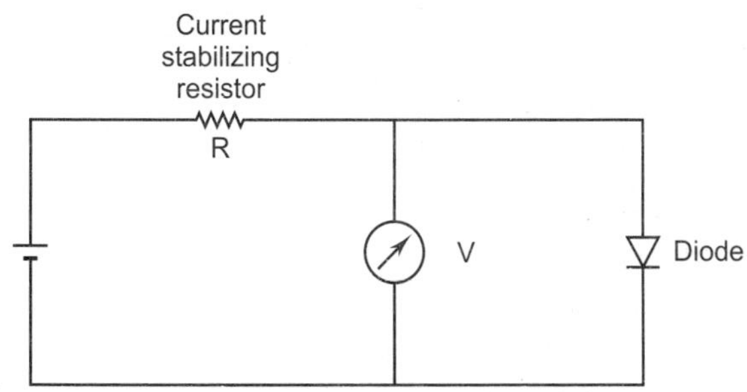

Fig. 6.50: A diode thermometer circuit

When a diode is replaced, a recalibration is necessary. Also, in some cases, size may be a factor. Advantages of using diodes in thermometers are their easy availability as also they are very cheap. Diode thermometry finds application in cryogenic temperature range but in that case linearisation of diode characteristic is a must.

6.7.7.2 Transistor Thermometers

Temperature sensitive thermometers, like diodes, find application in the range from absolute zero to 54°C.

The operating principle involves enhancing the base bias to maintain a constant collector current when the temperature decreases. Thus, the base bias can be calibrated in terms of temperature for a constant or fixed collector current. Temperature differential—as low as 0.056°C can be measured.

Transistor thermometers are much more sensitive than diode thermometers because of current amplification factor.

6.7.7.3 Integrated Circuit Thermometry

IC temperature sensors find widespread application in temperature measurements. LM 335 IC is one such type and its two packaged versions are shown in figure 6.51. It operates as a two terminal zener with an output of 10 mV/K. It operates within a current range of 400 μA to 5 mA – the performance of the IC remains invariant. It offers a dynamic impedance which is less than 1Ω. Its useful ranges are –10° to 100°C.

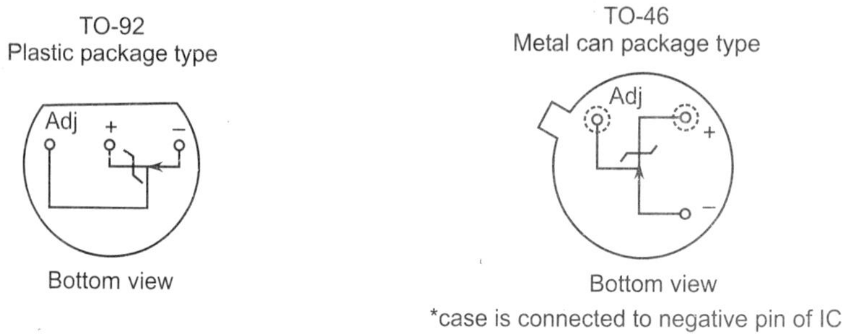

Fig. 6.51: The two packaged versions of LM 335 IC

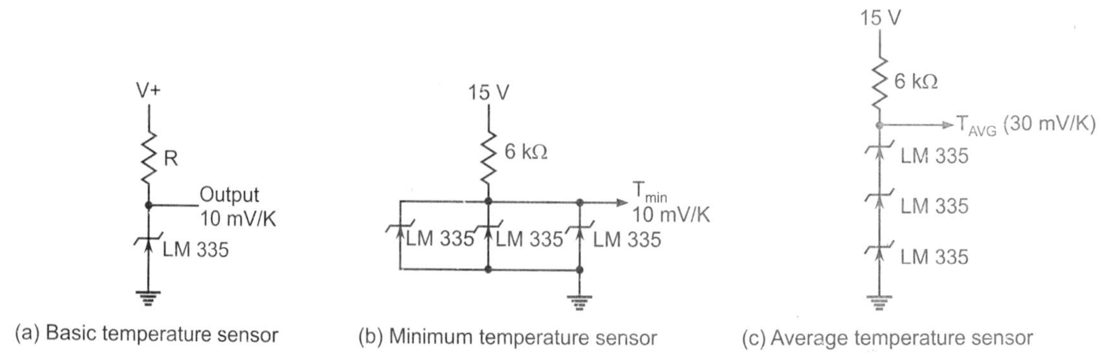

Fig. 6.52: Temperature measurement employing LM 335
(*a*) Basic temperature sensor (*b*) Minimum temperature sensor (*c*) Average temperature sensor

Figure 6.52 (*a*) shows a basis temperature sensor while figures 6.52 (*b*) and (*c*) are meant for minimum and average temperature sensing respectively. The sensor which is exposed to lowest temperature sets the output voltage in figure 6.52 (*b*) while figure 6.52 (*c*) gives an output which is the sum total of the individual IC sensors.

Self-heating errors of the IC sensors can be reduced by operating the IC at the minimum current.

6.8 RADIATION TEMPERATURE MEASUREMENT

6.8.1 Introduction

When the temperature measuring sensors can not be brought in physical contact with the process—may be because of (*a*) possible damage to sensors or (*b*) long distance involved whose temperature is sought or when the hot body is a movable one–radiation temperature measurements can be very successfully applied in such situations. The principles involved in such measurements are:

(*a*) to measure total radiation energy from a hot body, giving rise to a radiation pyrometer.

(*b*) to measure the spectral radiant energy from a hot body, giving rise to an optical pyrometer.

The radiation temperature measurement involves a range from 50°C to as high a temperature as 55,00°C, although it is mostly used for high temperature measurements not possible with other sensors like thermometers, RTDs, thermocouples, thermistors, etc. Radiation temperature measurements are very fast responding because of involvement of very small mass which is to be heated be radiation. Also the accuracy of measurement is very good.

6.8.2 Radiation Fundamentals

Radiant energy, which is electromagnetic in nature and whose wavelength lies in between 0.01 to 100 μ (1 μ = 10^{-4} cm) corresponds to thermal radiation.

A black body is a body which absorbs all radiations (*i.e.* of all wavelengths) that fall on it. Similarly, it is such a body which radiates energy of all wavelengths at a maximum rate corresponding to the temperature of the black body. The radiation laws are based on the concept of black body.

Kirchhoff's law states that a body, in thermal equilibrium with its surroundings, emits as much radiation as it receives per unit time for any given temperature and wavelength while Stefan - Boltzmann's law states that the amount of radiation energy of all wavelengths (or radiant flux) W_b from a black body is proportional to the fourth power of its absolute temperature. Thus, $W_b = \sigma.A.T^4$

where, W_b = radiant energy, ergs/ second

σ = Stefan Boltzmann constant, 5.77 × 10^{-5} erg/(sec)(cm^2)(deg^4).

A = area of the black body, cm^2

T = absolute temperature of the black body, K.

When two bodies are at absolute temperatures T_1 and T_2 (assume $T_2 > T_1$), then radiant heat energy, which is transferred from the hotter body, is given by:

$$(W_b)_{1,2} - \sigma\,(T_2^4 - T_1^4).$$

The emissive power of a radiating surface at a certain temperature is defined as the ratio of the amount of radiation emitted per second per unit area of the surface to that emitted per second by unit area of a perfectly black body at the same temperature.

Now, the emissive power of a black body is a function of the wavelength λ and is expressed by Planck's law: $W_{b\lambda} = \frac{C_1}{\lambda^5} \cdot \frac{1}{e^{c_2/\lambda T} - 1}$

where, $W_{b\lambda}$ is in ergs/sec/cm²/0.01μ

C_1 = 3.7403×10^9 ergs/sec/cm²/0.01μ

C_2 = 1.4384 cm K

T = absolute temperature, K

A plot of $W_{b\lambda}$ vs. λ is shown in figure 6.53 for several temperature. $W_{b\lambda}$ is also known as spectral intensity of radiation.

Two observations can be made from the figure:

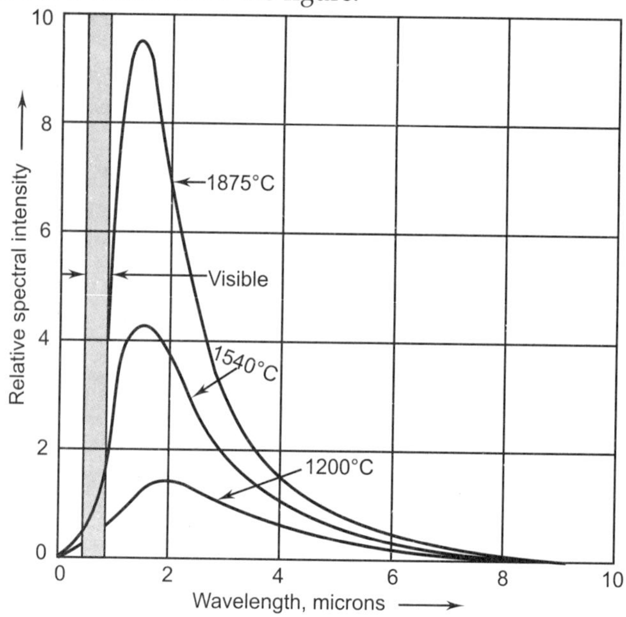

Fig. 6.53: Spectral distribution of radiant energy at different temperatures

The spectral intensity is wavelength dependent.

The points on the curves where maximum spectral intensity occur, shift to the shorter wavelengths as temperature increases more and more.

According to Wein's displacement law:

$$\lambda_m T = 2900$$

where λ_m is the wevelength at which the spectral intensity becomes maximum corresponding to the absolute temperature T. Here λ_m is in micron (μ) and T in K.

For low T values, λ_m is more and vice-versa.

Thus, when a body (say a metal) is gradually heated, the wavelength progressively decreases. And this is the reason for the changing colour of a gradually heated body.

6.8.3 Classification of Pyrometers

Pyrometers, also called infrared thermometers or radiation thermometers are normally classified according to their spectral response and also the operating method involved. This is shown in figure 6.54.

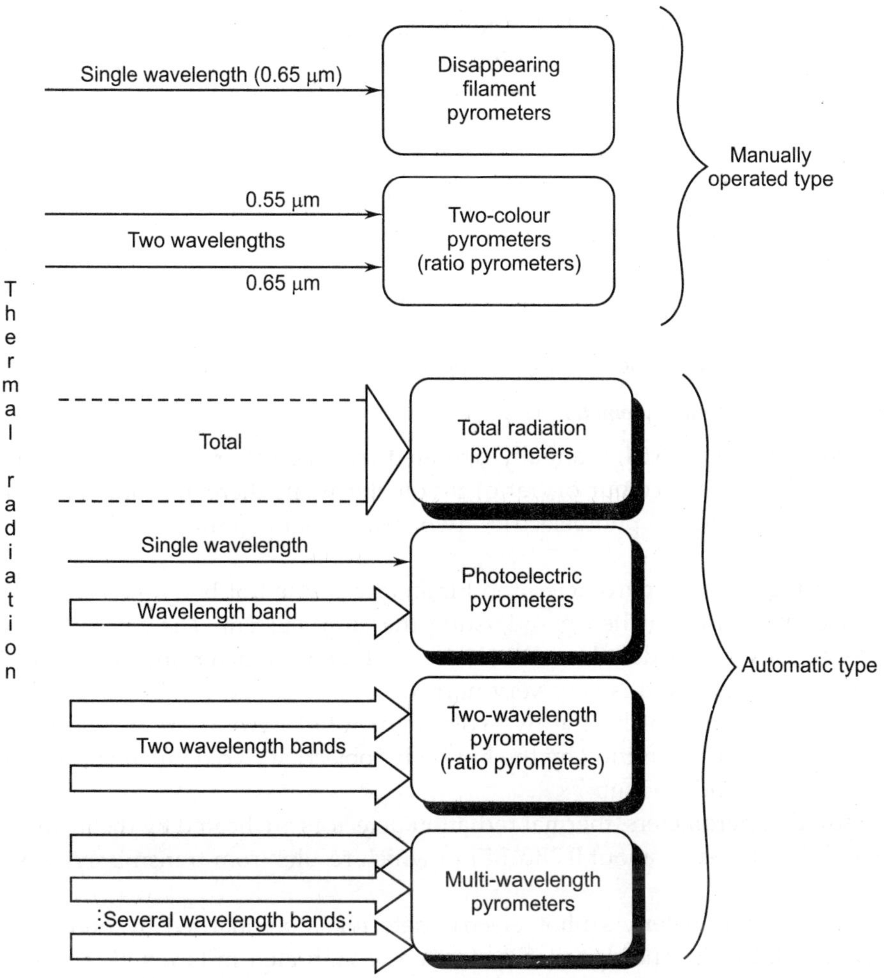

Fig. 6.54: Classification of pyrometers based on wavelength and different operating methods

Block diagrams of manual and automatic pyrometers are now shown. Figure 6.55 shows a manually operated pyrometer in a block diagram form.

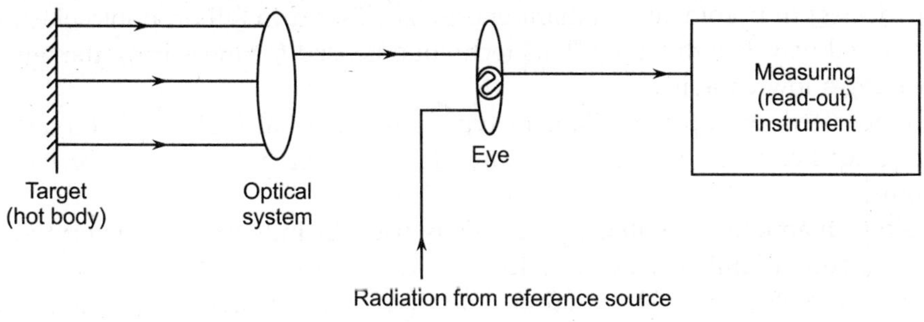

Fig. 6.55: Block schematic of a manually operated pyrometer

In this, radiation from the hot body (target) is guided to the eye via an optical system. Eye also receives radiation from a reference source. These two are compared by the eye of the operator who thus makes the adjustment of the measuring unit — calibrated in terms of temperature.

Figure 6.56 shows the block diagram of an automatic pyrometer. In this, the optical system guides the radiation from the hot body (target) on to a radiation detector which may either be a thermal or photoelectric sensor. The signal conversion unit conditions the output from the detector which is then fed to the measuring unit for display of temperature.

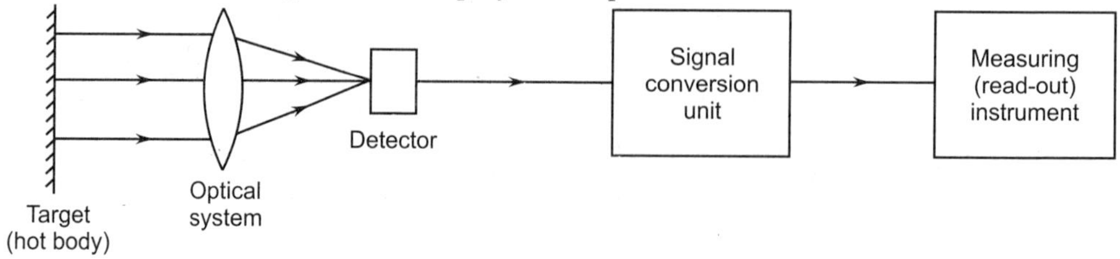

Fig. 6.56: Block schematic of an automatic pyrometer

6.8.3.1 Principles of Different Pyrometers

As already shown in figure 6.54, manually operated pyrometers are of two types: disappearing filament pyrometer and two colour or (ratio) pyrometer while there are four types of automatic pyrometers: total radiation, photoelectric, two wavelength (ratio) and multi wavelength pyrometers.

In disappearing filament pyrometer, the luminance of the hot body (object) and that of the filament (of the lamp) are matched by adjusting the lamp current. The observer's eye matches this comparison. This is considered a monochromatic type pyrometer operating at 0.65 μm since the operating wavelength band is very very narrow.

Temperature measurement by the two colour (or ratio) pyrometers is based upon finding the ratio of radiant intensity emitted by the hot body (object) at two defined spectral wavebands, which are mostly 0.55 and 0.65 μm.

In total radiation pyrometers, thermal radiation detectors are heated by the incident radiation. The wavelength band used is about 0.2 to 14 μm which results from transmissivity of the optical system.

In photoelectric pyrometers, a photoelectric detector is bombarded by photons from some selected wavelength bands. The signal so generated is calibrated in terms of temperature of the hot body.

In two wavelength (ratio) pyrometers, a ratio is found out of the radiant intensity in two selected wavelength bands. This ratio is found out by employing photoelectric detectors and is calibrated in terms of temperature of the hot body.

In multiwavelength pyrometers, radiant energy is allowed to fall on photo electric detectors from some selected wavelength bands. This technique is used for measuring the temperature of hot bodies having a low emissitivity.

Manually operated, disappering filament type pyrometer has high precision and often used for industrial reference measurement. They are also used for measurement of temperature of non-black bodies.

A two-colour, manually operated pyrometer is used in industries to determine the actual temperature of a body of unknown emissivity.

Automatic pyrometers of total radiation type that uses a quartz of glass lens are used for continuous temperature measurement in an industrial environment for targets approaching black bodies or non-black bodies. These pyrometers are used for measurement of temperature of electric chamber furnaces, glass tank furnaces, etc.

6.8.4 A Total Radiation Pyrometer

The total radiation pyrometer, shown schematically in figure 6.57 consists of a hot body (whose temperature is to be measured), a lens, a diaphragm, a sensor (a thermopile here) and an indicating or a recording instrument.

Radiation from the hot body is concentrated on the thermopile via the converging lens and the diaphragm. Both the position of the lens and the diaphragm are adjustable for calibration purposes.

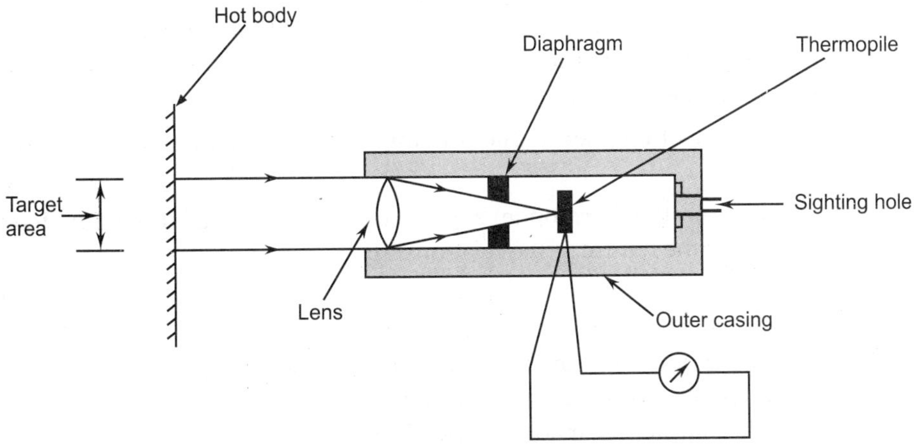

Fig. 6.57: Schematic of a total radiation pyrometer

Let the hot body temperature be T_h, detector temperature T_d and casing temperature T_c, then the radiant energy received by the receiving unit is given by $\delta A T_h^4$,

$$A = \text{area of the target.}$$

Since thermopile output (emf) is proportional to $(T_d - T_c)$, hence T_c should be maintained constant, since this emf is calibrated in terms of target temperature.

The above equation is true for black body radiation. In practice, the emissivity of the target surface is to be considered. Thus, amount of radiant energy received by the receiving unit is

$$E = e \sigma A T_h^4$$

where, e = emissivity or emission efficiency of the target surface. e, the emissivity is dependent on the condition of the surface. Closed furnaces very much approximate black body conditions whereas for molten metals, the emissivity value becomes less and also that its value does not remain constant. Any dust, dirt, smoke or gas present between the target and the thermopile also reduces the actual value of emissivity. If these kind of situations are confronted, it is better to calibrate the receiving radiation unit under actual operating conditions.

Thus, if correction is not applied, temperature recorded by the receiving unit will be less than the actual; it will thus record T_a, the apparent value of temperature. T_h and T_a are related empirically as:

$$T_h = K.\, T_a$$

where K is a constant, whose value is obviously less than unity. The value of K is dependent on the emissivity of the target surface as also on the condition existing between the target and the receiving unit (as already mentioned above).

Again, Planck's equation for black bodies, which has already been mentioned, can as well be applied in this case of gray bodies (*i.e.*, bodies which are not black bodies). In this case, the equation is

$$W = \frac{\in c_1}{\lambda^5 (e^{c_2/\lambda T} - 1)}$$

where, $\in$ = emissivity or emission efficiency of the gray body (here target source) surface and the other symbols have their usual meanings.

The emf obtained at the thermopile is either measured by a millivoltmeter or an automatically balanced potentiometer.

The pyrometer which has been discussed is not a total radiation pyrometer because the band of wavelengths emitted by the hot surface cannot be accomodated by the lens. This restricts to a given band of wavelengths determined by the particular lens.

The thermal capacitance of the thermopile is nominal giving rise to very fast response of this type of pyrometer.

The voltage output (from the thermopile) and the temperature of the hot body does not bear a linear relationship, rather the nature of the relationship is shown in figure 6.58.

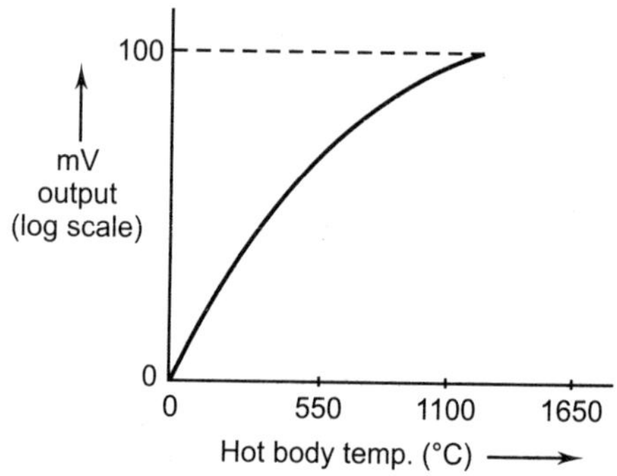

Fig. 6.58: mV vs. temperature curve

Different types of detectors can be used in this pyrometer. They are thermopile, thermistor, bolometer (all thermal type) or photo cell. The detectors are given a black coating and the surface is made rough. This ensures maximum absorption of incident radiation.

Of the different types of detectors, photo cell is fastest in response, followed by bolometer (resistance type) and the rest. Thermopile gives the maximum output and mostly used. Bolometers are costly, while compensation difficulty in case of thermistors restricts its use to a large extent. Lastly, the photo cells have large sensitivity but their spectral efficiency is poor, thus limiting its use.

6.8.5 The Optical Pyrometer

An optical pyrometer, a manually operated device, is used in the temperature range of 550° to 3300°C.

The optical pyrometer is operated with very nearly monochromatic light, the band of wave length is about 0.01 μ wide at 0.65 μ which lies in the red part of the visible spectum.

6.8.5.1 Disappearing Filament Type Optical Pyrometer

Radiant energy from the target area of the hot body is focussed on the filament via the lens and the adjustable diaphragm as shown in figure 6.59. The lamp filament appears against a background of the target surface. In the field of the eyepiece of the microscope, the filament is viewed, shown in figure 6.60. A red filter is used to make the radiation monochromatic in nature. The current through the filament can be adjusted by the rheostat. By adjusting this current, the brightness of the filament is made exactly identical to that of the target source. Under this condition, the filament disappears fully from the field of view, shown in figure 6.60, which also shows the nature of the field of view of the eyepiece when brightness of the filament is made more or less than the brightness due to the target source.

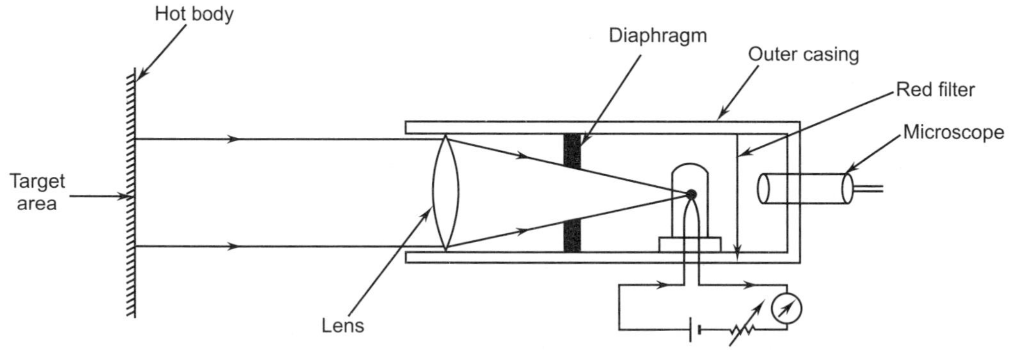

Fig. 6.59: Disappearing filament type optical pyrometer

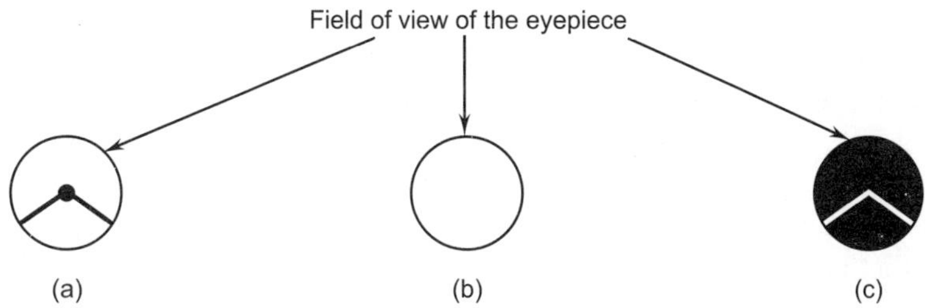

Fig. 6.60: (*a*) Filament brightness more than target surface brightness, (*b*) Filament brightness equal to target surface brightness, (*c*) Filament brightness less than target surface brightness

A variation of the above optical pyrometer can be made by inserting an optical wedge in the path of the rays from the target surface. Such an arrangement is shown in figure 6.61.

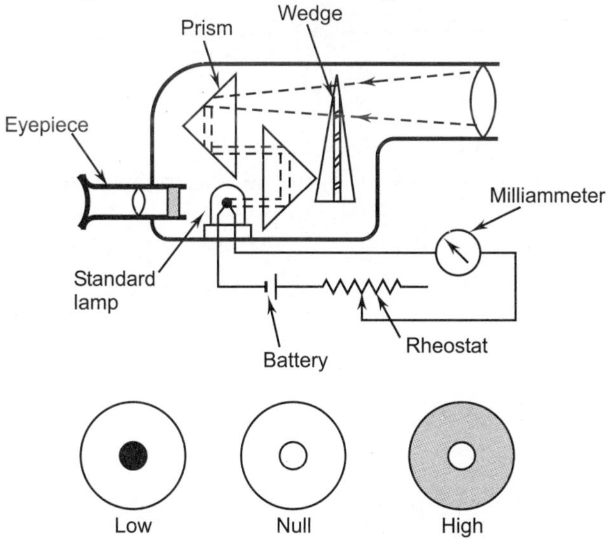

Fig. 6.61: Schematic arrangement—pyrometer instrument company optical pyrometer

Here, by means of a lead-screw arrangement, the variable thickness optical wedge may be made to move in a vertical manner. Thus, the wedge thickness which will be presented to the incoming radiation will be changed, thereby affecting the brightness of the image that will be formed in the field view of the eyepiece. In practice, the current though the filament is set to a standard value by altering the rheostat and the wedge is moved until a brightness match is obtained. If the wedge position is calibrated, it will read directly in terms of temperature of the unknown hot body.

In some instruments, provision is made to control both wedge position or the brightness of the lamp and the calibration made accordingly.

6.8.5.2 Factors in Optical Pyrometers

There are two influencing factors which must be taken into consideration with optical pyrometers. These are:

The presence of any absorbing medium lowers the reading of optical pyrometer. However, it should be borne in mind that the absorbing gases have greater effect in the low wavelength side but in the red band of the spectrum the effect is quite minimal. Ambient temperature changes almost do not have any effect on the accuracy of an optical pyrometer. The distance between the target surface and the pyrometer does not affect reading so long as the field of view of the eyepiece is filled up with radiation from the target area.

An absorbing glass is sometimes placed in the path between the target surface and the pyrometer. The reasons are two fold:

- to increase the range of the instrument.
- to protect the mechanism from damage.
- for ideal block body radiation at a wavelength λ, the simplified Planck's radiation law is

$$W_\lambda = \frac{c_1}{\lambda^5} \cdot \frac{1}{e^{c_2/\lambda T}}$$

where $e^{c_2/\lambda T} >> 1$ which is actually the case.

When the absorbing glass is inserted, then

$$W'_\lambda = \frac{c_1}{\lambda^5} \cdot \frac{1}{e^{c_2/\lambda T_a}}$$

where, T_a is the apparent temperature as measured by the pyrometer.

The transmittivity or transmission factor τ_λ of the absorbing glass is then

$$\tau_\lambda = \frac{W'_\lambda}{W_\lambda}$$

Obviously, τ_λ has a value less than 1.

Combining, we have:

$$T = \frac{1}{\frac{1}{T_a} + \frac{\lambda}{c_2} \ln \tau_\lambda}$$

6.8.6 Two Colour Pyrometer

Since emitted radiation energy vs. temperature relationship from a hot body is dependent on its emissivity, thus a radiation pyrometer needs recalibration if different types of hot body temperature measurements are done using the same pyrometer.

This problem is avoided in a two-colour pyrometer. Radiation from the hot body is splitted into two equal parts and applied to two narrow band filters. The outputs from the filters are applied to two detectors sensitive to the outcoming wavelengths of the two filters. The two detector output voltages are v_1 and v_2 respectively. The ratio of the output voltages (v_1/v_2) is found to be a function of temperature and independent of emissivity of the hot body, provided that the two wavelengths λ_1 and λ_2 are very close together.

6.9 FIBRE OPTIC THERMOMETERS

6.9.1 Introduction

Thermometers, based on the properties of optical fibres, have widespread fields of applications — like automobile, aircraft maintenance, medicine and biological fields, production of electronic components, high voltage and chemical industries. The temperature range of fibre optic thermometers roughly range from –200° to +2000°C. These thermometers are immune to electric and magnetic fields.

6.9.2 Classification of Fibre Optic Thermometers

It is grouped into two categories—extrinsic sensors and intrinsic sensors.

Extrinsic sensor, also known as indirect use thermometers, utilises the optical fibre as a medium for transmission of electromagnetic waves between the sensor and the medium whose temperature is desired.

In intrinsic sensor, also known as direct use thermometers, the optical fibre itself is being used as the temperature sensing element. In this, either the average temperature of the fibre or the temperature distribution along the length of the fibre is measured. Thermometers based on extrinsic sensors, are divided into the following categories:

- GaAs sensor absorption type,
- "Fabry-Perot" sensor optical interference type,
- Thermochromic reflection type,
- Flourescent type,
- Black body sensor thermal radiation type.

Thermometers, based on Intrinsic sensor, are divided into the following types:

- Raman scattering type,
- Change of refractive index type, and
- Optical coupling of two fibres type,

Some of the extrinsic and intrinsic types of thermometers are discussed below:

6.9.3 Thermometer Based on GaAs Semiconductor Sensor

The principle of operation of this kind of thermometer is based on the fact that the absorption coefficient A of a semiconductor material depends upon both its temperature and the frequency of radiation incident upon it. For GaAs, the empirical formula is like:

$$A = A_0 \exp [\alpha(h\nu - E_g/KT].$$

where A_0 is a constant, α is a material constant (also known as extinction coefficient), h is the Planck's constant, υ is the frequency of the incident radiation, E_g is the energy gap, K is Boltzmann constant and T is the temperature in K.

By varying υ, $h\upsilon$ can be made to approach E_g, in which case A becomes a function of temperature T.

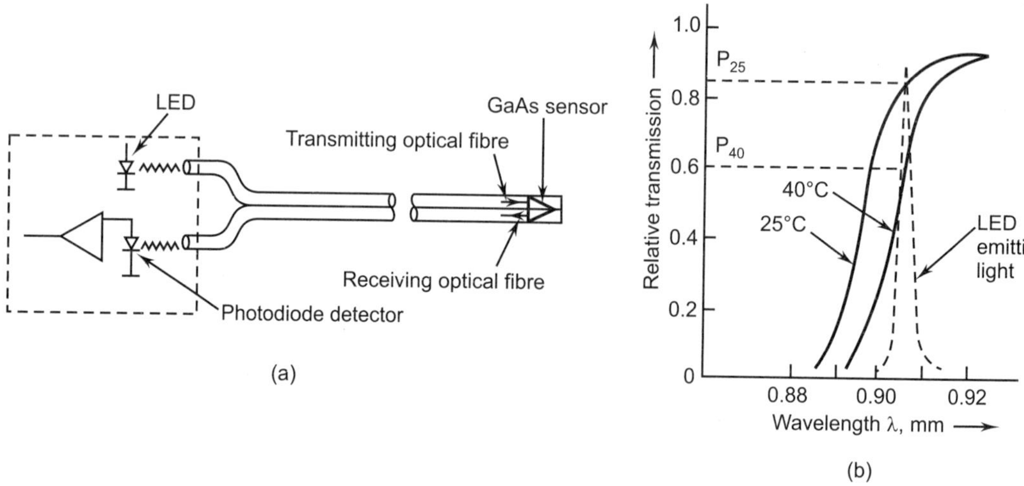

Fig. 6.62: A fibre optic thermometer with the basic operating structure of (*a*) use of a GaAs semiconductor sensor with the relative radiation transmitted through a 0.25 mm thick prism in (*b*)

In the experimental set-up shown in figure 6.62, a narrow-band LED acts as a light source which is guided to the GaAs prism sensor via the optical fibre. After passing through the sensor the light is received by a diode photo-detector via the receiving optical fibre. The sensor has the property of shifting the radiation band border and hence the output signal is dependent on the thermal dependence of the GaAs prism sensor.

This type of thermometer has a maximum range of 50°C and has applications in biology and medicine fields.

6.9.4 A fibre-optic Based Thermometer

A schematic of a fibre-optic based temperature measurement system is shown in figure 6.63. In the figure BS1 and BS2 are acting as beam splitters and M_1, M_2 are the two mirrors included at 45° to the horizontal.

BS1 acts as a beam splitter for the He-Ne laser source with one part falling on mirror M_1, gets a deflection by 90° and reaches BS2. The other part, after passing through BS1, falls on M_2. After reflection at M_2, the light ray passes through a optical fibre guide kept in a place whose temperature is desired. Here the optical fibre acts as a temperature sensitive element, such that the light ray which reaches the detector has a phase difference and detected by a Mach-Zhander interferometer. This phase difference is a function of temperature of the object and hence the detector can be calibrated in terms of the unknown temperature.

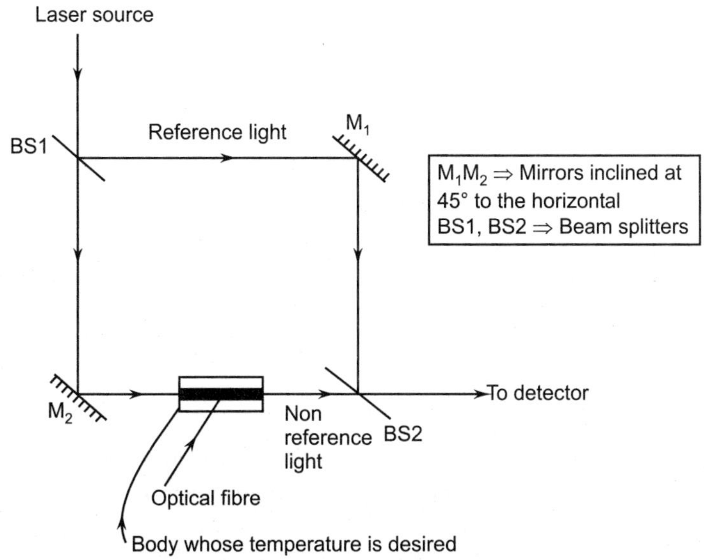

Fig. 6.63: Temperature measurement based on phase difference between light paths

6.9.5 Thermochromic Thermometers

Several liquid crystals change their colour with temperature. These are called thermochromic materials. These thermometers are applied in the temperature range of 25° to 75°C in the field of medicine and biology.

The reflection factor of these thermochromic materials is a function of both temperature and frequency of the incident visible radiation. In the experimental set-up by Brenci et al and Grattan shown in figure 6.64, cobalt chloride solution in water/alcohol is used as a thermochromic material. The cobalt chloride solution exhibits a remarkable colour change between 25°C and its spectral absorption with wavelength is also shown. The system uses a halogen lamp as the source of light. This is chopped and transmitted by a silica optical fibre to a temperature sensor. This sensor contains the thermochromic solution. The sensor has a mirror bottom so that the incident light is reflected from this portion to another receiving optical fibre. Here the received beam splits into two beams of wavelengths 0.655 μm and 0.800 μm. As shown in the figure, the light beam at 0.800 μm acts as a reference beam, since no absorption occurs at this wavelength. But the amount of absorption at wavelength 0.655 μm is temperature dependent. These two signals are then sent to photo-diodes, amplified and finally converted to dc form. The ratio of these two signals is a measure of temperature. This type of thermometric measurement is independent of (*a*) fluctuations in source intensity and (*b*) transmission losses, because both signals are made to pass through the same or common optical path.

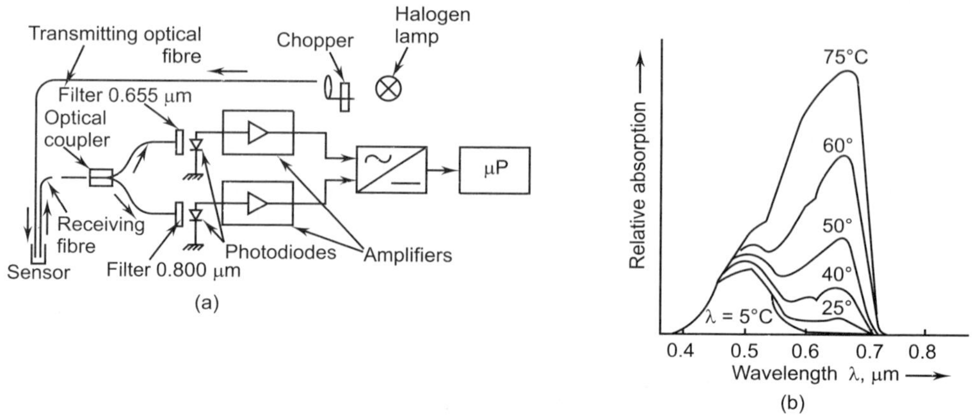

Fig. 6.64: (*a*) Block diagram of a thermochromic optical fibre thermometer, (*b*) Characteristic curves of spectral absorption of cobalt chloride solution in water-alcohol at different temperatures

6.9.6 Black Body Sensor Thermometers

The basic principle of operation of this type of thermometer is the dependence of temperature on the spectral radiant intensity emitted by a black body and is given by Planck's law.

The black body sensor consists of a cavity which is made up of platinum or rhodium or iridium—the type depending on the application range of the thermometer. The cavity is covered by a very thin film metal coating. This itself is covered by a ceramic protective film. The radiation from the sensor is sent via a high temperature optical fibre guide to a photo diode by means of a low temperature optical fibre, a lens and a narrow band filter. The output of the photodiode is sent to a signal processing circuitry which can be calibrated in terms of the temperature of the black body sensor.

This thermometer has a useful temperature range of 300° to 2000°C with a very high resolution of 0.01°C and can be used upto a frequency limit of 10 kHz.

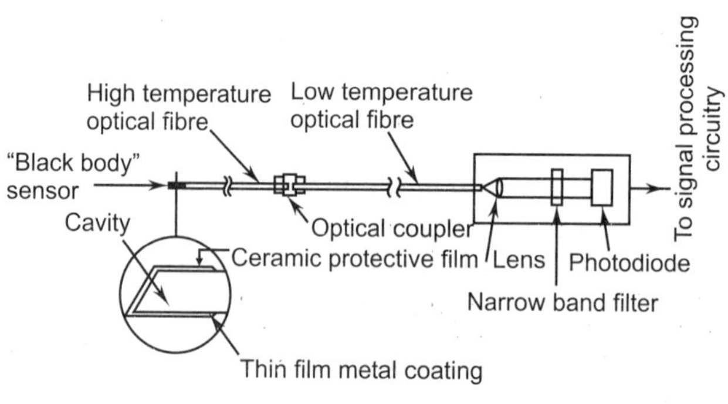

Fig. 6.65: A fibre optic thermometer with a 'black body' sensor

The application areas of this thermometer include (*a*) diamond film deposition (*b*) production of semiconductors (*c*) plasma etching and (*d*) sintering of ceramic products.

6.9.7 Thermometer Based on Fabry-Perot Sensor

The principle of operation of this type of thermometer is based on the fact that spectral reflection coefficient of Si monocrystal is a function of temperature. These thermometers have a temperature range from 0° to 400°C, have a very high resolution and immune to stray electromagnetic fields. The application areas include (*a*) food processing (*b*) microwave drying (*c*) a study of thermal effects of induced electromagnetic fields in aircrafts.

In the experimental set-up shown in figure 6.66, an LED acts as a radiation source which is fed to the Fabry-Perot sensor via an optical coupler and optical fibre. In the sensor, a thin Si monocrystal is enclosed between two glass layers. The sensor is placed in a light shield system. The light is reflected from the thin film Si sensor whose reflection coefficient is wavelength dependent. The reflected rays are guided via a lens to a semi-transparent mirror (a dichroic mirror) where it splits into two radiations of wavelengths λ_1 and λ_2. These two radiations are then fed to two photo detectors to be ultimately fed to a signal processor. The ratio of the electrical signals outputted by the photo detectors is a function of the temperature of the Si film sensor at the wavelengths λ_1 and λ_2.

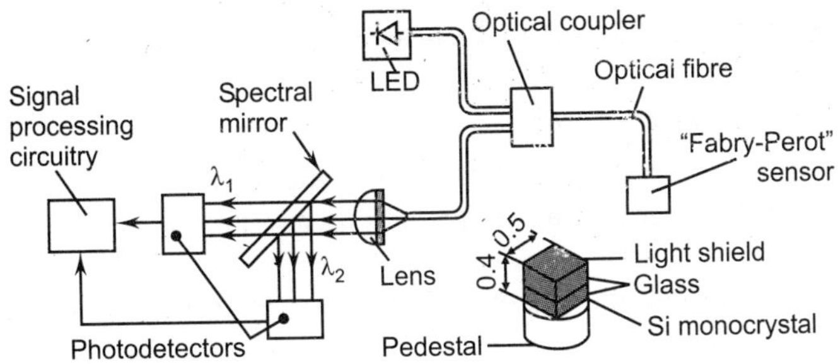

Fig. 6.66: A fibre optic thermometer with "Fabry-Perot" sensor

6.9.8 Raman Scattering Thermometer

These thermometers are based on the well-known Raman scattering effect; it is temperature dependent. When a light pulse is passed through an optical fibre, Raman scattering takes place. Due to this scattering, Stokes light, antistokes light and Rayleigh scattering light are observed. These three components are having differing wavelengths and different from the wavelength of the incident light.

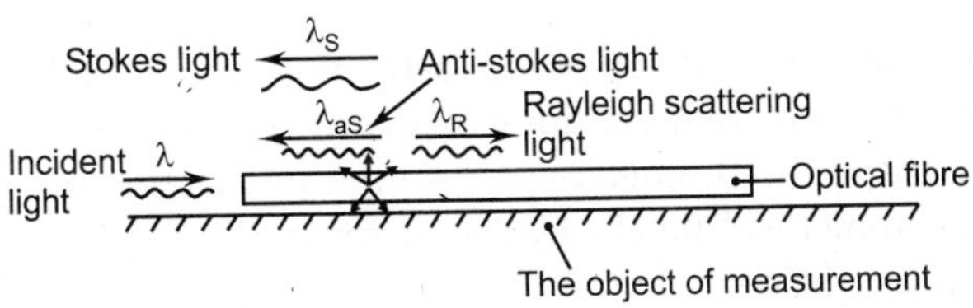

Fig. 6.67: Raman scattering in an optical fibre

This type of thermometer is normally used for temperature distribution or average temperature measurement along the length of an object. While figure 6.67 shows the Raman scattering in an optical fibre, figure 6.68 shows the block diagram of measurement of temperature distribution of an object, courtesy Lida et al 1992. An optical wavelength division multiplexer guides the Stokes and Anti-Stokes light to the respective detectors, the output from which is fed to an average circuit and finally to a signal processor. The signal processor output gives out the measured temperature distribution along the length of the object. A synchronising signal from the averager block guides the light rays from the semiconductor laser to the optical fibre.

The main application areas of this types of thermometer includes the temperature distribution and the average temperatures of large surfaces and along very long objects (like pipelines).

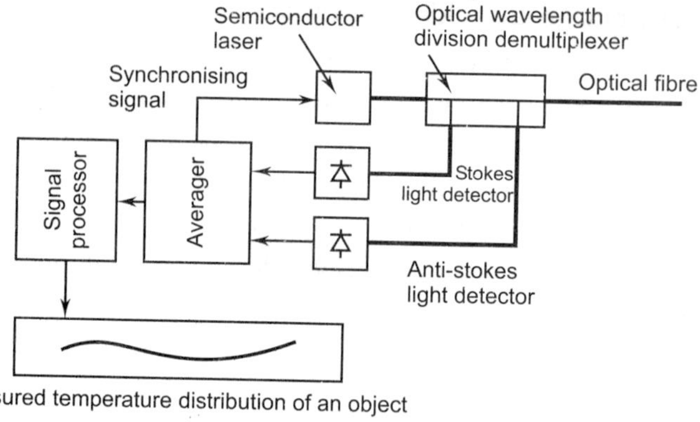

Fig. 6.68: Thermometer based on Raman scattering for measurement of temperature distribution

6.9.9 Refractive Index Thermometer

Refractive index of optical fibres is a function of temperature. This was the basis on which Gottlieb and Brandt (1979) described a method for measurement of average temperature along the length of a fibre.

Thermometers based on change of refractive index are normally applied for well defined problems.

Another application area of this type of thermometer is for detection of leaks in a low temperature liquid container (*Grattan*, 1887). The method employed here is to place the optical fibre below the low temperature liquid container. The fibre is made up of a glass core light guide and is covered by a plastic sheath.

If any leakage of cold liquid is there, it will prevent the light wave from being transmitted because at temperatures below 25°C, the refractive indices of both the core and the sheath become equal.

6.9.10 Conclusion

Thermometers based on fibre optic principles is a fastly developing field in the area of temperature measurement. It has its own inherent advantages over conventional temperature measurement methods, *viz*, commonly used thermometers, as also thermoelectric, resistive and semiconductor type thermometric measurements. The chief advantages of fibre optic thermometers are:

- Very low thermal inertia, resulting in less error in temperature measurement.
- Immune to both electric and magnetic fields – resulting in temperature measurements in induction, dielectric and microwave heading.
- Because of insulating properties, contact with "live" metallic elements possible, and does not present any danger.
- Immune to mechanical and chemical influences.
- Very wide temperature range roughly from – 200° to +2000°C.
- Temperature measurement of living organisms possible because of very small diameters of fibre optic cables.
- Multichannel temperature measurement possible, because several optic fibre cables can be coupled.

- Moderate to very high length of optic cables can be employed for temperature measurements, without impairing sensitivity of measurements.

6.9.11 Suitability of Temperature Transducers

A host of several temperature transducers have been discussed in this chapter. The choice of the appropriate temperature transducer for a given situation is guided by several factors like (*a*) numerical value of temperature to be measured (*b*) fastness (time cosntant) (*c*) cost involved (*d*) solid or liquid temperature (*e*) contact or non-contact type measurement. (*f*) accuracy, measurement sensitivity, etc.

To cover the entire range of industrial temperature measurements, different transducers are to be employed — on the lower side are RTDs, thermistors, in the mid range Hg-in-glass, bimetallic thermometers; thermocouples, pressure thermometers whereas in the higher temperature range pyrometers are used.

For fast changing temperatures, transducers having very fast response (i.e. low time constant) should be employed provided that the temperature to be measured falls under the purview of that transducer. For example, thermistors can be employed because of its low time constant on the lower temperature range. Narrow band radiation pyrometers are very suitable for high temperature measurements for temperatures changing as fast as 10 μs. For measurement of solid body temperatures, a very good physical contact is desired to get accurate temperature; hence RTDs, thermocouples, thermistors are to be used here. Of course, the non-contact type radiation pyrometers are employed for high temperature measurements of solid bodies. In case of fluids, any of the temperature transducers discussed can be used for temperature measurements.

Where accuracy of temperature measurements is at a premium, transducers having higher values of accuracy are to be employed. Some forms of radiation pyrometers give accuracy to the tune of ±0.05%. Quartz thermometers offer an accuracy limit of ± 0.1%, whereas for the case of RTDs it is ±0.5%. Base metal thermocouples have an accuracy of ±0.5% of full scale whereas it is ± 0.2% for noble metal thermocouples, although the latter type is much costlier.

While employing radiation pyrometers, correction for variable emissivity should be employed, although the problem is avoided in the two-colour pyrometer.

SOLVED EXAMPLES

1. A bimetal element is formed from stainless steel and invar, and the combined thickness is 0.075 cm. The brazing temperature is 40°C and it is raised to 100°C. Find the radius of curvature so formed.

   Given, $\alpha_s = 1.6 \times 10^{-5}/°C$ and $\alpha_i = 1.68 \times 10^{-6}/°C$

   Ans.
   $$r = \frac{2S}{3(\alpha_s - \alpha_i)(t - t_b)} = \frac{2 \times 0.075}{3(16 - 1.68) \times 10^{-6} \times 60}$$
   $$= 58.194 \text{ cms}$$

2. Find the emf developed across a thermocouple for which α = 40 μV/°C and the hot and cold junction temperatures are at 100°C and 27°C.

   Ans. The emf produced is given by
   $$e = \alpha(T_h - T_c)$$
   $$= 40 \times (100 - 27) \ \mu V$$
   $$= 2.92 \text{ mV}$$

3. A chromel—constantan thermocouple has its two junctions maintained at 400° and 0°C and gives a thermo-emf of approximately 30 mV. The resistance of the coil of the milliammeter is 50 Ω and

Chapter 6

a current of 0.1 mA gives full scale deflection. The resistance of the lead wires is 0.2 Ω and that of the thermocouple wires is 2 Ω. Calculate

(*a*) the resistance to be put in series with the coil to get full scale deflection at 400°C.

(*b*) the error due to a rise of temperature of the meter coil by 5°C. α of the meter coil = 0.00425/°C.

(*c*) The error of the lead resistance increases by 0.05 Ω.

Ans. We have,

$$e_{th} = i(R_{th} + R_{ld} + R_s + R_c)$$

where,

e_{th} = emf available from the thermocouple

i = current through the milliammeter

R_{th} = reistance of the termocouple wires

R_{ld} = resistance of the lead wires

R_c = resistance of the millivoltmeter coil

R_s = resistance in series with the coil

∴ From the above equation,

$$30 \text{ mV} = 0.1 \text{ mA } (2 + 0.2 + R_s + 50)$$

or

$$R_s = 247.8\ \Omega$$

(*b*) increase in resistance of the meter coil due to a rise of temperature by 5°C = 50 × 0.00425 × 5 = 1.0625 Ω

∴ The new meter current is

$$= 30/(2 + 0.2 + 247.8 + 51.0625)$$
$$= 0.099647 \text{ mA}$$

∴ Error in temperature measurement

$$= \frac{0.0999647 - 0.1}{0.1} \times 400$$
$$= -1.412°\text{C}$$

(*c*) The new current is

$$= 30/[2 + (2+0.05) + 247.8 + 50]$$
$$= 0.099983 \text{ mA}$$

∴ Error in temperature measurement

$$= \frac{0.099983 - 0.1}{0.1} \times 400$$
$$= -0.068°\text{C}$$

4. For a type *K* thermocouple, the hot and cold junctions are at temperatures of 1000°C and 40°C respectively. Find the emf across the thermocouple.

Ans. Referring to the table for type *K* thermocouple,

The millivolts corresponding to 1000°C and 40°C are 41.276 mV and 1.612 mV respectively.

Thus, the millivolt across the thermocouple is

$$= (41.276 - 1.612) \text{ mV}$$
$$= 39.664 \text{ mV}$$

5. For a type *J* thermocouple, the hot and cold junctions are at temperatures of 500°C and 50°C respectively. A portion of the two wires of the thermocouple are exposed at 200°C. Find the emf across the thermocouple.

Ans. Referring to the table for type *J* thermocouple, the millivolts corresponding to 500°C and 50°C are 27.393 mV and 2.585 mV respectively.

Thus, the millivolt across the thermocouple is

$$= (27.393 - 2.585)\ \text{mV}$$
$$= 24.808\ \text{mV}$$

To note at this point that exposure of the two wires of the thermocouple at 200°C will not alter the output mV, because the thermocouple mV output is independent of any temperature gradient existing along the length of the two thermocouple wires.

6. For a type R thermocouple (Platinum – Platinum plus 13% Rhodium) the outputs are 14.629 mV and 0 mV corresponding to 1300°C and 0°C respectively. Find the mV output when the cold junction is maintained at 30°C. With cold junction at 20°C and the mV output 10.5 mV in a particular measurement, determine the actual temperature of the measurement point.

Ans. 1st part: From the given data,

$$\text{sensitivity of the thermocouple} = \frac{14.629 - 0}{1300 - 0}\ \text{mV/°C}$$
$$= 0.0112531\ \text{mV/°C}$$
$$\text{emf corresponding to 30°C} = 0.0112531 \times 30\ \text{mV}$$
$$= 0.3376\ \text{mV}$$

Thus, the mV output when the cold junction temperature is maintained at 30°C = (14.629 – 0.3376) mV

$$= 14.2914\ \text{mV}$$

2nd part: Difference of temperature between the hot and cold junctions

$$= 10.5/0.0112531 = 933.08°\text{C}$$

Thus, the hot junction temperature is = 933.08C + 20 = 953.08°C

7. A platinum resistance thermometer has resistances of 100 Ω and 119.55 Ω at 0°C and 50°C respectively, the temperature coefficient of platinum being 0.00391°C^{-1}. Find the temperature when the resistance becomes 200 Ω.

Ans. We have:

$$t_2 = t_1 + \frac{R_2 - R_1}{\alpha R_o}$$
$$= 50 + \frac{200 - 119.55}{0.00391 \times 100} = 255.75°\text{C}$$

8. In a certain temperature measurements case with thermistor, β = 3140 and the reference temperature is 27°C at which resistance is 1000 Ω. Find the temperature when the resistance measured is 2300 Ω.

$$R_o = 1000\ \Omega$$
$$R = 2300\ \Omega$$
$$\beta = 3140$$
$$T_o = 273 + 27 = 300\ K$$
$$T = ?$$

From the relation

$$R = R_o \exp\left[\beta\left(\frac{1}{T} - \frac{1}{T_o}\right)\right]$$
$$\frac{1}{T} = \frac{\ln R - \ln R_o}{\beta} + \frac{1}{T_o}$$
$$= \frac{7.741 - 6.908}{3140} + \frac{1}{300}$$
$$= 0.0035987$$
$$T = 277.88\ \text{K}$$

$\therefore$ Unknown temperature = 4.878°C

REVIEW QUESTIONS

1. Mention the six basic fixed points on the Celsius scale, as provided by ITS-48. Discuss the revisions made to this later.
2. Mention the basis on which an unknown temperature measurement is undertaken and which falls under the category of non-electrical methods.
3. Mention the different changes on which temperature measurements are based. Classify the temperature measuring instruments/sensors by their temperature measuring ranges.
4. Mention the different types of expansion thermometers.
5. What are the two basic drawbacks of a mercury-in-glass thermometer? What is 'dip effect' in mercury-in-steel thermometer?
6. Why nitrogen is mostly used in gas thermometers? At what temperature it is filled?
7. What is the principle of operation of a vapour pressure thermometer? Draw the schematic of a vapour pressure thermometer and describe its working.
8. What type of thermometers fall under the category of filled-in systems? Discuss the errors normally associated with a filled-in system.
9. What are the characteristic requirements of a thermomcouple based system?
10. Mention the names of frequently used thermocouple in industries and their ANSI designations. Mention their useful temperature ranges.
11. Why 'cold junction' compensation is necessary in thermocouple-based systems? What are the methods followed to eliminate the problems?
12. Why compensating cables are used in thermocouple based temperature measurement systems?
13. Mention the available methods for measurement of thermocouple output.
14. Describe the potentiometer method of temperature measurement employing thermocouples.
15. Discuss the techniques used to linearise thermocouple outputs.
16. Mention the RTDs normally employed to measure temperature. Discuss their relative merits/demerits.
17. Describe Siemen's 3-lead RTD method of temperature measurement. Discuss the draw back of the circuit, if any.
18. Describe the pulse excitation technique to measure unknown temperature of several places employing RTDs. What is the advantage of 'pulsing'?
19. Describe a temperature compensator circuit employing thermistors and its usefulness.
20. How thermal conductivity is measured employing thermistors?
21. What is the working principle of an ultrasonic thermometer? Draw the sketch of temperature measurement using such a thermometer and list its advantages and disadvantages.
22. Describe an optical pyrometer for measurement of temperature of a distant object. Discuss the effects of (*a*) emissivity of the target surface (*b*) effect of absorbing madia.
23. What is the difference between intrinsic and extrinsic sensors?
24. Draw and describe a fibre-optic based thermometer based on phase difference between light paths.
25. Describe a fibre optic thermometer with a black body sensor.
26. Discuss the choice of appropriate temperature transducers, considering all aspects of temperature measurements.

Fill in the blanks:

1. The latest temperature scale is known as ....
2. A mercury-in-glass thermometer should not be employed to measure rapidly fluctuating temperature because it has a relatively high value of ....
3. Mercury-in-glass thermometers are categorised into .... grades.
4. The 'dip effect' is related to .... type of thermometer.
5. Thermometers based on expansion of liquids have a smaller 'dead zone' compared to thermometers based on expansion of ....

6. .... temperature variations do not influence the readings obtained from them.
7. A thermocouple is an .... type of transducer.
8. Thermocouple outputs are always in the range of ....
9. A thermocouple has a .... time constant when it is bare.
10. Increasing the diameter of a thermocouple increases its ....
11. A thermopile can measure .... temperature of a place.
12. Amongst the RTDs used, the metal which is having a decidedly non-linear temperature-resistance characteristics is ....
13. An RTD has, in general, a .... value for its time constant.
14. Platinum is used as a material for RTD because it is available in a very .... state.
15. Callender's 4-lead method of temperature measurement is particularly suitable for .... temperature measurement.
16. A .... circuit is required to measure temperature when the sensor employed is an RTD.
17. Power dissipated by a thermistor is very ....
18. A thermistor is .... sensitive.
19. Temperature coefficient of resistance of thermistor is about .... times that of platinum.
20. Specific heat of thermistor materials must have a .... value in order that the time constant of the thermistor to remain low.
21. A .... is used as a temperature compensator.
22. .... is most sensitive amongst RTD, thermistor and thermocouple.
23. A quartz thermometer operates in the range of .... to ....
24. Temperature measurement by a quartz thermometer is dependent on .... frequency.
25. The principle of operation of an ultrasonic thermometer is based on the fact that the .... of sound waves in a medium is dependent on its temperature.
26. Radiation temperature measurement is essentially a .... of measurement.
27. Out of the different types of detectors used in a total radiation pyrometer, ... is fastest in nature.
28. Fibre optic thermometers operate in the range of .... to ....
29. Fibre optic thermometers are immune to .... and .... fields.
30. Fibre optic thermometers are grouped into .... and .... sensors.

Tick the correct answer:

1. Measurement of temperature based on change of pressure of a gas is related to—
   (*a*) Pressure law (*b*) Charle's law (*c*) Change in Vapour pressure. ☐
2. In thermostats, the material which has lower temperature coefficient of resistance is—
   (*a*) Invar (*b*) Brass (*c*) Platinum ☐
3. Mercury-in-steel thermometer is used for—
   (*a*) domestic purpose (*b*) industrial purpose (*c*) both cases ☐
4. The bulb of a vapour pressure thermometer is filled with—
   (*a*) non-volatile liquid (*b*) volatile liquid (*c*) both ☐
5. The reference junction of a thermocouple is—
   (*a*) 0°C (*b*) 100°C (*c*) a variable temperature point. ☐
6. The important characteristic curves for thermistors are
   (*a*) Resistance-temperature characteristics
   (*b*) Voltage-current characteristics
   (*c*) Current-time characteristics
   (*d*) All three characteristics. ☐
7. The emf developed between two junctions of a thermocouple is (e_h = *emf* developed at the hot junction and e_c = emf developed at the cold junction)
   (*a*) $e_h - e_c$ (*b*) $e_h - e_c$ (*c*) $(e_h + e_c)/2$ ☐

Chapter 6

8. Amongst the three thermocouples, which is suitable to measure a temperature of 1000°C
   (*a*) Chromel-constantan (*b*) Iron-constantan (*c*) Chromel-Alumel ☐
9. Suitability of a thermocouple for a particular temperature measurement is dependent on
   (*a*) temperature range, (*b*) type of environment,
   (*c*) accuracy required, (*d*) all of these ☐
10. The emf developed by a thermocouple is dependent on
    (*a*) materials used, temperatures of the hot and cold junctions
    (*b*) diameters of materials, temperatures of the hot and cold junctions
    (*c*) the length of the wires of the materials used, temperature of the hot and cold junctions. ☐
11. Response of RTD to temperature in agitated water is fastest when it is used
    (*a*) bare
    (*b*) with stainless steel well
    (*c*) with stainless steel well and steel outer tube ☐
12. Resistance values of industrial RTDs are normally
    (*a*) 40 – 100 Ω (*b*) less than (*a*) (*c*) more than (*a*) ☐
13. A thermistor is not recommended to be used at around 1000°C because of reduced
    (*a*) stability (*b*) accuracy (*c*) both ☐
14. A thermistor has only
    (*a*) NTC (*b*) PTC (*c*) both (*a*) and (*b*) ☐
15. Radiation temperature measurement is essentially a
    (*a*) contact type (*b*) non-contact type (*c*) can be both (*a*) and (*b*) ☐
16. Radiation temperature measurement is used for
    (*a*) high temperature (*b*) low temperature (*c*) both (*a*) and (*b*) ☐
17. The temperature vs. mV output curve from a radiation pyrometer is:
    (*a*) linear (*b*) non-linear (*c*) both (*a*) and (*b*) ☐
18. In Raman scattering thermometer, the different types of light that are available are:
    (*a*) Stokes light (*b*) Antistokes light
    (*c*) Rayleigh scattering light (*d*) all (*a*), (*b*) and (*c*) ☐
19. A thermocouple is suddenly immersed in a medium of high temperature. The approximate time taken by the thermocouple to reach 98% of the steady-state value is
    (*a*) equal to the time constant of the thermocouple,
    (*b*) equal to twice the value of the time constant of the thermocouple,
    (*c*) equal to four times the values of the time constant of the thermocouple,
    (*d*) independent of the time constant. ☐
20. The sensing elements of a thermocouple at its hot junction is provided with a shield while this measurement is in a high temperature gas. The principal reason for providing the shield is
    (*a*) to reduce conduction and convection errors,
    (*b*) to reduce radiation error,
    (*c*) to provide temperature compensation to the Seebeck voltage, and
    (*d*) to improve air supply to the sensing element for better response. ☐
21. An RTD is assumed to be represented in the range 0°C to 100°C by a linear model as

    $$R_T = R_o\,(1 + 0.004\,T)$$

    where T is the temperature in °C. R_o = 100 Ω with a variation of ± 2Ω. The true model of the RTD is

    $$R_T = R_o\,(1 + 0.004\,T + 6 \times 10^{-7}\,T^2)$$

    The worst-case error magnitude in the resistance value that will be introduced, if the linear model is used over a range of 0 to 100°C, is
    (*a*) 2 Ω (*b*) 2.3 Ω (*c*) 3.0 Ω (*d*) 3.4 Ω ☐

[*GATE 2003*]

22. A temperature measuring instrument is modelled as a first order system with a time constant of 5s. The sensor of the instrument is placed inside an oil bath whose temperature has a sinusoidal variation with amplitude of 10°C and a period of 20s around an average temperature of 200°C. The sinusoidal component at the output of the instrument will have an amplitude of

(*a*) 0°C (*b*) 5.37°C (*c*) 8.57°C (*d*) 10°C ☐

23. The table provides themo-emf sensitivity of five materials with reference to platinum around 273K.

Material	Constantan	Nickel	Copper	Iron	Nichrome
Sensitivity (μK^{-1})	-35	-25	+6	+18.5	+25

(*a*) The thermocouple pair that gives the maximum sensitivity around 273K is

(*a*) Platinum-Constantan (*b*) Nichrome-Constantan

(*c*) Nickel-Constantan (*d*) Copper-Nickel ☐

(*b*) Two copper-constantan thermocouples are connected in such a way that the two constantan wires are joined together. The two copper wires are connected to the input of a low noise chopper stabilized differential amplifer having a gain of 1000. One of the thermocouple junctions is immersed in a flask containing ice and water in equal proportion. The other thermocouple junction is at a temperature *T*. If the output of the amplifier is 2.050 V, the temperature *T* is

(*a*) 205°C (*b*) 102.5°C (*c*) 51.25°C (*d*) 50°C ☐

[*GATE 2003*]

24. An hot body is radiating energy in the visible spectrum. The best method to measure the temperature is

(*a*) thermopile (*b*) thermocouple (*c*) bolometer (*d*) optical pyrometer ☐

25. A thermocouple can be considered as a single capacity system with a time constant of 100 ms and unity gain. It is immersed in a heat bath whose temperature is varying at 1°C/s from 0°C. When the actual temperature of the heat bath is 70°C, the reading of the thermocouple will be

(*a*) 70.1°C (*b*) 71°C (*c*) 69.9°C (*d*) 69°C ☐

26. An iron-constantan thermocouple is to be used to measure temperature between 0° and 300°C. With reference junction at 0°C, the emf induced at 100°C and 300°C are

$$E_{100} = 5268\ \mu V, \quad E_{300} = 16325\ \mu V$$

Then the nonlinearity at 100°C as a percentage of full scale is

(*a*) + 3.23% (*b*) – 1.06% (*c*) – 0.96% (*d*) + 3.23% ☐

27. A thermocouple at room temperature 30°C is dipped suddenly into a bath having water at 100°C. It takes 30 seconds to reach 96.5°C. The time required to reach a temperature of 98°C is

(*a*) 32.5 s (*b*) 34.6 s (*c*) 35.6 s (*d*) 38.6 s ☐

28. An iron-constantan thermocouple has its measuring junction in a bath whose temperature can vary between 100°C–200°C. The reference junction is maintained at ambient temperature. If the emf developed is to correspond to that with the reference junction at 0°C, the maximum error, expressed as a percentage of the emf developed, due to ambient temperature variation in the range 0°C – 45°C, is

(*a*) 8.182 (*b*) 22.5 (*c*) 45 (*d*) 81.82 ☐

[*GATE 2005*]

29. A thermocouple is made using a copper and an iron wire. Its measuring junction is kept at a temperature of 100°C and the reference junction is kept at the ambient temperature of 30°C. The emf developed would be

(*a*) 700 μV (*b*) 1000 μV (*c*) 3010 μV (*d*) 5300 μV ☐

30. A voltage generated using a AD590 sensor is used to compensate for ambient temperature error for the iron-constantan thermocouple in the circuit shown in the figure in such a way that the output

Chapter 6

voltage V_o is proportional to the measuring junction temperature T_m. The values of R_L and E for exact compensation, assuming ideal components, are

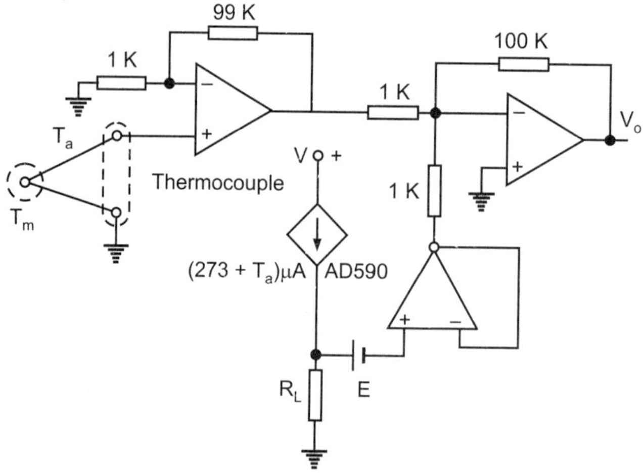

(*a*) 1 kΩ, 273 mV (*b*) 4.3 kΩ, 1.174 V (*c*) 5.3 kΩ, 1.44 V (*d*) 1 kΩ, 303 mV ☐

[*GATE 2005*]

31. In the amplifier circuit shown in the figure, a 10 kΩ resistance is connected in parallel to a negative temperature coefficient thermistor to reduce non-linearity in the range 298–232 K. The thermistor specifications are:

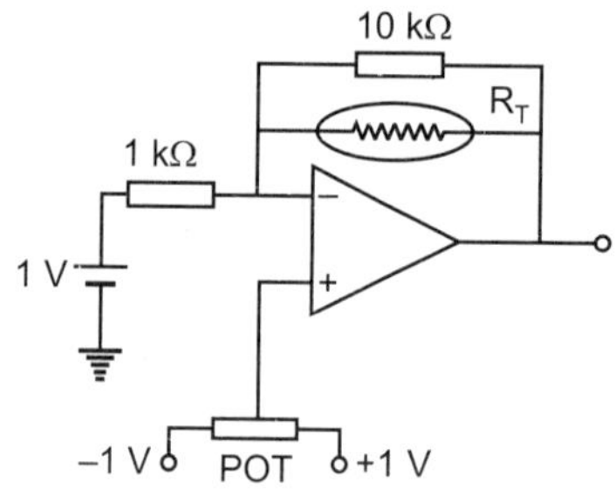

$$R_{298} = 10\text{ k}\Omega, \text{ the characteristic constant}$$
$$\beta = 4000\text{ K},$$
$$R_T = R_{298} \exp\left[\beta\left(\frac{1}{T} - \frac{1}{298}\right)\right]$$

(*a*) The magnitudes of the resistance change, between the temperatures 298 K and 323 K, when the thermistor is used alone and when it is used in parallel combination with the resistance, are

(*a*) 6.5 kΩ, 2.4 kΩ (*b*) 5.7 kΩ, 1.6 kΩ (*c*) 6.1 kΩ, 2.0 kΩ (*d*) 5.3 kΩ, 1.2 kΩ ☐

(*b*) In the given figure the potentiometer is adjusted so that the output voltage V_o is 0 at T = 298 K, the voltage output at 323 K is

(*a*) 500 mV (*b*) 400 mV (*c*) 2.4 V (*d*) 2.6 V ☐

[*GATE 2005*]

32. A type J (iron-constantan) thermocouple has a voltage sensitivity of 55 μV/°C

A digital voltmeter (DVM) is used to measure the voltage under the condition shown in the following figure.

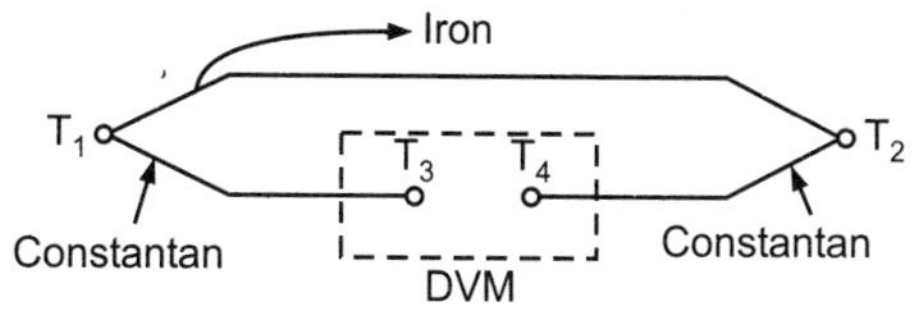

Given that $T_1 = 300°C$, $T_2 = 100°C$ and $T_3 = T_4 = 20°C$, the meter will indicate a voltage of

(*a*) 11.0 mV (*b*) 15.4 mV (*c*) 16.5 mV (*d*) 17.6 mV. ☐

[*GATE 2006*]

33. A thermometer with time constant τ, initially at the ambient temperature, is used to measure the temperature of a liquid in a bath. The excess temperature of the thermometer and the liquid over the ambient are $\theta(t)$ and $\theta_l(t)$, respectively, where t denotes the time. If $\theta_l(t) = kt$, where k is constant, the steady state error, defined as $\lim_{t \to \infty}[\theta(t) - \theta_l(t)]$, is

(*a*) $\propto$ (*b*) 0 (*c*) $-k$ (*d*) $-k\tau$ ☐

34. The temperature being sensed by a negative temperature coefficient (NTC) type thermistor is linearly increasing. Its resistance will

(*a*) linearly increase with temperature (*b*) exponentially increase with temperature
(*c*) linearly decrease with temperature (*d*) exponentially decrease with temperature ☐

35. A pair of identical thermocouples is employed for measuring the temperature of a specimen as shown below. The characteristic of the thermocouple is tabulated below.

The reference junction is at 2°C. The meter reads 48 μV. The correct temperature of the specimen is [*GATE 2008*]

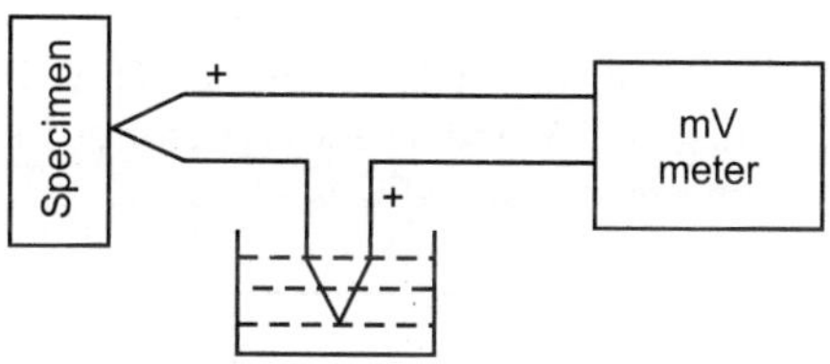

Temperature (°C)	0	10	20	30	40	50	60	70	80	90
Output (μV)	35	45	55	65	75	85	95	105	115	125

(*a*) 13°C (*b*) 46°C (*c*) 48°C (*d*) 50°C ☐

36. A bridge type rectifer meter and a thermocouple meter employ moving coil movement for indication. Both are calibrated on a 100 Hz sinusoidal waveform. If a 100 Hz rectangular waveform is applied to each, the ratio of their readings will be

(*a*) 2 : 1 (*b*) 1.11 : 1 (*c*) 1.41 : 1 (*d*) 1.21 : 1 ☐

37. Thermocouples are:

(*a*) passive transducers, (*b*) active transducers,
(*c*) both (*a*) and (*b*) above, and (*d*) output transducers. ☐

38. Three types of temperature transducers are compared as regards their sensitivity. The order in which they exhibit their sensitivities (highest to lowest) is

(*a*) RTDs, thermocouples, thermistors (*b*) RTDs, thermistors, thermocouples
(*c*) Thermocouples, RTDs, thermistors (*d*) Thermistors, RTDs, thermocouples ☐

Chapter 6

39. The temperature transducers exhibit non-linear behaviour. The order in which they exhibit non linearity (highest to lowest) is
   (*a*) Thermocouples, RTD, thermistors (*b*) Thermistors, thermocouples, RTDs
   (*c*) RTDs, thermocouples, thermistors (*d*) Thermistors, RTDs, thermocouples ☐
40. The resistance value of a thermistor is 5000 Ω at 25° C and its resistance temperature co-efficient is 0.041°C. A measurement with a lead resistance of 10 Ω will cause an error of
   (*a*) 0.02°C (*b*) 0.05°C (*c*) 0.1°C (*d*) none of these ☐
41. Some wire-wound resistors have bifilar winding. This type of winding is used to
   (*a*) increase the thermal stability, (*b*) reduce the tolerance,
   (*c*) reduce the inductance of winding, and (*d*) double the power rating of the resistance. ☐
42. A thermocouple ammeter gives full-scale deflection of 10A. When it reads one fifth of the scale, the current will be
   (*a*) 2 A (*b*) 4 A (*c*) 4.47 A (*d*) 5.78 A ☐

REVIEW PROBLEMS

1. A Chromel-Alumel thermocouple operates linearly up to 1100°C and gives 45.13 mV with reference junction temperature at 0°C. Find the emf if the hot junction temperature is 800°C and the reference junction temperature is 30°C. **[Ans.** 31.59 mV**]**
2. Power emitted from the outer surface of a furnace is 20.5 ± 0.5 kW/m^2, the surface emissivity being 0.85 ± 0.1. Find the surface temperature of the furnace alongwith the uncertainity. **[Ans.** (891.9 ± 26.5)K**]**
3. Find the resistance value at 100°C for a copper RTD if its value is 53 Ω at 0°C. α for copper = 0.0043/°C. **[Ans. 75.79 Ω]**
4. A thermistor has a resistance of 10 kΩ at 25°C and 1 kΩ at 100°C. The range of operation is 0°C–150°C. The excitation voltage is 5 V and a series resistor of 1 kΩ is connected to the thermistor. The power dissipated in the thermistor at 150°C is

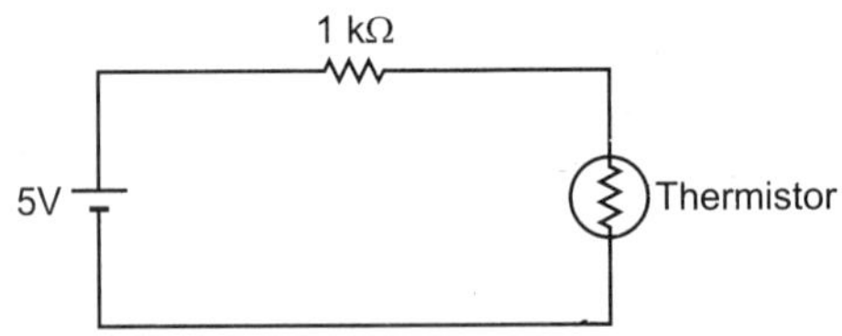

   (*a*) 4.0 mW (*b*) 4.7 mW (*c*) 5.4 mW (*d*) 6.1 mW
   [*GATE* 2007] **[Ans.** (*c*)**]**
5. The values of the material constant β for thermistors *P* and *Q* are 4000 K and 3000 K respectively. The resistance of each thermistor at 298 K is 2 kΩ. At 373 K, the ratio of the resistance of thermistor *P* to that of thermistor *Q* will be closest to
   (*a*) 1.33 (*b*) 1.00 (*c*) 0.75 (*d*) 0.50
   [*GATE* 2009] **[Ans.** (*d*)**]**
6. Consider a temperature measurement scheme shown in the figure. It was an RTD whose resistance at 0°C is 100 Ω and temperature coefficient of resistance (α) is 0.00392/°C.

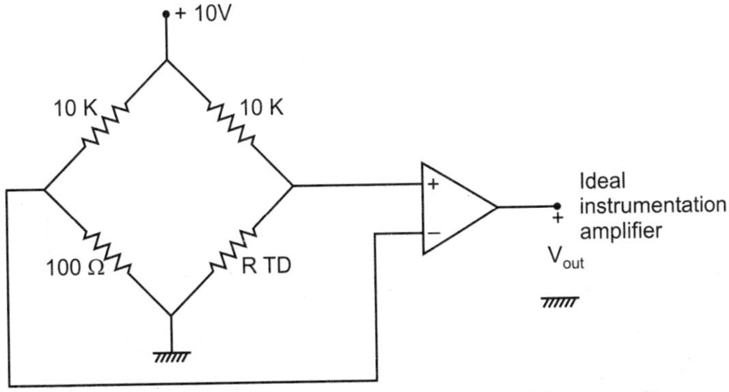

The differential gain of the instrumentation amplifier to achieve a voltage sensitivity of 10 mV/°C at 0°C should be approximately.

(*a*) 13.41 (*b*) 26.02 (*c*) 57.53 (*d*) 90.14 **[Ans. (*b*)]**

7. The RTD is placed in hot water bath of temperature 100°C. Based on the gain calculated from above the error in the measured value of temperature due to bridge non-linearity is

(*a*) – 0.1°C (*b*) – 0.4°C (*c*) – 0.9°C (*d*) + 1.2°C

[*GATE 2009*] **[Ans. (*c*)]**

8. A thermocouple based temperature measurement system is shown in the adjoining figure. Relevant thermocouple emf data (in mV) is given below. The cold junction is kept at 0°C. The temperature is 30°C in the other parts of the system.

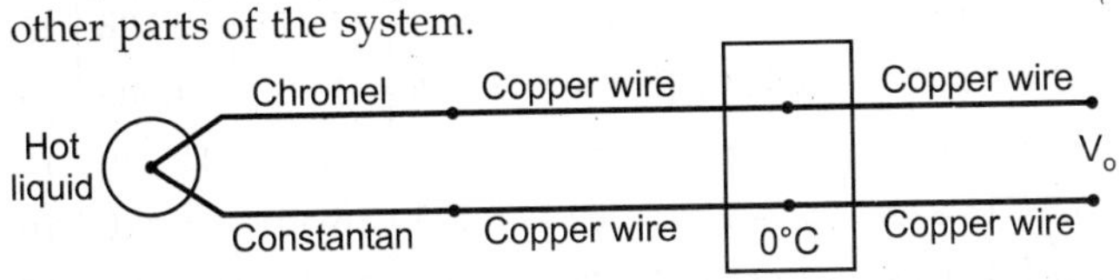

The emf V_o is measured to be 26.74 mV, the temperature of the hot liquid is

(*a*) 370.0°C (*b*) 372.4°C (*c*) 376.6°C (*d*) 380.0°C

Temperature	*Emf of Chromel-Constantan*	*Emf of Copper-Constantan*
10°C	0.591	0.391
20°C	1.192	0.789
30°C	1.801	1.196
370°C	26.549	19.027
380°C	27.345	19.638

[*GATE 2010*] **[Ans. (*d*)]**

9. The temperature of a furnace is monitored at a distance of 50 m away. The temperature transmitter has a range of 0° – 500°C and provides 4–20 mA current output. The measured temperature and the output current have a straightline relationship with positive slope. The temperature is determined from the voltage measured across a resistance of 500 Ω in the current loop. If the voltage measured across the resistance is 4 V, the temperature of the furnace is

(*a*) 100°C (*b*) 125°C (*c*) 150°C (*d*) 200°C

[*GATE 2011*] **[Ans. (*b*)]**

10. For a copper-constantan (Type *T*) thermocouple, the junction potential *E* (in μV) at θ°C is given by

$$E = 38.74\,\theta + 3.3 \times 10^{-2}\,\theta^2 + 2.07 \times 10^{-4}\,\theta^3 - 2.2 \times 10^{-6}\,\theta^4$$

The sensitivity of thermocouple at 100°C is approximately

(*a*) 45.34 μV/°C (*b*) 42.75 μV/°C (*c*) 38.74 μV/°C (*d*) 0.66 μV/°C

[*GATE 2011*] **[Ans. (*a*)]**

Table Tc 1: Reference calibration table for type-S thermocouple; conforms to IEC 584-1.1995. Reference temperature 0°C; emf in mV

°C	0	−10	−20	−30	−40	−50	−60	−70	−80	−90	−100
	0	-0.053	-0.103	-0.150	-0.194	-0.236					
°C	**0**	**10**	**20**	**30**	**40**	**50**	**60**	**70**	**80**	**90**	**100**
0	0	0.055	0.113	0.173	0.235	0.299	0.365	0.433	0.502	0.573	0.646
100	0.646	0.720	0.795	0.872	0.950	1.029	1.110	1.191	1.273	1.357	1.441
200	1.441	1.526	1.612	1.698	1.786	1.874	1.962	2.052	2.141	2.232	2.323
300	2.323	2.415	2.507	2.599	2.692	2.786	2.880	2.974	3.069	3.164	3.259
400	3.259	3.355	3.451	3.548	3.645	3.742	3.840	3.938	4.036	4.134	4.233
500	4.233	4.332	4.432	4.532	4.632	4.732	4.833	4.934	5.035	5.137	5.239
600	5.239	5.341	5.443	5.546	5.649	5.753	5.857	5.961	6.065	6.170	6.275
700	6.275	6.381	6.486	6.593	6.699	6.806	6.913	7.020	7.128	7.236	7.345
800	7.345	7.454	7.563	7.673	7.783	7.893	8.003	8.114	8.226	8.337	8.449
900	8.449	8.562	8.674	8.787	8.900	9.014	9.128	9.242	9.357	9.472	9.587
1000	9.587	9.703	9.819	9.935	10.051	10.168	10.285	10.403	10.520	10.638	10.757
1100	10.757	10.875	10.994	11.113	11.232	11.351	11.471	11.590	11.710	11.830	11.951
1200	11.951	12.071	12.191	12.312	12.433	12.554	12.675	12.796	12.917	13.038	13.159
1300	13.159	13.280	13.402	13.523	13.644	13.766	13.887	14.009	14.130	14.251	14.373
1400	14.373	14.494	14.615	14.736	14.857	14.978	15.099	15.220	15.341	15.461	15.582
1500	15.582	15.702	15.822	15.942	16.062	16.182	16.301	16.420	16.539	16.658	16.777
1600	16.777	16.895	17.013	17.131	17.249	17.366	17.483	17.600	17.717	17.832	17.947
1700	17.947	18.061	18.174	18.285	18.395	18.503	18.609				

Table Tc 2: Reference calibration table for type-R thermocouple; conforms to IEC 584-1.1995. Reference temperature 0°C; emf in mV

°C	0	−10	−20	−30	−40	−50	−60	−70	−80	−90	−100
	0	-0.051	-0.100	-0.145	-0.188	-0.266					
°C	**0**	**10**	**20**	**30**	**40**	**50**	**60**	**70**	**80**	**90**	**100**
0	0	0.054	0.111	0.171	0.232	0.296	0.363	0.431	0.501	0.573	0.647
100	0.647	0.723	0.800	0.879	0.959	1.041	1.124	1.208	1.294	1.381	1.469
200	1.469	1.588	1.648	1.739	1.831	1.923	2.017	2.112	2.207	2.304	2.401
300	2.401	2.497	2.597	2.696	2.796	2.896	2.997	3.099	3.201	3.304	3.408
400	3.408	3.512	3.616	3.721	3.827	3.933	4.040	4.147	4.255	4.363	4.471
500	4.471	4.580	4.690	4.800	4.910	5.021	5.133	5.245	4.357	5.470	5.583
600	5.583	5.697	5.812	5.926	6.041	6.157	6.273	6.390	6.507	6.625	6.743
700	6.743	6.861	6.980	7.100	7.220	7.340	7.461	7.583	7.705	7.827	7.950
800	7.950	8.073	8.197	8.321	8.446	8.571	8.697	8.823	8.950	9.077	9.205
900	9.205	9.333	9.461	9.590	9.720	9.850	9.980	10.111	10.242	10.374	10.506
1000	10.506	10.638	10.771	10.905	11.039	11.173	11.307	11.442	11.578	11.714	11.850
1100	11.850	11.986	12.123	12.260	12.397	12.535	12.673	12.812	12.950	13.089	13.228
1200	13.228	13.367	13.507	13.646	13.786	13.926	14.066	14.207	14.347	14.488	14.629
1300	14.629	14.770	14.911	15.052	15.193	15.334	15.475	15.616	15.758	15.899	16.040

°C	0	−10	−20	−30	−40	−50	−60	−70	−80	−90	−100
	0	−0.051	−0.100	−0.145	−0.188	−0.266					
°C	**0**	**10**	**20**	**30**	**40**	**50**	**60**	**70**	**80**	**90**	**100**
1400	16.040	16.181	16.323	16.464	16.605	16.746	16.887	17.028	17.169	17.310	17.451
1500	17.451	17.591	17.732	17.872	18.012	18.152	18.292	18.431	18.571	18.710	18.849
1600	18.849	18.988	19.126	19.264	19.402	19.540	19.677	19.814	19.951	20.087	20.222
1700	20.222	20.356	20.488	20.620	20.749	20.877	21.003				

Table Tc 3: Reference calibration table for type–B thermocouple; conforms to IEC 584-1.1995. Reference temperature 0°C; emf in mV

°C	0	10	20	30	40	50	60	70	80	90	100
0	0	−0.002	−0.033	−0.002	0.	0.002	0.006	0.011	0.017	0.025	0.033
100	0.033	0.043	0.053	0.065	0.078	0.092	0.107	0.123	0.141	0.159	0.178
200	0.178	0.199	0.220	0.243	0.267	0.291	0.317	0.344	0.372	0.401	0.431
300	0.431	0.462	0.494	0.527	0.561	0.596	0.632	0.669	0.707	0.746	0.787
400	0.787	0.828	0.870	0.913	0.957	1.002	1.048	1.095	1.143	1.192	1.242
500	1.242	1.293	1.344	1.397	1.451	1.505	1.561	1.617	1.675	1.733	1.792
600	1.792	1.852	1.913	1.975	2.037	2.101	2.165	2.230	2.296	2.363	2.431
700	2.431	2.499	2.569	2.639	2.710	2.782	2.854	2.928	3.002	3.078	3.154
800	3.154	3.230	3.308	3.386	3.466	3.546	3.626	3.708	3.790	3.873	3.957
900	3.957	4.041	4.127	4.213	4.299	4.387	4.475	4.564	4.653	4.743	4.834
1000	4.834	4.926	5.018	5.111	5.205	5.299	5.394	5.489	5.585	5.682	5.780
1100	5.780	5.878	5.976	6.075	6.175	6.276	6.377	6.478	6.580	6.683	6.786
1200	6.786	6.890	6.995	7.100	7.205	7.311	7.417	7.524	7.632	7.740	7.848
1300	7.848	7.957	8.066	8.176	8.286	8.397	8.508	8.620	8.731	8.844	8.956
1400	8.956	9.069	9.182	9.296	9.410	9.524	9.639	9.753	9.868	9.984	10.099
1500	10.099	10.215	10.331	10.447	10.563	10.679	10.796	10.913	11.029	11.146	11.263
1600	11.263	11.380	11.497	11.614	11.731	11.848	11.965	12.082	12.199	12.316	12.433
1700	12.433	12.549	12.666	12.782	12.898	13.014	13.130	13.246	13.361	13.476	13.591
1800	13.591	13.706	13.820								

Table Tc 4: Reference calibration table for type–J thermocouple; conforms to IEC 584-1.1995. Reference temperature 0°C; emf in mV

°C	0	−10	−20	−30	−40	−50	−60	−70	−80	−90	−100
−200	−7.890	−8.095									
−100	−4.633	−5.037	− 5.426	−5.801	−6.159	−6.500	−6.821	−7.123	−7.403	−7.659	−7.890
0	0	− 0.501	−0.995	−1.482	−1.961	−2.431	−2.893	−3.344	−3.786	−4.215	−4.633
°C	**0**	**10**	**20**	**30**	**40**	**50**	**60**	**70**	**80**	**90**	**100**
0	0	0.507	1.019	1.537	2.059	2.585	3.116	3.650	4.187	4.726	5.269

°C	0	10	20	30	40	50	60	70	80	90	100
100	5.269	5.814	6.360	6.909	7.459	8.010	8.562	9.115	9.669	10.224	10.779
200	10.779	11.334	11.889	12.445	13.000	13.555	14.110	14.665	15.219	15.773	16.327
300	16.327	16.881	17.434	17.986	18.538	19.090	19.642	20.194	20.745	21.297	21.848
400	21.848	22.400	22.952	23.504	24.057	24.610	25.164	25.720	26.276	26.834	27.393
500	27.393	27.953	28.516	29.080	29.647	30.216	30.788	31.362	31.939	32.519	33.102
600	33.102	33.689	34.279	34.873	35.470	36.071	36.675	37.284	37.896	38.512	39.132
700	39.132	39.755	40.382	41.012	41.645	42.281	42.919	43.559	44.203	44.848	45.494
800	45.949	46.141	46.786	47.431	48.074	48.715	49.353	49.989	50.622	51.251	51.877
900	51.877	52.500	53.119	53.735	54.347	54.956	55.561	56.164	56.763	57.360	57.953
1000	57.953	58.545	59.134	59.721	60.307	60.890	61.473	62.054	62.634	63.214	63.792
1100	63.792	64.370	64.948	65.525	66.102	66.679	67.255	67.831	68.406	68.980	69.553

Table Tc 5: Reference calibration table for type-T thermocouple; conforms to IEC 584–1.1995. Reference temperature 0°C; emf in mV

°C	0	–10	–20	–30	–40	–50	–60	–70	–80	–90	–100
–200	–5.603	–5.753	–5.888	–6.007	–6.105	–6.180	–6.232	–6.258			
100	–3.379	–3.657	–3.923	–4.177	–4.419	–4.648	–4.865	–5.070	–5.261	–5.439	–5.603
0	0	–0.383	–0.757	–1.121	–1.475	–1.819	–2.153	–2.476	–2.788	–3.089	–3.379
°C	**0**	**10**	**20**	**30**	**40**	**50**	**60**	**70**	**80**	**90**	**100**
0	0	0.391	0.790	1.196	1.612	2.036	2.468	2.909	3.358	3.814	4.279
100	4.279	4.750	5.228	5.714	6.206	6.704	7.209	7.720	8.237	8.759	9.288
200	9.288	9.822	10.362	10.907	11.458	12.013	12.574	13.139	13.709	14.283	14.862
300	14.862	15.445	16.032	16.624	17.219	17.819	18.422	19.030	19.641	20.225	20.872

Table Tc 6: Reference calibration table for type–E thermocouple; conforms to IEC 584-I.1995. Reference temperature 0°C; emf in mV

°C	0	–10	–20	–30	–40	–50	–60	–70	–80	–90	–100
–200	–8.825	–9.063	–9.274	–9.455	–9.604	–9.718	–9.797	–9.835			
–100	–5.237	–5.681	–6.107	–6.516	–6.907	–7.279	–7.632	–7.963	–8.273	–8.561	–8.825
0	0	–0.582	–1.152	–1.709	–2.255	–2.787	–3.306	–3.811	–4.302	–4.777	–5.237
°C	**0**	**10**	**20**	**30**	**40**	**50**	**60**	**70**	**80**	**90**	**100**
0	0	0.591	1.192	1.801	2.420	3. 048	3.685	4.330	4.985	5.648	6.319
100	6.319	6.998	7.685	8.379	9.081	9.789	10.503	11.224	11.951	12.684	13.421
200	13.421	14.164	14.912	15.664	16.420	17.181	17.945	18.713	19.484	20.259	21.036
300	21.036	21.817	22.600	23.386	24.174	24.964	25.757	26.552	27.348	28.146	28.946
400	28.946	29.747	30.550	31.354	32.159	32.965	33.772	34.579	35.387	36.196	37.005
500	37.005	37.815	38.624	39.434	40.243	41.053	41.862	42.671	43.479	44.286	45.093
600	45.093	45.900	46.705	47.509	48.313	49.116	49.917	50.718	51.517	52.315	53.112
700	53.112	53.908	54.703	55.497	56.289	57.080	57.870	58.659	59.446	60.232	61.017

°C	0	10	20	30	40	50	60	70	80	90	100
800	61.017	61.801	62.583	63.364	64.144	64.922	65.698	66.473	67.246	68.017	68.787
900	68.787	69.554	70.319	71.082	71.844	72.603	73.360	74.115	74.869	75.621	76.373
1000	76.373										

Table Tc 7: Reference calibration table for type–K thermocouple; conforms to IEC 584-1.1995. Reference temperature 0°C; emf in mV

°C	0	−10	−20	−30	−40	−50	−60	−70	−80	−90	−100
−200	-5.891	-6.035	-6.158	-6.262	-6.344	-6.404	-6.441	-6.458			
−100	-3.554	-3.852	-4.138	-4.411	-4.669	-4.913	-5.141	-5.354	-5.550	-5.730	-5.891
0	0	-0.392	-0.778	-1.156	-1.527	-1.889	-2.243	-2.587	-2.920	-3.243	-3.554
°C	**0**	**10**	**20**	**30**	**40**	**50**	**60**	**70**	**80**	**90**	**100**
0	0	0.397	0.798	1.203	1.612	2.023	2.436	2.851	3.267	3.682	4.096
100	4.096	4.509	4.920	5.328	5.735	6.138	6.540	6.941	7.340	7.739	8.138
200	8.138	8.539	8.940	9.343	9.747	10.153	10.561	10.971	11.382	11.795	12.209
300	12.209	12.624	13.040	13.457	13.874	14.293	14.713	15.133	15.554	15.975	16.397
400	16.397	16.820	17.243	17.667	18.091	18.516	18.941	19.366	19.792	20.218	20.644
500	20.644	21.071	21.497	21.924	22.350	22.776	23.203	23.629	24.055	24.480	24.905
600	24.905	25.330	25.755	26.179	26.602	27.025	27.447	27.869	28.289	28.710	29.129
700	29.129	29.548	29.965	30.382	30.798	31.213	31.628	32.041	32.453	32.865	33.275
800	33.275	33.685	34.093	34.501	34.908	35.313	35.718	36.121	36.524	36.925	37.326
900	37.326	37.725	38.124	38.522	38.918	39.314	39.708	40.101	40.494	40.885	41.276
1000	41.276	41.665	42.053	42.440	42.826	43.211	43.595	43.978	44.359	44.740	45.119
1100	45.119	45.497	45.873	46.249	46.623	46.995	47.367	47.737	48.105	48.473	48.838
1200	48.838	49.202	49.565	49.926	50.286	50.664	51.000	51.355	51.708	52.060	52.410
1300	52.410	52.759	53.106	53.451	53.795	54.138	54.479	54.819			

Table Tc 8: Reference calibration table for type–N thermocouple; conforms to IEC 548-I.1995. Reference temperature 0°C; emf in mV

°C	0	−10	−20	−30	−40	−50	−60	−70	−80	−90	−100
−200	-3.990	-4.083	-4.162	-4.226	-4.277	-4.313	-4.336	-4.345			
−100	-2.407	-2.612	-2.808	-2.994	-3.171	-3.336	-3.491	-3.634	-3.766	-3.884	-3.990
0	0	-0.260	-0.518	-0.772	-1.023	-1.269	-1.509	-1.744	-1.972	-2.193	-2.407
°C	**0**	**10**	**20**	**30**	**40**	**50**	**60**	**70**	**80**	**90**	**100**
0	0	0.261	0.525	0.793	1.065	1.340	1.619	1.902	2.189	2.480	2.774
100	2.774	3.072	3.374	3.680	3.989	4.302	4.618	4.937	5.259	5.585	5.913
200	5.913	6.245	6.579	6.916	7.255	7.592	7.941	8.288	8.637	8.988	9.341

°C	0	10	20	30	40	50	60	70	80	90	100
300	9.341	9.696	10.054	10.413	10.774	11.136	11.501	11.867	12.234	12.603	12.974
400	12.974	13.346	13.719	14.094	14.469	14.846	15.225	15.604	15.984	16.366	16.748
500	16.748	17.131	17.515	17.900	18.286	18.672	19.059	19.447	19.835	20.224	20.613
600	20.613	21.003	21.393	21.784	22.175	22.566	22.958	23.350	23.742	24.134	24.527
700	24.527	24.919	25.312	25.705	26.098	26.491	26.883	27.276	27.669	28.062	28.455
800	28.455	28.847	29.239	29.632	30.024	30.416	30.807	31.199	31.590	31.981	32.371
900	32.371	32.761	33.151	33.541	33.930	34.319	34.707	35.095	35.482	35.869	36.256
1000	36.256	36.641	37.027	37.411	37.795	38.179	38.562	38.944	39.326	39.706	40.087
1100	40.087	40.466	40.845	41.223	41.600	41.976	42.352	42.727	43.101	43.474	43.846
1200	43.846	44.218	44.588	44.958	45.326	45.694	46.060	46.425	46.789	47.152	47.513
1300	47.513										

Table RTD 1: Reference calibration table for Ni–100 Ω resistor and its tolerances. Resistance R in Ω versus temperature t in °C conforms to DIN 43760 (Not standardised by IEC)

°C	0	–5	–10	–15	–20	–25	–30	–35	–40	–45	ΔR/Δ*t* (Ω/°C)
–50	–74.3	71.9	69.5								0.47
0	100.0	97.3	94.6	91.9	89.3	86.7	84.1	81.6	79.1	76.7	0.52
°C	**0**	**+5**	**+10**	**+15**	**+20**	**+25**	**+30**	**+35**	**+40**	**+45**	
0	100.00	102.8	105.6	108.4	111.2	114.1	117.1	120.0	123.0	126.0	0.58
50	129.1	132.2	135.3	138.6	141.7	145.0	148.3	151.6	154.9	158.3	0.65
100	161.8	165.3	168.8	172.4	176.0	179.6	183.3	187.1	190.9	194.7	0.74
150	198.6	202.6	206.8	210.6	214.8	218.9	223.2	227.4	231.8	236.2	0.81

Temperature, t (°C)	Tolerances	
	± (°C)	± (Ω)
–60	2.1	1.0
0	0.4	0.2
100	1.1	0.8
200	1.8	1.6

Table RTD 2: Reference calibration table for Cu–100 Ω resistor and its tolerances. Resistance R in Ω versus temperature t in °C conforms to DIN 43760 (Not standardised by IEC)

°C	0	–5	–10	–15	–20	–25	–30	–35	–40	–45	ΔR/Δ*t* (Ω/°C)
–50	78.70										
0	100.00	97.87	95.74	93.61	91.48	89.35	87.22	85.09	82.96	80.83	0.43
°C	**0**	**+5**	**+10**	**+15**	**+20**	**+25**	**+30**	**+35**	**+40**	**+45**	
0	100.00	102.13	104.26	106.39	108.52	110.65	112.78	114.91	117.04	119.17	0.43
50	121.30	123.43	125.56	127.69	129.82	131.95	134.08	136.21	138.34	140.47	0.43
100	142.60	144.73	146.86	148.99	151.12	153.25	155.38	157.51	159.64	161.77	0.43
150	163.90	166.06	168.16	170.29	272.42	174.55	176.68				0.43

Temperature, t	Tolerances			
	Class 2		Class 3	
(°C)	± (°C)	± (Ω)	± (°C)	± (Ω)
-50	0.5	0.20	0.6	0.26
0	0.3	0.13	0.3	0.13
100	0.7	0.28	0.9	0.39
150	0.9	0.4	1.4	0.59

Table RTD 3: Reference calibration table for Cu–100 Ω resistor and its tolerances. Resistance R in Ω versus temperature t in °C conforms to IEC751: 1983 + A1: 1986 + A2: 1995. (t_{90} – ITS 90 temperature)

°C	0	−5	−10	−15	−20	−25	−30	−35	−40	−45	$\Delta R/\Delta t_{90}$ (Ω/°C)
−200	18.52										
−150	39.72	37.64	35.54	33.44	31.44	29.22	27.10	24.97	22.83	20.68	0.42
−100	60.26	58.23	56.19	54.15	52.11	50.06	48.00	45.94	43.88	41.80	0.41
50	80.31	78.32	76.33	74.33	72.33	70.33	68.33	66.31	64.30	62.28	0.40
0	100.00	98.04	96.09	94.12	92.16	90.19	88.22	86.25	84.27	82.29	0.39
°C	**0**	**+5**	**+10**	**+15**	**+20**	**+25**	**+30**	**+35**	**+40**	**+45**	
0	100.00	101.95	103.90	105.85	107.79	109.73	111.67	113.61	115.54	117.47	0.39
50	119.40	121.32	123.24	125.16	127.08	128.99	130.90	132.80	134.71	136.61	0.38
100	138.51	140.40	142.29	144.18	146.07	147.95	149.83	151.71	153.58	155.46	0.38
150	157.33	159.19	161.05	162.91	164.77	166.63	168.48	170.33	172.17	174.02	0.37
200	175.86	177.69	179.53	181.36	183.19	185.01	186.84	188.66	190.47	192.29	0.36
250	194.10	195.91	197.71	199.51	201.31	203.11	204.90	206.70	208.48	210.27	0.36
300	212.05	213.83	215.61	217.38	219.15	220.95	222.68	224.45	226.21	227.96	0.35
350	229.72	231.47	233.21	234.96	236.70	238.44	240.18	241.91	243.64	245.37	0.35
400	247.09	248.81	250.53	252.25	253.96	255.67	257.38	259.08	260.78	262.48	0.34
450	264.18	265.87	267.56	269.25	270.93	272.61	274.29	275.97	277.64	279.31	0.34
500	280.98	282.64	284.30	285.96	287.62	289.27	290.92	292.56	294.21	295.85	0.33
550	297.49	299.12	300.75	302.38	304.01	305.63	307.25	308.87	310.49	312.10	0.32
600	313.71	315.31	316.92	318.52	320.12	321.71	323.30	324.49	326.48	328.06	0.32
650	329.64	331.22	332.79	334.36	335.93	337.50	339.06	340.62	342.18	343.73	0.31
700	345.28	346.83	348.38	349.92	351.46	353.00	354.53	356.06	357.59	359.12	0.31
750	360.64	362.16	363.67	365.19	366.70	368.21	369.71	371.21	372.71	374.21	0.30
800	375.70	377.19	378.68	380.17	381.65	383.13	384.60	386.08	387.55	389.02	0.30
850	390.48										

Temperature, t t_{90} (°C)	Tolerances			
	Class A		Class B	
	± (°C)	± (Ω)	± (°C)	± (Ω)
-200	0.55	0.24	1.3	0.56
-100	0.35	0.14	0.8	0.32
0	0.15	0.06	0.3	0.12
100	0.35	0.13	0.8	0.30
200	0.55	0.20	1.3	0.48
300	0.75	0.27	1.8	0.64
400	0.95	0.33	2.3	0.79
500	1.15	0.38	2.8	0.93
600	1.35	0.43	3.3	1.06
650	1.45	0.46	3.6	1.13
700			3.8	1.17
800			4.3	1.28
850			4.6	1.34

CHAPTER 7

PRESSURE MEASUREMENT

7.1 INTRODUCTION

Pressure is a very important process parameter and its measurement in an accurate manner is critical to the quality of end product in any process industry. A very large variety of instruments varying widely in principle of operation, design, accuracy and purpose are employed for measurement of pressure. The physical and chemical nature of the substance whose pressure is required to be measured, influences the type of instrument/transducer to be employed. Again, the physical/chemical characteristics of the transducer must be known apriori before it is employed for a specific application. For example, in several situations, a manometer may be considered for pressure measurement because of its simplicity, very low cost and relatively high inherent accuracy but is a poor choice because its dynamic response is sluggish. As a second example, a Bourdon tube is a good choice for measurement of high static pressure but is not recommended when the pressure fluctuates rapidly.

Industrial pressure measurements range from below atmospheric to several hundred atmospheric pressures. Pressure as low as 10^{-11} mm of mercury can be measured very easily.

Both mechanical and electrical methods are employed for pressure measurements.

7.2 DIFFERENT TYPES OF PRESSURES

Pressure measurement falls under the following categories: absolute, gauge, vacuum and differential.

- **Absolute pressure:** This is the pressure of the fluid measured with respect to zero pressure (*i.e.* total vacuum).
- **Gauge pressure:** This is the pressure of the fluid measured with respect to standard atmospheric pressure.
- **Vacuum pressure or Simply vacuum** (also called **suction** or **negative pressure**): When gauge pressure is negative, it is known as vacuum, *i.e.* any pressure less than the atmospheric pressure is known as vacuum.
- **Differential pressure:** It is the difference between two pressures. For instance, in a flowing fluid, if an orifice is placed, a pressure difference will be there on either side of the orifice and is known as differential pressure. Gauge pressure is also an example of differential pressure in which one end is always atmospheric pressure. Figure 7.1 is self-explanatory and represents the different kinds of pressure measurements encountered in industries:

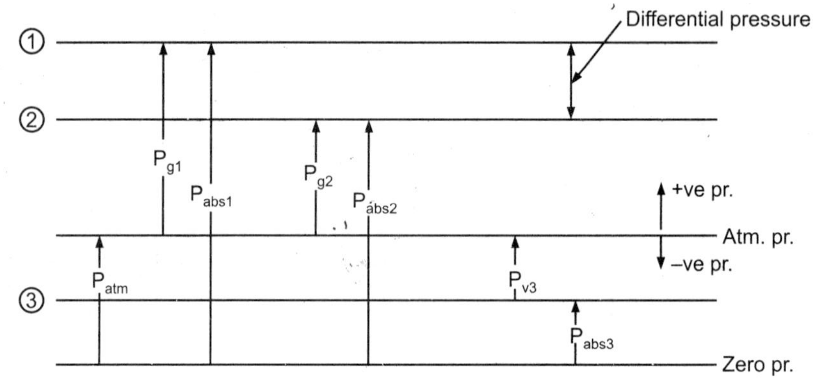

Fig. 7.1: Representation of gauge, absolute, vacuum and differential pressures

P_{g1} and P_{abs1} are the gauge and absolute pressures at level 1 respectively. The following relationship holds between P_g, P_{atm} and P_{abs}

$$P_{abs} = P_g + P_{atm}$$

or

$$P_g = P_{abs} - P_{atm}$$

If now, $P_{abs} > P_{atm}$, then P_g is +ve and is called gauge pressure while if $P_{abs} < P_{atm}$, then P_g is –ve and is called vacuum pressure or vacuum or suction pressure.

Industrial pressure measurement ranges vary from as low as 10^{-11} mm of mercury column to as high a pressure as 1,50,000 psi. Figure 7.2 indicates the approximate ranges and spans of various devices for pressure measurement.

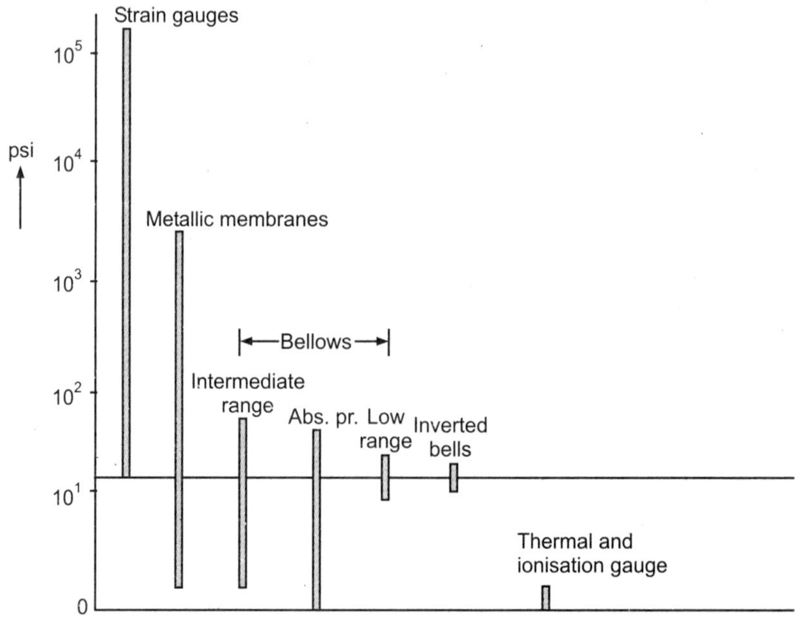

Fig. 7.2: Approximate ranges of different elements and gauges for pressure measurement

7.3 TYPES OF PRESSURE MEASURING INSTRUMENTS BASED ON RANGES

Instruments which measure atmospheric pressure are known as barometers, while the ones which measure absolute pressure of a process are called absolute pressure gauges.

Again, instruments which measure gauge pressures are known by the name of Pressure Gauges while Vacuum Gauges measure the vacuum existing at a place. Instruments capable of measuring both gauge pressure and vacuum are called Compound Pressure Gauges. Instruments which measure small differential pressure, are called manometers.

Lastly, instruments which measure low gauge pressure or vacuum are called Forced draft and Draft gauges respectively. And the instruments measuring both Forced-draft and Draft are known as Compound Draft Gauges. Table 7.1 shows the different instruments and the types of pressures measured by them.

Table 7.1: Different Instruments/Gauges and the Types of Pressures Measured by them

Instruments/Gauges	*Types of pressures measured*
Barometer	Atmospheric pressure
Absolute Pressure Gauge	Absolute pressure
Pressure Gauge	Gauge pressure
Vacuum Gauge	Vacuum
Compound Gauge	Gauge pressure and vacuum
Manometer	Small pressure differential
Draft Gauge	Low vacuum
Forced Draft Gauge	Low gauge pressure
Compound Draft Gauge	Low gauge pressure and low vacuum

Chapter 7

7.4 UNITS OF PRESSURE

The unit of pressure is not a primary quantity but is a derived unit. The MKS or SI unit of pressure is Pascal, abbreviated as Pa. It is defined as the force of one Newton per square metre, *i.e.*

1 Pa = 1 N/m^2 = 10 dyne/cm^2

and 1 MPa = 10^6 Pa.

In many cases, pressure is expressed in terms of mm of water column (mm wc) or mm of Hg column (mm Hg). When pressures are so expressed, reference should always be 4°C for water and 0°C for mercury and for acceleration due to gravity its standard value 9.80665 m/s^2 is taken. For meniscus type pressure gauges (manometers), units mm wc or mm Hg are so often used, while pressure measuring instruments are calibrated in terms of Kgf/m^2 or Kgf/cm^2 ('*f*' for force).

In FPS unit, absolute pressure is expressed in Pounds per in^2 absolute (psia) or simply Pounds absolute (pa), while the gauge pressure is expressed in Pounds per in^2 gauge (psig).

Lastly, another unit 'atmosphere' or simply 'atm' is used for expressing high pressures. Thus, a 'pressure of 30 atm' means that the pressure of the concerned place is 30 times the atmospheric pressure.

Some other units are given below to give an overview of the different units used.

1μ (micron) = 10^{-4} cm of Hg = 10^{-6} m of Hg

1 Pa = 10 dyne/cm^2 [1 Pa = 1 N/m^2 = $\frac{10^5 \text{ dyne}}{10^4 \text{ cm}^2}$ = 10 dyne/cm^2]

1 bar = 10^5 Pa = 1 Atm = 1.03 Kg/cm^2

1 Torr = 1 mm of Hg

1 Pa = 1 N/m^2 = 10μ bar

1 Atm = 76 × 13.6 gm.wt/cm^2 = 14.7 psi.

Conversion between different units is given in Table 7.2.

Table 7.2: Conversion Table between Pascal, Kgf/m², Kgf/cm², mm wc and mm Hg

Unit	*Pa*	*Kgf/m²*	*Kgf/cm²*	*mm wc*	*mm Hg*
1 Pa	1	0.10197	10.197×10^{-6}	0.101197	7.50×10^{-3}
1 Kgf/m²	9.80665	1	10^{-4}	1	73.56×10^{-3}
1 Kgf/cm²	98.0665×10^3	10^4	1	10^4	735.56
1 mm wc	9.80665	1	10^{-4}	1	73.56×10^{-3}
1 mm Hg	133.322	13.595	13.595×10^{-4}	13.595	1

7.5 PRESSURE STANDARDS

Several pressure standards are available against which all pressure measuring instruments are calibrated to ensure that the desired accuracy is always maintained. Some of the standards used are:

- Dead weight testers,
- Liquid column testers and
- Test gauges—which themselves are periodically checked and calibrated against a more accurate standard.

7.5.1 Dead Weight Tester

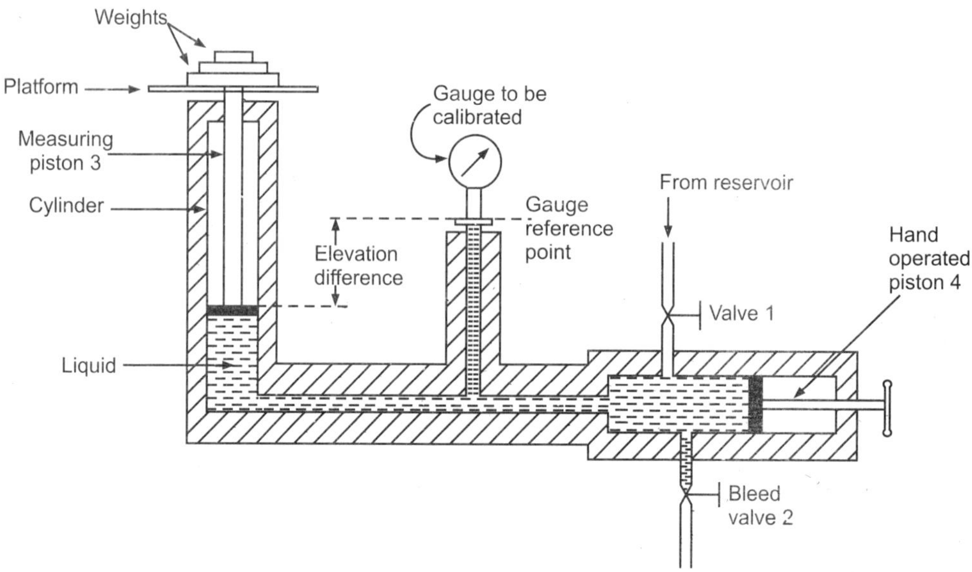

Fig. 7.3: A dead weight gauge calibrator

A dead weight or piston gauge is a primary pressure standard and is employed as a standard for calibration of less accurate gauges/transducers. Pascal's law is the basis for its operation. It is a sturdy instrument and should have a lifelong accuracy, if handled with proper care.

The schematic of a dead weight tester is shown in figure 7.3. Piston 4 is fully drawn out and the system is filled up with oil via valve 1, which is then closed. Weights are then placed over the platform of the measuring piston 3 and the piston 4 is turned clockwise to build up the pressure inside. When the inside pressure goes on increasing, a point is reached when the measuring piston 3 is just lifted which balances the weights alongwith the weight of the platform. At this point, the pressure indicated by the gauge (to be calibrated) equals the weights supported by the piston divided by the piston's area.

Several considerations should be borne in mind for accurate results to be expected from the system. First, a small clearance exists between the cylinder and piston 3 resulting in an axial flow of fluid from high pressure (below piston 3) to low pressure side (above piston 3). This thus results in a viscous drag (shear) force that supports a part of the weights. This effectively increases the cross-sectional area which is the average of the piston and the cylinder areas. Second, the local acceleration due to gravity influences the reading. Third, buoyancy of air acts on the platform and weights on it and also the buoyancy of oil on the piston, thereby reducing the weight.

Fourth, there is an effect of temperature on both the piston and weights. Fifth, there is a height difference between the lower end of the measuring piston and the gauge reference level.

Conventional dead weight testers cannot measure pressures below the 'tare' pressure. The tare pressure is obtained by dividing the piston assembly weight by its area. This is overcome in 'tilting' type piston gauges.

The readings of the dead weight tester is subject to corrections for extremely high pressures. This is due to variations in the diameter of the piston and the cylinder due to increase in pressure.

Due to the above mentioned errors, the following equation is utilised, for reduction of errors, which is applicable in the normal pressure range.

$$P = \left(\frac{m}{a} - \rho h - \rho h'\right)\frac{g}{g_s} \quad ...(7.1)$$

where P = pressure

m = mass of weights and platform.

a = effective piston area.

ρ = density of liquid.

h = difference in level between that of the gauge under test and the bottom position of the piston.

h' = correction due to buoyancy for the immersed part of the piston.

7.6 MANOMETERS

7.6.1 Introduction

Manometers, or more commonly known as liquid-column manometers, are used for measurement of low values of differential pressure measurements. There are many varieties of manometers which are in common use. These are: *U*-tube type, *U*-tube type with sealing liquid, inverted *U*-tube manometer, well type, seal pots type, enlarged leg float type, bell type, inclined tube type, ring Balance type, etc.

Manometers are simple in their operation, are direct reading local indication type and are very accurate. They are also sometimes used for reference purposes.

The most commonly used manometric liquids are water and mercury—the choice of liquid depends upon the type and character of the fluid whose pressure is to be determined. The bore of the *U*-tube must not be less than 8–10 mm and must remain constant along the tube. Lesser bore diameter tubes must not be used when the manometric fluid is water because of capillary effects.

The error in a *U*-tube manometer mainly depends upon the height of the liquid column and the accuracy with which this height is read. To increase accuracy, *U*-tube manometers are sometimes fitted with a mirror scale.

The accuracy of a manometer is inherent in nature since it has no moving parts, no gears, no rack and pinion arrangement, no bellows, no diaphragms, etc., which are subject to wear, fatigue etc. A manometer requires no calibration. In fact, a manometer is used as a standard against which other gauges are calibrated.

7.6.2 The *U*-Tube Manometer

A simple *U*-tube manometer is shown in figure 7.4. It consists of a glass *U*-tube partly filled with manometric liquid. The liquid may be mercury, water or alcohol. A scale attached to the two limbs can read off the difference in reading between the two columns.

Let ρ_l is the density of the manometric liquid. Since the system is balanced, thus pressure at *A* must be equal to the pressure at *B*. Thus,

$$P_1 = P_2 + h \rho_l g$$

or,

$$P_1 - P_2 = h \rho_l g$$

when P_1 and P_2 are connected to two processes, the manometer measures the differential pressure. When gauge pressure is desired, the left-leg is connected to the process while the right leg is exposed to the atmosphere. To measure vacuum, the right leg is connected to the process vacuum while the left leg is left open.

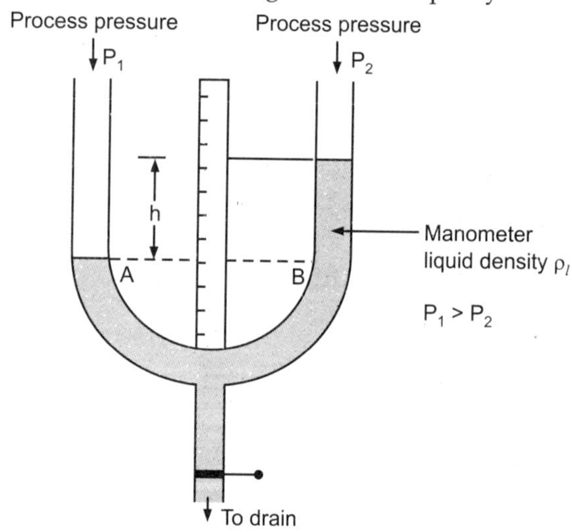

Fig. 7.4: A U-tube manometer

The bore of the *U*-tube should be constant and should be of such diameter that capillary action will not affect the measurement.

There is a provision at the bottom for draining the manometric liquid. If less amount of liquid is there in the manometer, a greater differential can be measured for a given manometer.

In case the process fluid reacts chemically with the manometer liquid, there is a provision for sealing liquids in the manometer. Such an arrangement is shown in figure 7.5. In this case, the process fluid is prevented from coming into direct contact with the manometer liquid by the sealing liquid. Equal volume of sealing liquid is put on either limb of the manometer. Under balance condition, since pressure at *A* is equal to the pressure at *B*, thus,

$$P_1 + l \rho_s g = P_2 + l \rho_s g + h \rho_l g$$

or,

$$P_1 - P_2 = h \rho_l g \qquad ...(7.2)$$

- This equation is the same one as before.

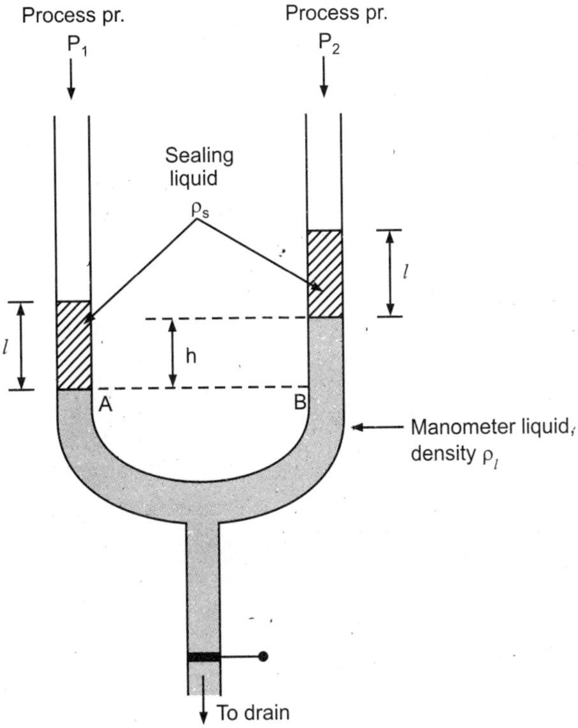

Fig. 7.5: U-tube Manometer with sealing liquid

On application of pressure, liquid in one limb goes down while in the other it moves up. Thus, no fixed reference is there in either of the limbs. This makes the measurement more error prone than it would if one limb could have been maintained at a fixed reference level.

7.6.3 Inverted *U*-tube Manometer

Sometimes, it is desired to measure the pressure differential across an orifice placed in a flowing fluid. It may so happen that employing the same fluid in the manometer as that the flowing fluid in the pipe is recommended. In such situations, an inverted *U*-tube manometer is employed to measure the said pressure differential and shown in figure 7.6.

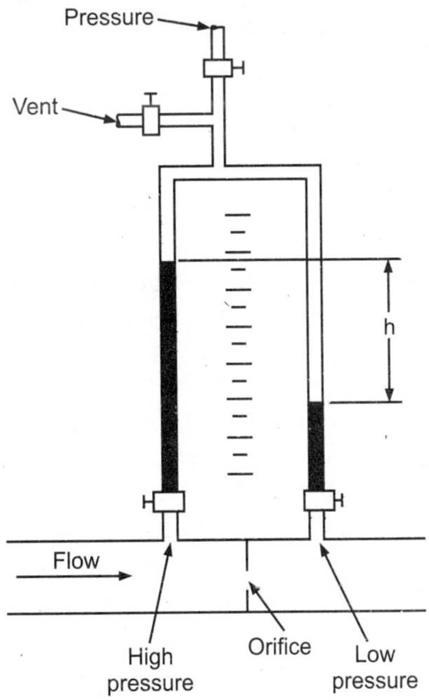

Fig. 7.6: Inverted-U-tube manometer

The pressure connections are made directly to manometer legs, while equal pressure is applied to the top of the liquid in each leg. A vent is provided in the pressure line applied from the top. This facilitates in keeping the differential pressure reading in the vicinity of central portion of the manometer. It should be noted at this point that venting does not alter the differential pressure reading; *h*– only alters its position to facilitate reading.

7.6.4 Well (Reservoir) Type Manometer

Two versions of well type manometer is shown in figure 7.7. Figure 7.7(*a*) consists of a large volume container, called the well, having cross-section A_1 and a glass tube, called the leg, having cross-section A_2. The system is filled with the manometric liquid (may be water or mercury, depending on pressure differential).

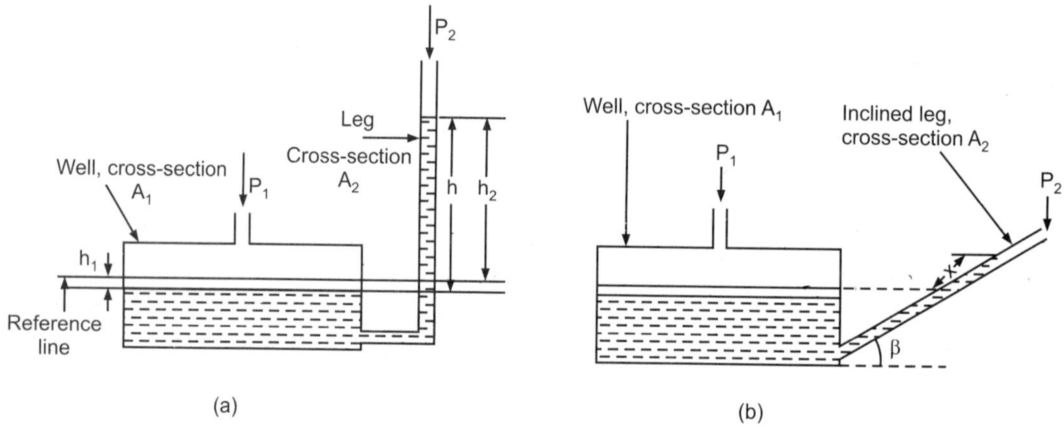

Fig. 7.7: Well type manometer (*a*) vertical leg (*b*) inclined leg

In measuring pressure differentials, the well is connected to the higher process pressure and the leg to the lower process pressure. In measuring gauge pressure, it is connected to the well and in measuring vacuum the leg is connected to the unknown vacuum pressure.

Let an unknown pressure differential lowers the well height by h_1 and increases the leg height by h_2 from the reference level. Thus, $h = h_1 + h_2$ is actually the pressure differential from the figure,

$$A_1 h_1 = A_2 h_2$$

$$\therefore \qquad h = h_1 + h_2 = h_2 \cdot \frac{A_2}{A_1} + h_2 = h_2\left(1 + \frac{A_2}{A_1}\right) = h_2\left(1 + \frac{d_2^2}{d_1^2}\right)$$

where d_1 and d_2 are the bore diameters of well and leg respectively.

Thus, if ρ_l is the density of the manometric liquid, then

$$\Delta P = P_1 - P_2 = \rho_l h = \rho_l h_2\left(1 + \frac{d_2^2}{d_1^2}\right) = \rho_l h_2\left(1 + \frac{A_2}{A_1}\right) \qquad ...(7.3)$$

If $A_2 << A_1$, then $\Delta P = \rho_l h_2$ and a single height measurement (h_2) from the reference level is suffice to find out the unknown pressure, unlike two height measurements as in a simple *U*-tube manometer. This is obviously an advantage of well type manometer. A_1/A_2 may be as much as 400 times so that error in measurement is greatly reduced. Figure 3.7(*b*) represents a well-type manometer with inclined tube. This increases sensitively of measurement in which the leg is inclined at an angle β with the horizontal.

In this case,

$$\Delta P = P_1 - P_2 = \rho_l\, x\left(1 + \frac{A_2}{A_1}\right). \sin \beta \qquad ...(7.4)$$

The instrument scale is obviously enlarged because if β is increased, $\cos\beta$ decreases, so that for a given unknown pressure difference x will increase — such that $x\cos\beta$ remain constant.

7.6.5 Enlarged Leg Float Type Manometer

An enlarged leg float type manometer is shown in figure 7.8. In this manometer, in the enlarged leg, *i.e.* well, there is a float, the position of which varies with pressure differences. The displacement of the float from the reference line due to change in differential pressure is taken to the pointer via the thread. The pointer is calibrated in terms of differential pressure. The pressure balance

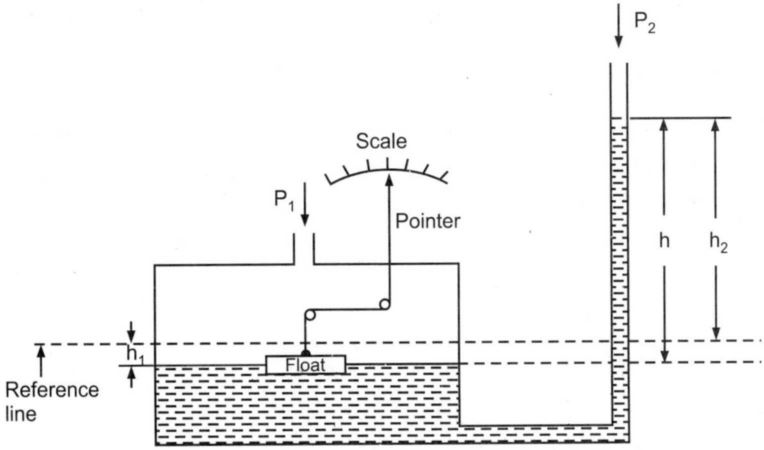

Fig. 7.8: An enlarged leg float type manometer

equation remains the same as in the case of well type manometer (vertical leg) and is given by:

$$\Delta P = P_1 - P_2 = \rho_l h_2\left(1 + \frac{A_2}{A_1}\right)$$

The tubing connecting the well and the leg can be changed. If provision is made for changing the leg then its cross-section A_2 will be changed. This is sometimes done to change the measurement span of the system.

7.6.6 Bell (Inverted Bell) Type Manometer

The bell type manometer shown in figure 7.9, can measure pressure differential, low gauge pressure or vacuum.

It consists of an inverted bell floating on the surface of the liquid. A change in pressure causes the bell to move up or down against spring tension which is attached to the top inside surface of the container, as shown. Thus, a scale attached to the bell can be calibrated in terms of differential pressure.

The two pressures P_1 and P_2 are led into the manometer as shown by means of connections to some tubing.

At balance, making use of force balance equation:

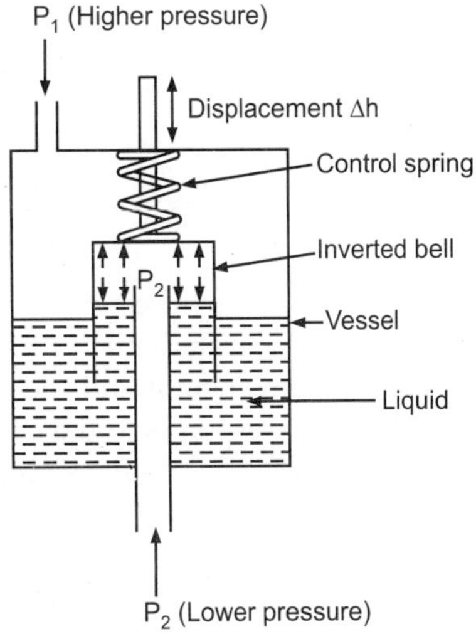

Fig. 7.9: Inverted bell-type manometer

$$(P_1 - P_2) \times A = K \cdot \Delta h \qquad \text{...(7.5)}$$

where A represents the bell cross-sectional area, K spring constant and Δh is the shift in the position of the spring.

For small pressure differential measurements the cross-section (A) of the bell is to be increased (thus the site of the system will increase—a disadvantage) or springs should be chosen with low spring constant (K) values.

7.6.7 Ring Balance Manometer

It is used for the measurement of low vacuum and low gauge pressure of gases. It is also used to measure the flow rate of gases based on the pressure difference across a restriction, as applied in head flow meters.

The arrangement of a ring balance manometer is shown in figure 7.10. A hollow ring having a circular cross-section (a tube of light material like polythene which is transparent) is separated at the top by a gas-tight partition and the bottom part is filled up with a sealing liquid like paraffin or transformer. A knife-edge supports the ring at the centre.

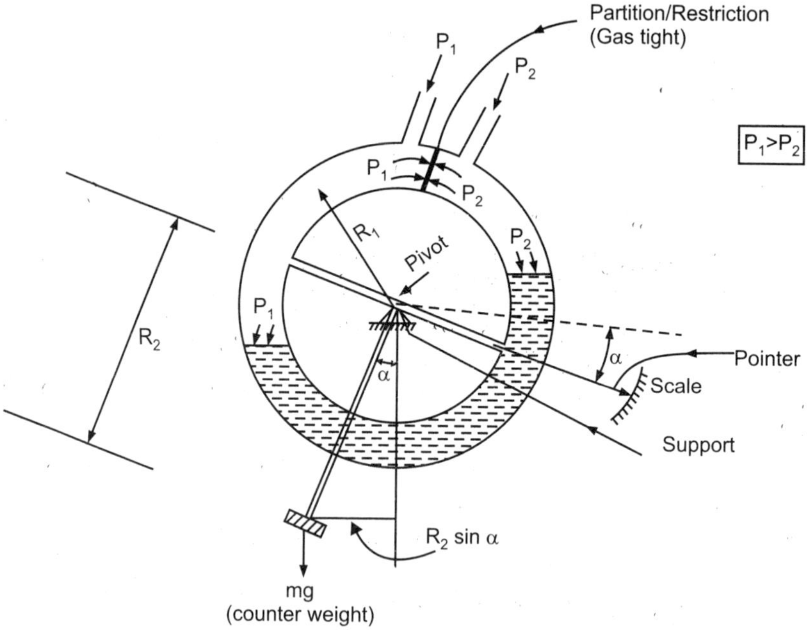

Fig. 7.10: A Ring Balance Manometer

Two pressure taps are provided at the top end of the ring on either side of the partition through which the differential pressure ($P_1 \sim P_2$) is applied. When $P_1 = P_2$, the liquid on either side will be at the same height and the counter weight (mg) will be in its bottom most position. The figure shown corresponds to the case of $P_1 > P_2$, so that the liquid level on the left is depressed a bit (due to higher pressure P_1 existing on that side) and the counter weight assuming the position as shown. Under balanced conditions, the controlling moment provided by the counter weight is balanced by the rotating moment arising out of the pressure differential across the partition. From the figure, it can be written as:

$$\text{Rotating moment} = A\, R_1\, (P_1 - P_2)$$

and, Controlling moment = $mg\, R_2 \sin \alpha$

Thus, the system balance equation is

$$A\, R_1\, (P_1 - P_2) = mg\, R_2 \sin \alpha$$

or

$$P_1 - P_2 = \frac{mg\, R_2 \sin \alpha}{A\, R_1} \quad ...(7.6)$$

Here, R_1 = mean radius of the ring

R_2 = length of the counter weight arm

A = cross-sectional area of the ring

α = angle through which the ring has turned due to the pressure difference $(P_1 - P_2)$

In the expression, no density term of the fluid is involved and hence the instrument is independent of any variation in density due to temperature changes.

7.6.8 Tilting Manometer

It utilises the same principle as the ring balance manometer and is shown in figure 7.11. Pressure is applied at the two ends of the tube containing mercury as sealing liquid. The tube is balanced against a knife-edge pivot. The spring, as shown in figure, provides the restoring force required for balancing. The angle of tilt may be calibrated in terms of differential pressure.

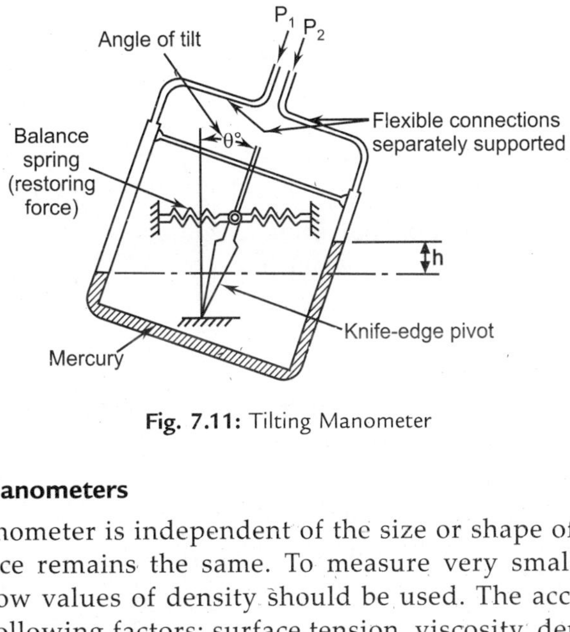

Fig. 7.11: Tilting Manometer

7.6.9 Accuracy of Manometers

The accuracy of a manometer is independent of the size or shape of the U-tube, provided the pressure difference remains the same. To measure very small pressure differences, liquids having very low values of density should be used. The accuracy of a manometer is dependent on the following factors: surface tension, viscosity, density, temperature and vapour pressure.

Surface tension effect is avoided by using large-bore tubes. Viscosity effect is not much of an importance in vertical type U-tube manometers, but it is not so in inclined gauges. Since density of a liquid is temperature dependent, hence pressure readings obtained from manometers would be erroneous, unless temperature effect is corrected for. Since manometers are increasingly applied to measure vacuum, hence vapour pressure of the manometric liquid should be given due consideration.

Chapter 7

7.7 MICROMANOMETERS

7.7.1 Introduction

To extend the range of manometers, micromanometers are used which have a range of 0.005 mm to 500 mm of water column. Several varieties of micromanometers are in use which eliminate the effects of meniscus and capillary effects. Micromanometers—both laboratory and industrial types—are used for measurement of low pressure, suction and pressure differential of air and noncorrosive gases in testing thermal power plants and similar installations. Micromanometers are classified into working and reference standards. Working micromanometers are further subdivided into engineering and laboratory types.

Several types of micromanometers are available which are mentioned below:

- Prandtl-type micromanometer,
- Micrometer type micromanometer and
- Air micromanometer.

These are now discussed below:

7.7.2 Prandtl-type Micromanometer

The construction of this type of micromanometer is shown in figure 7.12. When the pressure in both the limbs is same (*i.e.* $p_1 = p_2$), the meniscus in the inclined tube is adjusted by a fixed hairline to act as a reference point. For better viewing, a magnifier is used. The reading of the micrometer is noted. With the application of pressure. *i.e.*, when $p_1 \neq p_2$, the menisecus in the inclined tube moves off the hairline (either goes down or up, depending on whether $p_1 > p_2$ or otherwise). But the meniscus can be restored to touch the hairline again by adjusting the well along with the micrometer. This new reading of the micrometer is noted. The difference between the two micrometer readings is thus a measure of the pressure difference between p_1 and p_2.

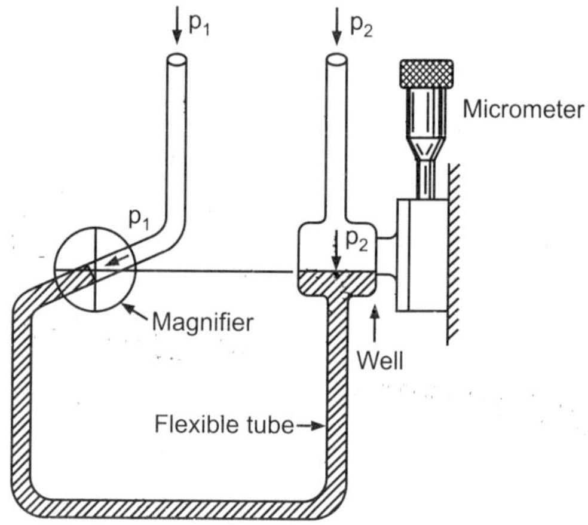

Fig. 7.12: A Prandtl-type manometer

When water is used as the working fluid, readings as low as 0.0254 mm of water can be read for an instrument range of 254 or 508 mm of water.

7.7.3 Micrometer Type Micromanometer

In this type of micromanometer, shown in figure 7.13, two large bore transparent tubes are connected as shown, forming the shape of a U-tube.

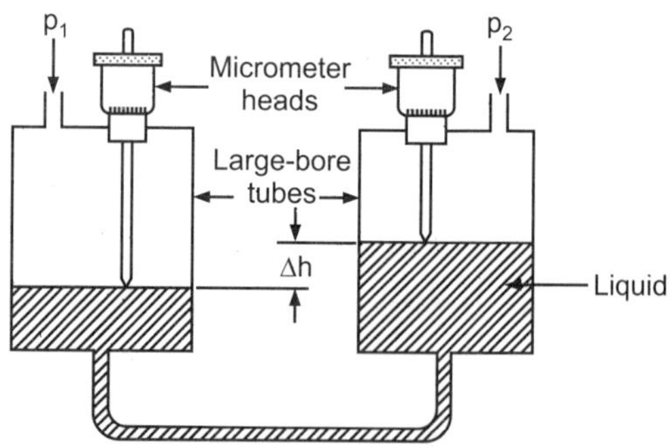

Fig. 7.13: A micrometer type manometer

The huge bore tubes eliminate the effect of capillary. Pressures p_1 and p_2 are applied as shown while measurements are done by adjustable sharp index points touching the central part of the two liquids. The contacts with the surface of the manometric liquids may be done either optically or electrically. The difference between the two micrometer readings is a measure of Δh and hence that of $p_1 \sim p_2$.

7.7.4 Air Micromanometer

An air micromanometer is shown in figure 7.14. It is an extremely sensitive one in which all capillary and meniscus effects are avoided by using air as the working fluid instead of customary liquid.

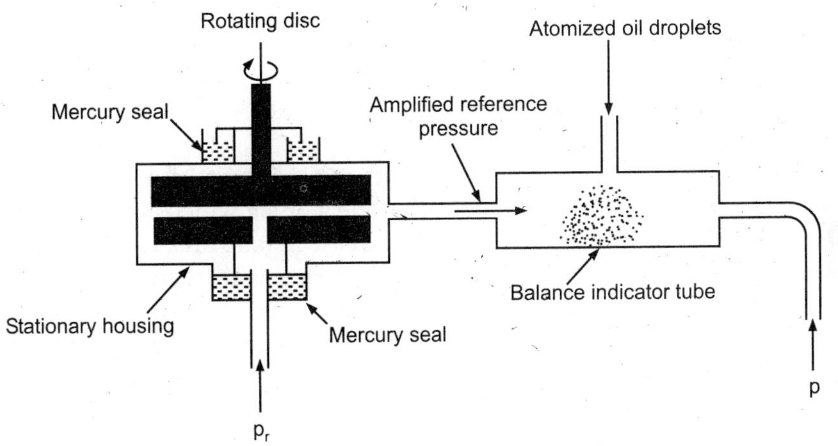

Fig. 7.14: An air micromanometer

The rotating disc in a stationary housing mechanically amplifies the reference pressure by the principle of centrifugal action. The disc speed is adjusted until the amplified reference pressure just balances the unknown pressure p. This balance is achieved when a lack of movement of oil droplets is observed, kept in a glass indicator tube.

Under balance condition, the pressure difference between p_r and p is given by:

$$\Delta p = p_r - p = \alpha \rho \omega^2 \quad ...(7.7)$$

where α is a constant term depending on the disc radius and the annular clearance between the disc and the housing, ρ is the reference air density and ω is the angular speed of the disc.

This air-type centrifugal micromanometer is in a position to measure a pressure as low as 0.005 mm of water column, with an uncertainty factor of 1% only.

7.8 ELASTIC PRESSURE SENSING ELEMENTS

7.8.1 Introduction

There are several varieties of elastic pressure sensing elements for measurement of pressure in the range of 50 Pa (5 kgf/cm^2) to 1000 MPa (10,000 kgf/cm^2). The elements normally used are diaphragms, bellows and Bourdon tubes (also called tubular springs). Based on these sensors, pressure gauge, vacuum gauge, draft gauge, forced draft gauge, compound pressure and vacuum gauges are available. These instruments are classified into *reference* and *working* standards. The variation of pressure produces a deformation or a bending moment in the sensing element. These variations are then calibrated in terms of pressure or pressure differential, as the case may be.

Chapter 7

The pressure-deflection or force-deflection characteristics of an elastic sensor should be linear so that a uniform scale for pressure is obtained, while its performance is decided by the two parameters stiffness and sensitivity. These two quantities are reciprocal of each other. Sensitivity to an applied pressure is given by $s = \lambda/p$, where λ is the deflection corresponding to the applied load. For an elastic element having a non-linear pressure-deflection characteristics, it is expressed as:

$$s = \frac{d\lambda}{dp}.$$

Elastic elements used as pressure sensor show a marked hysteresis loop in their behaviour *i.e.,* the deflection for an increasing or decreasing load will differ for a given applied pressure. The corresponding curve is shown in figure 7.15(*a*).

Accuracy of the instrument is dependent on the hysteresis – a thinner sized hysteresis loop will result in better accuracy.

The hysteresis loop of the elastic elements has its origin in non-ideal elastic behaviour. They also suffer from another drawback—the elastic after-effect or lag which arises due to the variation in their properties with time. This thus leads to an increase in loop area as shown in figure 7.15(*b*). This is represented by the outer curve in the figure. Performance variation of elastic elements may also be due to creep and excessive stress application beyond elastic limit. Thus, for reliable operation of an elastic element, the strains produced in it due to external and internal force must not exceed its elastic limit.

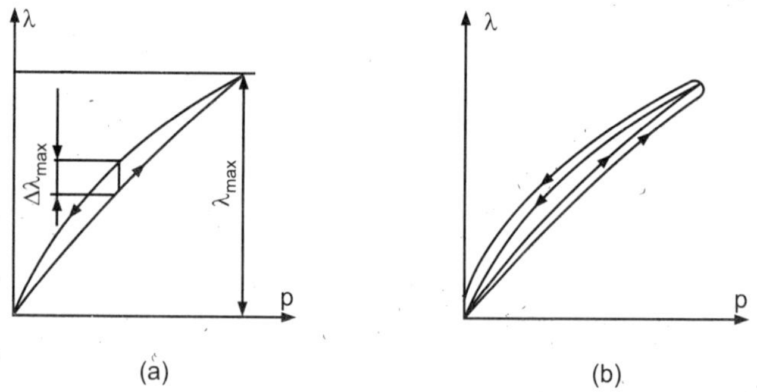

(a) (b)

Fig. 7.15: Pressure-deflection characteristics of an elastic element

The elastic pressure elements normally used in industries are age-hardened alloys like phosphor bronze, trumpet brass, beryllium copper, stainless steel, etc. They are sufficiently plastic and high amount of elasticity and strength are imparted to them during the process of formation.

Modulus of elasticity of elastic elements, as used in pressure-measuring instruments, vary slightly with temperature and this introduces a complementary (temperature) error. Sometimes complementary error may be introduced by variations in the linear dimensions of an elastic element as it is subjected to temperature variations.

Elastic pressure sensors, which convert applied pressures to forces and thence to forces, are collectively called *force summing* devices.

7.8.2 Diaphragm

7.8.2.1 Introduction

As pressure sensing elements, various types of diaphragms are available, namely flat type, convex type, corrugated type, catenary type. Diaphragms are mostly used for relatively low pressure measurements compared to bellows and Bourdon tubes.

7.8.2.2 Flat Type

Figure 7.16 shows a flat diaphragm alongwith its pressure-deflection characteristics.

The flat diaphragm is a circular disc of constant thickness, fixed around its periphery and made of steel or bronze. An applied pressure causes both bending and tension in the diaphragm, resulting in a non-linear pressure-deflection characteristics of the same. The linear portion of the curve should be utilised so that the scale calibration remains uniform throughout the range of use of the instrument.

Since time response of flat diaphragms are very fast, they can be used to measure pressures changing at several hundred or thousand hertz.

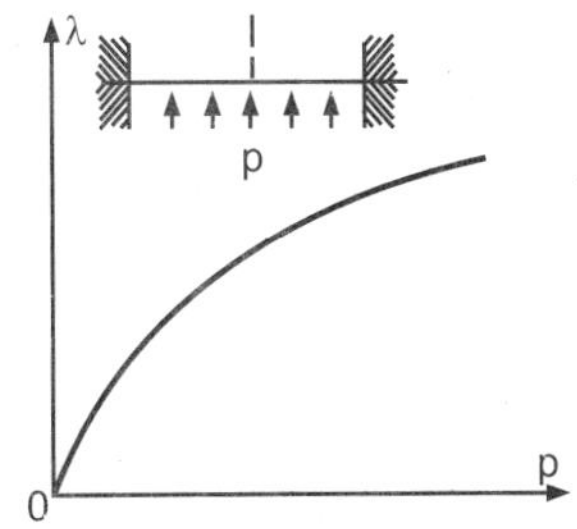

Fig. 7.16: Flat diaphragm and its pressure-deflection characteristic

Assuming pressures p_1 and p_2 being applied on either side of a flat diaphragm, D the diameter of the diaphragm, λ the deflection of the centre of diaphragm, t the thickness, μ the Poisson's ratio and Y the Young's modulus for the material of the diaphragm, the differential pressure $p_1 - p_2$ across the diaphragm is given by:

$$\Delta p = p_1 - p_2 = \frac{256Yt^4}{3D^4(1-\mu^2)}\left[\left(\frac{\lambda}{t}\right) + 0.488\left(\frac{\lambda}{t}\right)^3\right] \quad ...(7.8)$$

If, $0.488\left(\frac{\lambda}{t}\right)^3 << \left(\frac{\lambda}{t}\right)$, then $(p_1 - p_2) \propto \lambda$ and hence Δp becomes proportional to the deflection λ of the diaphragm.

Higher terms in the above empirical formula occurs due to the occurrence of tension alongwith bending at the same time.

It is recommended that the maximum value of λ should be less than $t/3$ for the diaphragm to remain within elastic limit always.

7.8.2.3 Convex (flapping) Type

A convex diaphragm alongwith its pressure deflection characteristics is shown in figure 7.17. As in the case of flat diaphragms, this also is made from steel or bronze and is used in alarm annunciations whenever a deviation from a preset pressure takes place. The nature of the curve can be explained in parts as follows:

- With increasing pressure (starting from zero), the deflection of the diaphragm follows an almost linear path along 'oea'.
- As the pressure is still increased, the diaphragm suddenly loses its stability and flexes in the opposite direction (shown dotted on the curve). In so doing, the diaphragm either opens or closes the contacts of the associated alarm.
- A further increase causes the characteristics to follow 'bc' path, again in an almost linear manner.
- When the applied pressure is decreased, at a value p_2 as shown, the diaphragm flexes back to its original shape and follows the path 'ge'. Any further decrease in applied pressure entails the curve to follow 'eo' path.

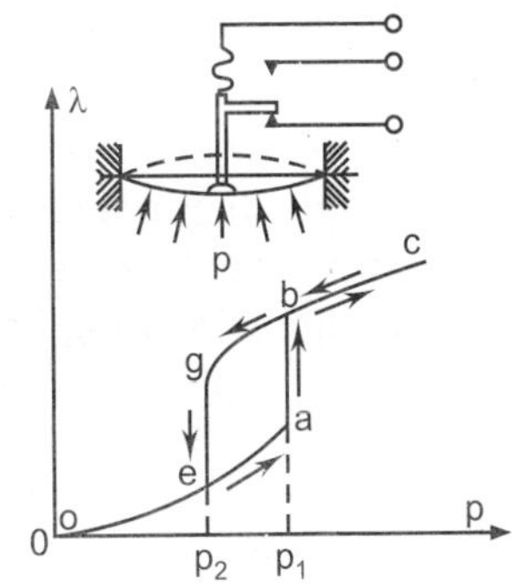

Fig. 7.17: A convex diaphragm and its pressure-deflection characteristic

7.8.2.4 Catenary Type

These are non-metallic (slack) diaphragms, least affected by temperature and are used for low-pressure measuring instruments and draft gauges. These are fabricated from large-mesh nylon or silk fabric impregnated with petrol or any oil resistant rubber or plastic.

Non-metallic (slack) diaphragms are shown in figure 7.18. They have a rigid centre. The annular part of the diaphragm is corrugated to enhance stability. The diaphragm shown in figure 7.18 (*a*) has a low stiffness which, however, can be increased by inserting a coiled spring, as shown in figure 7.18 (*b*).

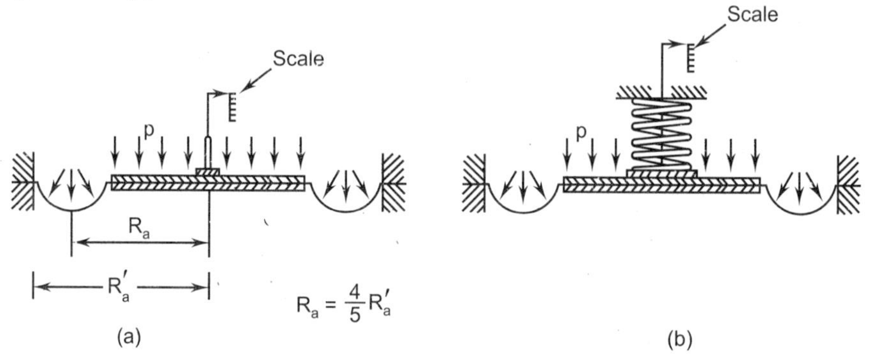

Fig. 7.18: Non-metallic (slack) diaphragms (*a*) without spring (*b*) with spring

7.8.2.5 Corrugated Diaphragms

Shell, Capsule and Element Types:

To improve reliability of operation and to linearise the pressure-deflection characteristics, corrugated diaphragms are used. Corrugated diaphragms can have different convolutions—sinusoidal, trapezoidal or sawtooth type and shown in figure 7.19. In this figure, all the diaphragms are of shell type.

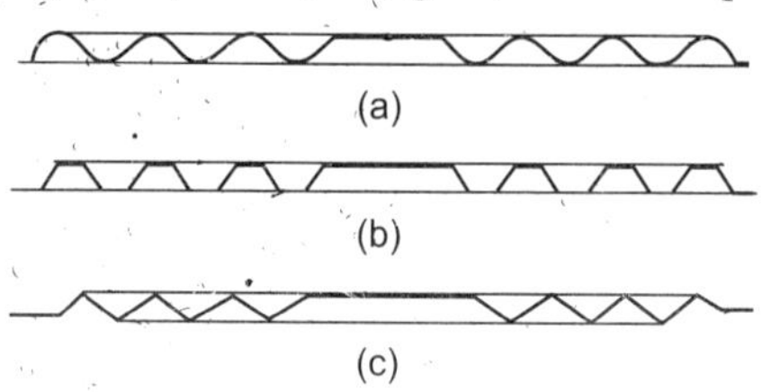

Fig. 7.19: Corrugated diaphragms: (*a*) sinusoidal convolution; (*b*) trapezoidal convolution; (*c*) sawtooth convolution

A single diaphragm is known as diaphragm shell. A diaphragm capsule is formed when two shells are brazed or welded together; while a diaphragm element is formed when two or more capsules are joined together. A capsule and an element are shown in figure 7.20.

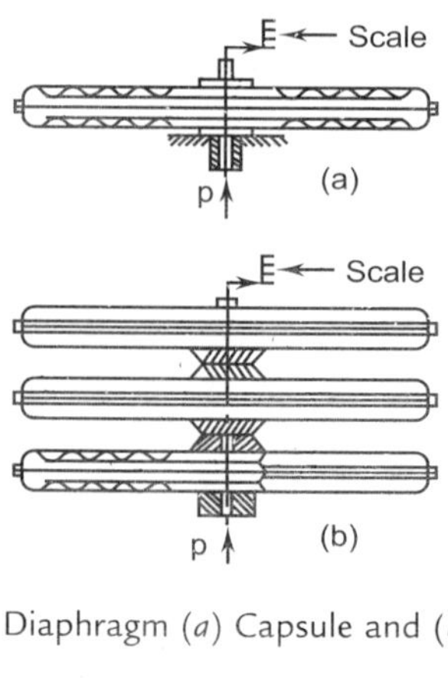

Fig. 7.20: Diaphragm (*a*) Capsule and (*b*) Element

Figure 7.21 shows a pressure element which is assembled from two collapsible liquid-filled diaphragm capsules.

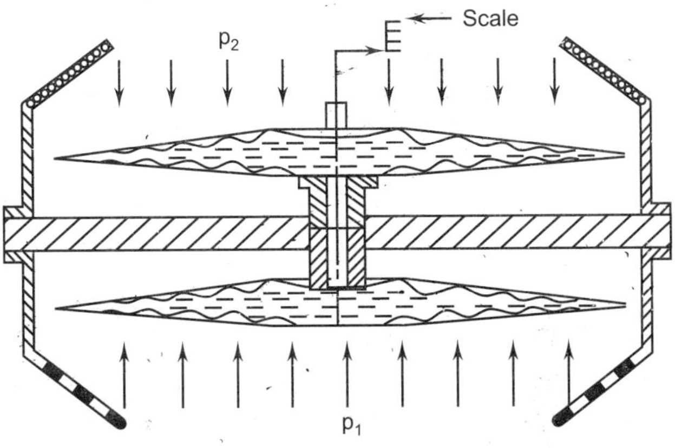

Fig. 7.21: A pressure element assembled from two collapsible liquid-filled diaphragm capsules

This kind of arrangement is usually used for measuring pressure differentials as in head or pressure-difference flow meters.

At this stage, several points need mentioning about corrugated diaphragms of different convolutions:

- Even a shallow convolution leads to an appreciable linearisation in characteristics—shown by different depths of convolution, H
- Increase in depth of convolution leads to a practically linear pressure-deflection characteristics. This is due to higher bending resistance.
- Stiffness of diaphragm increases with corrugation depth.
- Characteristics of a diaphragm is not significantly affected by corrugation profile.
- To control the characteristics of a corrugated diaphragm, stress should be put on (*a*) depth of corrugation and (*b*) material thickness.

Figure 7.22 shows the effect of convolution depth on the pressure-deflection characteristics of a diaphragm.

For diaphragms several terms need to be explained so that their characteristics can be rightly understood.

- Linearity: This is the relationship between the applied pressure and the motion, expressed as a percentage derivation from the best straight line.

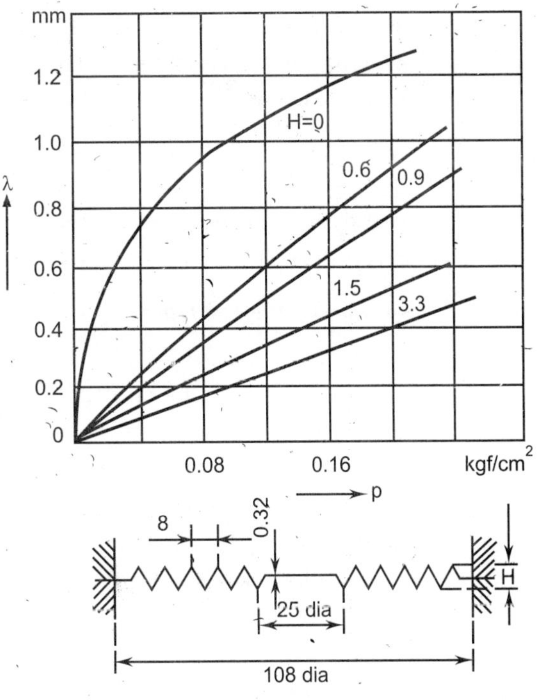

Fig. 7.22: Effect of convolution depth on the pressure-deflection characteristic of a diaphragm

- Maximum pressure: It is that pressure which will give rated motion in an unrestricted form.
- Travel: It is the useful distance the diaphragm would move when it expands without experiencing non-linear movement. Travel should be kept to a minimum if a high degree of linearity and repeatability are desired.

Some typical diaphragm pressure sensing elements are shown in figure 7.23.

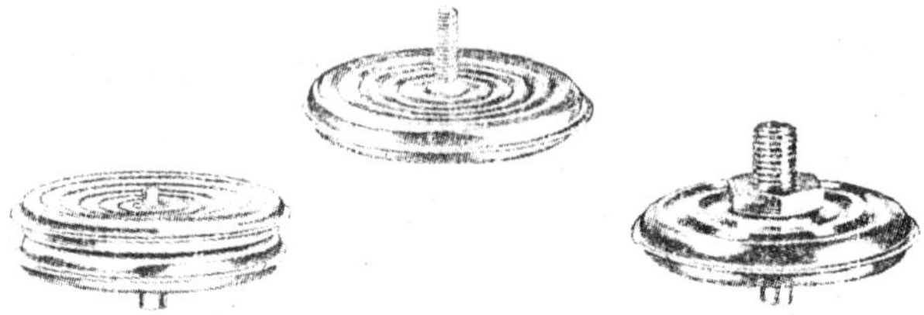

Fig. 7.23: Typical diaphragm pressure-sensing elements. (U.S. Gauge Co.)

7.8.3 Bellows

7.8.3.1 Introduction

A metallic bellow comprises a series of circular parts, brazed or welded together so as to give it a cylindrical like shape when viewed from the front. It is a single piece expansible, collapsible member in which corrugations are done at right angles to the long axis. The expansion/compression takes place along the long axis, but negligible in any other direction. The circular parts are called corrugations or convolutions or folds.

Special type metallic alloys are employed to obtain precision bellows for relatively high pressure measurements, while bellows made of plastic and rubber materials are utilised for lower pressure measurements.

Since the maximum travel of bellows should be limited, sometimes springs are used which acts in opposition to bellow movement. The bellow in conjunction with the spring incorporate stiffness to the system and ensure linear relationship between the applied pressure and the displacement of the system.

7.8.3.2 Seamless and Welded Types

Bellows can be of two types—seamless and welded. The former type can be fabricated by either hydraulic or mechano-hydraulic process from thin-walled seamless tubing. The welded type is fabricated from stamped diaphragm shells welded around internal and external peripheries. Two types of seamless bellows are shown in figure 7.24 (*a*) and (*b*) while an welded below is shown in figure 7.25.

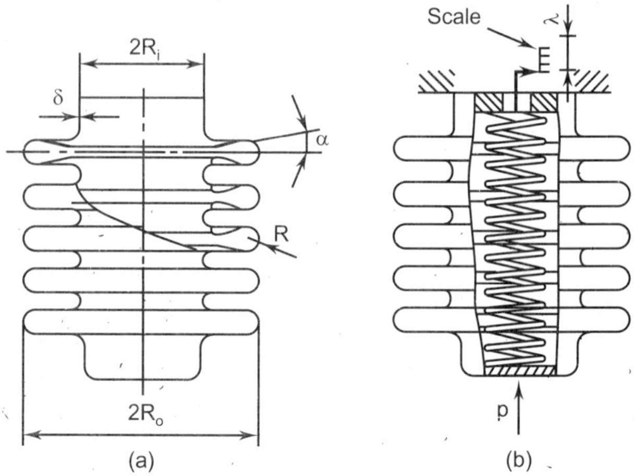

Fig. 7.24: Seamless bellows

The stiffness of a bellow element is dependent on the following factors:

- Dimensions of the bellow
- Wall thickness of the tubing, δ
- Elastic properties of the material
- Radius of corrugation, R
- Formation angle α and
- Whether a bellow is used with a spring or not

The effective area of a bellow element is given by the empirical formula:

$$A_{eff} = \pi\left(\frac{R_o + R_i}{2}\right)^2 \quad ...(7.9)$$

where, R_o and R_i are the outside and inside radii of a bellow element, respectively.

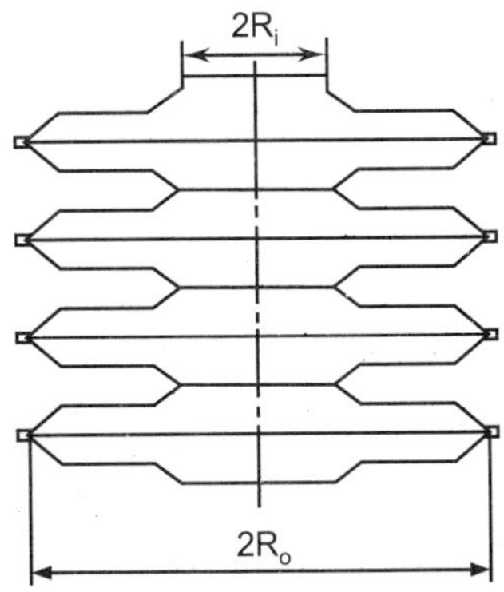

Fig. 7.25: An welded bellow

Depending on the application, bellows are available in vastly differing sizes—their diameters range from 5 mm to 150 mm and wall thickness (shown by δ in the diagram) from 0.08 mm to 3 mm.

A bellow element has both advantages and disadvantages while in use. On the advantage side lies robust and rugged construction, ability to measure absolute, gauge and differential pressures, reasonably accurate, moderately priced, produces large force when pressurised and most importantly a near linear deflection vs. pressure relationship, while on the disadvantage side are the hysteresis effect (although can be reduced largely by proper manufacturing techniques) and its sensitivity to fluctuating pressures.

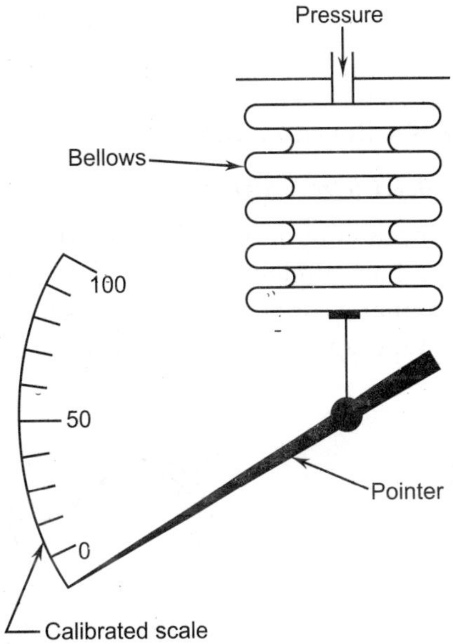

Fig. 7.26: A simple bellow measuring gauge pressure

7.8.3.3 Gauge Pressure Measurement by Bellows

A simple bellow element can measure gauge pressure, as shown in figure 7.26. Increasing pressure causes the bellow element to expand, enabling the pointer to move along the scale. The scale can therefore be calibrated to record gauge pressure.

7.8.3.4 Absolute Pressure Measurement by Double Bellows

Figure 7.27 shows a double bellow system, designed to measure absolute pressure. The two bellows act in opposition, so that any change of atmospheric pressure or temperature affects both the bellows identically and hence their effects are cancelled out. One bellow is evacuated in such a way that its pressure can be brought down in the region of 5 microns. Thus, the pressure of this bellow is as good as negligible and hence the pointer represents the absolute pressure of the system under measurement.

Chapter 7

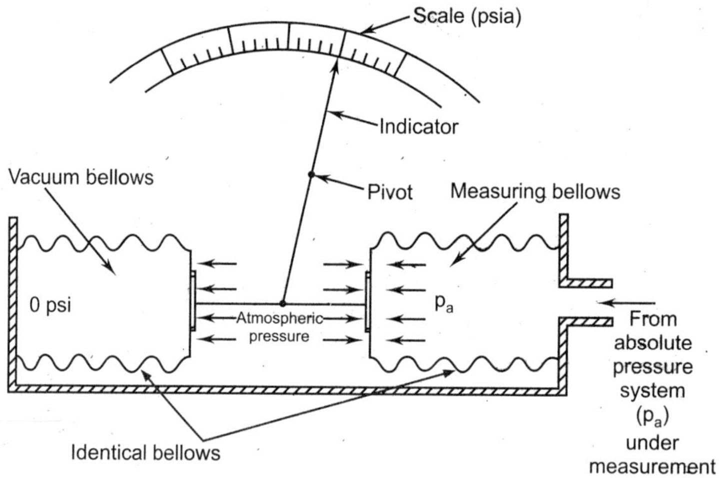

Fig. 7.27: A double-bellow unit to measure absolute pressure

7.8.4 Bourdon Tube

7.8.4.1 Introduction

Proposed by *E.* Bourdon in 1849 and patented in 1952, a Bourdon tube is the most sought-after mechanical element employed to measure pressure over a very wide range and having a non-circular cross-section. Four types of Bourdon tubes are available, as shown in figure 7.28 and are C-type, spiral type, twisted type and helix type.

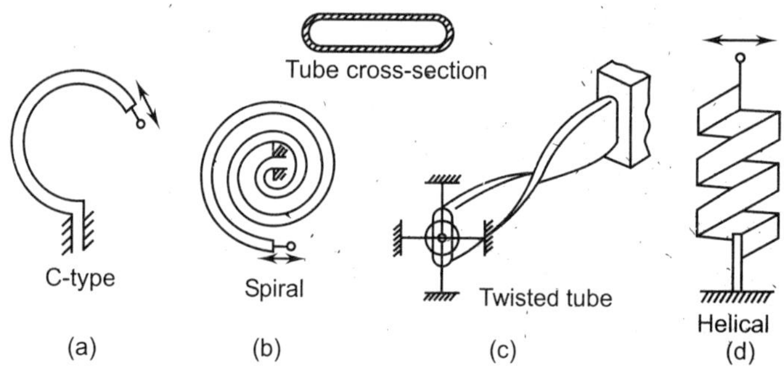

Fig. 7.28: The four Bourdon tube varieties

The most commonly used Bourdon tube is the C-type. Spiral or helix types are employed when more movement for a given pressure is desired, thereby increasing sensitivity and resolution. The spiral type Bourdon tube will have several turns, all of them remaining in the same plane. The helix type also has several turn, but with all turns having the same diameter. The increased sensitivity is, of course, at the expense of higher cost compared to C-type. Also, these two types can measure a much lesser maximum pressure than a C-type Bourdon tube.

The material of the Bourdon tube must have good elastic or spring characteristics for repeatable performance to be obtained. The sensitivity of the Bourdon tube increases with increase in radius of curvature and decrease in wall thickness of the tube.

7.8.4.2 The C-type Bourdon Tube and its Description

The *C*-type Bourdon tube is formed by bending the tube circularly so that it forms the arc of a circle of about 250°–270° angle. For this particular shape, it is given the name *C*-type. The Bourdon tube is welded to a socket at the base through which the external pressure is applied to the tube. The other end of the tube is sealed and known as 'tip'. The socket and the tip are welded, brazed or soldered according to the applied pressure range. A *C*-type Bourdon tube is shown in figure 7.29.

When a pressure difference occurs, with the inside of the tube having a higher pressure, the tube cross-section moves more and more towards a circular one, resulting in a curvilinear motion of the tip for '*c*', spiral and helical types and an angular rotation in the twisted type. This tip travel is suitably amplified by a combination of link (adjustable), a segment lever, sector and pinion arrangement. Thus, ultimately a rotational pointer movement is obtained, called 'multiplication', which can be adjusted by varying the length of the lever, The Bourdon tube with link, lever, pinion, scale, etc., is shown below in figure 7.29.

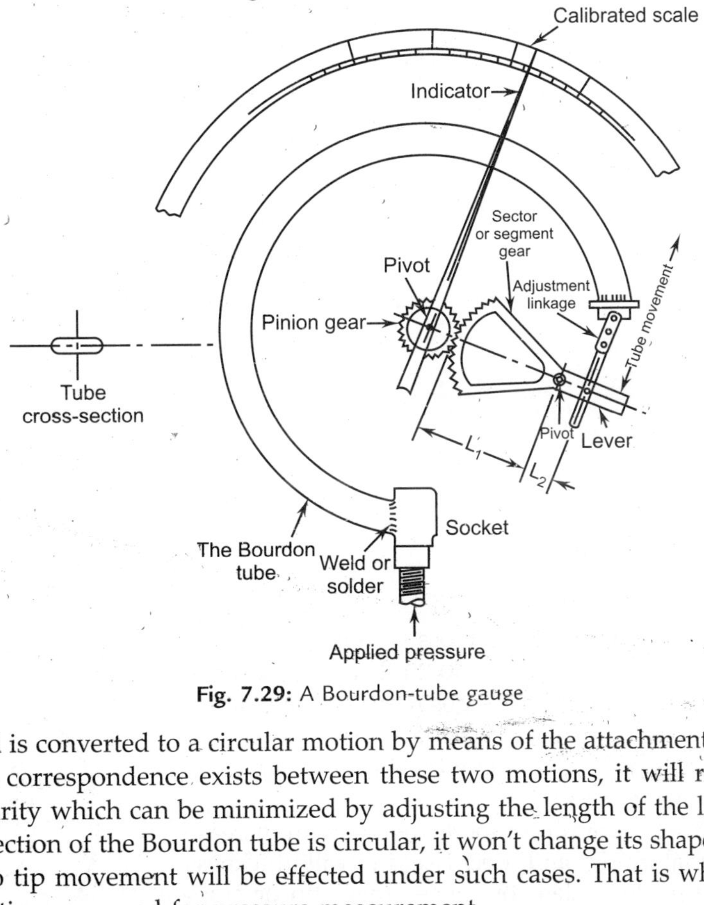

Fig. 7.29: A Bourdon-tube gauge

The tip travel is converted to a circular motion by means of the attachments described above. If no one to one correspondence exists between these two motions, it will result in distortion, known as angularity which can be minimized by adjusting the length of the link.

If the cross-section of the Bourdon tube is circular, it won't change its shape on application of pressure; thus no tip movement will be effected under such cases. That is why tubes with non-circular cross-sections are used for pressure measurement.

The materials which are normally used for construction of Bourdon tubes are phosphor bronze, beryllium copper, steel, stainless steel, etc. The deflection of the tip depends upon the total tube length, radius of the bend, the thickness of the wall, the major and minor axes of the cross-section and Young's modulus of the tube material.

The *C*-type Bourdon tube is formed by winding the tube circularly so that it forms the arc of a circle which subtends an angle $\gamma \simeq 200°$ to $270°$, as shown in figure 7.30 (*a*). The tube cross-section may be elliptical (figure 7.30 (*b*)), flat or oval (figure 7.30 (*c, d*)). For high pressure applications, figures (*b*) or (*d*) are employed while for a low value of pressure the cross-section shown in figure (*c*) is preferred over others. The major axis 2*a* is at right angles to the mean radius of curvature R_c. One end of the tube is stationary and is connected to the unknown pressure while the other end is brazed. This end is left free to move and is connected to a pointer which moves over a graduated pointer. Gauge pressure up to 60 Kgf/cm² (6 MPa) or vacuum down to 1 Kgf/cm² (0.1 MPa) can be measured by thin-walled Bourdon tubes while a thick walled oval Bourdon tube can measure gauge pressures up to 200–1600 Kgf/cm² (20–160 MPa).

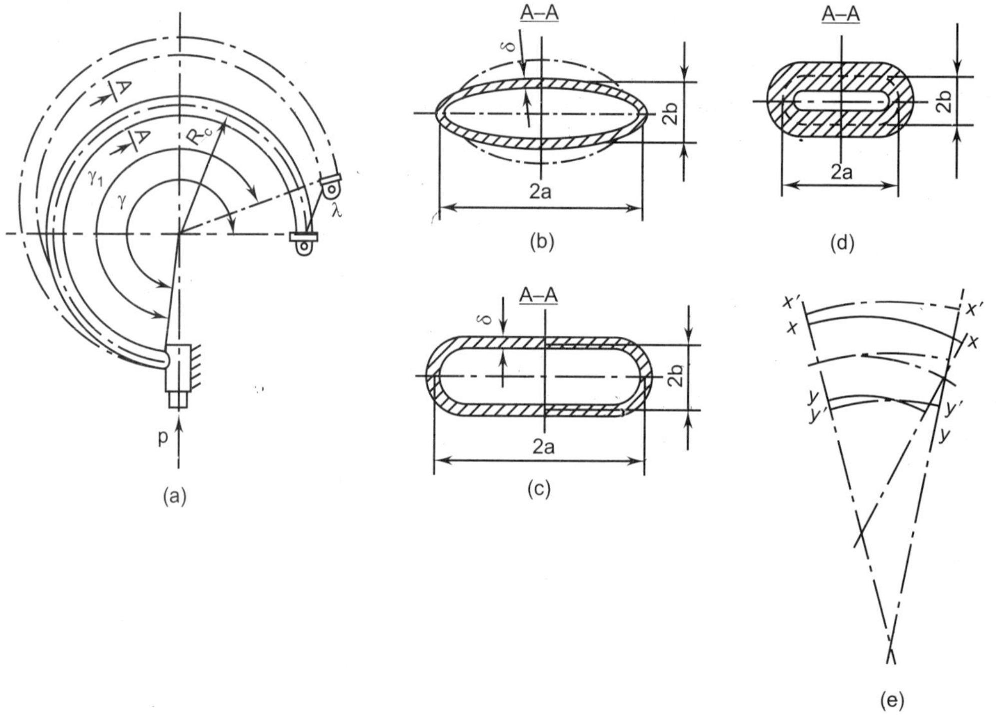

Fig. 7.30: Elliptical and flat *C*-type Bourdon tubes

When pressure is applied, the cross-section approaches the dot-dash line as shown in figure 7.30 (*b*). Thus, as shown in figure 7.30 (*e*), *xx* and *yy* assume the positions *x′x′* on a larger-radius arc and *y′y′* on a smaller-radius arc. As a result, the Bourdon tube rotates anti-clockwise with the tube curvature reducing by an angle $\Delta\gamma = \gamma - \gamma_1$.

The sensitivity of a Bourdon tube increases with increase in radius of curvature R_c and decrease in wall thickness δ. Sensitivity is also dependent on *a*/*b* — the ratio of cross-sectional axes and also on the shape of the section near the ends of the major axis.

The total deflection λ of the tip is a vector sum of λ_r and λ_t— the radial and tangential deflection respectively and shown in figure 7.31

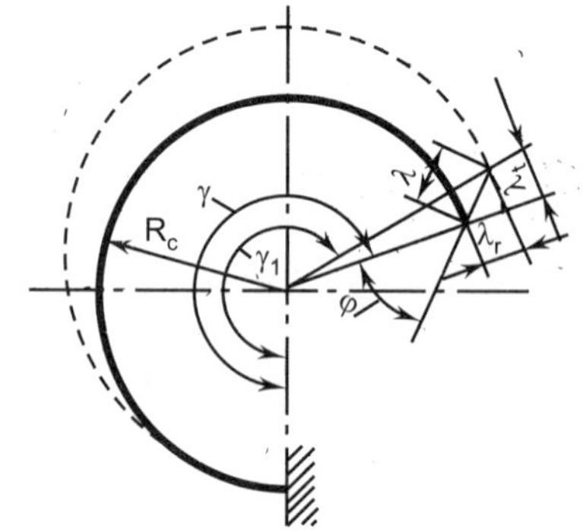

Fig. 7.31: Path described by the tip of a Bourdon tube

$$\therefore \qquad \lambda = \sqrt{\lambda_r^2 + \lambda_t^2} \qquad ...(7.10)$$

7.8.4.3 Bourdon Tube Movement Relations

The pointer deflection of a Bourdon tube is inherently non-linear in nature with respect to the applied pressure. This is shown pictorially in figure 7.32, where A is the link or travelling angle and *B* is, as shown.

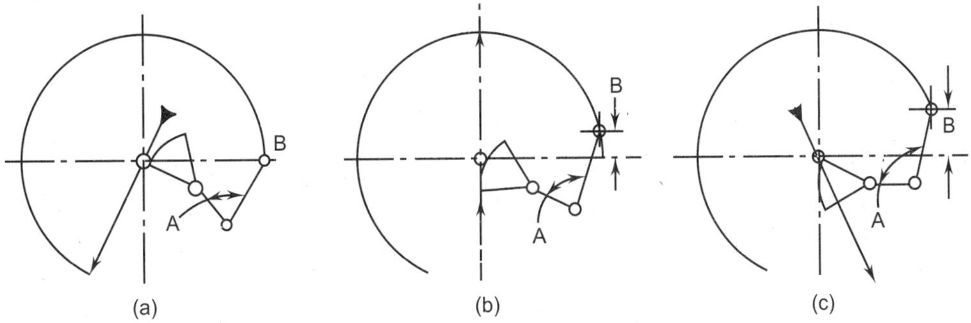

Scale	A	B
(a) 0%	70°	0 mm
(b) 50%	80.5°	6.6548 mm
(c) 100%	110.0°	9.525 mm

Fig. 7.32: Bourdon tube movement relations

Bourdon tubes respond to ambient temperature changes—for a rise in temperature, the reading is faster. Again, if the tube is to be used in a position other than vertical, it should be calibrated in that position. A Bourdon tube must not be subjected to excessive pressure, pressure shock or vibration. When a Bourdon tube is to be used for steam pressure measurement or else where higher temperatures are encountered, the tube must be protected by a water seal which is discussed later in this chapter.

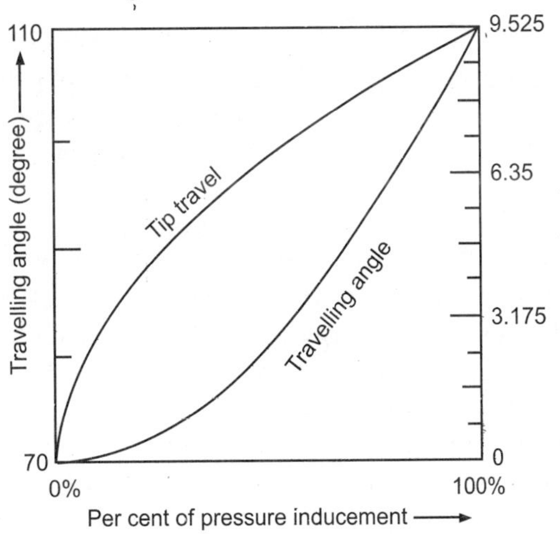

Fig. 7.33: Non-linear curves of the tip travel and travelling angle, A.

With variations in pressure, a Bourdon tube gives a non-linear output which must be converted to a linear rotational motion. This is achieved by a geared sector-and-pinion arrangement. Figure 7.33 shows the non-linear nature of tip travel and travelling angle with pressure variations. The nature of curves suggest that the two almost compensate each other, giving rise to a linear circular scale.

7.8.4.4 Accuracy and Error in Bourdon Tubes

C-type Bourdon tubes can measure up to 6000 bar. Measurement accuracy is typically ±1% of full-scale deflection. For helical and spiral types, more resolution at the same accuracy limit is obtained, but the maximum measurable pressure is in the region of 700 bar.

A major source of error with Bourdon tubes which, in the majority of cases, escapes attention, is concerned with the fluid with which the tube was calibrated and the fluid whose pressure is to be measured. The pointer of the Bourdon tube is normally set to zero using air as the calibrating medium. While using this calibrated Bourdon tube to measure some unknown liquid pressure, an error in reading in excess of 5% may result. One way out to nullify this is to calibrate the tube with the same fluid whose pressure is to be measured later on.

Lastly, it should be borne in mind that Bourdon tubes offer very good accuracy while measuring gaseous pressures. But, in the case of liquid pressure measurements, care must be taken to see to it that the entire tube is totally filled with liquid both during calibration and measurement, otherwise error will occur.

7.8.4.5 Quartz Bourdon Tubes

Rapid strides in quartz technology has resulted in the development of pressure sensitive helical Bourdon tube made of fused quartz. Bourdon tubes made of quartz have essentially no hysteresis, creep, relaxation or fatigue. Temperature effects are minimised in quartz Bourdon tubes because of the fact that thermal expansion coefficient of quartz is extremely low. Quartz Bourdon tubes are insensitive to seismic vibrations.

These are available in the sizes ranging from 762 mm of mercury to about 500 psi rated maximum pressure. The helical type fused-quartz Bourdon tubes rotate approximately 100 degrees for their rated pressure.

7.8.5 Choosing of Proper Elastic Pressure Sensing Materials

Proper thought should be given to the selection of material for diaphragm, bellow, Bourdon tube which act as pressure sensors as well as springs which are used for damping purposes.

Materials, which are to be chosen for the above purpose, should have a small value for its internal friction. It is the minimum force which must be overcome before elongation/contraction occurs. Again, materials should repeat their performance with either increasing or decreasing forces. Thus, the materials should have a small value for their hysteresis loop. Figure 7.34 shows the amount of energy required for a material to go through a complete stress cycle.

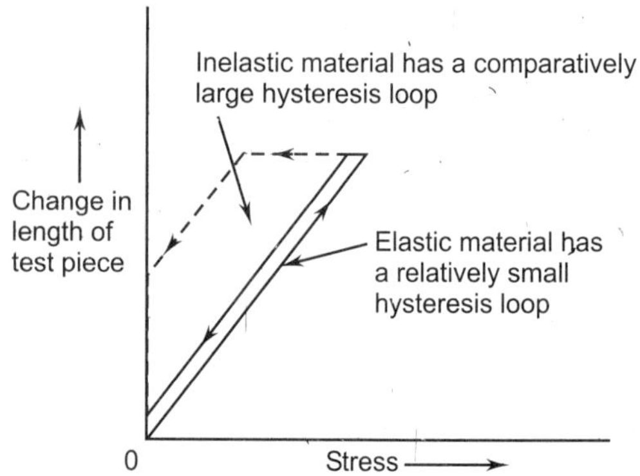

Fig. 7.34: Hysteresis loops of elastic and inelastic materials

The hysteresis loop represents the amount of energy that is to be spent while the material goes through a complete cycle of pressure or stress changes. Every material has a hysteresis loop. As ideal elastic material has no hysteresis loop, while an inelastic material has a large hysteresis loop and hence a large hysteresis loss per cycle. Materials possessing hysteresis loop will not exhibit the same change of length for the same stress when it goes through increasing and decreasing stress paths. Again, materials possessing high value of hysteresis loss are having a greater chance to fail under repeated operations. Hysteresis loss appears as heat in the material, thereby affecting instrument accuracy.

7.9 ELECTRICAL TYPE PRESSURE GAUGES

7.9.1 Introduction

The electrical type pressure transducers can be resistive inductive or capacitive type.

In the resistive type, a change of pressure changes the resistance of the sensor, in the inductive type it is the inductive coupling that changes with pressure whereas the value of capacitance is made to change with pressure in the capacitive type of transducers. These are now discussed below.

7.9.2 Resistive Type Pressure Gauge

Different types of resistive pressure gauges are shown in figure 7.35 (*a-d*). The unknown pressure to be measured is fed at the input to the Bourdon tube. The changed pressure causes the pointer to move along a resistive element, thereby changing the output voltage. This output voltage can be calibrated in terms of unknown pressure.

Figure 7.35 (*b*) shows another version of resistance pressure sensor in which a change of process pressure moves a wiper along a potentiometer, which is connected to one arm of a Wheatstone's bridge. The output of the bridge can be calibrated in terms of unknown process pressure. It is a low cost one having high output signal so that no amplification is necessary. But its mechanical wear and tear is certainly a disadvantage.

Figure 7.35 (*c*) represents a system in which the applied pressure differential ($P'_1 \sim P'_2$) is straightway calibrated in terms of galvanometer current.

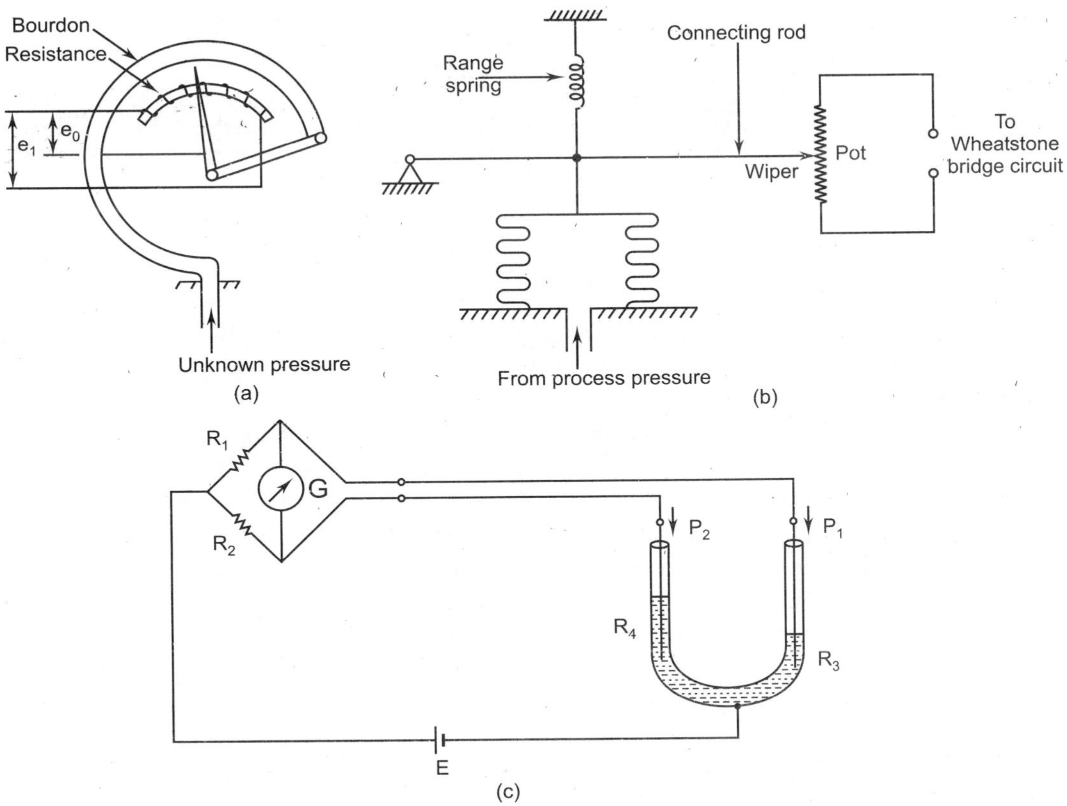

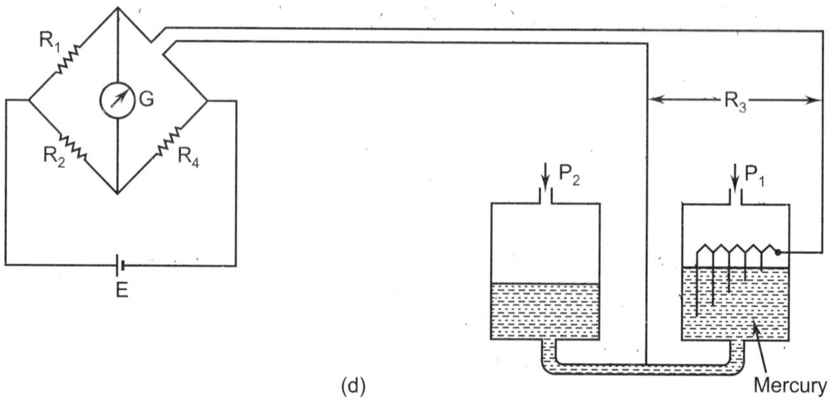

(d)

Fig. 7.35: Different types of resistive pressure gauges

Figure 7.35 (*d*) shows a resistance pressure gauge of liquid contact type and the unknown pressure differential can be measured in a manner similar to figure 7.35 (*c*).

7.9.3 Differential Pressure Measurement Using Strain Gauges

Figure 7.36 shows the technique of measuring the differential pressure ($P'_1 \sim P'_2$) by measuring the changes in resistance values of the strain gauge employing a Wheatstone's bridge. This change is a function of local strain which is a function of diaphragm deflection and hence differential pressure. When the diaphragm deflection is less than one third of its thickness, then only the deflection can be claimed to follow a linear relationship to the applied differential pressure.

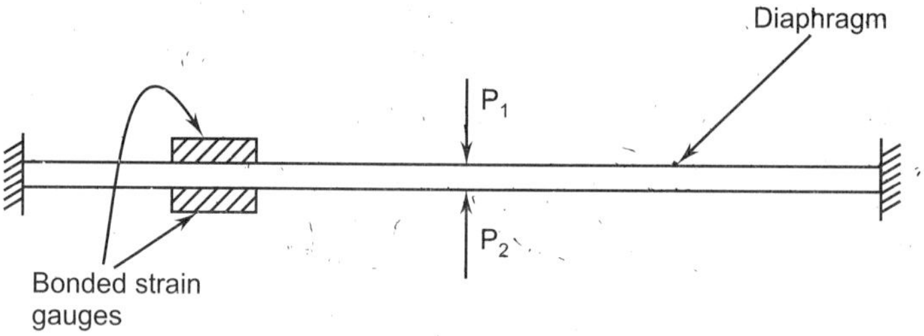

Fig. 7.36: Differential pressure measurement by strain gauges

The bonded strain gauge is a major advantage over an unbonded one because in the former the presence of mechanical frame by attaching the sensing wire directly to the strained surface is dispensed with.

When strain gauges are used as sensors for pressure measurements, process temperature variations must be compensated because temperature changes cause both the base material on which the strain gauge element is bonded and the element itself to either expand or contract. Several temperature compensation techniques are available and discussed below:

- By matching the thermal properties of the base material and the strain gauge element.
- The strain gauge element is initially calibrated for the expected temperature range variations. The readings thus can be corrected from this calibration curve, after noting the process temperature.
- By employing dual strain gauges—one having positive and the other negative response to temperature changes. Thus, they cancel each other's effect due to temperature changes.

- By employing dummy elements. The dummy strain gauge is mounted on the same surface as the active one and is exposed to identical temperature changes but is not subjected to the applied pressure. This dummy gauge is placed in an adjacent arm to the active one in a Wheatstone's bridge circuit, thereby eliminating the effect of temperature.

7.9.4 Diaphragm Type Bonded Strain Gauge Pressure Transducer

The cross-section of a diaphragm based pressure transducer with strain gauge being used as a sensor is shown in figure 7.37(*a*). It is a circular diaphragm having thickness *t*, and radius *a* and rigidly fixed around its periphery. In all, four resistance strain gauges are utilised—R_1 and R_3 are mounted diametrically opposite with radius a_r while R_2 and R_4 are placed very close to the centre with radius a_t. Application of pressure results in resistance changes in the strain gauges and measured in terms of unbalanced voltage in a Wheatstone's bridge configuration—with a strain gauge placed in each arm of the bridge.

Both tensile and compressive stresses are developed on the diaphragm and the arrangement of the strain gauge in the bridge circuit is such that their effects are added up—thus a large output is obtained for a given applied pressure resulting in higher sensitivity.

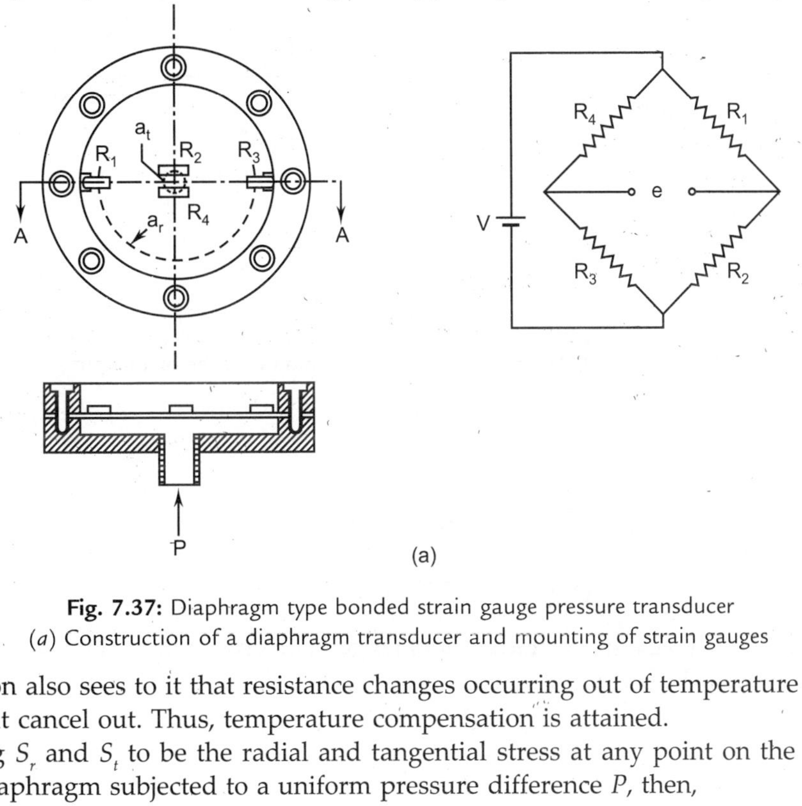

Fig. 7.37: Diaphragm type bonded strain gauge pressure transducer
(*a*) Construction of a diaphragm transducer and mounting of strain gauges

The connection also sees to it that resistance changes occurring out of temperature changes give an output that cancel out. Thus, temperature compensation is attained.

Assuming S_r and S_t to be the radial and tangential stress at any point on the low pressure side of the diaphragm subjected to a uniform pressure difference *P*, then,

$$S_r = A\left[\left(1+\frac{1}{\mu}\right)-\left(1+\frac{3}{\mu}\right)\frac{r^2}{a^2}\right] \quad ...(7.11)$$

and,

$$S_t = A\left[\left(1+\frac{1}{\mu}\right)-\left(3+\frac{1}{\mu}\right)\frac{r^2}{a^2}\right] \quad ...(7.12)$$

where, the constant A, for a given pressure P, is given by:

$$A = \frac{3\,Pa^2 \propto}{8t^2}, \quad \mu = \text{Poisson's ratio of the material of the diaphragm.}$$

while the deflection-pressure relationship for the diaphragm is as follows:

$$y = \frac{3\,P\,(1-\mu^2)}{16\,Et^3}(a^2 - r^2)^2 \qquad ...(7.13)$$

E being the modulus of elasticity of the diaphragm.

The above three expressions give rise to linear relationships between stress-pressure or deflection-pressure—these are valid only for small pressure differential across the diaphragm and shown in figure 7.37(*b*).

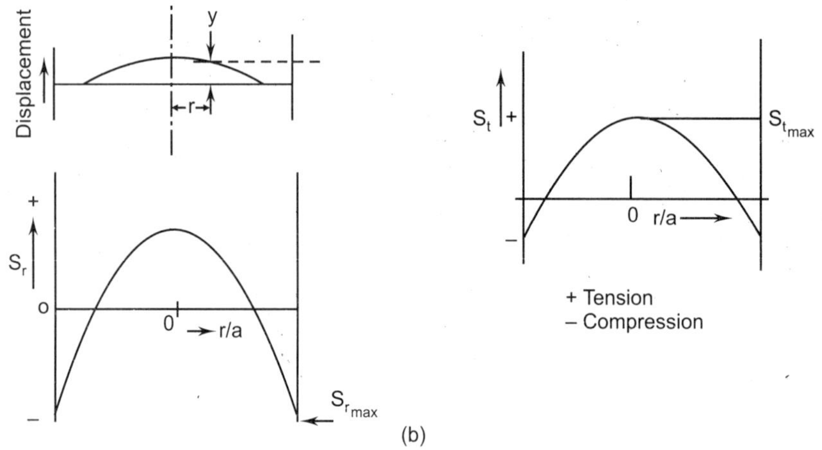

Fig. 7.37: (*b*) Deflection and stress distribution over the diaphragm

The maximum positive tangential stress occurs at the centre ($r = o$), while the maximum negative radial stress occurs at the edges ($r = a$). Their values are derived from the above and are given by

$$(S_t)_{max} = +A\left(1 + \frac{1}{\mu}\right)$$

and

$$(S_r)_{max} = -\frac{2\,A}{\mu}$$

The existence of biaxial stress (radial and tangential) gives rise to biaxial strain and the corresponding relationships are:

$$\in_r = \frac{S_r - \mu\, S_t}{E}$$

and,

$$\in_t = \frac{S_t - \mu\, S_r}{E}$$

S_r and S_t can be found out from the parameters of the diaphragm. Once S_r and S_t are known, $\in_r$ and $\in_t$ can be found out. The output of the bridge, from the bridge sensitivity equation, is as follows:

$$e = \frac{V}{4}\left[\frac{\Delta R_1}{R_1} - \frac{\Delta R_2}{R_2} + \frac{\Delta R_3}{R_3} - \frac{\Delta R_4}{R_4}\right] \qquad ...(7.14)$$

where V = excitation voltage

R's and ΔR's are the values of the gauge resistances and changes in gauge resistances due to application of pressure. $\in_t$ and $\in_r$ are measured by (R_2, R_4) and (R_1, R_3) respectively which produce an additive output. Thus,

$$e = \frac{V\,G_F}{4}\,(2\in_r + 2\in_t) \quad ...(7.15)$$

G_F being the gauge factor.

For a given pressure, therefore, the output e can be found out.

The dynamic response of a diaphragm type strain gauge pressure transducer is reasonably good. The diaphragm can be excited into an infinite number of modes, although the lowest frequency is the one of interest for understanding its response to dynamic input. For an edge-clamped diaphragm vibrating in vacuum (*i.e.* having no fluid inertia effects), the lowest natural frequency is given by:

$$w_n = \frac{10.21}{a^2}\left[\frac{Et^2}{12\rho\,(1-\mu^2)}\right]^{1/2} \text{ radian/second.}$$

where, E is in N/m², density ρ in Kg/m³ and 't' and 'a' are in meters.

The natural frequency w_n is dependent on the elastic properties of the material and the physical dimensions of the diaphragm. Factors, such as imperfect clamping, wrinkling of diaphragm, liquid inertia, etc. may render the actual frequency to differ from the calculated one. Wrinkling on the diaphragm stiffens the same and raises w_n.

7.9.5 Differential Pressure/Pressure Measurement by Bonded/Unbonded Strain Gauges

Figure 7.38 shows the differential pressure/pressure measurement techniques employing bonded/unbonded strain gauges.

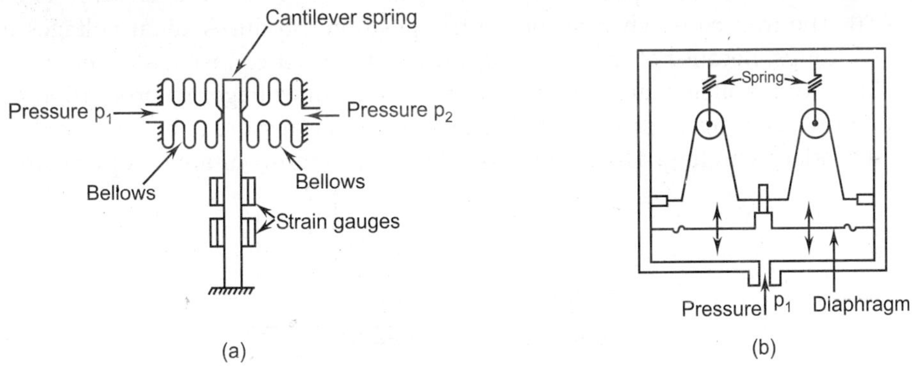

Fig. 7.38: (*a*) Differential pressure measurement by bonded strain gauges.
(*b*) Pressure measurement by unbonded strain gauge.

In figure 7.38 (*a*), the strain gauges are mounted on a cantilever spring. On either side of the cantilever, two bellow elements, one opposing the other, are mounted. The dimension of the cantilever is so chosen that it compensates for its change in Young's modulus value with temperature.

Figure 7.38 (*b*) shows a technique for measurement of pressure with unbonded strain gauge alongwith a diaphragm.

7.9.6 Inductive Type Pressure Gauge

This type of pressure gauge utilises the magnetic properties (say self or mutual coupling) for conversion of the movement of a sensor into an electrical signal. The schematic of an inductive type pressure gauge is shown in figure 7.39. Process pressure causes a change in the self-inductance of the coil, which is then fed to a standard measuring circuitry.

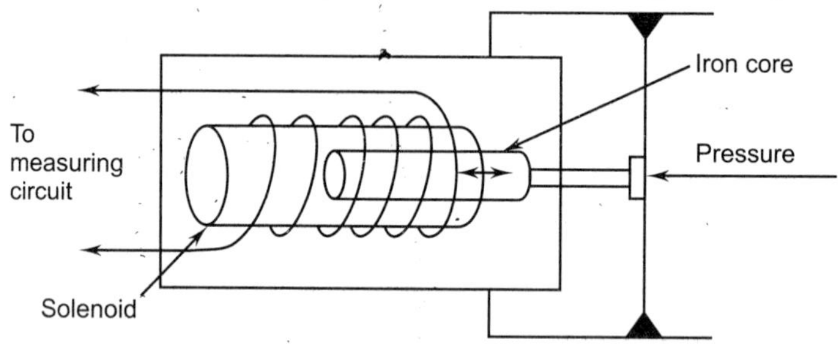

Fig. 7.39: An inductive type pressure gauge

The output of this bridge circuitry is then calibrated in terms of the process pressure. For example, when the process pressure increases, the mechanical arrangement ensures that the iron core is pushed into the solenoid, thereby increasing its self-inductance.

In the ratio inductance type of transducer, there are two solenoid coils. The movement of the iron core, with changes in the process pressure, changes the inductance ratio of the two coils. The sensitivity and linearity of this type of transducer is better than the self-inductance type.

7.9.7 Pressure Measurement by LVDT

A linear variable differential transformer (LVDT), having a primary winding and two secondary windings connected in series opposition, can be employed to measure pressure and is shown in figure 7.40. The iron core, when at the central position, produces equal voltages in the two secondary windings. Because of the series opposition nature of connection of the two secondary windings, these two voltages are cancelled out. An output voltage, in proportion to the core movement and which is again proportional to the unknown pressure to be measured, is obtained across the secondary winding. This voltage is calibrated in terms of applied pressure.

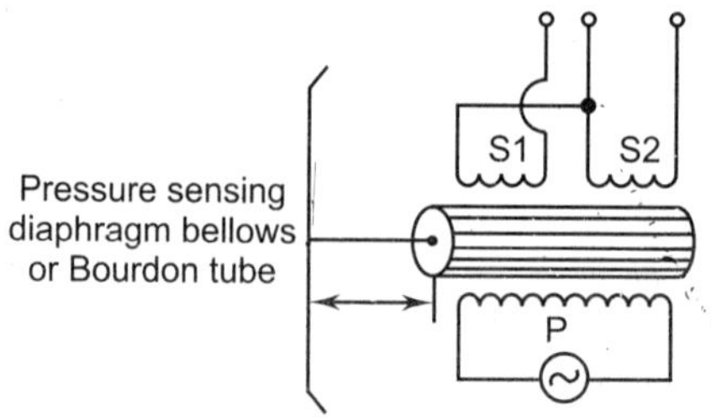

Fig. 7.40: Pressure measurement employing LVDT

7.9.8 Capacitive Type Pressure Gauge

The principle of operation of this type of pressure gauge involves the movement of an elastic element with changes in pressure which results in capacitance variation. This variation is measured electrically and calibrated in terms of unknown input pressure. A capacitance type pressure gauge can measure absolute, gauge or differential pressure, depending on the reference pressure used.

Figure 7.41 (*a*) shows the constructional details of how the capacitance value of the capacitor changes. The diaphragm is connected to the pressure bellows, which, in its turn, is connected to the input pressure point. The diaphragm acts as the variable plate. With increasing pressure, the diaphragm is pushed upwards and vice versa.

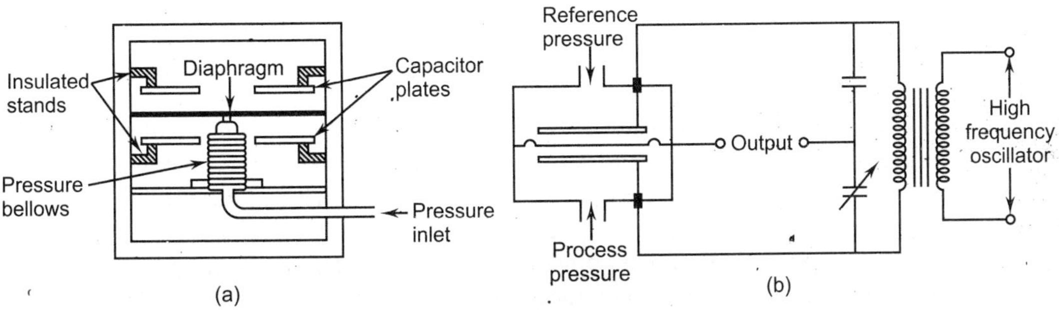

Fig. 7.41: (*a*) Construction of capacitor (*b*) Pressure measurement employing capacitor

Figure 7.41 (*b*) shows a capacitive arrangement, connected to an electrical circuit to measure an unknown pressure. The supply to the circuit is a high voltage high frequency oscillation. With process pressure variation, the output of the circuit varies. In the null detector method, the capacitor arms are varied to get balance. The capacitor can then be calibrated in terms of unknown pressure. In the unbalanced mode, the ratio between the output voltage and excitation voltage is a direct measure of the unknown pressure.

A capacitive type pressure transducer has both advantages and disadvantages. On the advantage side lies linearity, good accuracy, rangeability and a reasonable speed of response. On the disadvantage side are the very high output impedance, temperature dependence, stray capacitance effect, long lead lengths, etc.

This type of pressure gauges have an accuracy of ± 0.1 to ± 0.2% of instrument span. Pressure as high as 35 MPa or vacuum as low as in the micron range can be handled by such a gauge.

7.9.9 The Piezoelectric Effect

Certain crystals like quartz, tourmaline, barium titanate, Rochelle salt, etc., give rise to an emf when external pressure is applied along specific axes. Such a phenomenon is called piezoelectric effect and is used for measurement of dynamic pressure. Conversely, when an electric field is applied to such a crystal, a mechanical strain will be induced in it. This is known as inverse piezoelectric effect.

The most commonly used piezoelectric pressure gauge employs quartz—silicon dioxide (SiO_2) as the piezoelectric material. This is because quartz does not absorb moisture, almost temperature insensitive in the range of 20° – 400°C, has good insulating properties and has also good mechanical strength.

Figure 7.42 shows a rectangular slab cut out of a quartz crystal with its faces parallel to the three axes—optic, electric and mechanical. Applying a tensile force along the electric axis will result in electrostatic charges appearing on the faces perpendicular to the electric axis. If the same force changes from tensile to compressive, electrostatic charges of the same magnitude but opposite in sign will appear on the faces as it was in the first case.

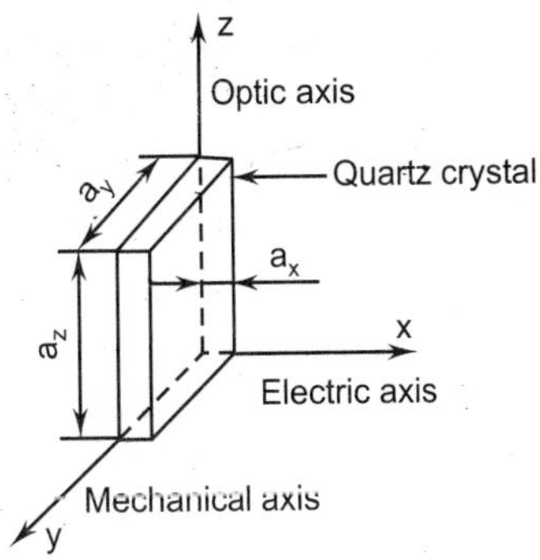

Fig. 7.42: A quartz crystal slab and its three axes

Applying a force F_x along the electric axis will result in a charge Q_x given by

$$Q_x = KF_x = kp_xA_x \quad ...(7.16)$$

where K is the piezoelectric constant or modulus of the crystal employed, expressed in *c/N*, $A_x = a_y\ a_z$ and p_x is the pressure (tensile/compressive) applied on either side of crystal lying along the electric axis.

K is independent of size of crystal but its value differs from crystal to crystal. A typical value for K is 2.1×10^{-12} c/N.

Applying a force F_y, of the same sign and magnitude as F_x along the mechanical axis, would produce the same charge but of opposite sign. Hence, tensile force applied along electric axis is equivalent to a compressive force applied along mechanical axis. Thus, longitudinal piezoelectric effect along x-axis is equivalent to a transverse piezoelectric effect along y-axis.

To note at the this point that a force applied along the optic axis does not produce a charge.

Piezoelectric pressure transducers can measure pressures up to 1000 Kgf/cm² (100 MPa) and is particularly suitable for situations of pulsating pressures. It is seen that accuracy increases with higher pulsations. The piezoelectric effect is instantaneous and is time-invariant in nature.

7.9.9.1 Charge and Voltage Sensitivity

As already mentioned, the piezoelectric effect is direction-sensitive, *i.e.*, tensile and compressive forces produce voltages of opposite polarities. If a charge q coulomb is produced for an applied force F Newton, then

$$q = dF$$

where d, the proportionality constant, is called the charge sensitivity of the crystal. The crystal capacitance is

$$c = \frac{\epsilon_r \epsilon_0 A}{t}$$

where, t and A are the thickness and area of plate of the crystal. Thus, the voltage produced is

$$q = Ec$$

or,

$$E = \frac{q}{c} = \frac{d}{\epsilon_0\ \epsilon_r} \cdot \frac{F}{A} t = gpt \quad ...(7.17)$$

where p = applied pressure = $\dfrac{F}{A}$

and, $g = \dfrac{d}{\epsilon_0\ \epsilon_r}$ is called the voltage sensitivity of the crystal. From the above,

$$g = \frac{E}{pt} = \frac{E/t}{p} \quad ...(7.18)$$

E/t is the electric field strength (in V/m). Hence, g is the electric field strength per unit pressure and is expressed in Vm/N.

7.9.9.2 A piezoelectric Pressure Transducer

Figure 7.43 shows a pressure gauge employing a piezoelectric transducer. The housing has two quartz slabs b and d with their same polarity facing each other (it is the negative charges of both slabs facing, as shown). The negative faces of both the slabs are made common by a metal support c, while the positive charges of the slabs fit into metal supports a and e. The top lid of the assembly is f while i is a

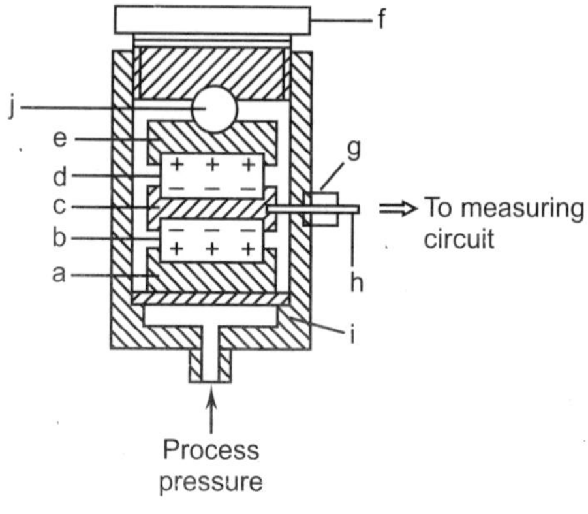

Fig. 7.43: A pressure gauge using a piezoelectric transducer

flat metal diaphragm. Between the lid f and upper support e is a ball, j. The ball distributes the applied pressure uniformly over the total quartz stack area. Through the bottom, the process pressure p is applied.

When an unknown pressure is applied at the bottom, the positive charges produced, as shown, are connected to the ground while the negative charges are collected and taken out via h and connected to a measuring device–calibrated in terms of unknown pressure. An insulator g ensures the delivery of negative charges to h.

If a measurement demands placing of the piezoelectric transducer system at a fairly high temperature—as in an internal combustion engine—the sensitivity of the transducer is to a certain extent affected although the piezoelectric constant K somewhat remains invariant. In such cases, cooling is arranged so that transducer temperature is maintained within acceptable limits.

Sensitivity of the transducer may be increased by increasing the number of quartz slabs, increasing the effective area of the diaphragm.

7.9.10 The Bridgeman Gauge-Measurement of Very High Pressure

Some materials like platinum, manganin, constantan, tungsten, bismuth, lithium, antimony etc. change their ohmic resistance value when subjected to external pressure. Studies have shown that manganin is the best choice for such a purpose. Manganin exhibits an approximately linear relationship between its resistance and the applied pressure and is given by:

$$R = R_0 (1 + \beta \, \Delta p) \quad \text{...(7.19)}$$

where, R_0 is the resistance at standard atmospheric condition, Δp is the increase in pressure and β is the pressure coefficient of resistance, given in cm^2/Kgf.

For manganin, the linearity between applied pressure and resistance is maintained up to a pressure of 30,000 Kgf/cm^2 (3000 MPa). It has a further advantage of very low temperature coefficient of resistance. Thus, ambient temperature variations will have negligible effect on the measurements. But manganin's low sensitivity limits its usability in the range of very high pressures (can be used ~ 100 MPa).

Again, value of β is not a constant for manganin—having variations for different batches of manganin. Thus, a recalibration of the system is needed whenever manganin is to be replaced with a new one.

Figure 7.44 shows a Bridgeman manganin-wire resistor transducer. The transducer is in the form of a resistor (2), held in place by metal supports (3) and is housed inside the gauge case (1).

The metal supports also act as output leads. The supports are insulated from the transducer case and are held in place by a nut (4). The coupling (5) connects the process pressure to the transducer, while the joint between the nut and the transducer case is sealed by two washers and insulating spacers.

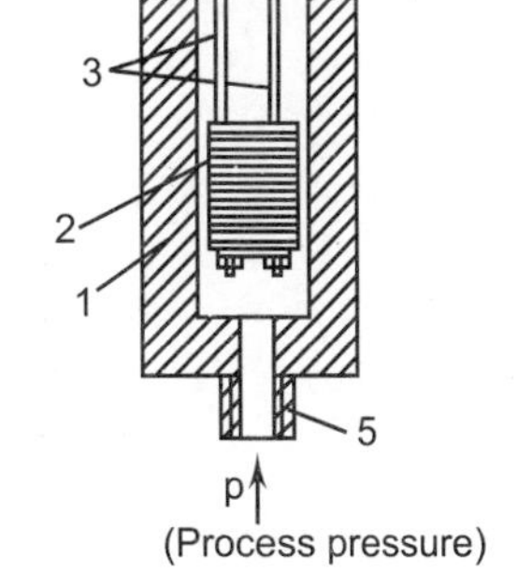

Fig. 7.44: Pressure gauge using a manganin-resistor transducer

In normal cases, the resistance of the manganin-resistor is measured by a bridge, while for more precision cases, potentiometer circuitry is used. The transducer accuracy is of the order of ± 1% of its range.

Resistance pressure sensors have a number of advantages like stability, small size, low cost, ease of manufacture and a conventional method of measurement using bridge circuitries.

Chapter 7

7.9.11 A Photoelectric Pressure Transducer

The principle of action of this type of pressure transducer uses the properties of a photo-tube. The photo-tube consists of a curved sheet of thin metal coated with a photo emissive material. At the centre of this sheet lies the anode rod. The voltage-current characteristics of such a tube is shown in figure 7.45.

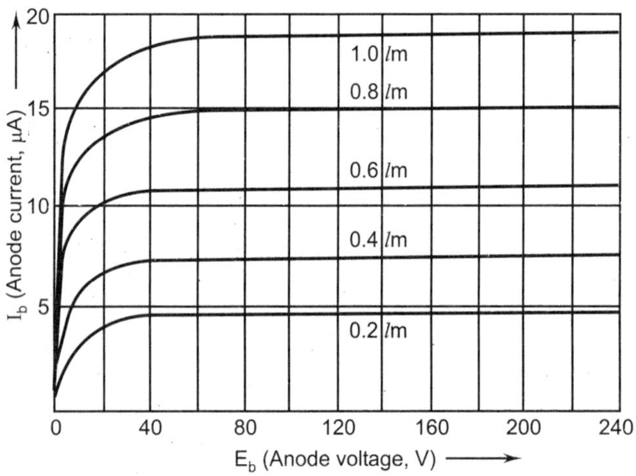

Fig. 7.45: Current vs. voltage characteristics of a photoemissive cell.

It is observed that the current through the photo-tube is dependent on the following:

- wavelength of incident light,
- voltage difference between anode and cathode and
- intensity of incident light.

It is also observed from the characteristics that for an anode voltage in excess of 40V, the output current becomes independent of anode voltage but in a function of the amount of incident light.

The photoelectric transducer used for measurement of pressure is shown in figure 7.46.

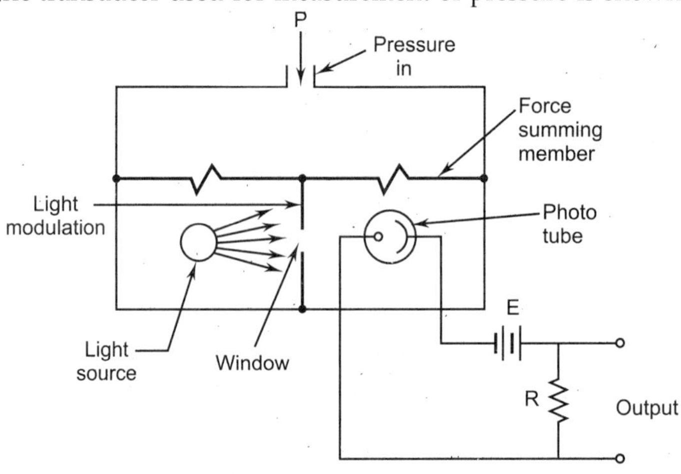

Fig. 7.46: A Photoelectric transducer for measurement of pressure

In the experimental setup, the amount of light, incident on the photo tube from the light source, is controlled by a window. The aperture of the window is controlled by a force summing member of the pressure transducer.

As the input pressure varies, a change in position of the force summing member changes the position of the window. It thus alters the amount of light falling on the photo tube. Thus, the drop across the resistance changes, which is then amplified suitably. This output can be calibrated in terms of the applied pressure.

The system can be applied for static and low-frequency dynamic inputs and has a high efficiency, although the long term stability is not good enough.

7.9.12 An Optical Pressure Transducer

This type of transducer utilises a diaphragm or a bellow element as a primary sensor; its movement with pressure variation is then detected by optical means.

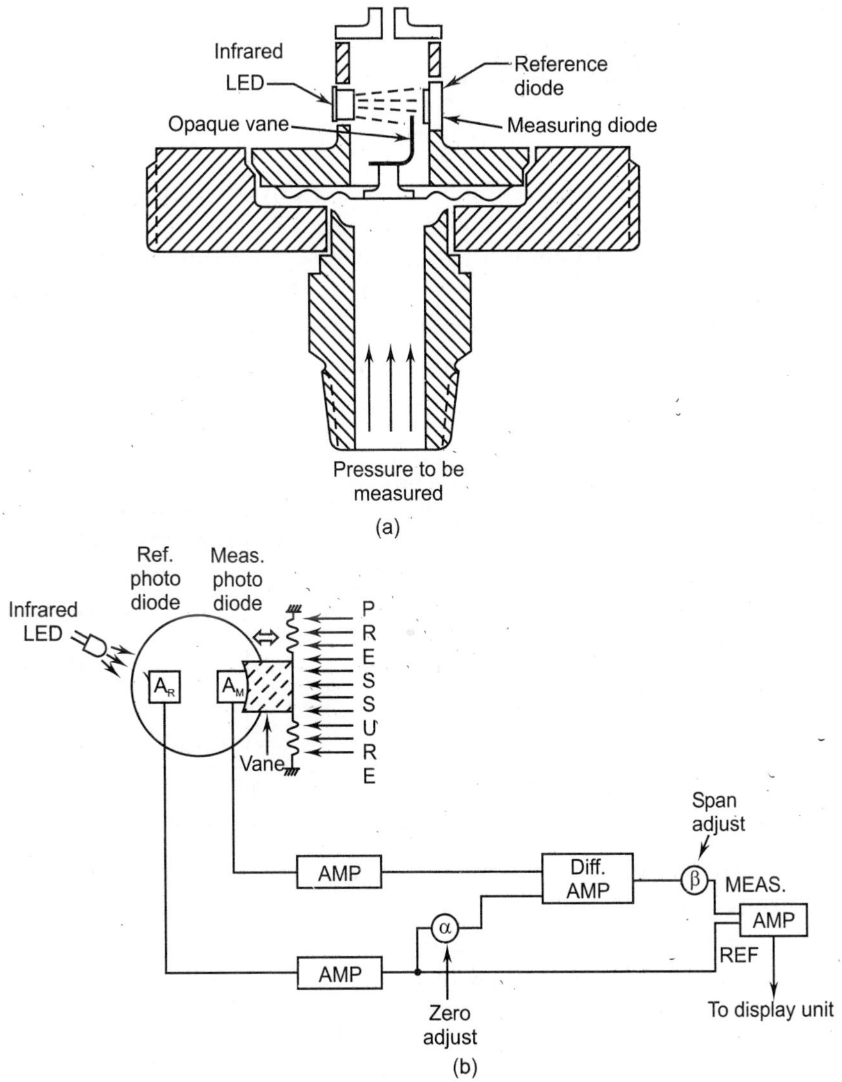

Fig. 7.47: (*a*) An optical pressure transducer,
(*b*) An arrangement to measure pressure by optical pressure transducer

In the diagrammatic arrangement shown in figure 7.47, the diaphragm position changes with changes in applied pressure. An opaque vane, which is connected to the diaphragm thus moves in a vertical plane thereby altering the amount of light received by the measuring photo-diode from the infrared LED. The measuring photo diode alongwith the reference photo diode are housed in a single chip thereby nullifying any change in LED output due to temperature variations or aging effect. The outputs of the diode illuminated areas A_R and A_M of the reference and measuring photo diodes (these outputs are in mV) are then amplified, zero and span adjusted by electronic circuits before finally getting the digital display readout which is calibrated in terms of applied pressure.

This optical pressure transducer eliminates the effect of any electromagnetic energy transfer that might take place through hazardous spaces. An example of such a hazardous space is in coal mines containing methane. Diode non-linearities can be taken care of by using PROMs.

7.10 MEASUREMENT OF VACUUM

7.10.1 Introduction

Vacuum can be thought of as a space entirely devoid of any kind of matter, *i.e.*, air has been totally exhausted from that space. It is just but impossible to achieve absolute vacuum. This has led to the concept of degree of vacuum in vacuum technology field with the different degrees of vacuum specified as follows:

medium high vacuum : 1 to 10^{-3} Torr

high vacuum : 10^{-3} to 10^{-7} Torr

very high vacuum : less than 10^{-7} Torr

with 1 Torr equal to 1 mm of Hg.

Vacuum of varied degree has widespread applications in fields like fabrication of television tubes, x-ray tubes, magnetrons, metallurgical fields and also in pharmaceutical industry for the manufacture and preservation of different drugs, blood plasma and serums.

Vacuum measurements can be made either by mechanical or electrical means.

Some of the absolute pressure measuring gauges with their useful ranges are shown in figure 7.48.

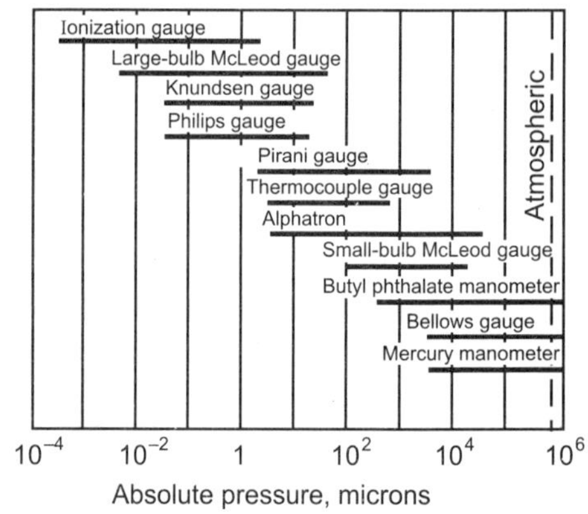

Fig. 7.48: Useful Ranges of Absolute Pressure-Measuring Gauges (Produced by Permission of Distillation Products Industries)

7.10.2 Mechanical Methods

Measurement of vacuum by mechanical means can be done by Knudsen's gauge, McLeod gauge or Philips gauge, of which the first two will be discussed.

The mechanical methods are simpler than their electrical counterparts although the electrical ones can measure lower absolute pressure and are also more suited for high frequency pressure variation situations.

7.10.2.1 Knudsen Gauge

It is a radiometer type instrument, the principle of operation of which is the development of a force between one heated surface and another surface (which is maintained at the temperature of the gas whose pressure is desired to be measured) by the molecules rebounding from the hotter surface with more kinetic energy than the surface at the temperature of the gas. This force is proportional to the absolute pressure. A schematic of a Knudsen gauge is shown in figure 7.49.

This gauge is insensitive to gas composition but is dependent on accommodation coefficient. Accommodation coefficient is a measure of the extent to which the rebounding molecules has attained the temperature of the hot surface. Another advantage is that the gauge can give a continuous output reading.

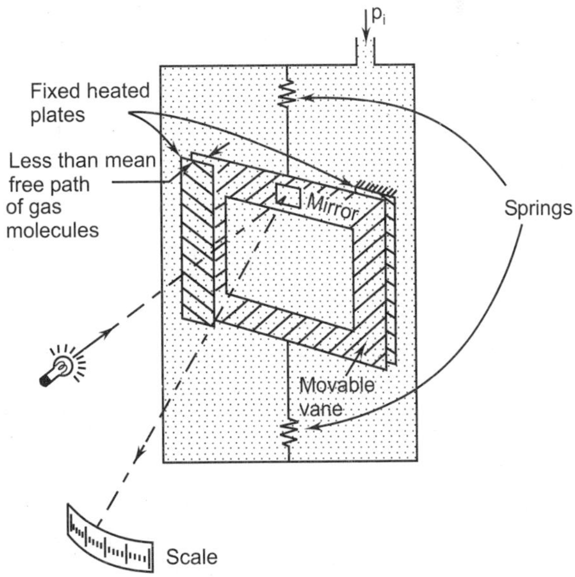

Fig. 7.49: Knudsen gauge

The experimental set up contains two fixed heated plates and a movable vane. The movement of the movable vane is restricted by springs attached at its both ends and supports the vane. A lamp and scale arrangement records the movement of the vane. The scale is calibrated in terms of absolute pressure. The pressure is applied into the system from the top. The movable vane is in between the two fixed plates. The gap between the movable and the fixed plates is less than the mean free path of the gas molecules. Since the temperature of the fixed heated surface (T_f) is higher than that of the vane (T_v), the gas molecules that rebound from the heated surface impinge on the vane with a net force which is proportional to the gas pressure (this is true for fixed values of T_f and T_v). It can be shown that the unknown gas pressure is given by the following formula:

$$p_i = \frac{KF}{\sqrt{T_f/T_v - 1}} \cong \frac{KFT_v}{T_f - T_v} \quad \text{(for small difference of temperature)} \qquad \text{... (7.20)}$$

where F is the net force on the vane and K is a constant, having a nominal value 4×10^{-4}. The Knudsen gauge is suitable for application in the pressure range of 10^{-2} to 10^{-8} Torr.

7.12.2 McLeod Gauge

It is modified type mercury manometer, which is described in figure 7.50.

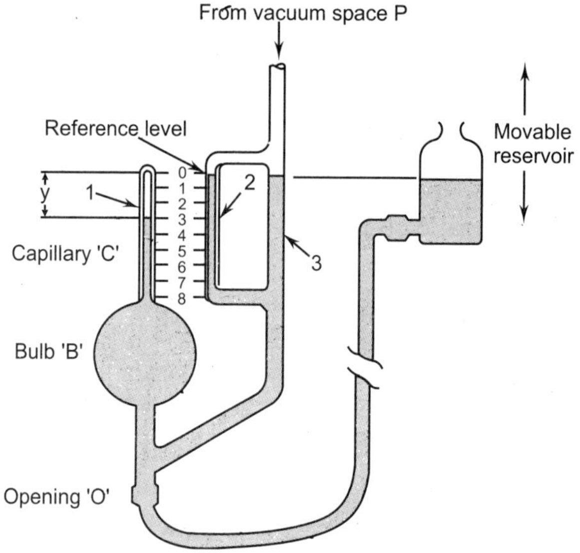

Fig. 7.50: The non-linear McLeod gauge

A capillary C is connected to a large diameter bulb B, the lower part of which has an opening 'O'. This opening is connected to a movable reservoir (R) and another portion, as shown in the figure, to which the unknown pressure point is connected. If the capillary C contains total vacuum as the reservoir is raised, the mercury level in 1 (*i.e.*, capillary C), 2 and 3 will rise and remain at the same level till mercury reaches the top end of the capillary C (*i.e.*, '1').

With the arrangement, as shown in the figure, the movable reservoir is moved down till the mercury level in the bulb goes just below the opening O. At this point, the pressure source is connected to the system via the top end of 3. Thus, the bulb and 1, 2, 3 are all at the unknown pressure P, which is to be measured. The movable reservoir R is now raised till the height of mercury in limb 2 touches the reference level. The unknown gas is now trapped in a small volume ya, where a = cross-sectional area of the bore of the capillary and y is as indicated in the figure.

Let now,

V = Volume of the capillary, bulb and upto the opening 'O' down below

P_c = Pressure in the capillary

P = Unknown pressure

Thus, applying Boyle's law, *i.e.*,

original (Pressure × Volume) = Final (Pressure × Volume)

or
$$P \times V = P_c \times ay$$

Now, $P_c > P$ and is given by:

$$P_c = P + y$$

$$\therefore \quad PV = ay\,(P + y) \quad \text{or} \quad P = \frac{a}{(v - ay)} \cdot y^2$$

Since, $ay << V$

$$\therefore \quad P = \frac{a}{V} \cdot y^2 = Ky^2 \qquad \text{...21)}$$

where $K = \frac{a}{V}$ is known as compression ratio or gauge constant.

$\therefore \quad P \propto f(y^2)$

Thus, the unknown pressure P maintains a non-linear relationship with y and hence sensitivity of the instrument is not constant. However, the pressure scale can be linearised, *i.e.*, a proportional relationship can be derived between y and P if measurements are done in the following way, as shown in figure 7.51.

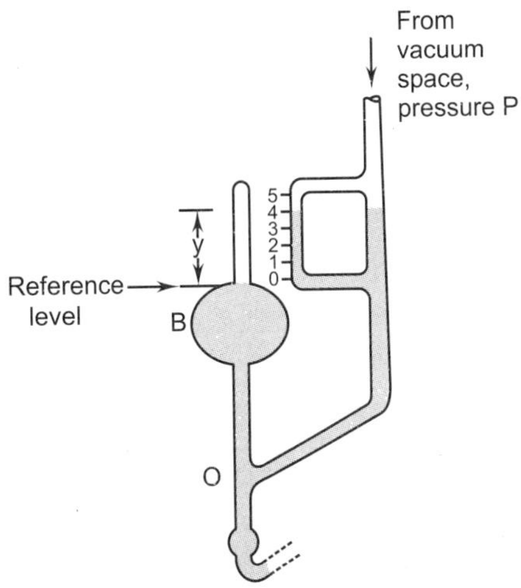

Fig. 7.51: McLeod Gauge (Linear type)

In this case, the reference level is shifted as shown, which is the junction of the top end of the bulb and the beginning of the capillary. Here the reservoir height is so adjusted that the entrapped gas always occupies the whole length of the capillary, *i.e.*, a constant volume. Let this volume be V_c, which is the internal volume of the capillary. Here,

$$\therefore \quad PV = P_c V_c = (P + y).\ V_c$$

$$\Rightarrow \quad P = \frac{V_c}{V - V_c} \cdot y$$

$$\simeq \frac{V_c}{V} \times y \quad (\because V >> V_c) \qquad ...(7.22)$$

$$\therefore \quad P \propto f(y)$$

Thus, the unknown pressure is linearly related to the height h and hence the scale becomes linear.

McLeod gauge can measure vacuum in the range of 1 mm to 10^{-6} mm of Hg and is used for the calibration of electrical pressure gauges like Pirani gauges.

In case the gas, whose pressure is to be measured, contains any vapour, compression may lead to condensation. In that case, pressure measurement will become erroneous.

The gauge cannot be used for continuous output reading, has limitations on the lowest measurable pressure and is bulky, but has the advantage that the gauge is not influenced by the composition of the gas.

When this gauge is used for calibration of other gauges, a liquid-air trap is used between the McLeod gauge and the gauge to be tested in order to prevent passage of mercury vapour. The bulb is placed in between the opening '*O*' and the capillary. The purpose of the bulb is to entrap a larger amount of gas (whose vacuum pressure is to be measured). The presence of the bulb enhances the accuracy of measurement and, at the same time, can measure very low vacuum pressures.

7.10.3 Electrical Methods

Pirani, thermocouple and thermionic gauges are used for the measurement of high vacuum where continuous measurement is of prime importance.

7.10.3.1 The Pirani Gauge

The Pirani Gauge, named after M.V. Pirani, was designed in 1906. It is also known as hot-wire pressure transducer. It is based on Wheatstone's bridge principle and the out-of-balance potential is balanced by a potentiometer, straightway calibrated in terms of unknown vacuum pressure. This unbalance may also be connected to a μA which may be calibrated to read vacuum directly. The arrangement is shown in figure 7.52.

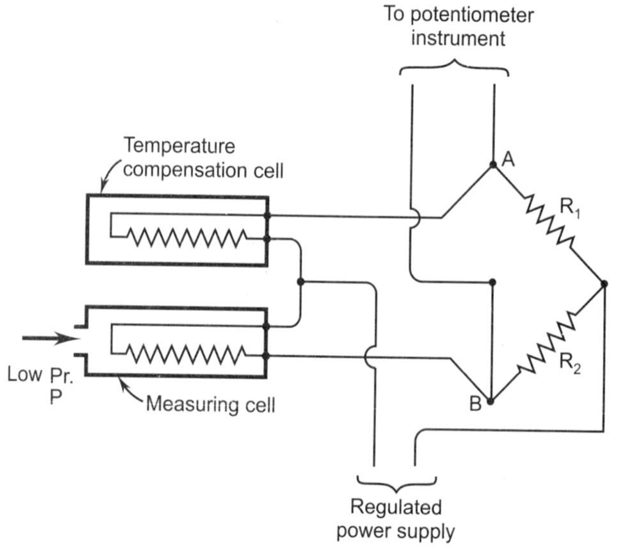

Fig. 7.52: Pirani vacuum gauge and circuit

The measuring cell and the temperature compensation cell, both made of hard glass, contain tungsten filaments. This gauge operates on the principle that the heat loss by conduction and convection from a **heated wire** is a function of the pressure of the gas surrounding the wire. The measuring cell is connected to the low pressure point while the pressure in the temperature compensation cell is maintained at around 1 micron. Thus, any change in environmental temperature would affect both the cells in a like manner and would thus cancel out. The current through the filaments is maintained constant by a regulated power supply system. The unbalance bridge voltage, measured by an automatic balance potentiometer is calibrated in terms of unknown pressure.

Pirani gauges have the useful operating range from 20 to 200 microns. The gauge is simple in design and is very easy to operate, although the calibration depends upon the gas (*i.e.*, composition dependent) whose pressure is being measured by the gauge.

The open end of the measuring cell is not recommended to be exposed to atmosphere because chances of corrosion of the filament is there, which may lead to a change in calibration. Again, the measuring cell filament should not exceed or value of 230°C roughly, otherwise organic vapours, if present, would deposit itself on the filament leading to a zero shift of calibration.

Figure 7.53 shows the calibration curve for air with bridge unbalance measured in μA and air pressure in terms of Torr (*i.e.*, mm of Hg column).

Pirani gauges are accurate to within ± 2% of the calibration pressure range. This gauge is recommended not to be used beyond 1.0 Torr, because at such a high pressure the thermal conductivity of gases does not decrease with decrease in pressure. Pirani gauge is linear between 10^{-3} Torr to roughly 4×10^{-2} Torr. Outside this range, it is logarithmic.

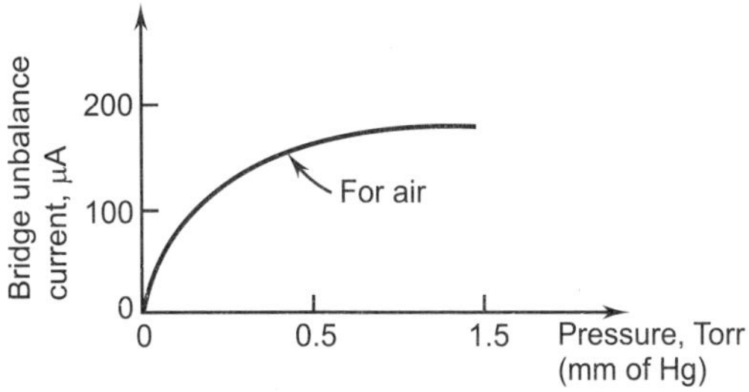

Fig. 7.53: Calibration curve

As already mentioned, since Pirani gauge is composition dependent, hence the gauge needs recalibration when it is used for different gases. Also, because of its composition dependence characteristics, such a gauge can also be used to detect leaks in vacuum experiments. The surface condition and emissivity of the heated element play an important part since the operation of the gauge is dependent on thermal conductivity.

7.10.3.2 The Thermocouple (Thermal) Vacuum Gauge

The thermocouple type vacuum gauge is shown in figure 7.54. The transformer primary is connected to a regulated power supply while the measuring cell and the temperature compensating cell are connected to the two secondaries. Two thermocouples, connected in series opposition, measure the temperature of the two cells. Since the emfs of the two thermocouples are in opposition, the instrument can be made a direct reading one.

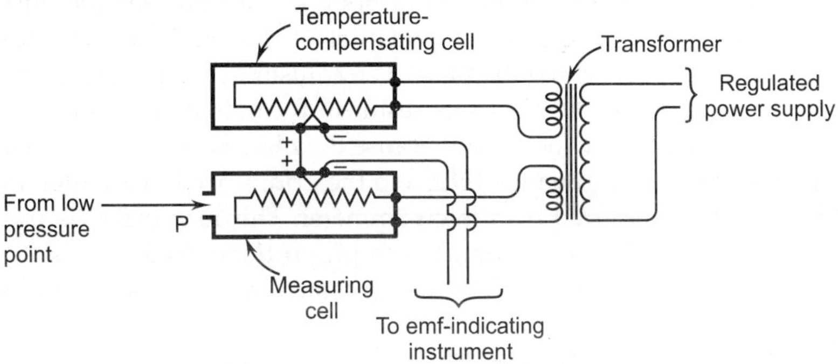

Fig. 7.54: Thermocouple vacuum gauge and circuit

The principle of operation of this gauge is identical to that of Pirani gauge and has the same features and handicaps as that of Pirani gauge. The output of this gauge is about 20 mV for a vacuum of 450 microns of mercury. This output can either be measured by a potentiometric arrangement or else by a μA and calibration done accordingly.

The use of transformer ensures that the source impedance is small resulting in electromagnetic or radio frequency interference.

The normal range of a thermocouple vacuum gauge is about 1 to 10^{-3} mm of Hg. The filament temperature should be kept below 200°C, otherwise there will be a possibility of sample gases getting decomposed and or forming a deposit on the heating element. The accuracy of the gauge is about ± 10% over the full range but is around ± 2% at the midscale.

The gauge may lose its calibration if the filament becomes dull or contaminated – in that case radiation losses would occur. One way to overcome this is to precoat the filament, but in that case sensitivity problem can be avoided by employing thermopile which provide a higher potential output.

7.10.3.3 The Hot Filament Ionisation (Thermionic) Gauge

This gauge, shown in figure 7.55, is in the form of a triode with a small hole for connection to the unknown low pressure point. It consists of a cathode, grid and a plate.

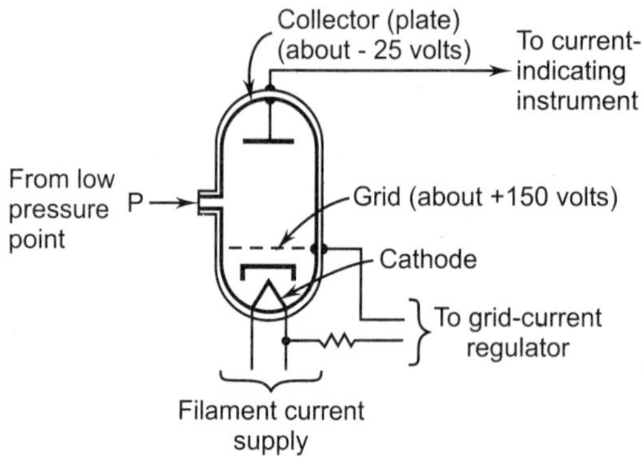

Fig. 7.55: The hot filament ionisation (thermionic) gauge

Here the grid is maintained at a high positive potential with respect to the cathode while the plate is maintained at a negative potential with respect to the grid. Those potentials are unlike the case in which a triode is normally connected. Normally, the grid is maintained at approximately +150 V and the plate at –25 V.

Electrons (thermions) emitted from the hot cathode surface proceed (because of high kinetic energy of the electrons and due to positive potential of the grid) towards positive grid, pass through the gird and start approaching the plate. The collision between the gas molecules (whose pressure is to be measured) and the electrons result in formation of positive and negative ions. These positive ions are collected by the plate (because the plate is at a negative potential) resulting in an ion current. The electrons ultimately return to the grid, since it has a high positive potential. The positive ion current is measured by a microammeter. This ion current is proportional to the number of molecules present in the gas and hence proportional to the absolute pressure of the gas, provided the electron current is maintained constant, which is achieved by a grid current regulator.

The vacuum pressure is given by: $p = \frac{1}{K}\left(\frac{i_p}{i_g}\right)$, where

K = sensitivity of the gauge

i_p = plate current

and i_g = grid current

Low pressures in the range of 10^{-3} to 10^{-11} mm of Hg can continuously be measured by this gauge. This gauge can be calibrated by using inert gases, but it is not reliable with oxygen, hydrogen and other gases which decompose at the hot filament. The accuracy of this device is low-about ± 10%.

For measurement of extremely low values of pressures $\left(\sim \frac{1}{100} \text{micron}\right)$, it is recommended to degas the tube, *i.e.* to increase momentarily the filament current to a high value with the tube at a low pressure. This heats up the tube appreciably and remove all occluded and adsorbed gas from the tube walls, the elements of the tube and the collector. The gauge is not suitable for measurement of absolute pressure above 2 microns owing to the possibility of burning out of the filament.

The high temperature of the cathode may result in:

- The cathode being attacked by the gases present or may undergo some chemical combination with them.
- The gases present may dissociate.
- The cathode itself may be oxidised if sufficient amount of oxygen is present in the gas.

This gauge provides the widest pressure measuring range and a fast response. It is not applicable for gases which are going to decompose at high filament temperatures (~2200°C). The ionisation tube is subject to mechanical damage and its utility is normally confined in laboratories. Because of high filament temperature, it deteriorates fast and is also subject to chemical attack.

7.10.3.4 Ionisation Gauge (Radioactive Type): An Alphatron

In this gauge, ionisation is produced by collision of molecules of the gas with the alpha-particle radiation from a very small radium or other similar source instead of by electron collision. The positive ions so produced are passed through a resistor R. The magnitude of the current is of the order of 10^{-10} A and the drop across R is fed to a very high input impedance voltmeter. The voltmeter reading can be calibrated in terms of the unknown vacuum pressure.

Since no filament is used in this case, there is no possibility of burning out of the same at higher pressure, as was the case for thermionic type, but in this case the gauge sensitivity is much less because of availability of less energy.

This gauge can monitor the decrease in pressure with higher fidelity. The low penetrating power of the α emitter makes the job of shielding very easy. A schematic of an alphatron is shown below in figure 7.56, alongwith its characteristics.

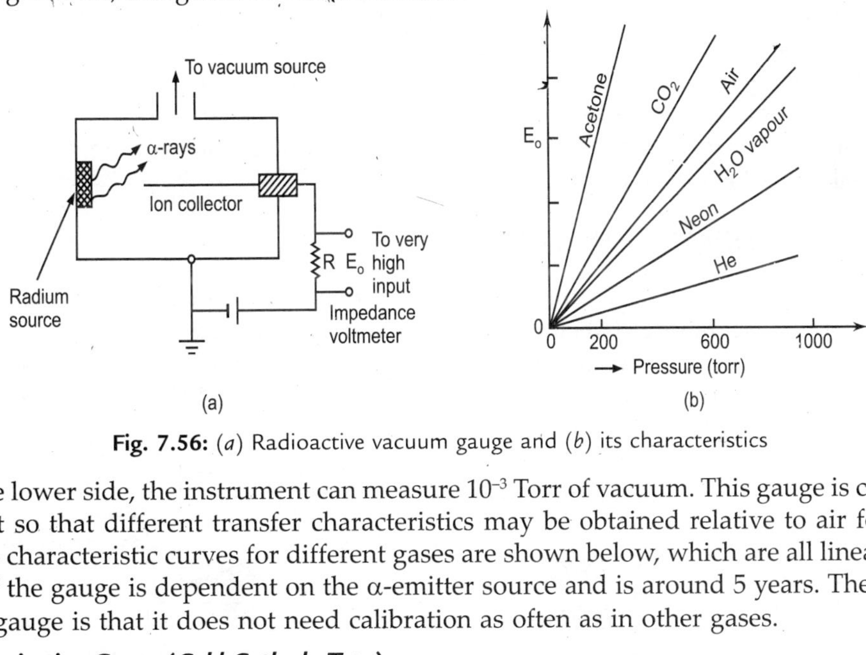

Fig. 7.56: (*a*) Radioactive vacuum gauge and (*b*) its characteristics

On the lower side, the instrument can measure 10^{-3} Torr of vacuum. This gauge is composition dependent so that different transfer characteristics may be obtained relative to air for different gases. The characteristic curves for different gases are shown below, which are all linear in nature. The life of the gauge is dependent on the α-emitter source and is around 5 years. The advantage with this gauge is that it does not need calibration as often as in other gases.

7.10.3.5 Ionisation Gauge (Cold Cathode Type)

F.M. Penning, in 1937, first explained its operating principle. The gauge is also known as a Philips detector. The basic difference between the hot and cold cathode types is the manner of obtaining the ions. In the hot filament type, electrons are ejected by thermionic principle while in cold cathode type, electrons are derived by employing a strong magnetic field. Since less number of electrons are emitted in the cold cathode variety, hence less ions will be formed by the collision

Chapter 7

of gas molecules with the electrons if they travel a straight path from cathode (*C*) to anode (*A*). Thus, sensitivity will be poor. To increase sensitivity, a magnetic field, as shown in figure 7.57, is applied which causes the electrons to spiral as they proceed from cathode to anode. Since, in this case, the electrons travel more path, they will ionise more molecules and hence more ions will be emanated and thus the sensitivity is enhanced.

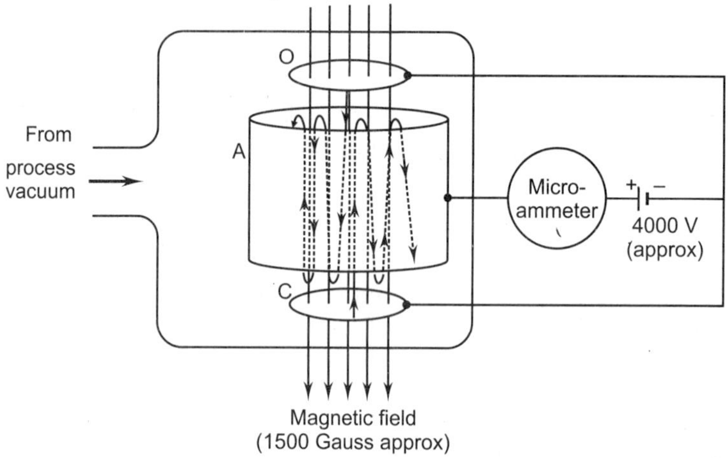

Fig. 7.57: Ionisation gauge (Cold cathode type)

The micro ammeter meter is calibrated in terms of process vacuum. The range of this gauge is approximately from 10^{-2} to 10^{-7} mm of Hg with accuracy of the order of $\pm$ 20% at an absolute pressure of around 10^{-5} mm of Hg. The sensitivity of the gauge is around 5 mA per micron pressure.

While the advantages of cold cathode type are its lower cost and very little chance of burn-out of the cathode, compared to the hot filament type, its disadvantages are mainly non-linear output signal and a low value of accuracy.

7.11 A FLAPPER NOZZLE SYSTEM

A flapper-nozzle system is a displacement to pressure transducer. Either very small displacements or else variation in clearance between two surfaces can be converted into air pressure variations by this transducer. It can be used as a precision gauging equipment as well as a basic component in pneumatic or hydraulic measurement, transmission and control. Figure 7.58 shows a basic schematic arrangement for such a flapper-nozzle system.

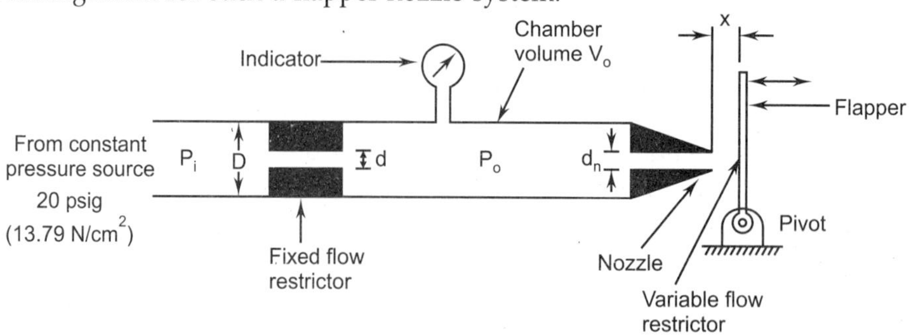

Fig. 7.58: The basic Flapper-Nozzle system

Fluid at a constant pressure (P_i) is applied to the fixed flow restrictor, which is in series with a variable flow restrictor. The fluid (normally air which should be dust free and filtered) is vented into the atmosphere via the nozzle, the rate of flow of which (*i.e.* air) is regulated by the flapper position. When the flapper is held tightly against the nozzle, no air can be rented into the atmosphere and the air pressure inside ultimately assumes the supply pressure, P_i. When the flapper is held a long distance away from the nozzle, air escapes out of the nozzle without any obstruction and the size and the shape of the nozzle governs the output flow rate.

The distance between the nozzle and the flapper should not be made excessively large such that the chamber pressure never falls to the atmospheric pressure.

A pressure measuring device measures the chamber pressure P_o, which is also the output pressure signal. When the flapper moves, changing the value of x, it, in its turn, changes the value of P_o. For small displacements, the pressure-displacement curve is linear and in this linear range, P_o is very sensitive to variations in x.

The response time of the system is determined by the compressibility of gases. The system dynamics is a linear, first-order system—for small changes in x. For finding out the transfer function of the system, the principle of conservation of mass is applied to the chamber—the difference between the mass entering the chamber and that leaving it must show up as an additional mass in the chamber, thereby altering its pressure.

With P_i maintained at a constant value, the ratio of output pressure to input pressure, *i.e.*, P_o/P_i can vary from zero to one theoretically—these two extreme values occur when the clearance between the nozzle and the flapper varies from a very large value to zero, respectively. Under steady state conditions, the mass of fluid entering the chamber is equal to the mass bleeding through the nozzle.

For finding out a relationship between output pressure P_o and input supply pressure P_i, it is assumed that flow through the restrictions is incompressible. Thus, by equating the flow through the fixed restrictor and the nozzle, we can write that:

$$c \times \frac{\pi d^2}{4} \times (P_i - P_o)^{1/2} = c \times (\pi d_n x) \times P_o^{1/2} \quad \text{...(7.23)}$$

where c is the discharge coefficient, assumed equal for both the cases. All other quantities are as shown in the diagram.

Thus,
$$\frac{P_o}{P_i} = \frac{1}{1 + 16\left(\frac{d_n x}{d^2}\right)^2}$$

as
$$P_n = \frac{1}{1 + 16x_n^2} \quad \text{...(7.24)}$$

where,
$$P_n = \frac{P_o}{P_i} \text{ and } x_n = \frac{d_n x}{d^2}$$

The displacement-pressure characteristic curve is shown in figure 7.59. For small and large values of x_n, the curve is non-linear and hence should be avoided. For large x_n values, P_n becomes almost constant and hence the system becomes insensitive. To avoid this, the system should be operated in the linear region. Thus, the operating point is chosen exactly at the middle of the linear region and the mean position of the flapper corresponds to this. From the curve, these two extreme values for P_o/P_i were found to be 0.15 to 0.75. The maximum slope (and hence maximum sensitivity) is derived by making $d^2P_n/dx_n^2 = 0$, which gives $P_n = 0.75$ at $x_n = 0.144$. the lower and upper limits of P_n are taken to be 0.15 and 0.75 and with $P_i = 20$ *psi*, P_o values becomes 3 and 15 *psi* respectively.

When worked in the linear region, a very small change in x_n (*i.e.*, x – flapper position) results in a large change in P_o, *i.e.*, why the system can be treated as a displacement-to-pressure converter.

As can be seen from the above, the industrial control pressure (*i.e.*, output pressure P_o) is kept in between 3 and 15 *psi* with a mean at 9 *psi*. The upper limit is governed by pipe thickness while the lower pressure limit (*i.e.*, 3 *psi*) is not brought down to zero (*a*) to avoid any leakage from atmosphere into the system (*b*) to eliminate poor sensitivity and also (*c*) to have a check on the control pressure.

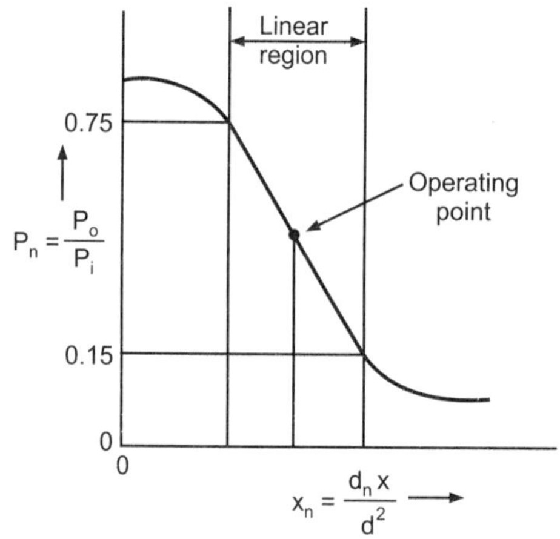

Fig. 7.59: Displacement-pressure characteristic curve of a flapper-nozzle system

The total displacement of the flapper which covers the total output pressure range (*i.e.*, 3 to 15 *psi*) is called *throttling range*. In most cases, it is approximately 0.1 mm. Hence, this displacement-to-pressure transducer has very high sensitivity and is particularly employed in situations prone to fire hazards. The volume of the chamber is kept at around 4–5 *cc* so that response of the transducer becomes fast and, as such, the time constant is kept at a low value of the order of ms. The time constant is determined by the compressibility of the gas (normally air).

Since the system is susceptible to vibrations, care should be taken while housing a flapper-nozzle assembly to avoid such effects.

The flapper-nozzle assembly forms the basis in many pneumatic and hydraulic instrumentation systems with the flapper actuated by bimetallic strip, etc.

7.11.1 A Reverse Acting Control Relay

For maximum sensitivity to be derived from the flapper-nozzle system, *i.e.*, to get a considerable change in the output pressure from a very small change in flapper movement, the system must be operated in the linear region. This is achieved by a reverse acting pneumatic relay, shown in figure 7.60.

The process pressure is applied to the bellow while the regulated pressure input of value 20 *psi* is applied as shown. The principle of operation of the relay is as follows:

If the flapper moves towards the nozzle, the intervening gap decreases, thereby increasing the pressure P, and indicated by the pressure meter P_1. This increased pressure is transmitted to the diaphragm G, thereby lifting the ball valve B upwards. This action partially closes the path of air that is entering via H, and increasing the clearance at E, thereby bleeding some air into the atmosphere. The ultimate effect is a reduction in the record of the pressure meter P_2 and is connected to the line taken to the remote destination for gauging purpose.

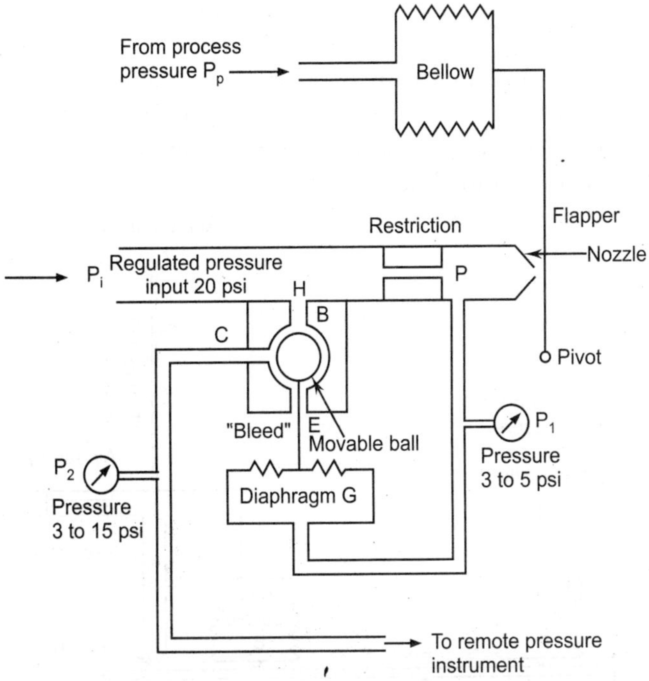

Fig. 7.60: A reverse acting control relay

It is seen that a small change of 2 *psi* in the nozzle pressure, as indicated by pressure meter P_1 (and hence corresponding to a very small change in flapper movement) results in a large change of about 12 *psi* at the output (indicated by pressure meter P_2). With the reverse acting relay being employed, the movement of the flapper is reduced to 1/6th of its former value and, as such, the flapper-nozzle system operates totally in the linear region.

Since an increase in pressure P_1 results in a decrease in pressure P_2, the relay is called reverse acting. Also, a very small change in nozzle pressure (P_1) results in a large change in output pressure (P_2); as such the relay may be called a pressure amplifier.

7.11.2 A Basic Pneumatic Pressure Transmitter

Apart from local indication/record of pressure of a line, sometimes it is desired to read/record pressure at a considerable distance. A pneumatic pressure transmitter can be employed for such a case where the employed system is totally non-electrical in nature. Such a scheme is shown in figure 7.61.

The pressure to be transmitted is applied at the left end. If the process pressure increases, it will result in expansion of bellow B_1. The left end of the beam would be depressed by this increased pressure about the pivot F causing the right end of the beam, which acts as a flapper to be lifted up a bit. Thus, the gap between flapper and the nozzle is decreased, effecting an increase in pressure in bellow B_2. The pressure in B_2 increases linearly with decreasing distance between the flapper and the nozzle. This bellow B_2 obviously acts as a control or balance bellow. This increase in pressure in B_2 causes it to expand, thereby moving the flapper away from the nozzle until a torque balance is obtained against the pivot *F*.

Hence, it is seen that an increase in pressure in B_1 also increases the pressure in B_2–albeit in a proportionate manner. This changed pressure in B_2 can be taken to a remote destination and measured.

The range (or restraining) spring helps B_2 in opposing the movement of B_1. By varying the position of pivot *F* and the flapper (*i.e.*, the right end of the beam), a very small flapper-nozzle movement will ensure torque balance-ensuring that the graph of flapper-nozzle displacement *vs.* pressure change (in B_2) works in the linear region.

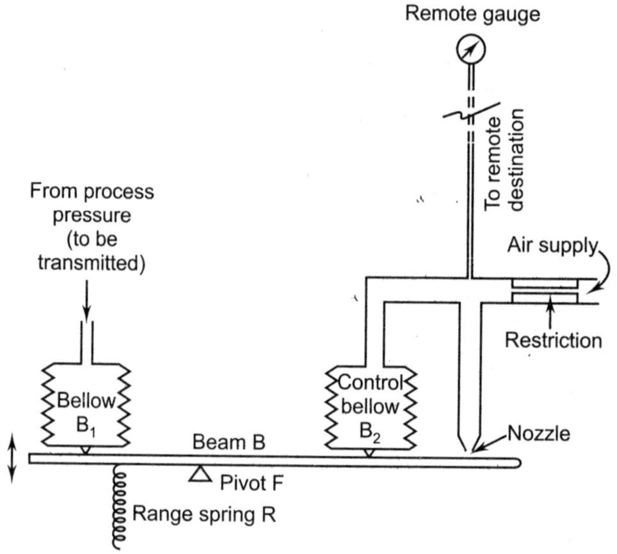

Fig. 7.61: Block schematic of a pneumatic pressure transmitter

For differential pneumatic pressure transmitters, the range spring *R* is discarded. A bellow, (say B_3) similar like, B_1, is placed on the bottom side of the beam *B* facing B_1. The two pressures are applied at bellows B_1 and B_3 for such a differential pneumatic pressure transmitter.

7.11.3 A Force Balance Pressure Transmitter

A force balance pressure transmitter, shown in figure 7.62 is much like a pressure balance pressure transmitter, although here the range (or restraining) spring is done away with. Here the output pressure signal (indicated by transmitted output) generates the necessary force for restoring the beam balance with the help of feedback bellows B_2. With increase in process pressure, the transmitter output signal decreases and vice-versa.

Although force balance transmitters are considered to be motionless, in reality a very small motion of the beam of the order of a few thousandths of an inch actually takes place.

Zero adjustment (zero shift error) is provided in the transmitter caused by the operating temperature and pressure being different from test conditions.

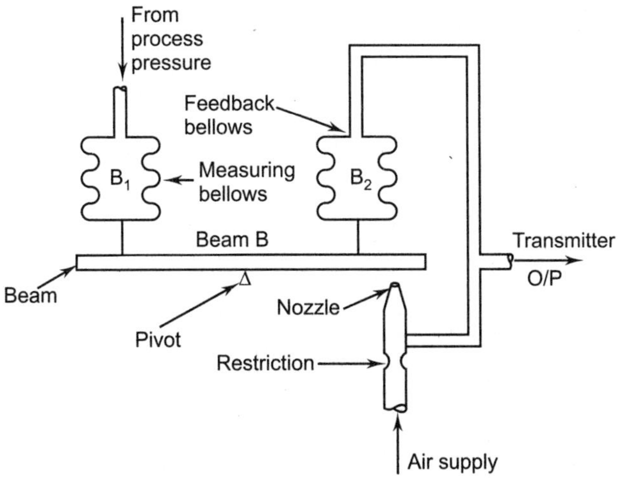

Fig. 7.62: A force balance pressure transmitter

7.11.4 An Electronic Type Force Balance Transmitter

An electronic force balance transmitter is shown in figure 7.63 in which the flapper-nozzle system is replaced by a detector coil and the feedback bellow by a 'force motor'. It is a typical pneumoelectric transmitter with pressure as input and DC current as output.

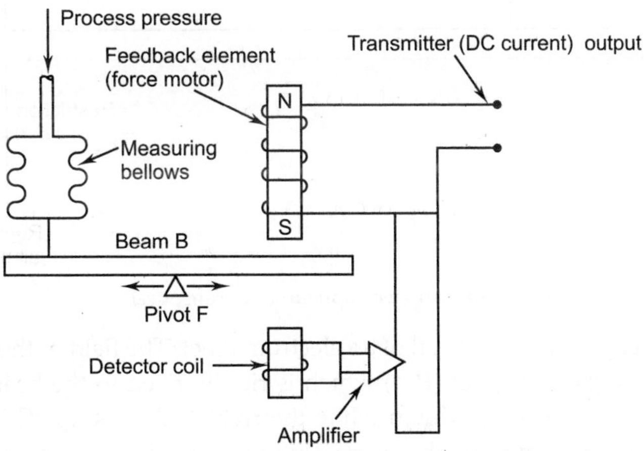

Fig. 7.63: An electronic type force balance transmitter

The detector coil's proximity (nearness or otherwise) changes with process pressure changes. This, thus, changes the inductive reactance of the coil. This change is converted into the transmitted *DC* current output, the force motor being in series with this output. The force motor consists of a permanent magnet on which is wound a coil consisting of several turns. The feedback force is generated by this force motor which thus counterbalances that generated by the process pressure.

The *DC* current output (*i.e.*, the transmitter output) can be changed by repositioning the pivot *F*. With this, a range adjustment of the order of 10 : 1 can be derived.

It is obvious that in this case the transmitter output current replaces the output air signal of the conventional force balance pressure transmitter.

7.11.5 An Electropneumatic Transducer

In a considerable number of cases in industries, valves are pneumatically operated. In such cases, the transmitted 4–20 mA *DC* current signal is to be reconverted back to pneumatic signal. Such an electropneumatic pressure transducer is shown in figure 7.64.

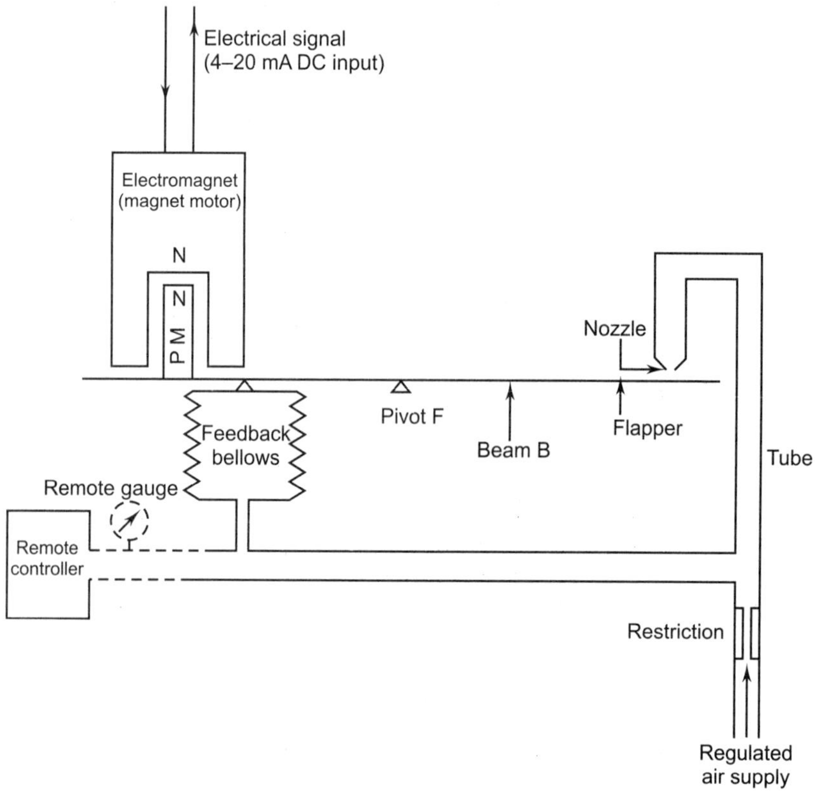

Fig. 7.64: An electropneumatic transducer

The 4–20 mA DC signal is fed to the coil of an electromagnet. The field of this electromagnet acts in opposition to the permanent magnet (*PM*) which is rigidly fixed to the beam *B*. The net action is that the left end of the beam comes down while the right end goes up. This right end (*i.e.,* the flapper) moves closer to the nozzle, thereby decreasing the 'bleeding' of the air to the surrounding. Thus the air pressure inside the tube increases and transmitted to the feedback bellows which thus acts in a manner to oppose the motion created by the electrical signal. A balance is ultimately achieved with the gauge pressure representative of the electrical output signal.

In a typical electropneumatic pressure transducer, the movement of the beam is effectively enhanced by a factor of around 50. In such a case, for the entire range of variation of input current signal, the beam movement becomes as small as 5.08×10^{-4} mm. Here, the force of repulsion between the electromagnet and the permanent magnet becomes a linear function of the current in the coil.

7.12 A DIFFERENTIAL PRESSURE GAUGE

A differential pressure gauge, which can either measure differential pressure or gauge pressure, is shown in figure 7.65.

Two diaphragm capsules —1 and 3, set up on either side of a base plate 2, measure the differential pressure. The two input pressures p_1 and p_2 are admitted into the diaphragm assembly via two input lines with p_1 acting as the high pressure and p_2 low pressure. The spaces inside the diaphragm capsules are filled with liquid.

4 is a plunger and connected with the upper capsule through a non-magnetic rod. Movement of 4 causes a change in output voltage. 5 is the differential transformer transducer which is housed in a non-magnetic stainless steel housing 6.

For zero setting of the instrument, the two shut-off valves (7 and 8) are closed and the equalising valve (9) opened. The plunger is then moved up/down for zero adjustment.

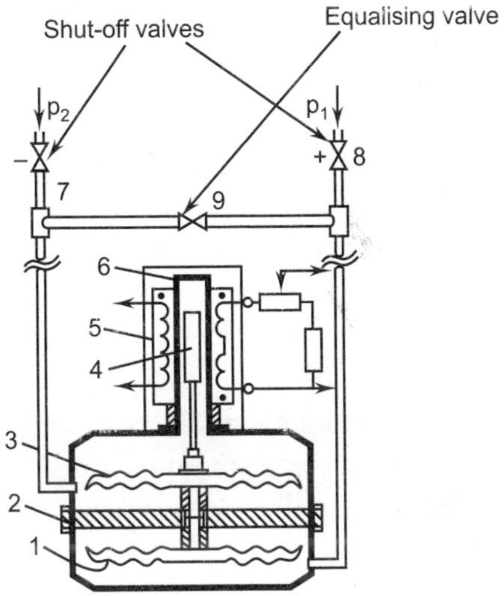

Fig. 7.65: A differential pressure gauge

A pressure differential is applied with p_1 acting as the higher pressure. This causes liquid from the lower capsule to be moved to the upper one and a consequent plunger movement. Thus, a change in output voltage will be obtained, which is proportional to the pressure difference p_1–p_2.

If the applied pressure differential exceeds the maximum value (maximum pressure differential for a particular DPT) or if one of the capsules is subjected to one-sided loading, the capsule would not fail. This is because the diaphragm shells of the capsule will collapse over its profile and the liquid in this capsule moves over into the other one.

Capsules of different stiffness are used for different maximum pressure differentials. Variations in ambient temperatures do not affect the instrument readings.

7.13 CONNECTING PRESSURE GAUGES TO LINES–A PRACTICAL APPROACH

7.13.1 Introduction

Numerous considerations must be borne in mind while connecting a pressure gauge to a process line whether it is 'in situ' (*i.e.*, local measurement) type of measurement or else measurement at a distance.

Apart from the metrological characteristics of the pressure gauges employed, another determining factor in the choice of instruments is their range, the minimum and maximum pressures or differential pressure to be encountered in a given situation.

The accuracy of pressure measuring instruments is normally stated as a percentage of the range used. Thus, the lower half of the instrument scale will give more error than the upper half. This thus calls for a judicious choice of the range of instrument to be employed to measure a given pressure.

Servicing a pressure gauge influences its location; a gauge needs occasional service and its reading needs to be observed by the operator from time to time. Also, the length of the line from the tapping point (*i.e.*, the point of measurement) to the instrument should not exceed 50 m — an increase in this line length reduces the bandwidth of the instrument.

When the pressure sensing element cannot be exposed directly to the hot/corrosive fluid whose pressure is to be measured, suitable safeguards must be provided.

The tapping point in the pipe should be so chosen that the process fluid at that point must be free from turbulences/disturbances. Also, the sensing element must not disturb the process flow. The edges of the hole in the pipeline/duct must be free from burrs or any kind of asperities – on the side facing the process fluid flow – because even a minute irregularity might affect the result.

Sometimes pressure fluctuations (pulsations) do take place in furnaces leading to oscillations in the elastic sensing elements like diaphragms, etc. Thus they pressure gauge will not record steady pressure. To avoid such oscillations, pulsation dampeners (dash-pots) are installed in the connecting line upstream of the instrument.

For bell and float type pressure gauges, dash-pots are not necessary because of the fact that the bell or float have an appreciable large volume compared to diaphragms. Thus, they inherently act pulsation dampeners.

When the line pressure is steady, the working gauge pressure range should be at least three-fourth of the maximum range while in the case of varying (fluctuating) pressure, they same should be at least two-thirds of the maximum range.

7.13.2 Installation of a Draft Gauge in a Gas/Air Pipeline

Figure 7.66 shows how a draft gauge can be connected to a pipeline carrying gas/air. The connecting line is provided with an incline in the direction of pressure tap. In case if this cannot be provided, a condensate pot must be installed at the lowest point on the line.

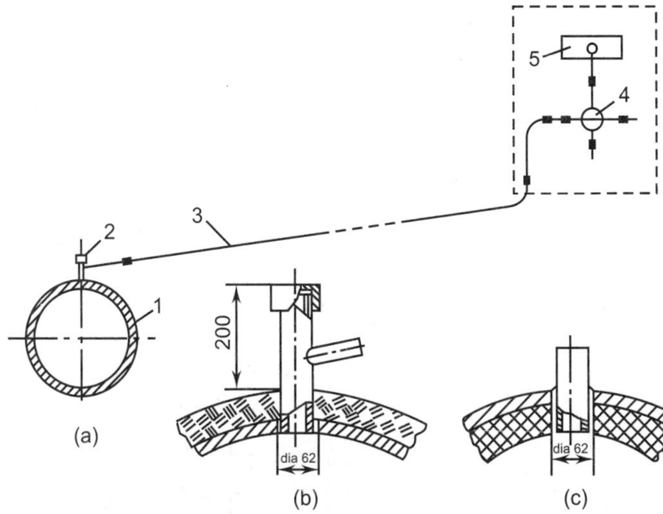

Fig. 7.66: Connection of a pressure-measuring instrument to the process: (*a*) For an indicating draft (or forced-draft) gauge; (*b*) and (*c*) Taps on gas and air flues with external and internal insulation (all dimensions are in mm)

In the figure, 1 represents gas/air pipeline, 2 is the pressure tap, 3 the connecting line, 4 is a 3-way switching cock and 5 is the draft gauge. Pressure taps must be located on the upper side of the pipeline containing gas/air.

7.13.3 Use of Cyclones in Pressure Measurements

Figure 7.67 shows the installation of taps with cyclones (dust catchers) with figures (*a*) and (*b*) representing upper and lower directions of gas/air flow. The dust catcher is employed when the gas/air stream contains dust/suspended solid particles. In case the dust catcher is not used, dust would have found its way into the pressure measuring instrument thus clogging it and ultimately leading to erroneous reading.

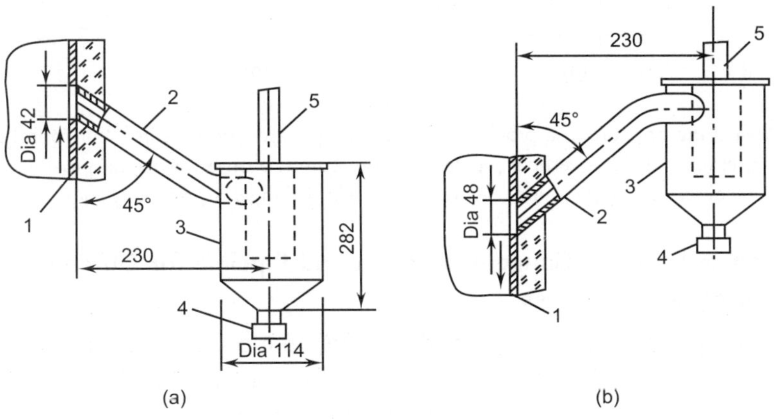

Fig. 7.67: Installation of taps with cyclones for different directions of flow: 1-externally insulated dust flue wall; 2-tube tap; 3-cyclone; 4-clean-out plug; 5-connecting tube

7.13.4 Liquid Pressure Measurement in a Horizontal Pipeline

Figure 7.68 (*a*) shows connections for measurement of a liquid, say water, in a pipeline with pressure ≤ 25 kgf/cm^2 and $t < 70$°C. 1 represents the horizontal pipeline, 2 is the pressure tap, 3 is a 3-way cock, 4 is a flange and 5 is the pressure gauge. With such a 3-way cock, the instrument can be put on/off the line, check its zero setting blow off the connecting line or else to test the instrument in situ. Sometimes, a reference pressure gauge can be connected to the flange 4 to calibrate the pressure gauge 5 in a 'in situ' condition. Figure 7.68 (*b*) and (*c*) show the tap arrangements on a vertical and horizontal line respectively.

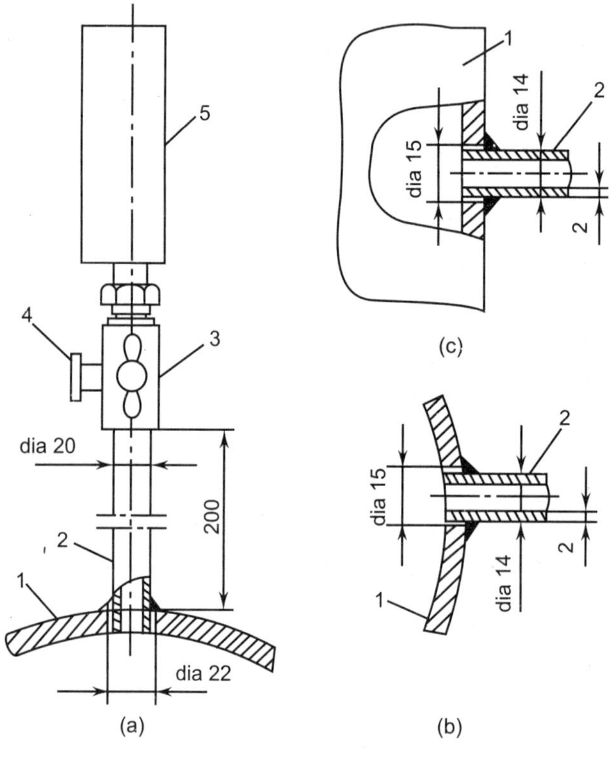

Fig. 7.68: Measuring pressure in a pipeline: (*a*) measuring water pressure in a horizontal pipeline; (*b*) pressure tap on a vertical pipeline; (*c*) pressure tap on a horizontal pipeline; 1-pipeline, 2-pipe union; 3-three way cock; 4-flange, 5-pressure gauge

7.13.5 Installation of Pressure Gauge in a Horizontal Pipeline for Different Temperatures and Pressures

When the test pressure is $p \leq 100$ kgf/cm^2 and $t > 70$°C, the connections are made as shown in figure 7.69. A piezometer ring 3 is connected in between the line 2 and the 3-way cock 4. This prevents direct exposure of hot process fluid to the pressure gauge 1.

Before the pressure gauge 1 is put in the line, the 3-way cock 4 is put in the blow-off position to blow off the connecting line. The piezometer rings 3 will thus be filled with steam (water) and condensate. The condensate is allowed to assume room temperature and then only the gauge is connected to the line.

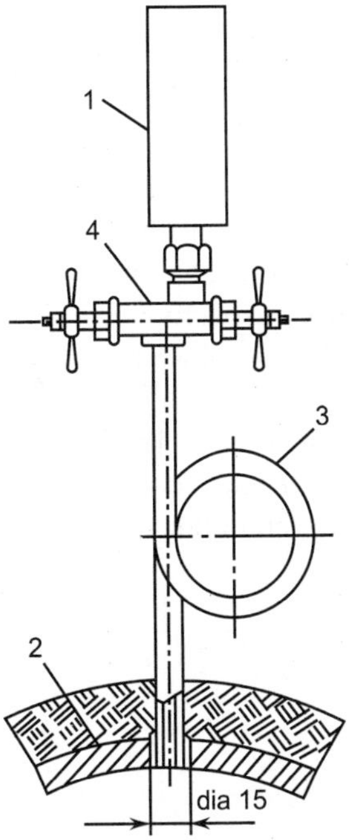

Fig. 7.69: Installation of a pressure gauge on a horizontal pipeline at $p \leq 100$ kgf/cm^2 and $t > 70$°C

For measurement of pressure in the range of $p = 100$ to 400 kgf/cm^2 and $t > 70$°C, the connections to be made are shown in figure 7.70, in which 4 is a shut-off value placed upstream of the piezometer ring 3. Downstream of the ring is a 3-way cock 2 which is connected to the pressure gauge 1.

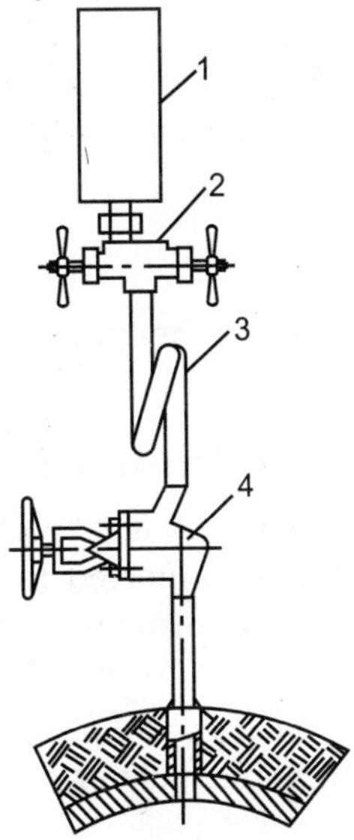

Fig. 7.70: Installation of a pressure gauge on a horizontal pipeline at p = 100 to 400 kgf/cm^2 and $t > 70$°C

If the gauge needs to be installed at a sufficient distance from the line, the piezometer ring is dispensed with, the pressure and temperature range remains the same, as in the above case. This is shown in figure 7.71.

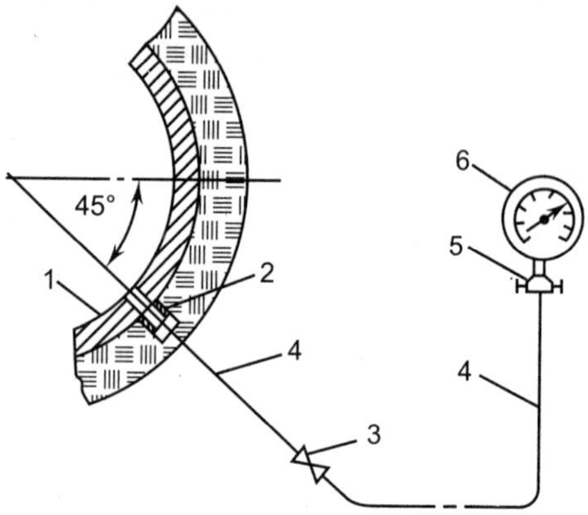

Fig. 7.71: Installation of a pressure gauge on a wall to measure pressure in a pipeline at p = 100 to 400 kgf/cm^2 and t > 70°C: 1— insulated pipeline; 2–pipe union: 3–shut-off valve; 4–connecting line; 5–three-way valve; 6–pressure gauge

7.13.6 Installation of Pressure Gauge in a Fuel-oil Line

The installation of pressure gauge in a fuel-oil line is shown in figure 7.72. 1 represents the fuel-oil line while 2, 3 and 4 represent the shut-off value, sealing chamber (liquid seal) and the pet-cock respectively. The sealing chamber 3 should be installed as close to the shut-off valve.

Normally, water is used as a sealing liquid. The chamber is half-filled with water. The pet-cock is used to check the level of water in the sealing chamber.

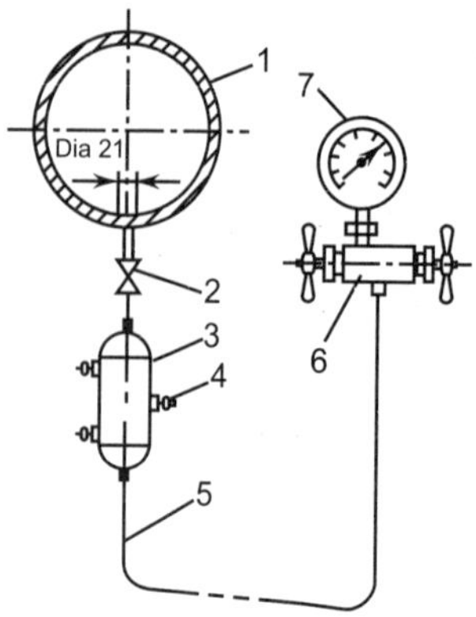

Fig. 7.72: Installation of a pressure gauge on a fuel-oil line at $p \leq 64$ kgf/ cm^2. 1–fuel-oil line; 2–shut-off valve; 3–sealing chamber; 4–pet-cock; 5–connecting line; 6–three-way valve; 7–pressure gauge

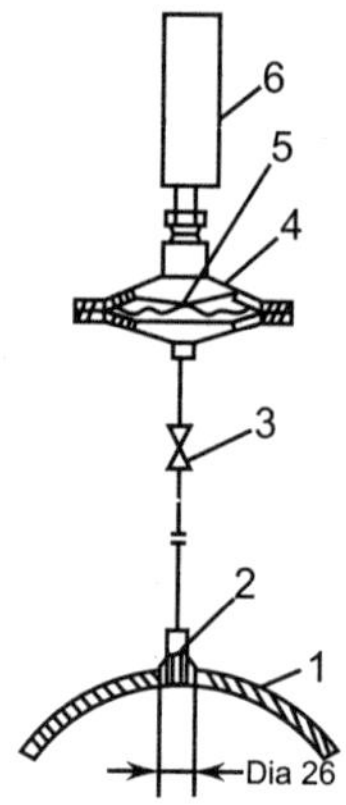

Fig. 7.73: Installation of a pressure gauge in the case of corrosive process fluid and 1–pipeline; 2–pipe union; 3–shut-off needle valve; 4–diaphragm seal; 5–diaphragm; 6–pressure gauge

A gas trap is provided at the top point of the connecting line when the gauge is installed above fuel-oil line.

7.13.7 Installation of Pressure Gauge in a Corrosive Fluid

Figure 7.73 shows the installation procedure for a pressure gauge when the process fluid is corrosive in nature. The pressure gauge 6 is connected to the line 1 via a diaphragm seal 4. 5 is the diaphragm which transmits the process pressure to the gauge 5 but does not allow the corrosive fluid from reaching the same (5).

It should be borne in mind that when the use of diaphragm is mandatory (because of corrosive nature of fluid), it itself introduces a complementary error.

7.14 PRESSURE SWITCH

A pressure switch has got a pressure sensor inside with electrical contacts which are activated at a preset pressure—called set point. Application of a pressure switch involves setting an alarm or for some form of control like operating a solenoid valve, to start/stop a pump-all at some prefixed set point.

The switch proper is either a microswitch or a mercury switch. This switch is activated by some form of pressure sensor like a diaphragm, a Bourdon tube or a bellow. The pressure switch, at the preset pressure, energises a relay coil. The contacts of the relay (either NO or NC—normally open or normally closed) are used for control as stated above.

Figure 7.74 shows a pressure switch with *D* the diaphragm, (*c*, *c*) the contacts, (*s*, *s*) the binding screws, *R* the relay and *K* its contact, *A* is a metal contact.

With increasing pressure, the pressure builds up in the diaphragms *D*. It expands upwards. Thus, the metal contact *A* touches the contacts *c*, *c* and hence the relay coil gets its supply and is thus energised.

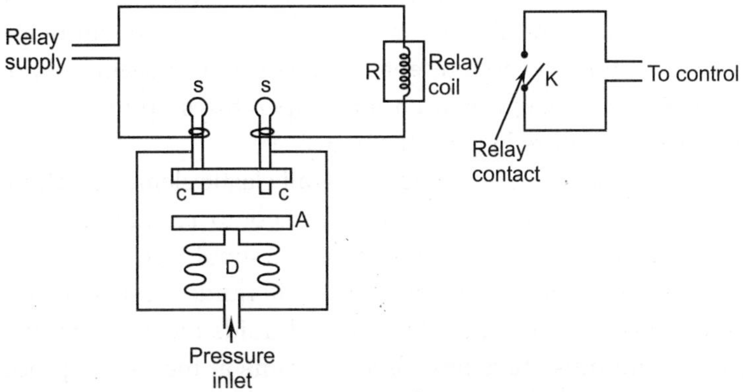

Fig. 7.74: Schematic of a pressure switch

Consequently, the relay contact closes which is used for control purposes. Normally, the set point pressure at which the contact closes is an adjustable one.

The factors which go in selecting a pressure switch for a particular application include electrical ratings of contacts (NO or NC), lifespan, proof pressure, housing, etc.

When a pressure falls and comes below the set point, the relay gets de-energised; the switch then resets itself to the normal position. The difference between the set point and reset point pressure is known as 'dead band'.

7.15 ACCESSORIES FOR PRESSURE MEASUREMENT

7.15.1 Introduction

The diverse field of pressure measurement often encounters difficulties in the form of pressure pulsations (fluctuating pressure), slurries, corrosive fluids, high or low temperatures of process fluids, etc. Several accessories are employed to remedy the situation depending on the kind of difficulty encountered.

Chapter 7

Accessories include chemical (diaphragm) seals. Volumetric seals, pulsation dampeners snubbers, pigtail siphons, freeze protection, valve manifolds, etc.

7.15.2 Chemical Seals

In many cases, chemical seals (or diaphragm protectors) are used with pressure sensors for the following reasons: (1) they prevent poisonous and corrosive fluids from reaching the pressure sensor (2) the process fluid is not allowed to freeze due to ambient or process temperature change and (3) they prevent slurries in the process fluid from clogging the sensing element.

A chemical seal is made up of a flexible diaphragm that separates the process fluid from the pressure sensor. The intervening gap between the diaphragm and the sensor is filled up with a temperature stable liquid. This filling liquid should have properties like very low thermal expansion, very low viscosity and in case of diaphragm rupture it must not contaminate the process fluid. Different filling liquids are recommended as chemical seals for use with process fluids-depending on the temperature of the process. In the –1° to 149°C temperature range, a 70–30 mixture of glycerine and water is recommended; in the range of –46° to 38°C the sealing liquid used may be kerosine, ethyl alcohol or toluene while a temperature range of 25° to 816°C necessitates the use of sodium-potassium eutectic mixture as a sealing fluid. Gaskets used are also temperature dependent-teflon; it is used for temperatures up to 205°C while metal gaskets are recommended for temperature beyond the above.

Chemical seals operate nearly satisfactorily with all elastic elements - although the volumetric capacity of the seal is an area of concern. This may be kept to a minimum to reduce errors caused by temperature changes while, on the other hand, it should be sufficiently large enough to activate the pressure sensor. Again, the stability of the fill fluids must be there for the temperature ranges to which it is expected to be subjected. The fill fluids must not decompose at the highest operating temperature and must have low viscosity value at the lowest process temperature.

Standard and volumetric seals are the two types of chemical seals to be described here.

7.15.2.1 Standard Seal

A standard seal is shown in figure 7.75. It consists of upper and lower housings and also the diaphragm assembly. The filling screw ensures that the seal and the sensor are filled up with the fill liquid under vacuum. This housing is made up of carbon steel or stainless steel 304.

The diaphragm assembly is in the form of a capsule or element and is welded to the upper housing.

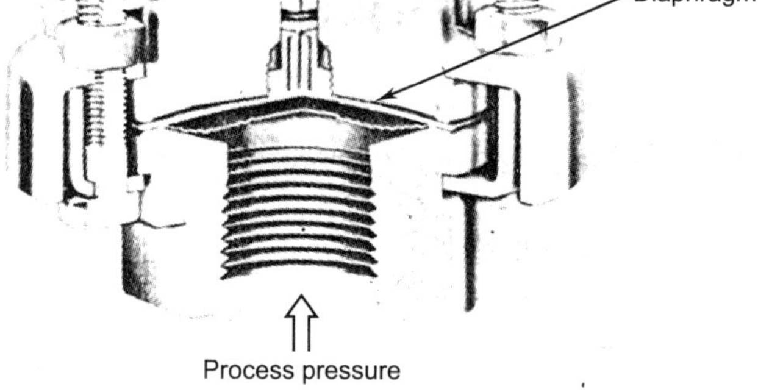

Fig. 7.75: A typical standard chemical seal (Courtesy of Dresser Industries, Inc.)

In case of any leakage occurring above the diaphragm capsule, it presses against an upper backup plate, which has corrugations. These corrugations prevent diaphragm distortion/collapse and seals it off the process.

The upper housing and the diaphragm assembly can be removed alongwith the instrument for maintenance purposes. The diaphragms are mostly rupture-proof and can withstand pressures as high as 2500 psig.

The lower housing is in contact with the process fluid. This housing is made up of corrosion-resistant materials. Some seals have flush connections which allow purging of the connecting line. This ensures removal of any material that is accumulated below the diaphragm.

There are other seals like off-line chemical seals, in-line chemical seals, large displacement seal and full-stream spool seal.

Off-line seals provide protection against corrosion and freezing. They have dead-ended cavities and, as such, not suitable to protect against plugging.

In-line (or flow-through) seals eliminate almost the difficulty forced due to dead-end cavities but the maintenance of these seals requires the draining of process pipes.

Displacements in the off-line and in-line seals are small and are used with small-displacement sensing elements like Bourdon tubes, small-diameter bellows, etc. Large diameter bellows produce large displacements and large displacement seals have to be used. In such a seal, a rolling diaphragm is used.

In-line seals cannot remove the dead-end cavities fully, which are eliminated in full-stream spool seal.

7.15.2.2 Volumetric Seal

The principle of operation of these seals is the same as other seals, but here the seal element is connected to the pressure sensor via a capillary tubing which is filled with a suitable sealing liquid. Two types of volumetric seals are shown in figure 7.76. These seals minimise/eliminate the dead-end cavity problem which is prevalent in off-line seals. The seal elements are normally the made of stainless steels.

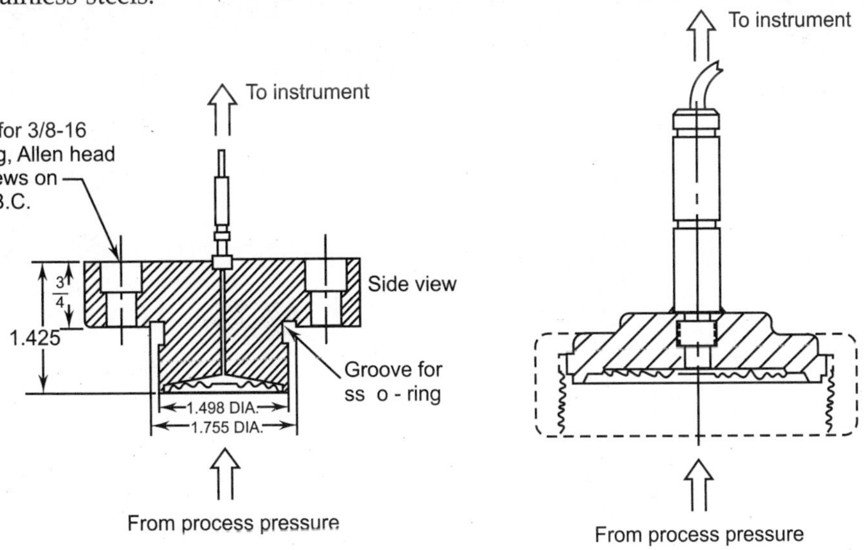

(a) (b)

Fig. 7.76: Volumetric seals

7.15.3 Pulsation Dampeners (Snubbers)

Such a device is used in pressure measurement systems when the pressure sensor is either subjected to rapidly fluctuating pressures or a great pressure surge—thus leading to wear and tear of the sensor or damage it permanently. Techniques involved to minimise the effects of such dampening/surging effect on the sensor include putting in a restriction that produces sufficient resistance to flow that pressure at the sensing part of the sensor changes in a slow manner. This, of course, results in a slow response of the indicator, sometimes to the tune of several seconds. Figure 7.77 shows some of the techniques employed in snubbers. Figure 7.77(*a*) employs a precision needle valve in-line going to the pressure gauge, figure 7.77(*b*) employs a fine bore plug that is screwed into the gauge socket, figure 7.77(*c*) employs a typical filled bulb pulsation dampener, while figure 7.77(*d*) employs putting a small piston in the inlet that rises and falls with pressure impulses.

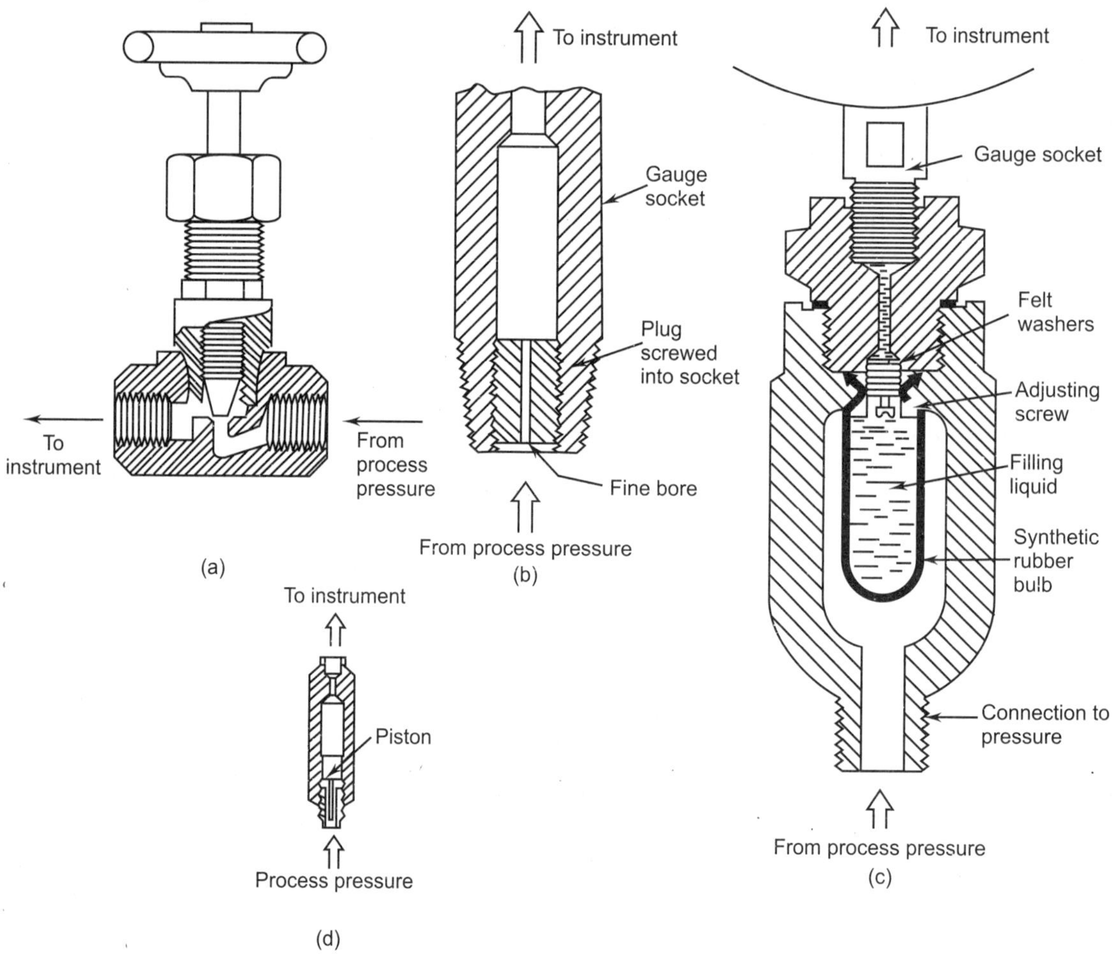

Fig. 7.77: Typical snubber designs employing a
(*a*) Precision needle valve in-line
(*b*) A fine-bore plug
(*c*) A liquid - filled bulb
(*d*) A piston filled in the inlet

7.15.4 Pigtail Siphon

Exposing a pressure sensing element like a Bourdon tube to steam or a hot fluid may damage it permanently. When a Bourdon tube is used for steam pressure measurement, water seals are used as shown in figure 7.78. The figure on the left is a pig-tail.

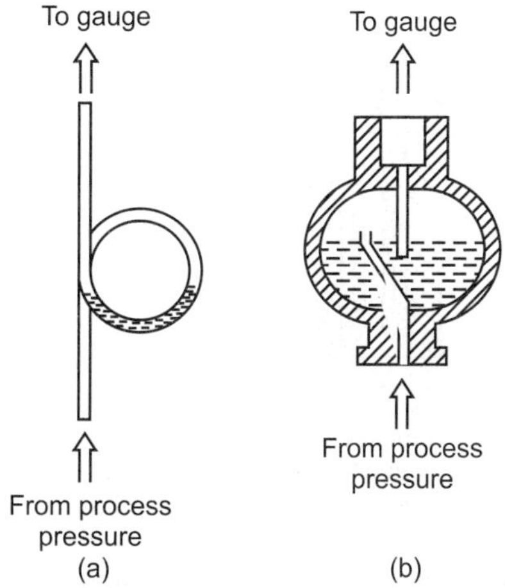

Fig. 7.78: (*a*) Pigtail; (*b*) Water seal

Figure 7.79 shows a pigtail connected to a Bourdon tube. From the bottom steam or hot vapour enters into the pigtail where it is condensed, but the steam pressure is made to reach the Bourdon tube with such an arrangement. The Bourdon tube is not coming into direct contact with steam and is thus saved from damage.

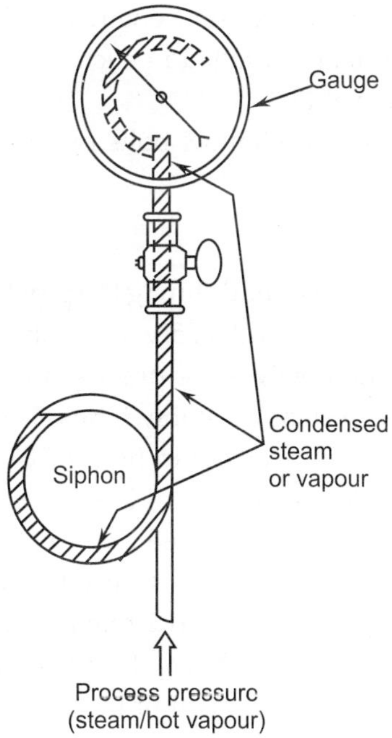

Fig. 7.79: A pigtail connected to a Bourdon tube

Chapter 7

Figure 7.80 shows another way of saving the sensing element which uses an internal siphon that prevents the element from coming into direct contact with steam.

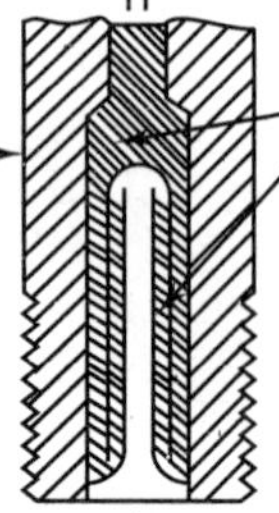

Fig. 7.80: Schematic of an internal siphon

7.15.5 Freeze Protection

Process fluids like steam or other liquids of metals, such as sodium, potassium may freeze inside pressure sensing element because of extreme environmental conditions.

In such cases, steam tracing or small elective heaters are installed inside the casings of large instruments. For instruments of lesser size, resistance wire type heating tracing is wrapped around the transmitter. A thermostat is normally installed in such cases so that the temperature does not overshoot the prescribed maximum value.

In some cases, manufacturers provide hollow bolts in the transmitter casing that connects it to a low-pressure steam supply.

7.15.6 Valve Manifolds

In pressure lines, valve manifolds are used for draining the pipelines or else for periodic calibration of pressure gauges/instruments. 2-valve, 3-valve, 5-valve manifolds are available serving specific needs. In a 3-valve manifold, the three valves are pre-assembled into a manufactured manifold. The use of manifolds saves space as well as assembly times. The chances of leak are also reduced. The 2, 3 and 5-valve manifolds are shown below in figure 7.81.

In the 2-valve manifold, when 1 remains open, 2 is closed and the instrument records the process pressure. For draining the line (necessitated due to clogging), 1 is closed and 2 is made to open. When the instrument needs calibration, 1 is closed while 2 is opened and connected to a standard test bench. In the 3-valve manifold, 1 and 2 remain open and 3 closed when differential pressure of the line is sought to be measured. Valve 3 is the equaliser valve which, when opened (then 1 and 2 are closed), is utilised for adjustment of the zero setting of the pressure instrument.

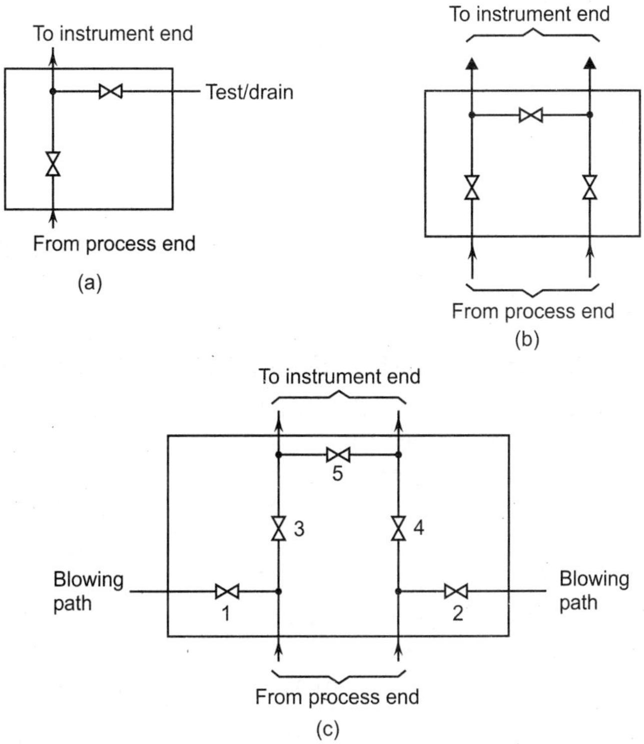

Fig. 7.81: Valve manifolds (*a*) A 2-valve manifold, (*b*) A 3-valve manifold, (*c*) A 5-valve manifold

7.16 SOME POINTS ABOUT PRESSURE MEASUREMENTS

Several methods and instruments have been described in this chapter for absolute pressure, gauge pressure and vacuum pressure measurements. These pressure measuring instruments are used in varied process control industries, as also in research. The accuracy of measurement depends on the method used the metrological characteristics of the instruments, the test conditions and also on the accuracy of the instrument itself. Thus, the choice of the instrument depends on the factors mentioned above. A point to be noted is that the accuracy of an instrument depends upon the existing conditions under which measurements are being done and not on the reference conditions.

Accuracy is expressed as a percentage of full scale deflection of the instrument. For example, with a 100 psi full scale indication pressure measuring instrument, let the pointer indicates pressures of 5 psi and 80 psi for two different measurements. If it is assumed that the instrument accuracy is 2% of f.s.d., then the magnitude of limiting error is

$$0.02 \times 100 = 2 \text{ psi}$$

Thus, error for 5 psi indication is $= \dfrac{2}{5} \times 100 = 40\%$

whereas, error for 80 psi indication is $= \dfrac{2}{80} \times 100 = 2.5\%$

Thus less the measured value, the more the possibility of error in measurement.

Primary pressure sensors should be so employed as to obviate the need for an intermediate transducer, otherwise this would add to uncertainty in measurement. Those sensors whose output give a standard dc signal are usually preferred over others because of the ease of subsequent processing.

Primary pressure sensors should be located at a place which is convenient for servicing and local indication. Again, the length of the line connecting the point of measurement and the instrument must be kept to a bare minimum. A primary pressure transducer should be protected against hot, humid and hostile process environment.

The pressure sensor should be installed at a place within the fluid which is free from process turbulences and disturbances. Again, the installation should be such that it does not disturb the process fluid flow.

SOLVED EXAMPLES

1. What will be the (*a*) gauge pressure and (*b*) absolute pressure at a depth of 20 meters under water? Provide answers in kg/cm² and in mm Hg.

Ans. (*a*) Pressure at a depth of 20 meters

$$= \rho h = 1000 \text{ kg/m}^3 \times 20 \text{ m} = 20{,}000 \text{ kg/m}^2$$
$$= 20{,}000 \times 10^{-4} \text{ kg/cm}^2 = 2 \text{ kg/cm}^2.$$

now, 10,000 mm of water column = 1 kg/cm²

∴ 20,000 mm of water column = 2 kg/cm²

∴ Required pressure = 2 kg/cm²

$$= 20{,}000 \text{ mm of water column}$$
$$= \frac{20{,}000}{13.546} \text{ mm of Hg column}$$
$$= 0.738 \times 10^3 \text{ mm of Hg column.}$$

(*b*) Absolute pressure = Gauge pressure + Atmospheric pressure

$$= 2 \text{ kg/cm}^2 + 1.03 \text{ kg/cm}^2$$
$$= 3.03 \text{ kg/cm}^2.$$
$$= 3.03 \times 10^4 \text{ mm of water column}$$
$$= (3.03 \times 10^4)/13.546 \text{ mm of Hg column}$$
$$= 2.237 \times 10^3 \text{ mm of Hg column}$$

2. A *U*-tube manometer is employed to measure a differential air pressure. The manometric fluid has a density of 800 kg/m³. Air is at a 300 kpa pressure and at a temperature of 20°C. Find the differential pressure if the difference in height in the manometric fluid is 100 mm.

Ans.

$$\rho_f = 800 \text{ kg/m}^3.$$
$$P_a = 300 \text{ kPa}$$
$$t = 20°\text{C}$$
$$h = 100 \text{ mm}$$
$$\rho_a = \frac{P_a}{RT} = \frac{300 \times 10^3}{287 \times 293} = 3.57 \text{ kg/m}^3.$$

∴ The required differential pressure

$$= gh\,(\rho_f - \rho_a)$$
$$= 9.81 \times 0.1 \times (800 - 3.57) = 0.78 \text{ kPa}$$

3. A differential pressure of 3 kg/cm² is applied across a metallic diaphragm of thickness 3 mm which gives a deflection of the centre of the diaphragm by 0.8 mm. Find the diameter of the diaphragm if its Young's modulus and Poison's ratio are 1,00, 000 kg/cm² and 0.35 respectively.

Ans. We have

$$\Delta p = p_1 - p_2 \cong \frac{256\,\gamma\, t^4}{3D^4\,(1-\mu^2)}\left(\frac{\lambda}{t}\right)$$

Here,

$$\Delta_p = p_1 - p_2 = 3 \text{ kg/ cm}^2$$
$$\gamma = 1{,}00{,}000 \text{ kg/ cm}^2$$
$$t = 3 \text{ mm}$$
$$D = ?$$
$$\mu = 0.35$$
$$\lambda = 0.8 \text{ mm}$$

$$D^4 = \frac{256\,\gamma\, t^4}{3(p_1-p_2)(1-\mu^2)}\left(\frac{\lambda}{t}\right) = \frac{256\times 10^5\times(0.3)^4}{3\times 3\times(1-0.35^2)}\times\left(\frac{0.08}{0.3}\right)$$

or, $$D^4 = 7001.71$$

or, $$D = 9.15 \text{ cm}$$

4. A Bridgeman gauge has manganin wire as its coil having a nominal resistance value of 100 Ω at a pressure of 1 MPa. The gauge is to be used to measure 50 MPa pressure with an uncertainty of 0.5%. Find the allowable uncertainty in the resistance measurement. β for manganin = 25 pΩ/Pa.

Ans. $$R_o = 100\ \Omega$$
$$\beta = 25\ \text{p}\Omega/\text{Pa}$$

uncertainty in pressure measurement = 0.5 %

Maximum uncertainty in differential pressure measurement

$$= \omega_{\Delta p} = 0.5\,(50-1)/100 = 0.245 \text{ MPa}$$

Therefore, maximum uncertainty in measuring resistance

$$= R_o\,\beta\,\omega_{\Delta p}$$
$$= 100\times(25\times 10^{-12})\times 0.245\times 10^6\ \Omega$$
$$= 0.0006125\ \Omega$$

5. A Knundsen gauge is designed for a maximum pressure of 3 μm of Hg. The spacing between the fixed plates and the movable vane is less than 0.3 mean free path at this pressure. Calculate the force acting on the vane when the inlet gas temperature is 27°C and the temperature difference (between the fixed plates and the movable vane) is 50 K.

Ans. We have

$$p_i = \frac{kFT_v}{T_f - T_v} \quad\text{or;}\quad F = \frac{p_i(T_f - T_v)}{kT_v}$$

here,

$$p_i = 3\times 10^{-6} \text{ m of Hg.}$$
$$T_f - T_v = 50 \text{ K}$$
$$K = 4\times 10^{-4}$$

and $$T_v = 300 \text{ K}$$

∴ $$f = \frac{3\times 10^{-6}\times 13{,}600\times 9.81\times 50}{4\times 10^{-4}\times 300}$$

$$= 166.77 \text{ N}$$

6. A McLeod gauge has a volume of capillary, bulb and tube down to its opening equal to 100 cm³, and a capillary diameter of 1 mm. Calculate the vacuum pressure when the capillary reads 1 cm.

Ans. We have

$$P = \frac{ay^2}{v - ay}$$

here, $v = 10^5\ \text{mm}^3,\ a = \frac{\pi}{4}(1)^2 = 0.785\ \text{mm}^2$

$y = 10\ \text{mm}$

$$\therefore \quad p = \frac{0.785 \times 10^2}{10^5 - 0.785 \times 10}$$

$$= 0.0007851 \text{ mm of Hg.}$$

7. For an ionisation gauge, the chamber pressure is 10^{-5} Torr. The plate current is 10^{-5} A. If the gauge sensitivity is 100/Torr, find the grid current.

Ans. We have,

$$p = \frac{1}{k}\left(\frac{i_p}{i_g}\right)$$

here, $p = 10^{-5}/\text{Torr}$

$k = 100/\text{Torr}$

$i_p = 10^{-5}\ \text{A}$

$$\therefore \quad i_g = \frac{1}{K} \cdot \frac{i_p}{p}.$$

$$= \frac{1}{100} \times \frac{10^{-5}}{10^{-5}}\ \text{A}$$

$$= 0.01\ \text{A}$$

REVIEW QUESTIONS

1. (*a*) Define the following terms: (*i*) absolute pressure (*ii*) gauge pressure (*iii*) differential pressure (*iv*) atmospheric pressure.

   (*b*) Describe the construction and operation of a Pirani gauge. Also, mention its advantages and disadvantages. [*WBUT (EI 501) 2003*]

2. Describe with a neat diagram the construction and operation of a dead weight tester. Why the tester is used as a laboratory standard for pressure gauge calibration? How the 'tare' pressure in a dead weight tester can be reduced? What precautions should be taken while calibrating a dead weight tester?

3. Mention the different manometers available for differential pressure measurements. Describe a well type manometer with vertical leg. What is the advantage of inclined leg manometer compared to its vertical counterpart?

4. (*a*) Explain, with a neat sketch, the most important advantage of a well manometer over a simple U-tube one.

   (*b*) What are over-range and under-range protections in a pressure measuring instrument? Explain with an example. [*WBUT (EI 501) 2005*]

5. Explain the working principle of a ring balance manometer. Also, mention its advantages and limitations. [*WBUT (EI 501) 2003*]

6. (*a*) Discuss the use of elastic diaphragm for pressure measurement. What secondary transducers are generally used with elastic diaphragms?

   (*b*) Describe the use of an inclined U-tube manometer. What are the characteristics of the liquids used in manometers. [*WBUT (EI 501) 2003*]

7. Describe a C-type Bourdon tube pressure sensing element. Draw a neat sketch of the same indicating its different parts. Explain the following two terms: multiplication and angularity.
8. Describe a capacitance type pressure gauge showing also the construction of the capacitor. What are its advantages and disadvantages? Indicate its accuracy.
9. Define *'d'* and *'g'* coefficients by which the sensitivity of a crystal is identified and obtain interrelationship between them. [*WBUT (EI 501) 2005*]
10. Describe a Bridgeman gauge for measurement of very high pressure. What are the advantages of this gauge? State the range of maximum pressure that can be measured.
11. Draw the neat sketch of an optical pressure transducer and also describe a circuit arrangement to measure pressure with its help.
12. (*a*) Why is a McLeod gauge considered to be a standard for measurement of pressure in the vacuum range?
    (*b*) What are the limitations of a McLeod gauge?
    (*c*) What modification is done to linearise the scale of McLeod gauge at higher pressures? [*WBUT (EI 501) 2005*]
13. Describe a basic flapper-nozzle system. Draw its characteristic curve. What is meant by 'throttling range'?
14. Draw the schematics of a pneumatic pressure transmitter. Explain its working.
15. Draw the installation diagram of a draft gauge in a gas/air pipeline.
16. What are the accessories used for measurement of pressure? Draw and describe a chemical seal.
17. Discuss (*a*) pulsation dampeners (*b*) valve manifolds.

Fill in the blanks:

1. When a pressure is measured with respect to zero pressure, it is called .... pressure.
2. Instruments which measure atmospheric pressure are called ....
3. A manometer requires .... calibration.
4. To extend range of manometers, ... are used.
5. Hysteresis effect of elastic pressure sensing elements should be ....
6. Time constant of a flat diaphragm is very............... .
7. For a corrugated diaphragm, linearity in the pressure-deflection characteristic curve improves with .... of convolution.
8. Sensitivity of Bourdon tube .... with increase in radius of curvature.
9. The cross-section of a Bourdon tube can never be ....
10. The motion of the tip of a bourdon tube is .... in nature.
11. When external pressure is applied along specific axes of a quartz crystal, an .... is generated.
12. Bridgeman gauge utilises the principle of variation of resistance with ....
13. In optical pressure transducer, an .... element acts as a primary sensor.
14. Very high vacuum is a pressure which is less than .... Torr.
15. McLeod gauge cannot be used when .... reading is required.
16. The standard pressure used in a flapper-nozzle system is always ....
17. The output pressure variation in a flapper-nozzle system is always maintained in between .... psi and .... psi.
18. The total displacement of flapper which covers the total output pressure range is called .... range.

19. A flapper-nozzle system is a .... to .... transducer.
20. A reverse acting relay employing flapper-nozzle ensures that the system always operates in the .... region.
21. Cyclones are used in pressure tap lines when the fluid contains ....
22. Chemical seals are used to prevent .... fluids from reaching the pressure sensor.
23. Pulsation dampeners are used when the line pressure is .... in nature.
24. When steam pressure measurement is desired, a .... seal is connected in the line prior to Bourdon tube.
25. Valve manifolds are used for .... pipelines as well as for .... pressure instruments.

Tick the correct answer :

1. A pressure lower than the atmospheric pressure is called
   (*a*) Gauge pressure (*b*) absolute pressure (*c*) vacuum pressure ☐
2. For gauge pressure measurement, the reference used is
   (*a*) atmospheric pressure
   (*b*) zero pressure
   (*c*) no standard reference is used ☐
3. A dead-weight tester is used as a
   (*a*) primary pressure standard (*b*) secondary pressure standard ☐
4. Micromanometers are used to
   (*a*) extend the range of manometers
   (*b*) reduce the range of manometers
   (*c*) both (*a*) and (*b*) ☐
5. 1 torr of pressure is equal to
   (*a*) 1 mm of Hg column
   (*b*) 760 cm of Hg column
   (*c*) 1 Pascal ☐
6. For relatively low pressure measurements by elastic pressure sensors, the one that should be used is
   (*a*) diaphragm (*b*) bellow (*c*) Bourdon tube ☐
7. The most commonly used Bourdon tube is
   (*a*) *c*-type (*b*) helix type (*c*) spiral type ☐
8. The deflection of a Bourdon tube is dependent on the cross-section of
   (*a*) major axis (*b*) minor axis (*c*) both (*a*) and (*b*) ☐
9. The tip travel of a Bourdon tube is
   (*a*) linear (*b*) circular (*c*) curvilinear ☐
10. A capacitive pressure transducer can measure
    (*a*) absolute pressure (*b*) gauge pressure
    (*c*) differential pressure (*d*) all (*a*), (*b*) and (*c*) ☐
11. In a photoelectric pressure transducer, the
    (*a*) wavelength of the incident light
    (*b*) voltage difference between the anode and cathode
    (*c*) amount of light falling on the photo tube is varied. ☐

12. A piezoelectric type 50 MPa full scale pressure sensor with built-in electronics has a sensitivity of 100 mV per MPa. If this sensor is subjected to a static pressure of 10 MPa, its output will be

(*a*) 0 mV (*b*) 100 mV (*c*) $1000/\sqrt{2}$ mV (*d*) 1000 mV ☐

[*GATE 2002*]

13. A quartz piezoelectric type pressure sensor has a built-in charge amplifier. The sensor has a sensitivity of 1μV/Pa. It is subjected to a constant pressure of 120 kPa. The output of the transducer at steady state is

(*a*) 0 mV (*b*) 100 μV (*c*) 120 μV (*d*) 120 mV ☐

[*GATE 2004*]

14. In Pirani gauge, calibration is

(*a*) dependent on the composition of the gas

(*b*) independent of the composition of the gas

(*c*) cannot say ☐

15. A Pirani gauge is used to measure pressure of the order of

(*a*) 10 MPa (*b*) 1 MPa (*c*) 100 Pa (*c*) 1 Pa ☐

[*GATE 2003*]

16. A Pirani gauge measures vacuum pressure and works on the principle of

(*a*) change in ionising potential (*b*) change in thermal conductivity

(*c*) deformation of elastic body (*d*) change in self-inductance ☐

[*GATE 2006*]

17. The ionisation gauge (cold cathode type) has

(*a*) non-linear output and low accuracy

(*b*) linear output and high accuracy

(*c*) non-linear output and high accuracy ☐

18. A flapper-nozzle system is a

(*a*) displacement to pressure transducer

(*b*) pressure to displacement transducer

(*c*) both (*a*) and (*b*) ☐

19. A pressure gauge is calibrated from 0–50 kN/m^2. It has a uniform scale with 100 scale divisions. One fifth of a scale division can be read with certainty. The gauge has a

(*a*) resolution of 0.1 kN/m^2 (*b*) threshold 0.1 kN/m^2

(*c*) dead zone of 0.2 kN/m^2 (*c*) a resolution of 0.5 kN/m^2 ☐

20. The operating point on the displacement-pressure characteristic curve of a flapper-nozzle system is

(*a*) middle part of the linear region

(*b*) upper part of the linear region

(*c*) lower part of the linear region ☐

21. A pressure sensor has the following specifications: sensitivity at the design temperature = 10 V/MPa, zero drift = 0.01 V/°C, sensitivity drift = 0.01 (V/MPa)/°C.

When the sensor is used in an ambient 20°C above the design temperature, the output from the device is 7.4 V. The true value of the pressure will be

(*a*) 0.71 MPa (*b*) 0.68 MPa (*c*) 0.65 MPa (*d*) 0.61 MPa. ☐

[*GATE 2004*]

REVIEW PROBLEMS

1. A manometer uses transformer oil of specific gravity 0.864 as the measuring liquid. The scale is graduated in mm of water. If one leg is a 2 mm bore tube and the other a 20 mm well, calculate the angle to the horizontal at which the tube and scale must be inclined to give 4 mm scale deflection for a pressure of 1 mm head of water. Assume, 1 mm of water = 9.81 Pa [*WBUT (EI 501) 2003*]

   [**Ans.** 16.22°]

2. A well-type manometer has its 'capillary diameter-to-well diameter, ratio as 1:20. It is required to measure a pressure differential of 1 Pascal. What should be the approximate height of the mercury column in the capillary? [*WBUT (EI 501) 2005*]

   [**Ans.** 0.0075 mm of Hg column]

3. A McLeod gauge has a capillary diameter of 1 mm and bulb volume 140 cc. Find the vacuum pressure when the reading is 10 mm. [**Ans.** 0.0005607 mm of Hg]

4. Manganin has a pressure coefficient of resistance 2.25×10^{-10} m²/kg. What should be the change in pressure for a 0.2% change in the value of resistance. [**Ans.** 0.89 × 107 kg/m2]

5. The closed limb (*i.e.* the capillary) of a McLeod gauge has a diameter of 1 mm and a volume V above the opening 'O' of 100 cc. If the difference in mercury levels is 3 mm, calculate the unknown pressure. Also, find the value of gauge constant.

   [Note: gauge constant = $\frac{a}{V}$] [**Ans.** 7.1 × 10–5 mm of Hg

   7.9 × 10–6 (mm)–1 of Hg]

6. Calculate the maximum allowable flapper movement from the following data:

$$P_o = 3 \text{ to } 15 \text{ psi}$$
$$P_i = 20 \text{ psi}$$
$$d = 0.5 \text{ mm}$$
$$d_n = 0.2 \text{ mm}$$

   [**Ans.** 0.5632 mm]

7. The diameter of an aluminium sphere is found to be 50 mm at an atmospheric pressure of 10^5 Pa. The bulk modulus for aluminium is 68.6 GPa. The change in the diameter of the sphere in microns, when placed in vacuum is

   (*a*) 0.018 (*b*) 0.024 (*c*) 0.032 (*d*) 0.038

   [*GATE 2005*] [**Ans.** (*b*)]

8. A mercury barometer reads *h* mm Hg with the temperature of the mercury at T°C. The barometer reading corrected for standard temperature 0°C with β denoting the volumetric expansion coefficient of mercury in $(°C)^{-1}$ is

   (*a*) $h/(1 + \beta T)$ (*b*) $h(\beta + T)$ (*c*) $h(1 + \beta T)$ (*d*) $h(\beta - T)$

   [*GATE 2005*] [**Ans.** (*a*)]

9. A well of cross-sectional area a_w is connected to an inclined tube of cross-sectional area a_t to form a differential pressure gauge as shown. When $P_1 = P_2$, the common liquid level is denoted by *A*. When $P_1 > P_2$, the liquid level in the well is depressed to *B*, and the level in the tube rises by *l* along its length such that the difference in the tube and well levels is h_d. The angle of inclination θ of the tube with the horizontal is

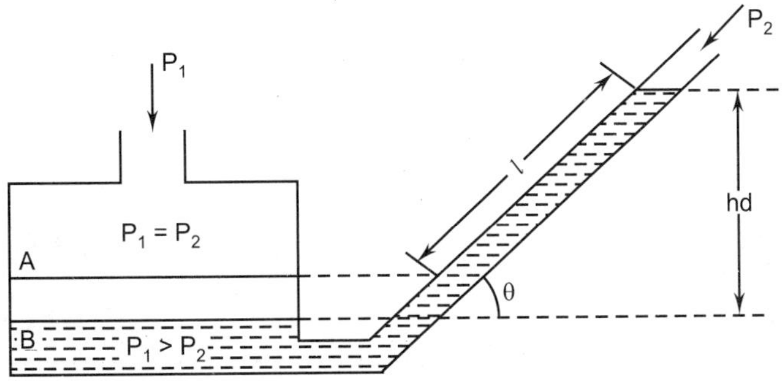

(*a*) $\sin^{-1}\left[\frac{l}{h_d} - \frac{a_w}{a_t}\right]$ (*b*) $\sin^{-1}\left[\frac{h_d}{l} + \frac{a_t}{a_w}\right]$ (*c*) $\sin^{-1}\left[\frac{h_d}{l} - \frac{a_t}{a_w}\right]$ (*d*) $\sin^{-1}\left[\frac{h_d}{l} + \frac{a_w}{a_t}\right]$

[*GATE 2007*] [**Ans.** (*c*)]

10. A differential pressure transmitter is used to measure the flow rate in a pipe. Due to aging, the sensitivity of the pressure transmitter is reduced by 5%. All other aspects of the flowmeter remaining constant, change in the sensitivity of the flow measurement is

(*a*) 10.0% (*b*) 5.0% (*c*) 2.5% (*d*) 2.2% [*GATE 2009*]

CHAPTER 8

SOUND PRESSURE LEVEL MEASUREMENT

8.1 INTRODUCTION

Sound pressure measurements play a very important part in the design of sound recording and sound reproducing equipments, for the development of less noisy equipments and machineries, study of water-borne sound (sonar) for underwater equipments, aero-dynamic noise, and for vibration problem diagnostics.

The sound pressure produced by the exhaust in jet engines may lead to fatigue failure in metals because of enormous vibration, normally termed as "accoustic fatigue". Again, prolonged exposure to heavy sound and noise (as encountered in steel plants) may lead to noise-induced hearing loss. Again, bursting of huge crackers may lead to damage of ear drums. Nowadays, laws are being enacted to curb sound pollution.

Sound is basically an elastic longitudinal wave which propagates by alternate compression and rarefaction of the medium through which it passes. A source of sound is always in a state of vibration. An elastic wave carries energy from one place to another due to vibrations of the medium.

8.2 SOUND PRESSURE LEVEL (SPL)

The threshold of hearing and the threshold of pain for human beings, measured at 1 kHz, are 0.0002 μ bar and 1 m bar respectively, where

$$1\mu \text{ bar} = 1 \text{ dyne/cm}^2 = 0.1 \text{ N/m}^2 = 1.45 \times 10^{-5} \text{ lb/m}^2$$

Thus, the threshold of pain to threshold of hearing represents a ratio of 5×10^6. Because of this large ratio value over which human ear can respond, sound pressure measurements involve a logarithmic scale, and expressed in decibel (dB), for convenience.

The value of 0.0002 μ bar is accepted as a standard reference value for pressure against which other pressures are compared. Thus, $P_{ref} = 0.0002$ μ bar.

The magnitude of fluctuations in pressure when sound passes through a fluid medium is a measure of its sound pressure level and is formally defined as the following ratio:

Sound pressure level (SPL) $= 20\log_{10}\dfrac{P}{P_{ref}} = 20\log_{10}\dfrac{P}{0.0002}$ decibel (dB) where P is the rms value of sound pressure, measured in μ bar. In the above, rms value for P is chosen for the fact that mostly sound waves are random signals and not pure sine waves. It should be remembered that at 1 kHz the SPL in dB is also termed as Phon.

When P = 0.0002 μ bar (*i.e.* the threshold of hearing) the corresponding SPL is 0 dB. Thus an SPL of 0 dB is the lowest pressure fluctuation that can be discerned by human beings (in quiet surroundings). On the other side, rocket engines generates an SPL of approximately 160 dB while one atmosphere corresponds to 194 dB.

Table 8.1 gives some of the sound pressure levels encountered in human surroundings:

Table 8.1: SPLs for Different Cases

Source of sound	*SPL in dB*	*Sound pressure in μ bar or dyne/cm²*
Human threshold of hearing	0	2×10^{-4}
Whisper	20	10^{-3}
Typewriters/normal conversational speech (at 1m distance)	60	10^{-1}
Road with heavy vehicles	120	10^{2}
Human tolerance level	150	–
Rocket engine	160	10^{4}
One atmosphere	194	–

8.3 SOUND POWER LEVEL AND VARIATION OF INTENSITY OF SOUND WITH DISTANCE

Total sound energy radiated by a source of sound in all directions is called sound power. Some of the sound power levels are shown in Table 8.2.

Table 8.2: Power Levels for Different Cases

Source of sound	*Power level*
Faint whisper	10^{-9} W
Normal conversational speech	10^{-5} W
Jet engine	50 kW
Space rocket	10 MW

Because of this wide power range of sound sources, a logarithmic scale is used for sound power. This is abbreviated as PWL and is defined as:

$$\text{PWL} = 10\log\frac{W}{W_{ref}}\text{ dB}$$

where W is the second (accoustic) power of the source and W_{ref} is the reference sound power, having a value of 10^{-12} W.

Now, intensity of sound is defined as the rate of flow of sound energy through unit area placed normal to the direction of propagation of sound in unit time. Two sounds of equal intensity may not appear to be equally loud to different observers if the two sounds differ greatly in their frequencies. Intensity of sound depends on many factors, namely: amplitude of vibration of the source, size of source, distance from the source, density of the medium, motion of the medium,

Chapter 8

presence of other bodies, etc. If d be the distance between the source of sound (assumed point source) and the listener and I be the intensity of the source, then,

$$I \propto \frac{1}{d^2}$$

Thus, intensity falls off rapidly with distance.

8.4 BASIC PRINCIPLE

For commonly encountered sound measurements, a sound pressure level meter is employed. It consists of a microphone which acts as a transducer. Firstly, the microphone converts the sound pressure into motion employing a thin diaphragm. Secondly, this motion is converted into a voltage by employing another transducer, usually a capacitive, piezoelectric or moving-coil type.

An "air leak" in the form of an air leak capillary is provided on either side of the microphone. This is done to equalize the (*a*) average pressure (atmospheric) and (*b*) to prevent the bursting of the diaphragm. This air leak is necessary (*a*) for the diaphragm not to respond to slow atmospheric pressure changes and (*b*), at the same time, it must respond to the actual sound pressure level fluctuations which is ultimately to be measured. Note that the eustachian tube of the human ear behaves in the same manner.

The microphone output is characterised by a voltage which is very low and also the output impedance is very high. To circumvent this problem, a FET-input amplifier with a high gain and high input impedance is employed. This amplifier is inbuilt in the microphone housing which eliminates stray capacitance effects.

8.5 SCHEMATIC OF A SOUND PRESSURE LEVEL METER

The basic schematic of a sound pressure level meter is shown in figure 8.1.

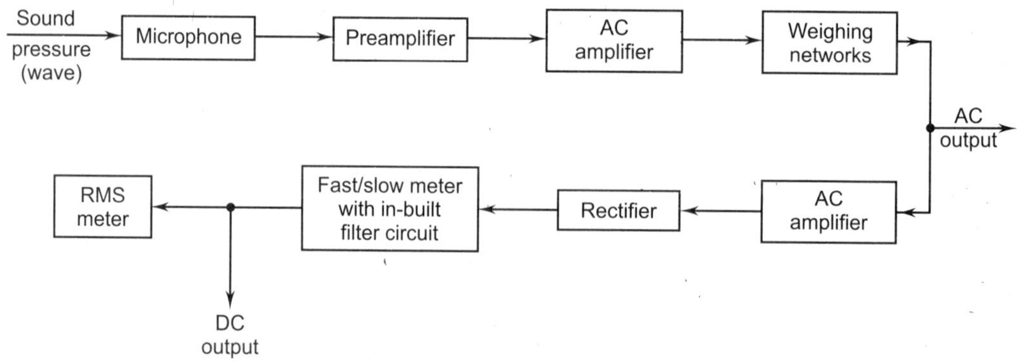

Fig. 8.1: A schematic of a sound pressure level meter

The sound pressure is applied to the microphone where it is transduced and subsequently amplified before feeding into a weighing network. The weighing network is basically an electrical filter whose frequency response characteristic is almost identical to that of the average human ear.

8.6 CHARACTERISTICS CURVES

Response of human beings to sound pressure level differs for varied frequencies. Figure 8.2 shows a typical representation consisting of five curves, each corresponding to a particular loudness. Each curve has been labelled with a typical loudness (40, 60, 80, 100, 120 Phons), the unit of which is Phon, zero (0) Phon corresponds to the loudness level of threshold of hearing. From the curves it is apparent that a higher SPL is required at low frequencies than that at higher frequencies to produce the same sensation of loudness. By having a look at the 60 Phon curve, it is apparent that at 1000 Hz, an SPL of 60 dB is required, but when the frequency is brought down to 100 Hz, the SPL requirement would be about 72 dB.

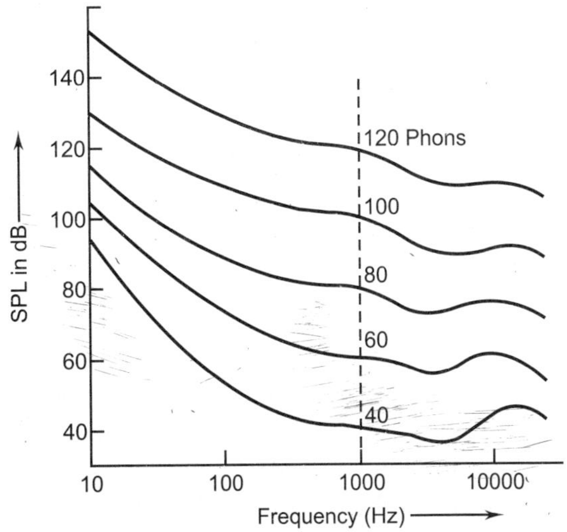

Fig. 8.2: A typical representation of equal loudness contours indicating SPL in dB at different frequencies (in Hz)

A sound level meter does not have a flat frequency response because the meter determines the loudness as perceived by human being rather than the measurement of actual sound pressure. The weighing networks are actually electrical filters and have three filters – *A*, *B* and *C* working at 40, 70, and 100 Phons respectively of human ear response.

Figure 8.3 shows the frequency response of these three filters (*A*, *B* and *C*) alongwith that of a linear one. Meters which are to measure the actual sound pressure utilise the linear or flat curve. The next best curve is the one corresponding to *C* filter. Meters which incorporate the filters (*i.e.* weighing networks) in them are actually sound level (SL) rather the sound pressure level (SPL) meters.

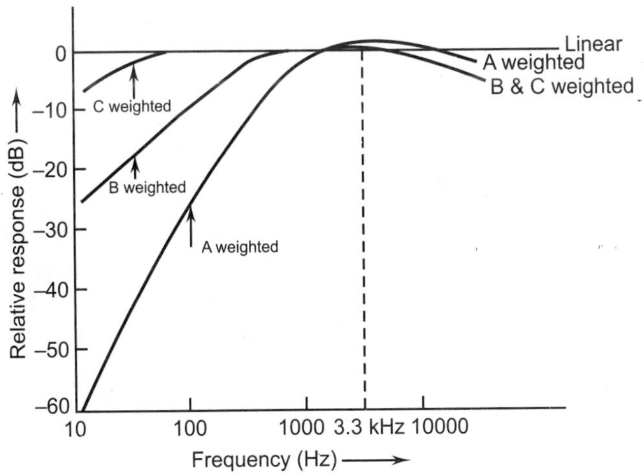

Fig. 8.3: Linear and A, B and C weighted response of sound-level meters

The slow-fast response switch in the meter has a definite purpose: The 'slow' position gives a steady reading neglecting any fast variations in the input signal while the 'fast' position can record fast changes in the input signal.

8.7 MICROPHONES

8.7.1 Introduction

While the design of microphones is a very complex one and falling outside the purview of the present discussions, due attention should a be paid to the response of microphones to varied frequency. Now, by pressure response of a microphone is meant the variation in output voltage when a constant sound pressure of varied frequency is applied across the diaphragm of a microphone.

Another very important aspect is the 'free-field response' of a microphone. This is the relationship between microphone output voltage and the sound pressure existing at the microphone location 'before' the microphone was introduced into the sound field. It should be borne in mind that the accoustical impedance of a microphone is totally different from the air medium in which the microphone is immersed.

The microphone, as a whole, is treated as a rigid body to the incoming sound pressure signal that falls on it. Sound wave, which falls on a microphone, is reflected in a complex manner; the nature of reflection depends on the shape and size of the microphone on the one hand and frequency and direction of propagation of the sound wave on the other. At low input sound frequencies, reflection from the microphone is independent of the angle of incidence; for such a case the pressure response and the free field response are identical. This is certainly not the case for high value of frequency of input sound wave. Here the diaphragm of the microphone will experience twice the pressure than if the microphone was considered absent for waves falling perpendicularly on the diaphragm. While for sound waves propagating parallel to the microphone, the average pressure experienced by the same would be zero.

A variety of microphones are used for sound pressure level meters—the type depending on sensitivity frequency response, linearity and dynamic range. The types normally available are:

1. Capacitor type,
2. Electrodynamic type,
3. Piezo-electrical crystal type, and
4. Electret type.

8.7.2 A Capacitive Type Microphone

A capacitive type microphone, designed to operate approximately between 20 Hz and 15 kHz, works on the principle of variation of capacitance with variation in sound pressure. As already mentioned, this transducer remains insensitive to atmospheric pressure variations but is sensitive to pressure fluctuations, although the changes in atmospheric pressure may be much greater than the latter. The transducer is so designed that the diaphragm does not burst due to variations in atmospheric pressure.

The basic construction is shown in figure 8.4. The diaphragm is kept under radial tension. A capillary leak ensures that both sides of the diaphragm are subjected to the same atmospheric pressure variations and hence the diaphragm becomes immune to such atmospheric pressure changes. The chamber and the capillary have very small volume so as to keep the pneumatic time constant to a low value. These two act as a low pass filter which allow the slow rate of change of atmospheric pressure variations but prevent the sound pressure variations from reaching the bottom surface of the diaphragm. The two plates of the capacitor are the diaphragm and the fixed plate with air in between acting as the dielectric. The fixed plate is provided with holes, called *damping holes,* which allow the air to pass to the bottom side of the chamber. In case the holes were not there, deflection of the diaphragm would have been resisted.

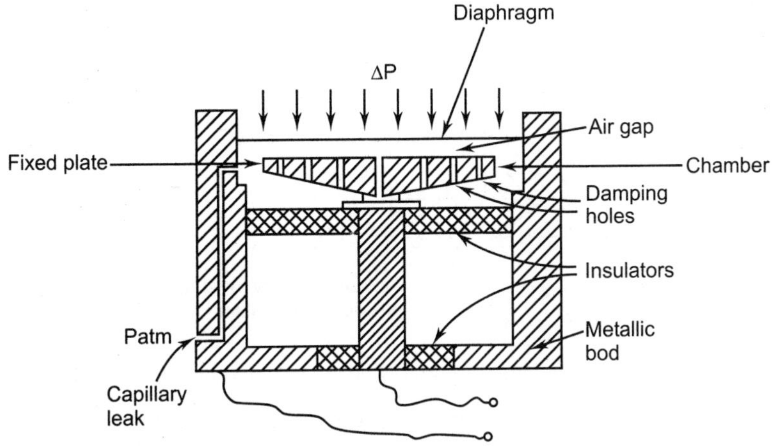

Fig. 8.4: Schematic of a capacitive microphone

The capacitive microphone gives a flat frequency response between 20 Hz to 20 kHz. The diaphragm thickness is about 5 mm to 25 mm and the spacing between the electrodes is about 20 μm.

Figure 8.5 gives the free field response and pressure response for a typical microphone for zero degree angle of incidence, *i.e.* $\theta = 0°$. Usually, microphone calibration charts are available with manufacturers for selective angles of incidence starting from 0° to 90°.

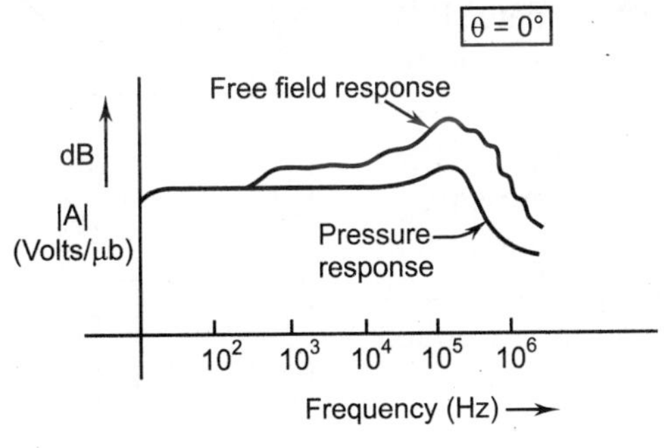

Fig. 8.5: Microphone response in free-field and non-free field conditions

8.7.3 Electrodynamic Type Microphone

A coil, moving in a magnetic field, gives rise to an emf across its terminals. This is the principle employed in an electrodynamic type microphone. The diaphragm, which senses the change of pressure, is attached to the coil, which, in its turn, is placed within the pole pieces of a permanent magnet. A soft iron core, shown in figure 8.6, is inserted to direct the magnetic field through the coil, thereby increasing sensitivity. As the diaphragm moves due to changing pressure, the coil moves in the magnetic field and a voltage is developed across the coil which is a measure of sound pressure.

Chapter 8

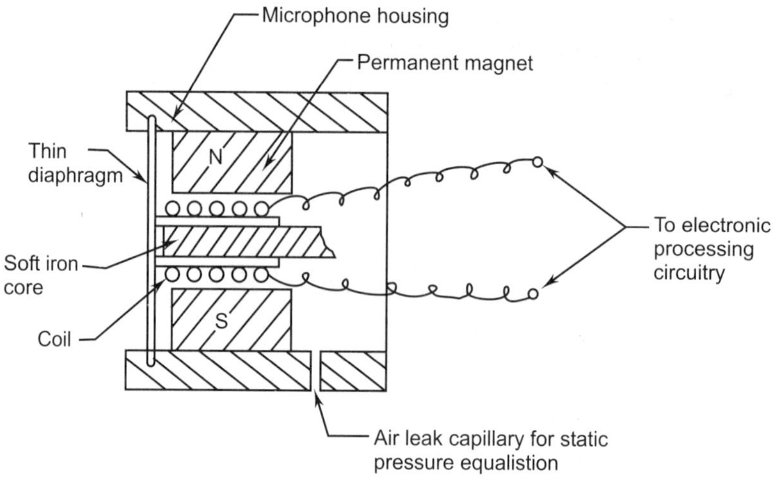

Fig. 8.6: Schematic of an electrodynamic type microphone

This microphone has advantage of being self-generating type but it is sluggish because of high inertia of the moving coil.

8.7.4 Piezoelectric Crystal Type Microphones

Some commonly employed materials, such as Rochelle's salt, quartz, barium titanate, etc. generate an electrical potential when it is subjected to mechanical strain.

The schematic of a piezoelectric crystal type microphone is shown in figure 8.7. A conical shaped diaphragm with holes in it is attached to a lead zirconate titanate (PZT) ceramic beam. The beam acts as a piezoelectric material. An electrical output is obtained across the beam when it is mechanically strained by the diaphragm to which the pressure is applied. Holes are provided in the diaphragm to prevent damage to the diaphragm in case a considerable pressure differential is applied across it. A back up plate prevents excessive sagging of the diaphragm.

This type of microphone is very sensitive at lower frequencies. But it is sensitive to vibrations and requires impedance matching between the microphone output and the circuitry that it drives.

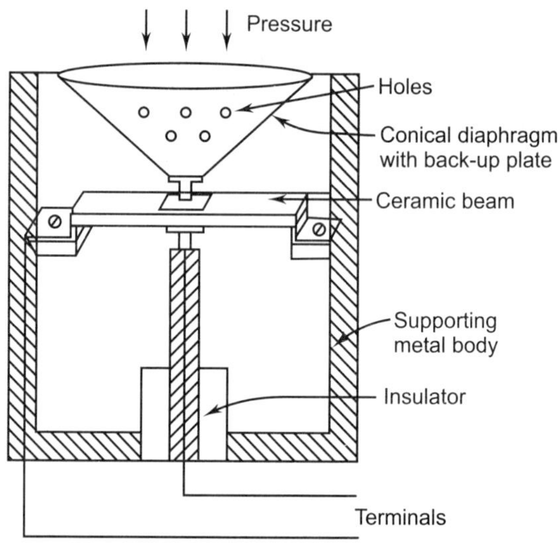

Fig. 8.7: Microphone using a lead-zirconate-titanate ceramic piece

8.7.5 Electret-type Microphone

In this type, the diaphragm consists of a polymer film on which charge is permanently built and, as such, no polarising voltage is needed. A back up plate is provided to prevent sagging of the polymer film.

The electret type microphone is less costly, can be used in high humidity conditions, robust in nature and has an excellent stability.

On the other hand, its electrical characteristics vary with temperature and also its vibration sensitivity is considerable so that care should be taken to minimise these effects.

8.8 ACCOUSTIC INTENSITY

Recent studies and developments have made accoustic intensity measurements a more preferred one than sound pressure measurements for location of source of noise, sound absorption and sound transmission loss.

Sound intensity is a vector quantity and is defined as the quantity and direction of time-averaged flow of accoustic power per unit area at a given position. Thus,

$$I_r = \underset{T \to \infty}{Lt} \frac{1}{T} \int_0^T p(t).v_r(t)\,dt \quad (\text{W/m}^2)$$

where, I_r = accoustic intensity, in the selected direction r

$p(t)$ = pressure at the given position

T = time for which averaging is done

$v_r(t)$ = particle velocity in the selected direction and position

Both p and v_r are functions of time.

Sound pressure p can be measured by a microphone but measurement of v_r using hot-wire anemometers has some difficulties which can be overcome by Euler's equation for zero mean flow, which is

$$\rho \frac{\partial v_r}{dt} + \frac{\partial p}{dr} = 0 \qquad \rho \text{ being the fluid mass density.}$$

By replacing $\frac{\partial p}{dr}$ by $\frac{\Delta p}{\Delta r}$, we can write:

$$v_r \simeq \frac{1}{\rho \Delta r} \int_0^t (p_2 - p_1)\,dt$$

where, p_1 and p_2 are the pressures measured by two microphones placed in the direction of r at distances r_1 and r_2 ($r_2 > r_1$). Approximating p as $(p_1 + p_2)/2$ gives:

$$I_r \simeq \frac{1}{2\rho \Delta r} \frac{1}{T} \int_0^T \left(\frac{p_1 + p_2}{2}\right)(p_2 - p_1)\,dt$$

An intensity probe can be designed in many ways; the two-microphone method is implemented in several ways: (*i*) face to face (*ii*) side by side (*iii*) back to back, but the face to face method with a solid spacer in between is superior than the other two with $r_1 \sim r_2$ is kept at a small value than the accoustic wavelength. The spacer prevents the reflection between the two microphones which would have otherwise occurred, but at, the same time, allows the pressures to reach the diaphragms of the microphones. To cover different frequency range, different spacers alongwith different microphones are used. Figure 8.8 shows the different microphone configurations and errors for accoustic intensity probes.

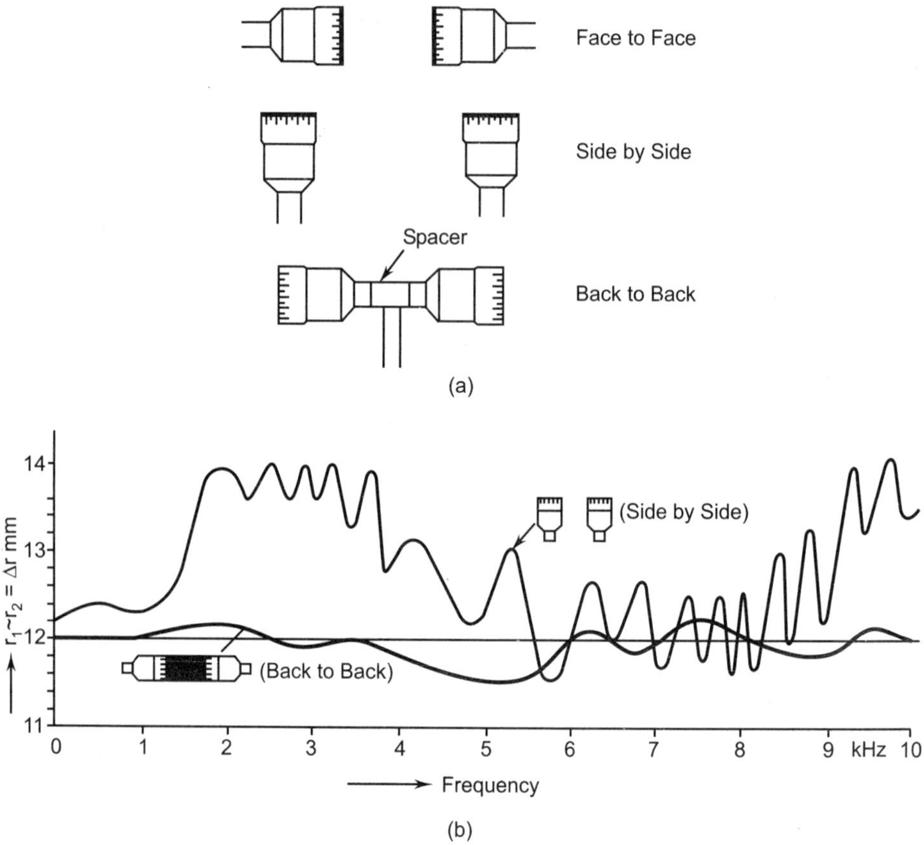

Fig. 8.8: Microphone configurations and the respective errors for accoustic intensity probes

8.9 MULTI-SOUND SOURCES: ADDITION OF SPLs

Let two sound sources having SPLs $X_1\, dB$ and $X_2\, dB$ be acting at the same point and at the same time.

Then,
$$X_1 = 10 \log \frac{I_1}{I_0} \qquad \therefore\ I_1 = I_0 \text{ antilog} \frac{X_1}{10}$$

and
$$X_2 = 10 \log \frac{I_2}{I_0} \qquad \therefore\ I_2 = I_0 \text{ antilog} \frac{X_2}{10}$$

where, I_1 and I_2 are the intensities of the two sound sources and I_0 is the intensity of the reference source.

Resultant sound intensity is $I_1 + I_2$ and hence the resultant SPL is

$$= 10 \log \frac{I_1 + I_2}{I_0}$$

$$= 10 \log \left[\text{antilog} \frac{X_1}{10} + \text{antilog} \frac{X_2}{10}\right]$$

Thus, in the general case of n different sound sources having SPLs $X_1, X_2, ..., X_n$ the resultant SPL is given by:

$$= 10 \log \left[\sum_{n=1}^{n} \text{antilog}\,(x_n/10) \right]$$

SOLVED EXAMPLES

1. If the sound pressure from a source is 110 dyne/cm^2, determine the sound pressure level (SPL). Assume $P_{ref} = 0.0002\ \mu$ bar.

Ans. By definition

$$\text{SPL} = 20 \log_{10} \frac{P}{P_{ref}}$$

both, P and P_{ref} are in μ bar (1 μ bar = 1 dyne/cm^2)

$$= 20 \log_{10} \frac{110}{0.0002}$$

$$= 114.81 \text{ dB}$$

2. SPL at a distance of 20 m from a source of sound is 60 dB. Find the SPL at 40 m distance.

Ans. We have, $I \propto \frac{1}{d^2}$

Thus, doubling the value of d reduces I by one-fourth at 40 m,

$$\text{SPL} = 60 + 10 \log \frac{1}{4}$$

$$= 60 - 6$$

$$= 54 \text{ dB.}$$

3. A source of sound having SPL of 95 dB is working in presence of noise having its own SPL of 65 dB. Find the combined SPL of the system.

Ans.

$$\text{Combined SPL} = 10 \log \left[\text{antilog}\left(\frac{95}{10}\right) + \text{antilog}\left(\frac{65}{10}\right) \right]$$

$$= 10 \log\,[3.1623 \times 10^9 + 3.162277 \times 10^6]$$

$$= 10 \log\,[3.1654 \times 10^9]$$

$$= 95.0044 \text{ dB. } \textbf{(Ans.)}$$

REVIEW QUESTIONS

1. Define sound pressure level. Why it is expressed in dB unit? Draw and describe a sound pressure level meter.
2. Describe a capacitive type microphone as used in sound pressure level meters.
3. Discuss the relevant points about sound pressure level measurement techniques.

Fill in the blanks:

1. The threshold of pain to threshold of hearing represents a ratio of ....
2. SPL of human tolerance level in dB is around ....
3. Zero Phon corresponds to the loudness level of .... of hearing.
4. Normally, ... type of microphones are used for sound pressure level meters.
5. A capacitive microphone gives a flat frequency response between .... Hz to .... kHz.
6. Decibel (dB) is the unit of ....
7. An SPL of 0 (zero) dB represents ....

Tick the correct answer:

1. SPL of rocket engines (in dB) is around

(*a*) 60 (*b*) 100 (*c*) 160 ☐

2. The type of microphone to be used in SPL meters depends upon

(*a*) sensitivity (*b*) frequency response

(*c*) linearity and dynamic range (*d*) all of *a*, *b* and *c* ☐

3. Two automobile horns of 100 dB are blown at the same time. The resultant intensity is

(*a*) 200 dB (*b*) 103 dB (*c*) 50 dB (*d*) 110 dB ☐

4. The SPL at double the original distance would become

(*a*) $\frac{1}{2}$ (*b*) $\frac{1}{4}$ (*c*) $\frac{1}{8}$ (*d*) $\frac{1}{16}$ ☐

5. Audio frequency range, for human beings, lies between:

(*a*) 20 Hz to 20 kHz (*b*) 0 Hz to 40 kHz

(*c*) 10 kHz to 50 kHz (*d*) 20 kHz to 1 MHz ☐

REVIEW PROBLEMS

1. Sound pressure at a point is 100 dyne/cm^2. Determine the SPL. Given, $P_{ref} = 0.0002$ μbar.

[**Ans.** 113.98 dB]

2. Determine the SPL at a distance of 80 m, if the SPL at 10 m distance from a source of sound is 110 dB.

[**Ans.** 91.9 dB}

CHAPTER 9

FLOW MEASUREMENT

9.1 INTRODUCTION

Apart from temperature and pressure, measurement of flow rate and total flow (over a period of time) is another very important process parameter in the field of instrumentation.

Flow measurements are made in numerous chemical plants, refineries, power plants, steel plants and a host of other industries as well where the quality of end product or the performance of the plant depends upon its precise measurement. Domestic water and gas measurements and intake of petrol into cars from petrol stations are some of the examples of our daily association with flow measurements.

Industrial flow measurements include measuring the flow rates of fluids (liquids and gases) and solids. Fluid measurements are done either on the basis of volume or weight while for solids it is weight (or mass) only. For gas flow, any measurements made from the process are always converted to NTP.

9.2 UNITS OF FLOW

Several units are now commonly used dependent on the needs of the process.

Solids: Tonnes/hr, kg/min, etc.

Liquids: expressed either in volume rate or weight rate

Volume rate: Litre/hr, litre/min, m^3/hr, gallons/hr

Weight rate: Tonnes/hr., kg/min

Gases: m^3/hr, Nm^3/hr (at NTP)

Steam: Tonnes/hr, kg/min

[Since steam density changes with temperature and pressure, measurements made are converted into weight rate of water.]

9.3 FUNDAMENTALS FOR FLOW OF LIQUIDS

9.3.1 Introduction

When a liquid passes through a pipeline, resistance in the form of friction and viscosity are encountered by the flowing liquid. If the velocity of the liquid is low, flow advances in parallel lines. Flow is then said to be laminar or viscous or streamline. Beyond a certain velocity value called critical velocity, eddies start forming in the liquid. The laminar flow pattern is then destroyed and flow thereafter is said to be turbulent.

9.3.2 Pattern of Flow: Reynold's Number

Reynold's number—a dimensionless quantity, determines the pattern of flow (*i.e.* streamline or otherwise) in a pipeline. It is expressed as:

$$R_n = \frac{D\rho v}{\mu} = \frac{Dv}{\nu}$$

where, R_n = Reynolds number
D = diameter of pipe
ρ = mass density of liquid
v = velocity of liquid
μ = dynamic viscosity
and ν = kinematic viscosity = μ/ρ

It is found that if $R_n < 2000$, flow is laminar, viscous or streamline. If $R_n > 4000$, flow is turbulent and if $2000 < R_n < 4000$, then flow is in a critical state–changing itself from streamline to turbulent pattern as the value of R_n continues to rise from 2000.

Figure 9.1 shows the pattern of flow under streamline and turbulent flow conditions.

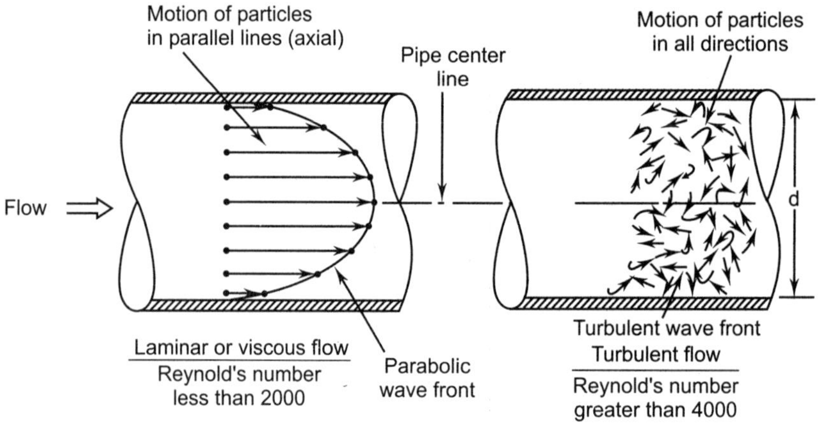

Fig. 9.1: Flow pattern under laminar and turbulent conditions

The wave front of velocity for streamline (or laminar or viscous) and turbulent flow conditions are shown below in figure 9.2, the velocity distribution curve is paraboloid in nature for streamline flow and the average velocity is half the maximum.

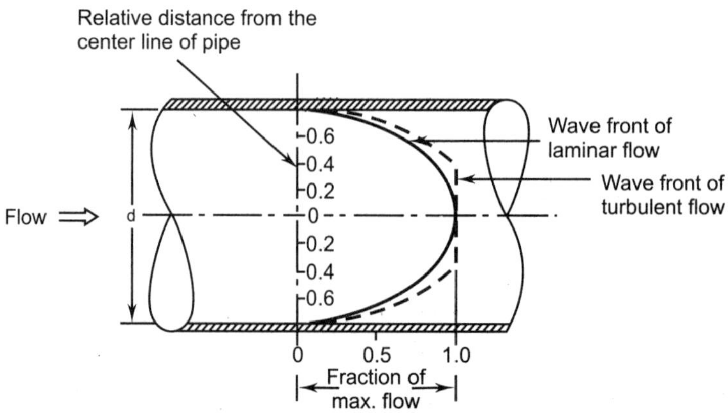

Fig. 9.2: Velocity distribution in laminar and turbulent flow

It should be noted that frictional resistance to streamline motion is proportional to velocity and surface area of contact, while it is independent of pressure value and nature of the surface in contact and greatly affected by temperature fluctuations.

In case of turbulent motion, frictional resistance is proportional to square of velocity, mass density and surface area of contact, while remaining independent of pressure value and temperature effects (almost negligible).

Figure 9.3 shows separately the velocity profiles for streamline or laminar or viscous and turbulent flow conditions, where r is the radius at a point, r_p is the pipe radius, V_{max} is the maximum velocity (along the pipe axis), while V represents the velocity corresponding to radius r.

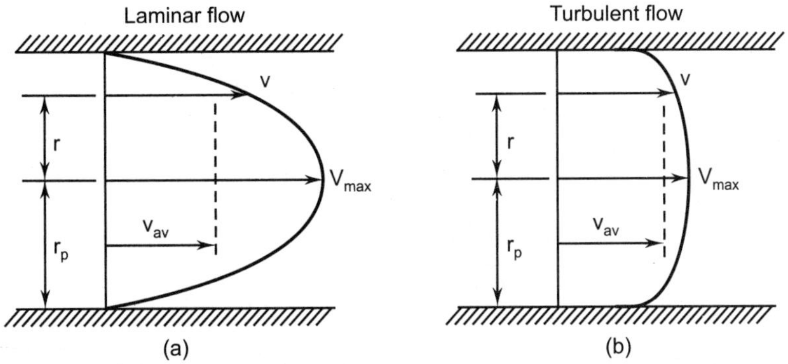

Fig. 9.3: Velocity profiles, maximum and average velocity for (*a*) streamline and (*b*) turbulent flow

The velocity for laminar and turbulent flow are given by

$$v = V_{max}\left(1 - \frac{r^2}{r_p^2}\right) \quad \text{for laminar flow}$$

and

$$v = V_{max}\left(1 - \frac{r^2}{r_p}\right)^{1/n} \quad \text{for turbulent flow}$$

n is an exponent and depends on Reynolds number R_n. Figure 9.4 shows n as a function of R_n alongwith the corresponding velocity ratio $V_{av}/V_{max.}$

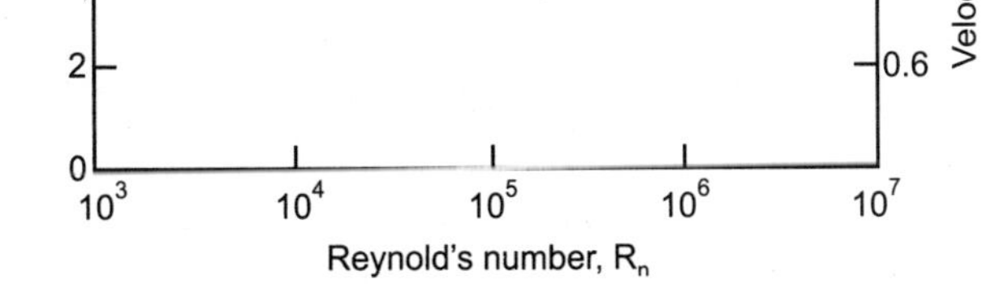

Fig. 9.4: Exponent n and velocity ratio V_{av}/V_{max} as a function of Reynold's number

9.3.3 Flow Through Pipes: Reynold's Experiment

Reynold's performed an experiment for flow of liquid through a pipe. He measured the head loss (h) between two sections of the pipe separated by a distance l. Also, velocity of flow was found out by measuring the discharge Q and dividing it by the cross-section A.

He then plotted a graph of head loss per unit length of pipe against velocity, which is shown in figure 9.5.

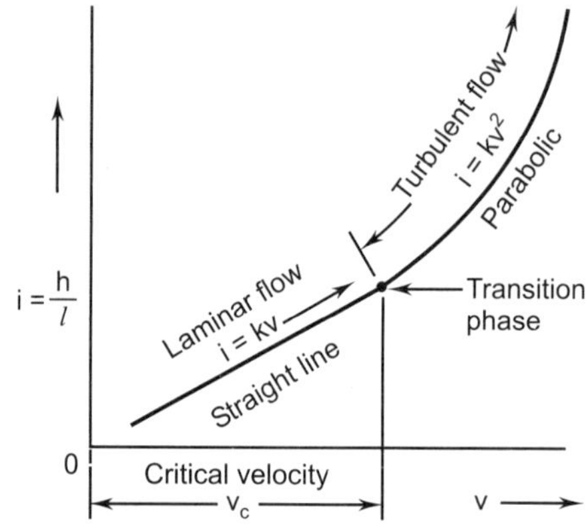

Fig. 9.5: Graph of head loss per unit length of pipe against velocity

where, in the figure i = head loss per unit length = h/l and $v = Q/A$.

Upto the point of critical velocity when the flow is laminar or streamline, the relationship between i (= head loss per unit length of pipe) and velocity of flow is given by $i = kv$, that is a straight line, while it is $i = kv^2$ for turbulent flow. The curve is parabolic in nature in this region.

9.3.4 Determination of Critical Velocity: Colour Band Method

The value of critical velocity in case of water may be found by passing water through a glass pipe and injecting a colored liquid (say potassium permanganate solution) into the centre of the passing stream. The colour band remains in a straight line fashion along the centre of the stream for velocities less than the critical value. When turbulence sets in because of higher velocity, the colour band is mixed up and the band is broken. Here no band is observed and the whole water becomes uniformly coloured.

9.3.5 Flow of Incompressible Fluid in Pipes: Bernoulli's Equation

It states that for an incompressible, frictionless and adaptive flow in a pipe, each particle in the system has the same total energy, regardless of its position.

If the symbols p, v and z stand for static pressure, fluid stream velocity and elevation above some datum line, Bernoulli's equation is as follows:

$$\frac{p_2}{\rho} + \frac{v_2^2}{2g} + z_2 = \frac{p_1}{\rho} + \frac{v_1^2}{2g} + z_1 \quad ...(9.1)$$

ρ represents the fluid weight density and g the local acceleration due to gravity.

Suffixes 1 and 2 correspond to positions 1 and 2 respectively, as shown in figure 9.6.

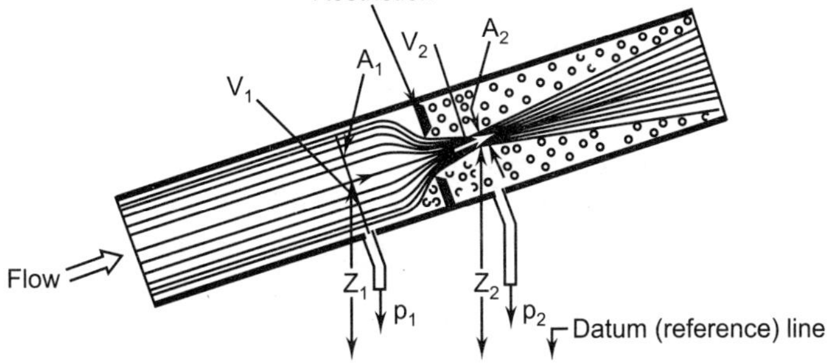

Fig. 9.6: Flow through a restriction in a closed pipe

Assuming $z_1 = z_2$, the above equation can be written as:

$$v_2^2 - v_1^2 = \frac{2g}{\rho}(p_1 - p_2) \quad \text{...(9.2)}$$

Now, from the equation of continuity,

$$q = A_2V_2 = A_1V_1 \quad \text{...(9.3)}$$

where, q = volume flow rate

Area A_2 is not always constant. A_2 corresponds to the area of the stream at which *vena contracta* occurs. Now, this varies with flow rate. Thus, combining equations (9.2) and (9.3).

$$q = KA_2\sqrt{\frac{2g}{\rho}(p_1 - p_2)} \quad \text{...(9.4)}$$

where K is a flow coefficient and represents the coefficient of contraction and coefficient of velocity taken together.

$$K = c/\sqrt{1-\alpha^4}, \quad \text{where } c = \text{discharging coefficient}$$

$$\alpha = \sqrt{A_2/A_1} = d/D.$$

where d = diameter of orifice

D = internal diameter of the pipe

and, $(p_1 - p_2)/\rho = h$ = differential head, expressed in units of length and measured by a manometer.

Equation (9.4) can be written in a more convenient way as

$$q = \frac{\pi k\alpha^2 D^2}{4}\sqrt{\frac{2g}{\rho}(p_1 - p_2)} = \frac{\pi kd^2}{4}\sqrt{\frac{2g}{\rho}(p_1 - p_2)} \quad \text{...(9.5)}$$

In equation (9.5), $(p_1 - p_2)$ is a variable quantity and is derived for an incompressible, ideal fluid.

In practice the differential head h is measured by a manometer, the manometric fluid density being ρ_m. Another fluid of density ρ_f, rests on either limb of the manometric fluid. This fluid (of density ρ_f) normally differs from the flowing fluid because of varied temperature conditions. When flow rate is to be converted to some standard temperature (say control room temperature) at which the flowing fluid density is ρ_s, the flow rate would be given by:

$$q = \frac{\pi kd^2}{4}\sqrt{\rho(\rho_m - \rho_f)/\rho_s} \cdot \sqrt{2gh} \quad \text{...(9.6)}$$

9.3.6 Vena Contracta: Different Coefficients Pertaining to Flow

Let the level in a water tank be maintained constant by means of a tap and a hole or orifice fitted at the side bottom. The situation is shown in figure 9.7.

Amount of water coming out of the orifice per unit time. *i.e.*, the flow rate is dependent on the shape, size and the type of orifice employed.

There will be frictional resistance to the flow of fluid via the orifice, which can be reduced by making the orifice sharp edged, thereby reducing the contact area.

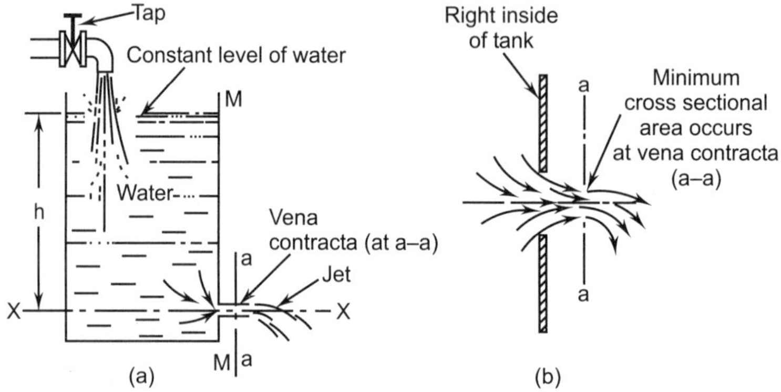

Fig. 9.7: (*a*) Jet from orifice; (*b*) Magnified view at vena contracta

The section (shown in the figure as a-a) at which the flow pattern becomes parallel is called vena contracta. At this, the flow velocity is maximum and the jet cross-section becomes minimum. The section a-a depends on the flow rate. For flow rates through an orifice, vena contracta plays the determining factor.

9.3.6.1 Coefficient of Contraction (C_c)

It is defined as

$$C_c = \frac{\text{Area of jet at vena contracta}}{\text{Area of orifice}}$$

Slight variation in the value of C_c is noticed for variations in head, shape, size and the type of orifice. A typical value is 0.65 for small, sharp-edged orifices.

9.3.6.2 Coefficient of Velocity (C_v)

It is defined as

$$C_v = \frac{\text{Actual velocity at vena contracta}}{\text{Theoretical value of velocity}}$$

For a head H, the denominator is $\sqrt{2gH}$, and if v_a is the actual velocity at vena contracta then

$$C_v = \frac{v_a}{\sqrt{2gH}}$$

A typical value for C_v is around 0.97. Actual velocity value is diminished because of frictional effects to flow at orifice edge.

9.3.6.3 Coefficient of Discharge (C_d)

Amongst the three coefficients, C_d is most important and is defined as:

$$C_d = \frac{\text{Actual discharge (or flow rate)}}{\text{Theoretical discharge (or flow rate)}}$$

Again, due to frictional effects at the orifice edge, the actual discharge is less than the theoretical discharge.

C_d, the co-efficient of discharge corrects the following:

- influence of velocity profile (Reynolds number)
- assumption of no energy loss between the taps, and
- pressure tap locations.

9.3.7 Flow of Compressible Fluids in Pipes

For compressible fluids, law of conservation of energy as applied in thermodynamics is employed to find out the relationship between pressure and velocity of flow through a pipe.

Assuming adiabatic (*i.e.* neither heat accepted nor rejected by the flowing fluid) flow of fluid through the pipe and also assuming no external work done on or by the fluid, it can be written as

$$p_1 v_1 + \frac{V_1^2}{2g} + JE_1 = p_2 v_2 + \frac{V_1^2}{2g} + JE_2 \quad ...(9.7)$$

In the above equation, the terms z_1 and z_2 are omitted from the two sides. Since their difference ($z_1 \sim z_2$) is very small and hence is neglected.

where, E = internal molecular energy of the fluid

J = work equivalent of heat

v = specific volume of fluid

Again, $V_2^2 - V_1^2 = 2gJ(H_1 - H_2)$ ...(9.8)

where H stands for enthalpy.

For an ideal gas, if specific heats remain constant,

$$H_1 - H_2 = \frac{\gamma R}{J(\gamma - 1)} T_1 \left[1 - \frac{p_2}{p_1}^{(\gamma-1)/\gamma}\right] \quad ...(9.9)$$

where, γ = ratio of the two specific heats of the gas = C_p/C_v

R = Gas constant

and T = absolute temperature

Applying the conservation of mass principle:

$$\omega = \frac{A_1 V_1}{v_1} = \frac{A_2 V_2}{v_2} \quad ...(9.10)$$

where, ω is the weight rate of flow. Combining the above equations:

$$\omega = \frac{\pi C\alpha^2 D^2}{4} \sqrt{\frac{2g\gamma p_1}{(\gamma - 1)v_1}} \sqrt{\frac{\left(\frac{p_2}{p_1}\right)^{2/\gamma} - \left(\frac{p_2}{p_1}\right)^{(\gamma+1)/\gamma}}{1 - \alpha^4 \left(\frac{p_2}{p_1}\right)^{2/\gamma}}} \quad ...(9.11)$$

where, C stands for discharge coefficient.

Equation (9.11) shows that the flow rate is dependent on p_2/p_1. Since manometer measures the pressure differential $(p_1 - p_2)$, hence it is advisable to convert the term p_2/p_1 in equation (9.11) in the form of $(p_1 - p_2)$.

For this, we assume

$$\frac{p_2}{p_1} = 1 - x$$

Since, p_2 is very close to p_1 thus, x is very close to zero.

or
$$\left(\frac{p_2}{p_1}\right)^{2/\gamma} \cong (1-x)^{2/\gamma}$$
$$= 1-\left(\frac{2}{\gamma}\right)\times x$$
$$= 1-\frac{2}{\gamma}\left(1-\frac{p_2}{p_1}\right) \qquad ...(9.12)$$

and
$$\left(\frac{p_2}{p_1}\right)^{\left(\gamma+\frac{1}{\gamma}\right)} \cong 1-\left(\gamma+\frac{1}{\gamma}\right)+\left(\gamma+\frac{1}{\gamma}\right)\left(\frac{p_2}{p_1}\right)$$
$$= 1-\left(\frac{\gamma+1}{\gamma}\right)\left(1-\frac{p_2}{p_1}\right) \qquad ...(9.13)$$

Substituting (9.12) and (9.13) in (9.11), we have the expression for weight flow rate as:

$$\omega = \frac{\pi C \alpha^2 D^2}{4}\sqrt{\frac{2g(p_1-p_2)}{v_1\left[1-\alpha^4\left(\frac{p_2}{p_1}\right)^{2/\gamma}\right]}} \qquad ...(9.14)$$

In equation (9.14), the term $\left[1-\alpha^4\left(\frac{p_2}{p_1}\right)^{2/\gamma}\right]$ is the velocity of approach factor. Since flow coefficient is not much affected because of a reasonable variation in the pressure ratio p_2/p_1, the velocity of approach factor can be taken as $(1 - \alpha^4)$. Hence, equation (9.14) is reduced to

$$\omega = \frac{\pi\alpha^2 D^2}{4}\cdot\frac{C}{\sqrt{1-\alpha^4}}\sqrt{\frac{2g(p_1-p_2)}{v_1}} \qquad ...(9.15)$$

An additional parameter, ϕ, termed as rational expansion factor, is now introduced

$$\phi = \frac{\text{weight of compressible flow rate}}{\text{weight of incompressible flow rate}}$$

Thus, by finding out the weight of incompressible flow rate and multiplying by ϕ, the weight of compressible flow rate can be found out. The term ϕ is given by:

$$\phi = \sqrt{\frac{(1-\alpha^4)\left[\left(\frac{p_2}{p_1}\right)^{2/\gamma}-\left(\frac{p_2}{p_1}\right)^{(\gamma+1)/\gamma}\right]}{\left(\frac{\gamma-1}{\gamma}\right)\left(1-\frac{p_2}{p_1}\right)\left[1-\alpha^4\left(\frac{p_2}{p_1}\right)^{2/\gamma}\right]}} \qquad ...(9.16)$$

For orifices using flange taps or vena contracta, an empirical expansion factor exists, which is as follows:

$$\phi_1 = 1-[0.41+0.35\,\alpha^4]\left(\frac{p_1-p_2}{p_1}\right)\left(\frac{1}{\gamma}\right) \qquad ...(9.17)$$

for orifices using pipe taps, it is

$$\phi_2 = 1-[0.333+1.145\,(\alpha^2+0.7\alpha^5+12\alpha^{13}]\left(\frac{p_1-p_2}{p_1}\right)\left(\frac{1}{\gamma}\right) \qquad ...(9.18)$$

For gases containing moisture, a still further correction is needed to account for the density of vapour.

The moisture factor M is

$$M = 1+\frac{p_v}{p}\left(\frac{G_v}{G}-1\right) \quad ...(9.19)$$

where p_v = Vapour pressure

G_v = Specific gravity of vapour referred to air at the same condition of pressure and temperature.

The specific volume of the gas may be found from the relationship:

$$v = \frac{yRT}{p} \quad ...(9.20)$$

where, y = compressibility factor

The flow equation for gases is given by:

$$Q = \frac{\pi}{4} k \alpha^2 D^2 Y \frac{v_b}{M_b} \sqrt{\frac{d_m M_1}{v_1}} \sqrt{2gh} \quad ...(9.21)$$

where Q = flow rate at the base temperature T_b and base pressure p_b
k = flow coefficient
d_m = density of manometer fluid
M_1 = moisture factor at upstream conditions
M_b = moisture factor under base conditions
v_b = specific volume of gas under base condition
v_1 = specific volume of gas at upstream conditions
h = difference in height between the two limbs of the manometer
Y = expansion factor

9.4 FACTORS INFLUENCING FLOWMETER ACCURACIES

9.4.1 Introduction

Accuracy of flowmeters is determined for a steady flow of a homogeneous, single-phase Newtonian fluid for which the coefficient remains fixed in a long, straight pipe run. In reality, the above ideal conditions cannot be met. Factors which affect accuracy are called "influence quantities" and are velocity profile distribution, nonhomogenous flow, pulsating flow and cavitation.

The error associated with a particular influence quantity depends on (*a*) the sensitivity of the flow meter to the disturbance and (*b*) whether a correction can be incorporated or not. If the fluid is Newtonian in nature, velocity profiles can be brought within desirable limits by (*a*) increasing straight length of pipe both upstream and downstream of the flow or (*b*) for shorter pipe lengths, by introducing flow conditioners. Again, less sensitive flowmeter or pulsation dampers may be used to offset the effects of influence quantities.

9.4.2 Velocity Profile

Amongst the four influence quantities, velocity profile is the most important one. It is affected by swirl, non Newtonian fluids, non axisymmetric flow etc.

9.4.3 Newtonian Fluids

A newtonian fluid is a fluid which when acted upon by a shear stress, produces a velocity gradient which is proportional to the applied shear stress. This constant of proportionality is the absolute viscosity of the fluid. Examples of newtonian fluids are all gases, mostly all liquids and very fine mixture of spherical particles in a liquid. The velocity profile obtained for a Newtonian fluid is a reference one and from this reference, corrections can be made.

In the absence of viscosity, the velocity of a flowing fluid would be the same across any pipe cross-section. In presence of viscosity, shearing action would be present between two adjacent layers. In such a case, velocity would be maximum on the pipe progressively reducing to zero at the pipe wall. Osborne Reynolds observed that, in the absence of swirl, a newtonian fluid has two distinctly different velocity profiles.

Viscous drag between two adjacent layers forces the fluid into parallel-layer motion when the velocity is low or viscosity is high. On the other hand, if the velocity is high or viscosity is low, inertial forces overcome viscous forces and the fluid flows in a turbulent manner. The laminar profile is easily analysed, but the turbulent one is complex and cannot be represented mathematically in a simple manner.

For Reynolds number less than 2000, the laminar flow profile is parabolic in nature. This is unaffected by wall roughness. If r *be* the radius of a point and r_p the pipe radius, the relationship between the velocity at the point and the maximum velocity (occurring on the centre line) is

$$v_p = V_{\max}\left[1 - \left(\frac{r}{r_p}\right)^2\right]$$

from which the average velocity is

$$v_{av} = \frac{1}{2} V_{\max}$$

The fully developed parabolic profile from a blunt profile takes a developed length L, given by,

$$L = 0.028\ DR_n$$

where, D = internal pipe diameter
and R_n = Reynolds number

Length L is independent of pipe roughness but pipe-entrance conditions, like sharp edge versus rounded inlet, may require a higher value of L.

A transition in the nature of the parabolic profile occurs for Reynold's number in between 2000 and 4000. At Reynold's number of 4000 or more, the profile becomes more blunt and with increasing number, the profile becomes more and more flatter until all the particles across any cross-section, except at the walls, will be moving with the same velocity. This flat profile also goes by the name of 'infinite Reynolds number profile', the 'rectilinear profile' or 'plug-flow profile'.

Turbulent profile does not have a fixed geometry. It is dependent on the pipe roughness and Reynolds number. The velocity at a point for turbulent motion is given by

$$V_p = V_{\max}\left(1 - \frac{r}{r_p}\right)^{1/n}$$

where, $n = 1.66 \log R_n$

9.4.4 Non-Newtonian Fluids

For non-Newtonian fluids, the ratio between the shear stress and shear rate is not constant. This ratio changes with stress and is called the apparent viscosity. Non-Newtonian fluids are subdivided into two classes—time dependent and time independent. The apparent viscosity of the former depends on time while for the latter it is not.

The different types of non-Newtonian fluids are shown in figure 9.8.

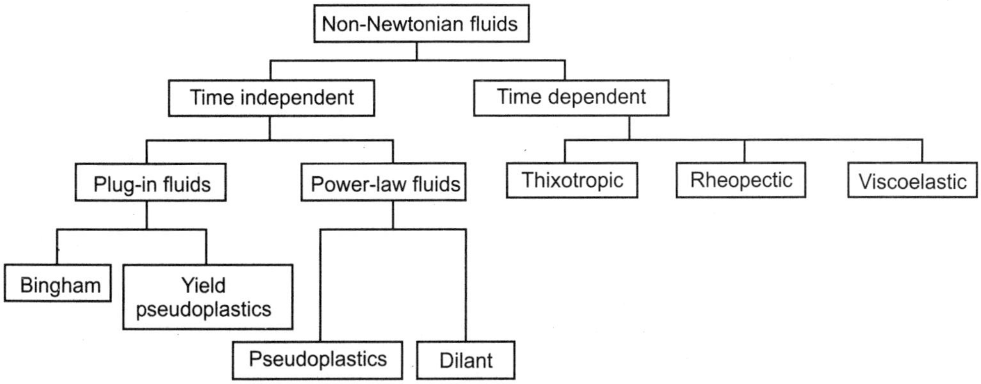

Fig. 9.8: Different types of non-Newtonian fluids

9.4.5 Profile Distortion

Velocity profiles are affected by elbows, strainers, pipe fittings, reducers, expanders, etc. These are necessary for any plant piping system. The velocity profile is a locus of all the axial velocity vectors across a given cross-section.

The axial velocity vector is affected by:

- A pure swirl motion that causes mction about the pipe axis.
- An asymmetrical profile peaks near the wall.
- A symmetrical profile that has a high core velocity which results from a very sudden reduction in pipe cross-section.
- Secondary flows, like two or more counter rotational vortices in a plane perpendicular to the pipe axis.

Straight lengths of pipe required to eliminate or reduce swirl within acceptable limits is different for liquids and gases. For liquids with higher viscosities, this length is obviously smaller.

Fittings are a must for flow systems. Upstream fittings are divided into two broad categories—those that distort profile but produce little swirl, and the second that distort profile and causes swirl also. A distorted profile can be brought within acceptable limits by a sufficiently long upstream pipe run or else by using flow conditioners or straighteners in which case straight run of pipe is reduced.

Flow conditioners are used where sufficient straight length of upstream pipe is not available to remove or reduce swirl within acceptable limits. Swirls are generated when the flowing liquid passes through a number of bends, *i.e.*, change in direction of flow takes place. Three flow conditioners are shown in figure 9.9. Different flow conditioner types are available and some of them are Etoile, AGA, ASME, AMCA, Zanker, Mitsubishi, Sprenkle, etc. Three concepts in flow

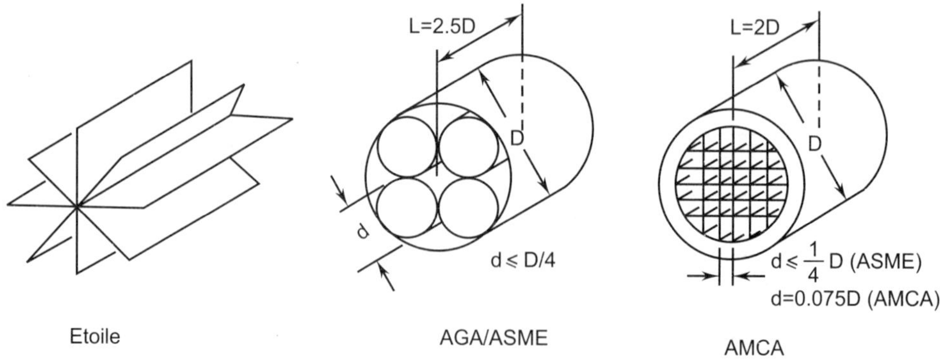

Fig. 9.9: Different types of flow conditioners to reduce swirl

conditioner/straightener are employed. The first one counteracts swirl by splitting the conduit into a large number of parallel tubes. Perforated plates/grids are employed in the second that removes uneven distribution and the third one tries to establish a given flow pattern. Examples of the first two are Etoile, Sprenkle and the third one is standardised for use with orifice plates in ISO 5167.

The number of tubes in a flow conditioner is not a fixed one. Usually, four to eight tubes are employed to remove distortion. It is to be noted that increasing the number of tubes in the conditioner does not, in general, improve profile restoring ability.

9.4.6 Non-homogeneous Flow

The theoretical and practical flow equations applied for flow meters are valid only for homogeneous flow. A non-homogeneous flow can have a small or significant bearing on a flowmeter's operating principles.

Non-homogeneous fluids are grouped as per the following:

- Two phase fluids
- Fluids with two or more components that are individually homogeneous in nature.

Homogeneous and non-homogeneous flow differ on two counts: for the latter density cannot easily be determined and secondly the different phases comprising the non-homogeneous flow may not have the same velocity as the main flow.

In some cases, for non homogeneous flow, a component may flow along the bottom of the pipe. This is known as '*hold up*' or '*slip*'. This is a function of viscosity, density difference, particle size, surface tension of superficial velocity of each component. Gravity effect also changes flow pattern in horizontal vertical or incline pipelines.

Some of the two phase fluids encountered in industries are: air in water, steam and ammonia vapour, pulp in water.

9.4.7 Pulsating Flow

While discussing pulsating flow, it is pertinent to understand what steady flow means. ISO Technical Report 3313 (1974) states that "flow condition in a measuring section in which the flow does not vary with time." ISO 5167 (1980) notes that "this international standard applies only to pressure differential devices in which the flow remains subsonic throughout the measurement section, is steady or varies only slowly with time."

Flow is always a dynamic process. Vortex formation, pressure fluctuations within lead lines, turbine-rotor area are some of the influencing factors leading to pulsations.

A threshold pulsation index I_p is defined by Head (1956) when pulsations exist in flow. For the index remaining below a certain threshold, reference condition accuracy applies. If otherwise, passive-type filters (*e.g.*, restrictors), increased volume, or a combination of the two are employed to bring the index below the threshold.

Pulsation index I_p was defined by Head (1956) as the ratio of the difference between maximum and minimum flow rate to the average flow rate.

Flow pulsations of higher magnitude would be apparent from flowmeter signal or a sustained hum or an audio steady rumble. A high frequency response velocity measuring device, placed in close proximity of the primary device, is used to ascertain pulsating flow. For such cases, the threshold pulsation index is exceeded and low-pass in-line filter is installed to cut off frequencies beyond a certain minimum value. Such filter design is dependent on whether fluid is itself a gas or a liquid. Pulsation dampers are used to contain pulsation when it causes unacceptable pipeline vibration. Figure 9.10 shows typical filters for gas and liquid flows.

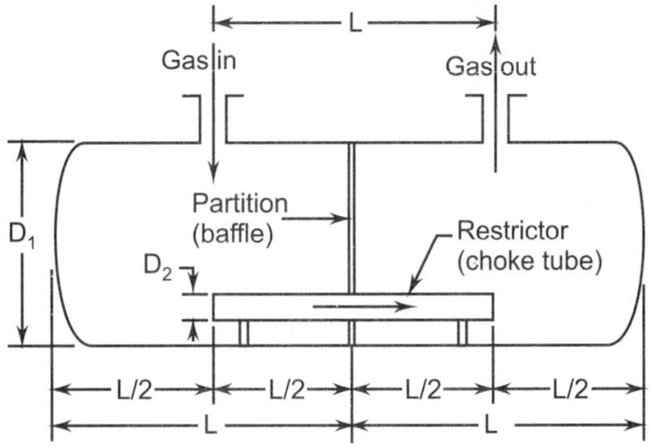

Flow rate $\times 10^3 =$ (CC/hr)	Tank dia. D_1 (cm)	Choke dia. D_2 (cm)	Length L (cm)
3.5×10^6	40.5	3.8	96.5
$3.5 \times 10^6 - 12.7 \times 10^6$	61	5	86.4
$12.7 \times 10^6 - 19.8 \times 10^6$	76	7.6	102

(a)

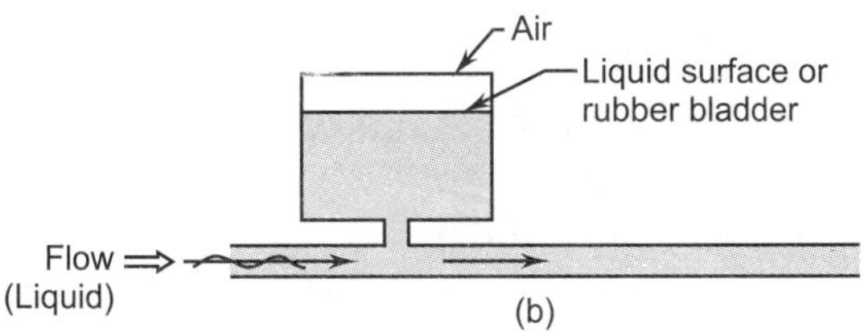

(b)

Fig. 9.10: Typical passive filters for (*a*) gas (*b*) liquid flows

9.4.8 Cavitation

Cavitation is the process of boiling a liquid as its pressure decreases. When an obstruction is placed in a pipeline, the downstream line pressure of the liquid may approach its vapour pressure leading to cavitation. Cavitation, in essence, is the formation and collapse (implosion) of vapour cavities. Implosion is associated with generation of audible noise. Extensive cavitation may lead to damage of pipe, restricts flow, produces audible noise or may damage turbine blades.

Sudden huge pressure drop caused by flow restrictors such as valves or vortex elements or friction along the flow path, flow separation etc. are some of the causes of cavitation. Liquids with dissolved gases or gas bubbles in it supports formation of cavitation. It is observed that gas concentration in the range of 40 ppm will give rise to cavitation in fluids at higher static pressures.

A dimensionless cavitation number is defined as follows:

$$n_c = \frac{2g_c (P_{f2} - P_{v2})}{e_f V_{f2}^2}$$

with suffix 2 referring to the location of maximum velocity within the flowmeter. P_{f2} is the static pressure and P_{v2} is the vapour pressure and their difference is seen as the force required to collapse the vapour bubbles. The term V_{f2}^2 in the denominator is the dynamic pressure required to initiate bubble formation. Thus, the dimensionless number n_c is the ratio between the collapsing and forming forces.

For n_c below a certain value, called the *incipient cavitation number*, cavitation becomes more and more volatile and destructive to piping. It was shown by Oba et al (1978) that addition of very small amount of polymer reduces both the incipient cavitation number and audible noise caused by cavitation.

9.5 DIFFERENT TYPES OF FLOWMETERS

Head type
- Orifice
- Venturi
- Flow nozzle
- Dall tube
- Pitot tube

Inferential type
- Rotameter
- Turbine
- Target

Positive displacement meter
- Nutating disc
- Reciprocating piston
- Sealed drum
- Rotary vane
- Lobed impeller
- Fluted spiral rotor
- Oval gear

Mass flowmeter
- Impeller turbine
- Twin turbine
- Gyroscopic
- Coriolis
- Bridge circuit
- Radiation type
- Thermal

Electrical type
- Electrical manometer type
- Linear resistance element type
- Electromagnetic
- Ultrasonic
- Anemometer

Oscillatory type
- Vortex shedding
- Fluidic
- Swirl

Miscellaneous type
- V-cone
- Wedge
- Gas injection
- Flow marker
- Compound
- Cross-correlation
- Centrifugal force element

Solid flowmeters
- Conveyor weighing
- Servo drive
- Nuclear

Open channel
- Weirs
- Flumes

9.6 FLOWMETER SELECTION TABLE

Table 9.1

Flowmeter	*Pipe size, in (mm)*	*Gases (Vapours)*		*Liquids*					
								Slurries	
		Clean	*Dirty*	*Clean*	*Viscous*	*Dirty*	*Corrosive*	*Fibrous*	*Abrasive*
SQUARE ROOT SCALE: MAXIMUM SINGLE RANGE 4:1									
Orifice		□	□	□	□	□	□	□	□
Square-edged	> 1.5(40)	■	□	■	□	▒	▒	□	□
Honed meter run	0.5 - 1.5 (12–40)	■	□	■	▒	□	▒	□	□
Integral	< 0.5 (12)	■	□	■	■	□	▒	□	□
Quadrant /conic edge	> 1.5 (40)	□	□	■	■	▒	▒	□	□
Eccentric	> 2 (50)	▒	■	▒	□	■	▒	□	□
Segmental	> 4 (100)	▒	■	▒	□	■	▒	□	□
Annular	> 4 (100)	▒	■	▒	□	■	▒	□	□
Target	> 0.5 - 4 (12 - 100)	■	■	■	■	■	▒	□	□
Venturi	> 2 (50)	■	▒	■	▒	▒	▒	▒	▒
Flow nozzle	> 2 (50)	■	▒	■	▒	▒	▒	□	□
Lo-Loss	> 3 (75)	■	□	■	▒	□	■	□	□
Pitot	> 3 (75)	■	□	■	□	▒	▒	□	□
Annubar	> 1 (25)	■	□	■	□	□	▒	□	□
Elbow	> 2 (50)	■	▒	■	□	▒	▒	▒	▒
LINEAR SCALE: TYPICAL RANGE 10:1									
Magnetic	0.1 - 72 (25 - 1800)	□	□	■	■	■	■	■	■
Positive-displacement	< 12 (300)	■	□	■	□	□	▒	□	□
Turbine	0.25 - 24 (6 - 600)	■	□	■	□	□	▒	□	□
Ultrasonic		□	□	□	□	□	□	□	□
Time of flight	> 0.5 (12)	□	□	■	▒	□	■	□	□
Doppler	> 0.5 (12)	□	□	□	▒	■	■	■	■
Variable-area	≤ 3 (75)	■	□	■	■	□	▒	□	□
Vortex	1.5 - 16 (40 - 400)	■	▒	■	□	▒	▒	□	□

■ = designed for this application; ▒ = normally applicable; □ = not designed for this application.

URV = Upper range value

Temperature, °F (°C)	*Pressure psig (kPa)*	*Accuracy, uncalibrated (including transmitter)*	*Reynolds number*
Process temperature to 1000°F (540°C): transmitter limited to–20°– 250°F (–30° – 120°C)	To 6000 psig (41,000 kPa)		
		± 1–2% URV	$R_n > 2000$
		± 1% URV	$R_n > 1000$
		± 2–5% URV	$R_n > 100$
		± 2% URV	$R_n > 200$
		± 2% URV	$R_n > 10,000$
		± 2% URV	$R_n > 10,000$
		± 2% URV	$R_n > 10,000$
		± 1.5 – 5% URV	$R_n > 100$
		± 1 – ± 2% URV	$R_n > 75,000$
		± 1 – ± 2% URV	$R_n > 10,000$
		± 1.25% URV	$R_n > 12,500$
		± 5% URV	No limit
		± 1.25% URV	$R_n > 10,000$
		± 4.25% URV	$R_n > 1000$
360 (180)	≤ 1500 (10,500)	± 0.5% of rate to ± 1% URV	No limit
Gases: 250 (120) Liquids: 600 (315)	≤ 1400 (10,000)	Gases: ± 1% URV Liquids: ± 0.5% of rate	≤ 8000 cSt
– 450 – 500 (– 268 – 260)	≤ 3000 (21,000)	Gases: ± 0.5% of rate Liquids: ± 1% of rate	≤ 2 – 15 cSt
– 450 – 500 (– 268 – 260)			
– 300 – 500 (180 – 260)	Pipe rating	± 1% of rate to ± 5% of URV	No limit
– 300 – 250 (– 180 – 120)	Pipe rating	± 5% URV	No limit
Glass: ≤ 400 (200) Metal: ≤ 1000 (540)	Glass: 350 (2400) Metal: 720 (5000)	± 0.5% of rate to ± 1% URV	To highly viscous fluids
≤ 400 (200)	≤ 1500 (10,500)	± 0.75 – 1.5% of rate	> 10,000

9.7 DIFFERENTIAL HEAD METHOD OF FLOW MEASUREMENT

9.7.1 Introduction

The underlying principle of all differential-producing flowmeters, also sometimes called *head-class flowmeters,* is Bernoulli's streamline energy equation. Flow measurements come under the category of head type and includes different types of standard restrictions that are put across the flow. These include orifice plate, venturi tube, flow nozzle, Pitot tube, Dall tube, etc. One advantage of such standard restrictions is that they can be interchanged in their use with various differential manometers in horizontal/sloping/circular pipelines on the basis of design data alone, without any need to calibrate them individually. However, certain requirements with regard to their installation must be adhered to. Details about orifice plates now follow.

9.7.2 Orifice Plate

A variety of orifice plates are available, which have industrial applications. The types are: Concentric square-edged orifice, small-bore honed-orifice, quadrant edged orifice, conical-entrance orifice, eccentric and segmental orifice.

9.7.2.1 Concentric Square-edged Orifice

A concentric square-edged orifice and its pressure profile is shown is figure 9.11.

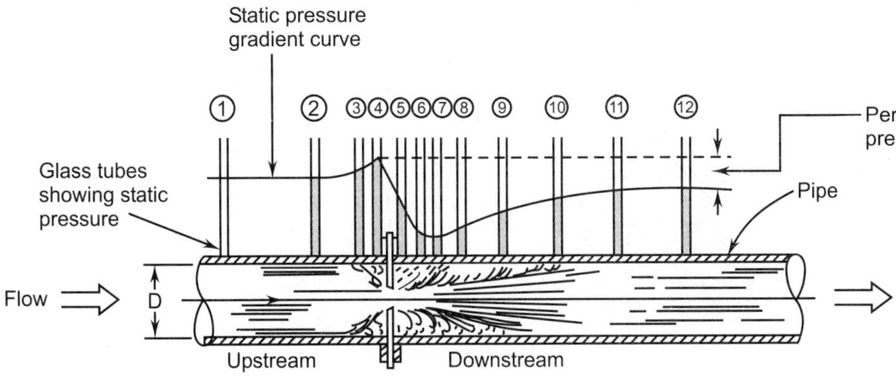

Fig. 9.11: Static-pressure gradient in an orifice pipe

The spacing for pressure taps for (*a*) flange (*b*) D and D/2 and (c) 2.5 D taps are shown in figure 9.12.

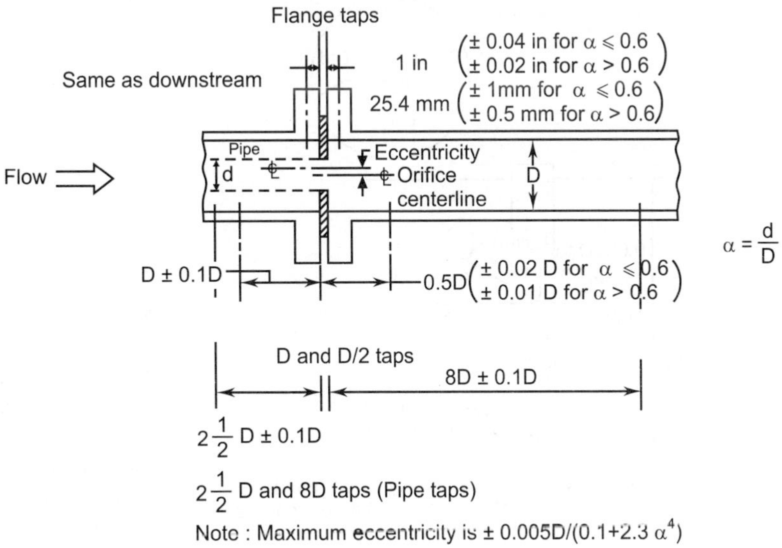

Fig. 9.12: Pressure-tap spacing

The designs for corner-tap for (*a*) individual and (*b*) annular slot types are shown in figure 9.13.

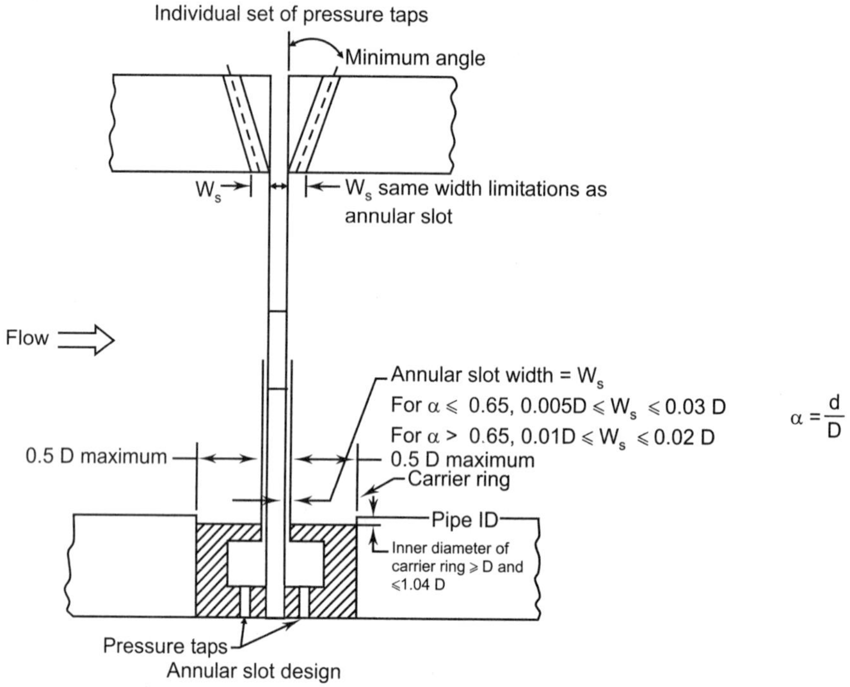

Fig. 9.13: Corner taps

The design requirement for a concentric square-edged orifice is shown in figure 9.14. Discharge coefficient and accuracy of installation are dependent on concentricity, edge sharpness, surface finish and plate bending. If reverse flow is accommodated, the plate should not be beveled and its thickness should be in between 0.005 D and 0.02 D.

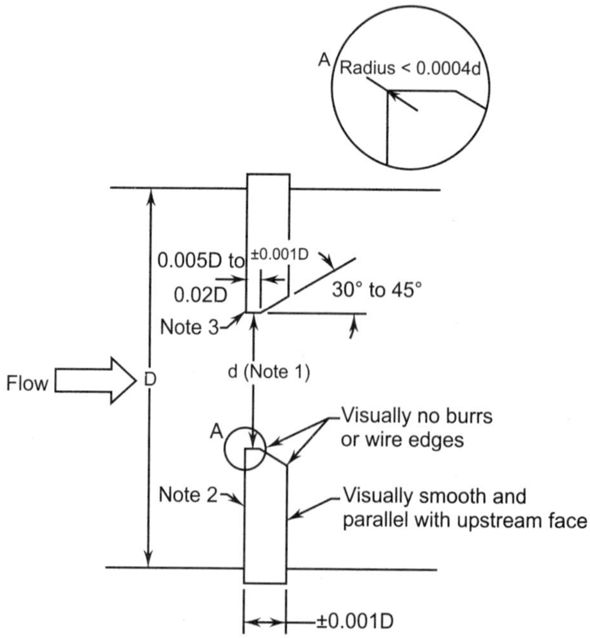

Notes. (1) Mean of four diameters, no diameter > 0.05% of mean diameter, (2) Maximum slope less than 1% from perpendicular; relative roughness < 10^{-4} *d* over a circle not less than 1.5 *d.* (3) Visually does not reflect a beam of light finish with a fine radial cut from center outward.

Fig. 9.14: Concentric square-edged orifice

9.7.2.2 Small Bore Honed Orifice

As the pipe diameter decreases the effects of plate eccentricity, inside pipe roughness, pressure-tap geometry, etc. become more glaring. For such cases, orifices appropriate to small-line-size pipes are installed in honed meter runs. The taps across the orifice are annularly grooved to ensure manufacturing uniformity. Such a situation is shown in figure 9.15.

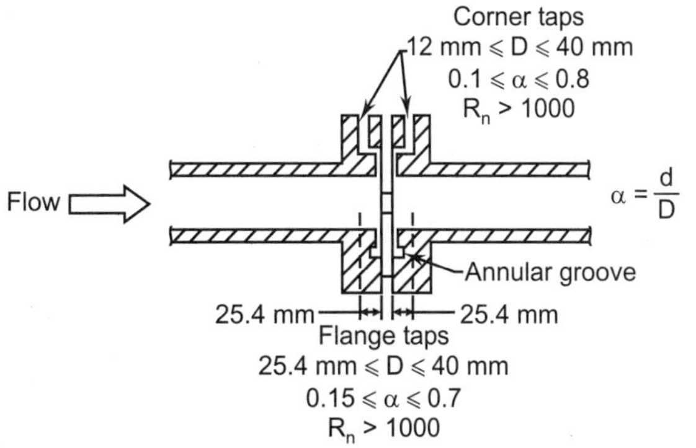

Note. Threaded or flanged tube ends, assembled and then bared and honed

Fig. 9.15: Honed-orifice meter run

9.7.2.3 Quadrant-edged and Conical-entrance Orifices

These orifices are used essentially for Reynolds numbers below 10,000. Their geometries ensure a constant discharge coefficient over the laminar flow regime.

9.7.2.3.1 Quadrant-edged Orifice

The quadrant edged orifice, also called quarter-circle orifice, is shown in figure 9.16 and are manufactured from standard orifice plates. Deposition of solids on the orifice surface, corrosion etc. don't have a significant bearing on the performance of the orifice.

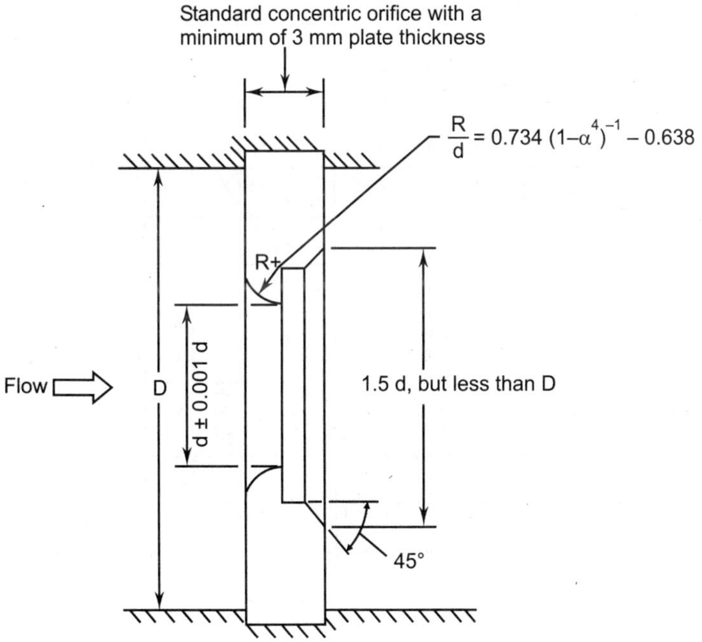

Fig. 9.16: Quadrant (quarter-circle) concentric orifice

9.7.2.3.2 Conical Entrance Orifice

This is shown in figure 9.17. Here a conical inlet replaces the quarter-circle of the quadrant-edged orifice.

It is difficult to manufacture. Figure 9.18 shows the discharge coefficient for quadrant and conical orifices. The figure suggests that the discharge coefficient is assigned two values depending on the Reynolds number range.

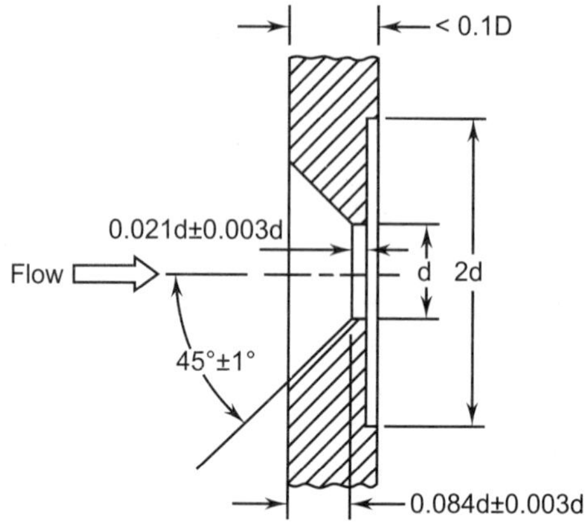

Fig. 9.17: Conical-entrance orifice

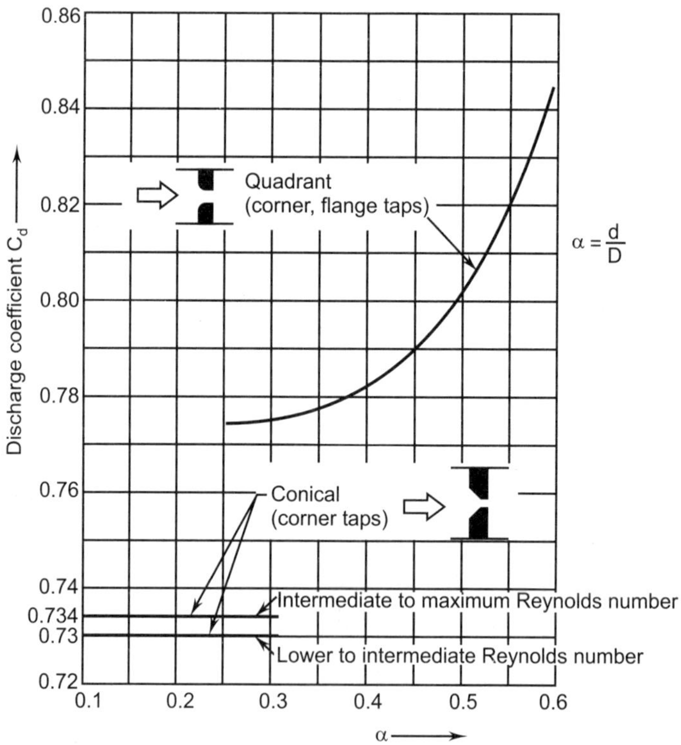

Fig. 9.18: Discharge coefficient for quadrant and conical orifices

9.7.2.4 Segmental Orifice

This is shown in figure 9.19 in which the opening is a segment of a circle with a diameter 98% of the pipe diameter. The circular part of the opening is concentric with the pipe and is accurately centered. This ensures that the opening is not covered by inlet pipe or gaskets or pipe eccentricity.

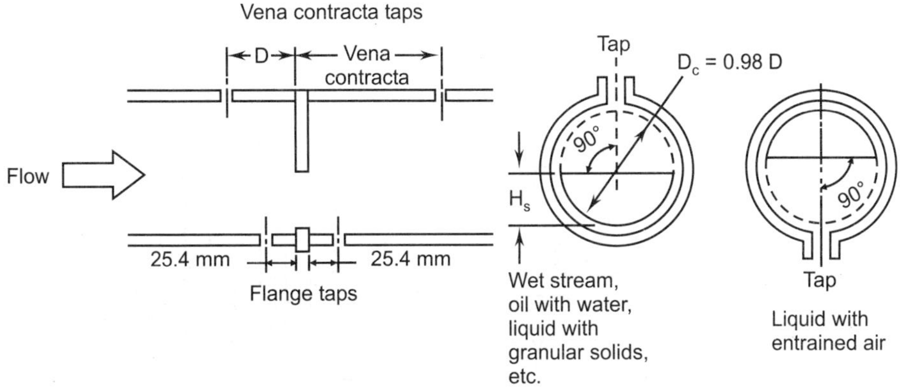

Fig. 9.19: Segmental orifice

The segmental orifice eliminates damming of foreign particles at the side of the pipe at which the opening is located. This property makes segmental orifice a suitable choice for a horizontal pipe when the fluid is a liquid containing granular solid, wet steam, oils having water in it etc. However, because of better discharge coefficient accuracy of a concentric orifice, it is preferred when the flow is vertically downwards.

A segmental orifice with its opening at the top of the pipe is used when the liquid contains gas or vapour. However, if the flow is upwards, a concentric orifice is preferred. For liquids containing sticky solids or solids having a density near to that of the flowing liquid, a segmental orifice is not recommended.

9.7.2.5 Eccentric Orifice

A two tap arrangement for an eccentric orifice is shown in figure 9.20. No part of flange or gasket should cover any part of the opening of the orifice.

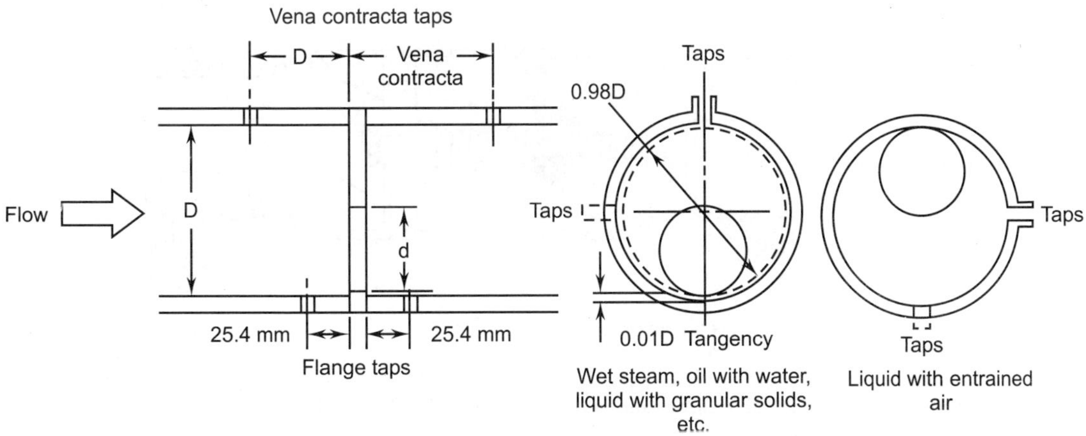

Fig. 9.20: Eccentric orifice

An eccentric orifice has the advantage of a segmental orifice, but does not allow free drainage. On other counts, an eccentric orifice is better than a segmental orifice.

Figure 9.21 shows the location of minimum pressure (vena contracta) for segmental, eccentric and concentric type of orifice plates.

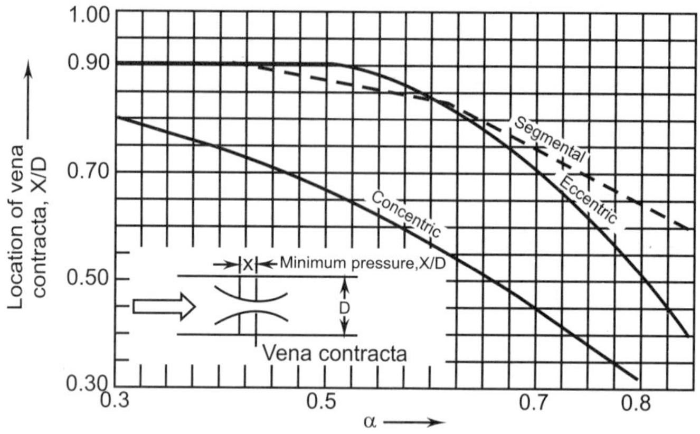

Fig. 9.21: Location of minimum pressure (vena contracta)

9.7.2.6 A Variable Orifice Meter

The Gilflo orifice meter type B is a variable orifice meter and is shown in figure 9.22. It was patented by Hayward.

Such an orifice meter comprises an annular orifice whose area varies as per the fluid flow rate. This is so adjusted that the differential pressure generated across the flow is proportional to the flow rate over a considerable part of the operating range. The rangeability of such an instrument can go up to 100 to 1 compared to a maximum of 5 to 1 for conventional square-edged orifice plates.

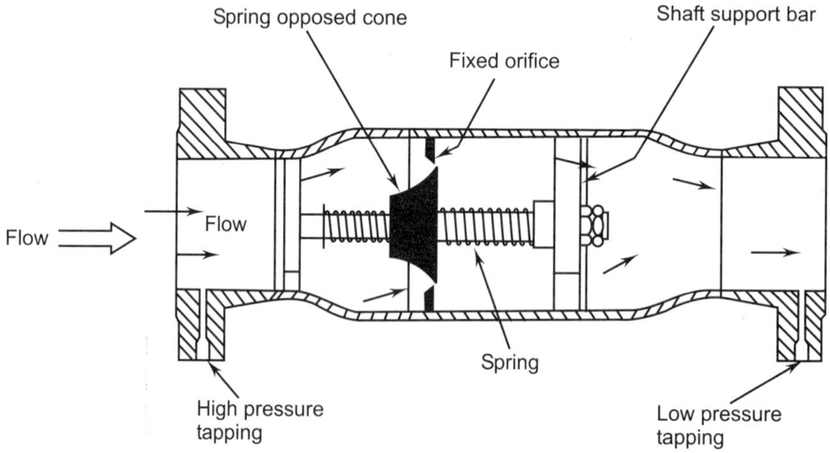

Fig. 9.22: Gilflo variable orifice meter, type B, (*Courtesy:* Gervase Instruments Ltd.)

A fixed orifice is attached to the body of the meter. A specially designed contoured cone, fixed on a shaft with spring attachment allows a variable flow space. With an increase in flow rate, the cone movement produces an increased annulus and results in a higher differential pressure.

The meter is applicable for a wide range of fluids in the temperature range from –200° to 300°C. Instrument hysteresis can be avoided by a combination of proper selection of materials of the orifice, over-range stops and a relatively large cone area.

9.7.2.6.1 Advantages

- Applicable to fluid and steam
- For standard designs, calibration not required
- First cost of plate lowest
- Robust primary elements
- No moving parts
- Reliable over a considerable period of time

9.7.2.6.2 Disadvantages

- High pressure loss
- Rangeability 3 – 4 : 1
- Output nonlinearly related to flow rate
- Performance changes with plate damage/coating with time
- Swirls/poor flow profiles affect measurements
- Requires regular checking/maintenance
- An orifice meter should be flow calibrated so that uncertainty in measurements can be limited (as prescribed in BS 1042)
- Installation requirement—stringent and critical
- Poor turn-down ratio

9.7.2.7 Different Curves for Orifice Plates

For values $\alpha(= d/D)$ remaining the same, it is seen that flow coefficient k [refer equation (4)] depends only on the Reynolds number *i.e.* $k = f(R_n)$. For lower values of Reynolds number, this dependence is more pronounced. As R_n increases, it affects k to a lesser degree. Thus, for a considerably high value of R_n (which is different for different restrictions), the flow coefficient may be assumed to be constant, as shown in figure 9.23.

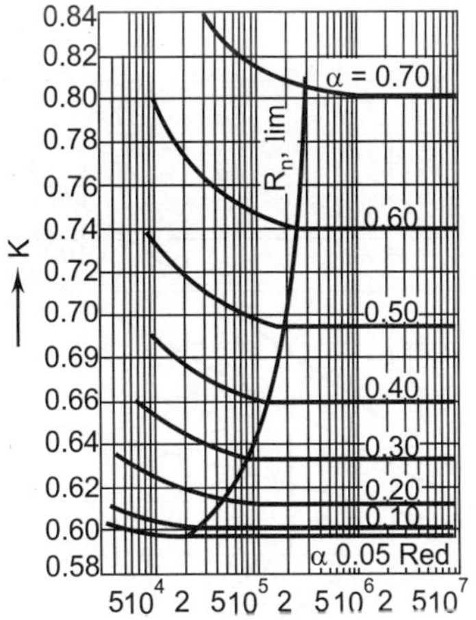

Fig. 9.23: Flow coefficient of standard orifice plates as a function of the Reynolds number at various fixed values of α

The value of Reynolds number, above which the flow coefficient k remains constant for different α values is called its limit and is symbolized by R_n lim. This value is dependent on the value of α, and the respective flow coefficient is termed as basic and designated as k_b. The curve of basic flow coefficient as a function of α in this case α_b is shown in figure 9.24.

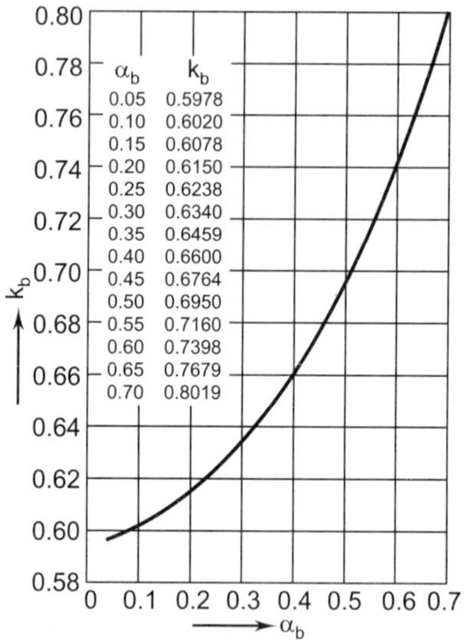

Fig. 9.24: Basic flow coefficient of standard orifice plates as a function of α_b

The value of flow coefficient discussed so far is a function of smoothness of the internal pipe surface as well as the upstream edge sharpness of the orifice. In figures 9.25 and 9.26 are shown two curves—the first one showing the pipeline *roughness correction factor* k_r with different diameters of orifice plates and the second one showing the *edge sharpness correction factor* k_e with the different diameters of the orifice plates.

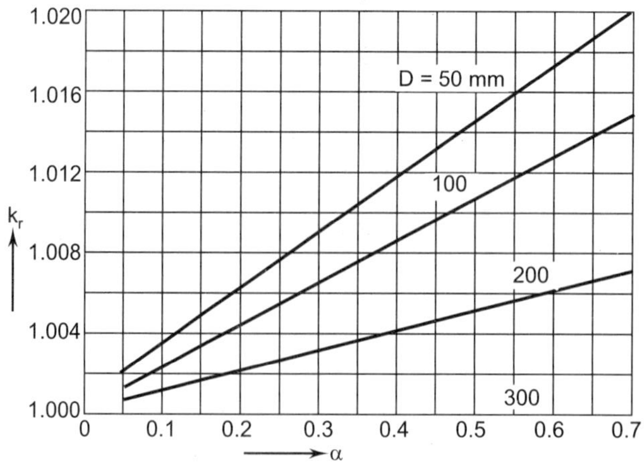

Fig. 9.25: Pipewall roughness correction factor for orifice plates

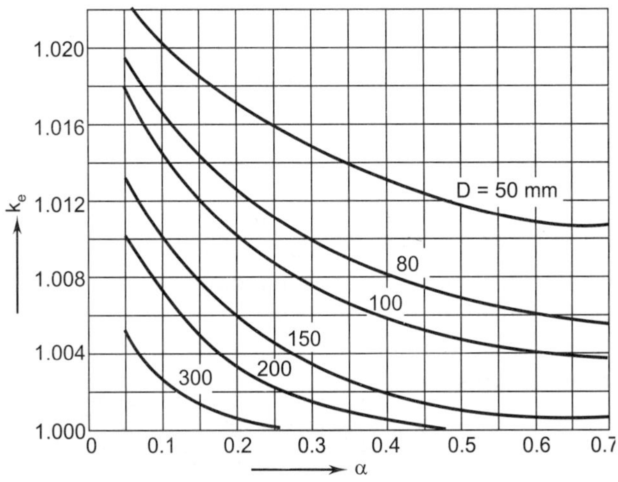

Fig. 9.26: Edge sharpness correction factor, k_e, for orifice plates

9.7.3 Venturi Tube

9.7.3.1 Introduction

A venturi tube is a head type flow measurement device and its operating principle is based on Bernoulli's theorem. Venturi tubes are available in two forms—short venturi and long venturi, also known as Herschel standard venturi designed by C. Herschel. A venturi meter is normally employed in water distribution system and its standardised version, as per BS 1042, is shown in figure 9.27.

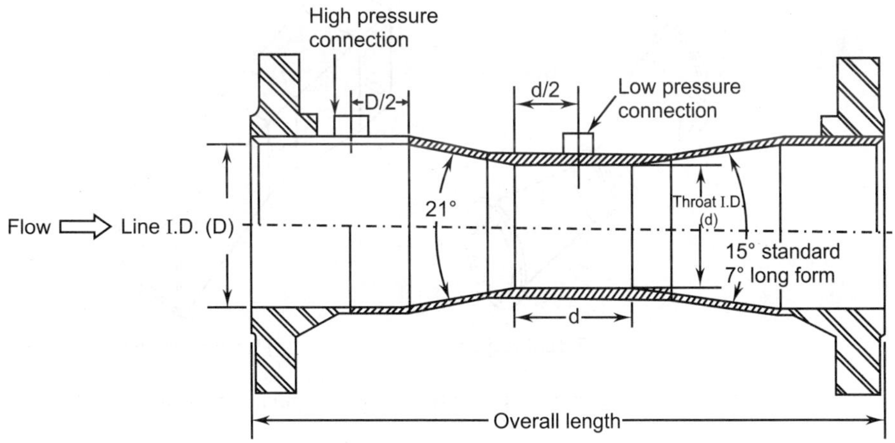

Fig. 9.27: Classical venturi tube

A venturi tube is usually made of cast iron or steel. The throat (*i.e.* portion having smallest area) is sometimes made from a separate assembly of bronze. This is done for easy replacement purposes. The entrance section to the venturi is 21° included angle and the exit section is 7° included angle for long venturi and 15° for short venturi. The throat diameter to tube diameter varies between 0.25 to 0.5. The pressure taps are located one-quarter to one-half pipe diameter upstream of the inlet cone and at the middle of the throat.

For a venturi tube, the equation for flow rate of a liquid is given by:

$$q = \frac{\pi}{4} \times \frac{C_d \alpha^2 D^2}{\sqrt{1-\alpha^2}} \sqrt{\frac{2g(\rho_m - \rho_f)}{\rho} \times h} \quad ...(9.22)$$

where,
q = flow rate
C_d = discharge coefficient
α = d/D = throat dia/pipe dia
ρ_m = weight density of manometer fluid
ρ_f = density of fluid over manometer fluid
ρ = density of flowing fluid
h = pressure differential

It can also be expressed in this form:

$$q = \frac{\pi}{4} \times C_d \times \frac{d^2}{\sqrt{1-\alpha^2}} \sqrt{\frac{2g\,\Delta p}{\rho}} \quad ...(9.22a)$$

where, Δp is the pressure differential between the pipe and the venturi tube.

9.7.3.2 Other Forms of Venturi

Two other forms of venturi are available. These are:

Eccentric type

Rectangular type

These are shown in figure 9.28.

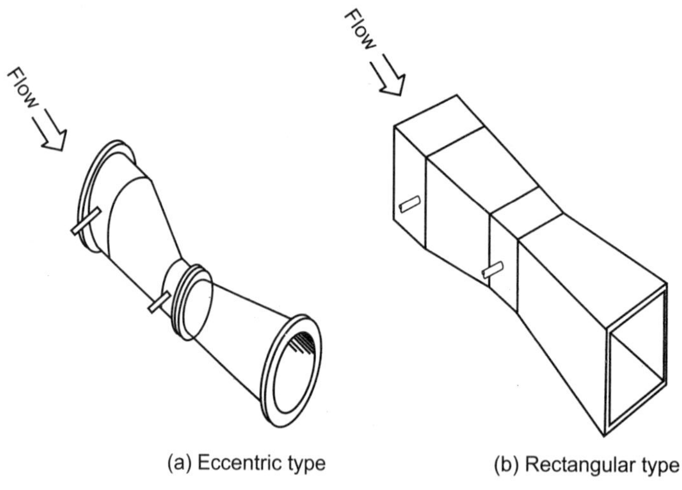

Fig. 9.28: (*a*) Eccentric type venturi (*b*) Rectangular type venturi

9.7.3.3 Advantages

- Can measure very high flow rates ~ 5 × 10^6 gallons/hr
- Very high efficiency and good pressure recovery
- Best choice for DP meters over 1000 mm bore
- Performance enhanced as ratio of bore diameter to internal pipe diameter increases
- Robust and resistant to internal erosion-long working life
- Unaffected by solid particles/gaseous bubbles and can handle dilute slurries

- Reproducibility is very high
- No moving parts
- Highest accuracy of all the pressure differential pressure elements, if properly installed and calibrated
- Permanent pressure loss (head loss) is less than an orifice plate of same capacity.

9.7.3.4 Disadvantages

- Occupies considerable length of pipe
- Very expensive to manufacture and install
- Large size poses difficulties in handling
- Pressure-velocity relationship is square root in nature
- Poor turn-down ratio.

9.7.4 Flow Nozzles

9.7.4.1 Types

It is a design in between the venturi tube and the orifice plate. Figure 9.29 show an venturi nozzle, an ISA nozzle and a flow nozzle.

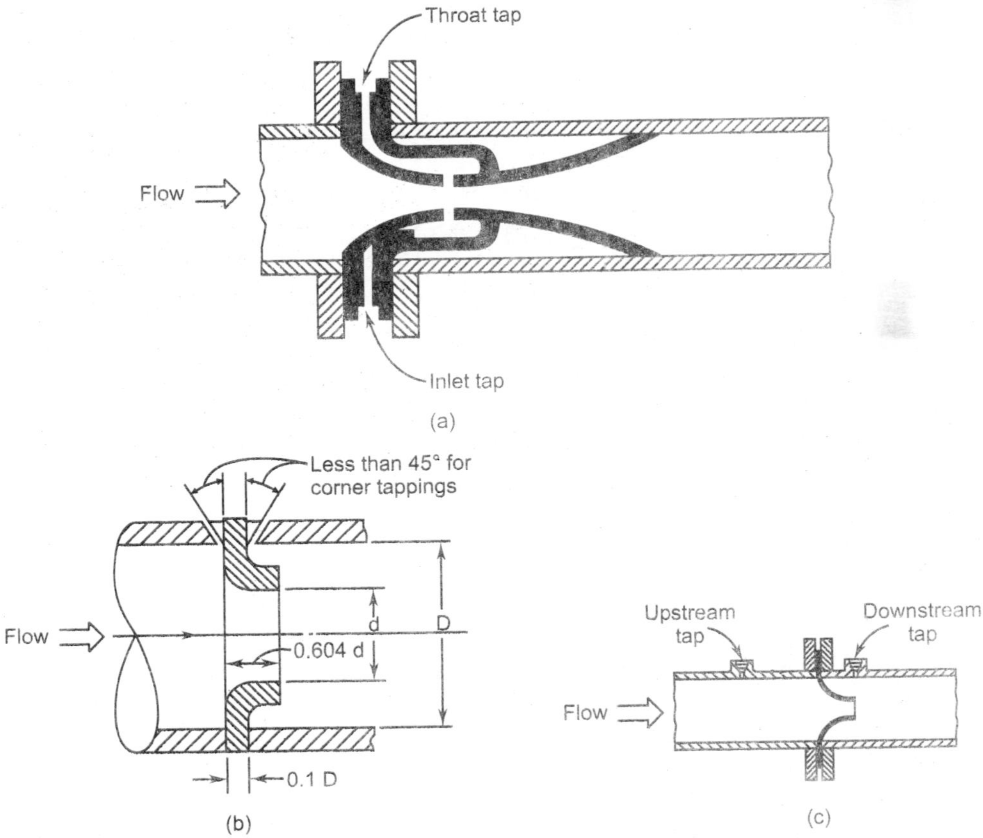

Fig. 9.29: Different nozzle types (*a*) Venturi nozzle (*b*) ISA nozzle (c) Flow nozzle

The entry and exit sections of a venturi nozzle are shortened compared to a venturi tube and can be fitted easily into pipe lines.

The ISA nozzle with corner tappings are generally employed for testing the discharge from compressors.

In a flow nozzle, the exit section (as in a venturi tube) is simply omitted-thereby; it resembles an orifice with a well-rounded upstream edge. The upstream tap is roughly about one pipe diameter from the entrance to the nozzle while the downstream tap is chosen directly above the straight portion of the nozzle.

The flow coefficients of flow nozzles are either supplied by the manufacturer or must be determined by test.

The flow nozzles find their applications to measure flow rate of high pressure superheated steam in thermal and nuclear power plants.

9.7.4.2 Advantages

- Less sensitive to corrosive effects and dirts
- Flow coefficient less affected as compared to orifice plates in presence of dirt
- More accurate to measure flowrates of steam/gases compared to liquids
- Sweeps solids present in flow through the throat because of its streamlined contour
- Can meter fluids with high viscosities; can be applied for Reynolds number as low as 50
- Nozzles with throat diameter of 0.6 mm can meter flowrates as low as 0.002 m^3/hr

9.7.4.3 Disadvantages

- Initial cost of installation many times higher than orifice plate
- Cost of maintaining a flow nozzle is more because it is necessary to remove a section of pipe to inspect it
- More difficult to replace than an orifice plate. This is due to its width.

9.7.5 The Dall Tube

9.7.5.1 Description

The Dall tube is shown in figure 9.30 and was designed by H.E. Dall of George Kent Company.

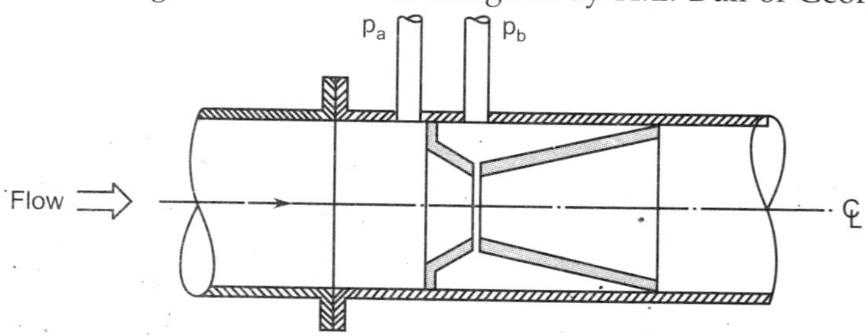

Fig. 9.30: Dall tube

It produces a larger differential pressure with low pressure loss when compared to a long type venturi tube. Again, the Dall tube is smaller and lighter than an equivalent short type venturi tube. Thus, a Dall tube is better than any venturi design, although it can be used only for fluids containing no solid particles which may clog the small opening.

Constructionally, it consists of a converging upstream cone followed by a diverging downstream cone with a circumferential slot (gap) in between. The two pressure taps are shown in the figure. An abrupt change in the flow contour results in a larger differential pressure. The discharge coefficient of such a tube is in the region of 0.6–0.75.

Dall orifice or insert is a shortened version of Dall tube and possesses all the features of Dall tube. Lengthwise, it is only 0.3 pipe diameter.

9.7.5.2 Advantages

- Short length
- Low head loss
- Has no upper line size limit
- Can be made from various materials
- Can be used for fluids with no solid particles
- Large differential pressure produced, compared to venturi tube
- Smaller than a venturi.

9.7.5.3 Disadvantages

- Pressure differential is sensitive to upstream disturbances
- Very costly due to machining problems
- Not suitable for measuring flowrate of hot feed water.

9.7.6 Pitot Tube

9.7.6.1 Introduction

A Pitot tube is an obstruction type primary element used to measure (*i*) local or point velocities (*ii*) pressure in a stream (*iii*) velocities in boundary layers (*iv*) flow rates of fluids.

A Pitot tube operates on measuring the pressure difference between the impact (total) pressure and static pressure. The difference between the two pressures is called the kinetic pressure and related to the velocity of flowing fluid by Bernoulli's theorem as

$$p_t - p_s = \frac{\rho_f V_f^2}{2} \quad \Rightarrow \quad V_f = \sqrt{\frac{2}{\rho_f}(p_t - p_s)}$$

where, p_t and p_s are impact (or total) pressure and static (or hydrostatic) pressure respectively. V_f and ρ_f are the velocity and density of the flowing fluid.

For compressible fluids, assuming the flow to be adiabatic, the equation is:

$$V_f = \sqrt{\left[\frac{2g\gamma}{\gamma - 1} p_s v_f \left\{\left(\frac{p_t}{p_s}\right)^{\gamma-1/\gamma} - 1\right\}\right]}$$

Expanding $(p_t/p_s)^{\gamma-1/\gamma}$ and neglecting higher order terms for $[1 - (p_t/p_s)]$ and employing a manometer to measure the pressure difference, it can be written as:

$$V_f = C_v \sqrt{2gh(\rho_m - \rho_f) v_f}$$

where, h = manometric head and $v_f = RT/p$

A Pitot tube is a very simple device, easy to install and can measure velocity in both closed pipes and open channels. It is not suitable for fluids containing solid suspensions and has a low pressure loss.

Consider a spherical object placed in a fluid stream. As fluid approaches the solid, its velocity decreases and becomes zero when it impinges on the object. This is called stagnation point, as shown in figure 9.31.

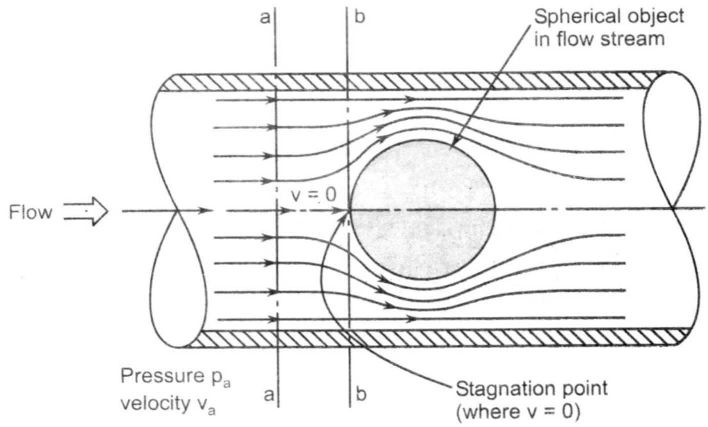

Fig. 9.31: Pitot tube—stagnation point

Figure 9.32 shows a Pitot tube with the spherical object being replaced by a tube with a small opening facing the flow. At the stagnation point, velocity head is converted to pressure head. If the two pressures, as shown in the figure, are connected to a sensitive differential pressure gauge, it can be calibrated in terms of velocity of the stream. The tube, used to measure static pressure, is also called piezometer tube.

In closed pipes and open channels, it is sometimes desirable to measure flow rate of liquids at a point to determine mean velocity or flow profile. Apart from Pitot tube, some others are used to measure the above, e.g. Doppler anemometer, hot-wire anemometer, Doppler velocity probe, insertion vortex, insertion turbine and propeller-type current meter.

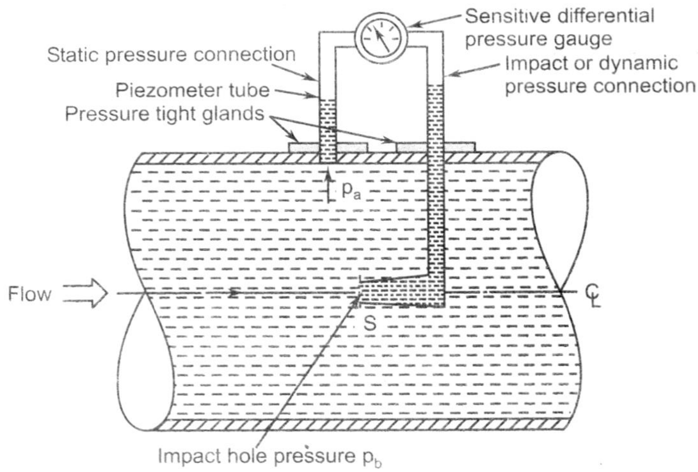

Fig. 9.32: A Pitot tube

9.7.6.2 *Short Single Opening Type*

A Pitot tube of short single opening type is shown in figure 9.33. The impact opening is about 3.175 to 6.35 mm in diameter with the opening facing the incoming fluid. Differential pressure is obtained from static pressure tap and impact pressure taps. This is calibrated in terms of the velocity of stream.

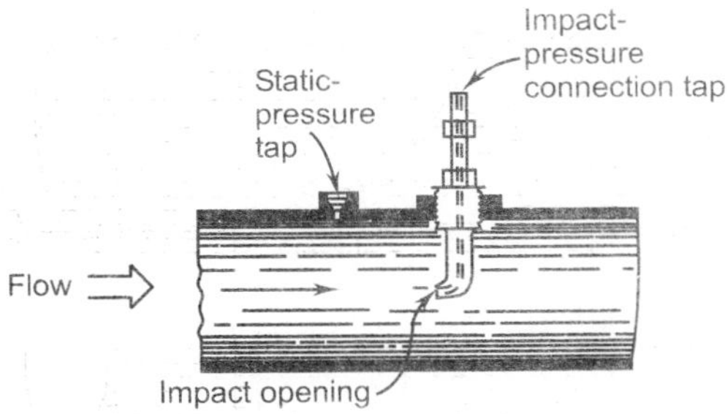

Fig. 9.33: Pitot tube: short single-opening type

9.7.6.3 A Combined Pitot Tube

Figure 9.34 shows a combined Pitot tube. The impact pressure connection and static pressure connection are housed in the Pitot tube assembly. One disadvantage with this type is the formation of eddies around the assembly which makes the static pressure measurement a bit erroneous.

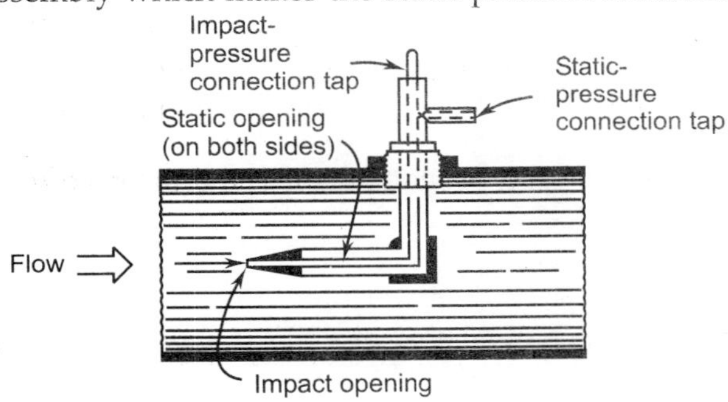

Fig. 9.34: A Combined Pitot tube

9.7.6.4 A Multiplying Pitot Tube

A further variant of a Pitot tube is the *"Multiplying Pitot tube"* shown in figure 9.35. As usual, the 'impact' opening faces upstream, while the 'static' opening faces downstream. For a given flow, the pressure differential (*i.e.*, pressure difference between impact pressure and static pressure) is more in this case because of the 'suction'. Head is added with static head leading to a greater static head. The accuracy of measurement is thus increased.

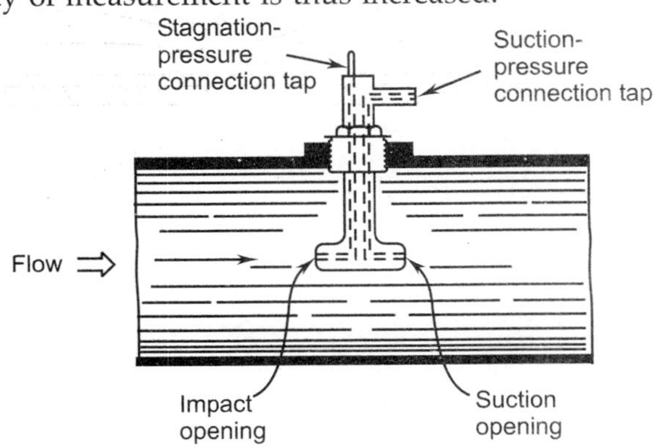

Fig. 9.35: A Multiplying Pitot tube

9.7.6.5 *Piezometer Ring*

Measurement of static pressure requires a high degree of accuracy. The shape of the pressure tap, such as a rounded corner or a machining burr causes significant errors in static pressure measurements. To overcome such an error, a piezometer ring is used. Multiple holes around the periphery enhances the accuracy of such a static pressure measurement. Such a piezometer ring is shown in figure 9.36.

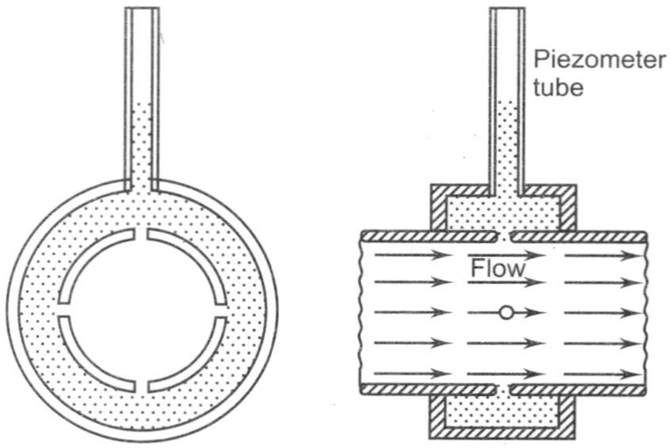

Fig. 9.36: A piezometer ring for static pressure measurements

9.7.6.6 *Advantages*

- Tube installation is rather simple
- Readily adaptable in very large diameter pipes/ducts
- Can be employed in open-channel streams
- Produces no appreciable pressure loss
- Can be used to plot a velocity distribution along a cross-section of the tube/duct
- A low-cost device
- For noticeable differential pressure to be obtained, fluid velocity should be appreciable
- For a maximum misalignment up to 5° between the axis of the head of the tube and the direction of incoming flow, the velocity values obtained are within acceptable limits. For a misalignment of around 20°, the error in velocity determination is around 2%. This misalignment is known as '*angle of attack*' or '*yaw*'.

9.7.6.7 *Disadvantages*

- Is used where very high accuracy is not required
- For high velocity fluids, tubes must have high stiffness and strength
- For high velocity fluids, tubes may get broken because of vibrations
- Applicable for clean fluids
- Sensitive to swirl and distorted profiles
- Axial alignment critical.

9.7.6.8 *Annubar–An Averaging Pitot Tube*

It is employed to obtain an average value of flow across a cross-section of the pipe by measuring the average value of several flow measurements across the cross-section.

A Pitot static tube measures a single-point flow-rate which can be calibrated in terms of volume flow rate. But such volume flow rate measurements become erroneous if the existing velocity profile differs from the calibrated conditions.

To overcome this difficulty, several spatially dispersed sensing ports are housed in a fixed probe as shown in figure 9.37. The individual port readings are averaged to produce a single 'impact' pressure and a single 'static' pressure. The difference between these two pressures Δp is related to the volume flow rate by the square root relationship.

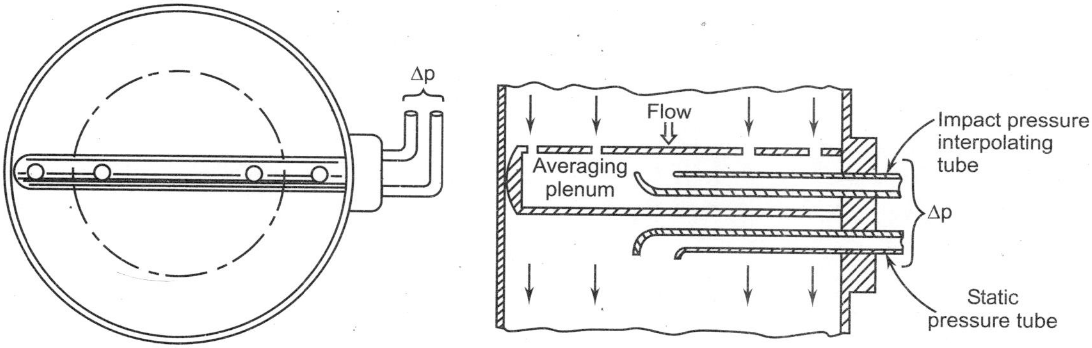

Fig. 9.37: Averaging Pitot tube

9.7.6.8.1 Advantages

- No pressure loss
- Ready adaptability for measurement of flow in large pipes/duct

9.7.6.8.2 Disadvantages

- Sensitive to upstream disturbances
- Accuracy is poor
- Not suitable for sticky fluids or fluids containing solid particles.

9.7.6.9 Point Velocity Measurement by Pitot Tube

Two Pitot tubes are employed at the same depth around a cross-section in the fluid stream for point velocity measurement. The scheme is shown in figure 9.38. The tube facing the flow measures $p + v^2/2g$, where p is the fluid pressure and $v^2/2g$ represents velocity head. The second tube, placed at the same depth as the first one, measures only p. This tube is placed normal to the flow. The difference of the two readings is the pressure differential that represents the fluid velocity at the depth at which the two measurements were carried out.

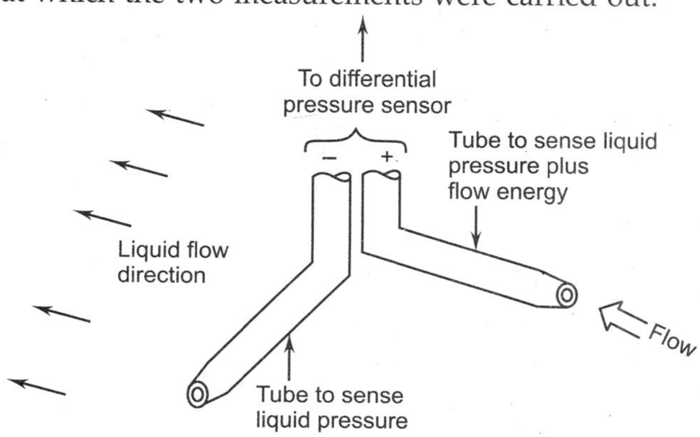

Fig. 9.38: Point velocity measurement by Pitot tube

Chapter 9

This method is a crude one and gives a rough estimate of point velocity. It is better suited at low flow velocities. It is mostly employed in the laboratory. Several pairs of tubes may be employed at different depths of a cross-section to get an idea of average velocity at that section.

9.7.7 Pressure Loss Curves for Orifice, Venturi and Nozzle

The primary elements, viz., orifice plate, venturi tube, nozzle, etc., develop the pressure differential which is subsequently measured by a secondary device. Calibration is then done to infer flow rates.

These primary elements, when placed in a flow line, cause an irrecoverable pressure loss of varying degree. Obviously, the lower this loss, the greater the advantage. The magnitude of this loss is a major factor in the selection process of a primary element. The pressure loss curves of the different primary elements is shown in figure 9.39.

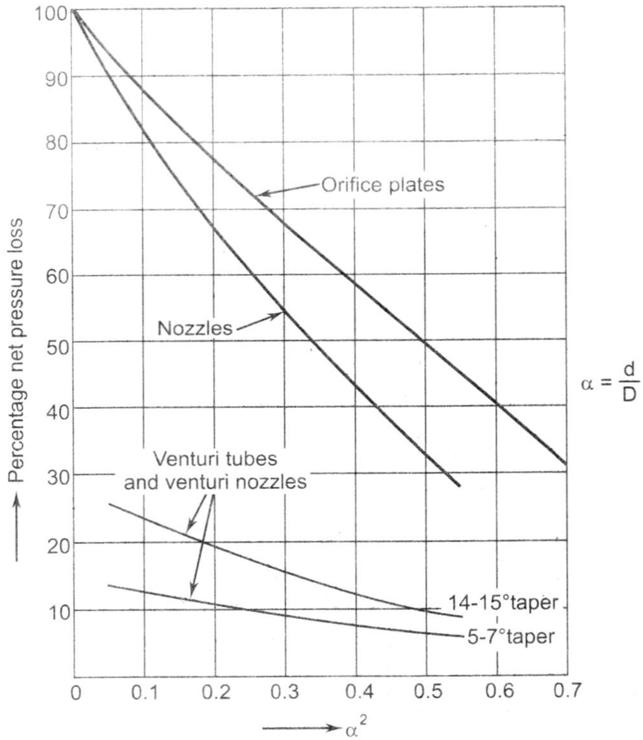

Fig. 9.39: Net pressure loss as a percentage of pressure difference. *Courtesy*: British Standards Institution

9.7.8 Installation of Head (Differential Pressure) Flowmeters—a Practical Approach

9.7.8.1 Introduction

Several differential pressure manometers, like diaphragm gauge, bellow gauge, U-tube manometer, bell gauge, ring-type gauge, enlarged leg manometers, are utilized for flow metering purposes. The accuracy with which the flow rate of a fluid can be measured utilising the pressure drop across a p.d. device not only depends on the type of p.d. device used but also on the following factors:

- The fluid must fill in the entire cross-sectional area—both upstream and downstream of the installed p.d. device.
- The fluid must be in a steady state.
- Steam should be super heated.
- The phase state of the fluid must not be altered on passing through the p.d. device—for example a liquid should not vapourise, steam shall remain in the superheated condition, etc.
- P.d. device must be so installed as to be convenient to the attending personnel.
- Provision shall be there for cleaning a p.d. device (by liquid or air purging) in case of deposits in the p.d. device.
- In case a fluid (gas/steam) produces a condensate or liberates dust or if a liquid forms fumes/sediments – these must not be allowed to accumulate near to the p.d. device.
- If shock/vibrations are present, the device should be installed away from such places-or else shock-mounts or shock-absorbers must be used.
- For corrosive fluids, active substances or petroleum products, the differential manometers must be separated from the flow line by employing sealing chambers.
- Leaks around valves, unions, glands must be prevented—because leaks influence instrument readings in a considerable way.
- Purge valves/vents must be fitted to the connecting lines for occasional removal of foreign matters/particles.
- The gauge should be installed at the same height as the tapping position—otherwise '*zero-elevation*' or '*zero-suppression*' error would occur. With the gauge placed higher than the tap position, zero elevation error occurs. Zero suppression error occurs in the other case—*i.e.* gauge placed at a lower position than the position of the tap.
- A sufficient straight length of pipe must be provided—both upstream and downstream of a restriction where a p.d. device is installed. Presence of valves, elbows, pipe fittings, etc. causes distortion in fluid flow. The straight run of the pipe should be such as to straighten any distortion due to the above. It should be borne in mind that flow disturbances upstream of a p.d. device are more troublesome than downstream.
- Any valve installed upstream of a p.d. device, particularly which remain open partially, cause much disturbance. Thus, it is advisable to install a p.d. device upstream of a valve.
- The pipe upstream and downstream of a p.d. device must have any change in cross-section and must not have rivets, welds, etc.
- When flanges are used for installation of a p.d. device, care must be taken for the joints to be made as tight as possible. The gaskets between the device and the flange must be installed, flush with the inside surface of the flow line.

9.7.8.2 *Layout for Connecting Lines for Flow Measurements: Some Considerations*

The most commonly used locations for p.d. devices relative to differential manometers and metering accessories as also, the layout for connecting lines for flow measurement purposes will be discussed here.

- The length of connecting lines is a maximum 15 m, although it can be extended to 50 m, if necessary.
- The substance entering the differential manometer should be at the room temperature.
- The length and diameter of the connecting lines greatly influence the dynamic error that a differential manometer introduces when used to measure a non-steady state flow.

- Connecting lines should be laid along the shortest route, vertically or having a gradient of at least 1 in 10. The curves/bends along a line must be smooth and must not be abrupt in nature. The joints must be hermetically sealed and the lines must be protected against heat/cold.
- The inside diameter of the connecting lines must be atleast 8 mm.
- The inside diameter of the tubes connecting piezometer rings or pressure tap holes of a p.d. device to equalising or sealing chambers should never be less than 12 mm.
- The diameters of the connecting lines should be of large size while measuring fluid flow at low pressures.
- When a single p.d. device output is to be connected to several differential manometers, separate lines should connect each manometer and terminate in separate holes of a piezometer ring.

Now will be shown of how to connect a p.d. device with a differential manometer by means of connecting lines for (*a*) liquid (*b*) steam (*c*) gas flow measurements.

9.7.8.2.1 Layout for Liquid Flow Measurement

Figure 9.40 shows the layout for connecting lines with a differential manometer installed (*a*) below (*b*) above a p.d. device for liquid flow measurement.

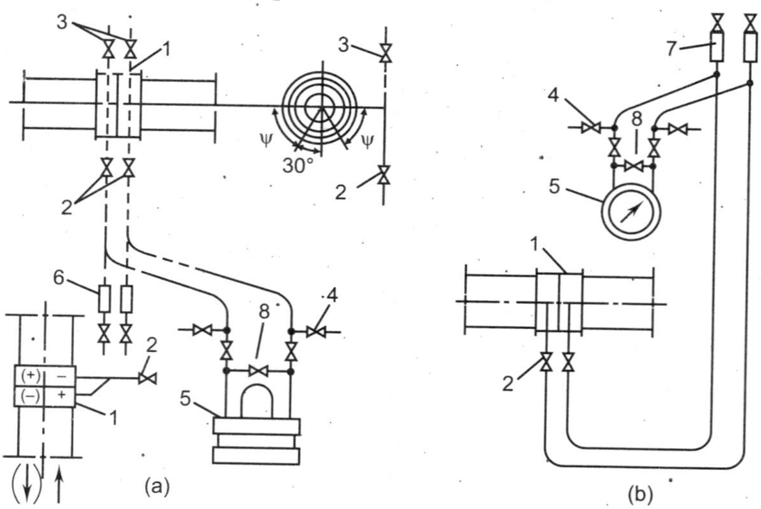

Fig. 9.40: Layout for connecting lines with a differential manometer installed (*a*) below and (*b*) above a p.d. device for liquid flow measurement:
1. p.d. device; 2. shut-off valve; 3. vent or purge valve (installed if necessary); 4. vent or purge valve on connecting line; 5. differential manometer; 6. settler and purge valve or drain (installed if necessary); 7. gas trap and vent; 8. equalising valve

With the manometer below the p.d. device, any gas that might be liberated by the liquid cannot accumulate either in the instrument (manometer) or the connecting lines. The slopes of the connecting lines must be same and in the same direction for the entire length of their run.

When pipelines are either horizontal or sloping in nature, the connecting lines must be connected to the lower half of a p.d. device. The angle being ψ = 30 to 90° from the vertical. The connecting lines must not be connected to the upper half or vertically to the lower half lest a gas enters the connecting lines in the former case and sediments in the latter.

A gas trap and purge valve (vent) is provided when the manometer is to be installed above the p.d. device. In case the connecting lines here have elevated portion or if the connecting lines cannot be laid out with the same gradient or if they are to bypass a wall, the same procedure of attaching gas traps (or vents) must be adhered to. The amount of gas entering the connecting lines can be reduced by having a U-bend extending for at least 0.7 m below the pipeline.

9.7.8.2.2 Layout for Steam Flow Measurement

The scheme for layout of connecting lines for steam flow measurement with a differential manometer installed either below or above or p.d. device is shown in figure. 9.41.

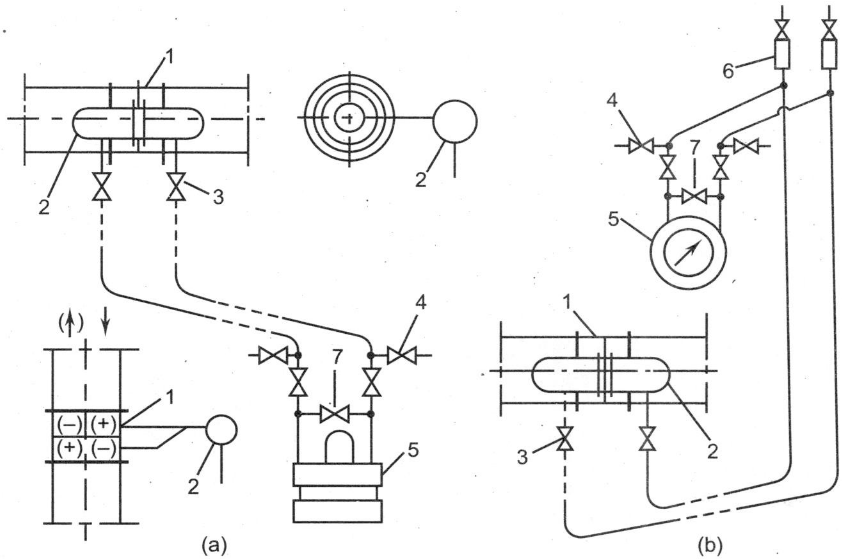

Fig. 9.41: Layout of connection lines for steam flow measurement with a differential manometer installed (*a*) below and (*b*) above the p.d. device: 1. p.d. device, 2. condensing chamber, 3. shut-off valve; 4. purge valve or vent; 5. differential manometer, 6. gas trap and vent; 7. equalising valve

For measurement of flow rate of steam, the condensing chambers in the connecting lines are placed very closely to the p.d. device. This arrangement ensures that the condensate in both the connecting lines are kept at the same and constant upper level.

When steam flow rate changes very fast, the levels of the condensate in the two condensing chambers may differ, giving rise to *complementary error*. This is reduced by making the condensing chambers liberally proportioned.

When the differential manometer is below the p.d. device, entrapped air, if any, can be vented by vent valve.

Again, when the differential manometer is to be placed above the p.d. device, gas traps or vents are provided at the highest points in the connecting lines.

9.7.8.2.3 Layout for Gas Flow Measurement

Figure 9.42 shows the layout diagram of the connecting lines with a differential manometer for gas flow measurements.

When the manometer is placed above the p.d. device like the one shown in figure 9.42(*a*), any condensate forming in the connecting lines will go back in the pipeline. For horizontal and sloping pipelines, the connecting lines must be attached to the bottom half of the pipeline. This would prevent any condensate accumulation in the connecting lines.

If a situation warrants attaching elbows in the connecting lines they should have condensate pots and drains at their lowest points.

When the differential manometer is to be installed below the p.d. device, a vent or purge valve is installed at the highest point while the settler and drain are placed at the lowest point as shown in figure 9.42(*b*).

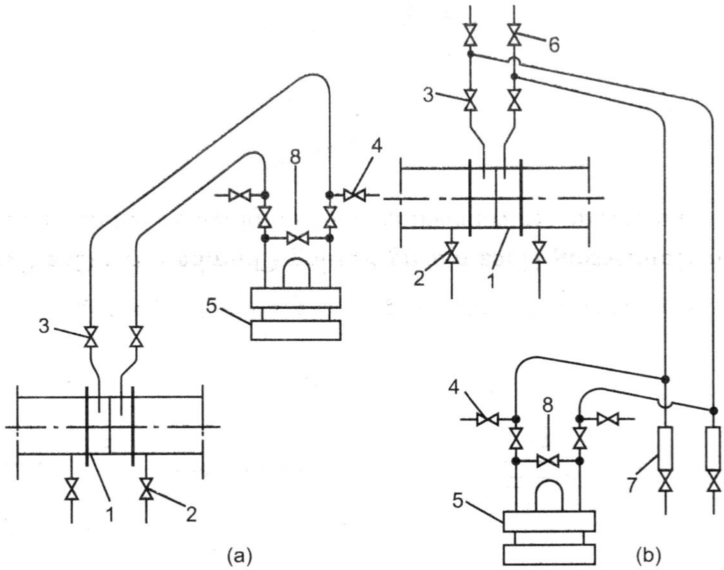

Fig. 9.42: Layout of connection lines on a line carrying a gas, with a differential manometer installed (*a*) above and (*b*) below the p.d. device: 1. p.d. device, 2. purge valve or drain (installed if necessary), 3. shut-off valve, 4. purge valve, 5. differential manometer, 6. additional vent or purge valve, 7. settler and drain, 8. equalising valve

9.8 SEALING CHAMBERS

9.8.1 Introduction

Sealing chambers (or pots) are used to protect differential manometers from chemically aggressive mediums. It is also used for oils or tarry liquids, which are less viscous at higher process temperatures but become viscous in the relatively cool connecting lines. As such, meter response becomes sluggish in nature.

9.8.2 Layout for Connecting Lines having Sealing Chambers for Liquid Flow Measurements

Figure 9.43 gives the layout of connection lines employing sealing chambers for flow measurement of liquids with density lower/higher than that of the sealing liquid.

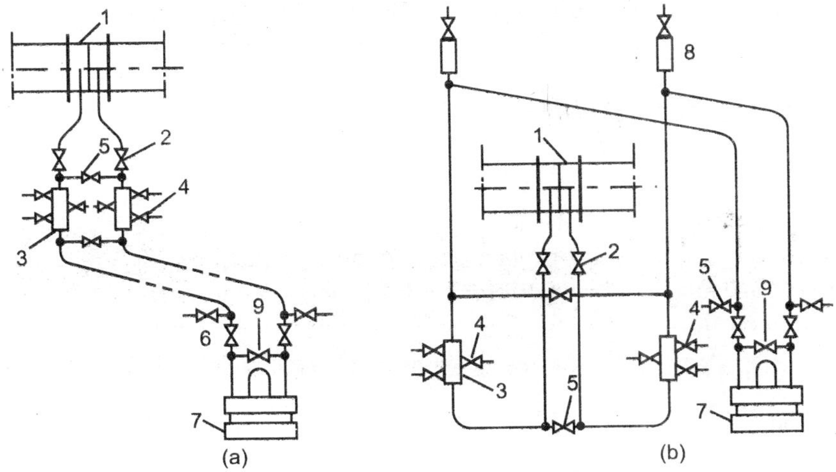

Fig. 9.43: Layout of connection lines containing sealing chambers for flow measurement of liquids with a density: (*a*) lower and (*b*) higher than that of the sealing liquid: 1. p.d. device, 2. shut-off valve, 3. sealing chamber, 4. pet cock, 5. auxiliary valve, 6. purge valve, 7. differential manometer, 8. gas trap and vent, 9. equalising valve

9.8.3 Layout for Connecting Lines having Sealing Chambers for Gas Flow Measurements

Figure 9.44 shows the layout for connecting lines employing sealing chambers for gas flow measurements.

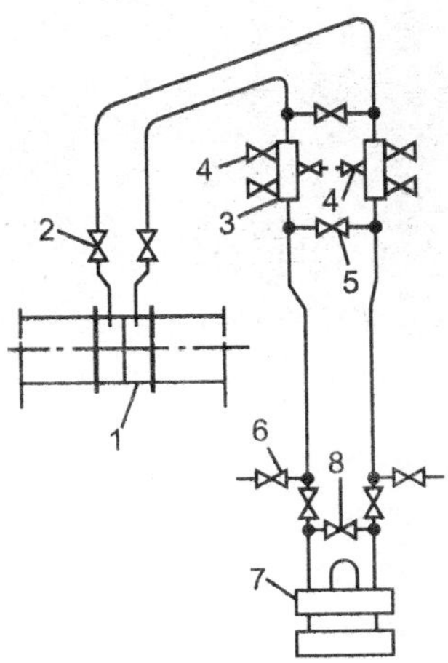

Fig. 9.44: Layout of connection lines containing sealing chambers for flow measurement of gas: 1. p.d. device, 2. shut-off valve, 3. sealing chamber, 4. pet cock, 5. auxiliary valve, 6. purge valve, 7. differential manometer, 8. equalising valve

When pressure differential across the p.d. device is zero, the levels in the two sealing chambers must assume the same height. The sealing liquid should not react chemically or mix with the process fluid and form deposits or attack material of which the sealing chamber is made, pressure leads, etc.

The commonly used sealing liquids are water, glycerine, a mixture of water and glycerine, light mineral oils, ethylene glycol, etc.

The pressure differential across the pressure difference device comes a change in level in the two sealing chambers. It thus introduces a complementary metering error, which must be taken care of.

9.9 INFERENTIAL FLOWMETERS

9.9.1 Introduction

Variable area flowmeters (commonly termed as rotameters), turbine flowmeters and target flowmeters fall under the purview of inferential flowmeters. In these flowmeters, the flow rate is inferred from a particular characteristics effect of a phenomenon associated with each of the above three mentioned flowmeters.

9.9.2 Variable Area Flowmeter

9.9.2.1 Introduction

A variable area flowmeter is commonly referred to as a rotameter or a flowrator. These names are, however, trade names for the most commonly used pattern. The term rotameter derives its name from the fact that initially the floats in a rotameter were made with slots which made them rotate with flow. The rotation was effected for the purpose of centering and stabilising the float against a variable flow rate.

9.9.2.2 Principle of Operation

The basic equation for flow through orifice, venturi or nozzle is given by:

$$Q = AC_d\sqrt{2gh}$$

where, Q = flow rate
A = orifice or throat area
C_d = discharge coefficient
and, h = differential head across the restriction

Thus, if h is kept constant, Q will linearly vary with A. This is the operating principle of a rotameter in which a float is made to move inside a tapered or conical glass tube, as shown in figure 9.45.

Thus, while in a head flowmeter, the flow rate is made proportional to the pressure differential across a fixed restriction, in the case of a area flowmeter (rotameter) the flow rate is made proportional to a variable area maintaining a fixed pressure differential across a restriction (float).

9.9.2.3 Description and Theory

A rotameter is based on drag principles in which the drag force on a body in a uniform flow is related to fluid velocity.

It employs a float kept in a fluid system whose velocity (flow rate) need to be measured. The float obstructs the flow and fluid passes through the annular space between the float and the inner wall of the glass tube. This is shown in figure 9.46. When the different forces acting on the float balance each other, the float is in equilibrium and assumes a particular position within the tube. With increasing/decreasing flow rate, the float takes up a higher/lower position in the tube. The position of the float can be calibrated on a scale attached to the outside of the tube.

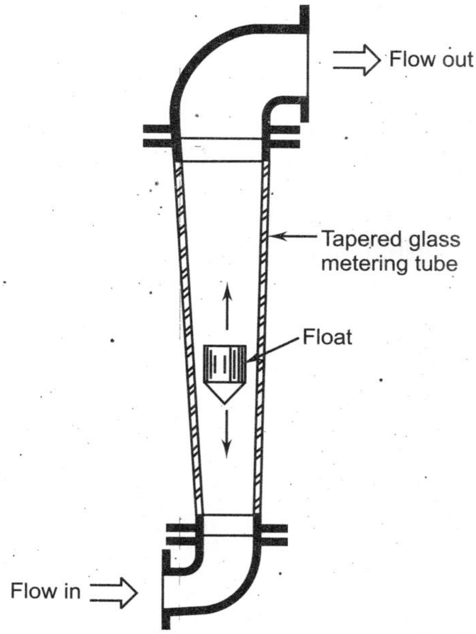

Fig. 9.45: Area meter (rotameter)

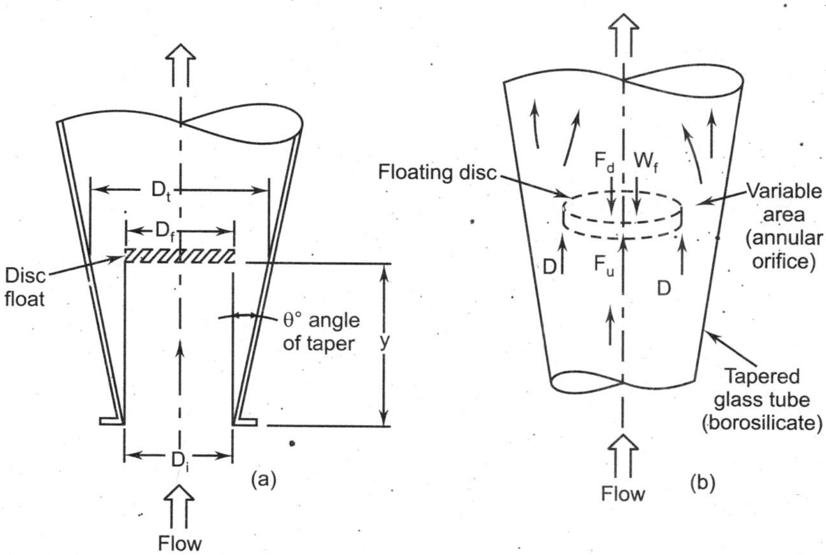

Fig. 9.46: Theory of rotameters

The weight of the float W_f, acting in the downward direction, is

$$W_f = V_f(\rho_f - \rho)g\downarrow$$

where, V_f = volume of the float

ρ_f = density of the float

ρ = density of the fluid

Now, the downward force F_d and upward force f_u, acting on the upper and lower force of the float respectively, are given by

$$F_d = p_d \cdot A_f \downarrow$$

and
$$F_u = p_u \cdot A_f \uparrow$$

where, p_d and p_u are the pressure acting on the top and bottom face of the float, having an effective area A_f.

A drag force D (viscous drag) also acts on the float. Its value is dependent on the shape of the float and the flow condition–turbulent or viscous. Its value is given by

$$D = C\mu VL\uparrow$$

where, C = a constant

μ = absolute viscosity of the fluid

V = velocity of fluid

and, L = a dimensional function (length)

Neglecting viscous drag, when the float is in equilibrium, it can be written as:

Net downward forces = Net upward forces

or,
$$W_f + F_d = F_u$$

or,
$$V_f(\rho_f - \rho)g + p_d A_f = p_u A_f$$

or,
$$p_u - p_d = \frac{gV_f}{A_f}(\rho_f - \rho) = \Delta p$$

Thus, the volume flow rate is

$$Q = AC_d\sqrt{\frac{2gV_f(\rho_f - \rho)}{A_f\rho}}$$

Now, from the figure,

$$D_t = D_i + 2y \tan\theta$$

where the symbols are as shown in figure (*a*). D_t is a variable diameter and its value depends on the position of the float. Its minimum value is D_i when $y = 0$.

If now D_f represents the diameter of the float, the variable annular area (this area depends on the position of the float, *i.e.* dependent on y)

$$\therefore \quad A = \frac{\pi}{4}\left(D_t^2 - D_f^2\right)$$

For simplicity sake, $D_f \simeq D_i$, although in reality D_i is slightly greater than D_f. Thus,

$$\therefore \quad A = \frac{\pi}{4}(D_t^2 - D_i^2)$$

$$\therefore \quad = \frac{\pi}{4}\left[(D_i + 2y\tan\theta)^2 - D_i^2\right]$$

$$\therefore \quad = \pi D_i y \tan\theta + \pi y^2 \tan^2\theta$$

This equation suggests that A would have a linear relationship with y if angle of taper, *i.e.*, θ is fairly small.

In that case $A = ky$, k = a constant.

$$\therefore \quad Q = kyC_d\sqrt{\frac{2gV_f(\rho_f - \rho)}{A_f\rho}}$$

or
$$Q = k'y$$

where k' is another constant.

Hence, the flow rate Q is linearly related to the float position y.

Thus, a vertical scale can be calibrated in terms of flow rate with the zero of the scale coinciding with the bottom part of the tube at which tapering begins. The position of the float is read off the scale and represents y.

The mass flow rate is

$$\omega = Q\rho = k'y\rho$$

The ratio ω/y is called *metering ratio* and is represented by R.

Density ρ of the fluid changes with temperature. This, thus, affects the flow rate. For the flow rate to be linearly related to y, Q must be independent of ρ. Hence, differentiating the expression for Q with respect to ρ and putting it to zero, it is seen that $\rho_f >> \rho$. But, in practice, the mass density of float (ρ_f) is made several times the fluid density ρ. When weight flow is desired, it can be shown that float density should be twice the fluid density for weight flow rate to become independent of fluid density.

To ensure that drag force does not affect float movement, its length is made small and Reynolds number is limited to within 2000.

The accuracy of a rotameter is typically ± 2% of full scale (± 1 per cent, if calibrated) and repeatability of around 0.25 per cent of reading.

The rotameter tube is normally made of high strength glass. If even greater strength is required, glass is replaced by metal. Float position in such a case is detected magnetically. If electrical or pneumatic output signal is required, float motion is measured with a suitable displacement transducer.

9.9.2.4 *A Rotameter for Horizontal Flow*

For horizontal flow, obviously the tube is horizontal in nature, along with the float inside. In such a case, the float is backed by a spring and is shown in figure 9.47. The displacement of the float and the force on the spring are related in a non-linear manner. Hence, flow rate is also non-linearly related to float displacement. The transducer circuitry and associated electronics can, however, take care of this non-linearity. In such a case, the signal would be proportional to velocity.

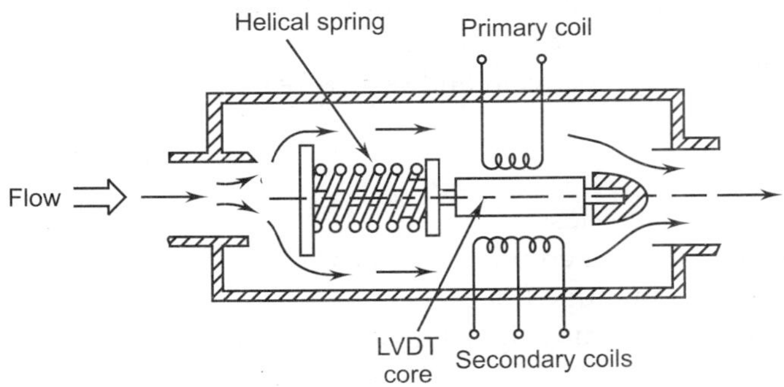

Fig. 9.47: A rotameter for horizontal flow

9.9.2.5 *Float Designs*

Industrial rotameter floats can assume a variety of shapes like ball float, viscosity immune float, viscosity non-immune floats, floats for low pressure losses. Some of the float designs, courtesy Fischer and Porter, are shown in figure 9.48. Shape, weight and material of a rotameter vary-depending on individual applications. Some very light weight floats are used for low pressure drops and applied for gas flow rate metering.

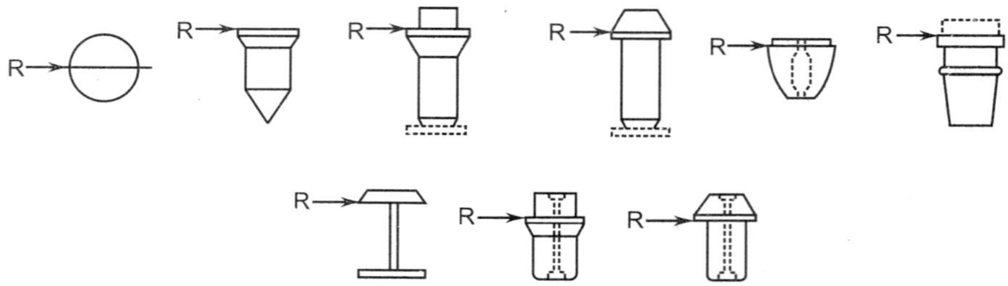

Fig. 9.48: Float designs (*courtesy* Fischer and Porter) R = reading edge of float. All floats are shown in the correct metering position

Rotameters are usually calibrated for a particular fluid. A correction factor is necessitated if the same meter is to be used for a fluid having either different viscosity or density. Some float designs incorporate a *'viscosity immunity ceiling'*, below which no viscosity correction is needed. For high viscous fluids like tar etc. cylinder-and plug type variable area meters are recommended.

A variable area flowmeter reads correctly under design conditions for a specific liquid. If, however, the design conditions vary or else if some other liquid is used, a correction factor is necessary. Manufacturers' data sheets must be referred to negate the effects of such deviations.

9.9.2.6 *Industrial Variable Area Flowmeters*

Such flowmeters vary from a simple ball floating in a glass tube to high pressure piston type meters operating up to a pressure of 4000 kPa. Industrial flowmeters can be direct viewing dial type or a standard digital output type.

Industrial piston type meters are made of cast iron, steel, etc. They are applied for high-pressure liquids or gases and their upper limit of pressure is adjustable using either springs or varying weights. These meters are very rugged and are used for tar, kerosine, gasoline and are less sensitive to vertical installation effects (at least within ± 5° of vertical). The fluids to be metered must be clean in nature or else a strainer may be necessary. In fact, periodic cleaning and an upstream strainer is recommended for dirty fluids.

When shutdown is not acceptable, a leak-type bypass manifold is suggested. Meter performance is least affected by upstream and downstream piping configurations.

9.9.2.7 *Advantages*

- Low cost
- Low and constant pressure loss
- Good accuracy at low and medium flow rates
- Availability of viscous-immune floats
- Applicable for fluids containing light slurries can measure volume flow rates as low as 0.1 cc/min
- Quite rugged
- Normal range of a rotameter is 10:1 and can be increased to 100:1 using sharp edged floats
- Can be applied to any kind of corrosive fluid

9.9.2.8 *Disadvantages*

- Care in constructing the tube

- Not suitable for pulsating flow
- Normally limited to small pipe diameter situations
- Mounting mostly vertical in nature

9.9.3 Turbine Flowmeter

9.9.3.1 Description

Such a flowmeter has a freely rotating turbine rotor, mounted axially with the centre line of the pipe and in the direction of flow of fluid. The basic turbine flowmeter with associated blocks is shown in figure 9.49.

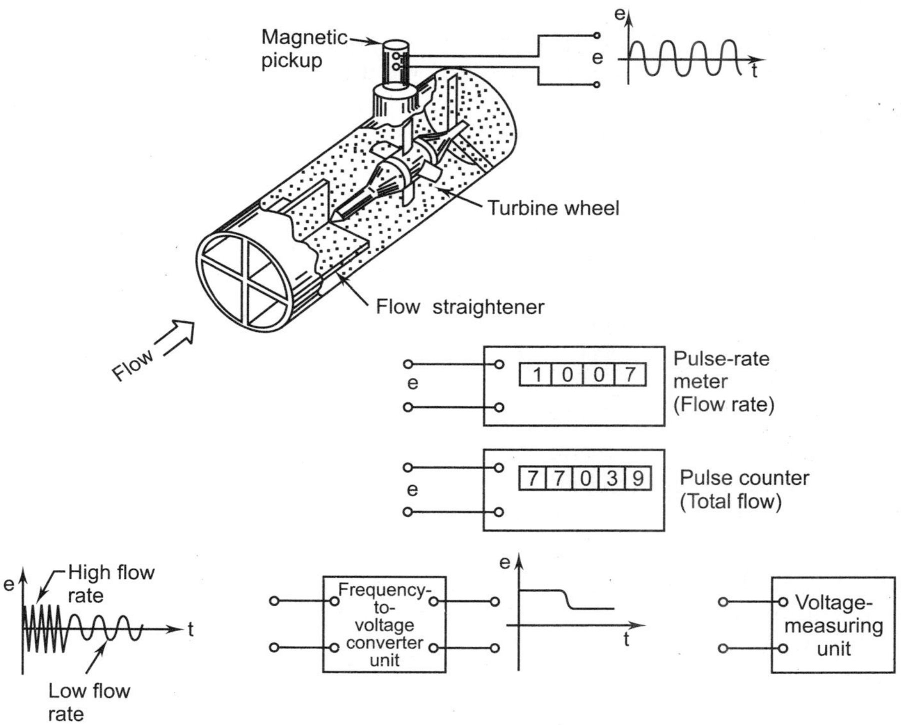

Fig. 9.49: A turbine flowmeter

When the fluid is in motion, it impinges on the rotor blades resulting in a rotating torque. The rotational speed of the rotor can be made to have a proportional relationship with fluid velocity over a specified flow range. Proper care should be taken to keep bearing friction and other mechanical losses to a minimum. The system gives a frequency output signal that varies linearly with volumetric flow rate–frequency obviously increasing with increasing flow rates.

A turbine meter consists of three parts–housing, rotor assembly and a magnetic pick-up coil. The metering technique employs the magnetic pick-up that produces voltage pulses as the rotor blades pass a given point of the pick-up. The rotor blades are normally made of magnetic steel to produce a pulse of sufficient strength. These pulses can be fed to a pulse-rate counter to measure flow rate. Integrating (summing the pulses) over a given time span gives the total flow. The output can be obtained in either analog or digital form.

The accuracy of turbine meters ranges from ± 0.25 to ± 0.5%, while the repeatability varies between ± 0.25% to ± 0.02%. Rangeabilities of turbine meters lie between 10 : 1 and 20 : 1. For low values of flow rates, it is less than 10:1, while for the better designed meters, this figure may attain a value of 100 : 1.

Pressure drop across a turbine meter varies as the square of the flow rate. Thus, a turbine meter is normally operated much below its rated capacity to have a check on the pressure drop and thereby saving pumping costs.

The size of turbine meters vary from 6 mm to 600 mm and liquid flow rates from 0.1 to more than 50,000 gpm.

9.9.3.2 Theory

For one dimensional frictionless flow past one dimensional blades, no retarding torque on the rotor will be felt. The rotational speed of the rotor is then related to the pipe velocity by the following expression:

$$f = \frac{N.v \tan \alpha}{2\,\pi r}$$

where,
N = number of blades
v = mean velocity
α = angle of attack
r = elemental radius

While the meter factor is defined as:

$$K = \frac{f}{v} = \frac{N.\tan \alpha}{2\,\pi r}$$

Rotor blades are of two types—helical and straight type. The above two expressions are true for helical rotors. For straight type blade rotors, the above equations are only true for one radial position. At this point, the retarding and driving forces balance each other. The above equations do not include the effects of viscosity, flow profile and bearing design etc. But they serve as a guide to predict meter constant to within 10 per cent of the value obtained in the linear region.

Two other theories have been proposed to overcome the limitations of the above. The first one is the momentum approach in which the driving torque is expressed as a function of the change of angular momentum of the fluid. The second is the air foil approach in which the forces exerted on an element of blade are integrated over the whole blade to get the driving torque. Hence, by increasing the number or size of the blades gives a corresponding increase in driving torque.

If the fluid density is ρ and the volume flow rate is q, then the force on the rotor in the rotational direction is $\rho q v \tan \alpha$ and the torque on the rotor is

$$T = k'\rho q(v \tan \alpha - \omega r)r$$

where, ω is the angular velocity and k' is a constant.

Thus, $$k'\rho q(v \tan \alpha - \omega r)r = T_b + T_\mu$$

where,
T_b = frictional counter torque on the bearing.
T_μ = counter torque due to fluid viscosity.

Hence, $$\omega/q = v \tan \alpha/(rq) - (T_b + T_\mu)/(k'r^2q^2)$$

Now, $$q = av, \text{ where } a = \text{pipe cross-sectional area}$$

$$\therefore \quad \frac{\omega}{q} = \tan \alpha/(ra) - (T_b + T_\mu)/(k'r^2q^2)$$

If torques due to bearing and viscosity are neglected,

Then, $$q = \omega/\{\tan\alpha/(ra)\}$$

The above equation does not hold true for low flow rates, since then $T_b \neq 0$ and $T_\mu \neq 0$.

9.9.3.3 Typical Calibration Curves

Typical turbine meter calibration curves are shown in figure 9.50. The characteristics are dependent on the nature of fluid as also its rate of flow. Thus, the meter factor, pressure loss, voltage magnitude and frequency of the output signal are all dependent on the above mentioned two factors. Curves have been shown for a low viscosity fluid. The meter factor curve is divided into non-linear and linear portions. If the meter is operated in the linear zone, an average meter factor can be assumed.

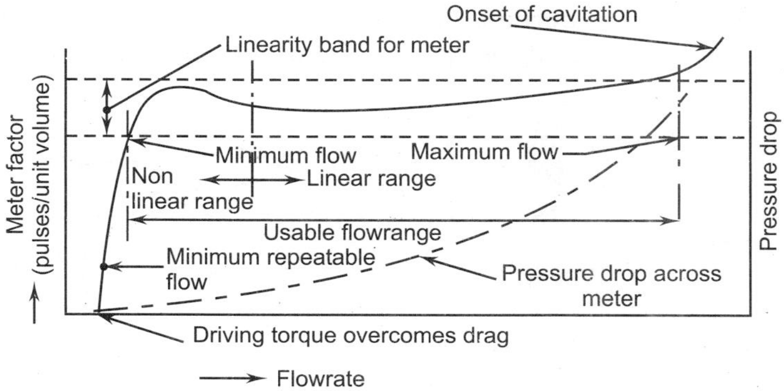

Fig. 9.50: Typical turbine meter calibration curves

With decreasing flow, the meter factor decreases. The non-linear portion is because of friction of bearings, magnetic drag and the velocity profile. Below a minimum flowrate, the turbine meter does not respond. When the flow rate is very high, if there is sufficient pressure to overcome cavitation, twice the recommended maximum flow rate is allowed for a short period, but in such a case the meter accuracy as also the life of the bearings would be impaired.

The response of a turbine meter to a positive and negative step change is shown in figure 9.51. This is usually expressed as the time for the meter output to reach 63 per cent of the final value for a step change in flowrate. A typical value for this time is 5–15 ms which varies with blade design, meter size or mass of the rotor.

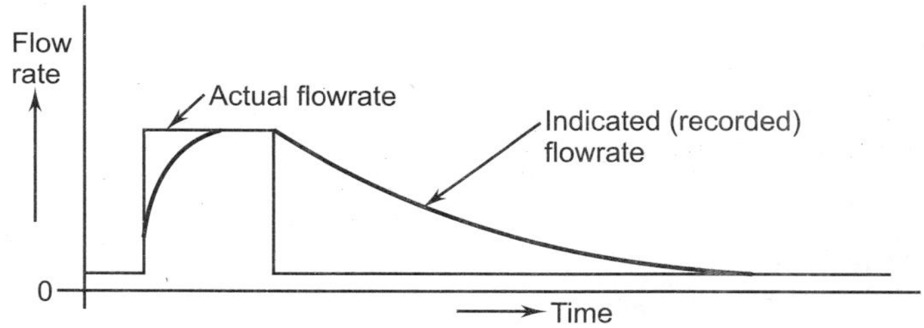

Fig. 9.51: The response of a turbine flowmeter to a positive and a negative step in flowrate

The nature of the curve suggests that the time taken to adjust to a positive step in flowrate is smaller than the time taken to adjust (to reach the final value) to a negative step change. It will lead to an average value when the meter is subjected to pulsating flow. In such a case, the meter indication is more than the actual one. This error would depend on both the frequency of pulsation and also on the waveform of pulsations.

For zero flowrate, the time constant is infinite. In such a case, if there is a negative step change, the rotor would spin indefinitely, in absence of friction.

9.9.3.4 *Effect of Viscous Fluids*

Calibration of a turbine meter changes when the meter is used for a fluid that is different from the one for which it was originally calibrated. This is where the viscosity effect comes in, which contributes to the change in calibration.

Figure 9.52 shows curves for different viscous fluids relating flowrate and meter factor. It is simply apparent that the linear range of the meter is more sensitive to viscosity than any other single factor. The transmission from non-linear to linear part occurs at progressively higher flowrates as viscosity increases. Experiments have confirmed that fluids having viscosity above 100 cSt do not possess any linear portion. Although linearity is affected by change in viscosity, repeatability is not. Viscosity effect affects smaller size meters more than the larger ones.

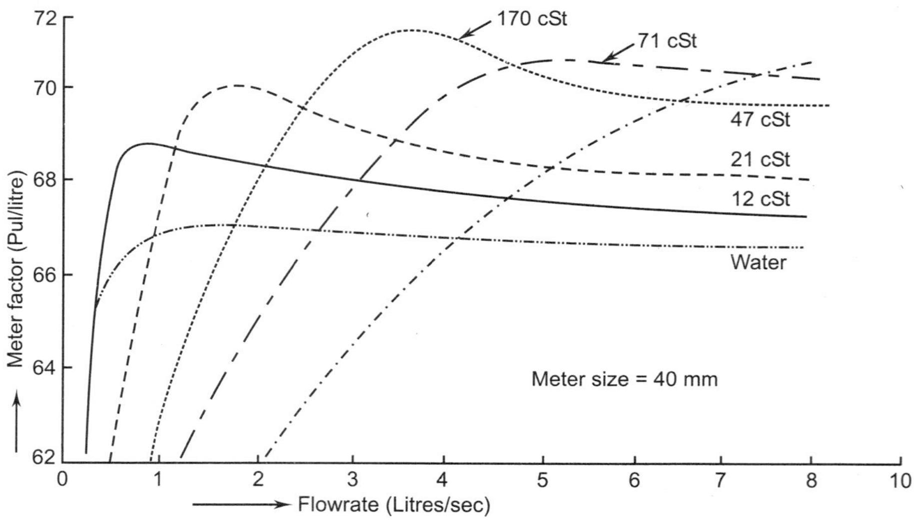

Fig. 9.52: Effect of viscosity on a constant angle turbine meter

Dimensional analysis carried out by Hochreiter has shown the following relationship between meter factor and viscosity as follows:

$$K = \frac{f}{Q} = \phi\left(\frac{f}{v}\right)$$

where,

f = frequency
Q = volume flow rate
ϕ = function
v = kinematic viscosity

Effect of viscosity at constant volumetric flow rate was studied extensively by Grey. The profiles for low, medium and highly viscous fluids are shown in figure 9.53. From the figure, it is evident that boundary layers become thicker as viscosity rises, leading to higher meter factor.

Lee and Karlby studied extensively the effect of viscosity, an axial turbine meter, and designed a viscosity compensator.

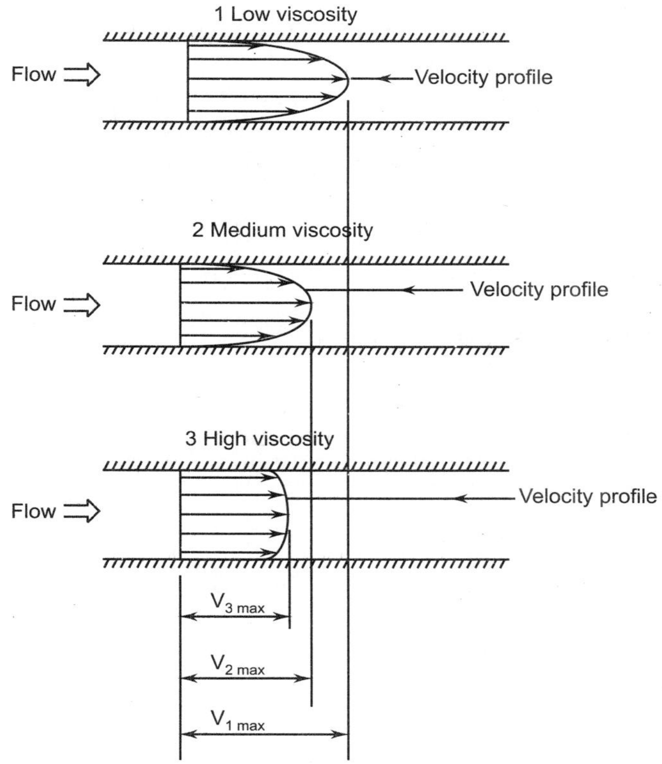

Fig. 9.53: Effect of viscosity at constant volumetric flowrate

9.9.3.5 Installation

Poor installation and variations in piping arrangements upstream of a turbine meter has a large bearing on the accuracy in flow measurements. Non-adherence to standard and recommended fitting arrangements, such as upstream valves, T-junctions, etc, affect accuracy. Figure 9.54 shows a typical turbine meter installation indicating the minimum upstream and downstream lengths within which no fittings can be installed.

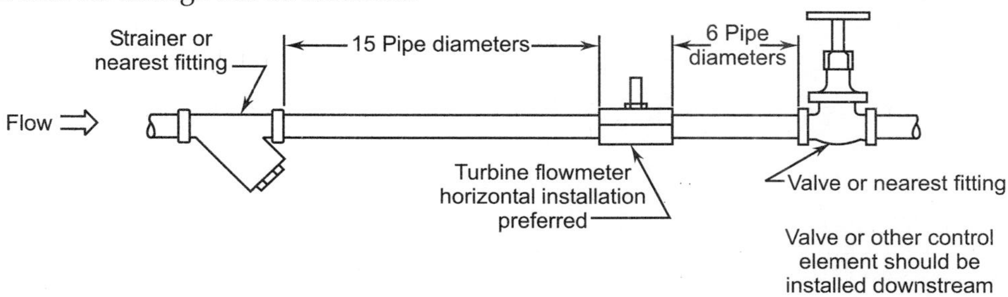

Fig. 9.54: A typical turbine meter installation (Courtesy of Brooks Instruments)

9.9.3.6 Industrial Applications

An industrial turbine meter offers a host of advantages like resolution, accuracy, intrinsically safe electrical output, corrosion resistance, case of installation, pressure rating, etc. Initially, turbine meter was used in aerospace industry as an accepted standard. This was followed by application areas in hydraulic, automotive and mechanical engineering industries. Oil industry is another application area for turbine meters. When such a meter is used with complex ball prover calibration system, it can be used for duty calculations to be payable for petroleum and allied products.

Turbine meters are extensively used for cost calculation purposes for many costly liquid materials because of high accuracy and repeatability. It is also used in mixing and batch operations. Another application area for turbine meters is food and chemical industries.

9.9.3.7 Advantages

- Certified for gas flow
- Suitable for cryogemic fluids
- Good accuracy
- Excellent repeatability
- Ease of installation and maintenance
- Suitable for extreme temperature and pressure
- Linear relationship between volume flow rate and angular velocity of the rotor

9.9.3.8 Disadvantages

- Applicable for low viscous fluids only
- Sensitive to contamination
- Conditioning sections required—both upstream and downstream
- High bearing cost
- Calibration changes with time
- Sensitive to flow disturbances
- Moving parts wear with time
- Limited material selection

9.9.4 Target Flowmeter

9.9.4.1 Description

A target flowmeter measures the flow rate of a fluid and operates on the principle that the force developed on a 'target' or disc centred in the pipe and at right angles to flow direction is proportional to the square of the flow rate.

A target flowmeter is shown in figure 9.55. The drag force on the disc is transmitted to the bar. Strain gauges attached to the bar can measure the force from which flow rate can be inferred. Here the disc acts as the primary element.

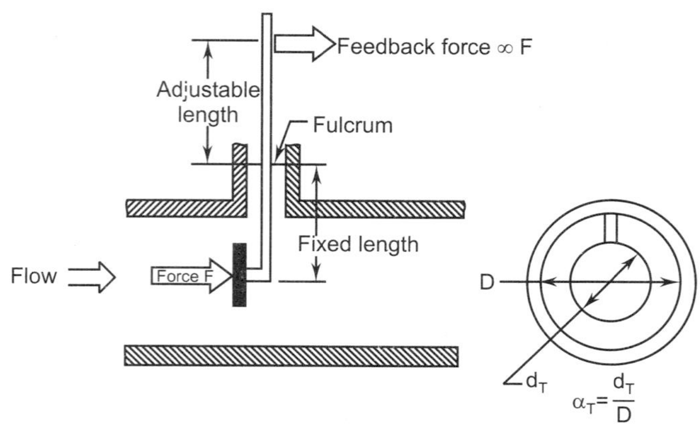

Fig. 9.55: Operating principle of target meter

For a fluid velocity V, the following relationship holds:

$$F = \frac{c_d \rho V^2 A}{2}$$

where, F = force on the target
c_d = drag coefficient
ρ = density of fluid
A = cross-sectional area of obstruction or target

The above relationship is valid for turbulent flow with R_n>4000. For turbulent flow, F is independent of Reynolds number and viscosity of fluid and has a square law relationship with velocity. For lower R_n values, F is proportional to velocity, but then F becomes dependent on coefficient of viscosity which again is dependent on temperature. For this reason, a target flowmeter is seldom applied for low flow velocities.

Figure 9.56 shows the schematic of a target flowmeter. The target is symmetrically placed relative to the pipe. Several strain gauges are connected, as shown. As the flow rate changes, the force changes, so also the resistance of the strain gauges. These are configured in a bridge circuit whose output is calibrated in terms of flow rate.

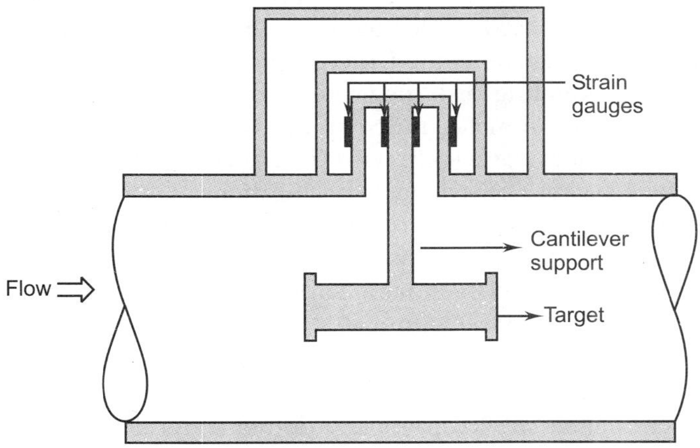

Fig. 9.56: Schematic of a target flowmeter

The accuracy of a target flometer varies in between ± 0.5 to ± 0.3%. The ratio of target to pipe size (α_T) varies in between 0.5 to 0.8. The target meter is a fast response meter with a response time of around 0.3 sec. The accuracy limit varies between 2% to 6% and its active range 3.5 : 1. It is a low cost device with moderate accuracy.

9.9.4.2 Advantages

- No pressure tap needed
- Inexpensive
- Applicable for Reynolds number as low as 100
- Suited for dirty, corrosive and high viscous fluids
- Can handle pressures upto 1500 psig and temperature upto 400°C
- Repeatability is good
- Good accuracy
- For a symmetric target, the flowmeter can be used as a bi-directional meter.

9.9.4.3 Disadvantages

- In-line mounting needed
- Pressure drop in high
- Limited flowrange
- Target susceptible to corrosion and clogging
- Maintenance cost increases with increase in suspended solids in fluid

9.10 POSITIVE DISPLACEMENT METERS

9.10.1 Introduction

Positive displacement meters are so called because they displace and then carry a fixed volume of liquid through the instrument. These meters are essentially of two types, depending on whether the quantity is expressed in terms of weight or volume.

Containers of either known weight or volumetric capacity are successively filled in and are emptied by the flowing fluid for a fixed number of times. The quantity (weight or volume wise) is then calculated by multiplying the container's capacity with that number.

Weighing meters are seldom used while volumetric meters of positive displacement type are used for liquid or gas flow measurements. Different positive displacement meters for volume measurement of fluids are: Nutating disk meter, reciprocating piston meter, rotary vane meter, lobed impeller meter, sealed drum meter, fluted spiral rotor meter, oval gear meter, etc.

Positive displacement meters are widely used in chemical, petroleum industries but seldom find use in power stations because of stringent specifications.

Positive displacement meters are low-friction, low maintenance and long-life devices and are used in applications where high accuracy and repeatability under steady flow conditions are desired. However, accuracy is affected by moving parts and, as such, needs recalibration. Accuracy of positive displacement meters may be within ±0.1% of the span of the meter. These meters are almost insensitive to viscosity, because of decreased leakage. Accuracy is also dependent on rate of flow.

Positive displacement meters are totalisers and do not show flow rates. Energy required for operation of these meters are derived from flow which results in a pressure loss. However, this energy requirement is very small.

Error in a positive displacement meter is defined as the difference between the recorded quantity and true quantity and recorded as a percentage of true quantity. This is dependent on the following factors:

- Amount of clearance between the rotor and the measuring chamber. The more the clearance, the more the error.
- Torque required to drive the register. Greater the torque, greater would be the pressure drop across the measuring element. This is turn determines the leakage rate past the rotor.
- Viscosity of the liquid. More the viscosity, more would be the pressure drop across the measuring element. But this is compensated by the reduction in flow through the rotor clearances for a given pressure drop.

9.10.2 Principle of Operation and Characteristic Curves

As liquid flows through the meter, it moves a measuring element which seals off the measuring chamber into a series of measuring compartments which are successively filled and then emptied.

Thus, for each complete cycle, a fixed quantity of liquid is allowed to pass through from the inlet to the outlet side of the meter. A pointer indicates the number of cycles for which the measurement takes place. Thus multiplying this number by the measuring chamber volume gives the total volumetric flow during the measurement period.

A positive displacement meter operates on the pressure drop across it. The pressure drop occurs due to fluid friction, friction of the metering element etc. The general characteristic of such a meter is shown in figure 9.57.

The pressure drop across the meter causes a certain amount of fluid to pass unmetered through the small clearance between the piston and the cylinder. The fluid so passed through the meter is referred to as *slip*.

If there is no slip (*i.e.* no unmetered fluid), then the actual quantity (volume) passed is equal to the displacement of the metering element per cycle times the total number of cycles. This is, thus, a straight line, represented by (1). With slip occurring, actual quantitiy passed would be more and represented by (2). Amount of slip is more at low flow rates, and this region is called the region of slip. Slip is compensated by meter manufacturers by changing the gear position in the gear train. The compensated curve is shown in (3).

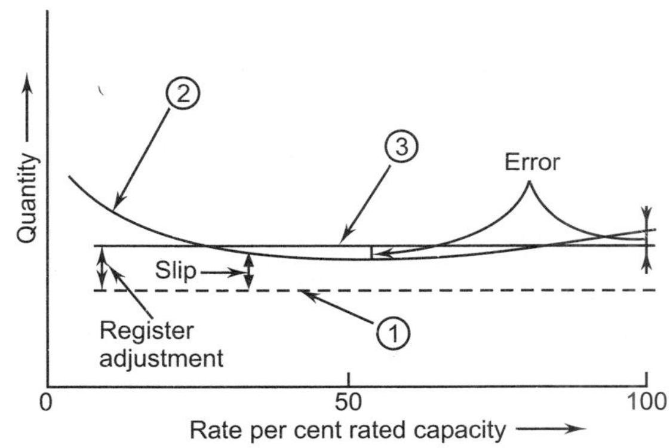

Fig. 9.57: General characteristics of a positive displacement meter

9.10.3 Comparison of Positive Displacement Meters

Table 9.2 gives a comparative study of the different positive displacement meters with regard to their size, accuracy %, temperature range in °C and application areas.

Table 9.2: Comparative Study of Different Positive Displacement Meters

Serial No.	*Meters*	*Size (cm)*	*Accuracy %*	*Temperature range in °C*	*Application areas*
1.	Nutating disc	1.25 to 5.25	± 1	- 150 to 120	Commercial as well as domestic water supply
2.	Reciprocating Piston	1.25 to 20	± 0.5	upto 150	common liquid
3.	Rotary vane	4 to 60	± 0.2	upto 200	Both liquids and gases including viscous fluids and slurries
4.	Lobed Impeller	upto 40	± 0.2	upto 400	-do-

9.10.4 Nutating Disc Meter

As seen from figure 9.58, fluid enters the meter via the inlet and leaves the outlet on the right side. It consists of a nutating disc which is eccentrically mounted and attached to a spherical ball. As fluid enters via the inlet a differential pressure is created and the disc wobbles or nutates giving rise to its rotational as well as sweeping motion.

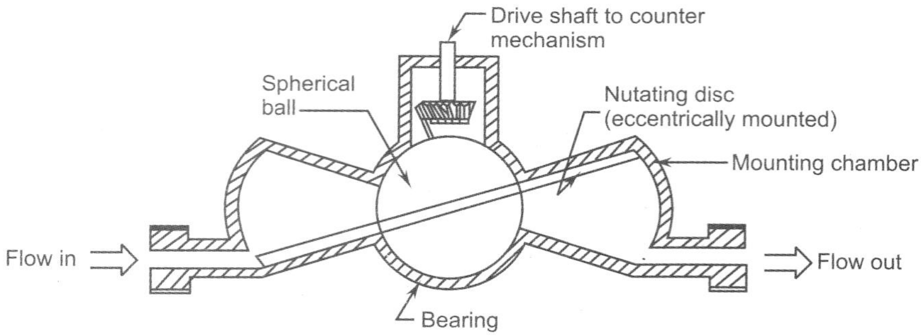

Fig. 9.58: Nutating disc meter

The disc maintains a tangential contact with the mounting chamber. The inlet and outlet chambers are separated by a partition, not shown in the figure. Each cycle of nutation of the disc results in shifting a precise volume of fluid to the outlet and a counter mechanism is activated. This can be calibrated in terms of volume.

9.10.4.1 Features

- Accuracy of the meter unaffected by variations in density and viscosity
- Static accuracy less than 1%
- Low cost, rugged construction and a low pressure drop device
- Widespread application as domestic meters
- Applied for flowrates between 15 to 500 gpm
- Limited in pipe size and capacity
- Applicable for clean fluids only

9.10.5 Reciprocating Piston Meter

Figure 9.59 shows reciprocating piston meter. It has a slide valve and a piston that reciprocates and allows the liquid alternately to pass through each end of the cylinder from the inlet to the outlet. The pistons operates the slide valve and also operates a counter mechanism. A number of pistons operating on a centre crank are generally incorporated in this type of piston meter. The crank arm operates a register to indicate the total quantity of fluid. The pressure differential for low flow rates is kept small to ensure operation of the system against its own friction. This is done by selecting large diameter pistons and operating them over a short stroke.

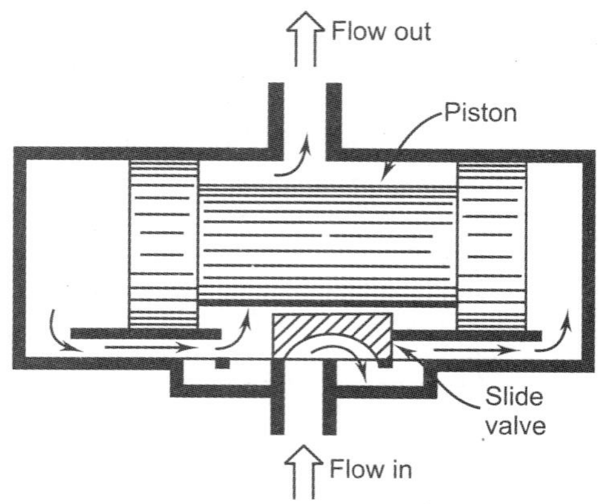

Fig. 9.59: Operating principle of a piston meter

9.10.5.1 Features

- Accuracy reasonably good—around 0.2 to 0.3%
- Variation in density and viscosity affects volumetric flow in a negligible way
- Applied for flowrates between 10 to 1000 gpm
- Suitable for measuring flow of fluids
- Suitable to measure flow rates of non-corrosive and light (low viscosity) liquids
- Cost is relatively high
- High maintenance cost
- Subject to leakage

9.10.6 Sealed Drum Meter

A sealed drum meter for gas flow is shown in figure 9.60. It consists of a horizontal cylindrical drum containing a motor with spiral-like vanes. The meter is filled with water to a height which is slightly over the central hub. The gas enters through the inlet into the rotor hub, passes through water and then to the chamber above formed by adjacent spiral vanes. The gas comes out via the outlet when the vane emerges from water.

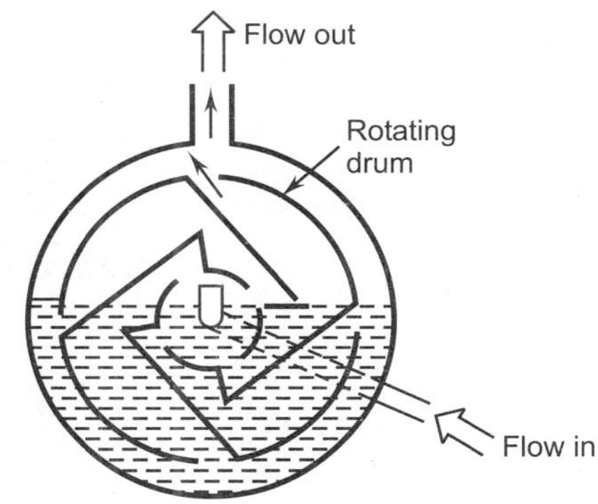

Fig. 9.60: A sealed-drum meter for gas

9.10.6.1 Features

- Used for domestic gas metering purposes
- Accuracy around 0.5%.
- Accuracy of the meter is dependent on (*i*) temperature and pressure of the gas (*ii*) improper water level and (*iii*) water should be at gas temperature and saturated with gas.

9.10.7 Rotary Vane Meter

The schematic of a rotary vane meter is shown in figure 9.61. A drum rotates which is placed eccentric to the meter casing. A number of spring loaded vanes also rotate at the same time making contact with casing and forming separate chambers. Because of the rotation of the vanes due to differential pressure, a fixed quantity of fluid is trapped inside each chamber which ultimately comes out via the outlet. A counter is attached to the eccentric drum. Thus, the total flow can be measured and registered.

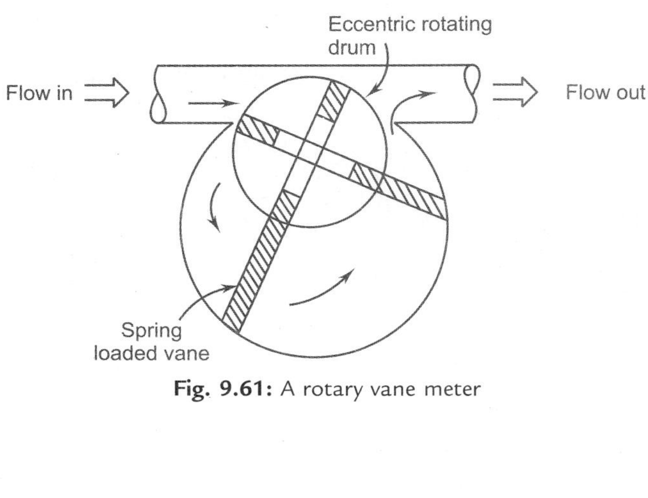

Fig. 9.61: A rotary vane meter

9.10.7.1 Features

- Density and viscosity variations of fluid affects the meter in a very small way
- Accuracy of the meter is around 0.2 to 0.3%.

9.10.8 Lobed Impeller Meter

The schematic of a lobed impeller meter is shown in figure 9.62. It consists of two rotors or impellers mounted on separate parallel shafts. The rotors rotate in opposite directions inside a closely fitting chamber for a full rotation of each rotor, a volume twice that of chamber *A* passes through the meter. Hence, rotation of rotor can be calibrated in terms of volume flow of the fluid.

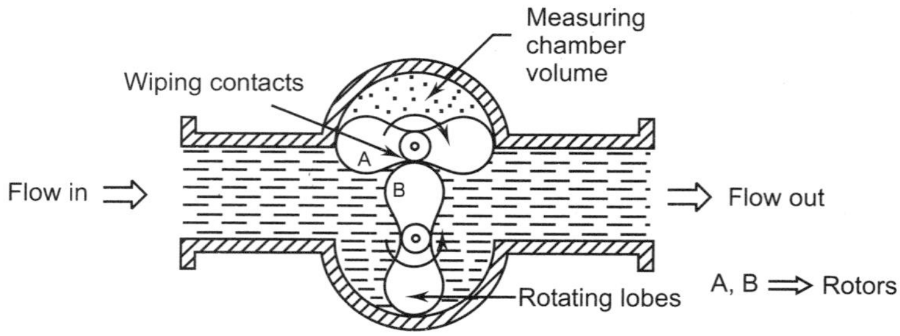

Fig. 9.62: A lobed impeller type flowmeter

9.10.8.1 Features

- Can be used for both liquid and gas flow
- Has a low pressure drop
- Accuracy of the order of ± 0.1%
- For increased accuracy, clearance (gap) between the rotors and the casing as well as between the two rotors must be as small as possible
- Applied to corrosive liquids
- Capacity range is good
- Cost is high
- Recurring maintenance required.

9.10.9 Fluted Spiral Rotor Meter

Figure 9.63 shows a fluted spiral rotor type meter. It consists of two fluted rotors housed in sleeve-type bearings. The whole set-up is mounted, so that it rotates like gears in a liquid tight case. When liquid enters the measuring chamber, the rotors divide the liquid into precisely defined volume chambers. The rotation of the rotors is made proportional to the amount of liquid that passes through. The index of a gear is rotated accordingly to record the flow.

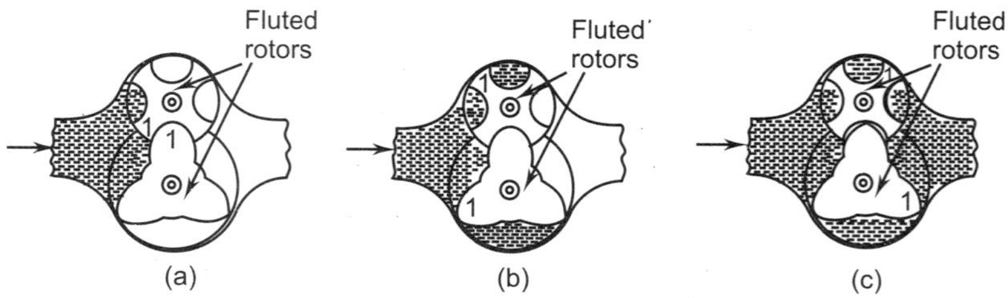

Fig. 9.63: Fluted spiral rotor type of meter

9.10.9.1 Features

- Mainly employed for flow of either crude or refined petroleum products at a maximum flowrate of upto 3000 m^3/hr and a pressure of upto 80 bar.
- Clearance between rotors and measuring chambers is kept to a minimum to reduce errors in measurements.

9.10.10 Oval Gear Meter

Figure 9.64 shows an oval gear meter consisting of two oval type gear wheels which are set in rotation as liquid passes the meter. The different positions of the two oval gears are shown from (*a*) through (*e*). One complete revolution of the two gears correspond to a fixed amount of liquid passing through the meter and measured by means of a spindle connected to one of the two gears and a suitable mechanical measuring device.

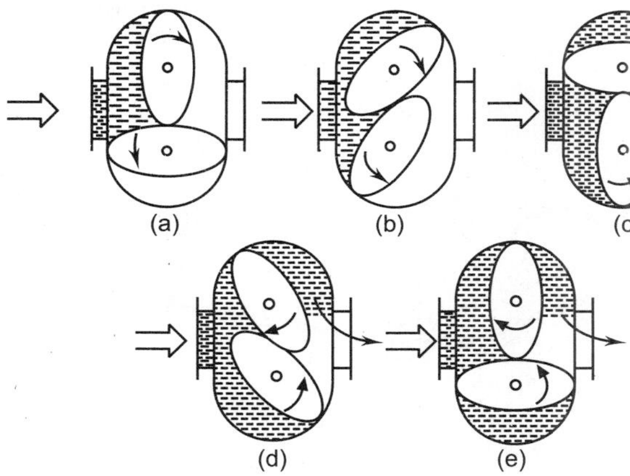

Fig. 9.64: Oval-gear meter

9.10.10.1 Features

- Available is sizes from 10 mm to 400 mm.
- Flow rate upto 1200 m^3/hr for pressures upto 60 bar.
- Accuracy around ± 0.25%.
- Available in a wide range of materials.

9.11 MASS FLOWMETERS

9.11.1 Introduction

In certain applications like finding out the range capability of an aircraft or liquid-fuel rockets, determination of mass flowrate is much more important than volume flowrate calculation because the range or velocity of rockets is dependent on the rate at which mass of fuel is consumed. In several chemical processes like pulp and paper, food processing, etc, mass flowrate is more important than volume flowrate.

Measurement of mass flowrates involve two approaches. In the first one, the volume flow rate is multiplied by density (also known as mass flow measurement) while the second method involves flowmetering that is dependent on mass flowrate.

In case of gases, flowmetering technique involves correcting the volume flowrate measurements at certain standard conditions and then multiplying the same with the gas density at those standard conditions. In some cases, air is taken of standard tables to take care of gas temperature and pressure variations.

Difficulty is encountered in mass flowmeters for two-phase (like gas-liquid, gas-solid, etc) fluid flow. The difficulty arises when the two phases have different velocities in different directions. The difficulty is overcome either by ensuring homogeneous flow (either by proper pipe sizing or meter positioning) or metering the two phases separately.

9.11.2 Impeller Turbine Mass Flowmeter

A commercial variety of impeller turbine mass flowmeter is shown below in figure 9.65.

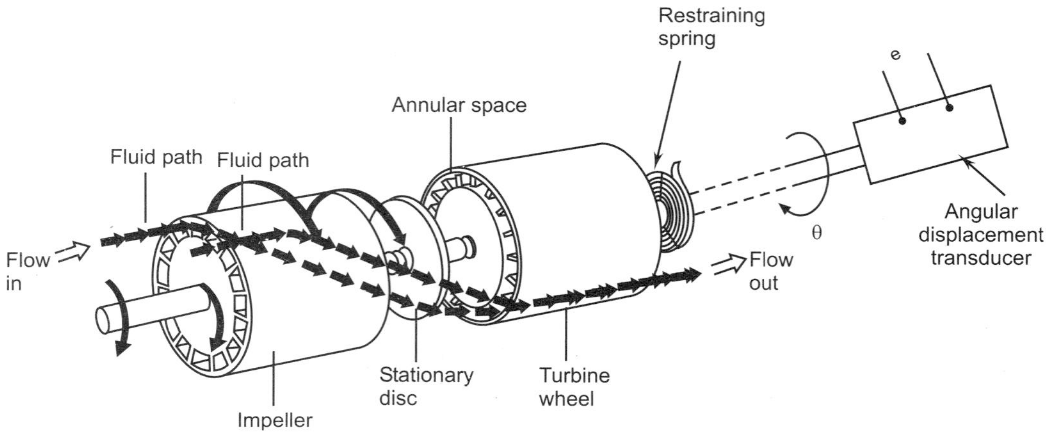

Fig. 9.65: A mass type flowmeter

If the flow input to the impeller has no rotational component and if the axial length of the impeller is long enough, the driving torque on the impeller, when it is rotated by a constant speed motor given by

$$T = r^2 \omega G$$

where, r is the radius of gyration, ω is the angular speed and G is the mass flow rate through the impeller wheel. It is to be noted that when $G = 0$ (zero mass flow condition), torque (T) will not be zero owing to frictional effects. Torque under such condition is also dependent on viscosity variations of the fluid.

The turbine wheel, shown on the right side, is held from turning by a restraining spring. Thus, the angular momentum of the fluid is removed in the turbine (considered long enough axially) and it experiences a torque proportional to the angular momentum.

If the restraining force applied by the spring is a linear one, the deflection is a linear measure of G. There can also be an output electrical signal by means of an angular displacement transducer.

The stationary decoupling disk ensures a minimum viscous coupling between the impeller wheel and the turbine at zero flow rate.

Since the momentum and, in consequence, the torque is mass flow rate dependent, that is why it is termed a mass flowmeter.

A number of meters are available in the market using angular momentum principles. These are twin turbine meter, vibrating gyroscope meter, the rotating gyroscope meter, Coriolis meter and the *S*-tube meter.

9.11.3 Twin-turbine Type Mass Flowmeter

The schematic of a twin-turbine mass flowmeter is shown in figure 9.66. Each turbine has a reluctance type pick-up and a strong magnet, while they are mounted on a common shaft and are connected by a torsion member.

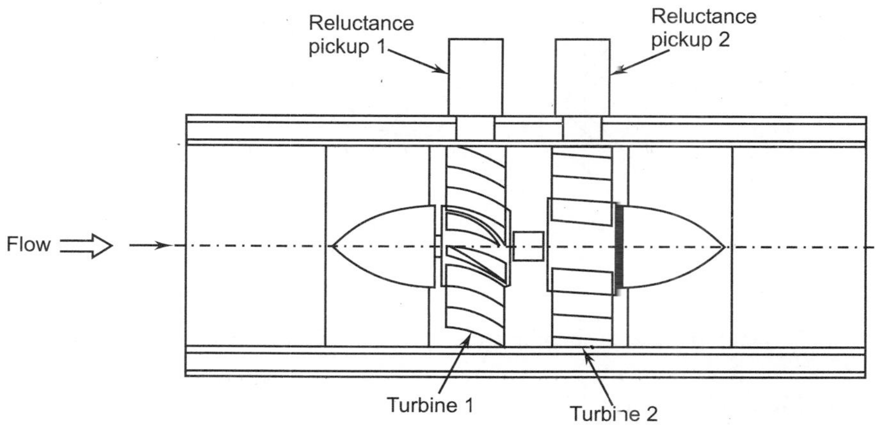

Fig. 9.66: A twin-turbine mass flowmeter

The blade angle of the two turbines are different and hence they would have different angular velocities. But since they are connected by a torsion member, they will have a common angular velocity with, an angular phase shift between the two turbines which is a function of angular momentum and hence, in its turn, a function of mass flow rate. In this case, no restraining spring is used and the torsion member holds the two turbines in twist. This twist or torque is thus a function of the angle between the two turbines.

The angle is measured in the following manner. With the rotation of the upstream turbine 1, the magnet associated with this turbine produces a pulse when the magnet passes underneath the pick-up coil. This pulse thus produced opens an electronic gate which is subsequently closed by the pulse generated by the pickup coil associated with turbine 2. A standard oscillator counts the number of pulses during this time interval which is thus a function of the torque developed.

9.11.4 Gyroscopic Mass Flowmeter

It operates on the principle of gyroscope. It is shown in figure 9.67. There are, in all, three axes—the input axis (*a-a*), the output axis (*b-b*) and the spin axis (*c-c*). Fluid motion takes place perpendicular to the spin axis. Thus, fluid motion is about *a-a* axis about which the assembly rotates. The shape of the pipe may be circular or square.

A torque is produced about the *b*-axis because of flow. The centrifugal force is corrected by the loop occurring along the spin axis. A deflection is produced about the output (*b*) axis and is measured by a transducer. This deflection is a measure of the mass flowrate.

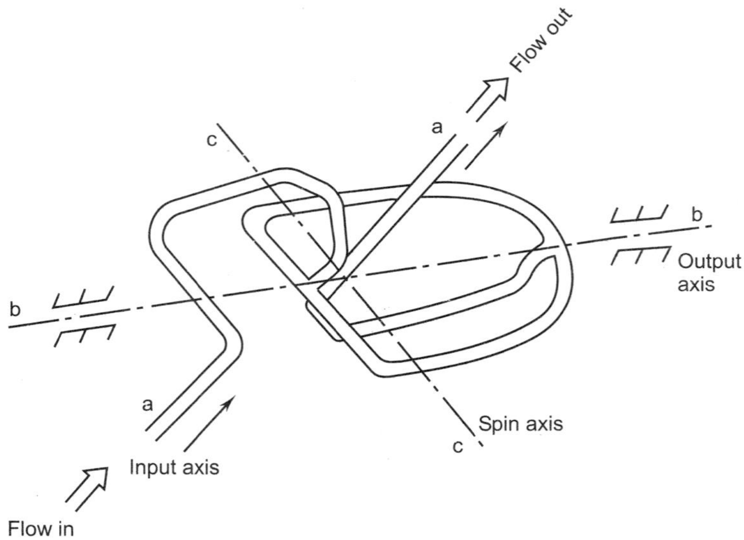

Fig. 9.67: Sketch to explain the operation of gyroscopic flowmeter

9.11.5 Coriolis Mass Flowmeter

9.11.5.1 Description

A body moving with a constant velocity v on a surface rotating with a constant angular velocity ω is subjected to a force, known as Coriolis force, which is perpendicular to both v and ω. The acceleration developed in the direction of the force is known as Coriolis acceleration f_c.

A Coriolis type mass flowmeter is shown in figure 9.68.

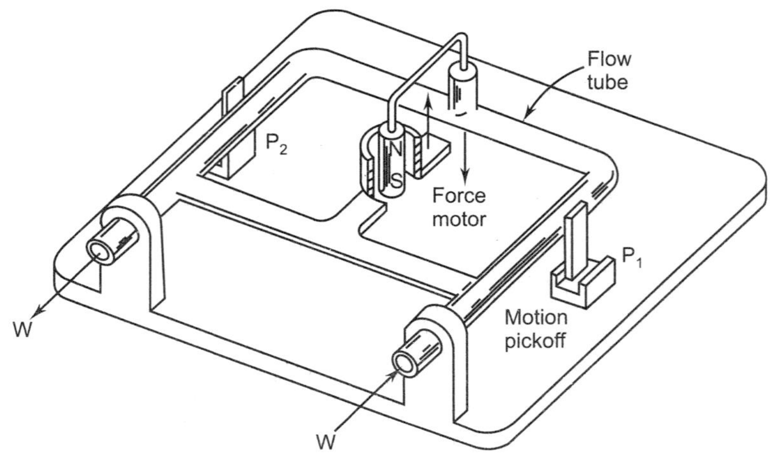

(a)

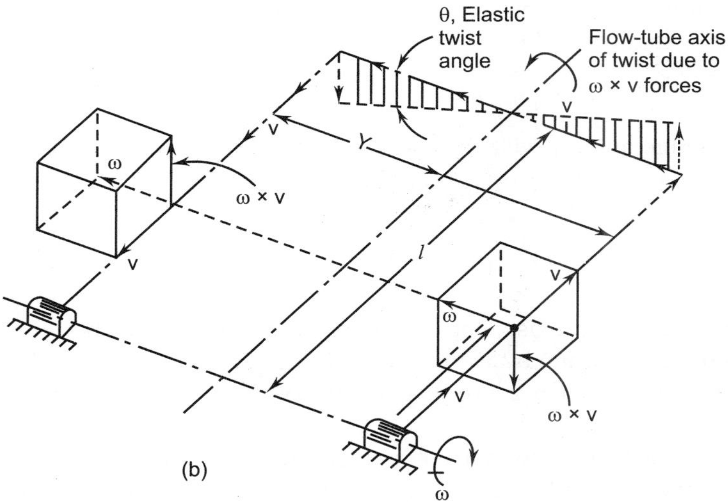

Fig. 9.68: Coriolis-type mass flowmeter

The fluid passes through a C-shaped pipe which is cantilevered by means of two supporting brackets, as shown. The drive system consists of two coils—a sensor coil and a driver coil. The force motor gets its drive signal from the sensor coil. The two coils share the same permanent magnet core. The sensor coil picks-up the frequency variation due to velocity change of fluid. As a result, the resonant frequency changes and the feedback ensures restoration of original resonant condition.

9.11.5.2 Theory

The Coriolis acceleration f_c is given by $2\vec{w} \times \vec{v}$ and hence the inertial force acting on an elemental length dl of the pipe in a direction perpendicular to fluid flow direction is

$$d\vec{F} = d_m \vec{F_c} = (\rho.\alpha.dl)(2\vec{\omega} \times \vec{v})$$

where, ρ and α are the density of the fluid and cross-section of the pipe respectively.

or $$d\vec{F} = 2\vec{w}(\rho\,\alpha\vec{v})\,dl = (2\,\omega W dl)$$

where, W = mass flow rate in kg/sec.

Hence, the total force is

$$F = \int dF = \int 2\omega W\,dl = 2\omega\,Wl$$

and hence, the inertial torque is given by

$$T = 2\omega\,Wl \times r$$

where r is as shown in the figure.

The angular velocity ω varies in a sinusoidal manner and hence torque T varies likewise also. This driving torque T tends to twist the C tube. Now since the torsional natural frequency of the C tube is greater than driving frequency, hence it can be thought of as a spring of spring constant K. Thus

$$K \times \theta = 2\omega\,Wlr$$

or, $$\theta = \frac{2\omega Wlr}{K}$$

Because of twist θ, both the pick-offs would not respond at the same time, rather one will respond Δ*t* time later than the other. Thus,

$$\theta = \frac{(l\omega_{av})\Delta t}{r} \simeq \frac{l\omega\Delta t}{r}$$

where, ω_{av} is the average angular velocity in time interval Δ*t* and it is almost equal to ω.

Thus,
$$\frac{2\omega Wlr}{K} = \frac{l\omega\Delta t}{r}$$

or,
$$W = \frac{k\Delta t}{2r^2}$$

Thus, a linear relationship exists between mass flow rate *W* and the time interval Δ*t*.

Δ*t* measurement can be done by PWM method and a gated oscillator along with a up/down counter. Again, total flow over a prescribed length of time can be done by digital integration method.

The meter can measure the flow rates of foams, slurries, mixtures, sticky fluids, peanut butter, liquids with gases in addition to clean liquid flow measurements. Accuracy is of the order of ± 0.15% for liquids and ± 0.5% for gases.

9.11.5.3 Advantages

- Obstruction less type
- Insensitive to viscosity, pressure and temperature of the flowing fluid
- Suitable for changing flows
- Can measure flow rates of clean liquids, liquids with gases, slurries, etc.

9.11.6 Bridge Circuit Mass Flowmeter

A bridge circuit for measurement of mass flow rate of liquids is shown in figure 9.69. Four identical orifices (1 to 4) are placed in a 'bridge' configuration. The pressure drop across an orifice is given by ρQ^2. A positive displacement pump is employed which runs at a constant speed and a volume flow rate *q*. From symmetry considerations, it is seen that flow rates through the orifices on the diagonally opposite arms will be identical.

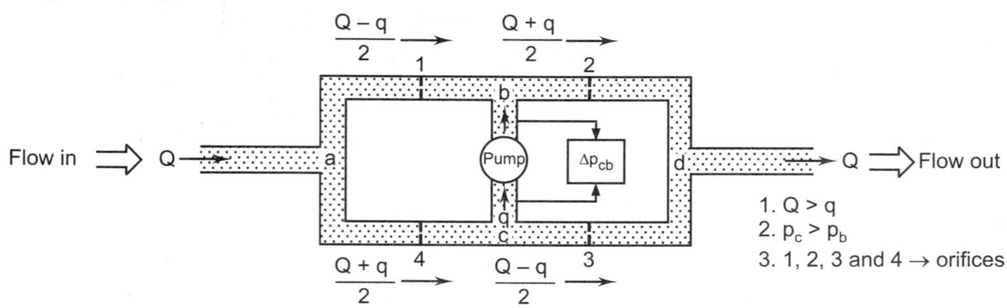

Fig. 9.69: Bridge-circuit flowmeter

Employing now the relationship for flow rates across an orifice plate,

$$p_a - p_b = k\rho\left(\frac{Q-q}{2}\right)^2$$

and
$$p_a - p_c = k\rho\left(\frac{Q+q}{2}\right)^2$$

Combining,

$$\Delta p_{cb} = p_c - p_b = \frac{k\rho}{4}[(Q-q)^2 - (Q+q)^2] = \frac{k\rho}{4}[4\,Qq]$$

$$= k\rho Qq$$

$$= kWq$$

$$\therefore \quad W = k'\Delta p_{cb}$$

Thus, mass flow rate can be found out by measuring the pressure drop and knowing the constant k'. The conversion factor k' is dependent on the pump discharge rate (q) and the orifice flow coefficient (k). Flow rates must be maintained so that Reynolds number is of considerable value and hence under such condition discharge coefficients of all the four orifices are equal and don't vary with varying Q.

9.11.7 Radiation Type Mass Flowmeter

A schematic of a radiation type mass flowmeter is shown in figure 9.70. It is a combination of two sensors packaged into a single unit—a magnetic flowmeter measures the volumetric flow rate and a gamma radiation based densitometer which measures the density of the fluid. The output of these two sensors are then fed to a microcomputer which then evaluates the mass flow rate by software programming.

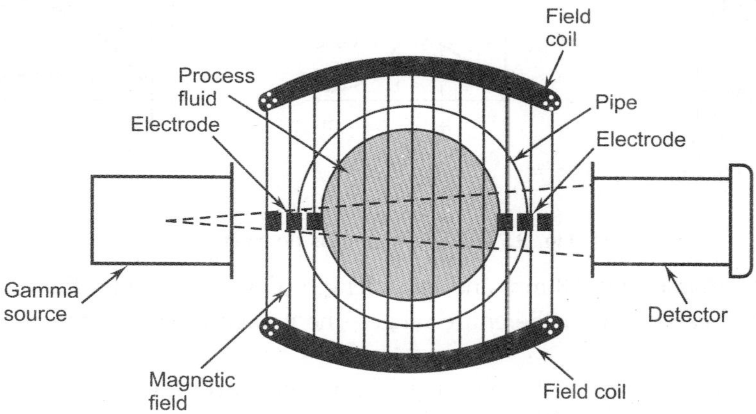

Fig. 9.70: Mass flowmeter combining a magnetic flowmeter and a radiation-type densitometer in a single unit

It is an obstructionless mass flowmeter which does not require any compensation for changes in process variables. It is a bulky and expensive flowmeter and is applicable to process fluids having at least 3.0 μS/cm conductivity.

9.11.8 Thermal Mass Flowmeter

A mass flowmeter, based on thermal principles, is shown in figure 9.71. In this, the fluid flow in the main pipe and the bypass pipe are assumed to be laminar with a range from 10 cm³/min to approximately 5.66 cm³/min for gas flow. A known fraction of the total flow is allowed through the bypass tube. Measurement is done on the bypass tube and total flow found out by multiplying with the flow-ratio factor. The use of bypass path for flow measurement has the advantage of measuring flows from small to very large values of flow rate conditions. This is achieved by simply altering the flow-ratio factor. It is the ratio of total flow to bypass flow and ranges from 1 to 10^5. Both the main and the bypass flow must have laminar flow condition in order to ensure a fixed flow ratio condition.

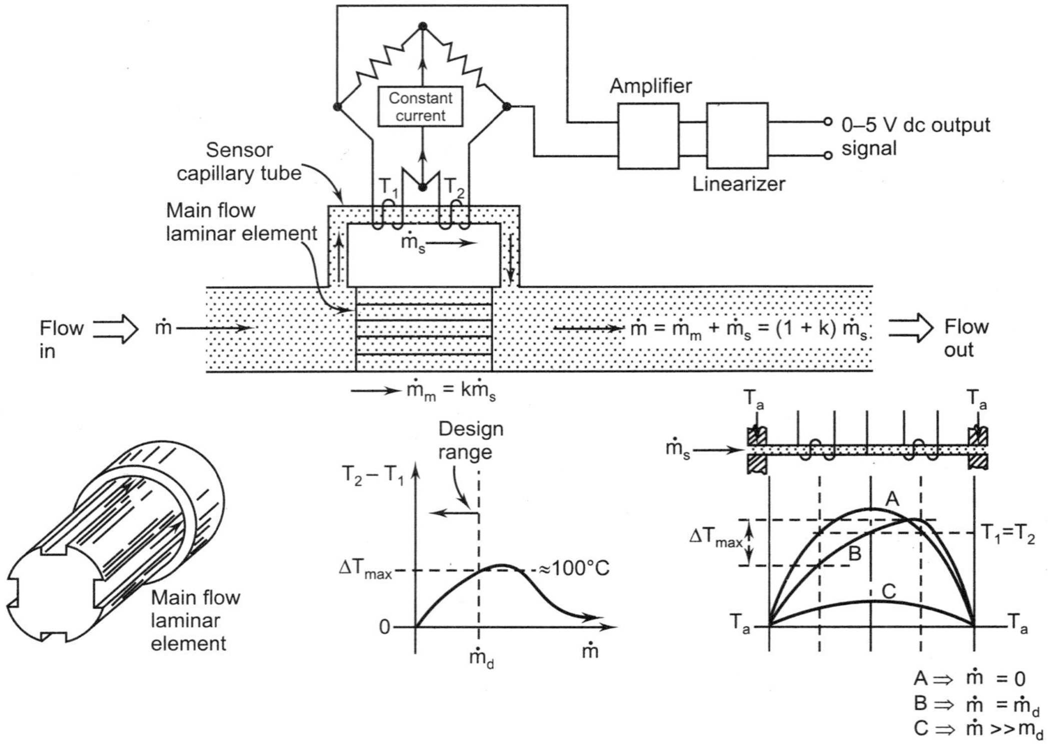

Fig. 9.71: A thermal mass flowmeter

The sensor tube diameter is 0.8 mm and has a length of 50 to 100 times the tube diameter. This ensures laminar flow condition for the entire range—the maximum sensor-tube flow being around 20 cm³/min.

In the figure, the flow rate to be measured is $\dot{m}$, that through the bypass tube is $\dot{m}_s$ with $\dot{m}_m = k\ \dot{m}_s$, where k is the flow-ratio factor and $\dot{m}_m$ is as shown in the figure. Two windings are wound around the bypass tube and arranged in a bridge configuration. They provide for a constant heat input to the fluid for all flow rates. The bridge resistances are so selected that the bridge output is zero for zero-flow rate (*i.e.*, $\dot{m}_s = 0$) and it corresponds to curve (A), which is thermally symmetric in nature. Under this condition ΔT or $T_2 - T_1$ is zero, heat in the coils is used up to balance heat losses from the tube to the surroundings, assumed to be at a temperature T_a.

For any mass flow rate $\dot{m}_s$ and assuming new thermal equilibrium, the temperature profile corresponds to curve (B) with T_1 dropping by a large amount while T_2 value stays nearly constant. T_1 temperature drops because of heat transfer at the rate of $\dot{m}_s\ c_p\ T_1$ from that region. Thus heat transfer rate from tube to fluid increases to support this energy loss. Thus ΔT increases requiring a drop in T_1 proportional to $\dot{m}_s$.

T_1 and T_2 are affected equally for changes in the ambient temperature and thus ΔT remains invariant for such changes. The instrument accuracy is of the order of ±1 per cent of the full scale.

The bypass arrangement ensures employment of a small, low power sensing tube. The flowmeter along-with control is widely used in electronic manufacturing units.

9.12 ELECTRICAL FLOWMETERS

9.12.1 Electrical Manometer Type Flowmeter

It is a type of flowmeter for which the flow rate varies directly with the conductance of the fluid and the scale is calibrated directly into flow units. A schematic of such a flowmeter is shown in figure 9.72.

A flow restrictor is put in the tube through which the fluid flows. Two taps—one upstream and another downstream—are taken and are joined in a manometric configuration. The downstream side contains a unit having a large number of vertical conducting rods arranged in a spiral form. These rods are connected to a resistance element, as shown, located within the unit. With increase in flow rates, the mercury level of the right chamber increases, thereby making contact with the next rod and decreasing the value of resistance that is put in series with the receiving instrument. The opposite happens for a decrease in flow rates, since the pressure differential across a restriction to flow bears a square relationship with the flow rate. Thus, if the resistance element can be physically made to have a parabolic curve, the meter reading can be made a linear one with flow rate variation. The oil above mercury prevents the rods from possible contaminations with the fluid. For the present case, the flow rate is

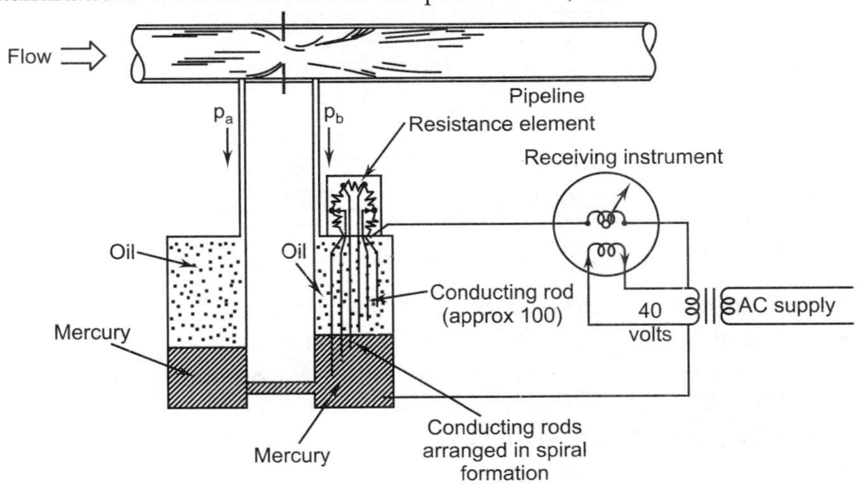

Fig. 9.72: An electrical manometer. (Republic Flowmeters Division of Rockwell)

$$Q = k\sqrt{p_a - p_b}$$

or,

$$Q = k_1 \sqrt{\Delta h}$$

where, Δh is the level difference of mercury in the two chambers.

Now, conductance, which is reciprocal of resistance is made proportional to $\sqrt{\Delta h}$ by the parabolic contour of the connecting rods. Hence,

$$Q = K_2 G, \quad G = \text{conductance}$$

and hence it is seen that flow rate and conductance are having a proportional relationship.

Care should be taken to see that instrument reading is independent of supply variations. This type of instrument is suitable for telemetring purposes.

9.12.2 Linear Resistance Element Flowmeter (Capillary Flowmeter)

The capillary flowmeter is shown in figure 9.73 and is suitable for small flow rates or for fluids having high viscosity. The Hagen-Poisemille equation for laminar flow is applied here since the Reynolds number is less than 2000. The equation is

$$Q = \frac{\pi D^4}{128\,\mu L}(P_1 - P_2)$$

where, Q is the flow rate
D is the inside diameter of the tube through which flow takes place
L is the length of the tube
$P_1 - P_2$ is the pressure drop across the length L, as shown
μ = coefficient of viscosity

Thus, the flow rate is proportional to the pressure drop across L, since other terms are constants. Hence, it is called a linear resistance element flowmeter.

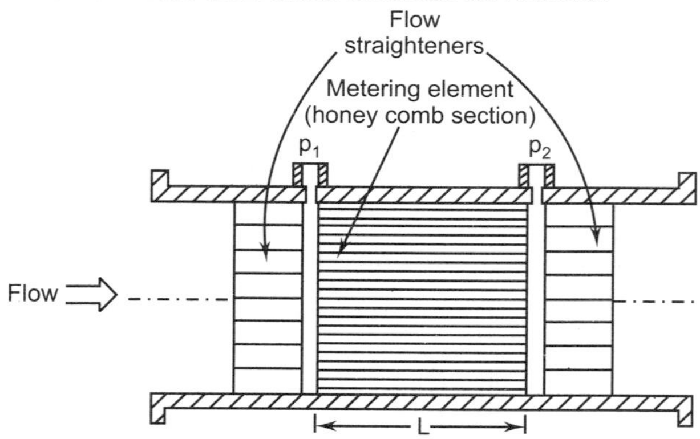

Fig. 9.73: Linear resistance element flowmeter

Fluid first passes through the flow straighteners before it faces the metering element. This is actually a bundle of capillary tubes in a honeycomb configuration.

Since the flowmeter offers good damping, thus an accurate average flow rate measurement of pulsating flow is possible. The flowmeter can also be used for reverse flow.

If the metered fluid is not clear, the metering element would be subject to plugging. The meter is expensive and a high pressure loss occurs of the metered flow.

9.12.3 Electromagnetic Flowmeter

9.12.3.1 Introduction

It is used to measure the rate of flow of electrically conducting fluids. Its principle of operation is based on Faraday's law of electromagnetic induction which states that if a conductor of length l moves with a velocity v in a direction perpendicular to a magnetic field of strength B, then the emf e (which is perpendicular to both B and v) induced across the conductor is given by $e = Blv$.
where

B = magnetic flux density, Wb/m^2

v = velocity of the conductor, m/sec

and, l = length of the conductor, m

This is shown in figure 9.74.

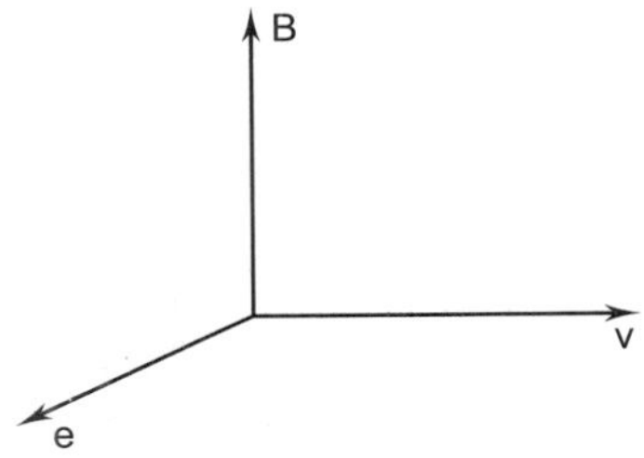

Fig. 9.74: Faraday's law of electromagnetic induction

A reference is made at this point to BS 5792 1980 which states, "If the magnetic field is perpendicular to an electrically insulating tube through which a conducting liquid is flowing, a maximum potential difference may be measured between two electrodes positioned on the wall of the tube such that the diameter joining the electrodes is orthogonal to the magnetic field. The potential difference is proportional to the magnetic field strength, the axial velocity and the distance between the electrodes". Thus, for a fixed magnetic field and fixed distance between the electrodes, p.d. becomes proportional to axial velocity. Thus, axial velocity and rate of flow can be determined.

9.12.3.2 Operating Principle and Operation

The pipe portion at which the two electrodes are placed must be non-magnetic (so that the magnetic field can penetrate the fluid) as well as non-conducting in nature, otherwise it will short-circuit the available emf across the electrodes. The schematic of a flow meter is shown in figure 9.75.

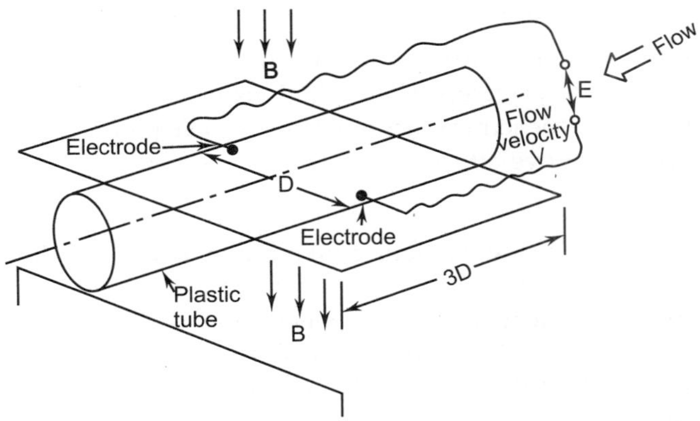

Fig. 9.75: Principle of operation of an electromagnetic flowmeter

Usually, a plastic pipe is used with the two electrodes placed at the points of maximum potential difference. On either side of the plastic pipe, the usual metal pipe is there. The inside of the plastic pipe is coated with teflon (PTFE), polyurethane etc.—the type of coating depends on corrositivity and temperature of fluid. The electrodes are normally of stainless steel type mounted flush with the inside face of the tube.

Figure 9.76 shows the transverse (figure *a*) and axial (figure *b*) potential distribution, with maximum potential obtained across a horizontal diameter of value BDv, where D is the diameter of the tube and v is the flow rate. The magnetic field applied is of limited length, roughly $3D$

length is sufficient. The strength of potential in the axial direction decreases on either side of the central field position, as shown in figure (*b*).

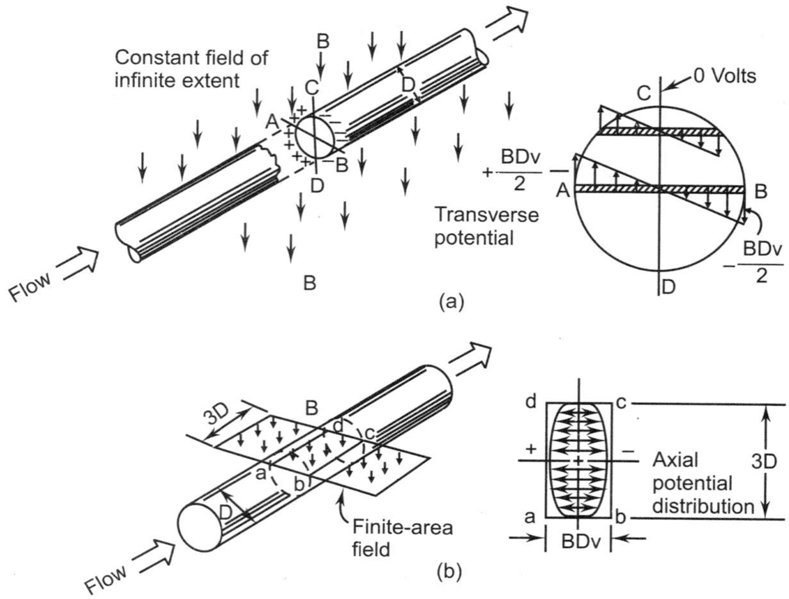

Fig. 9.76: Electromagnetic flowmeter

The electrical excitation needed to produce the magnetic field may either be dc or ac in nature giving rise to dc or ac output respectively. DC excitations give rise to polarization effect on the electrodes and subsequent electrochemical action. Thus, ac excitation (50 Hz or 60 Hz) systems came into being, reducing the above effect.

9.12.3.3 Primary Element

It consists of the metering tube, inside insulating tube lining, coil to produce field, flanged ends and electrodes.

The tube must be non-magnetic to allow field penetration, its diameter ranging from 3 mm to 1830 mm. The inside tube surface is insulated to prevent shorting of signals. Some of the lining materials are shown in table 9.3. The material is sometimes moulded around the tube ends forming a composite with the flanges.

Table 9.3: Typical Materials for Insulation

Material	*Application*	*Temperature limits (°C)*
Natural rubber	Wear and chemical resistance	-20 to 70
Neoprene	Good chemical and wear resistance in presence of oil and grease	0 to 100
Ebonite	Abrasion and chemical resistance	0 to 90
Polyurethane	Wear and impact resistance	-50 to 70
PTFE	Wear resistant, chemically inert. good for foodstuffs	-50 to 200

The electrodes penetrate through the tube and tube lining, so that they make contacts with liquid. Some of the electrode materials are: non-magnetic stainless steel (for non-agressive liquids), tantalum, monel, hastelloy, platinum-iridium, zirconium (for aggressive liquids).

The meter tube is earthed to the adjacent piping network. The flowmeter is connected to the earth via several earth connections, forming earth-loops. Earthing of the meter tube ensures the electrode output signal not to be affected by such spurious signals.

9.12.3.4 Conductivity of Fluids

If the conductivity of the fluid remains constant around the flowmeter housing, the operation of the meter would not be affected by any variation of fluid connectivity. This will be true if the output resistance of the primary device is at least two or more order-of-magnitude less than the input resistance of the secondary device. The output resistance of the primary device is given by:

$$R \approx \frac{1}{d\sigma}$$

where, d is the diameter of the electrode and σ the conductivity of the fluid.

Fluids having entrained gas or a tube running partially filled would result in variation in conductivity. These conditions should thus best be avoided. A secondary device with a typical input resistance of the order of 20 MΩ would suffice with most fluid types, except for pure water.

9.12.3.5 AC Electromagnetic Flowmeter

9.12.3.5.1 *Description and Operation*

The schematic arrangement of an AC magnetic flowmeter is shown in figure 9.77. The output signal (flow signal) and magnetic flux are in phase with each other. With changes in supply voltage output signal voltage would also vary, thus the supply voltage should remain constant.

The AC excitation scheme employs a frequency of 50–60 Hz.

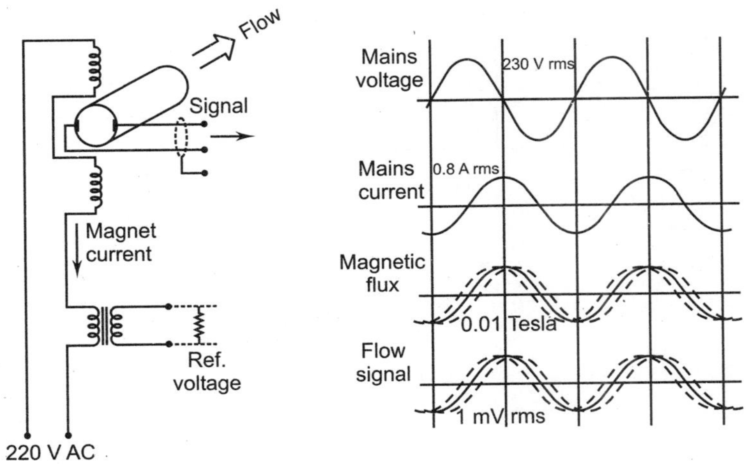

Fig. 9.77: Schematic arrangement of an AC-powered magnetic meter

Source: **Fluid Flow Measurement**, R A Furness (Longman, UK 1989/p:43). (Reproduced with kind permission of the author). *Courtesy*: Fischer & Porter Ltd.

9.12.3.5.2 Advantages

- Absence of polarisation effect present in a DC flowmeter
- High gain AC amplifiers and high pass filters eliminate slow and spurious voltage drifts which result from thermocouple and galvanic actions
- Not significantly affected by moderate flow rate variations
- Linear output variation with flow rate changes

9.12.3.5.3 Disadvantages

- Spurious AC signals may go into the measurement circuit
- High power consumption
- Zero drift caused by interfering voltage and electrode contamination
- Low output signal
- Cannot be applied for gas flow
- Performance is moderate compared to turbine or PD meters

9.12.3.6 Pulsed DC Type Electromagnetic Flowmeter

9.12.3.6.1 Description and Operation

The schematic of a pulsed DC electromagnetic flowmeter is shown in figure 9.78. It can be of two types: unipolar and bipolar. In the unipolar scheme, excitation is switched on or off while in bipolar scheme the excitation current is alternately reversed. Again, excitation scheme can be constant voltage or constant current type. The figure represents a unipolar scheme.

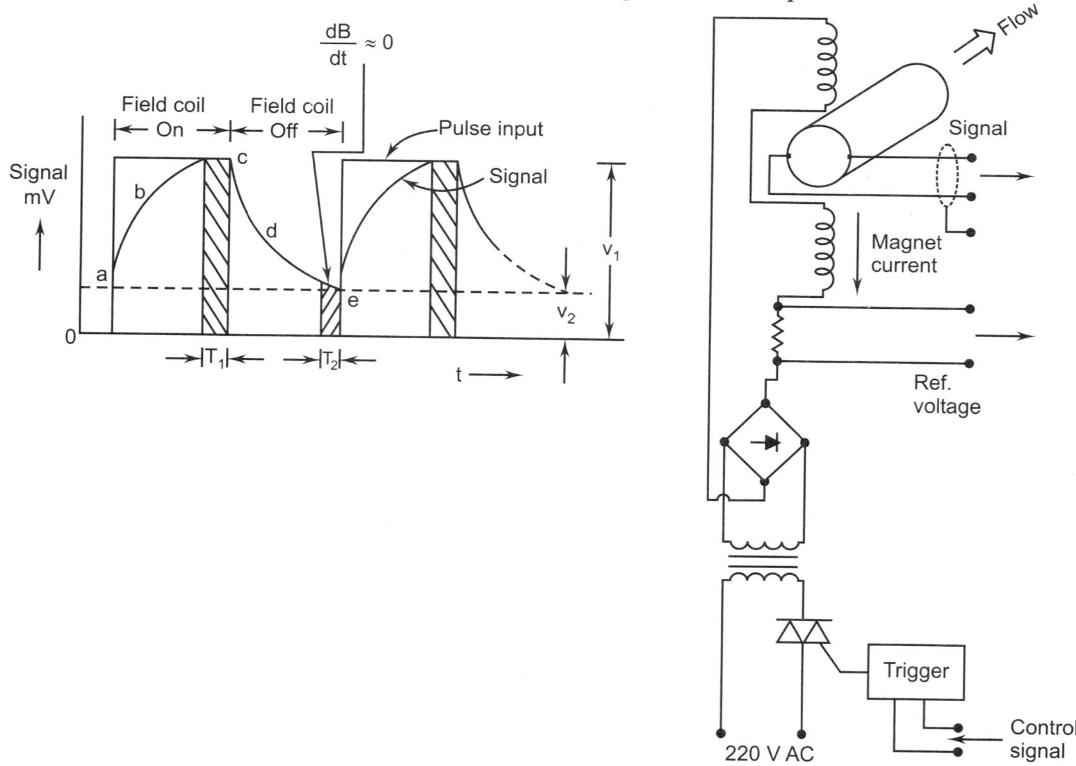

Fig. 9.78: Circuit design of pulsed DC magnetic meter

Source: **Fluid Flow Measurement,** R A Furness (Longman. UK 1989/p:45). (Reproduced with kind permission of the author). *Courtesy*: Fischer & Porter Ltd.

The pulsed DC supplied to the field coils is having a frequency of 2–7 Hz. The field coils are powered for a few cycles and kept off for a few cycles.

The output signal builds up when the field coils are excited and ultimately reaches a saturation along *abc*. This signal value v_1 is measured when $dB/dt \approx 0$ (T_1 being the sample time as shown in the figure), at which time the bridge rectifier circuit is switched off. The signal now decays along *cde* and a second sampling of signal voltage V_2 is made (again when $dB/dt \approx 0$), this time the sample time is T_2.

V_1 corresponds to flow-related signal and also noise while V_2 relates to noise only. Thus, v_1-v_2 is the flow related signal which could be calibrated for flow rate. Magnetic induction B changes with supply voltage variation which is compensated for by the reference voltage obtained across the precision standard resistor.

The lower and upper range of this type of flowmeter are approximately 1 m/sec and 10 m/sec respectively. The type of liquid, as also the flow rate, determine the flow tube size and the output emf.

9.12.3.6.2 *Advantages*

- About 75% power saving compared to its AC counterpart
- Cost around 35% less than its AC counterpart
- Ease of installation
- Simpler field wiring
- Better suited for pulsating flows
- Applicable for highly conducting fluids like molten metals, mercury, etc
- Since the frequency of excitation is very low, inductive effect is almost negligible
- More accuracy
- Very good zero stability

9.12.3.6.3 *Disadvantages*

- Not an automatic choice for low-conductivity, ion-conducting fluids—in these cases time-variant polarisation occurs
- Speed of response is slow (0.2–2 seconds)
- Noise does not remain constant for fluids containing slurries/low-conducting fluids

9.12.3.7 *A Dual Frequency Excitation Scheme*

9.12.3.7.1 *Description*

A dual frequency excitation scheme is sometimes undertaken for electromagnetic flowmeters that enjoys the advantages of AC (50 Hz) and pulsed DC (2–7 Hz) systems, *i.e.* higher accuracy and zero stability of pulsed DC type of excitation and good speed of response and insensitivity to measurement noise of AC systems.

Here the field coils of the electromagnetic flowmeter is excited with a compound waveform—a low-frequency waveform on which the high frequency composed is superimposed and is shown in figure 9.79. The output from the electrodes is subjected to both low and high frequency sampling. They are respectively integrated and differentiated. These two are then summed up in a summer circuit to obtain the output. This output signal is then properly calibrated which enjoys the advantages of both the individual schemes.

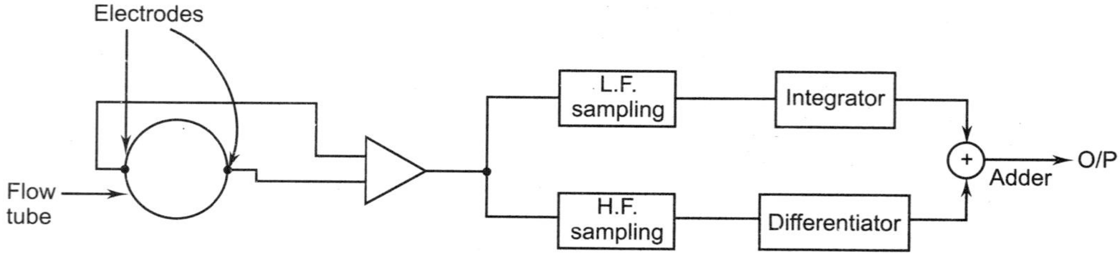

Fig. 9.79: Scheme for dual frequency excitation

9.12.3.7.2 Advantages

- No obstruction to the liquid flow in the pipe
- Particularly suited for fluids with solid matter, sewage, paper pulp, cement slurries, detergents, greasy fluids, etc.
- Bidirectional flow measurements possible
- Has a wide linear range
- Good transient response
- Very little maintenance required
- Not appreciably affected by upstream flow disturbances
- Practically not affected by variations in density, viscosity, pressure and temperature
- No head loss other than the length of the straight pipe which the meter occupies
- Blood flow in the vessels of living specimens can be measured
- Pure DC type electromagnetic flowmeters can measure the flow of liquid metals, such as mercury.
- Conductivity as low as 5 μS/cm (S for Siemens) can be measured by standard EM flowmeters—while it is 0.1 μS/cm for special meters. Conductivity as low as 0.1 ∞μS/cm can be measured with specially designed flowmeters.

9.12.3.7.3 Disadvantages

- An expensive device
- Cannot be applied to measure flow of fluids which are insulating in nature
- Accuracy in the range of ± 1% over a flow rate range of 5%
- Zero check can be done by stopping the flow
- Effective resistance of fluid between the electrodes should be limited to 1% of the impedance of the external circuit in order that the meter becomes insensitive to variations of the resistance of the fluid
- Cannot be applied for gas flow
- Output signal low
- Performance moderate w.r.t. turbine or PD meters

9.12.3.8 Electrode Cleaning

An electrode cleaning system by ultrasonic method is shown in figure 9.80 and introduced in recent times. This method of cleaning has been introduced by Fischer & Fischer.

Previously, a *'burn-off'* method consisting of passing a high current between the electrodes for a short time was used for large electrodes.

The ultrasonic method has a piezo-electric crystal, followed by a cone, washer and the electrode. The cone intensifies the vibrations of high frequency signal induced in the crystal when it is excited. The electrode heads thus experience a very violent vibration. This results in cavitation of the electrode surface resulting in an effective cleaning mechanism.

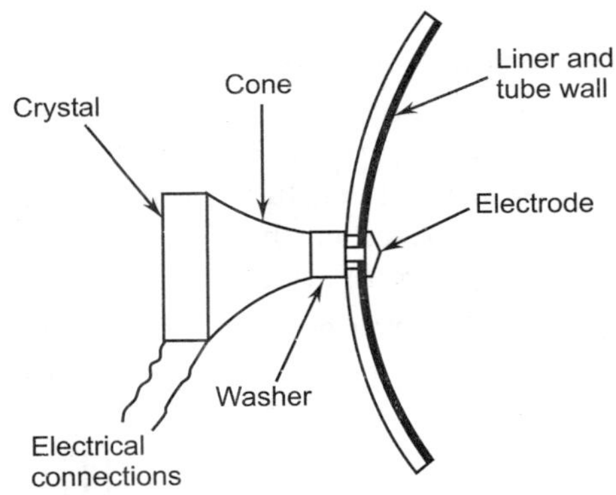

Fig. 9.80: Ultrasonic electrode cleaning

9.12.3.9 Electromagnetic Flowmeter in an Invert

There are certain instances in which a closed pipeline does not always run full of fluid always. But in such situations, the pipeline size is determined to accommodate occasional high flowrates, such as rain storm.

To measure flowrates in such situations, an electromagnetic flowmeter placed in an invert is used as shown in figure 9.81. At the place of invert, the line always runs full and hence a correct measurement can be made of the flowrate.

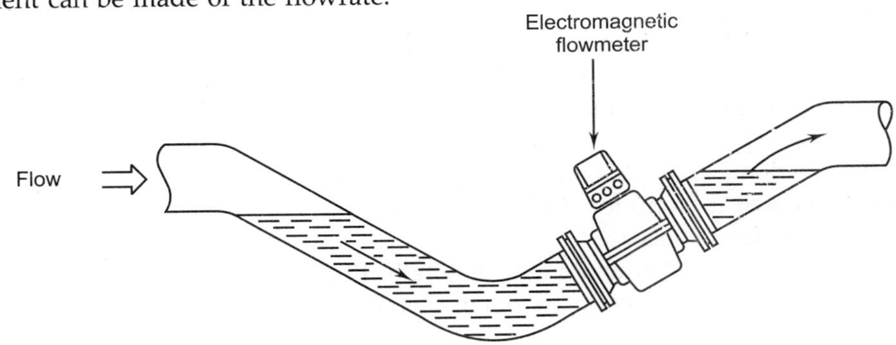

Fig. 9.81: Electromagnetic flowmeter in an invert

Source: **Guide to Flow Measurements**—W Hogrefe et. al. *Courtesy*: Bailey Fischer & Porter GmbH Göttingen, Germany.

However, there is a possibility of deposition of solids, particularly in waste water applications. During high flow, this deposit is either flushed out or a separate flush-water line is provided to eliminate the problem.

9.12.3.10 Electromagnetic Flowmeters for Partially Full Pipelines

Electromagnetic flowmeters calculate the flow rate by multiplying flowrate with the cross-sectional area of the pipe. This demands that the pipeline must always run full of fluid. For partially filled pipelines or free-surface flows, the cross-sectional area obviously becomes a varying quantity and the above relationship does not hold true.

Parti-MAG is a proprietary electromagnetic flowmeter which is designed specially for the above. Here two measurements are made as shown in (*a*) and (*b*) of figure 9.82. Figure (*a*) represents the case when the pipeline is totally filled up with fluid in which case the emf obtained across the electrodes is proportional to velocity. In figure (*b*) the emf obtained would be zero when the pipeline is empty.

When the pipeline is partially but more than half-filled, the emf obtained would be less than the case when it was full of fluid for case (*a*) while it would be non-zero for figure (*a*). In this case; it has been observed that for such a partially filled pipeline, the emfs obtained corresponding to figures (*a*) and (*b*) are a function of both the flow velocity and the fill-level. These two emfs are fed to a processor and the flow rate is computed via software.

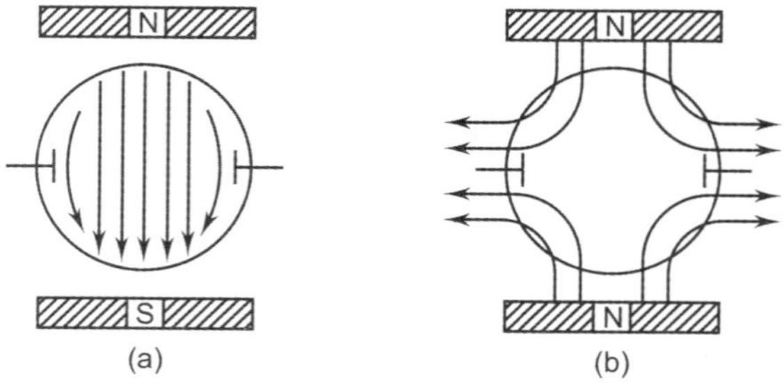

Fig. 9.82: (*a*) Magnetic field distribution when pipeline full of fluid (*b*) pipeline more than half-filled: *Source*: **Guide to Flow Measurements**, W Hogrefe *et. al. Courtesy*: Bailey Fischer & Porter GmbH. Göttingen, Germany

For cases where the pipeline is less than half-filled, an arrangement, as shown in figure 9.83 is employed. Here a total of three electrode pairs are employed with the bottom–most electrode pair capable of measuring a fill level of 5% of the total cross-sectional area. This meter can measure flow rate with a high degree of accuracy even at very low flow velocities and at very low levels.

An accuracy better than 1% and 1.5% for totally filled and partially filled pipes respectively are obtained for such a particular design.

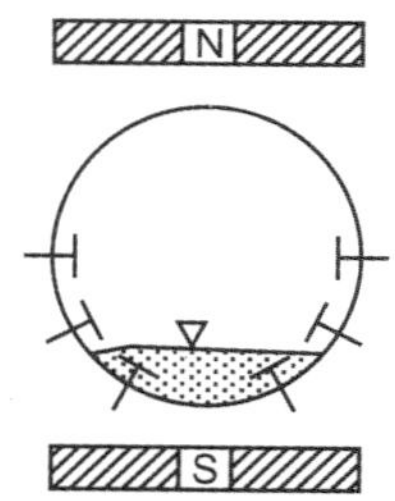

Fig. 9.83: Electrode arrangement

Source: **Guide Flow Measurements,** W Hogrefe *et al. Courtesy*: Bailey Fischer & Porter GmbH, Göttingen, Germany.

9.12.4 Ultrasonic Flowmeters

9.12.4.1 Introduction

Velocity of fluid affects the sonic speed of a pressure wave when it is transmitted through such a fluid. The effect is known as Doppler effect after the name of its discoverer. A typical frequency of 10 MHz is utilised for the above. There are two types of liquid ultrasonic flowmeters.

- Time of flight flowmeter (or transit time or counter propagating type) and Doppler frequency shift flowmeter (or Reflection type)

9.12.4.2 Time of Flight (Transit Time) Flowmeter

9.12.4.2.1 Description and Operating Principle

Different versions of time of flight or transit time flowmeters are shown in figure 9.84 (*a-d*). While figures (*a*) and (*b*) represent single and twin transmitter-receiver system respectively, figure (*c*) is an oscillating loop method and figure (*d*) represents a transit time flowmeter in which both transducers have a transmitter and receiver embedded into it.

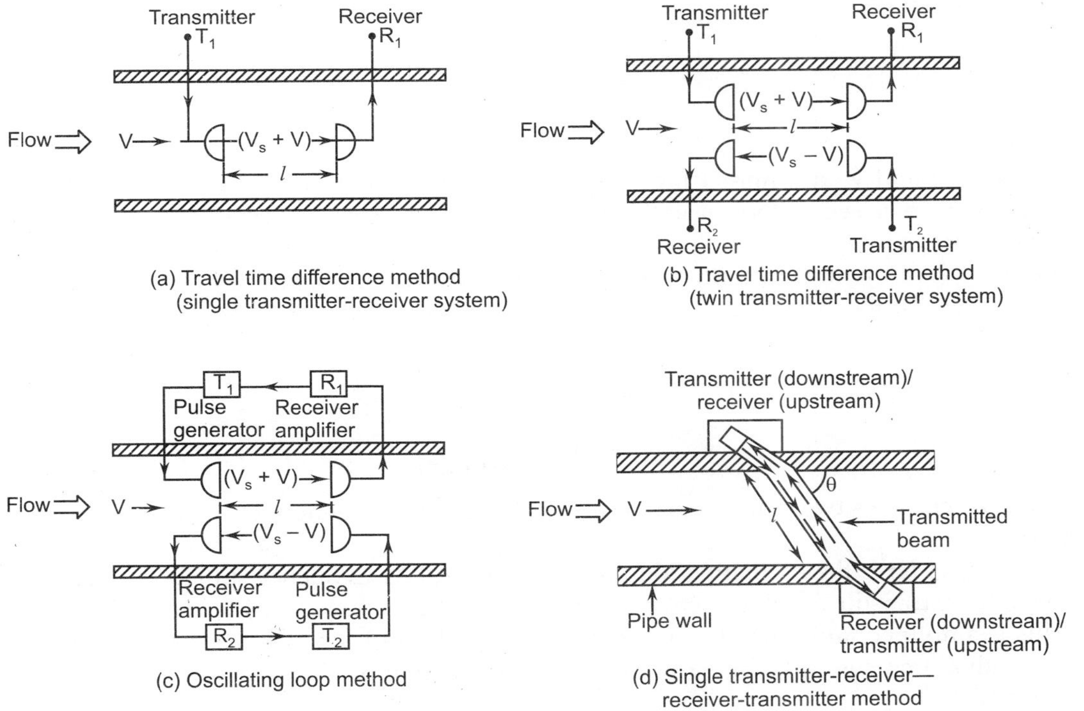

Fig. 9.84: Ultrasonic flowmeters of transit time type

For the transmitter and receiver functions, piezoelectric crystal transducers are used. In the transmitter part, a short burst high frequency (~ 10 MHz) is applied to the crystal which thereby starts to vibrate. Since the crystal is in physical contact with the fluid, this crystal vibration is transmitted to the fluid and propagate through it. The crystal in the receiver receives this pressure fluctuation and it starts vibrating accordingly. This vibration in the crystal produces an output electrical signal. The sonic or accoustic beam should be fairly narrow and well-defined.

The crystal diameter should be large compared to the wavelength of oscillation. This would ensure the crystal to be an efficient transmitter of accoustic energy.

We assume l to be the distance between transmitter and receiver, V the velocity of fluid and V_S to be the velocity of sound in the fluid.

Assuming a static fluid, a pulse from the transmitter reaches the receiver in time $t_o = \dfrac{l}{V_S}$.

If now the pulse travels in the direction of fluid flow, as in figure (*a*), then the transit time is

$$\begin{aligned} t &= \frac{l}{V_s + V} = \frac{l(V_S - V)}{(V_s + V)(V_s - V)} \\ &= \frac{l(V_S - V)}{V_S^2 - V^2} \\ &\simeq \frac{l}{V_S}\left(1 - \frac{V}{V_S}\right), \text{ assuming } V << V_s. \end{aligned}$$

$$= t_o \left(1 - \frac{V}{V_s}\right)$$

Thus, the difference in the travel time, Δt, is

$$\Delta t = t_o - t$$
$$= t_o \cdot \frac{V}{V_s}$$
$$= \frac{lV}{V_s^2}$$

Knowing l and V_s and measuring Δt, V can be found out. Now, V_s varies with temperature and it appears in its second power in the equation. It will thus lead to erroneous measurement of V. In a practical situation, Δt is of the order of µsec and hence it must be measured very precisely. Figure 9.84 (*b*) is a twin transmitter-receiver system. We assume t_1 and t_2 to be the transit times along and against the flow respectively. The travel time difference Δt here is:

$$\Delta t = t_2 - t_1$$
$$= \frac{l}{V_s - V} - \frac{l}{V_s + V} = \frac{2l\,V}{V_s^2 - V^2}$$
$$\simeq \frac{2lV}{V_s^2}, \quad \text{assuming } V << V_s.$$

Here, Δt becomes twice the previous case. But Δt is still dependent on V_s identically to figure 9.84 (*a*).

Figure 9.84 (*c*) is the oscillating loop method, also called frequency difference method. T_1 is a transmitting transducer which sends out a pulse. This is received by the receiving transducer R_1. This is amplified and immediately fed back to T_1. Thus a train of pules circulate between T_1 and R_1 with a time period t_1. Similarly the T_2–R_2 combination produces another train of pulses of time period t_2.

Thus,
$$f_1 = \frac{1}{t_1} = \frac{V_s + V}{l}$$

and,
$$f_2 = \frac{1}{t_2} = \frac{V_s - V}{l}$$

Thus, the difference in frequency, also called beat frequency, becomes:

$$\Delta f = f_1 - f_2 = \frac{V_s + V}{l} - \frac{V_s - V}{l} = \frac{2V}{l}.$$

This expression is independent of V_s—the velocity of sound in fluid. Normally, the frequency difference is very small. Thus, for a reasonably good resolution to be obtained, a long counting interval is needed. But this results in a delayed response.

Figure 9.84 (*b*) and (*c*) employ in all four transducers. But, now-a-days, mostly two are used instead of four. This reduces cost as well as errors due to length of path changes caused by deposit on the transducer faces.

A transit time flowmeter that employs two transducers instead of four is shown in figure 9.84 (*d*). This is a 'clamp-on' transducer type flowmeter unlike the 'wetted' transducer flowmeters shown in figure 9.84 (*a–c*).

Each of the two transducers act both as transmitter and receiver. Let the upper transducer acts as a transmitter and sends a pulse train of ultrasonic sound of certain frequency through the flowing fluid. The lower transducer then acts as a receiver of pulses. Let it receives the pulses from the transmitter at a repetition frequency f_1. In this case, the ultrasonic sound travels downstream. When the lower transducer acts as a transmitter of sound waves and the upper one as a receiver, sound waves travel upstream. In this case, let the repetition frequency be f_2. Thus,

$$f_1 = \frac{V_s + V\cos\theta}{l}$$

and

$$f_2 = \frac{V_s - V\cos\theta}{l}$$

hence,

$$\Delta f = f_1 - f_2 = \frac{2V\cos\theta}{l}$$

Thus, the transit time difference is

$$\Delta t = 1/\Delta f = \frac{l}{2V\cos\theta}$$

This expression is again independent of ultrasound velocity V_s.

In this 'clamp on' type meter, the two transducers are clamped on outside the pipe. Thus, installation is totally hassle free and no obstruction to fluid path takes place. But they have a problem of 'accoustic short circuiting'. That is, the receiving transducer receives both liquid-path and pipe-path signals. Again, the sonic path may be changed due to clamp slippage or temperature variation.

The transit time ultrasonic meters are sensitive to flow velocity profile. They give different readings for axisymmetric profiles of different shapes but the same average velocity.

9.12.4.2.2 Features

- Accuracies range from ± 1% to ± 5% of rate over flow ranges of 10:1 to 40:1.
- Transit time difference is of the order of ns. This, thus, requires very complex electronic circuit.
- Head loss is essentially zero.
- Wetted transducers are more accurate than clamp-on type.
- The sonic transmission frequency is in the range of 1 to 5 MHz for liquids and 100 to 500 MHz for gases.
- Applicable for relatively clean fluids. This minimizes signal attenuation and dispersion. Bubbles cause more attenuation than solids.
- Selection of transducer material is dependent on process fluid. The fluid must not react with the transducer material.
- Process temperature must be kept in mind while selecting the flowmeter.

9.12.4.3 Doppler Frequency Shift Flowmeter

9.12.4.3.1 Description and Operating Principle

Doppler ultrasonic frequency shift flowmeters operate by reflection of sonic energy from scatterers present in the fluid. The scatterers may be particulate solids, bubbles or strong eddies. As an example, for a 20 cm pipe, 10 per cent by volume of 40 μm reflectors or 0.2 per cent of 100 μm reflectors are needed for proper scattering to take place.

An ultrasonic Doppler frequency shift type flowmeter is shown in figure 9.85. A continuous wave ultrasonic signal (~ 0.5 to 10 MHz) is generated with the help of a piezoelectric crystal oscillator. This propagates through the fluid having a uniform velocity v. It is assumed that the scatterers also move in the fluid with the same velocity v. The sonic signal is scattered by suspended particles (scatterers) in the fluid. This producs a change in signal frequency which is detected by the receiver. Δf, the Doppler shift in frequency, is the difference between transmitted

(f_T) and received (f_r) frequencies. It can be shown that Δf is proportional to the volume flow rate of fluid. From analysis, it can be shown

$$\Delta f = f_T - f_R = \frac{2f_T \cos\theta}{V_s} V$$

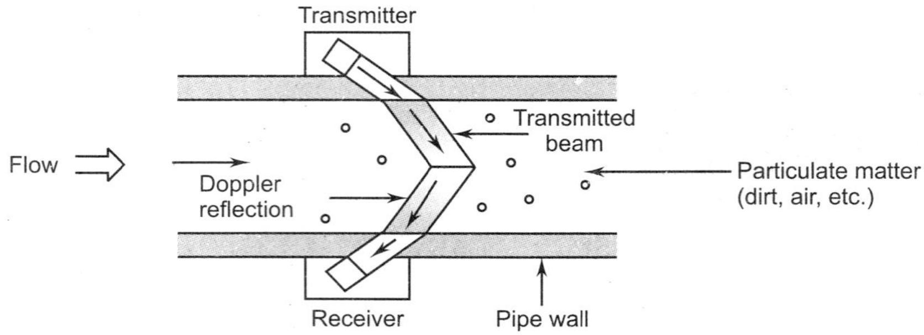

Fig. 9.85: Doppler frequency shift type flowmeter

V_s is the sonic velocity and θ is the angle that the transmitted and reflected waves make with the axis of flow in the pipe.

A change in V_s causes a change in θ. This makes the flowmeter to be independent of temperature effects.

Thus, $$V = k.\Delta f$$

where, $$k = \frac{V_s}{2f_T \cos\theta}, \text{ a constant.}$$

Thus, flowing fluid velocity can be found out by measuring Δf. At the receiver, instead of receiving a signal of frequency f_R, beat frequencies are received. The beat frequencies are the sum and difference of f_T and f_R. f_R can be extracted from this by FFT analysis. It is also possible to determine the peak value of Δf, which is a measure of fluid flow velocity.

An improvement in the basic method just described uses two different transmit frequencies. A special FFT technique enables measurement of 'point' velocity at the intersection of transmitted and reflected beams. This point velocity can be calibrated to give total volume flow rate. But for this, the flow profile must be reproducible.

There are several designs for the sound path to travel through the fluid. These can be chordal, diametric or along the axis of flow of fluid, as shown in figure 9.86.

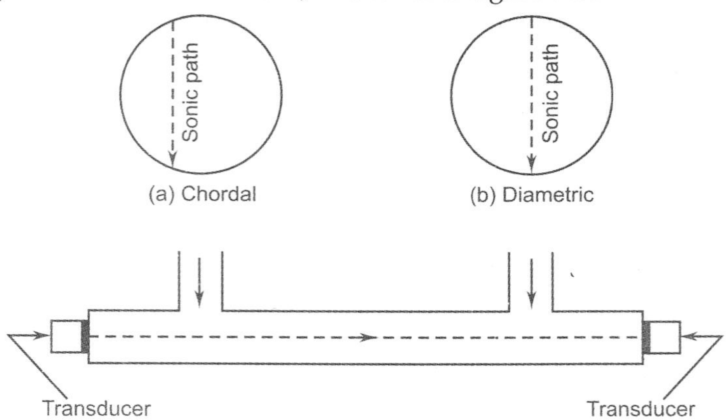

Fig. 9.86: Acoustic paths

9.12.4.3.2 Features

- Accuracy range ± 2% to ± 5% of flow rate.
- No restriction on flow rate or Reynolds number, but the flow velocity must be such that scatterers must be in suspension.
- Good rangeability
- Convenient installation for clamp-on type.
- Mainly used for liquids, but also applied for gases and steam.
- High temperature application possible.
- No head loss
- Applicable for both open channel flow and pipeflow.
- Application specific
- Long term stability not very good.
- Sensitive to velocity profiles.
- Applicable to varied pipe sizes.
- Any pipe material which does not absorb ultrasonic energy is suitable. For this reason, clay, concrete or very porus cast iron is unsuitable.
- Operating temperature limit is roughly 100°C.

9.12.5 Anemometers

9.12.5.1 Introduction

Anemometers find extensive application for the measurement of flow rate of fluids. Flow rate of non-conducting fluids in open channels and closed pipes and of gases in closed pipes can be very easily measured by such instruments.

9.12.5.2 Probe Types, Construction and Principle of Operation

Hot-wire and hot-film type resistance transducers are employed as anemometer probes in anemometric measurements. The two types are shown in figure 9.87.

The probes (either hot-wire or hot-film) are generally made of tungsten, platinum or platinum-iridium alloy. The wire diameter may vary from 0.0005 to 0.03 cm and a length of 0.1 to 1 cm. The length is normally half the diameter of the pipe through which fluid flows.

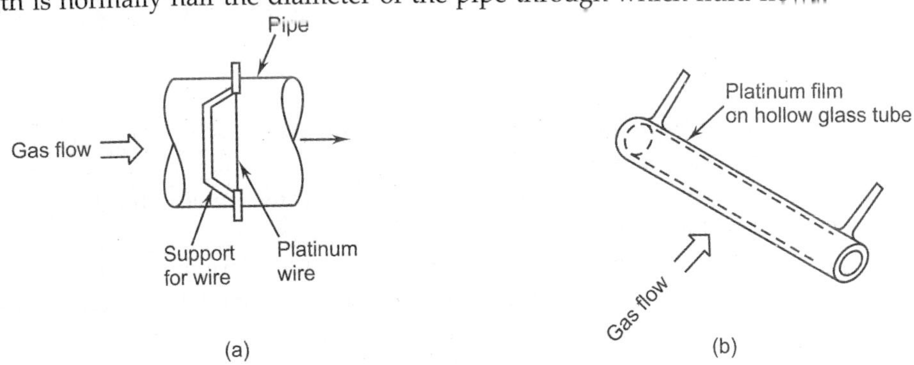

Fig. 9.87: Anemometer probes: (*a*) Hot-wire (*b*) Hot-film

The hot-wire filament is centrally placed inside the pipe and the axis of the wire is normal to the flow direction. The wire will snap if the flowing fluid contains particulate matter or if the flow rate is very high. In such cases, the filament must be rugged. A wire filament is tested under such actual hostile test conditions before its use. A smaller diameter wire gives a larger resistance per unit length and less thermal inertia (*i.e.,* faster response). But it may snap under large fluid pressure, *i.e.,* large flow rates. Hot film element, shown in figure 9.87 (*b*) is used when the fluid contains particulate matter or its velocity is very high. The film provides the required ruggedness and is made by depositing a film of platinum as a glass or ceramic base. The film element provides greater mechanical strength and can be used at a very high temperature by providing internal cooling.

The flow sensing element (hot wire or hot filament) is electrically heated by connecting it in a Wheatstone's bridge configuration. When such a probe is introduced in a flowing fluid, it is cooled and consequently there is a drop in resistance of the probe. The rate of cooling depends on several factors:

- Physical properties of the fluid
- Difference in temperature between the hot wire or film and fluid
- Shape, size and physical properties of the probe
- Velocity of fluid

Thus, an instrument can be designed to measure fluid flow rate provided the first three conditions are effectively made constant.

The flowing fluid should be non-conducting in nature so that the resistance of the wire or filament remains unaffected by the surrounding medium.

Heat loss by convection of the wire redness by a factor of two when it is oriented along the flow direction compared to the situation when the wire is placed normal to flow direction. Sometimes a single probe contains two matched wire elements with their axes at 90° to each other. This is employed to find the true flow direction. When such a probe is placed in a flowing fluid and both the wires indicate the same resistance value, it is concluded that the flow direction lies at 45° to either of the filament axes.

When the wire is placed in the flowing fluid, loss of heat due to convection is balanced by heat generated in the current carrying wire. It is assumed that radiation loss from the hot wire and conduction loss through the leads of the wire are neglected. The energy balance equation can be written as:

$$I^2R_W = KhA(t_w - t_f)$$

where,

I = current through the hot wire

R_w = resistance of the wire at temperature t_w.

t_w = temperature of the wire.

t_f = temperature of the fluid.

A = Surface area of the wire exposed to the flowing fluid.

h = heat transfer coefficient

and k = a conversion factor (from electrical to thermal power)

Now, hot wire anemometers are available in two forms: Constant temperature (resistance) type and constant current type. These are now discussed as under:

9.12.5.3 Constant Temperature (Resistance) Type Hot Wire Anemometer

This mode is used to measure steady velocity and is shown in figure 9.88. The hot wire R_w is placed in series with an ammeter in one of the arms of a Wheatstone's bridge. R_w and R_1 are very low resistances (~1Ω). There are two resistances R, R in the rest two arms with $R >> R_1$.

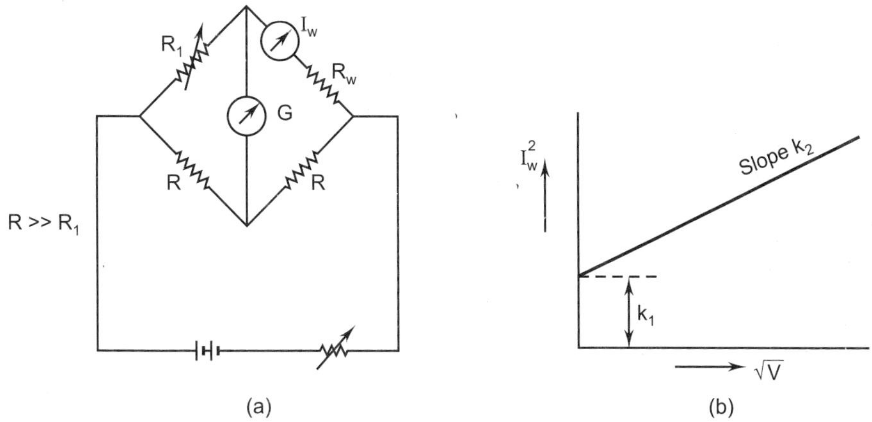

Fig. 9.88: (*a*) Circuit for constant temperature type anemometer; (*b*) Calibration curve

Before putting the circuit for measuring unknown velocities, it is calibrated. For this, the hot wire is exposed to known velocities over its intended range using the fluid which would be ultimately used. The temperature and pressure of the fluid should also be the same during calibration and measurement. Velocity measurements during calibration must be done with adequate accuracy so that a standard calibration curve is obtained for the fluid. As the flow rate increases, the hot wire resistance decreases and a higher amount of current is required through it in such a way that its temperature is maintained constant.

For a low fluid velocity, the resistance (~ 2*k*) in series with the battery is adjusted initially for the hot wire current I_w. This current should be high enough for adequate sensitivity of measurement but low enough to prevent burn-out of the filament. The bridge is balanced by R_1 after thermal equilibrium is reached. Each time the fluid velocity is changed, time is allowed for thermal stability, and R_1 again adjusted for bridge balance. For each such balance, corresponding I_w is noted.

For this case, the relationship between current I_w and flow rate is:

$$I_w^2 = k_1 + k_2\sqrt{V}$$

where, k_1 and k_2 are constants.

Once the calibration curve is drawn, it can be used to determine any unknown flow rate of the same fluid under the same temperature and pressure conditions of flow. Balance condition cannot be obtained if the velocity of fluid fluctuates.

A linear output from a constant temperature anemometer system allows accurate measurements of large amplitude, low frequency velocity fluctuations. It is also a preferred choice for studies of steady flow velocities or flow reversals. A constant current type anemometer system is, however, preferred for low-turbulence and high accuracy measurements.

9.12.5.4 Constant Current Type Hot Wire Anemometer

Figure 9.89 represents a constant current type hot wire anemometer. Current through the hot wire is maintained at a constant, suitable value I_w. Operating current I_w is set to such a value so as to derive the same advantages as in constant temperature case.

Chapter 9

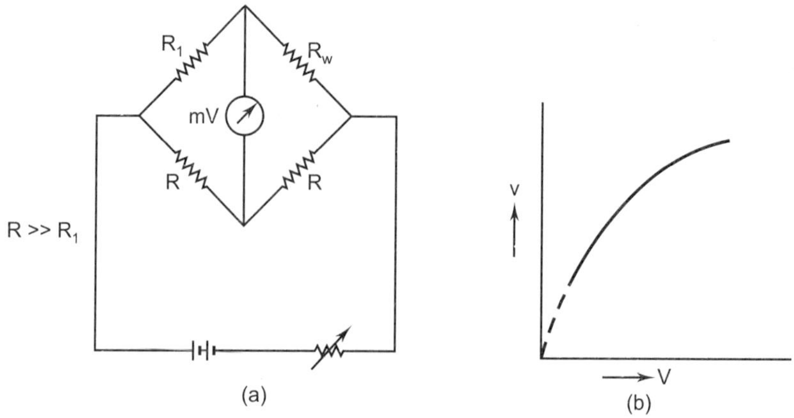

Fig. 9.89: (*a*) Circuit for constant current type anemometer; (*b*) Calibration curve

A calibration curve is first drawn for different flow rates. For each flow rate, resistance of hot-wire varies and consequently I_w varies also. By adjusting R_1, I_w is restored to its initially adjusted value. The millivoltmeter reading for each case is noted.

After calibration is over, it can be put to measure an unknown flow rate; it should be maintained at the same value as it was during calibration.

There is a possibility of the hot wire burn-out under no flow condition. Choice of lower operating value for I_w may prevent this, but at the cost of reduced sensitivity.

A constant current anemometer can be used to measure steady as well as fluctuating component of velocity.

For the case of liquids, the frequency of fluctuating components is very low, although it may be as high as 100 kHz for wind turbulences. Thermal inertia of the hot-wire element plays a major part in these cases. Such thermal inertia or time constant can be reduced by a proper choice of size and configuration of the wire. But there is a practical limit to such reduction leading to non-availability of flat frequency response upto 100 kHz. The time constant is also dependent on temperature coefficient of resistance of the wire material, flow velocity and wire current.

A value of 1ms for time constant is a realistic figure. This would ensure flatness in frequency response upto a few hundred Hz. Beyond this, dynamic compensation of measuring system is necessary.

Two outstanding features of a constant current anemometer are its very low noise level and very good sensitivity.

9.12.5.5 Dynamic Compensation Technique

Dynamic compensation by employing lead networks can be utilised for constant current type anemometers. The unbalance bridge voltage is amplified by a DC amplifier. It is then fed to the compensating network. Its output would indicate both the average and the fluctuating components of the flow rate.

The measuring circuit is modified, as shown in figure 9.90, if only the fluctuating component of velocity is to be displayed on a CRO. The AC amplifier removes the DC component and only the AC part is allowed to pass through. Passive networks are used to realise dynamic compensation.

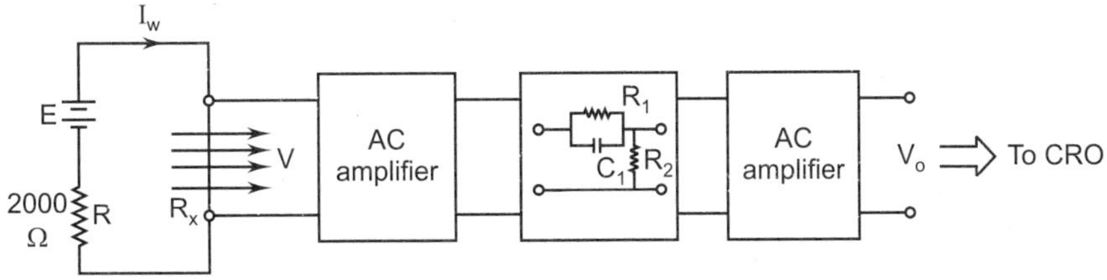

Fig. 9.90: Dynamic flow rate measurement

A compensation network reduces the signal amplitude considerably. Correct compensation depends on the time constant of the fluctuating flow field. This is not known with certainty and varies with flow conditions.

Figure 9.91 shows the frequency response characteristic of a constant current bridge circuit showing three outputs. Above a cut-off frequency f_c, which depends on the sensor material, diameter and coating, the sensor acts as an integrator when dynamic compensation is applied. In such a case, fluctuations in fluid flow are attenuated. With dynamic compensation, response curve becomes flat upto a maximum of 120 kHz. Flatness of response can be achieved only above 1–2 Hz, as is evident from the figure.

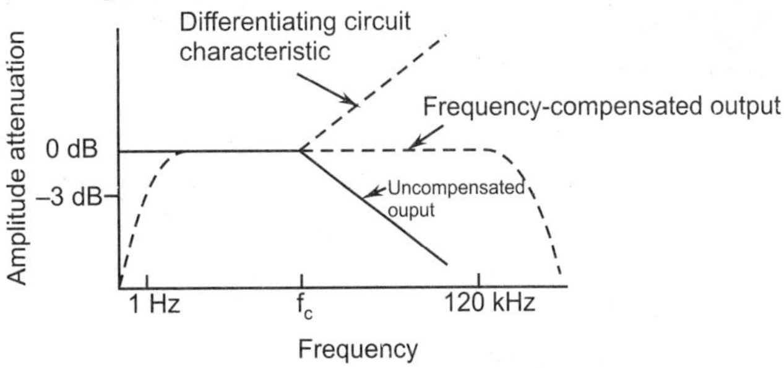

Fig. 9.91: Frequency response characteristics of a constant-current bridge anemometer system

A constant temperature type anemometer, with self-balancing, is shown in figure 9.92. It employs negative feedback.

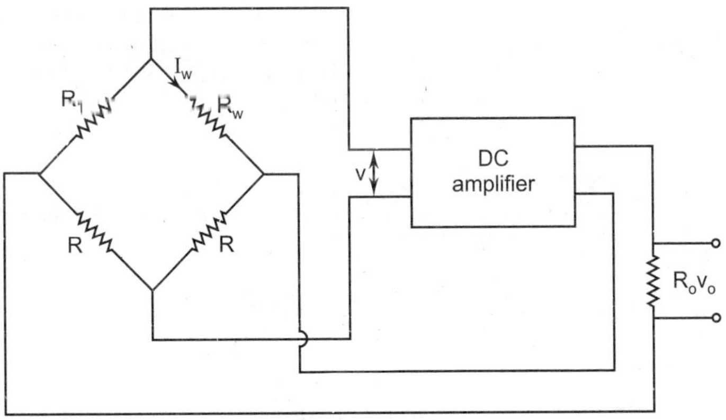

Fig. 9.92: Constant temperature anemometer, self-balancing type

A high gain DC amplifier amplifies the unbalance voltage v. It is then converted into a proportional current which is used to supply the bridge. The bridge current required for balance is a measure of flow velocity. Negative feedback reduces the time constant by a factor of 100. Flatness in frequency response upto 50 kHz is guaranteed. The circuit can record both steady and fluctuating components of velocity.

9.12.5.6 *Laser Doppler Anemometer (LDA)*

9.12.5.6.1 Introduction

It is an optical type fluid velocity meter and is sometimes known by the name Laser Doppler Velocimeter (LDV). It can measure the instantaneous velocity of gas or liquid flowing in a transparent channel.

Principle

The operating principle of an LDA is based on the Doppler shift in frequency encountered by light when it is scattered by an object moving relative to the source of light. A laser beam is focussed at a point in the fluid stream. The light beam is scattered by particles in the fluid and sensed by a photodetector. The output of the photodetector is processed which gives the magnitude of the Doppler frequency shift. This is proportional to the instantaneous fluid flow rate.

9.12.5.6.2 Description and Operating Principle

An arrangement for a Laser Doppler Anemometer is shown in figure 9.93. A He-Ne laser of 5 mW–15 mW power of wavelength 632.8 nm is used a source of light for fluid flow measurements in the vicinity. When the LDA is to be installed at a greater distance, laser power is to be increased. A beam splitter splits the laser beam into two equal parts and is fed to a focussing lens. The two beams are focussed at the focal point inside the fluid where the fluid flow rate is to be measured. An interference fringe pattern of alternate high and low intensity is formed at the focal point, shown in figure 9.93 (*b*). When the suspended particles in the fluid pass through the region of high intensity, they would scatter the light. The scattered light would have a Doppler shift in frequency. The scattered light is received by the receiver lens and directed to a photodetector. An electronic signal processing circuitry processes the detector output. This would be a varying electrical signal whose frequency is proportional to the speed at which the suspended particles cross the interference fringes.

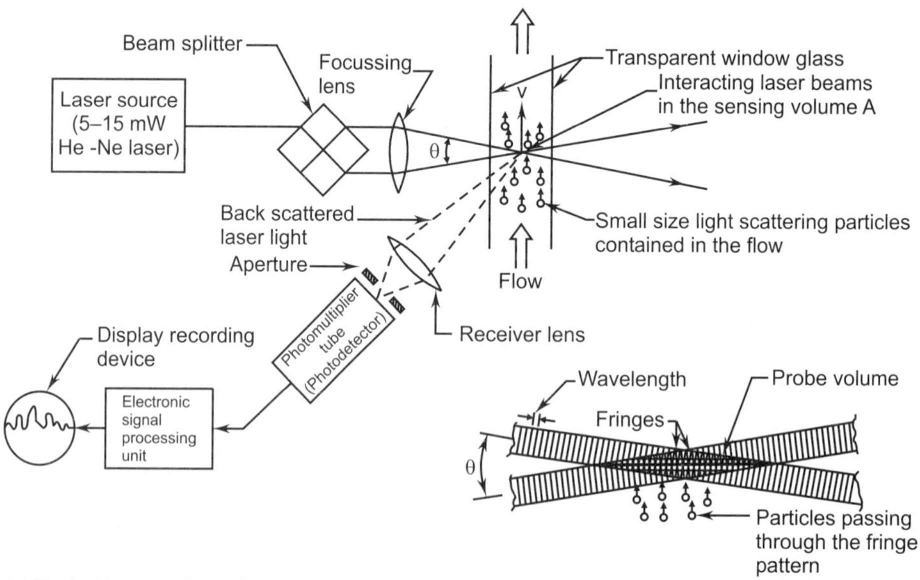

(a) Typical layout of laser Doppler anemometer; (b) Interference fringes in probe volume A

Fig. 9.93: Details of laser Doppler anemometer in dual beam or fringe mode

The gap between the fringes is given by:

$$y = \frac{\lambda}{2} \sin\left(\frac{\theta}{2}\right)$$

where, λ is the wavelength of the laser beam and θ is the angle between the two converging beams.

The frequency f of the electrical signal produced by a particle in the fluid moving across the light and dark fringe patterns with a velocity component v normal to the fringes is

$$f = \frac{2v \sin(\theta/2)}{\lambda}$$

When the laser source and the photo detector are in the vicinity, as it is in the figure, it is back-scattered mode. If the photodetector is located on the side opposite the flowing fluid, it is called forward-scattered mode.

Velocity components in different directions at a point can be achieved by either polarisation method using a single laser or two lasers of different wavelengths. Signal processors can be frequency tracking type or counter type. Frequency trackers work nicely with high particle concentrations while counter type is used for high flow rates. Hot wire anemometers can measure turbulence upto 150 kHz while the same is roughly 3 kHz for an LDA.

9.12.5.6.3 Advantages

- Non-contact type measurement
- Very high accuracy – of the order of ± 0.2%
- Sensing volume very small – a cube 0.2 mm side will do. This is volume *A* as referred to in the figure.
- Velocity measurement is direct in nature – unlike inference from pressure as in a Pitot tube or heat transfer, as in hot wire anemometer.
- Suitable for both liquid and gas flow measurements.
- Suitable for hydro dynamically and aero dynamical velocity measurements.

9.12.5.6.4 Disadvantages

- A transparent pipe portion is needed where velocity is to be measured.
- Expensive, requires skill to operate.
- Not suitable for clean flow.

9.13 OSCILLATORY FLOWMETERS

9.13.1 Types

There are three types of oscillatory flowmeters. They work on the principle of vortex shedding or COANDA effect to infer flow rates of fluids. These are:

- Vortex shedding meter
- Fluidic meter and
- Swirlmeter

9.13.2 Vortex Shedding Flowmeter

9.13.2.1 Introduction

When a blunt unstreamlined obstacle (known as bluffbody) is inserted in a flowing fluid, perpendicular to the flow direction, vortices in opposite directions are produced alternately downstream of the body on either side. The phenomenon is known as 'Von Karman (1912) vortex street'. The vortex generation phenomenon is shown in figure 9.94.

9.13.2.2 Principle of Operation

Since the bluffbody is not streamlined in nature, flow on the downstream side cannot follow the contours of the body under turbulent flow conditions. In this region, boundary layers of slow moving fluid are formed along the outer surfaces of the body and detached from it. A low pressure area, just behind the object, is formed. These layers, separated from the main flow stream, is rolled into eddies or vortices. A vortex grows and sheds itself off the tail part of the bluffbody. This, thus, gives rise to 'vortex shedding'. It results in velocity and pressure changes around and downstream of the vortex element. Thermal, pressure or ultrasonic detectors are used downstream to measure this vortex shedding frequency. The vortex-shedding frequency is a function of width and length of the barrier and also the ratio of barrier width to meter bore.

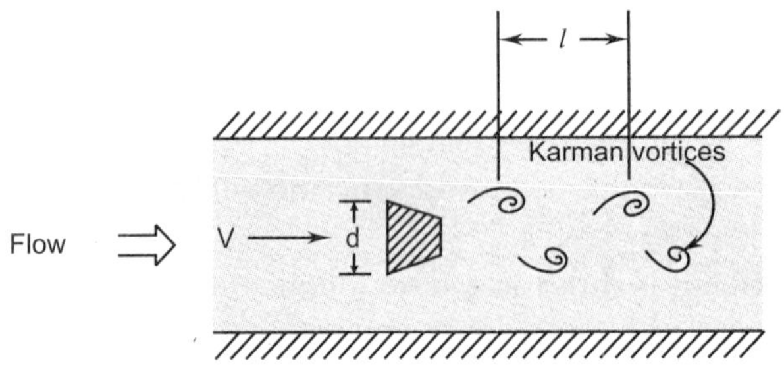

Fig. 9.94: Vortex generation by bluffbody

It should be noted that the distance l between two consecutive vortices is independent of flow velocity. This is known as Karman's principle. The vortex shedding principle is reliable for a Reynolds Number (NR) exceeding 10,000 and under this condition, the vortex shedding frequency f is:

$$f = \frac{N_{st} \cdot V}{d}$$

where,

V = fluid velocity, m/sec

d = characteristic dimension of the bluffbody, m

N_{st} = Strouhal number

The Strouhal number essentially remains constant over a wide velocity range and is also independent of fluid density. The velocity profile, however, influences the constancy of the Strouhal number. Since d is constant for a particular bluffbody, hence f can be made proportional to V provided N_{st} is maintained constant over the flow range for which the meter is used. Figure 9.95 shows the linear range over which the above relation holds for a particular bluffbody. X-axis denotes the Reynolds number (R_n) and consequently flow rate also.

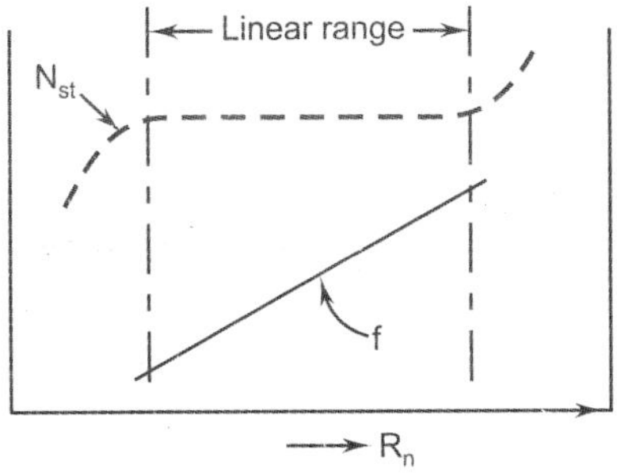

Fig. 9.95

Again, figure 9.96 shows how Strouhal number varies for two different bluffbodies.

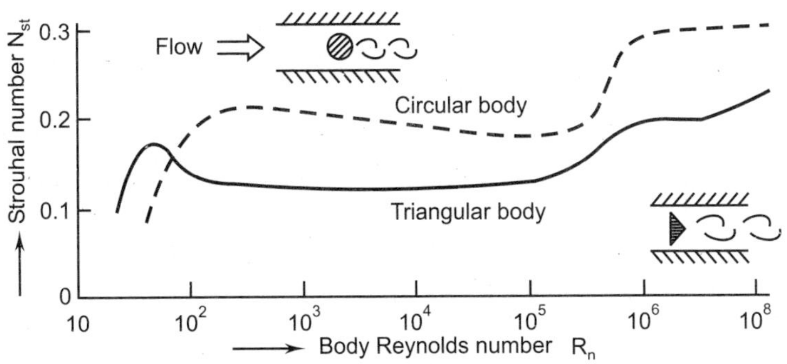

Fig. 9.96: Effect of Geometric shape & Flowrate on Vortex Shedding

Source: **Fluid Flow Measurement,** R A Furness (Longman, UK 1989/p:36). (Reproduced with kind permission of the author).

9.13.2.3 Bluffbody Designs

Figure 9.97 shows some typical bluffbody designs. These are Fischer & Porter, Eastech, Kent, Yokogawa and J-tec. Performance does not change significantly with different designs. Bluffbodies function properly with Reynolds number N_R lying between 10,000 to 40,000. Very recently, twin-body designs have been introduced in the market.

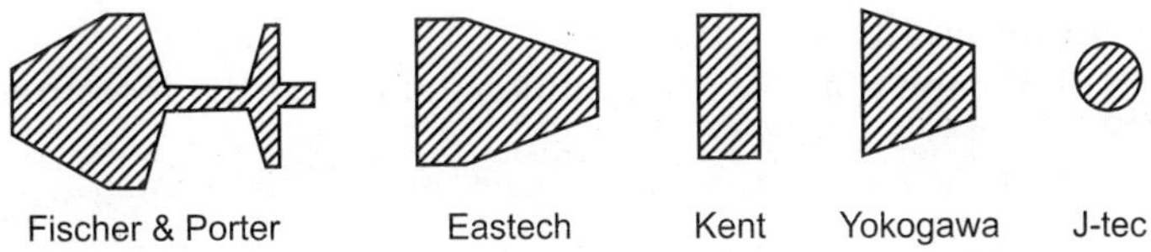

Fig. 9.97: Varied shapes of Bluffbodies used in typical Vortex Meter

Source: **Fluid Flow Measurement,** R A Furness (Longman, UK 1989/p : 37). (Reproduced with kind permission of the author).

9.13.2.4 Methods to Find Vortex Shedding Frequency f

Different frequency sensing schemes have been developed by manufacturers. Once *f* is determined, fluid flow rate can be found out.

Vortices cause alternating forces, *i.e.*, pressure across the bluffbody. Piezoelectric or strain gauge principles can be employed to detect these. The two sensors give two output signals which differ is phase by 180°. The two signal add up and the common noise signal is nullified leading to a large signal to noise ratio. Sometimes hot film anemometers are buried in the bluffbody to detect the periodic flow-velocity variations. Differential pressure created by the vortex will cause oscillations of a small caged ball whose motion can be detected by means of a magnetic proximity pick-up. Vortices would interrupt ultrasonic beams. These interruptions can be counted to determine the flow rate. A thermistor embedded in a bluffbody can be used as a flow sensitive detector.

9.13.2.5 Response

At the maximum flow rate, vortex frequency is in the region of 200–500 Hz and it responds to within 1 cycle to changing flow rates. If a F to V converter is used, then the system response would depend on the *LPF* connected at the output of the converter. This response is very fast but must be brought down to an acceptable value for lower flow rates.

By employing a Ono Sokki type F to V converter (having a zero time constant), velocity measurements can be done for each pulse. If fastness in measurement is not the sole criterion, several pulses can be averaged to have a higher accuracy.

9.13.2.6 Advantages

- A very good long term stability of calibration and repeatability better than ± 0.15% of rate.
- Rangeability approximately 10:1 for liquids and 20:1 for gases.
- Calibration is almost independent of operating conditions like pressure, temperature, density, etc.
- A low installed cost.
- Meter accuracy is very good over a wide flow range.
- Provides a linear digital output.
- No moving parts, low power consumption, little maintenance.
- Applicable for both liquid and gas flow with pipe sizes 15 mm or more.

9.13.2.7 Disadvantages

- Insensitive to low flow rates
- Vibration sensitivity
- Cannot be applied to measure flow rates of viscous fluids because of minimum Reynold's number constraint
- Sensitive to velocity profile distortions.
- A straight run of 50 pipe diameters recommended before the measurement point.

9.13.3 Fluidic Flowmeter

9.13.3.1 Principle and Working

The operating principle of a fluidic oscillatory type flowmeter is shown in figure 9.98. The central channel is filled up by the fluid in figure *a*. A part of this stream is diverted into the side opening which loops back to the inlet, shown in figure *b*. Figure *c* represents the main flow to be pushed into the opposite side wall. In figure *d*, the second feedback is shown. Thus, a complete cycle of oscillation is made and the flow oscillates from one side to another by mechanical means at a frequency which is proportional to the flow rate. The oscillating frequency can be measured by employing a thermistor or a strain gauge transducer. This is put in one of the feedback passages. For measurement purposes, a pulse is created for one oscillation.

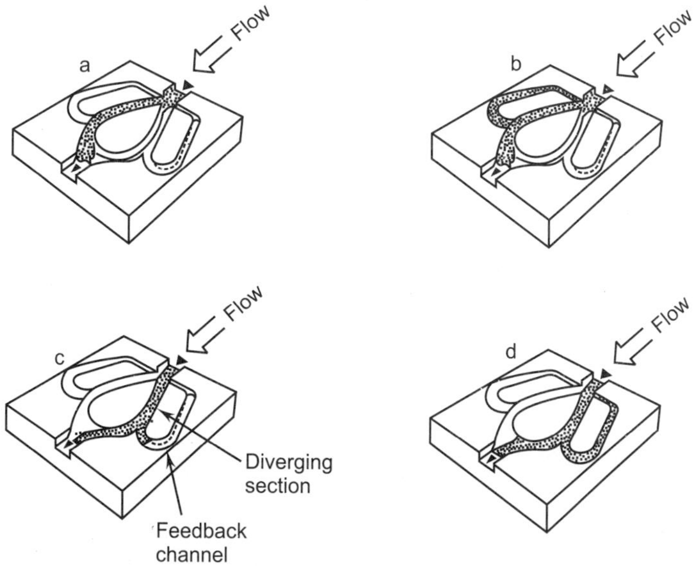

Fig. 9.98: Fluidic oscillatory type flowmeter

Source: **Fluid Flow Measurement**—R A Furness (Longman, UK 1989/p : 38). (Reproduced with kind permission of the author).

9.13.3.2 Features

- Can measure liquid flow rates of lesser Reynolds number
- Linearity about 1% of flow rate
- Has lower operating temperature and pressure
- Restricted to liquids primarily.

9.13.4 Swirlmeter

9.13.4.1 Principle and Working

It is a special form of oscillatory meter. It has no moving parts. Its output is in the form of pulses whose frequency is proportional to flow rate of fluids. It is generally restricted to gas flows. Figure 9.99 shows the schematic of a swirlmeter.

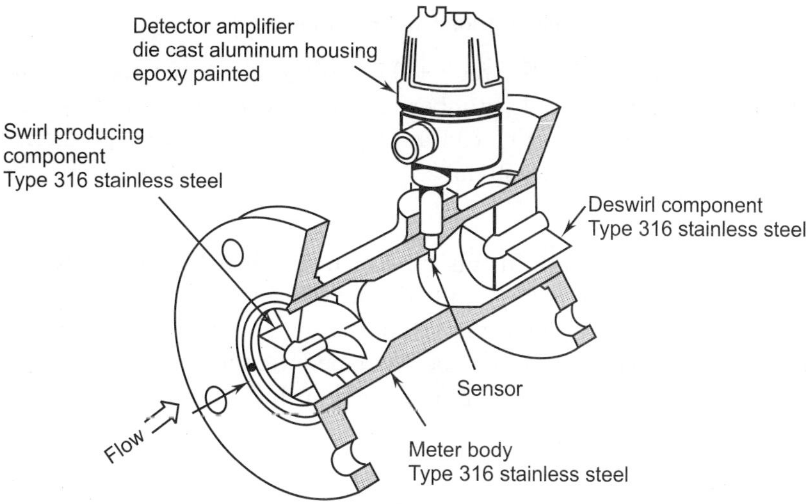

Fig. 9.99: Schematic of a swirlmeter

As the fluid passes through a fixed set of swirl blades, a swirling motion in the fluid is generated downstream of these blades. The fluid is then allowed to pass through a venturi like section where it is accelerated. Lastly, the fluid passes through the deswirler when it expands and consequently is decelerated. At this places, flow oscillates at a frequency proportional to the flow rate. Temperature variation occurs which can be measured by placing a thermistor at the deswirler. This is then converted into voltage pulses. These pulses are amplified, filtered and transformed into square waveform. This can be measured very easily by an electronic counter.

The deswirler straightens the flow before it leaves the meter.

9.13.4.2 Features

- Applicable for Reynolds number approximately between 10,000 to 1,000,000.
- Repeatable to within ± 0.25% of rate in the linear region
- Accuracy of ± 0.75% in the linear operating range of ± 1%

9.14 MISCELLANEOUS FLOWMETERS

9.14.1 Introduction

Several flowmeters are available which don't fall under any of the categories of flowmeters already discussed. Some of the flowmeters which are to be discussed here are *V*-cone flowmeter, wedge meter, gas-injection method, Flow marker method, Compound flowmeter GFO3.

9.14.2 V-cone Flowmeter

A *V*-cone flowmeter is shown in figure 9.100. It is placed at the central part of a constant diameter, precision machined metering tube. A lower pressure, just downstream of the cone, is created. The pressure differential in the upstream portion at the tube wall and downstream at the V-cone is measured and calibrated in terms of flow.

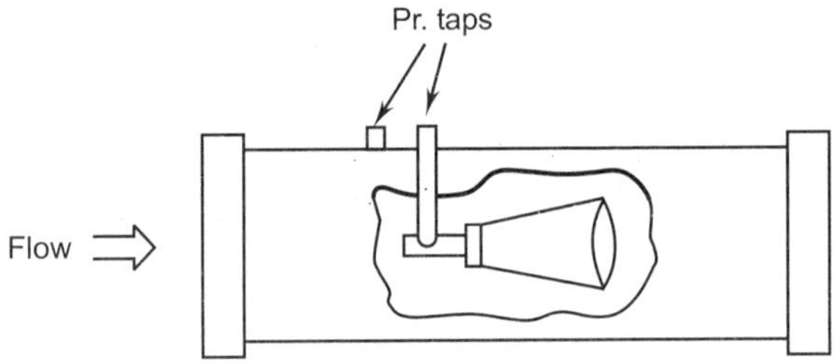

Fig. 9.100: A V-cone flowmeter

Accuracy of V-cone meters lies between ± 0.75% to ± 1%, repeatability of 0.1%. They can measure gas flow rates with Reynolds number less than 8000. These meters require very little straight pipe run where the *V*-cone is placed in the flow – one to two pipe diameters upstream and three to five pipe diameters downstream.

9.14.3 Wedge Flowmeter

A wedge is created in the flow path in this type of meter. It responds to bidirectional flow. A wedge meter is applicable to a whole gamut of flow like highly viscous fluids, gas or stream, slurries, clean, corrosive flow, multiphase flow, etc.

A differential pressure created across the V-shaped wedge is measured by the DP meter and the meter maintains an excellent square root relationship between flow rate and differential pressure. A wedge meter is shown in figure 9.101.

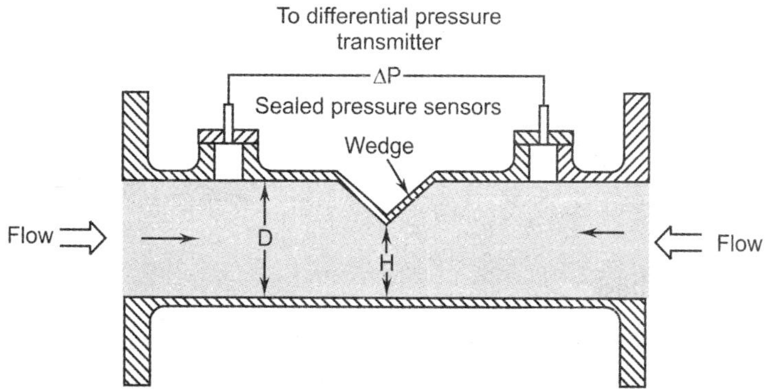

Fig. 9.101: A wedge meter

A wedge meter can be applied to measure flow rates of diverse flow types, as coal tar, fuel oil, iron ore, black liquor, etc. It is equally applicable for either very low Reynolds number (~ 500) or high viscous fluids (~ 300 centipoise).

The advantages associated with a wedge meter are:

- Very low maintenance.
- Operating cost low.
- Installation cost low.
- Applicable for bidirectional flow.
- Can measure flow rates of low conductivity fluids.
- Applicable to various types of fluids, such as abrasive, viscous, non-Newtonian, gummy fluids, etc.
- Generates a scouring action at the wedge which keeps it clean.

9.14.4 Gas-injection Flowmeter

A gas-injection flowmeter is shown in figure 9.102. It has no moving parts and was developed at Brennstoffinstitut, Germany.

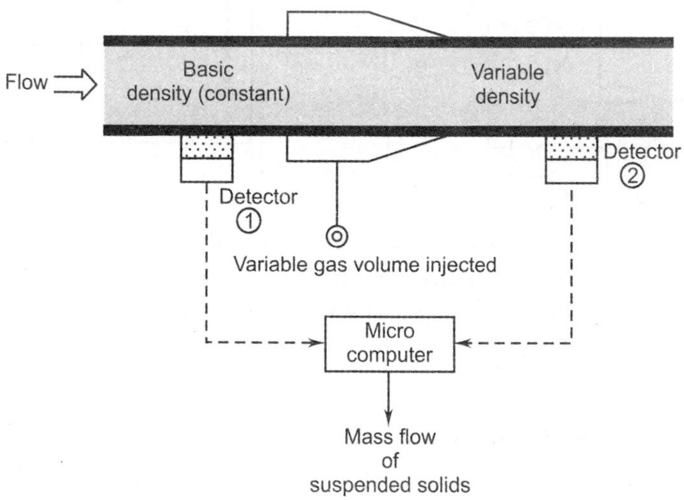

Fig. 9.102: A gas-injection flowmeter

A variable gas volume is injected to the flowing fluid whose flow rate is to be measured. Two detectors are placed—one upstream and another downstream of the place where gas is injected. The outputs of the two detectors are fed to a microcomputer which computes the two variations in density. This is calibrated in terms of flow rate.

Source: **1. Power,** September 1986;
2. Power, June 1995
3. Power, August 1995

9.14.5 Flow Marker Method

Flow markers in the form of ionic, thermal, optical radio active disturbances are introduced in the flow pipe. Some characteristics of flow is changed because of the above. This change is measured and calibrated in terms of the flow rate.

An ionic type flow marker is shown in figure 9.103, which is also known as electrolytic tracer method. Along the flow, a salt solution is introduced which leads to an increased conductivity. Two pairs of electrodes, *c-c* and *d-d*, are placed downstream, separated by a distance l. i_1 and i_2 currents are recorded in the ammeters when the electrolyte reaches electrodes *c-c* and *d-d*. The time difference t between the two records is noted. Then from the formula, flow rate is given by

$$q = \frac{\neq d^2 l}{4} \times \frac{1}{t}$$

where d is the pipe inside diameter.

If the instant of salt injection is known, measurements can then be made with a single pair of electrodes.

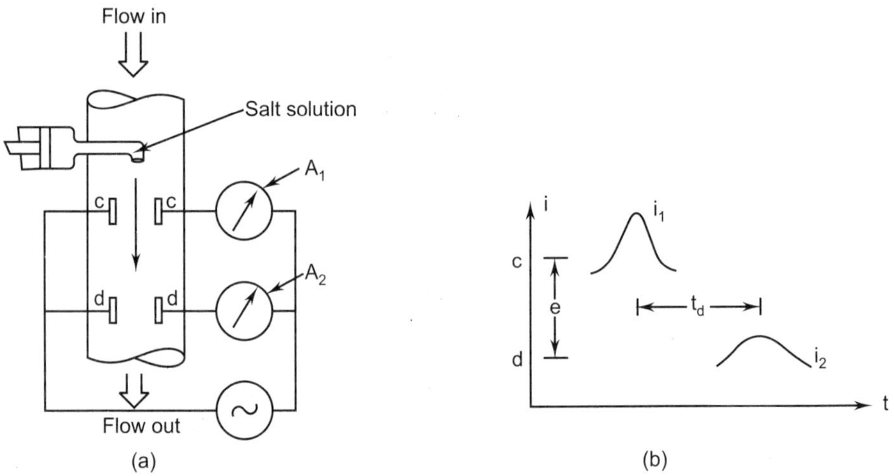

Fig. 9.103: Flow marker method

Errors in measurement by flow marker method are due to

- Current profiles i_1, i_2 are not sharp and spread over some area.
- Uncertainty in time over sensing by electrodes and their subsequent indication by ammeters.

9.14.6 Compound Flowmeter GFO3

In many process industries, flaring of gases is a common phenomenon. As an example, excess blast furnace gas produced in blast furnace is flared from excess gas bleeder station. Both the flow rate and composition of the gas to be flared are measured so that the burnt-out gases escaping into the atmosphere meet stringent clean air regulations.

Time-of-flight ultrasonic flowmeters with molecular weight sensing are employed at present for the above. Each such flowmeter has two ultrasonic sensors: a pressure and a temperature transmitter.

Very recently, Fluid Components International (FCI), San Marcos, California, USA has marketed a flowmeter called the GFO3 and shown in figure 9.104. This instrument has a matched pair RTD (for measuring gas flow rate). It also has a gas compensator that monitors heat-transfer properties of the gas. The instrument is software based and adjusts the flare rate as per the gas contents.

The instrument is designed to operate for temperature ranges 233 to 450 *K*, pressure upto 0.7 MPa. g. The advantages associated with GFO3 meter are:

- single-tap installation
- low maintenance
- considerable cost savings.

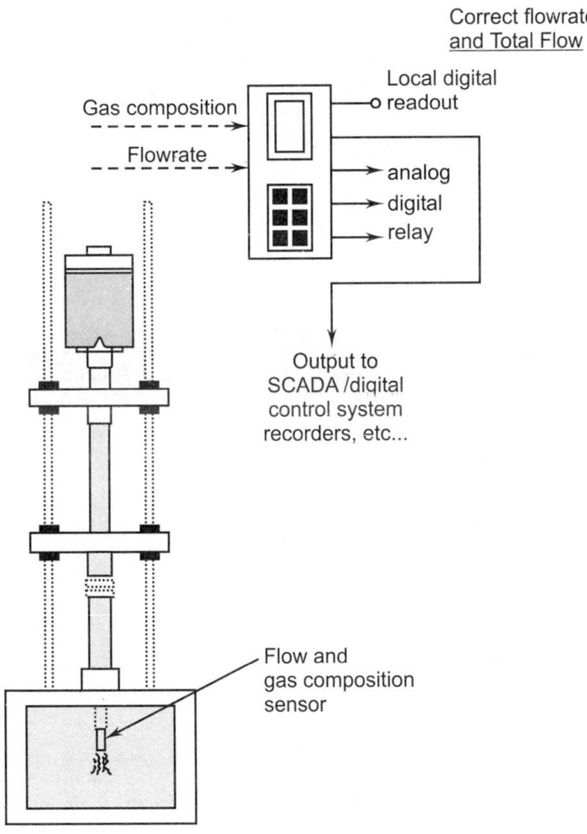

Fig. 9.104: The GFO3 flowmeter

9.14.7 Cross-correlation Flowmeter

A flow stream, which is noisy in nature, can be used to build a cross-correlation flowmeter. The noise pattern must persist for long enough to be detected by the two identical detectors D_1 and D_2. Flow rate is obtained by dividing the known distance between the two detectors by the transit time taken by the noise to travel from D_1 to D_2.

Cross-correlation flowmeters can be applied to a host of process variables like pressure, temperature, density, conductivity, ultrasonics, etc. Such meters have been developed to measure heavy water flow rate using ultrasonics, paper pulp flow rate using photometric sensors, etc.

A cross-correlation flowmeter is shown in figure 9.105, $m(t)$ and $n(t)$ are the upstream and downstream signals obtained from detectors D_1 and D_2 placed at positions A and B respectively.

Advantages of a cross-correlation flowmeter include its ability to measure (*a*) flow rate of very hostile fluids, (*b*) velocity of one component in a multi-directional flow without requiring calibration. Disadvantage of a such a flowmeter is that the two detectors must be identical.

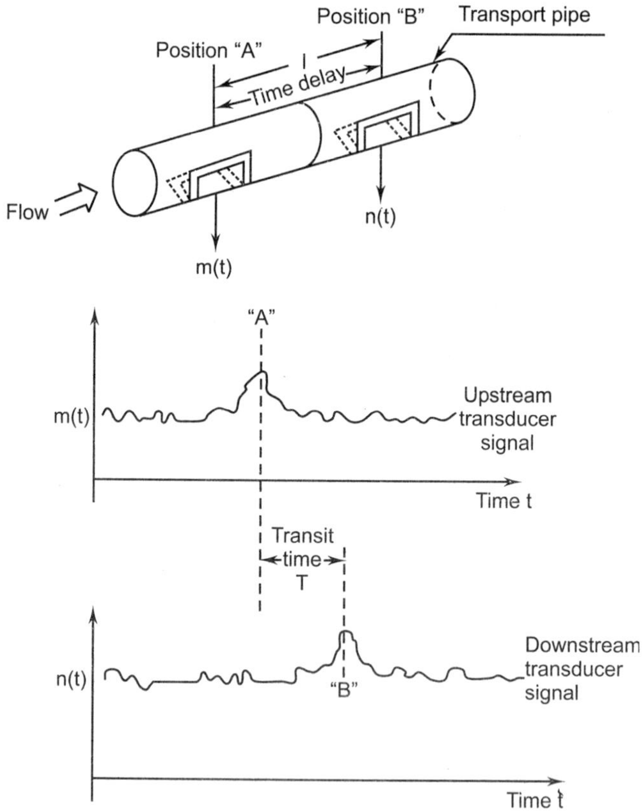

Fig. 9.105: Cross-correlation flowmetering

9.14.8 An Elbow Flowmeter

It is shown in figure 9.106. A horizontally laid pipeline is given a right angled bend in a manner to form a smooth arc of a circle of radius R in a horizontal plane. Centrifugal forces would be created at the bend by the flowing fluid and the difference in pressure $(P_1 - P_2)$ is measured by a pressure transducer. This difference in pressure depends on the velocity of flow and density of fluid. Assuming the elbow to be in the horizontal plane, it can be written as:

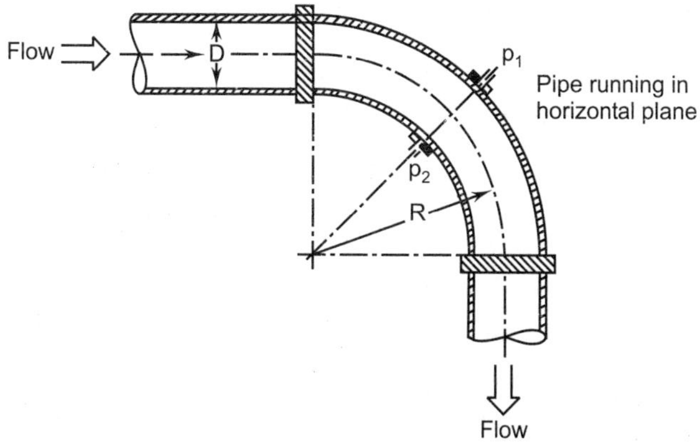

Fig. 9.106: An elbow flowmeter

$$v^2 = \frac{R(p_1 - p_2)}{\rho D}$$

where v is the mean velocity, R the radius of the bend, ρ the density of fluid, D the pipe diameter. The flowmeter does not require any separate installation, is very simple in nature and has an attainable accuracy P of the order of 4%.

The mass flow rate m is given by:

$$m = c_d A \sqrt{\rho(p_1 - p_2)}$$

where, c_d = discharge coefficient
A = cross-section of the pipe

Advantages associated with this flowmeter are:

- No obstruction to flow
- Pressure difference $(p_1 - p_2)$ is small.

While the disadvantages are:

- Pipe should run horizontally
- Horizontal pipe run must at least be 25 pipe diameters on either side of the bend.

9.15 MEASUREMENT OF SOLID FLOW

9.15.1 Introduction

In many industrial processes, such as coal, steel, cement industries, it is essential to determine the flow rate of dry materials. Meters for finding solid flow are essentially of the weighing type in that materials passing through a conveyor belt, say, are weighed for a definite length of time/continuously to ultimately find out the solid flow rate.

A few weighing schemes are now discussed for determination of solid flow.

9.15.2 A Conveyor Weighing System

A method for measuring the flow of dry materials by weighing is shown in figure 9.107.

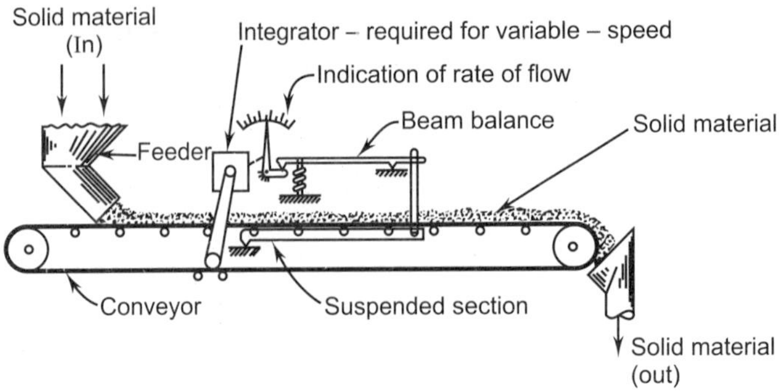

Fig. 9.107: Conveyor weighing mechanism for measuring flow of dry materials

In this case, material is fed on the conveyor via the feeder. The conveyor runs at a constant speed by electric motors. The suspended section on the conveyor, via the beam balance and spring mechanism, transmits the weight of that portion to the painter. This is calibrated in terms of rate of flow which can be tonnes/hr. The accuracy of measurement lies within 1%.

9.15.3 A Servo Drive Weighing System

A servo drive balance weighing system is shown in figure 9.108. A part of the conveyor is utilised for weighing purpose and that it should be flush with rest of the part. Weighing can be done by employing load cells having strain gauges as sensors. The electrical output so obtained is straightway related to the load measured. This output is utilised to control the motor speed.

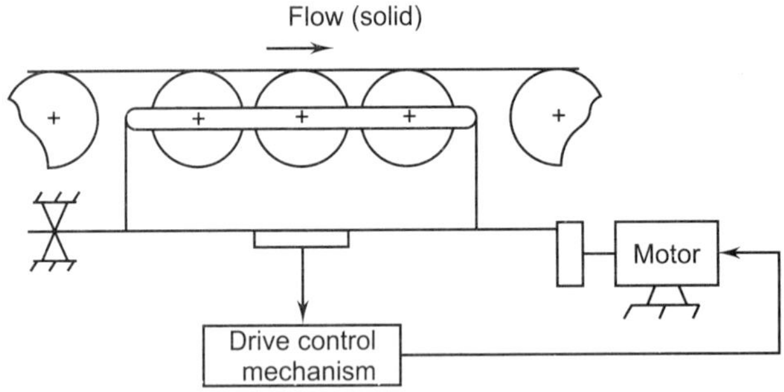

Fig. 9.108: A servo drive weighing system

9.15.4 A Nuclear Method for Solid Flow Rate Measurement

Nuclear weighing method to determine solid flow rate measurement is a proven one and applied to many process industries, such as paper, fertilizer, mining, steel, food processing, etc.

Figure 9.109 shows the method of solid flow rate measurement employing nuclear detector.

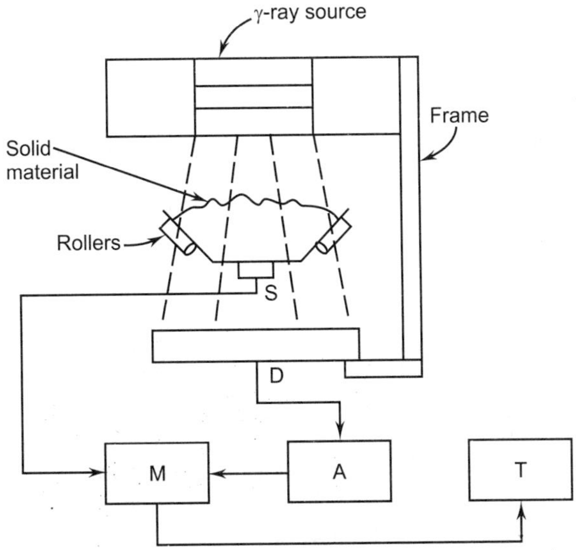

Fig. 9.109: Nuclear method of solid flow rate measurement

The γ-ray source passes through a particular length of material being carried on a conveyor. The conveyor speed is determined by the speed detector *S*. *D* is a radiation detector that gives the mass of material carried by the particular stretch on the belt. Its output is amplified. It acts as one input to the multiplier (*M*), the other input being from *S*. Thus, the multiplier output indicates the mass per unit time over that length. *T* is a totaliser which gives the total mass over any period of time.

The γ-rays source is normally ^{137}Cs while the detector may be Thallium activated sodium iodide crystal. A part of γ-radiation is absorbed when it passes through the solid material on the conveyor. Let the absorption coefficient of the material to falling radiation is γ, belt width *w* and average mass per unit length β, then the intensity of radiation received and consequently detected (*I*) by the detector is

$$I = I_o \exp(-\mu\beta/\omega)$$

I_o being the radiation intensity with no material on the conveyor.

A minimum amount of loading and a minimum percentage change in radiation intensity are required for measurements to be effective. A minimum 70% of full loading is needed for error-free operation. Conveyor loading is given by

$$\chi = \frac{kA}{V\omega}$$

where, *k* is a conversion constant, *A* the maximum mass flow rate, *V* the maximum speed.

Particle size variation of the product is taken into account by correcting the repeatability factor by adding the term R_p as

$$R_p = 1 - \left(\frac{\rho_{min}}{\rho_{max}}\right)_{bulk}$$

where ρ_{min} and ρ_{max} are respectively the minimum and maximum bulk density of the product.

The nuclear type solid flowmeter has a range from 0 to 100 tonnes/hour, its accuracy range being from ± 0.5 to ± 0.75 of full scale.

9.16 OPEN CHANNEL FLOWMETERING

9.16.1 Objective

An assortment of collection of river flow data provides an important information bank regarding the flow characteristics at any point along the flow path. The data so collected are used for water supply systems, design of highway bridges and culverts, pollution control purpose, flood management and development of reservoirs, energy production, irrigation, requirement on maintenance for navigation facilities, etc.

9.16.2 Methods

A number of methods are there to measure discharge in open channels and is known as open channel metering.

- Velocity-area method
- Ultrasonic method
- Slope-area method
- Dilution method
- Weirs and flumes
- Electromagnetic method
- Fall-discharge method
- Floats.

9.16.3 A Typical Velocity Profile for a River

Figure 9.110 shows a typical velocity profile provided by U.S. Geological Survey (USGS) for a large river. This is used as a reference for establishing flow rates in rivers, canals, etc. The profile is parabolic in nature with maximum velocity occurring at around 0.20 h from the top, where h is the depth of the stream. The velocity at the surface is approximately 1.18 times the average velocity, while the mean velocity occurs at around 0.6 h.

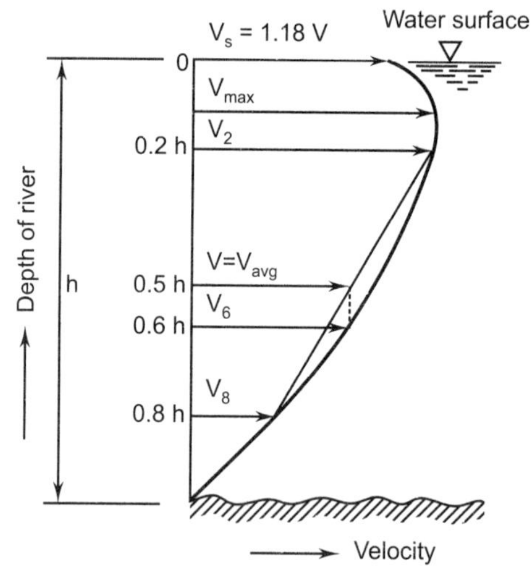

Fig. 9.110: A typical USGS velocity profile for a large river

9.16.4 Weirs

Variable area and variable head flowmeters are used for measurement of large flow rates in open channels. Weirs, nozzles and flumes are used for such a purpose. Their application areas include waterworks, sewage discharge, river flows, water-irrigation projects for measuring flow of water, semifluids, chemical wastes, etc.

A weir is an obstruction over which fluid is made to pass. They operate on the principle that if a restriction (obstruction) of a given shape and form is placed in the path of flow, a rise in the upstream fluid level would occur. This rise is a function of the rate of flow through the obstructed part. The rise in the fluid level can be measured by float, static head, bubbler system or echo sounder system.

The commonly used weirs are rectangular, triangular or *v*-notch or Thomson weir and Cipolletti or trapezoidal types. They are shown in figure 9.111.

Rectangular design is employed for measuring large flow rates, while the *v*-notch weir is used for small flow rates.

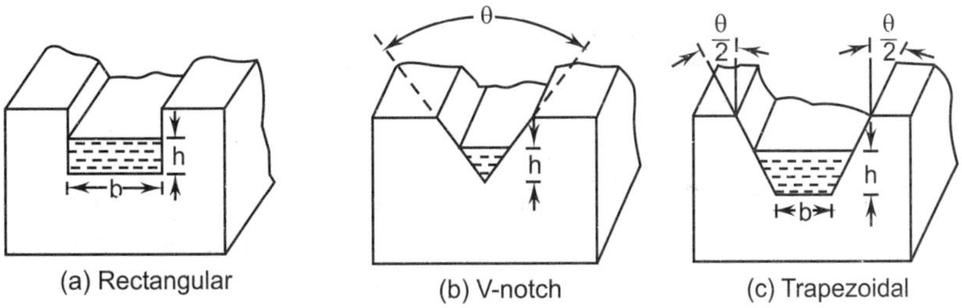

Fig. 9.111: Types of weirs

The flow rates for the three cases are given as under:

$$Q_R = \frac{2}{3} b C_R \sqrt{2gh^3}$$

$$Q_V = \frac{8}{15} C_V \tan\frac{\theta}{2} \sqrt{2gh^5}$$

and,

$$Q_T = \frac{2}{3} b C_T \sqrt{2gh^3}$$

C_R, C_V and C_T are the discharge coefficients for rectangular, *v*-notch and trapezoidal type weirs respectively. Their values are always less than 1 and depends on the flow rate. It is better to calibrate a weir after its installation to get the value of discharge coefficient in a given situation. The calibration can be done either by a weigh test or by another flow measuring device. However, manufacturers provide reference tables for the same.

Weirs are quite accurate devices, easy to instal and very simple in construction. Their main disadvantage is the non-linear flow characteristics, for example, for rectangular/Cipolletti weirs, 3/2 power law and for triangular weir, 5/2 power law are followed.

To derive adequate accuracy in measurements, limitations on head, width and depth of a weir are imposed during its construction. The upstream part of weir must have a straight, uniform cross-section channel to ensure uniform velocity distribution. Sometimes, baffle plates and screens are put across the channel to reduce disturbance. The downstream edge of a weir must be beveled to ensure that the stream breaks free of the edge and not fall down the face of the weir plate. The downstream face of a rectangular weir without side contraction, must be aerated to provide breakaway flow. Flow over a sharp-crested rectangular weir is shown in figure 9.112.

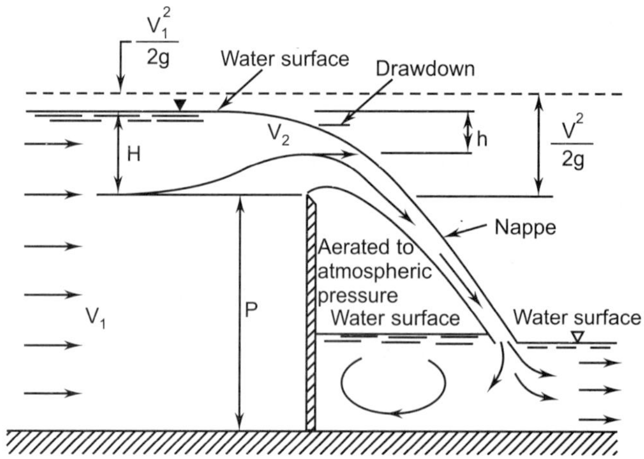

Fig. 9.112: Flow profile over a sharp-crested rectangular weir

9.16.5 Flumes

These are also used for open channel metering. As already discussed, water must be dammed upstream of a weir. This may cause changes in the inflow region under certain conditions. This drawback is absent in flumes. Flumes are of two types—Venturi flume and Parshall Venturi flume.

A Venturi flume operates on the same principle of a venturi nozzle. Like a venturi nozzle, a venturi flume puts a constriction in the flow path. This, thus, accelerates the fluid in the region of constriction.

9.16.5.1 Venturi Flume

A Venturi flume is shown in figure 9.113. A Venturi flume responds to the smallest flow rates. Normally, the constrictions are at the sides. Some elevated floor sections are provided.

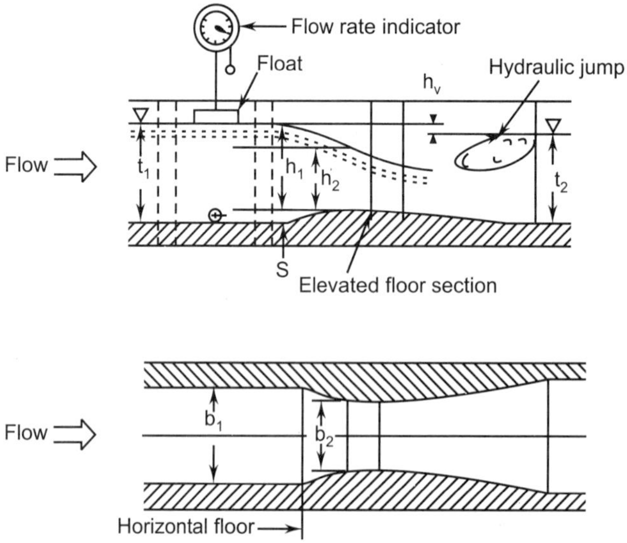

Fig. 9.113: Venturi flume

Source: **Guide to Flow Measurement**: *Courtesy :* Bailey Fischer & Porter

The volumetric flow rate through a Venturi flume is

$$Q_V = \frac{2}{3} C_v . b_2 . \sqrt{2gh^3}$$

where C_v is the discharge coefficient for the venturi flume.

At the flume inlet (*i.e.* upstream side), the water level is called 'headwater'. Water at the inlet side is in the subcritical range because of damming action. However, flow must be supercritical to establish a relationship between water level in the headwater and the flow rate. This is done by hydraulic decoupling of tailwater flow from headwater flow. A free outflow from the venturi flume must exist to avoid backflow. Otherwise, performance of the flume would be affected.

9.16.5.2 Parshall Venturi Flume

A Parshall Venturi flume is shown in figure 9.114. In this, in addition to side restrictions, there is a step in the bottom. This ensures smooth transition from subcritical to supercritical flow. The volumetric flow rate in this case is

$Q_{PV} = kh^n$, k is a constant while n is the constriction

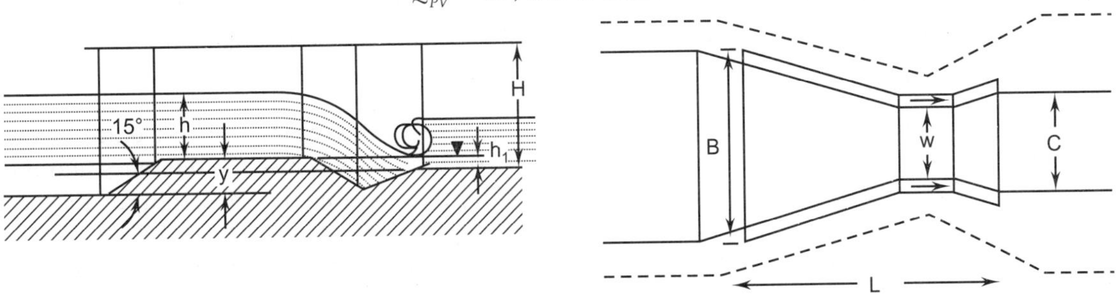

Fig. 9.114: Parshall venturi flume

Source: **Guide to Flow Measurement.** *Courtesy* : Bailey Fischer & Porter

While using a Parshall venturi flume, care should be taken of the following:

- Height measurement should be done at a place which is $3h$ upstream from inlet flange of the venturi flume.
- For proper accuracy to be obtained, the disturbance free upstream and downstream lengths must be a minimum of 10–20 times the flumes width B and $10B$ respectively.
- The ratio of headwater (h) to tailwater (h_1) height must be limited to a maximum of 1.5.
- Meter installation should be at a place of constant cross-section and slope.

9.17 FLOW SWITCHES

9.17.1 Introduction

When the flow rate is above or below a certain value, a flow switch can be activated which, in its turn, can activate an electrical or pneumatic circuit for control purpose for bringing back the flow rate within the specified limits.

The set points (both upper and lower) at which the flow switch is to be activated is set by the user. The switch remains activated until the flow rate comes back within the limits. The difference between the flow rate corresponding to the set point and the actual flow rate at which the flow switch is deactivated is called switch 'differential' - it can either be fixed or else adjustable. A small differential is desired, otherwise for flow rates operating around the set points, the flow switch will be activated more frequently.

In some cases, a manual reset feature is incorporated in the flow switch, *i.e.* once the flow switch is activated, it will not return to its previous state unless reset manually. It is done for cases where it is essential to fix up the causes for abnormal flow conditions.

9.17.2 Folding Paddle Type

In various forms, paddle flow switches are available. A paddle hangs in a loose manner when there is no flow in the pipeline. As flow starts in the pipeline, the paddle moves upwards which is translated into mechanical motion by various means like a bellows assembly, a pivoting can etc. This mechanical motion actuates a flow switch which can be mercury switch or a micro switch. By varying the length of the paddle of the paddle switch, the range and the point of actuation can be varied/adjusted.

Paddle switches are least expensive but are sensitive to pipeline disturbances, purturbations, vibrations, etc. For pipeline vibrations or pulsating flows, dampers are used. Paddle switches should be installed at least 10 pipe diameter downstream and that the pipe must have a straight run at that place. Paddle switches cannot distinguish between low flow velocities from no-flow condition. For such cases, folding type circular paddle switches are recommended.

9.17.3 Thermal Type

The principle of operation of thermal flow switch depends on the change of resistance of the thermal sensor when the flow rate varies. There are two thermal sensors—one reference and the other active. These are connected in a Wheatstone bridge configuration whose output may be used for control.

This type of flow switch is able to detect very low flow velocities. Activation of the flow switch can easily be adjusted from the output of the Wheatstone bridge. The major disadvantage of this flow switch is its relatively large time constant of the order of several seconds.

9.17.4 Bypass Type

In this flow switch type, an externally adjustable vane is placed in the flow path. In the vicinity of the vane, a bypass path is there. The presence of the vane produces a differential pressure in the flow stream, which forces a proportional flow through the bypass line. A piston supported by a spring is placed in the bypass line. The piston moves laterally with variation in flow rate. For a particular position of the piston, a switch can be activated.

This type of flow switch is employed for relatively low flow rates. Also, the flow rate at which the switch activation is desired can be externally adjusted.

9.18 FLOWMETER CALIBRATION

9.18.1 Introduction

Calibration of a flowmeter depends on how precisely volume or weight of fluid and the corresponding time of flow are measured. Conditions existing during the use of a flowmeter must match with those during calibration for the calibration to remain valid. Fluid properties like temperature, pressure, density, viscosity and flow disturbances, particularly upstream like valves, elbows, etc., affect flowmeter calibration. Temperature and pressure variations of a gas affect flowmeter calibration in a bigger way.

A flowmeter may be calibrated first by establishing a steady flow through it and measuring either mass or volume through it in a precise time interval. A flowmeter, so calibrated, can itself become a secondary standard against which a less accurate flowmeter may be calibrated.

9.18.2 A Liquid Flowmeter Calibrator

A commercial liquid flowmeter calibrator is shown in figure 9.115 (*a–f*). It follows a dynamic weighing scheme for calibration. It covers a range from 0.2 to 70,000 kg/hr with an overall accuracy of ± 0.1 per cent.

(*a*) *Running operation before test*: It is shown in figure 9.115 (*a*) Liquid from the reservoir is pumped. It enters filter first and then the heat exchanger which controls the temperature within. It then passes through the control valves, flowmeter under test and weigh tank before returning to the reservoir.

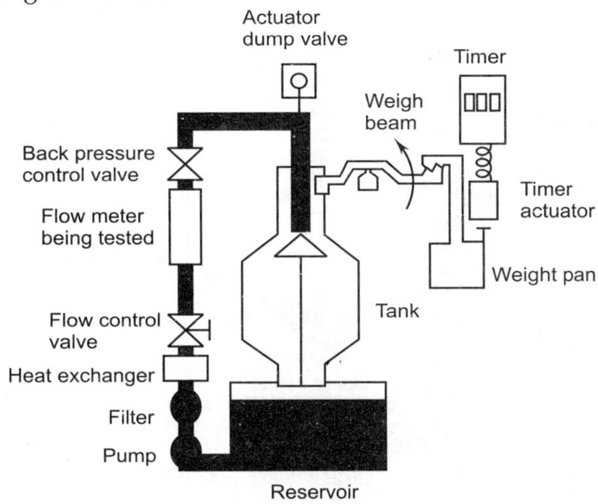

Fig. 9.115 (*a*)

(*b*) *Start of preliminary fill (tare time)*: It is shown in figure 9.115 (*b*). At the beginning of this cycle, the control valves have been adjusted for desired flow. A tare weight is placed on the weight pan. It is thus detached from the timer actuator. Start button is now pressed which closes the dump valve. Liquid thus starts to fill the weight tank.

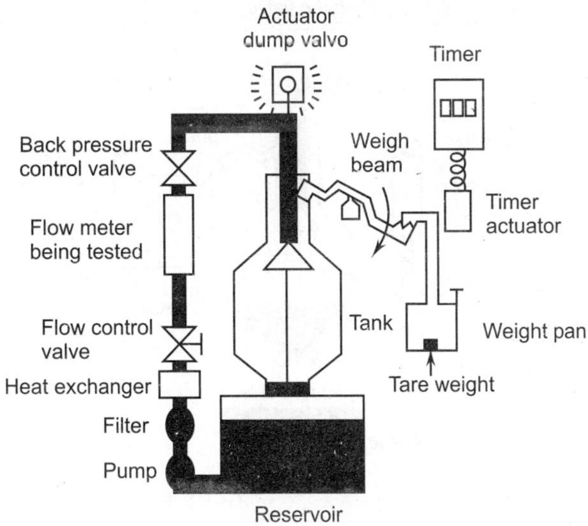

Fig. 9.115 (*b*)

(*c*) *End of preliminary fill and start of weighing cycle*: As the weight tank begins to fill, the weigh beam starts raising itself. Ultimately, the weigh pan touches the timer actuator and it is tripped. The electronic counter begins to count signalling the start of weigh cycle.

Before this, tare weight is balanced by the initial preliminary fill. The preliminary fill method eliminates mechanical errors during start and stop portions and allows dynamic error to be self-cancelling. The situation is explained in figure 9.115 (*c*).

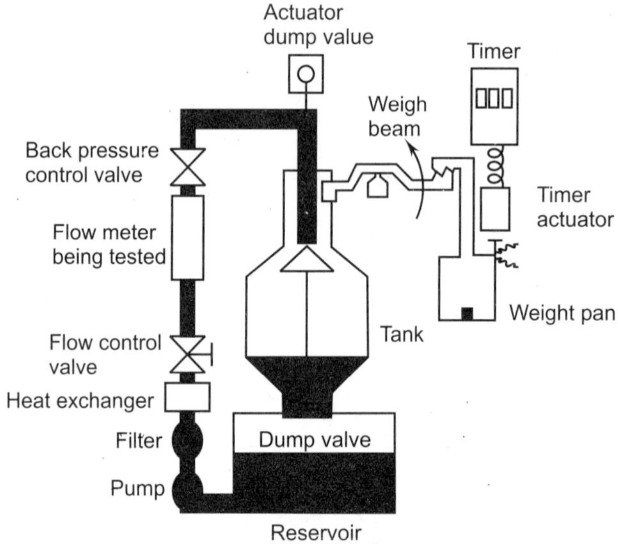

Fig. 9.115 (*c*)

(*d*) *Weighing cycle in operation*: This is shown in figure 9.115 (*d*). A precision weight placed on the pan allows weighing cycle to continue. The cone shaped deflector at the inlet of the weigh tank allows even distribution of liquid.

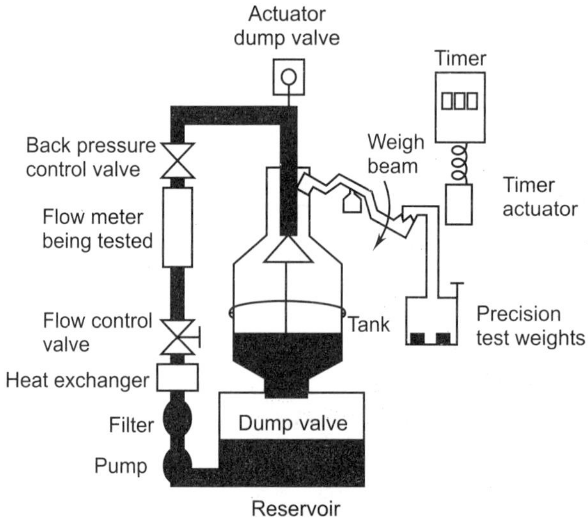

Fig. 9.115 (*d*)

(*e*) *End of weighing cycle*: This is shown in figure 9.115 (*e*). As the weigh tank continues to be filled in, the pan rises, ultimately tripping the timer actuator. It stops the timer. Flow rate is thus found out accurately by dividing the precision weight by the time interval.

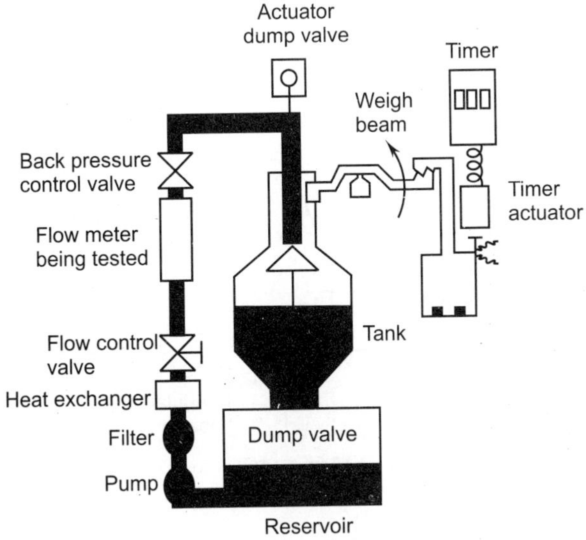

Fig. 9.115 (*e*)

(*f*) *Emptying for recycling*: This is shown in figure 9.115 (*f*). The weigh tank automatically empties after tripping of the timer acuator. The calibrator is now ready for the next flow rate setting.

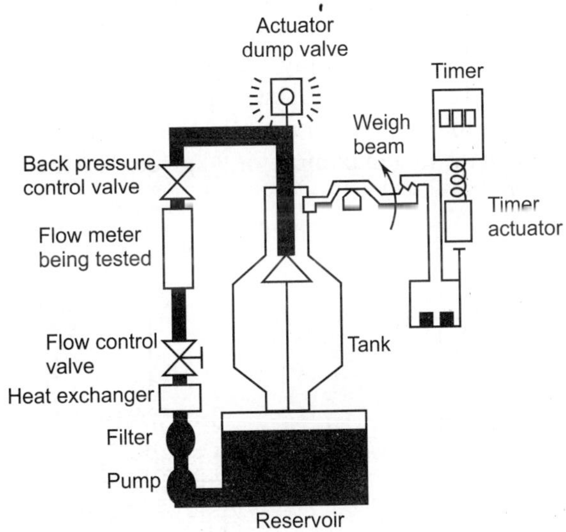

Fig. 9.115 (*f*)

9.18.3 A PVT System

A PVT (pressure-volume-temperature) system is basically a gas flow calibrator and is shown in figure 9.116. The initial temperature and pressure of the storage tank of known volume is measured. The gas is allowed to pass through the sonic nozzle and the meter under test. Temperature and pressure of the tank at the end of the measured period are noted.

The gas flow rate is thus calculated during this period. By regulating the valves and nozzle, a second test is done on the meter. The process is continued so that the meter is calibrated over a wide range of flow rate. Accuracies of the order of ± 0.5% to ± 0.2% can be achieved by this method.

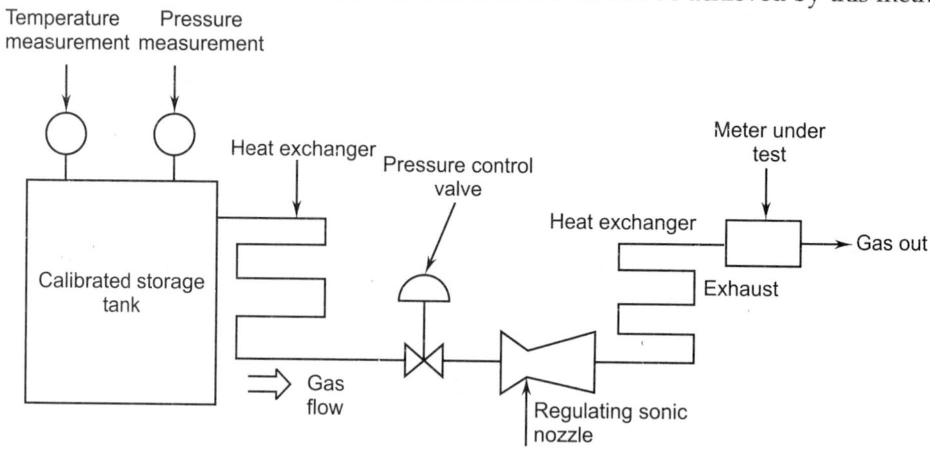

Fig. 9.116: A PVT system

9.18.4 A Bell Prover System

Figure 9.117 shows a bell prover system employed for calibration of a flowmeter by a gas.

Here the gas flows through the flowmeter during a known time interval and is trapped by the gasometer bell. Its volume is measured. By knowing the temperature and pressure of the gas, mass calculation is carried out for the volume entrapped. A counter weight ensures dynamic balancing of the entrapped gas. The bell can be raised to the top by adding appropriate weights. It can then be used as a source of gas supply which comes out of the flowmeter as the bell gradually comes down at a measured rate. Using a precision balance allows the measurement of mass of gas accumulated in the bell over a period of time. Accuracies of the order of ± 0.02% can be derived with flow rates up to a maximum of 9 kg/sec. The meter can also be calibrated in terms of volumetric flow rate.

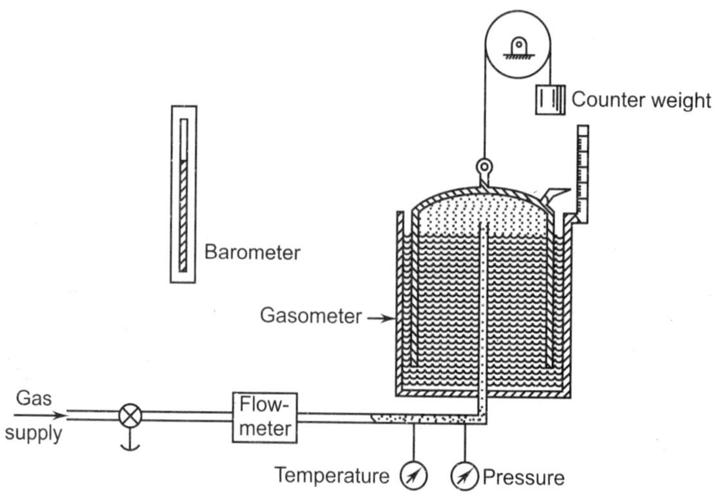

Fig. 9.117: Schematic of a bell prover system

9.18.5 A Soap-film Burette Method

This method is used to calibrate a meter for gas flow. This is shown in figure 9.118. Gas is allowed to pass through the meter under test and a burette. At the lower end of the burette, a soap solution is there. As gas passes through the soap solution, a soap film is formed which travels up the burette with the same velocity as the gas. By measuring the transist time of the soap film between two graduations on the burette, flow rate of the gas can be determined.

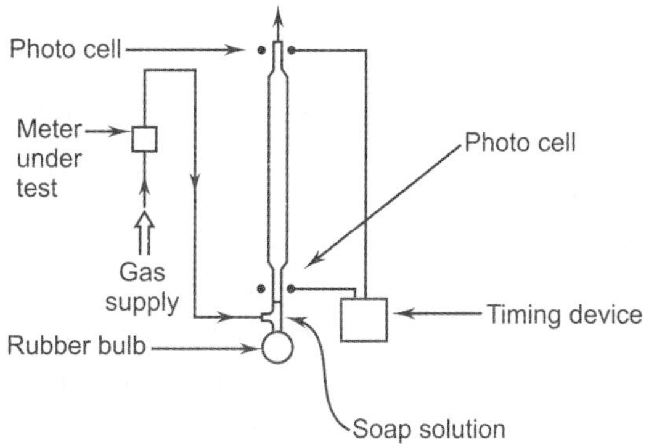

Fig. 9.118: Gas flowmeter calibration by soap-film burette method

9.19 SELECTION OF FLOWMETERS

9.19.1 Introduction

A multitude of flowmeters are available in the market from different makers. Choice of a flowmeter for a particular application is really a tough call, with a number of so many factors influencing the final selection.

British Standard (BS7405) has classified the different flowmeter types, dealt in details the procedure for flowmeter selection, broad areas of flowmeter applications and various data tables for the major areas of concern.

9.19.2 Selection Procedure

Figure 9.119 shows the flowmeter selection procedure. There is a table of data for each of the boxes shown in the figure. The data tables can be used to eliminate meters unsuitable for a specific application. The parameters to be considered are:

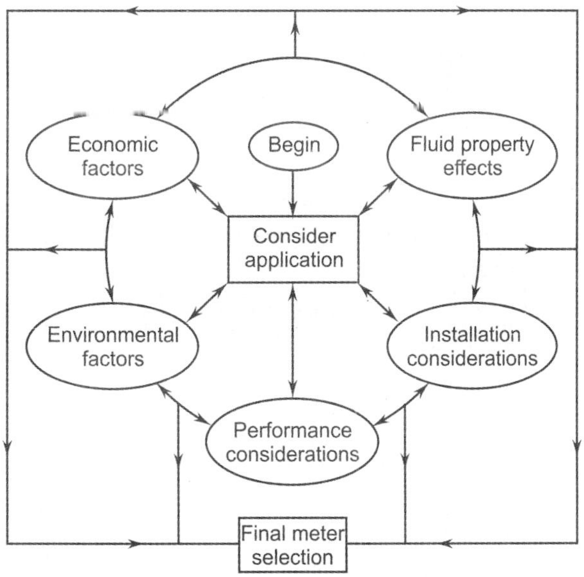

Fig. 9.119: A guide to flowmeter selection

- Performance
- Fluid property
- Installation
- Environmental
- Economic

Table 9.4 gives an elaboration on each of the five factors, namely performance, fluid property, economies, installation and environmental.

Table 9.4: Different Factors in Flowmeter Selection

Performance Factors	***Installation Factors***
Accuracy	Orientation
Repeatability	Flow direction
Linearity	Upstream and Downstream pipework
Rangeability (turndown)	Line size
Pressure drop	Location for servicing
Output signal characteristics	Effects of local vibration
Response time	Location of valves
Uncertainty	Electrical connections
	Provision of accessories
	Hazardous atmosphere
	Effect of pulsations/unsteady flow
Fluid Property Factors	***Environmental Factors***
Liquid or gas	Ambient temperature effects
Temperature and pressure	Humidity effects
Density	Safety factors
Specific gravity	Pressure effects
Viscosity	Electrical interference
Lubricity	
Chemical properties	
Surface tension	
Compressibility	
Abrasiveness	
Pressure of other phases	
Presence of other components	
Real gas effects	
Economic Factors	
Purchase price	
Installation costs	
Operation costs	
Maintenance costs	
Calibration costs	
Meter life	
Spares cost and availability	
Pumping power and headloss	
Technical optimization	

Table 9.5 shows broad areas of flowmeter applications.
Liquids (See note 1) Gases (See note 2) Miscellaneous (See note 3)

Table 9.5: Flowmeter Application Areas

Group	*Type*	*A*	*B*	*C*	*D*	*E*	*F*	*G*	*H*	*J*	*K*	*L*	*M*	*N*	*P*	*Q*	*R*	*S*	*T*
1.	Orifice	■	?	■	■	■	■	■		■	?	■	■	■	?	■	?	?	?
	Venturi	■		■	■	■	■			■	?		?	?	?	■	?	?	?
	Nozzle	■		■	■	■	■			■	■	?	?	■	?	■	?	?	?
2.	VA	■	■			#	?		■	■	■							?	
	Target	■				#				■				■		■	?		
	Averaging pitot	■		■	■	■	?	■		■		■	■	■		■	?	?	?
	Sonic nozzle									■	■	?	?						
3.	Sliding vane	■		#			■		■							?			
	Oval gear	■	■	#		#	■		■							?		■	
	Rotary piston	■	?			#	■		■							?		#	
	Gas diaphragm									■	■								
	Rotary gas									■	■								
4.	Turbine	■		■	#	■	?	■	■	■		■				?	?	#	
	Pelton	■	■			■		■		?	?					?		?	
	Mechanical meter	■													#	?			
	Insertion turbine	■		■	■	■		■		■		■	■	?		?	?		
5.	Vortex	■				■		■	■	■		?	■	■		?			
	Swirlmeter	■								■									
	Insertion Vortex	■		■	■	?		?	?	■		■	?	■		■		?	?
6.	Electromagnetic	■	■	■	■	#	?		■						■	■	?	■	
	Insertion electromagnetic	■		■	■	?			■						?	■	?	■	
7.	Doppler	■		?	?	#			?						■	■	?	?	
	Transit time	■	?	■	■	#	?	#	■	#					?	?		#	#
8.	Coriolis (direct)	■				#	■		■	?					?	?		#	
	Twin rotor (indirect)	■																	
9.	Anemometer	■		?	?	#				■									
	Thermal mass		#							■	■								
10.	Tracer	■	#	■	■	■	■	■		#		#	■	■		?	■	#	#
	Laser	■		?	?														

Key

■ is suitable: generally applicable

? is worth considering, sometimes applicable

is worth considering, limited availability or tends to be expensive

A blank indicates unsuitable: not applicable

Principles of Flowmeter Selection (British Standard BS7405)

Note: 1. Liquid applications are indicated by the following:

A. Central liquid application (< 50 cP) (< 0.05 Pd S)

B. Low liquid flows (< 0.12 m^3/h) (< 2 L/min)

C. Large liquid flows (> 1000 m^3/h) (> 1.7×10^4 L/min)

Note 2. Gas applications are indicated by the following:

J. General gas application

K. Low gas flows (< 150 m^3/h)

L. Large gas flows (> 5000 m^3/h)

Chapter 9

D. Large water pipes (> 500 mm bore)

E. Hot liquids (temperatures > 200°C

F. Viscous liquids (> 50 cP) (> 0.05 Pa.S)

G. Cryogenic liquids

H. Hygienic liquids

M Hot gases (temperatures < 200°C)

N Steam

Note 3. Miscellaneous applications are indicated by the following:

P Slurries and particle flows **Q** Liquid/liquid mixtures

R Liquid/gas mixtures **S** Corrosive liquids

T Corrosive gases

9.19.3 Influences of Fluid Properties

9.19.3.1 Viscosity

For laminar flow when Reynolds number (R_n) is low, viscosity effect is predominant and it influences varying effect on meter size, depending on the method. For gas flowmetering, viscosity effect does not make much of a difference on meter sizing.

For gas flow metering, viscosity affects meter sizing for lower flow ranges in case of variable area meters.

For liquid flow metering, mass meters and electromagnetic meters are totally viscosity independent. But ultrasonic meters have difficulty with meter sizing when flow crosses over from laminar to turbulent. Increased fluid viscosity shifts the minimum flow rate to a higher value in case of vortex, swirl and turbine meters. Oval gear and oscillating piston meters are more accurate at higher viscosity values.

9.19.3.2 Solids in Fluid

Gas flow rates are, in general, high and hence suspended particles in a gas may have sufficient kinetic energy to cause damage due to erosion. Oscillating piston meters require a clean gas and the gas is filtered before it is fed to the meter.

The sharp metering edges of variable area and DP meters may suffer permanent damage due to dust particles. Mass meters suffer from increased inaccuracy with increase in mass of suspended particles.

9.19.3.3 Corrosion

Corrosion effects by aggressive fluids on the meters must be attended to, otherwise the instruments would break down too often.

Corrosion effects are better attended to in electromagnetic and ultrasonic meters. For electromagnetic flowmeters, good quality corrosion resistant electrode materials are available like, SS, Hastelloy, Monel metal, platinum, etc. Corrosion resistant lining like PTFE can be deployed in case of ultrasonic flowmeters. The orifice and nozzles, pipe fittings and the pressure lines are also made of corrosion resistant materials in case of DP meters.

9.19.3.4 Gas Bubbles in a Liquid

Error in a liquid flowmeter increases in proportion to the increase in the proportion of gas in the liquid. In fact, a Coriolis mass meter is also affected by gas present in liquid.

Error in ultrasonic flowmeter, which depend on transit time principle, is noticeable even in the presence of small amount of gas bubbles in the liquid. Turbine meters suffer from overspeeding, while vortex and swirl meters suffer from cavitation in the presence of gas bubbles.

9.19.3.5 *Flow Stream Effects*

Depending on the Reynolds number, velocity profile of a fluid influences the flowmetering instrument performances.

All flowmetering instruments operate correctly with a turbulent axisymmetric velocity profile. With laminar axisymmetric flow profiles, these instruments operate likewise, but their flow ranges decrease with increasing viscosity.

Variable area, swirl meters, Coriolis mass meters operate independently of velocity profiles while DP meters are very sensitive. As such, they require long straight runs before metering. Ultrasonic flowmeters also require flow straightening run upstream for proper metering.

9.20 SITE INSTALLATION

9.20.1 Introduction

Installation site considerations go a long way in the process of choosing/selecting a meter. It may so happen that a flowmeter, otherwise suitable, cannot be installed because of the hostilities or some other considerations present at the meter site. For example, an oscillating piston meter cannot be installed in a pipeline carrying solids hydraulically.

While choosing a flowmeter, due consideration should be accorded to vibration, pulsation, reverse flow, temperature and pressure, as well as power supply.

9.20.2 Vibration

Vibration, if not contained, would degrade meter performance. Vibration destabilizes meter supporting system and would eventually lead to wear and tear of the moving parts, bearings, etc. It would lead to decalibration of meter.

The floats in the variable area meters are sensitive to vibrations. To reduce this, vibration dampeners should be used. Again, in the oscillating parts of vortex and mass flowmeters, the vibrating frequencies are superposed on the measurement frequencies, leading to erroneous reading. In case resonance occurs, the parts may fracture.

9.20.3 Pulsation

Inertia of a flowmeter influences its pulsation effects. If the metering element follows pulsations without any lag, measurement will be unaffected by such pulsations. Dampers are sometimes used to overcome such effects.

A damping device is employed to reduce pulsations in case of DP meters and variable area meters.

In the case of lobed impeller meters, pipeline connected to the meter may vibrate because of pulsations. If resonance occurs, loud noise and huge pressure drop may result. Thus, pulsation damper is a must for this. Oval gear and oscillating piston meters have considerable inertia which act as damper.

9.20.4 Reverse Flow

This occurs with extreme pulsation, when reverse flow takes place, volumetric totalisers reverse their direction and totalise backwards. Electromagnetic and mass meters can take care of flow in both directions automatically.

9.20.5 Temperature and Pressure

When temperature and pressure are very high, meter selection is governed by:

- Mechanical tolerances to be kept for temperature expansion.
- Strength of materials of meters
- Gasket types

When variable area meters are to be employed, the glass version is employed for lower temperature and pressure applications while the metal ones for higher limits.

Pressure lines, fittings and converters need to be taken into consideration for DP meters. It is the converter which is the most important one for which the upper temperature limit is around 120°C. Isolators in the pressure lines are provided to reduce temperature well below the above stated limit.

SOLVED EXAMPLES

1. Determine the volumetric flow rate of water in m³/hr in a pipe having diameter of 40 mm and orifice diameter of 10 mm. The manometer reads 20 cms of Hg across the tappings. Assume $k = 0.75$

Ans. Differential pressure across the tappings

$$\Delta h = (13.6 - 1) \times (20 \times 10) \text{ mm of water}$$

$$= 2520 \text{ mm of water} = \frac{p_1 - p_2}{\rho}$$

$$\alpha = \frac{d}{D} = \frac{10 \text{ mm}}{40 \text{ mm}} = 1/4.$$

Thus, from equation (9.5),

$$Q = \frac{\pi k d^2}{4}\sqrt{\frac{2g}{\rho}(p_1 - p_2)}$$

$$= \frac{3.14 \times 0.75 \times \left(\frac{10}{1000}\right)^2}{4} \times \sqrt{2 \times 9.8 \times 2520 \times 10^{-3}} \text{ m}^3/\text{sec}$$

$$= \frac{3.14 \times 0.75 \times \left(\frac{10}{1000}\right)^2}{4} \times \sqrt{2 \times 9.8 \times 2520 \times 10^{-3}} \times 3600 \text{ m}^3/\text{hr.}$$

$$= 1.48957 \text{ m}^3/\text{hr.}$$

2. Calculate the flow rate at the orifice for a pipe of diameter 50 mm and orifice diameter 25mm. Reynolds number is 10^5. Density of water = 1000 kg/m³ and kinematic viscosity = 10^{-2} stokes (cm²/second)

Ans. We know that Reynolds number

$$R_n = \frac{\rho v D}{\mu}$$

or,

$$v = \frac{R_n \mu}{\rho D} = \frac{R_n \cdot \nu}{D}, \qquad \nu = \text{kinematic viscosity} = 10^{-2} \text{ stokes} = 10^{-2} \text{ cm}^2/\text{sec} = 10^{-6} \text{ m}^2/\text{sec.}$$

∴ For the pipe

$$v_1 = \frac{R_n.\nu}{D} = \frac{10^5 \times 10^{-6}}{50 \times 10^{-3}} = \text{m/sec}$$

Applying equation of continuity,

$$A_1v_1 = A_2v_2$$

$\therefore$ v_2 = Flow rate at the orifice

$$= \frac{A_1}{A_2}.v_1 = \left(\frac{50}{25}\right)^2 \times \frac{10^5 \times 10^{-6}}{50 \times 10^{-3}}$$

$$= 8.0 \text{ m/sec.}$$

3. A venturi tube having a throat diameter of 10 cms is placed in a pipe having diameter 25 cms. The combination is used to measure flow rate. The flow rate through the venturi tube is 0.025 m³/sec. Viscosity of water is 10^{-3} Pa.S. Determine the Reynolds number for the venturi tube and the upstream to throat differential pressure developed. Assume $c_d = 0.98$

Ans. Velocity through the venturi tube $= \dfrac{\text{flow}}{\text{area}} = \dfrac{0.025}{\frac{\pi}{4}(10 \times 10^{-2})^2}$

$$= 3.185 \text{ m/sec}$$

Reynolds number for the venturi tube $R_n = \dfrac{\rho v d}{\mu} = \dfrac{1000 \times 3.185 \times (10 \times 10^{-2})}{10^{-3}} = 3.185 \times 10^5$

Again, from equation (9.22 *a*),

$$q = \frac{\pi}{4} \times C_d \times \frac{d^2}{\sqrt{1-\alpha^2}} \sqrt{\frac{2g\Delta P}{\rho}}$$

or,

$$0.025 = \frac{3.14}{4} \times 0.98 \times \frac{(10 \times 10^{-2})^2}{\sqrt{1-\left(\frac{10}{25}\right)^2}} \times \sqrt{\frac{2 \times 9.8 \times \Delta P}{1000}}$$

$\therefore$

$$\Delta P = 452.6 \text{ kg/m}^2$$

4. The flow rate through a Venturi tube ($C_d = 0.98$) of throat diameter 10 cms is 0.95 m³/sec, the pressure differential being 10 kPa. Calculate the flow rate through an orifice of 8 cm ($C_d = 0.65$) when installed in the same pipe with the same pressure differential.

Ans. We have,

$$\frac{q_{\text{orifice}}}{q_{\text{venturi}}} = \frac{C_{d_{\text{orifice}}}}{C_{d_{\text{venturi}}}}$$

or,

$$q_{\text{orifice}} = \frac{0.65}{0.98} \times 0.95$$

$$= 0.63 \text{ m}^3\text{/sec.}$$

5. Determine air velocity in a pipeline if the manometric head is 1.5 cm of mercury when a Pitot tube is installed in the line. The line pressure is 10 kg/cm², the temperature being 80°C. It is assumed that velocity coefficient $C_v = 1$ and manometric fluid is at 18°C.

Ans. In this case, the flow rate equation is $V_f = C_v\sqrt{2gh(\rho_m - \rho_f)v_f}$

$$v_f = \frac{RT}{p} = \frac{2880 \times 353}{8000} = 127.1 \text{ cm}^3\text{/gm}$$

Chapter 9

$$\rho_m = 13.5 \text{ gm/cc (at 18°C)}, h = 1.5 \text{ cm of Hg.}$$

It can be safely assumed that: $\rho_f << \rho_m$.

$$\therefore \quad V_f = \sqrt{2gh\,\rho_m v_f}$$

or,

$$V_f = \sqrt{2 \times 980 \times 1.5 \times 13.5 \times 127.1} = 2246 \text{ cm/sec}$$

6. Determine the fluid flow velocity when the measured pressure differential is observed to be 5 kPa by employing a Pitot tube. Fluid density = 1000 kg/m³.

Ans. We have the following fluid flow velocity when a Pitot tube is used:

$$v = \sqrt{\frac{2}{\rho}(p_t - p_s)}$$

$$= \sqrt{\frac{2 \times 5 \times 10^3}{1000}} \text{ m/sec}$$

$$= 3.16 \text{ m/sec.}$$

7. A cylindrical rotameter is having the following specifications: density = 4000 kg/m³, height = 4 cms, diameter = 4 cms. Maximum inside tube diameter is 6 cms. Assume fluid as water with $C_d = 0.6$, determine the maximum flow rate.

Ans. $$v_f = \text{Float volume} = \frac{\pi}{4}(4 \times 10^{-2})^2 \times (4 \times 10^{-2}) = 16\pi \times 10^{-6} \text{ m}^3$$

From the relation, $$Q = A\,C_d \sqrt{\frac{2gv_f(\rho_f - \rho)}{A_f\,\rho}}$$

where $$A = \frac{\pi}{4}\left(D_t^2 - D_f^2\right)$$

For maximum flow rate determination, $D_t = D_p$ (pipe diameter)

Then $$Q = C_d \cdot \frac{D_p^2 - D_f^2}{D_f} \sqrt{\frac{\pi g v_f(\rho_f - \rho)}{2\rho}}$$

$$Q = 0.6 \times \frac{0.06^2 - 0.04^2}{0.04} \sqrt{\frac{3.14 \times 9.8 \times 16\pi \times 10^{-6} \times (4000 - 1000)}{2 \times 1000}}$$

$$= 0.0014447 \text{ m}^3\text{/sec}$$

$$= 1.4447 \times 10^{-3} \text{ m}^3\text{/sec.}$$

8. A rotameter, calibrated for water, has a scale range from 1 to 500 litres/min and the float used has a density of 2500 kg/m³. This rotameter is to be used for a gas flow ranging between 20 to 10000 litres/min, the gas density being 1.15 kg/m³. Find the density of the used float (*i.e.* which is used for gas), its shape and volume remaining constant. Assume C_d to be constant for both the cases.

Ans. Since, $$\frac{10000}{500} = \frac{20}{1} = 20$$

$$\therefore \quad Q_2 = 20\,Q_1$$

where, Q_2 and Q_1 are the gas and water flow rates respectively.

From the general equation $$Q = C_d \cdot \frac{D_p^2 - D_f^2}{D_f} \cdot \sqrt{\frac{\pi g v_f(\rho_f - \rho)}{2\rho}}.$$

We can write $$Q_2 = K\sqrt{(\rho_{f_2} - \rho_2)/\rho_2} \text{ for gas}$$

and $Q_1 = K\sqrt{(\rho_{f_1}-\rho_1)/\rho_1}$ for water.

$$400 = \frac{(\rho_{f_2}-\rho_2)/\rho_2}{(\rho_{f_1}-\rho_1)/\rho_1} = \frac{(\rho_{f_2}-1.15)/1.15}{(2500-1000)/1000}$$

from, which, $\rho_{f2} = 691.15 \text{ kg/m}^3$

9. Find the angle of taper of a rotameter from the following data: maximum flow rate = 50 litres/minute, float volume = 10 cc, float diameter = inlet diameter = 2 cms, length of tapered part of the tube = 30 cms. $C_d = 1$. Also the float is fully compensated for fluid density.

Ans. The equation for flow rate for a rotameter is

$$Q = C_d \times \frac{D_p^2 - D_f^2}{D_f} \times \sqrt{\frac{\pi g v_f (\rho_f - \rho)}{2\rho}}$$

Since the rotameter float is fully compensated for fluid density, hence, $\rho_f = 2\rho$, also $C_d = 1$.

$$\therefore \quad Q = \frac{D_p^2 - D_f^2}{D_f}\sqrt{\frac{\pi g v_f}{2}} = \frac{(D_f + 2l\theta)^2 - D_f^2}{D_f}\sqrt{\frac{\pi g v_f}{2}}$$

$$= \frac{4D_f l\theta + 4l^2\theta^2}{D_f}\sqrt{\frac{\pi g\, v_f}{2}}.$$

here, $Q = \frac{50\times 10^3}{60}$ cc/sec, $D_f = 2$ cms, $l = 30$ cms, $\theta = ?$ $v_f = 10$ cc

putting the values above,

$$Q = 0.03627 \text{ radian}$$
$$= 2°$$

10. For a rotameter, the following data are provided:

float volume = 1cc, angle of taper = 5°, float diameter = 15 mm, total float movement = 30 cm, density of float = 2700 kg/m³, pipe diameter at the bottom most position of the float = 20 cm.

Calculate the range of flow rate when the fluid density is 0.7. Assume $C_d = 1$.

Ans. $v_f = 1\text{cc} = 1 \times 10^{-6}\ \text{m}^3$. $D_f = 15 \times 10^{-3}$ m; $\theta = 5°$.

$\rho_f = 2700 \text{ kg/m}^3$. $\rho = 700 \text{ kg/m}^3$.

$$A_f = \frac{\pi}{4} \times (15\times 10^{-3})^2 = 0.1766 \times 10^{-3}\ \text{m}^2.$$

Pipe diameter at the top most position of the float

$$= 20 + 2 \times 300 \times \tan 5° = 72.5 \text{ mm}.$$

The equation for flow rate for a rotameter is

$$Q = C_d \times \frac{D_p^2 - D_f^2}{D_f} \times \sqrt{\frac{\pi g v_f(\rho_f - \rho)}{2\rho}}$$

1st case: When float is at bottom most position:

$$Q|_{\min} = \frac{(20\times 10^{-3})^2 - (15\times 10^{-3})^2}{15\times 10^{-3}}\sqrt{\frac{3.14\times 9.8\times 1\times 10^{-6}\times(2700-700)}{2\times 700}}$$

$$= 77.35 \times 10^{-6}\ \text{m}^3/\text{sec}.$$

2nd case: When float is at top most position:

$$Q|_{\max} = \frac{(72.5\times 10^{-3})^2 - (15\times 10^{-3})^2}{15\times 10^{-3}}\sqrt{\frac{3.14\times 9.8\times 1\times 10^{-6}\times(2700-700)}{2\times 700}}$$

$$= 2\,223.8 \times 10^{-6}\ \text{m}^3/\text{sec}$$

11. Calculate the emf induced in an electromagnetic flowmeter when a conductive fluid passes through the pipe having inner diameter 3 cm.

Flux density = 45 weber/m² and flow rate = 200 cm³/min.

Ans.

$$\text{velocity} = \frac{\text{flow rate}}{\text{area}} = \frac{2000}{\frac{\pi}{4}(3)^2 \times 60} = 4.72 \text{ cm/sec} = 0.0472 \text{ cm/ sec}$$

$$\therefore \quad \text{induced emf } e = Blv = 45 \times (3 \times 10^{-2}) \times 0.0472$$

$$= 0.06372\ V = 63.72 \text{ mV}$$

12. For an ultrasonic flowmeter, the beat frequency Δf is 110 cps and the angle between the transmitter and receiver is 45°, the length of the sound path being 250 mm. Find the velocity of fluid. Assume that the transmitter is of single transducer assembly type.

Ans. The equation is

$$\Delta f = \frac{2V\cos\theta}{l}$$

$$\therefore \quad V = \frac{\Delta f \,.\, l}{2\cos\theta} = \frac{110 \times (250 \times 10^{-3})}{2 \times \cos 45°} = 19.45 \text{ m/sec}$$

REVIEW QUESTIONS

1. Draw and discuss the velocity distribution curves for laminar and turbulent flow. On what factors does frictional resistance depend for the two flow types?
2. Discuss vena contracta and the different coefficients pertaining to flow.
3. What are the different types of orifice plates used in industries? Mention their application areas. Draw the static pressure gradient in an orifice plate and discuss the same. Draw the different types of taps for orifice plate.
4. What are the different application areas of a Pitot tube? Draw and discuss a short single opening type Pitot tube.
5. Draw the layout of pipeline connections with a differential manometer when it is installed (*a*) below (*b*) above a p.d. device for liquid flow measurement.
6. Describe the basic theory of variable area meters with proper diagrams, showing the various forces acting on the float. Mention the advantages and disadvantages of a rotameter.
7. Explain the advantages or otherwise of the following types of restriction devices:
   (*a*) Pitot tube, (*b*) Nozzle, (*c*) Dall tube, (*d*) Elbow tap.
8. (*a*) Describe the construction and working of turbine flowmeters. Explain how the output is obtained in digital form for both flow rate and total flow.
   (*b*) What are the advantages and disadvantages of venturi, orifice and flow nozzle meter?
   [*WBUT (EI 501) 2003*]
9. List the different positive displacement meters. Why they are so called? Describe a reciprocating piston meter and state its features.
10. (*a*) Explain the difference between volume flow rate and mass flow rate.
    (*b*) Explain the working principle of a Coriolis mass flowmeter.
    (*c*) A flowmeter measures the flow of a liquid as follows:
        (*i*) constant at 10 kg/min for 30 minutes
        (*ii*) constant at 30 kg/min for 20 minutes.

What would be the measurement indicated by a mass-flow totaliser for the above period of 50 minutes? [*WBUT (EI 503) 2004*]

11. (*a*) Explain the working principle of an electromagnetic flowmeter.
    (*b*) "Electromagnetic flowmeter is not useful for organic fluids"—explain.
    (*c*) What are the advantages of pulsed DC excitation for electromagnetic flowmeters?
    (*d*) Explain why electromagnetic flowmeters are of 4-wire type instead of 2-wire type? [*WBUT (EI 501) 2004*]
12. Describe the working of ultrasonic flowmeters. What are the advantages and disadvantages of this type of flowmeter? [*WBUT (EI 501) 2003*]
13. Name some of the commonly used methods for solid flow measurement. Describe a conveyor weighing system for solid weight flow measurement.
14. Describe a Parshall Venturi flume for open channel flow measurement.
15. Describe a PVT system for gas flowmeter calibration.

Fill in the blanks:

1. For turbulent motion, frictional resistance is proportional to .... of velocity.
2. Approximate value of Reynolds number is .... when flow changes from stream line to turbulent motion.
3. The position of vena contracta is dependent on .... rate.
4. Eccentric type of orifice is used when .... is present in gas flow measurements.
5. A Pitot tube can measure velocity at a ....
6. A rotameter is an .... type flowmeter.
7. The tube material of a rotameter is normally made of ....
8. An orifice plate has no .... part.
9. Initial cost of installation of a flow nozzle is ....
10. Flow nozzles have a very .... coefficient of discharge.
11. A pitot is applicable for .... fluids.
12. Mass flowmeters utilise the principle of ....
13. A lobed impeller meter has a .... pressure drop.
14. A Coriolis mass flowmeter is .... type.
15. In an electrical manometer type flowmeter, the flow rate varies directly with .... of the fluid.
16. An electromagnetic flowmeter has a very wide .... range.
17. The tube material at which the sensor is placed in an electromagnetic flowmeter is .... in nature.
18. Sonic speed of a pressure wave transmitted through a fluid changes with changes in the .... of fluid.
19. The transit time between the transmitter and receiver for an ultrasonic flowmeter is of the order of ....
20. A laser Doppler anemometer can measure the .... velocity of a gas or liquid.
21. In a vortex shedding meter, Reynolds number should be of the order of ....
22. A swirlmeter is applied when the Reynolds numer is very ....
23. A .... flowmeter is useful to measure flow rate of multidirectional flow.
24. For solid flow rate measurement employing nuclear method, the source used is a .... ray.

Tick the correct answer:

1. Reynolds number
   (*a*) Increases with increase in average velocity of liquid
   (*b*) Increases with increase in density of flowing liquid
   (*c*) Decreases with increase in absolute viscosity of liquid
   (*d*) All of the above. ☐
2. Which of the following is a flow totaliser?
   (*a*) Venturimeter (*b*) Turbine flowmeter (*c*) Rotameter (*d*) Orifice plate ☐
3. A pitot-static tube is used for measuring velocity of a gas, flowing in a duct. The velocity is proportional to,
   (*a*) Square root of the total pressure measured by the tube
   (*b*) The total pressure measured by the tube
   (*c*) Difference between total and static pressures.
   (*d*) Square root of the difference between total and static pressures. ☐

   [*GATE 2002*]
4. In a rotameter, used for measuring flow rate of a fluid,
   P_a: Pressure above the float, p_b : pressure below the float, A: area of float,
   V: Volume of float, d_1: density of float material, d_2: density of the fluid, g: acceleration due to gravity. The following equation describes the equilibrium of the float.
   (*a*) $(p_b - p_a) A = vg(d_2 - d_1)$ (*b*) $(p_a + p_b) A = vg(d_2 + d_1)$
   (*c*) $(p_a - p_b) A = vg(d_2 - d_1)$ (*d*) $(p_b - p_a) A = vg(d_2 + d_1)$ ☐

   [*GATE 2002*]
5. Which of the following flowmeters has the lowest pressure drop for a given range of flow?
   (*a*) Orifice meter (*b*) Venturimeter (*c*) Flow nozzle (*d*) Rotameter

   [*GATE 2003*]
6. For a steady flow of liquid, the float of the rotameter will remain in a particular position when
   (*a*) Drag force is balanced by the weight of the float and buoyancy force
   (*b*) Weight of the float is balanced by the drag force and buoyancy force
   (*c*) Buoyancy force is balanced by the drag force and weight of the float
   (*d*) Drag force and buoyancy together is slightly greater than the weight of the float. ☐
7. A single-sensor contact-type ultrasonic flow detector uses a frequency of 330 kHz. When testing a specimen, an echo from a flaw is recorded 0.05 second after the transmitted pulse. If the velocity of sound in the test object is 6.0 km/s, the flaw is at a depth of
   (*a*) 120 cm (*b*) 60 cm (*c*) 30 cm (*d*) 15 cm ☐

   [*GATE 2004*]
8. An electromagnetic flowmeter is used to measure the average flow rate of a liquid in a pipe diameter 50 mm. The flux density in the liquid has a peak value of 0.1T. The output from the flowmeter is taken to an amplifier of gain 1000 and input impedance of 1MΩ. If the impedance of the liquid between the electrodes is 200 kΩ, the peak value of the amplifier output voltage for an average flow velocity of 20 mm/s is
   (*a*) 1.365 V (*b*) 1 V (*c*) 0.1 V (*d*) 0.083 V ☐

   [*GATE 2004*]

9. Figure given below shows an ultrasound flowmeter with the transducers separated by a distance $L = 25$ mm with the line joining the transducers making an angle 60° with the direction of flow. The transit time difference between upstream and downstream measurements is 10 ns with the sound velocity in the medium being 1000 m/s. Assuming that the size of the transducers is very small as compared to the diameter of the pipe, the flow velocity is

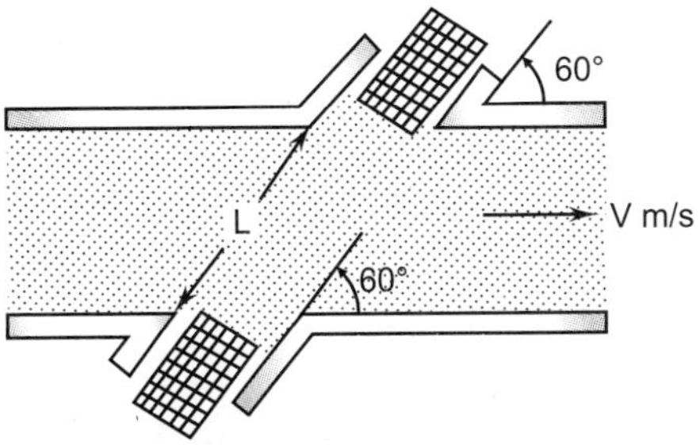

(*a*) 0.2 m/s (*b*) 2 m/s (*c*) 0.4 m/s (*d*) 40 m/s ☐

[*GATE 2005*]

10. "Vena contracta" is the cross-section where the flow area is a minimum for a restriction type flowmeter. For an orifice meter, if d is the diameter of the orifice opening, the area of the vena contracta is approximately,

(*a*) $\pi d^2/4$ (*b*) $0.99\ \pi\ d^2/4$ (*c*) $0.8\ \pi\ d^2/4$ (*d*) $0.6\ \pi d^2/4$ ☐

[*GATE 2006*]

11. The accompanying figure shows a VERTICAL venturi meter with upward water flow. When the measured static pressure difference, $p_1 - p_2$, between the inlet and the float is 30 kPa, the flow rate is found to be 50 liters per second. Assume that the coefficient of discharge remains the same. When $p_1 - p_2 = 20$ kPa, the flow rate, in litres per second, is

(*a*) 33.3 (*b*) 39.3 (*c*) 40.8 (*d*) 54.2 ☐

[*GATE 2007*]

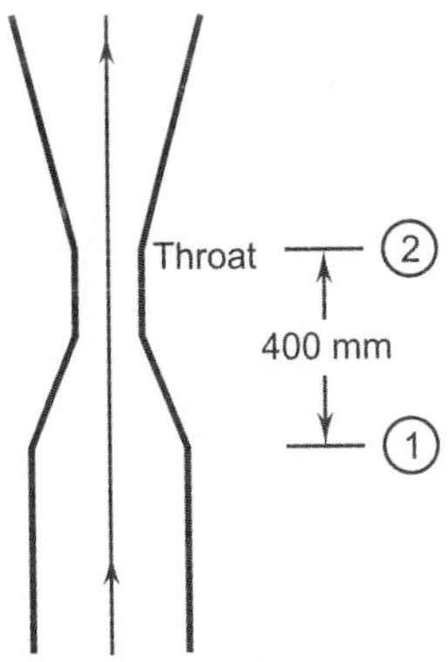

12. In a laminar flow experiment, fluid A is pumped through a straight tube and the volumetric flow rate and pressure drop per unit length are recorded. In a second straight tube having twice the

internal diameter of the first one, fluid B records the same pressure drop per unit length at the same volumetric flow rate. Assuming fully developed flow conditions in the tubes, the ratio of the dynamic viscosity of fluid B to that of fluid A is

(*a*) 16 (*b*) 32 (*c*) 64 (*d*) 128 ☐

[*GATE 2007*]

13. A rotameter can be used when the pipe line is

(*a*) Vertical (*b*) Horizontal (*c*) Incline (*d*) Both (*a*) and (*b*). ☐

14. A positive displacement flowmeter is a

(*a*) Quantity flowmeter (*b*) A variable area flowmeter

(*c*) A d.p. flowmeter (*d*) None of these. ☐

15. Hot wire anemometers are used to measure

(*a*) Liquid velocity (*b*) Gas velocity

(*c*) Pressure of a gas (*d*) None of the these. ☐

16. For clean fluids, the meter to be used is

(*a*) Laser doppler anemometer (*b*) Hot wire anemometer

(*c*) Ultrasonic flowmeter (*d*) Turbine flowmeter. ☐

17. Nutating disc flowmeters are mainly used for

(*a*) Industrial flow measurements (*b*) Residential flow measurements

(*c*) Both (*a*) and (*b*) (*d*) None of the these. ☐

18. A Pitot tube

(*a*) Has a poor accuracy (*b*) Has a high accuracy

(*c*) Is economical to instal (*d*) Both (*a*) and (*c*). ☐

19. Thermal flowmeters are mainly used for the measurement of

(*a*) Steady flow of gases (*b*) Unsteady flow of gases

(*c*) Unsteady flow of fluids (*d*) None of the these. ☐

20. The principle of operation of an electromagnetic flowmeter is based on

(*a*) Lenz's law

(*b*) Faraday's law of electromagnetic induction

(*c*) Both (*a*) and (*b*)

(*d*) None of the above. ☐

21. For a hot wire anemometers, the rate of cooling of the wire depends on

(*a*) Velocity of the fluid stream

(*b*) Physical properties of the flowing fluid

(*c*) Physical properties of the hot wire, its shape and size

(*d*) Difference in temperature between the hot wire and the fluid stream

(*e*) All the above. ☐

22. During intravascular measurement of arterial blood pressure, catheters may be introduced in different configurations, as shown. The static pressure will be measured correctly in the configuration(s)

(*a*) P and Q but not in R (*b*) R only

(*c*) Q only (*d*) P and R but not in Q **[Ans:** (*b*)**]**

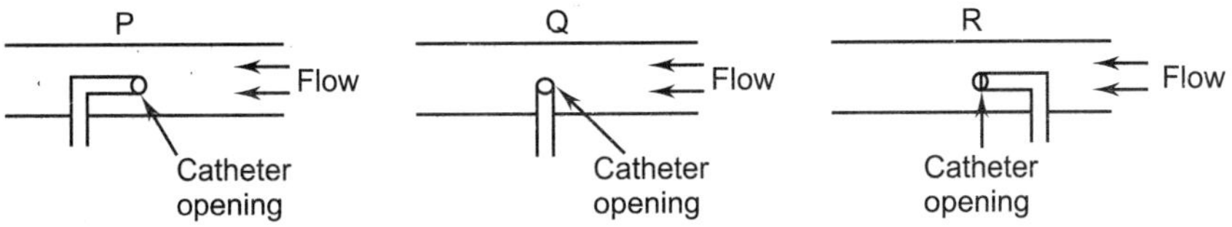

REVIEW PROBLEMS

1. An orifice plate of diameter 60 mm is placed in a pipeline of diameter 120 mm. Determine the volume flow rate of water when the pressure differentials is 250 Pa. Assume $C_d = 0.6$.

   [Ans. 3.88 litres/sec**]**

2. A gas with density 0.52×10^{-3} gm/cc flows through a pipe having a radius of 4.0 mm. A venturi tube of throat radius 2 cm is inserted in the pipe to measure the flow rate. Express the same in litre/sec when the manometer, containing Hg, records 10 cms. Assume $C_d = 0.95$.

   [Ans. 279.3 litre/sec**]**

3. In a pitot static velocity measurement, the differential head produced is 10 kPa. Calculate the rate of flow when the fluid is water. (ρ water = 1000 kg/m^3) **[Ans.** 4.472 m/sec**]**

4. A cylindrical float of 3 cm height and diameter and a density of 3000 kg/m^3 is used in a rotameter. Find the maximum value of flow rate when the maximum diameter of the metering tube is 5 cms. Assume $C_d = 0.6$. **[Ans.** 0.82×10^{-3} m^3/sec**]**

5. An electromagnetic flowmeter employs a pipe of diameter 80 mm and a flux density of 0.1 Wb/m^2 to measure the flow rate of a fluid. The impedance between the two electrodes is 300 kΩ. The flowmeter output is applied to gain amplifier having a gain of 1000, the input impedance of the amplifier being 300 kΩ. If the peak-to-peak amplifier output is 0.5 V, find the average velocity of flow. **[Ans.** 6.25 m/sec**]**

6. A calibrated orifice meter is used in a pipeline of 103 mm ID to calibrate the probe of a constant temperature hot wire anemometer (CTA). The orifice meter readings are recorded in mm of Hg and the CTA readings in volts. It is independently found that the average velocity in the pipeline is exactly equal to the velocity at its axis. The volume flow rate of the fluid (Q) can be measured from the orifice meter calibration equation $Q = 6.311 \times 10^{-4}\sqrt{h}$, where h is in mm of Hg and Q is in m^3/s. The readings of the CTA are correlated in the form (volt)$^2 = a + b$ (velocity)$^{1/2}$. Determine the constants 'a' and 'b' in this equation if the voltage readings are 0.284 and 0.323 V respectively when the corresponding orifice meter readings are 77 and 154 mm Hg.

   [*GATE 2002*] **[Ans:** $a = 0.079$; $b = 0.252$**]**

7. The figure shows the cross-sectional diagram of an orifice flow meter with an orifice radius R. Point 'a' is 30 R upstream while points 'b' and 'c' are 0.8 R and 30 R downstream from the orifice respectively. The pressure at points a, b and c are P_a, P_b and P_c respectively. Then

   (*a*) $P_c > P_b > P_a$ (*b*) $P_b > P_c > P_a$

   (*c*) $P_a > P_b > P_c$ (*d*) $P_a > P_c > P_b$ [*GATE 2009*] **[Ans:** (*c*)**]**

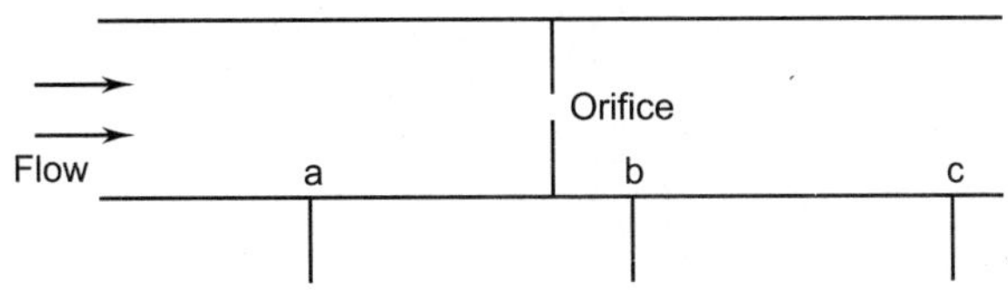

CHAPTER 10

LEVEL MEASUREMENT

10.1 INTRODUCTION

Level measurement and its control plays a very important and crucial part in process plants. For example, maintaining water at a predetermined level in a boiler drum or of liquids in storage tanks is very vital for the safety of plant and personnel.

Level gauges and meters continuously monitor the level of liquid while some level gauges operate within a predetermined range. Such level gauges actuate an alarm when either the upper or lower limit is reached. Level limit switches are fitted with contacts which is either broken or closed on reaching a predetermined level.

There are numerous ways to measure level like mechanical, electrical, ultrasonic, acoustical or radioisotope methods.

The different heads under which level measurement can be categorised are given below:

10.2 MECHANICAL TYPE LEVEL MEASUREMENT

10.2.1 Dipstick

Dipsticks are of two types—ordinary dipstick and optical dipstick.

10.2.1.1 Ordinary Dipstick

It is a very simple and cheap device to measure the liquid level. It is a graduated metal bar which can be inserted into the liquid up to a reference point or the bottom of the vessel. It is shown in figure 10.1. The dipstick is removed from the liquid and the level of liquid can be read off the scale.

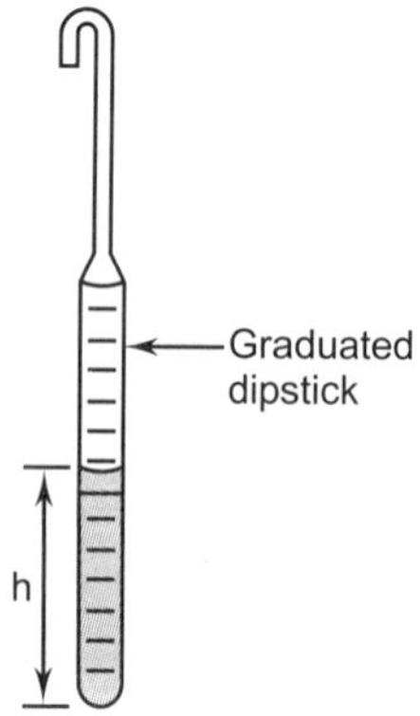

Fig. 10.1: A simple dipstick

The method is applicable for relatively small vessels. It is a rough method to measure liquid level. The method is normally applied for measurement of level of petrol/diesel tankers.

10.2.1.2 Optical Dipstick

It is shown in figure 10.2. It is applied to tanks having considerable depths and the dipstick need not be removed from the vessel. The dipstick can be moved in a vertical plane by a mechanical drive mechanism. When the lower end of the dipstick does not touch the liquid level, light enters into the detector. When the lower chamfered end of the dipstick comes in contact with the liquid, light cannot proceed towards the detector. Thus, adjusting the position of the dipstick, its position can accurately be determined.

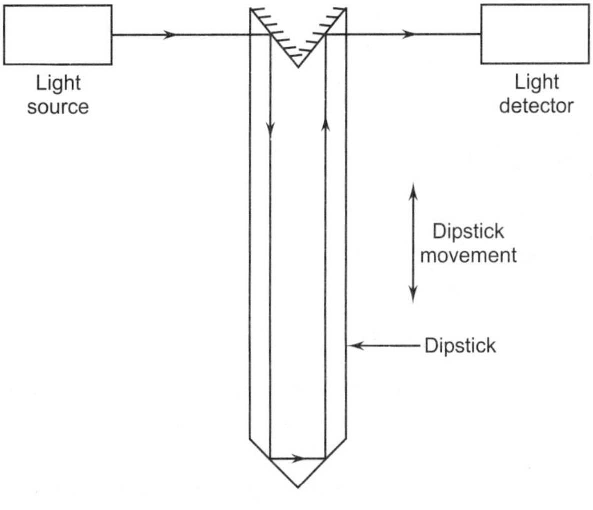

Fig. 10.2: An optical dipstick

The main advantage of this method is that the dipstick need not have to be removed from the liquid.

10.2.2 Sight Glass Method

This is a very simple and direct method of level measurement. It is shown in figure 10.3. The sight tube is a graduated glass tube mounted just besides the tank. As the liquid level changes inside the tank, the same variation takes place in the tube. The reading of the glass tube is read off from the graduated scale.

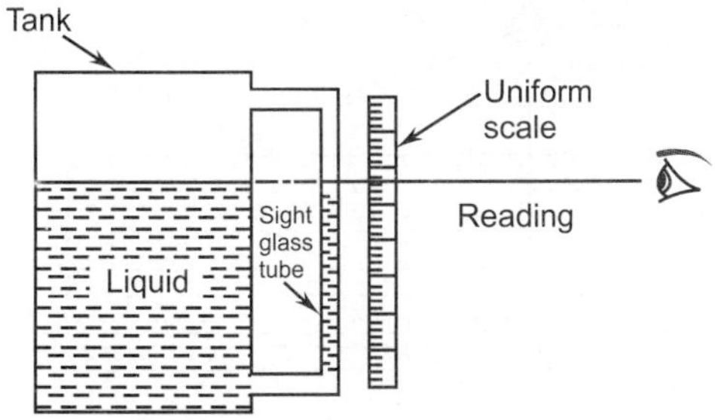

Fig. 10.3: Sight-glass level gauge

If the tank is at a very high temperature, a correction will be needed because of density variation of liquid inside the tank and the sight tube. The density of liquid inside the tube will be higher because it is at a much lower temperature.

10.2.3 Float Gauge Method

The float gauge method of measurement of level is shown in figure 10.4.

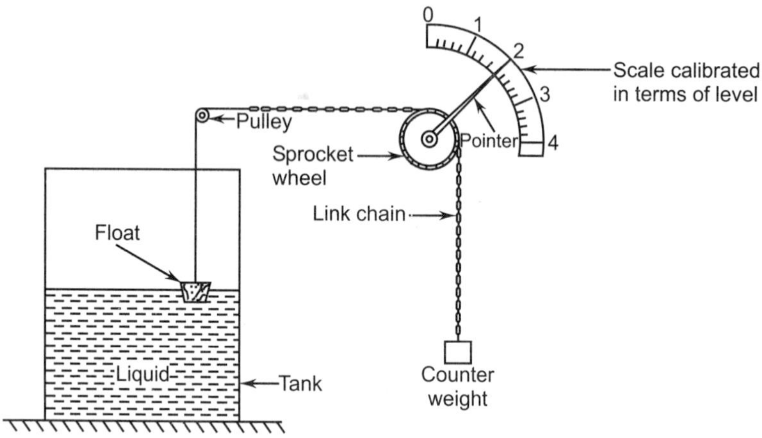

Fig. 10.4: Float level gauge

Its operation is based on Archimede's principle as applied to floating bodies.

It consists of a float, pulley, sprocket wheel, link chain, pointer, a counterweight and a scale. As the liquid level changes, the float changes its position accordingly. This change is reflected in the pointer movement along a scale graduated in terms of level of the liquid. The counterweight keeps the chain under tension and does not allow the chain to become slack with changing float position.

10.2.4 Rotating Shaft Level Gauge

Figure 10.5 shows a rotating shaft level gauge. Here the float is attached to a shaft. As the level changes, the float changes its position. The same is transmitted to the pointer via the shaft mechanism. The scale is calibrated in terms of liquid level.

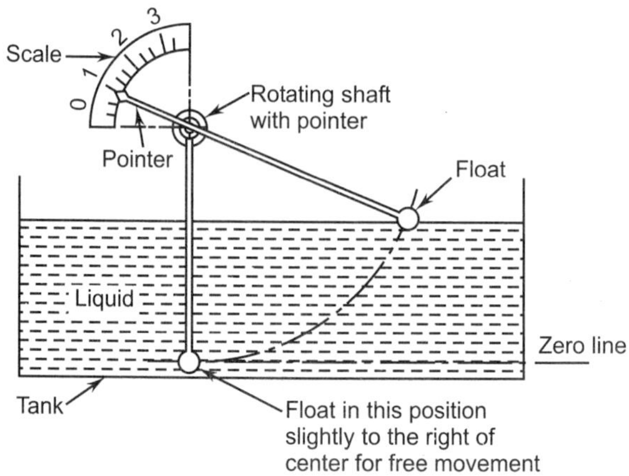

Fig. 10.5: Rotating shaft level gauge

10.2.5 Level Measurement Using Reed Switches

A method for measurement of level employing reed switches is shown in figure 10.6. Reed switches alongwith resistances are placed in a vertical plane, spaced apart from each other as per convenience. A float with a magnet moves up or down, very close to the reed switches, as level in the tank changes. When the magnet comes in close contact with a particular reed switch, it is closed. The magnitude of current through the ammeter is dependent on the particular reed switch which is closed. Ammeter current can thus be calibrated in terms of level of liquid in the tank. Current path may be completed either *via.,* 1, 2, 3 or 4, as per the level.

This method of level measurement is not a continuous one, rather it is discrete. A 5 mm difference between two consecutive readings is a practical proposition for this.

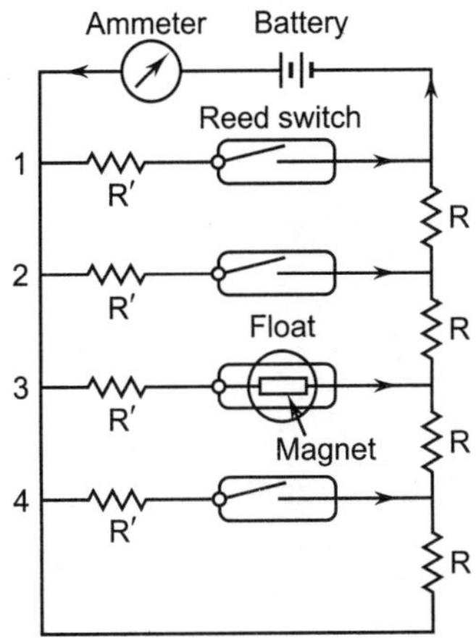

Fig. 10.6: A level indicator using reed switches

10.2.6 Level Measurement Using Magnetic Coupling

This is shown in figure 10.7. A float with a magnet moves up or down the liquid as level changes. A follower magnet with a tie rod attached to it is placed inside the guide tube.

If the level increases, the float magnet moves up. It pulls the follower magnet accordingly in a vertical plane. This movement of the follower magnet is sensed by a level indicator which is precalibrated.

The device finds its application in motor vehicles.

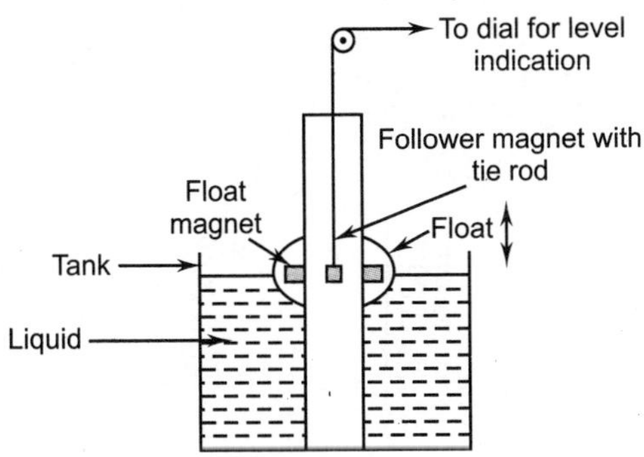

Fig. 10.7: Magnetically-coupled level indicator

10.2.7 Displacer Type

The schematic of a displacer is shown in figure 10.8. Its specific gravity is made slightly higher than the liquid whose level measurement is desired. Unlike a float, immersed portion of the displacer varies as liquid level changes. The displacer is cylindrical in nature having a fixed cross-sectional area. The displacer is suspended into the liquid with the help of a spring.

With rise in liquid level, the displacer is immersed more inside the liquid. This results in increased buoyancy on the displacer, resulting in more upward thrust. The spring is thus contracted in such a way that the system remains taut. The spring contraction is very small and the change in tension in the spring is a measure in the change of liquid level.

In the next section, a torque-tube displacer, based on the above principle, is discussed.

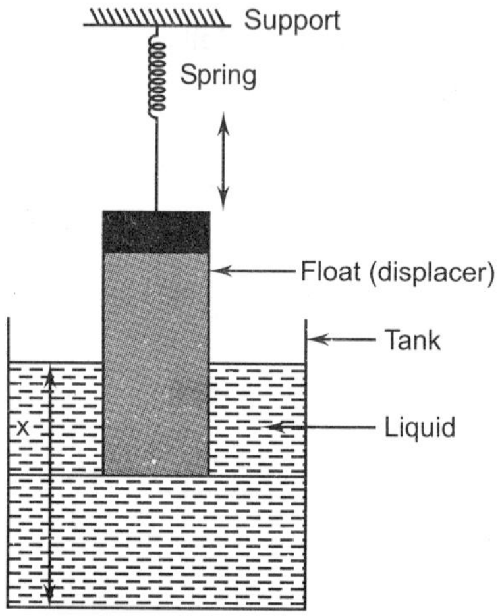

Fig. 10.8: Level measurement by displacer

10.2.8 Torque-Tube Assembly

The basic displacer type level indicator has been replaced by a torque-tube assembly shown in figure 10.9. The displacer is partially immersed in the liquid and is connected to the bearing assembly via the support rod. The rotary shaft is welded to the female socket *via* the torque rod. The rod is housed inside the torque-tube. The flange is bolted to the cage of the tank housing. Before bolting, an initial twist is imparted to the torque-tube which is balanced by the weight of the displacer at the zero setting of the scale.

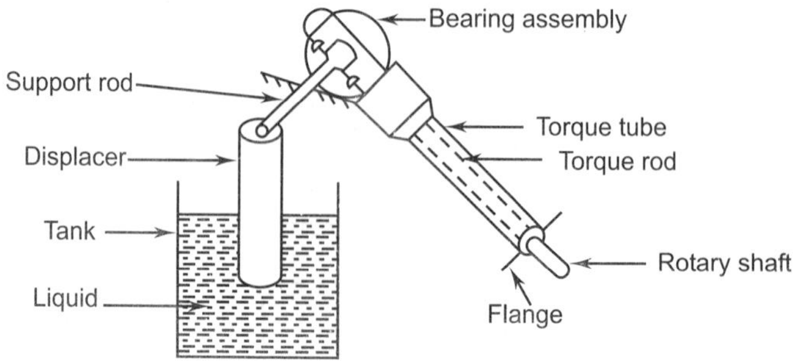

Fig. 10.9: The torque tube assembly

With rising liquid level, more of the displacer goes inside the liquid and buoyancy increases. This causes a rotation of the rotary shaft *via* the torque-tube. The angular rotation varies in between 2° to 6°. This may be used to vary the position of the flapper of a flapper-nozzle assembly. This can be used in conjunction with a pneumatic amplifier for either level measurement or its control.

The displacer movement can be mechanically coupled by a lever assembly and connected to a (*a*) strain gauge for a voltage variation proportional to displacement or a 4–20 mA signal (*b*) magnets and a proximity switch to derive an electrical signal (*c*) core of a rotary variable differential transformer which would cause consequent changes in the flux linkage of the secondary and thus variation in electrical output proportional to level changes.

This type of displacer system finds application in level measurements of tanks, high pressure/low pressure heaters etc.

10.3 HYDROSTATIC TYPES

10.3.1 Introduction

Hydrostatic types of level measurements fall under the purview of indirect method. It involves connecting a pressure gauge or a differential pressure gauge appropriately to the tank whose level measurement is desired. A rise or fall of liquid in the tank causes a change in hydrostatic pressure. The scale of the pressure gauge is calibrated in terms of level.

A change in temperature of liquid changes its density, thereby affecting calibration. Again, a pressure gauge, calibrated for a particular type of liquid level measurement, needs recalibration when another liquid is used.

If the nature of the liquid is such that it must not come into direct contact with the gauge, a sealing liquid is used in between the two. Sometimes air is also used for transmitting the pressure change caused by changes in level.

10.3.2 A Simple Hydrostatic Level Indicator

Figure 10.10 shows a simple arrangement of a level indicator. With rise or fall of level of liquid inside the container, an increase or decrease of pressure is effected. A connection to the pressure gauge is done as shown. The gauge is calibrated in terms of liquid level with respect to a reference level.

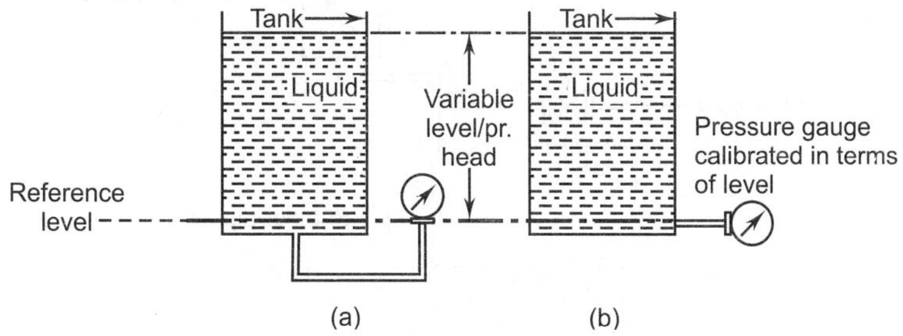

Fig. 10.10: Simple arrangement for level-pressure measurement. (*a*) Bottom connection (*b*) Side connection

Figure 10.10(*a*) connects the gauge from bottom while figure 10.10(*b*) is for side connection. For liquids which cannot be directly connected to a pressure gauge, an interface liquid is used to protect the gauge from damage.

10.3.3 Diaphragm Type Level Indicator

Figure 10.11 shows a level indicator with a diaphragm as the sensing element. It may be made of rubber or neoprene. With change in liquid level in the container, hydrostatic pressure changes. This is applied to the outside of the diaphragm. Air inside the diaphragm alongwith the capillary is compressed. The pressure instrument connected at the end of the capillary is calibrated in terms of level of liquid.

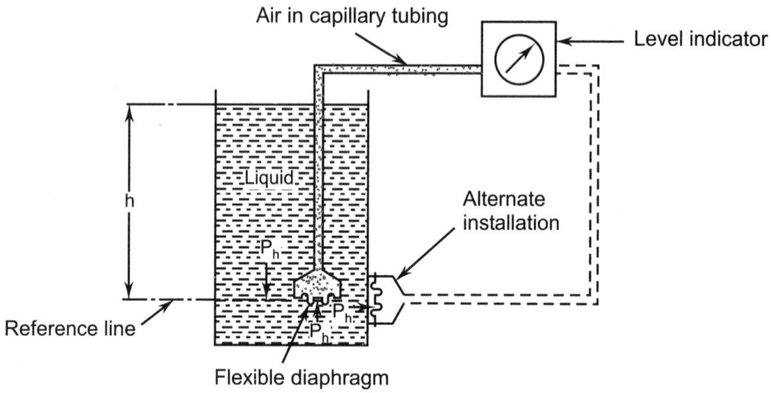

Fig. 10.11: Diaphragm type level indicator

Initially the diaphragm is placed at the reference level, before calibration is carried out. The system must be leak-proof. Any air leaking outside from the capillary would render the system inoperative. The diaphragm may also be affected by extreme temperature conditions of the liquid.

10.3.4 A Bubbler Level Indicator

Figure 10.12 shows a bubbler level indicator, also known as air purge or dip tube.

The bubbler tube or dip tube is installed vertically in the tank with its open end coinciding with the reference level. The other end of the tube is connected to air supply (*via* a regulating valve) and a pressure gauge.

When the air supply is blocked, water level in the bubbler tube would be equal to that in the tank. Air supply is regulated by the regulating valve. Bubbles will start coming out of the

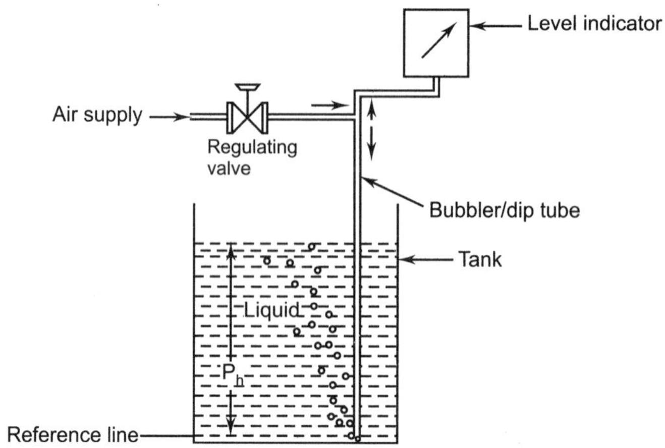

Fig. 10.12: Principle of bubbler level gauge

bottom end of the bubbler tube when air pressure just exceeds the hydrostatic column of the liquid level in the tank. The pressure gauge thus records the air pressure required to overcome the liquid column head above the open end of the tube. The pressure gauge is calibrated in terms of liquid level. The calibration is valid at a particular temperature.

An advantage of bubbler system is the fact that the measuring gauge can be placed at any elevation with respect to the tank. This is particularly suitable for measurement of level of underground tanks. The instrument requires recalibration if the liquid changes.

10.4 LEVEL MEASUREMENT IN CLOSED TANKS

The bubbler system just described pertains to open-tank measurement. In this, pressure on the liquid surface and on the reference side of the pressure sensing element are equal. Usually, this is atmospheric pressure. With change in atmospheric pressure, the above two changes are equally affected, resulting in cancellation. Thus, level measurement is not affected.

In the case of a closed tank, the pressure on the liquid surface may be variable. This may affect the pressure measurement, which is calibrated in terms of level. Thus, a differential pressure element is utilised for level measurement in closed tanks. This is shown in figure 10.13.

The variation of pressure on the liquid surface affects the high and low side of the *d/p* element identically. This results in an output signal only proportional to the liquid level.

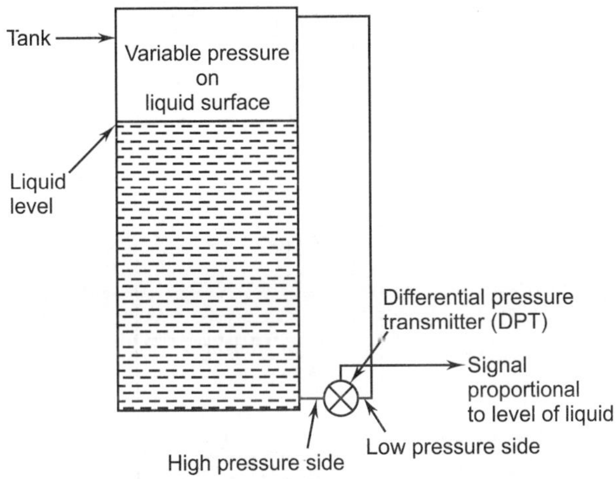

Fig. 10.13: Level measurement in closed tank

10.5 ZERO SUPPRESSION IN OPEN AND CLOSED TANKS

A level measuring instrument should be mounted at the reference level of the tank. Otherwise, the head pressure caused by the liquid, when the instrument is mounted at another location, must be taken care of during calibration of level measurement.

It is, however, not possible always to take the pressure tap at the reference level. The mounting of the Differential Pressure Transmitter (DPT) may also not match the reference level. In such cases, the pressure head caused by the liquid above or below the reference level must be properly compensated.

Figure 10.14 shows zero suppression in open and closed tanks. Figure 10.14(*a*) shows the case for an open tank. Here the pressure tap is taken at a depth *Y* below the minimum (reference) level, while the DPT is placed at a depth *Z* below the level of pressure tap. The hydrostatic pressure corresponding to $(Y + Z)$ would be recorded in the DPT when the liquid is at the reference level. This signal must be suppressed to make the output represent a zero value under such condition. This is zero suppression. This can be achieved by an adjustment served within the instrument or else electronically. Thus, with the level of liquid matching the reference level, the instrument output is made to represent the zero level value. For transmitters, this would correspond to 3 psi or 4 mA.

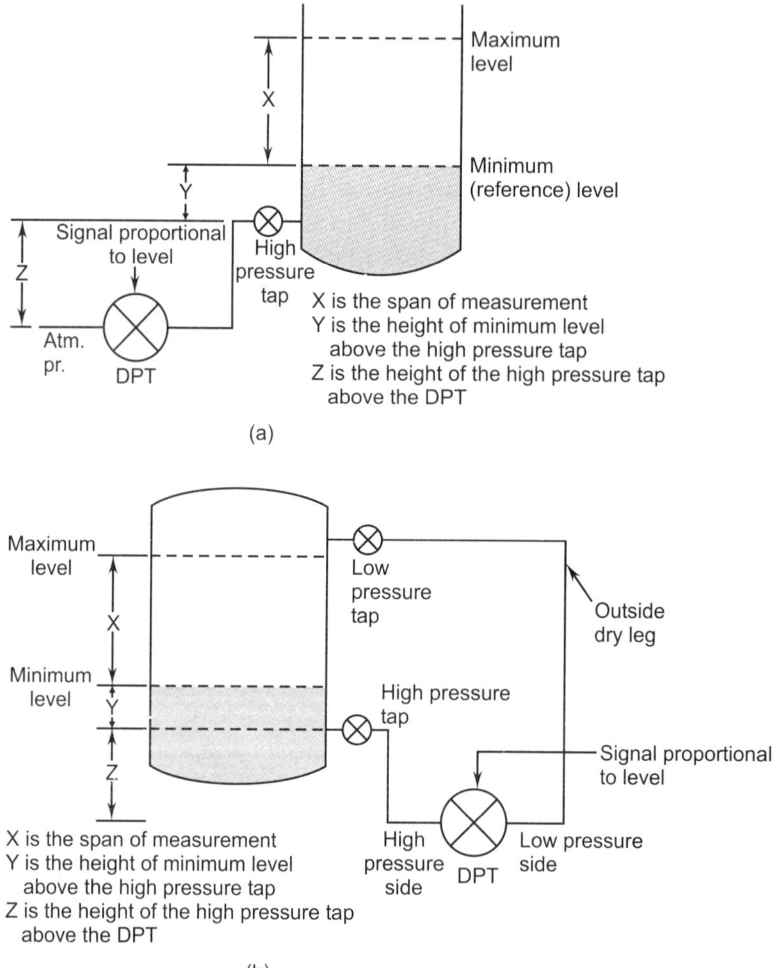

Fig. 10.14: Zero suppression in (*a*) open tank (*b*) closed tank

Zero suppression for the case of a closed tank and shown in figure 10.14(*b*) follows the same methodology.

10.6 ZERO ELEVATION IN CLOSED TANKS

Figure 10.15 shows the case of a closed tank in which the space above the liquid surface is filled with its vapour. It may so happen that the vapour on the right hand side of the low pressure tap condenses into liquid. This condensed liquid is sometimes called a wet leg and it produces a hydrostatic head on the low pressure side.

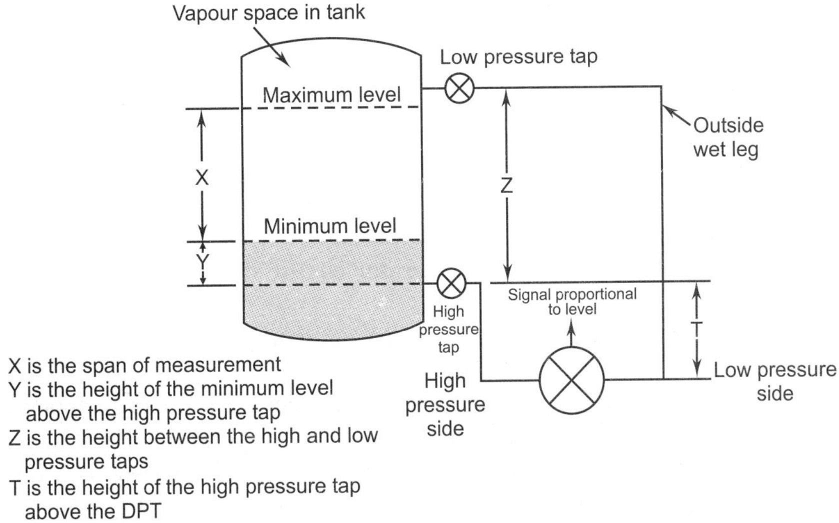

Fig. 10.15: Zero elevation in a closed tank

This would result in a negative reading when the level on the higher side corresponds to reference level. This is to be compensated and it is called zero elevation.

It must be understood that, in this case the problem of zero suppression is present also. Thus, when the level corresponds to the reference level, the instrument reading must give zero output, after taking into consideration both zero suppression and zero elevation errors.

10.7 USE OF PRESSURE REPEATERS IN CLOSED TANKS

Some process liquids whose levels are desired to be measured, are non-viscous at process temperature but become thick in nature or may even solidify at ambient wet leg temperature. In such cases, pressure repeaters are mounted in the vapour space above the process liquid in the tank and another tapping is taken out from the reference level. The vapour pressure is replicated by the repeater and transmitted to the low pressure side of the *d/p* transmitter. Thus, problem of wet leg on the low pressure side is avoided. A varying vapour pressure above the liquid level would not influence the level transmitter output.

10.8 ULTRASONIC LEVEL INDICATORS

10.8.1 Introduction

Sound waves of frequency 20 kHz and above are known as ultrasonic sound. When such a sound wave travels through a medium and strikes another medium, such as a wall, a particle in the liquid or a liquid surface, a small portion of the sound energy of the wave penetrates the barrier and the major portion is reflected, called an echo. Surfaces that reflect sound waves are called *live media* and those that reflect very poorly or not at all are called *dead media*. Examples of dead media are dust (in loose form), rust, etc. In this context, the sound absorption coefficient of a material is defined as:

$$c_a = \frac{\text{sound energy absorbed}}{\text{sound energy incident on the surface}}$$

Value of c_a is dependent on frequency of sound and also on such properties of the medium as porosity, material thickness and rigidity.

10.8.2 An Ultrasonic Level Gauge

The block diagram of a typical ultrasonic level gauge is shown in figure 10.16. Sound wave is produced by the generator and transmitter block. The transducer sends out the sound. It falls and is reflected from the level to be sensed/measured. The transducer in the reflected path receives this sound and converts it into an electrical signal. This is subsequently wave shaped and displayed in terms of level. The timing generator circuit synchronises the operations of all the blocks shown in the figure.

The time difference between the transmitted and received pulses is directly proportional to the distance between the transducers and the level being sensed.

The speed of ultrasonic sound is utilised to measure level. This speed is dependent on both the medium through which the pulse is passing as also its temperature. Thus, any temperature variation must be accounted/compensated for. As an example, the speed of ultrasonic sound through air varies with temperature at the rate of 0.607 metre per second per °C.

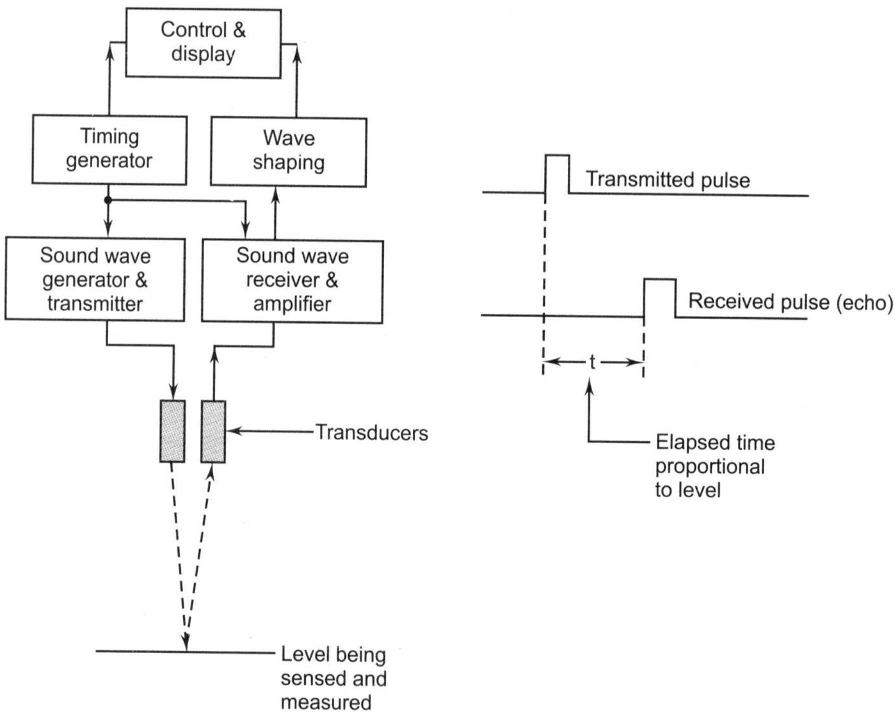

Fig. 10.16: Block diagram of an ultrasonic level measurement system

10.8.3 Different Types of Level Sensor Assembly

Figure 10.17 shows different variations in the ultrasonic level sensor assembly. Figures (*a*) and (*b*) are top mounted while figures (*c*) and (*d*) corresponds to bottom mounted assembly. In figure (*a*), the transmitter and receiver are separate while in figure (*b*), they are housed together. The advantages of top mounted ultrasonic level sensors are: (*i*) the sensor assembly does not come in contact with the fluid and (*ii*) the velocity of sound pulses is not fluid-specific. But the disadvantage lies in the scattering of some sound energy owing to scattering by liquid vapour that is invariably present at the top of the liquid. When the height of the liquid column is to be measured directly, the transducers are bottom mounted. Figure (*c*) exposes the transducer assembly to the process fluid and limits its accessibility to maintenance. Figure (*d*) circumvents these two drawbacks but at the cost of the ultrasonic pulse being weakened substantially by absorptive and dispersive effects of the bottom of the tank.

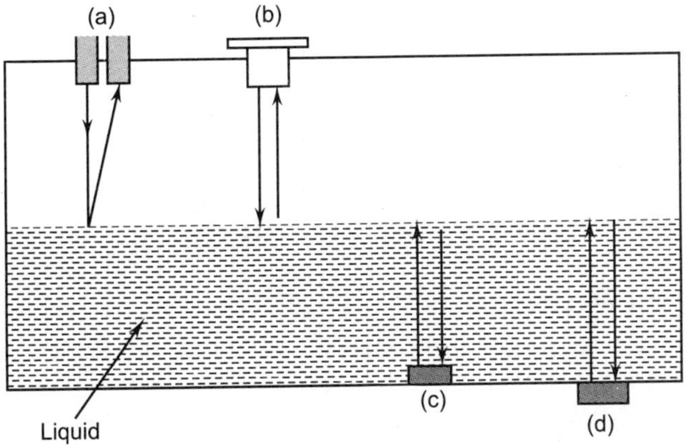

Fig. 10.17: Different types of ultrasonic level sensor assembly

10.8.4 Interface Level Measurement

Ultrasonic techniques can also be equally applied in measuring the position of interface between two immiscible liquids or measuring the sludge or precipitate level at the bottom of a liquid-filled tank. These are shown in figure 10.18.

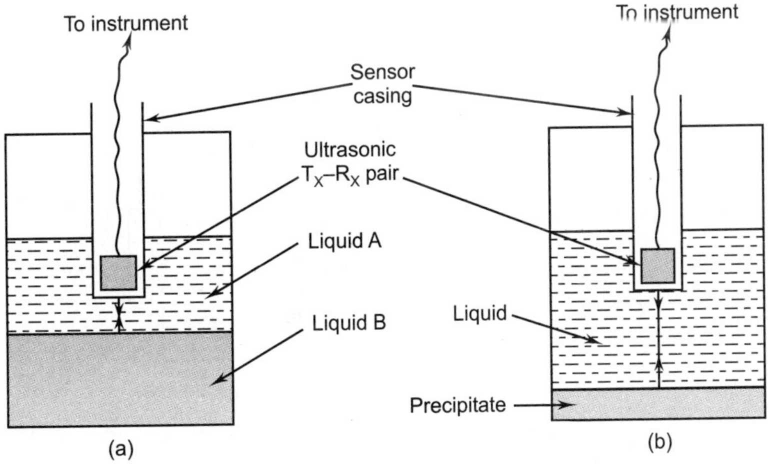

Fig. 10.18: Ultrasonic method of measuring (*a*) liquid/liquid interface; (*b*) liquid/precipitate interface

Commercial ultrasonic level detectors have a standard range of about 15 m with an accuracy limit of 0.1% of FSD.

Some points about ultrasonic level measurements:

- Stagnant liquids and solids having large and hard particles are good reflectors of ultrasonic waves.
- Foam, fluff and loose dirt are poor reflectors of ultrasonic waves.
- Dust, mist and humid environment in the vicinity of liquid surface tend to absorb ultrasonic waves.
- For measuring considerable depths, power of ultrasonic waves must be strong enough, because it attenuates with distance. In such cases, the pulses generated must be well-focussed.
- Surface whose depth is required to be measured must be flat. The pulses must strike the surface perpendicularly.
- For agitated liquids and sloping solids, the echo produced at the liquid surface has a tendency to disperse.
- It is a non-invasive type of measurement.

10.9 RADIATION TYPE LEVEL MEASUREMENT

10.9.1 A Nuclear Level Detector

Radiation method of level measurement normally uses a low-level γ-ray as a radiation source on one side and a detector system on the other side in a liquid filled tank. Figure 10.19 shows the schematic of the measurement system employed in which fig. (*b*) is a more accurate method which employs multiple sources.

Caesium-137 is used as the γ-ray source. The radiation level measured by the detector is related to the length of liquid as follows:

$$I = I_o \exp(-\mu \rho x)$$

where, I_o = intensity of radiation to be received by the receiver in absence of any liquid
I = intensity of radiation actually received by the receiver
μ = mass absorption coefficient
ρ = mass density of liquid
x = distance travelled through the medium.

The detector used for the purpose is a Geiger-Muller (GM) counter, an ionisation chamber or else a scintillation counter.

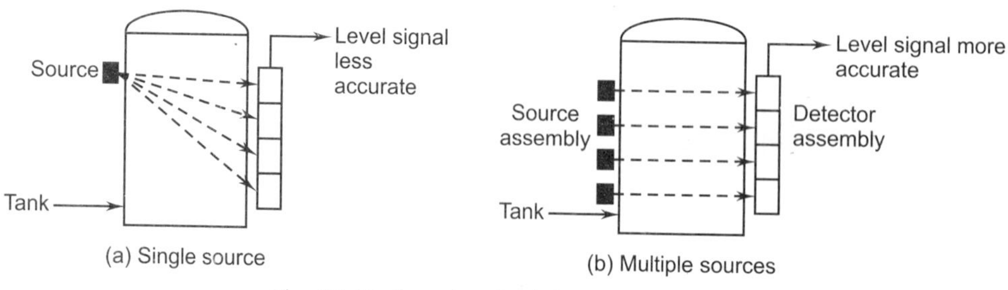

Fig. 10.19: A nuclear level detection method

Gamma (γ) rays, emitted by the source, pass through the walls of the container and the material in the tank. The strength of γ-ray reaching the detector would be a function of the level of material in the tank—the more the level the less the strength as in figure (*a*). The detector can then be calibrated in terms of level of material in the tank. The detector reading will be a non-linear function of level in this case and accuracy would be low.

When multiple sources and multiple detectors are used, as in figure (*b*), accuracy would be more. As level continues to rise, readings in the detectors would start decreasing, starting from the lowest detector. In this case, the accuracy of measurement will be more.

10.9.2 Salient Points

- A non-intrusive method.
- It is relatively expensive.
- Maintenance relatively easier, since source and detector assemblies are housed outside the tank.
- Operator safety is required because γ-radiation is involved. Generally, level measurements are carried out with a low permissible value of γ-radiation.
- Recalibration of detector (*s*) is required when material in the tank is changed with a new one or the material density changes.
- Can be used for density or interface level (between two liquids) measurement.
- Level indicators can have a range from 25 mm to 7 m with a precision of approximately 1% of the span.
- For high-pressure processes, wall thickness is significant. For such cases, more absorption will take place at the walls on both the source and detector sides. This would lead to erroneous measurements.
- Radiation source and detector(*s*) are placed outside the tank for (*a*) furnaces at very high pressure or temperature (*b*) highly corrosive fluid (*c*) coarse and abrasive type bulk materials, such as coal, limestone or ores, etc.

10.10 RADAR (MICROWAVE) LEVEL INDICATOR

A radar level sensor, consisting of a transmitter, an antenna, a receiver with signal processing circuitry in-built and an operator interface is shown in figure 10.20

The technique involves sending a constant amplitude frequency modulated microwave signal (a selected carrier frequency (*FMCW*) and an waveform) towards the surface of the process fluid. This signal is generated either by a parabolic dish antenna or a horn antenna. The reflected signal from the liquid surface is routed to the receiver. Time of flight is the time gap between the transmitted radar pulse and the received (echo) pulse. The frequency modulated wave frequency varies from 0 to 200 Hz for process fluid surface variation from 0 to 60 m.

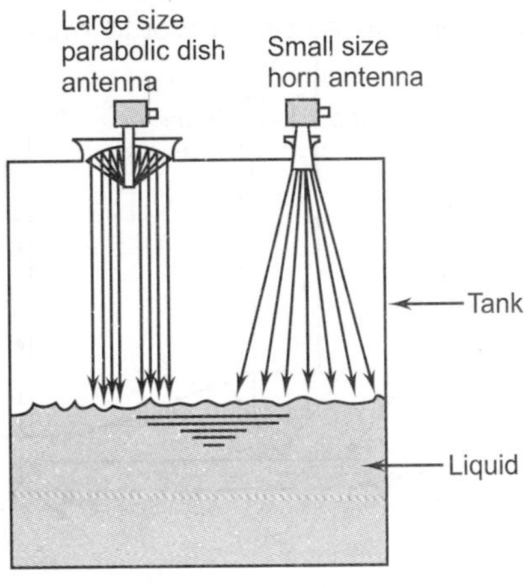

Fig. 10.20: A radar level sensor

Advantages

- Microwave velocity is about a million times more than sound velocity. Thus, this method is least affected by changes in process temperature or pressure or presence of foam, dust, mist in the process tank.
- Frequency domain analysis almost eliminates noise interfering effects.
- Non-intrusive type.

Disadvantages

- The dielectric constant of the liquid has to be substantial or else a very weak echo signal would result.
- Tanks lined with glass or tank nozzles functioning like waveguides would produce spurious echo effects.

10.11 OPTICAL METHODS

There are several optical methods of level measurements which are described below:

10.11.1 Reflection Based Level Measurement

There are three methods under this category. The first one is the time-of-flight measurement in which a single transmitter detector system detects the time gap between the emitted ray (from the transmitter) and the reflected ray (by the detector). In this case, the path of light travel is vertical. Knowing the time and the speed of light, the level can simply be determined. The second method involves measuring the angle between incident and reflected beams and calculating the level through simple trigonometry.

In the third method, light falls on the surface at a fixed angle of incidence. It is reflected as per the law of reflection. The detector detects the reflected light if it happens to be along the path of the reflected ray. This method thus is meant for detecting a fixed level.

10.11.2 Refraction-Based Level Measurement

Figure 10.21 shows the principle involved in the refraction based level measurement. So long as rays from the transmitter, incident on the left edge of the prism, is above the liquid level, there will be total internal reflection, as in figure (*a*). The reflected light intensity at the receiver will be maximum. In figure (*b*), the rays incident on the left edge of the prism corresponds to the liquid level. Thus, refraction of light at this point would occur, resulting in a reduced light intensity at the receiver. The vertical displacement of the prism would thus indicate the liquid level.

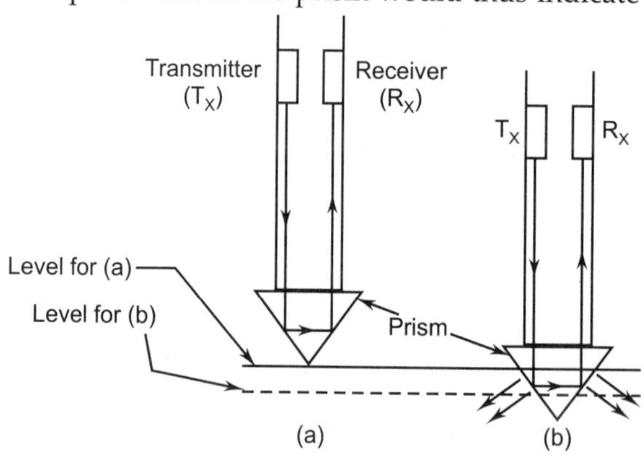

Fig. 10.21: Level measurement by refraction when the tip (*a*) does not touch the liquid (*b*) touches the same

This method is prone to error if the prism gets wet by the liquid and a few drops of it cling to the prism surface even when it is not in contact with the liquid.

10.11.3 Optical Fibre-Based Level Measurement

Single and multi-level detection is possible using step index multimode fibres.

Light rays pass through the core and cladding in multimode fibres. If the cladding is surrounded by a material having refractive index lower than that the cladding, propagation of light rays through the cladding remains unaffected. However, if the surrounding material has a higher refractive index, propagation of light rays through the cladding is affected. Light rays then leaks through this interface and the light rays reaching the detector would be of lesser power. In such a case, the propagation mode is said to be stripped off.

Figure 10.22 shows a single level detection by a step index fibre. A portion of the jacketed fibre is peeled off. This exposes the cladding. When this part is surrounded either by air or vapour of a liquid, the propagation of light rays through the cladding is unaffected because they have a refractive index less than that of cladding. When the liquid level rises to a height *h*, there would be loss of light rays because refractive index of liquid is more than that of cladding. The receiver (photo detector) would thus record lesser intensity.

Chapter 10

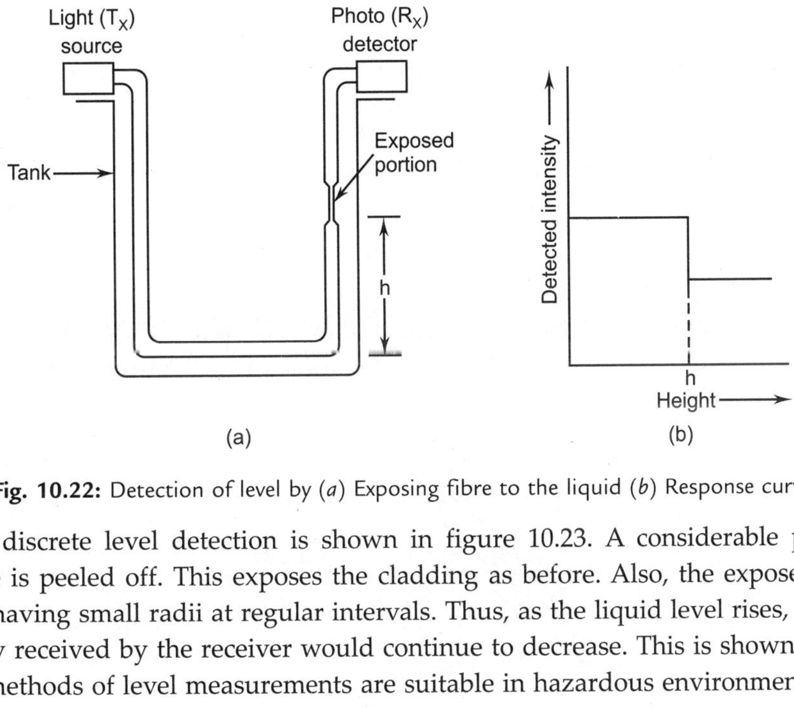

Fig. 10.22: Detection of level by (*a*) Exposing fibre to the liquid (*b*) Response curve

Multiple discrete level detection is shown in figure 10.23. A considerable portion of the jacketed fibre is peeled off. This exposes the cladding as before. Also, the exposed portion has small bends having small radii at regular intervals. Thus, as the liquid level rises, the amount of light intensity received by the receiver would continue to decrease. This is shown in figure (*b*).

Optical methods of level measurements are suitable in hazardous environments or electrical signals which would result in fires.

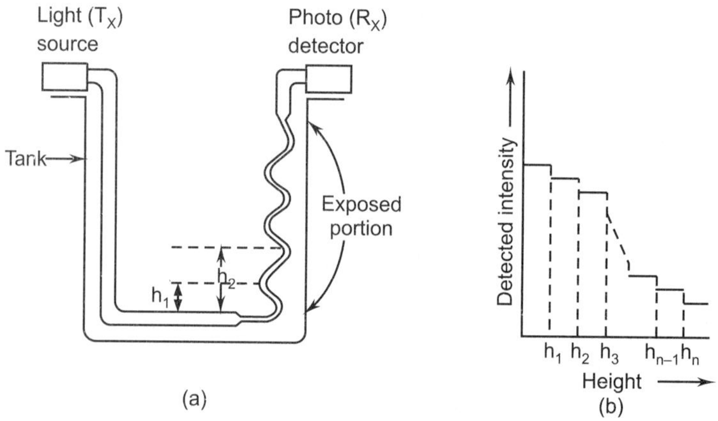

Fig. 10.23: Multi liquid level detection (*a*) methodology (*b*) response curve

10.12 SOLID LEVEL MEASUREMENT

10.12.1 Introduction

Measurement of solid level in several processes is an important parameter which need constant monitoring. In process industries, tanks, bins, sylos contain solids in bulk form the level of which are required to be measured. If dimensions of the container is known, the total weight in it can be determined. Depending on the process requirements, the following are required: (*a*) continuous monitoring and control of level and (*b*) point control at a preset level with level indication. Techniques employed for liquid level measurements can be capacitance type, weight measurement type, radiation type, conventional ultrasonic type, diaphragm type, etc. Some of the methods are described below:

10.12.2 Diaphragm Type

A diaphragm type solid level detection system is shown in figure 10.24. This can be used for point control. The diaphragm *D* is fixed to the container with a micro switch SW remaining in the chamber. As the level of solid in the container rises, diaphragm *D* is deflected to the left and the micro switch is closed. This closing of the switch can be used for controlling the level.

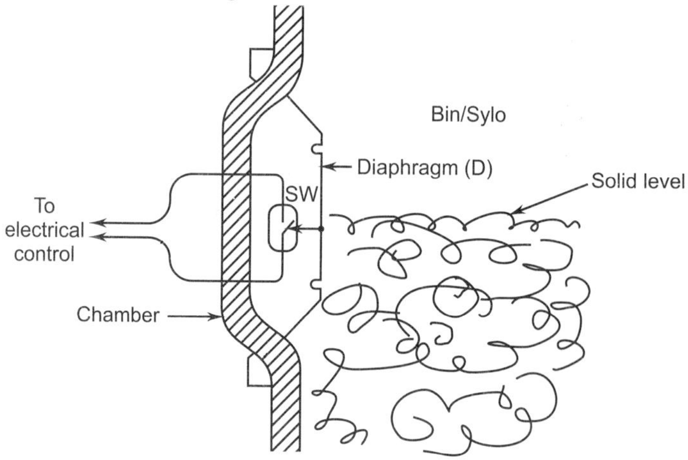

Fig. 10.24: Diaphragm type solid level detection and control system

10.12.3 Blast Furnace Solid Level Measurement

Huge tonnage of iron ore, coke and additives are poured into blast furnace by opening the big bell. The top level of solid material inside the blast furnace is measured by lowering a rod (called 'stock rod') which would rest on the solid. The measurement principle involved is shown in figure 10.25.

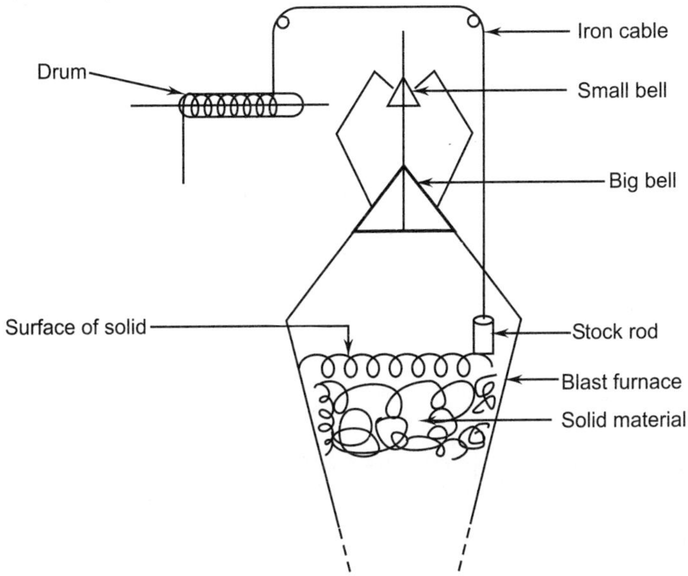

Fig. 10.25: Blast furnace solid level measurement

The measurement of level provides a continuous guide to the movement of material through the furnace and the rate at which the furnace is consuming raw materials and producing hot metal. From a knowledge of this measurement, the operator in the control room can adjust the intermittent charging rate of raw materials. This would prevent *'hanging'* of the blast furnace.

The level can be measured by dropping a sturdy float (rod) connected to a winch and cable assembly, until it rests on the solid level. Then by reference to the amount of cable unreeled from the drum, the level of solid inside the furnace can be determined accurately. When the furnace is charged (*i.e.*, opening of big bell), the rod is withdrawn to avoid its burying and damaging. The position of rod can easily be telemetered to remote locations by transmitters mechanically coupled to the cable drum. It should be mentioned here that now-a-days, 'bell-less' blast furnaces are used.

10.12.4 Load Cell Method

Figure 10.26 shows a storage tank or bin, in which the solid level can be measured by employing load cells. It is an indirect method of solid level measurement. As level increases, so also the weight. The tank should have a uniform diameter.

The load cells have strain gauges inside them. As the pressure on the load cells due to variations in the weight of the material inside the tank changes, the resistances of the strain gauges also change. This change in resistance is calibrated in terms of the level of solid in the tank by a simple bridge circuitry. For accurate level measurement of solid by this method, the density and granular or particle size of the solid will have to be constant. Also, moisture content should be fairly constant. For unevenly-sized solid, say coke, coal or iron ore, this method will not yield accurate result.

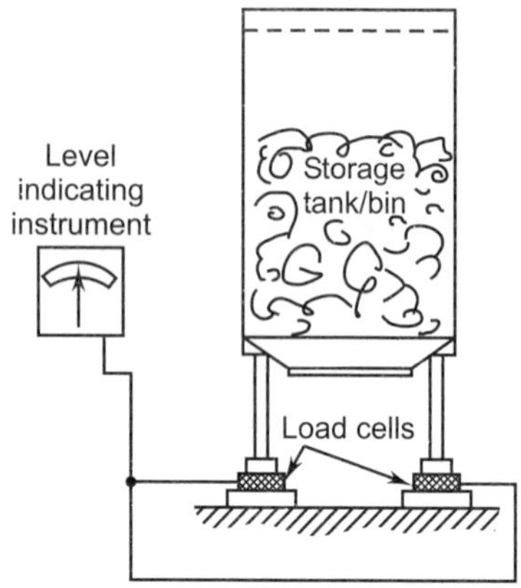

Fig. 10.26: Solid level measurement employing load cells

10.13 ELECTRICAL LEVEL MEASUREMENT

10.13.1 Introduction

Electrical level indicators can be of three types: resistive, capacitive and inductive. In all the three methods, a change in liquid level results in a corresponding change in electrical signal. This is subsequently measured and calibrated in terms of level.

10.13.2 Discrete-step Type Resistive Level Measurement

A discrete-step type level indicator is shown in figure 10.27. A mercury manometer is attached to the tank as shown. Resistors, each of value R, are placed at equal intervals. These resistances are in parallel to each other. A battery and an ammeter are connected as shown.

As the level of liquid in the tank increases, so also the mercury in the tube. Consequently, the current through the ammeter increases also, related to the liquid level as:

$$h_1\rho_1 = h_{Hg}\,\rho_{Hg}$$

ρ_1 and ρ_{Hg} being the density of the liquid and mercury respectively.

$$h_1 = h_{Hg}\cdot\frac{\rho_{Hg}}{\rho_1}.$$

$$= \frac{\rho_{Hg}}{\rho_1}\times ki,\quad k \text{ being a constant}$$

$$= k_1 . i,\quad k_1 \text{ being another constant}$$

It is obvious that the calibration curve relating liquid level h and current i is liquid specific.

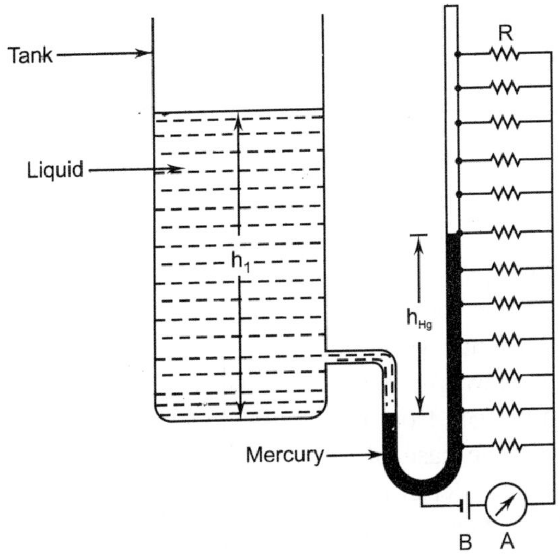

Fig. 10.27: Discrete step type level indicator

10.13.3 Tape Type Resistive Level Measurement

The resistance level sensor is in the form of a flat, coilable tape. The sensor length ranges from 1 to 30 m, width around 6-7 cm and thickness approximately 5 mm. There is a turn per cm and a resistance value of 10 ohm per helixal turn. The helix is spaced away from a gold-plated contact (shorting strip) and these two are held inside a sealed flexible plastic sheath. The schematic arrangement of the sensor is shown in figure 10.28.

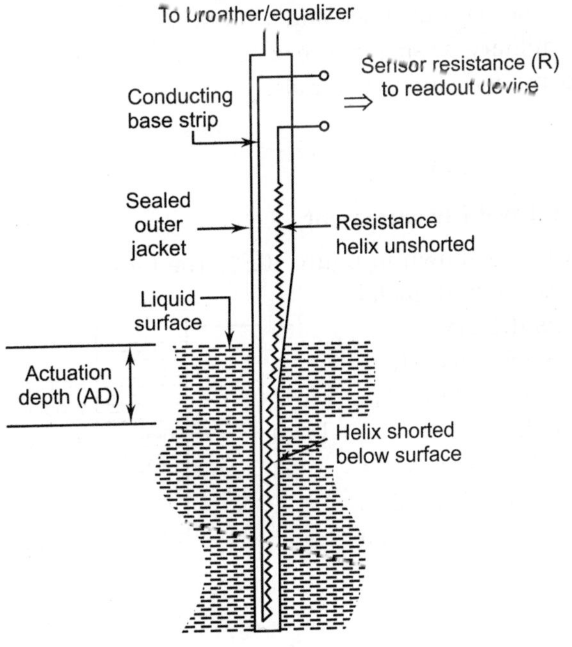

Fig. 10.28: Schematic of a resistance tape sensor for level measurement

The sensor alongwith the plastic sheath is suspended within the process vessel. With no process liquid in the vessel, full sensor resistance appears across the two leads. With liquid level inside the tank or vessel, the hydrostatic pressure pushes the sensor tape against the gold-plated contact. This lowers the resistance in proportion to the rise of liquid level in the tank. This change in resistance forms one arm of a Wheatstone's bridge. The unbalance bridge voltage is calibrated in terms of liquid level.

A minimum pressure is required to compress the outer flexible jacket which would result in shorting of the conducting base strip with the helical resistance. The tape is thus never shorted out all the way up to the surface of the liquid. It can be seen from the figure that shorting between the base plate and the resistance ends at some depth from the liquid surface, called the '*actuation depth*'. For water this depth is approximately 100 mm. The actuation depth occurs due to the built-in spring action of both the resistance and the jacket. The relationship between the actuation depth and specific gravity of the liquid is shown in figure 10.29.

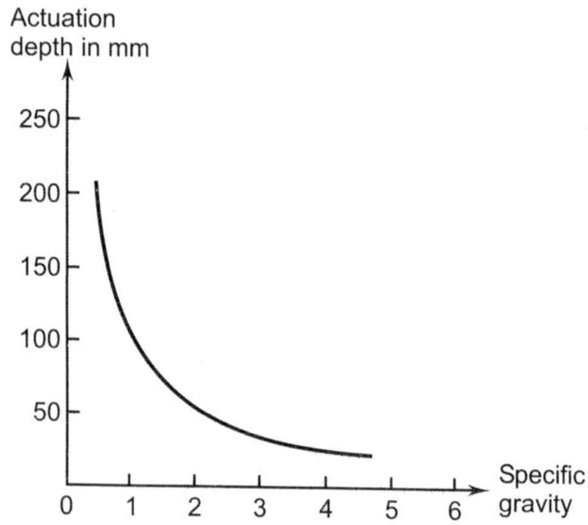

Fig. 10.29: Relation between actuation depth and specific gravity

This method can measure a level up to a precision of 3 mm for liquids (including slurries), the temperature range being from –30° to 100°C.

Advantages

- Highly resistive to mechanical shock and vibration.
- Thus suitable for military, seismic and hazardous applications.
- Can handle corrosive and slurry materials.

Disadvantages

- Suffers from actuation depth.

10.13.4 Inductive Type Level Measurement

An inductive level indicator is shown in figure 10.30. The tank contains a liquid having a good magnetic permeability, *i.e.*, liquid metal type. A side tube connects the main tank. The vertical limb contains mercury. There is a primary and two secondaries in opposition wound on the limb.

An AC voltage is applied across the primary. The voltage across the secondaries is detected by a phase-sensitive detection circuitry.

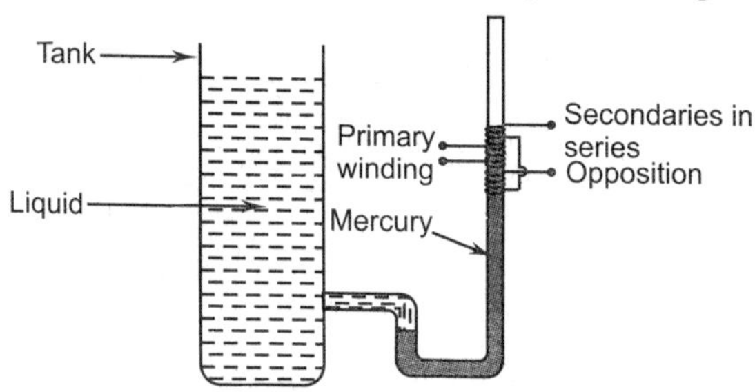

Fig. 10.30: An inductive level sensor

Advantages

- Accurate level measurement possible.

Disadvantages

- Applicable for the level range of the liquid for which a change in voltage across the secondaries is available.

10.13.5 Capacitance Type Level Measurement

Electrical capacitance level gauges are used in chemical, petrochemical and allied industries to measure remotely the level of homogeneous liquids in the temperature range –40° to 200°C and pressures upto 25–60 kgf/cm². Quality to thermal insulation decides the temperature limits of application of capacitance transducers. Capacitance level gauges are also referred to as Radio Frequency (RF) or admittance level sensors.

There is a variation of capacitance with level which is utilised for level measurements. Levels of both conducting and non-conducting liquids can be measured by this method. The two electrodes forming the two plates of the capacitor are electrically insulated for a conductive (conductivity of the order of 100 micro mhos/cm or more) liquid and those for a dielectric, they are not.

Figure 10.31 shows the construction of the capacitor for a non-conductive fluid. C_1, C_2 and C_3 in the figure represent gland capacitance, vapour phase capacitance and liquid phase capacitance respectively. *L* and *l* are vessel height and height of the liquid level while *D* and *d* are the diameters of the vessel and probe respectively.

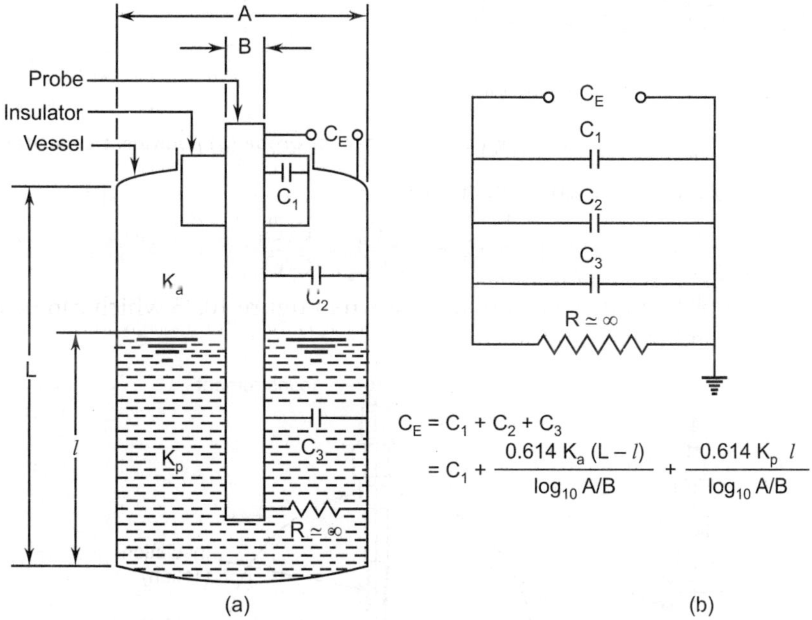

Fig. 10.31: Probe for a non-conductive fluid; (*a*) Capacitor probe; (*b*) Equivalent capacitance

Gland capacitance is the capacitance of the insulator between the probe and the tank for the bare probes used with non-conductive fluids. For a conductive fluid, the probe is insulated and then C_1 is the capacitance of the probe insulation. The equivalent capacitance C_E is

$$C_E = C_1 + C_2 + C_3 \text{ where,}$$

$$C_2 = \frac{0.614\ K_a\ (L - l)}{\log_{10}(D/d)} \text{ and } C_3 = \frac{0.614\ K_p\ l}{\log_{10}(D/d)}$$

K_a and K_p are the dielectric constants of the vapour and liquid phases respectively. When these two constants do not vary, the equivalent capacitance becomes a function of the liquid level height, l.

For a conductive fluid, the situation is shown in figure 10.32.

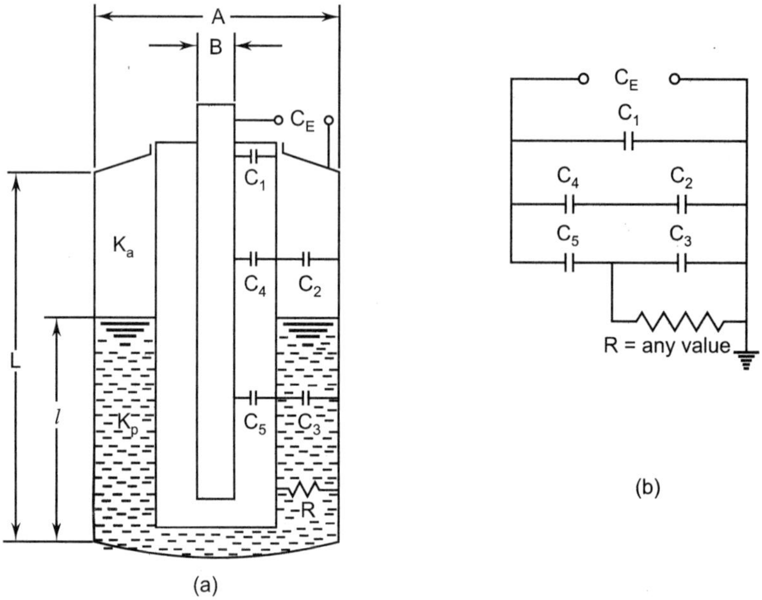

Fig. 10.32: Probe for a conducting fluid (*a*) capacitor probe (*b*) equivalent capacitance

Thus, in this case, the equivalent capacitance is

$$C_E = C_1 + \frac{C_2 C_4}{C_2 + C_4} + \frac{C_3 C_5}{C_3 C_5}$$

A teflon coated probe and its installation is shown in figure 10.33 which can be used for both conductive and non-conductive fluids.

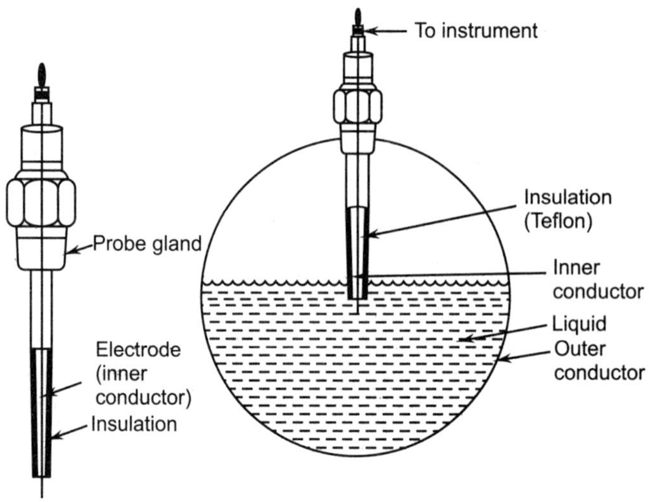

Fig. 10.33: Teflon coated probe and its installation

Increase of temperature of liquid decreases the dielectric constant of the insulator separating the plates of a capacitor. This results in measurement errors which must be compensated for if temperature variations are encountered.

Capacitance level measurements are used in on/off applications, alarms, switches and control applications. The following must be considered while selecting a capacitance probe:

- Conductivity of the process fluid.
- Metallic or non-metallic tank.
- Whether process fluid clings to the probe or not.

10.13.6 Cryogenic Type Level Measurement

For a cryogenic process fluid like nitrogen, the tank is always surrounded by a thermally insulated, evacuated '*cold box*'. The vapour space above the liquid is connected to the low pressure side of a DPT, while a connection taken out of the bottom portion of the liquid is connected to high pressure side of the same. When the bottom pipe connection, containing liquid nitrogen, comes out of the cold box, its temperature attains room temperature. Thus, liquid nitrogen is turned into vapour form. This would cause noise in level measurement. Thus, the pipe is provided with a slope towards the tank and also a vertical run, as shown in figure 10.34 to overcome the noise problem. Also, this pipe diameter is made big enough (about 2.5 – 3 mm) to decrease the turbulence caused by simultaneous boiling and condensation occurring at the liquid-vapour interface.

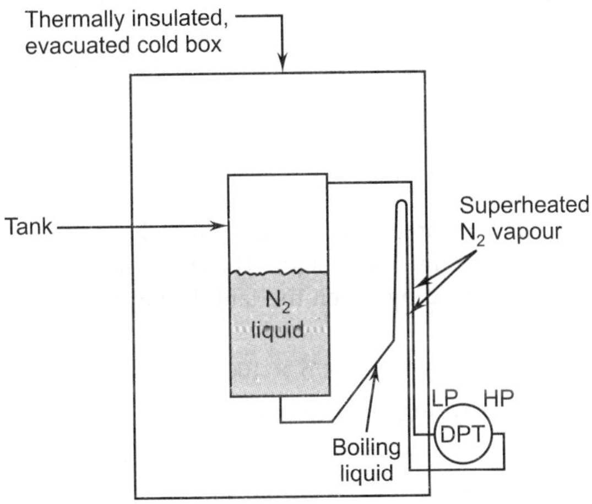

Fig. 10.34: Measurement of level of cryogenic fluid

SOLVED EXAMPLES

1. Level of water in a tank is measured by a two-wire pressure transmitter of range 0 – 8 kg/cm². The space above the water level is filled with vapour of a liquid of value 2 kg/cm². Find the output current of the transmitter if the level of water in the tank is 20 metres when the pressure transmitter is installed

   (*i*) at the bottom of the tank.

   (*ii*) 10 metres above the bottom of the tank.

   (*iii*) 10 metres below the bottom of the tank.

   Assume g = 10 m/sec².

6. Radar type level indicators are least/most affected by changes in process temperature. ☐
7. Level measurement using load cells is a direct/indirect one. ☐
8. Capacitive type level sensors for level measurements are dependent/independent of temperature variations of the liquid. ☐
9. Displacer type level gauges are superior/inferior to float type level gauges. ☐

REVIEW PROBLEMS

1. A DP transmitter is used for level measurement in an open tank. The specific gravity of the liquid is 1.1. The transmitter is connected to the tank 200 mm below the minimum level. The maximum level is 2 metres above the minimum level (refer to the figure below for dimensions).
   (*i*) Find the calibrated range and span of the transmitter in mm of water column (mmwc).
   (*ii*) What can you say about the zero of the transmitter? [*WBUT (EI 501) 2005*]
   [**Ans.** (*i*) Range: 0 mmwc to 2420 mmwc; Span: 2420 mmwc
   (*ii*) Zero of the transmitter corresponds to 200 mm below minimum level]
2. The air supply in a air-bubbler system is so adjusted that it can measure height in between 10 and 100 cms of water. (*a*) Find the readings of the mercury manometer for such a system. (*b*) Also indicate the minimum value of the supply pressure.
   [**Ans.** (*a*) 0.7353 mmwc, 7.353 mmwc (*b*) 0.7353 mmwc]

FURTHER READING

1. Instrumentation for Engineering Measurements 2nd Ed., J.W. Dally, W.F. Riley, K.G. Mcconnell, 1993. John Wiley and Sons, Inc. Wiley Student Edition.
2. Flowmeters and Flow Measurement, P. Chattopadhyay, Asian Books Pvt. Ltd., 2006.
3. Industrial Instrumentation: Principle and Design, T.R. Padmanabhan, Springer–Verlag Ltd., 2000, London..
4. Developments in Flow Measurement-1, Edited by R.W.W. Scott, Applied Science Publishers Ltd., 1982.
5. Instrumentation Measurement and Analysis., B.C. Nakra and K.K. Chaudhry, Tata McGraw Hill Pub. Co. Ltd., 2005.
6. Introduction to Measurements and Instrumentation, 2nd Ed., A.K. Ghosh, PHI Pvt. Ltd., 2007.
7. Basic Instrumentation Industrial Measurement, P.J.O'Higgins, McGraw Hill Book Company, 1966.
8. Flow Measurement Engineering Handbook, R.W. Miller, McGraw Hill Book Company, 1983.
9. Industrial Instrumentation and Control, S.K. Singh, Tata McGraw Hill Pub. Co. Ltd., 3rd Ed., 2009.
10. Mechanical Measurement and Instrumentation, E.E. Ambrosius, R.D. Felows and A.D. Brickman, The Ronald Press Co., N.Y. 1966.
11. Principles of Industrial Instrumentation, D. Patranabis, Tata McGraw Hill Pub. Co. Ltd., 2nd Ed., 8th Reprint, 2002.
12. Measurements and Instrumentation in Heat Engineering, Vol. 2: Mir Publishers. Moscow 1980.
13. Jone's Instrument Technology, Vol. 1: Mechanical Measurements, Edited by B.E. Noltingk, 4th Ed. ELBS, 1986.
14. Transducers and Instrumentation, D.V.S. Murty, PHI, 2006, 13th reprint.
15. Measurement Systems, Application and Design, E.O. Doeblin and D.N. Manik 5th Ed. Tata McGraw Hill Pub. Co. Ltd. 2007.
16. Principles of Measurement and Instrumentation, A.S. Morris, 2nd Ed., PHI Pvt. Ltd., 2003.
17. Applied Instrumentation in the Process Industries: A Survery, Vol. 1: 2nd Ed., W.G. Audrew and H.B. Williams, Gulf Pub. Co., Book Div., 1988.

18. Industrial Instrumentation, D.P. Eckman, Wiley Eastern Ltd., 16th Reprint, 1991.
19. Handbook of Instrumentation and Controls, H.P. Kallen, Editor, McGraw Hill Book Company, 1961.
20. Instrument Engineer's Handbook, Process Software and Digital Networks 3rd Ed., B.G. Lipta'k Editor-in-chief, CRC Press, 2002.
21. Principles of Measurement System, J.P. Bentley (Pearson Edu), 3rd Ed.
22. Industrial Level Measurement, D.R. Gillum, ISA, 1981.
23. Temperature Measurement, L. Michalski, K. Eckersdorf, J. Kucharski and J. McGhee, 2nd Ed. New York John Wiley, 2001.

INDEX

Q

R

S